THE FAMILIES AND GENERA OF VASCULAR PLANTS

Edited by K. Kubitzki

For further volumes see list at the end of the book and:
http://www.springer.com/series/1306

The Families and Genera of Vascular Plants

Edited by K. Kubitzki

XIV

Flowering Plants · Eudicots

Aquifoliales, Boraginales, Bruniales, Dipsacales, Escalloniales, Garryales, Paracryphiales, Solanales (except Convolvulaceae), Icacinaceae, Metteniusaceae, Vahliaceae

Volume Editors:
Joachim W. Kadereit · Volker Bittrich

With 76 Figures

Springer

Editors
Joachim W. Kadereit
Johannes Gutenberg
Universität Mainz
Mainz
Germany

Volker Bittrich
Campinas
Brazil

Series Editor

Prof. Dr. Klaus Kubitzki
Universität Hamburg
Biozentrum Klein-Flottbek und Botanischer Garten
22609 Hamburg
Germany

ISBN 978-3-319-28532-0 ISBN 978-3-319-28534-4 (eBook)
DOI 10.1007/978-3-319-28534-4

Library of Congress Control Number: 2016937409

Printed on acid-free paper

This Springer imprint is published by Springer Nature
The registered company is Springer International Publishing AG Switzerland

Preface

It is with great pleasure that I present volume XIV of the Families and Genera of Vascular Plants, a further step forward in the treatment of asterid (sympetalous) families of the eudicots. Recent advances in molecular systematics have necessitated an extensive rearrangement of the families, including the lumping, splitting or revival of well-known family concepts or, in some cases, the creation of completely new ones. To cite one example, the position of the Aquifoliaceae has long been unclear—in previous classifications the family has peregrinated through at least eight orders, whereas now it is placed in a new order at the base of the asterids, the Aquifoliales, which also comprise a few other presumably related families. The Icacinaceae, which always have been a problematic group for taxonomists, can now be presented in a substantially revised circumscription, although their closest relatives still remain unknown. It is worth remembering that early naturalists, who based their classifications on visible traits and geographical patterns, achieved classifications that have proven useful for integrating findings from both non-molecular and molecular data. Thus, the five families of the Dipsacales as treated in this volume are today confirmed as a monophylum—already in the 19th century, they had been recognised as a natural group but, despite much effort, their interrelationships still are not understood in all details.

It is with pride that we can include in this volume the monumental treatments of the two very large families Solanaceae and Boraginaceae, resulting from year- (if not life-) long involvement of their two main authors accompanied by numerous co-workers dealing with specific aspects, so that their treatments rest on broad factual bases. These include fully revised classifications of the two families as well as a wealth of biological data. Boraginaceae, which are predominantly herbaceous, comprise a wide range of growth forms such as rhizomatous herbs and other geophytes, and their flowers, fruits and seeds are diverse in terms of size, structure and functional adaptations. Solanaceae exploit habitats worldwide from coastal areas to high-montane regions, preferably in the tropics. Thus, their considerable variation of life-forms and reproductive structures is no surprise; almost every known pollination syndrome has been uncovered in this family and, in addition to nectar and pollen, pollinator attractants include curious substances such as liquid perfume and oil.

I am stressing all these biological traits included into the family treatments in this series from the very beginning, following the conviction that they are important for an understanding of the life history of plants. Very often, groups of closely related plant species differ in a single trait, their theme of diversification, while other traits may remain unchanged. Understanding adaptive variation may help understand life history evolution.

I would like to acknowledge the hard work and commitment of all authors and the volume editors; I am deeply indebted for their scholarly contributions and careful editorial work. I am also grateful to the copyright holders of the illustrations published under their responsibility. The artist Bobbi Angell, New York, is thanked for the generosity with which she authorized the use of her artwork.

Finally, I have great pleasure in thanking the copy editor of the present volume, Dr. Monique Delafontaine, for her dedicated editorial work. I also wish to acknowledge the very pleasant collaboration with Dr. Andrea Schlitzberger from Springer Verlag and with Mr. V. Anand from Spi Technologies India Pvt Limited.

Hamburg
April 2016

Klaus Kubitzki

List of Contributors

M.C.E. Amaral	Depto. de Biologia Vegetal, Instituto de Biologia, Universidade Estadual de Campinas, Campinas, SP, Brazil. mceamaral2009@gmail.com
S. Andrews	Kew Gardens, Royal Botanic Gardens, Kew, Richmond, Surrey, UK. s.andrews@kew.org
A. Anton	Instituto Multidisciplinario de Biología Vegetal, Córdoba, Argentina. anton@imbiv.unc.edu.ar
S. Arroyo	Botanischer Garten und Botanisches Museum Berlin-Dahlem, Freie Universität Berlin, Berlin, Germany. leu.arro@gmail.com
A. Backlund	Division of Pharmacognosy, Dept. of Medicinal Chemistry, Uppsala University, BMC - Biomedical Center, Uppsala, Sweden. anders.backlund@fkog.uu.se
G.E. Barboza	Instituto Multidisciplinario de Biología Vegetal, Córdoba, Argentina. gbarboza@imbiv.unc.edu.ar
G. Barriera	Conservatoire et Jardin botaniques de la Ville de Genève, Chambésy/Genève, Suisse. gabrielle.barriera@ville-ge.ch
G. Bernardello	Instituto Multidisciplinario de Biología Vegetal, Córdoba, Argentina. gabyberna@gmail.com
V. Bittrich	Rua Dr. Mário de Nucci 500, 13083-290 Campinas, SP, Brazil. folcar2007@gmail.com
C. Carrizo Garcia	Instituto Multidisciplinario de Biología Vegetal, Córdoba, Argentina. ccarrizo@imbiv.unc.edu.ar
R. Claßen-Bockhoff	Institut für Spezielle Botanik, Johannes Gutenberg-Universität Mainz, Mainz, Germany. classenb@uni-mainz.de
A.A. Cocucci	Instituto Multidisciplinario de Biología Vegetal, Córdoba, Argentina. aacocucci@gmail.com

J.G. Conran	ACEBB & SGC, School of Earth and Environmental Sciences, The University of Adelaide, North Terrace, SA, Australia. john.conran@adelaide.edu.au
M.T. Cosa	Instituto Multidisciplinario de Biología Vegetal, Córdoba, Argentina. cosa@imbiv.unc.edu.ar
L.P. Ronse De Craene	Royal Botanic Garden Edinburgh, Edinburgh, UK. l.ronsedecraene@rbge.org.uk
N. Diane	Milowstraße 10, 14195 Berlin, Germany. info@schulbuchredaktion.de
W.C. Dickison (deceased)	Formerly at University of North Carolina, Chapel Hill, NC, USA.
M.O. Dillon	Botany Department, The Field Museum, Chicago, IL, USA. mdillon@fieldmusem.org
R. Duno	Centro de Investigación Científica de Yucatán, Merida, Yucatan, Mexico. roduno@cicy.mx
H. Förther	Heilmannstr. 5, D-82049 Pullach, Germany.
V. Fuentes (deceased)	Formerly at Instituto de Investigaciones de Fruticultura Tropical, La Habana, Cuba.
M. Gottschling	Department Biologie I, Ludwig-Maximilians-Universität München, München, Germany. gottschling@biologie.uni-muenchen.de
H.H. Hilger	Systematische Botanik und Pflanzengeographie, Institut für Biologie, Freie Universität Berlin, Berlin, Germany. hahilger@zedat.fu-berlin.de
M. Hofmann	Formerly at Systematische Botanik und Pflanzengeographie, Institut für Biologie, Freie Universität Berlin, Berlin, Germany.
U. Hofmann (deceased)	Formerly at Systematic Geobotanical Institute, Georg August Universität Göttingen, 37073 Göttingen, Germany.
A.T. Hunziker (deceased)	Formerly at Instituto Multidisciplinario de Biología Vegetal, Córdoba, Argentina.
J.W. Kadereit	Institut für Spezielle Botanik und Botanischer Garten, Johannes Gutenberg-Universität Mainz, Mainz, Germany. kadereit@uni-mainz.de
T.G. Lammers	Department of Biology and Microbiology, University of Wisconsin Oshkosh, Oshkosh, WI, USA. lammers@uwosh.edu
A. Liston	Department of Botany & Plant Pathology, Oregon State University, Corvallis, OR, USA. listona@science.oregonstate.edu
P.-A. Loizeau	Conservatoire et Jardin botaniques de la Ville de Genève, Chambésy/Genève, Suisse. pierre-andre.loizeau@ville-ge.ch
F. Luebert	Nees-Institut für Biodiversität der Pflanzen, Rheinische Friedrich-Wilhelms-Universität, Bonn, Germany. fluebert@uni-bonn.de

J. Lundberg — Department of Botany, The Swedish Museum of Natural History, Stockholm, Sweden.
Johannes.Lundberg@nrm.se

T.D. Macfarlane — Western Australian Herbarium, Science and Conservation Division, Department of Parks and Wildlife, Manjimup, WA, Australia.
Terry.Macfarlane@dpaw.wa.gov.au

V. Mayer — Department of Botany and Biodiversity Research, Faculty of Life Sciences, University of Vienna, Vienna, Austria.
veronika.mayer@univie.ac.at

A.E. Moscone — Instituto Multidisciplinario de Biología Vegetal, Córdoba, Argentina.

M.J. Potgieter — Department of Biodiversity, University of Limpopo, Polokwane, South Africa.
martin.potgieter@ul.ac.za

A. Romanutti — Instituto Multidisciplinario de Biología Vegetal, Córdoba, Argentina.
romanut@imbiv.unc.edu.ar

V. Savolainen — Imperial College London, Ascot, Berkshire, UK.
v.savolainen@imperial.ac.uk

M. Schori — Department of Environmental & Plant Biology, Ohio University, Athens, OH, USA.
ms335704@ohio.edu

F. Selvi — Dipartimento di Scienze delle Produzioni Agroalimentari e dell'Ambiente (DISPAA), Sez. Scienza del Suolo e della Pianta, Laboratori di Botanica, Università di Firenze, Firenze, Italy.
federico.selvi@unifi.it

R. Spichiger — Conservatoire et Jardin botaniques de la Ville de Genève, Chambésy/Genève, Suisse.
rodolphe.spichiger@ville-ge.ch

R. Subils — Instituto Multidisciplinario de Biología Vegetal, Córdoba, Argentina.

M. Thiv — Staatliches Museum für Naturkunde Stuttgart, Stuttgart, Germany.
mike.thiv@smns-bw.de

D.C. Thomas — School of Biological Sciences, The University of Hong Kong, Hong Kong, China.
dthomas@hku.hk

T.M.A. Utteridge — Herbarium, Library, Art and Archives, Royal Botanic Gardens, Kew, Kew, Surrey, UK.
t.utteridge@kew.org

G.K. Walden — Department of Integrative Biology, University of California, Berkeley, CA, USA.
gkwalden@berkeley.edu

F. Weberling (deceased) — Formerly at Abteilung f. Spezielle Botanik, Universität Ulm, Ulm, Germany.

M. Weigend	Nees-Institut für Biodiversität der Pflanzen, Rheinische Friedrich-Wilhelms-Universität, Bonn, Germany. mweigend@uni-bonn.de
Q.Y. Xiang	Department of Plant and Microbial Biology, North Carolina State University, Raleigh, NC, USA. jenny_xiang@ncsu.edu
L.-B. Zhang	Science & Conservation Division, Missouri Botanical Garden, St. Louis, MO, USA. Libing.Zhang@mobot.org

Contents

Introduction to the Orders and Families of Uncertain Placement of this Volume

V. Bittrich and J.W. Kadereit

The present volume of this book series is the first of two to (almost) complete the treatment of the Asterids which started with Vol. VI (Cornales, Ericales, 2004), Vol. VII (Lamiales, 2004) and Vol. VIII (Asterales, 2007). It contains the orders Boraginales, Garryales and Solanales of the Lamiids (Asterids I) as well as three unplaced families of that clade, i.e. Vahliaceae, Icacinaceae and Metteniusaceae, and the orders Aquifoliales, Escalloniales, Bruniales, Dipsacales and Paracryphiales of the Campanulids (Asterids II).

Asterids I (Lamiids)

Conspectus of the Families of Garryales

1. Deciduous trees with alternate leaves. 1 genus/1 species, China **Eucommiaceae**
- Evergreen shrubs with opposite leaves 2
2. Inflorescence catkin-like, fruit a berry with 2 or 3 seeds. 1/15, N and C America, Greater Antilles **Garryaceae**
- Inflorescence not catkin-like, fruit a 1-seeded drupe. 1/8–11, E Asia **Aucubaceae**

As treated here, **Garryales** contain Garryaceae, Aucubaceae and Eucommiaceae with altogether three genera and 24 to 27 species. In APG III (2009), Aucubaceae was included in Garryaceae. Whereas a close relationship between Garryaceae and Aucubaceae had already been suspected by Baillon (1877), the relationship between Garryaceae/Aucubaceae and Eucommiaceae had not been suspected until revealed by molecular analyses (Soltis et al. 2000). Garryales at present are resolved as sister to all other Lamiids except Icacinaceae and Metteniusaceae (incl. Oncothecaceae; Refulio-Rodriguez and Olmstead 2014). All taxa in the order are dioecious woody plants with exstipulate leaves containing the iridoid aucubin. Beyond that, no obvious synapomorphy can be recognized, although the close relationship between *Garrya* and *Eucommia* (but not *Aucuba*) may be supported by the possession of gutta-percha (trans-1, 4-polyisoprene). The order is distributed in East Asia (*Aucuba*, *Eucommia*) and western North America (*Garrya*, with one species in the Greater Antilles). The stem group age of Garryales has been estimated at ca. 97 million years ago (Ma; Magallón and Castillo 2009), and the split between Garryaceae/Aucubaceae and Eucommiaceae has been dated at 20 ±8.6 Ma (Janssens et al. 2009) or 49.8 Ma (Magallón and Castillo 2009). The two families may represent remnants of a formerly more widespread north hemispherical lineage of presumably forest plants.

Three Families of Uncertain Placement within the Lamiids

Vahliaceae, **Icacinaceae** and **Metteniusaceae** (incl. Oncothecaceae) have all been associated with Garryales. However, new evidence indicates a relationship of Vahliaceae to Solanales, and Icacinaceae plus Metteniusaceae (incl. Oncothecaceae) appear to be sister to all other Lamiids (Refulio-Rodriguez and Olmstead 2014). No character is known which could support the relationship between Vahliaceae and Solanales. Whereas Vahliaceae with one genus and five species are annual or perennial herbaceous to subshrubby plants, Metteniusaceae with two genera and nine species and Icacinaceae with 35 genera

Boraginales by M. Weigend

J.W. Kadereit and V. Bittrich (eds.), *Flowering Plants. Eudicots*, The Families and Genera of Vascular Plants 14,
DOI 10.1007/978-3-319-28534-4_1,

and ca. 245 species are woody, sometimes lianaceous in Icacinaceae. All three families are largely tropical in distribution. Thus, Vahliaceae are found mainly in Africa (but also on the Indian subcontinent), Metteniusaceae grow on New Caledonia (*Oncotheca*) and Central to South America (*Metteniusa*), and Icacinaceae are pantropical with centres of diversity in Africa and Southeast Asia. The woody *Dendrobangia* of essentially unknown affinity with two species from Central to South America is here treated in Metteniusaceae, and this appears to be supported by new data (Greg Stull, pers. comm.). The circumscription of Icacinaceae presented here differs from past treatments of the family, and further changes in circumscription can be expected. Thus, several genera of Icacinaceae have been included in Cardiopteridaceae and Stemonuraceae of Aquifoliales and Pennantiaceae of Apiales (Kårehed 2001), and new data (Greg Stull, pers. comm.) seem to require the transfer of a number of other genera of Icacinaceae as treated here into an expanded Metteniusaceae.

If the relationships of Icacinaceae with Metteniusaceae (incl. Oncothecaceae) can be confirmed, they would represent a tropical to south hemispherical counterpart of the north hemispherical Garryaceae/Aucubaceae and Eucommiaceae. The ages of Icacinaceae and Metteniusaceae (and Oncothecaceae) have been estimated at ca. 97 Ma, and that of Vahliaceae at ca. 83.5 Ma (Magallón and Castillo 2009).

Conspectus of the Families of Solanales

1. Herbs of swampy areas, cortex with large air-spaces. Flowers small in dense terminal spikes, petal lobes folded inwards. 1/1, tropical and subtropical regions of the Old World, naturalized in the Americas **Sphenocleaceae**
– Characters not as above 2
2. Plants dioecious. Leaf axils with tuft of hairs. Ovary inferior. 3/5, equatorial and southern Africa and Madagascar **Montiniaceae**
– Plants rarely dioecious. Leaf axils without tuft of hairs. Ovary superior 3
3. Plants with milky sap (sometimes inconspicuous), sepals large, free. About 57/1600, cosmopolitan **Convolvulaceae** (not in this volume)
– Plants without milky sap. Sepals of various size, united at base 4
4. Bases of filaments conspicuously broadened. Stylodia 2, spreading. 1/11, cosmopolitan, mainly tropical **Hydroleaceae**
– Bases of filaments not conspicuously broadened. Style simple. 96/2400, cosmopolitan **Solanaceae**

The **Solanales** are composed of five families: Solanaceae (94 genera; incl. Duckeodendraceae, Goetzeaceae, Nolanaceae and Sclerophylacaceae), Convolvulaceae (57 genera; incl. Cuscutaceae), Sphenocleaceae (1 genus), Hydroleaceae (1 genus) and Montiniaceae (3 genera). Vahliaceae probably should also be included in the order as they appear to be sister to Solanales (Refulio-Rodríguez and Olmstead 2014). This latter relationship, however, had high support only in the Bayesian tree and not in the ML and MP trees. The close relationship of Convolvulaceae and Solanaceae, which are sister to each other, has long been accepted (e.g. Jussieu 1789) but the presently accepted composition of the order has been proposed only rather recently. In the past, Solanales often were considered part of a larger suprafamiliar group, the "Tubiflorae", which also included Polemoniales, Boraginales (Bartling 1830; Bessey 1915 as Polemoniales) and sometimes (Engler system, e.g. Engler 1903) also Lamiales. Until recently, Boraginaceae s.l. and Polemoniaceae (the latter now part of Ericales near the base of the asterids) were considered families closely related to Convolvulaceae/Solanaceae. The first large angiosperm phylogeny DNA sequence study (Chase et al. 1993) suggested that Montiniaceae (formerly included in Saxifragaceae, Cornaceae or Grossulariaceae) belong to Solanales. *Sphenoclea* was generally considered part of or close to Campanulaceae until Cosner et al. (1994), using *rbc*L sequence data, showed that it belonged into Solanales. *Hydrolea* had always been included in Hydrophyllaceae or in Hydroleaceae s.l., comprising various genera of Hydrophyllaceae s.l. until Di Fulvio de Basso (1990), based mainly on embryological data, showed that Hydroleaceae should include only *Hydrolea*. The *rbc*L sequence data of Cosner et al. (1994) later demonstrated that such monogeneric Hydroleaceae belong in Solanales instead of Boraginales. The placement of Polemoniaceae in Ericales, first suggested by Olmstead et al. (1992) based on *rbc*L sequence data, was unexpected (e.g. Cronquist 1981; Dahlgren 1983; Thorne 1983). This family shares with Solanales characters such as usually alternate leaves, 5-merous and usually actinomorphic perianth and androecium, plicate sympetalous corolla with adnate stamens, superior ovary, annular nectary disk frequent, and persistent calyx. On the other hand, non-DNA differences to typical Solanales are few: lack of alkaloids, internal phloem typical of Solanaceae and Convolvulaceae absent, sepals with broad membranous margins, and endosperm development nuclear (cellular in most Solanales with the exception of Convolvulaceae, but data lacking for Montiniaceae). Hufford (1992) suggested that the 3-carpellate gynoecia and loculicidal capsules of Polemoniaceae fit better in Ericales than in (core) asterids which tend to have bicarpellate gynoecia and septicidal capsules. Convolvulaceae, however, also often have loculicidal capsules, and 3-carpellate gynoecia occasionally occur in Solanales. Of the two possible morphological synapomorphies of Ericales discussed by Judd et al. (2008), i.e. theoid leaf teeth and protruding axile placentas, none is found in Polemoniaceae.

*rbc*L sequence data (Chase et al. 1993), *ndh*F sequence data (Olmstead et al. 2000), and sequence data of the 25,000 bp plastid inverted repeat (IR) region (Moore et al. 2011) weakly supported Boraginales as sister of Solanales. Soltis et al. (2011), however, identified Gentianales as sister of Solanales, but without bootstrap support. The phylogeny of Qiu et al. (2010), based on sequence data of mitochondrial genes, found Boraginales, Solanales and Gentianales as successive branches close to the base of Lamiidae. Corolla tubes with folding lines was suggested as a possible synapomorphy of Boraginales+Solanales (Judd and Olmstead 2004), but knowledge of this character, rarely mentioned in modern taxonomic literature, is rather incomplete. Alternate leaves, terminal inflorescences, and a persistent calyx are other morphological characters common to Boraginales and Solanales. Lack of iridoids has been suggested as a possible synapomorphy of Solanales and Boraginales by Judd et al. (2008), but iridoids are not completely absent from Solanales (see below). Pollen tubes with callose is another character possibly supporting a close relationship of Solanales and Boraginales. It occurs in Convolvulaceae, Solanaceae, Boraginaceae and part of Hydrophyllaceae (tribe Hydrophylleae), but not in Ehretiaceae, Heliotropiaceae and another part of Hydrophyllaceae (tribe Romanzoffieae; Prósperi and Cocucci 1979). In summary, a sister group relationship of Boraginales and Solanales seems best supported by the data currently available. The partly incomplete knowledge of morphological characters and lack of resolution at the base of Lamiidae, however, considerably complicate the evaluation of synapomorphies.

The members of Solanales are mainly herbs to shrubs or twining herbaceous or woody plants; trees, especially large trees, are rare (e.g. *Duckeodendron*, *Humbertia*). Leaves are alternate, usually simple and exstipulate. Flowers are arranged in terminal inflorescences, are normally bisexual, frequently 5-merous in perianth and androecium, with stamens adnate to the corolla tube; the ovary normally is superior and, with few exceptions (e.g. *Jaborosa*, *Nicandra*, *Nolana*), bicarpellate. The corolla is usually actinomorphic, and sympetaly is late, early only in *Sphenoclea* (Erbar 1995). A nectary often forming a disk around the ovary (always of gynoecial origin?) is rarely absent (e.g. *Solanum*, *Sphenoclea*). The endosperm is cellular (Hydroleaceae, Solanaceae, Sphenocleaceae) or nuclear (Convolvulaceae, probably an autapomorphy), unknown for Montiniaceae. Fruit types are various, but most common fruits are capsules (septicidal or septicidal-loculicidal, more rarely loculicidal or circumscissile) or berries, and the calyx is persistent.

Phytochemical data are abundant for Solanaceae and Convolvulaceae, but much less so for the three smaller families. In his detailed review, Eich (2008) characterized the former two families as follows:

General secondary metabolites: phenolics such as simple cinnamic acid derivatives, hydroxycoumarins, hydroxycinnamate conjugates, flavonoids, lignans, sterols, carotenoids, fats/oils and fatty acids.

Specific secondary metabolites: ornithine-derived alkaloids such as hygrines, N-acylpyrrolidines, nicotinoids, tropanes, calystegines. Sesquiterpenoid phytoalexins.

Absent: iridoids; essential oils, complex mixtures of lipophilic mono-/sesquiterpenes and/or phenylpropanoids secreted in oil cells, secretion ducts/cavities or glandular hairs; condensed tannins (proanthocyanidins).

While iridoids seem to be completely absent from Solanaceae and Convolvulaceae, the presence of the allochemical secoiridoid-glycosides, able to inhibit root growth of rice seedlings, has been reported by Hirai et al. (2000) for the weedy *Sphenoclea zeylanica*. An iridoid compound (Montinioside) has also been reported for *Montinia caryophyllacea* (Dahlgren et al. 1977). Alkaloids appear to be absent from *Sphenoclea*.

Phylogenetic relationships within Solanales as currently accepted are (((HYDROL-SPHENO)(MONTIN))(CONVOLV-SOLAN)). Embryological data support the sister group relationship of Hydroleaceae and Sphenocleaceae as both have endosperm haustoria at both the chalazal and micropylar ends; vestured pits occur in Montiniaceae and Hydroleaceae, but have not yet been reported from Sphenocleaceae. Convolvulaceae and Solanaceae share various supposedly apomorphic characters such as an intraxylary phloem (also common in Gentianales), lack of iridoids, similar alkaloids, flowers with oblique symmetry and similar corolla aestivation (Stevens 2001 onwards).

The stem age of Solanales was estimated to between about 80 Ma and 106 Ma (cf. Olmstead 2013). A historical biogeographical analysis based on current phylogenies suggested that the HYDROL-SPHENO-MONTIN clade arose in Africa, the Convolvulaceae in SE Asia, India and Madagascar, and the Solanaceae in South America. "Thus Solanales may present a classic case of vicariance following Gondwanan breakup, with its descendants on each of the three initial continental fragments giving rise to the three extant clades" (Olmstead 2013).

Conspectus of the families of Boraginales

1. Plants without chlorophyll, leaves reduced to scale-shaped cataphylls; ovary plurilocular by secondary subdivision, fruit with numerous "pits" (endocarpids—individual seeds enclosed separately in lignescent inner pericarp) at maturity. 2/4, SW USA and Mexico **Lennoaceae**
– Plants with chlorophyll, leaves green and with normal lamina; ovary uni- to tetralocular, fruit with only 1–4 endocarpids, or without endocarpids (nutlets, capsules) 2
2. Ovary subdivided into four mericarpids in flower, developing into (1–)4 individual nutlets, style insertion gynobasic. Ca. 85/1600–1700, subcosmopolitan, centres of diversity in the Mediterranean, western Asia and western North America, largely absent from the wet tropics **Boraginaceae** s.str.
– Ovary entire in flower, developing into a capsule or (dry or succulent) drupe or separating into 2–4 individual nutlets at maturity, style insertion apical 3
3. Stems and leaves spinose with stiff, white spines; flowers 10–20-merous, calyx segments strongly unequal. 1/2, SW Africa **Codonaceae**
– Stems and leaves not spinose, but sometimes setose or with spinescent brachyblasts (*Rochefortia*); flowers usually 4–5-merous [if more, then fruit drupaceous]; calyx segments equal or subequal 4
4. Fruit a laterally flattened capsule, obcordate, with 1–2 seeds; flowers always tetramerous, corolla with very short tube. 1/6, SW and NE Africa **Wellstediaceae**
– Fruit never laterally flattened, ovoid, with 4–? seeds if capsular; flowers usually pentamerous, corolla tube usually distinct and tubular 5
5. Fruit capsular with 4–? seeds. 15/280–300, mainly W North America, also C America and W South America, West Indies and Hawaiian Islands **Hydrophyllaceae**
– Fruit drupaceous or dry, often subdivided into mericarpids, 4-seeded, sometimes fewer by abortion, never capsular 6
6. Style undivided with a ring-shaped subterminal stigma (style-stigma complex); flowers usually in dense, scorpioid monochasia. 4/450, pantropical, also subtropical regions, some extending into warm temperate Eurasia, South America, Australia and North America **Heliotropiaceae**
– Style entire or divided, stigma(-ta) terminal; flowers variously arranged, not in dense, scorpioid, ebracteose monochasia, often in repeatedly dichasial or strongly congested inflorescences. 10/500, pantropical, also subtropical regions, very few in warm temperate E Asia **Ehretiaceae** s.l.

As treated here, **Boraginales** contain Boraginaceae s.str. (= Boraginaceae subfam. Boraginoideae), Codonaceae (= Boraginaceae subfam. Codonoideae), Ehretiaceae s.l. (incl. Boraginaceae subfam. Ehretioideae and Cordioideae, Hoplestigmataceae, Lennoaceae), Heliotropiaceae (= Boraginaceae subfam. Heliotropioideae), Wellstediaceae (= Boraginaceae subfam. Wellstedioideae) and Hydrophyllaceae, corresponding to a total of ca. 115 genera and 2950 species. The current treatment deviates from most classical treatments which generally accepted the four families Boraginaceae s.l. (subfam. Boraginoideae, subfam. Ehretioideae, subfam. Cordioideae, subfam. Heliotropioideae; Bentham and Hooker

1883; Gürke 1893), Hydrophyllaceae (incl. *Codon*), and Lennoaceae and Hoplestigmataceae as distinct families. Baillon (1891) recognized Boraginaceae in a slightly wider definition and included Hydrophyllaceae as another subfamily. The classification provided by Takhtajan (2009) circumscribed the order Boraginales with the families Hydrophyllaceae including subfam. Hydrophylloideae and Codonoideae, Boraginaceae including subfam. Ehretioideae, Cordioideae, Heliotropioideae, Boraginoideae and Wellstedioideae, and Hoplestigmataceae plus Lennoaceae. Some genera associated with Boraginaceae/Boraginales in the past have been shown not to be close relatives of the order/family—e.g. *Vahlia* (Vahliaceae, close to Lamiales; Albach et al. 2001; Weigend et al. 2014), Tetrachondraceae (Lamiales; Wagstaff et al. 2000), pantropical *Hydrolea* L. (Solanales; Cosner et al. 1994; Soltis et al. 2000) and *Pteleocarpa* Oliv. (Geraniales; Riedl 1997; Brummitt 2011).

Detailed molecular data have revealed that Boraginales consist of two main clades, i.e. Boraginales I with Codonaceae as sister to Wellstediaceae and Boraginaceae s.str. (= subfam. Boraginoideae), and Boraginales II with doubtfully monophyletic Hydrophyllaceae as sister (when monophyletic) or successive sisters (when not monophyletic) to Heliotropiaceae and Ehretiaceae s.l. (incl. subfam. Cordioideae, Ehretiaceae, Hoplestigmataceae, Lennoaceae; Ferguson 1999, Weigend et al. 2013, 2014).

Boraginales I are a predominantly Old World group, with Codonaceae and Wellstediaceae restricted to Africa (incl. Socotra) and Boraginaceae s.str. with the bulk of its lineages, including most basal lineages, restricted to the Old World. However, all four major clades of Boraginaceae s. str. (Echiochileae, Boragineae, Lithospermeae, Cynoglosseae) have dispersed one to several times to the New World, where the latter two subgroups (Lithospermeae, Cynoglosseae) underwent massive radiations.

Codonaceae and Wellstediaceae are monogeneric families of annual to perennial herbs or small shrublets, often with an "open" life span, i.e. growing and flowering until killed by drought. They are plants from deserts and semi-deserts. Both families as here recognized have some characters which are unique in Boraginales: the two species of Codonaceae differ mainly in details of floral morphology, but share several characters such as the possession of stiff, white spines, 10–20 calyx segments of markedly differing size, 10–20 corolla lobes, red, glandular structures on the corolla and many-seeded capsules with seeds with a very irregular, reticulate testa (Weigend and Hilger 2010). The six species of Wellstediaceae differ mainly in habit and indumentum, but share very small, tetramerous flowers with a weakly developed corolla tube, laterally flattened, 1–2-seeded, obcordate capsules and asymmetrically ovoid seeds often with a ring of longer hairs near the funicular pole (Thulin and Johansson, 1996).

All Boraginaceae s.str. have an ovary subdivided into four mericarpids with gynobasic style insertion. With a total of ca. 1600—1700 species in ca. 85 genera, the family is highly variable in its vegetative and generative morphology. However, Echiochileae, sister to the remainder of Boraginaceae s.str., are similar in ecology to Codonaceae and Wellstediaceae, with the bulk of species being shrubs from semi-desert and desert habitats; they are more species-rich (ca. 30 spp.) and more widespread than the former two families and range from western Asia to north-western Africa and are distributed with one genus with 10–15 spp. in the Americas. Core Boraginaceae—tribes Boragineae, Cynoglosseae and Lithospermeae—are the most species-rich clade in the order. They are predominantly herbaceous, with shrubs largely restricted to some subgroups of Lithospermeae, and are the only group of Boraginales where relatively complex flowers with both basal and faucal corolla scales occur. The three tribes of core Boraginaceae are particularly diverse in the warm-temperate and Mediterranean ecosystems of the northern hemisphere.

In contrast to Boraginales I, Boraginales II are predominantly New World in distribution, with Hydrophyllaceae restricted to the Americas. However, several isolated genera (*Hoplestigma*, *Coldenia*) und numerous species in Heliotropiaceae and Ehretiaceae (in *Euploca*, *Heliotropium*, *Cordia* and *Ehretia*) are found in the Old World. Hydrophyllaceae are a vegetatively and ecologically heterogeneous group, with a centre of diversity in south-western North America. They comprise 280–300 spp. in 17 genera. The majority of species are herbs and shrubs from seasonally arid habitats. Only a handful of taxa (e.g.

Hydrophyllum) are found as perennials in mesic forest habitats. The bulk of the genera are restricted to western North America, and only *Nama*, *Phacelia* and *Wigandia* extend into western South America. Ehretiaceae s.l. (500 spp. in 10 genera) and Heliotropiaceae (450 spp. in 4 genera) are larger and much more widespread groups than Hydrophyllaceae. Both families have their centres of diversity in seasonally arid subtropical habitats and both groups range into the humid tropics with only a handful of species. Ehretiaceae s.l. have their centre of diversity in the New World, but some isolated genera (*Hoplestigma*, *Coldenia*) and many species in the larger genera (e.g. *Cordia*, *Ehretia*) are restricted to the Old World. The family is almost exclusively woody, with some genera (*Coldenia*, *Tiquilia*) forming dwarf shrubs in dry, sandy habitats and the majority of taxa (*Cordia*, *Ehretia*, *Hoplestigma*) forming large, sometimes scandent shrubs or even tall trees. The only notable exception to this are the American former Lennoaceae (*Lennoa*, *Pholisma*), a group of parasitic herbs. Heliotropiaceae have all their basal lineages (*Myriopus*, *Ixorhea*) restricted to the Americas, but *Heliotropium* (incl. *Tournefortia*) has numerous species in the Old World tropics and subtropics. Unlike Ehretiaceae, Heliotropiaceae are vegetatively highly diverse and include a wide range of growth forms, ranging from ephemeral and perennial herbs (sometimes with stolons, rhizomes or root tubers), sub-shrubs and shrubs to small trees and lianas (Luebert et al. 2011a, 2011b). Superficially, Heliotropiaceae are similar to Boraginaceae, with mostly infundibuliform corollas with narrow tubes uncommon in the other groups. However, they differ profoundly in ovary (ovary entire, style apical) and corolla morphology (basal and faucal scales absent).

There are several characters which are widespread in Boraginales and have been used to characterize individual subgroups, but characters uniting the entire order or its two subgroups Boraginales I and Boraginales II so far have not been identified. Phytochemically, the presence of pyrrolizidine alkaloids (PAs) had been considered as characteristic for Boraginaceae s.str. and Heliotropiaceae (Hegnauer 1966) and they have since also been reported from some Ehretiaceae s. l. (Suri et al. 1980; Fu et al. 2002). However, PAs are absent from the Hydrophyllaceae studied so far and also from most Ehretiaceae, and detailed phytochemical data have not been published for either Codonaceae or Wellstediaceae. The available evidence thus fails to clarify the evolutionary history of PAs in Boraginales, but documents their absence in several groups. Also, PAs occur outside Boraginales and are not a unique substance class for the order.

Fruit morphology traditionally has been used to diagnose the constituent families, with four one-seeded mericarpids in the Boraginaceae s.l., multiple mericarpids in Lennoaceae and capsular fruits in Hydrophyllaceae. Recent data clearly retrieve clades with capsular fruits at the base of both Boraginales I (Codonaceae, Wellstediaceae) and Boraginales II (Hydrophyllaceae), implying the independent origin of fruits with four one-seeded mericarpids in Boraginales I and II (Weigend et al. 2014). A more detailed view further shows that the formation of "nutlets" is morphologically different between Boraginales I (Boraginaceae s.str. with nutlets separating from a persistent central column (= gynobase) and Boraginales II (mericarpids usually without a central column). This underscores the non-homology of the four one-seeded mericarpids found in these two clades. Moreover, representatives of Boraginales II with mericarpids (Ehretiaceae, Heliotropiaceae) have been shown to share a complex, derived pericarp anatomy with a multi-layered endocarp and specialized transfer cells in the seed coat which are absent in Hydrophyllaceae and Boraginales I (Diane et al. 2002).

Flower morphology provides relatively few informative characters at family level in Boraginales. Flowers are essentially tetracyclic, heterochlamydeous, synsepalous and sympetalous and with epipetalous filament insertion. The ovary is superior and consists of two united carpels with an apical (most) or gynobasic style (only Boraginaceae). The bulk of Boraginales have campanulate or infundibuliform corollas without conspicuous modifications of the corolla. Secondary increase of the number of perianth elements, irregular calyx aestivation, and an increased number of anthers (10 or more) are restricted to Lennoaceae, *Hoplestigma*, and some *Cordia* in Ehretiaceae s.l. Derivations from the pentamerous perianth are also found in Codonaceae (10–20), Lennoaceae (4–10) and Wellstediaceae (4). Boraginales I and

Hydrophyllaceae share a trend towards corolla tube modifications in the form of scale-shaped basal ingrowths, sometimes united into a ring of 5–10 distinct scales arching over the disk nectary. These basal scales are indistinct in Codonaceae and Wellstediaceae, variously present or absent in Hydrophyllaceae and absent in individual groups of Boraginaceae s.str. (e.g. Echiochileae), but very well developed in most groups of the latter family. Most Boraginaceae s.str. additionally have faucal scales, i.e. petal intrusions near the mouth of the tube, which are absent or indistinct in some taxa (e.g. hair tufts in Echiochileae) but very prominent in the majority of species (e.g. Boragineae, most Cynoglosseae with hypocrateriform flowers). The presence of some type of corolla tube invagination may thus be the ancestral character state for Boraginales, with several losses in Boraginales I (individual genera or species of Boraginaceae s.str.) and the crown group of Boraginales II.

An unequivocal fossil record of Boraginales is limited to Boraginaceae, Heliotropiaceae and Ehretiaceae s.l., and consists mainly of fruit remnants of Ehretiaceae s.l. from the Eocene and Miocene (Eocene of Europe: Chandler 1964; Gottschling et al. 2002; Miocene of Tasmania: Ettingshausen 1888), and of Boraginaceae s.str. from the Late Miocene (Texas to South Dakota in North America: Thomasson 1979, 1987). Putative fossil pollen has been recorded for Heliotropiaceae (Oligocene to Pliocene of Mexico and Puerto Rico: Graham and Jarzen 1969; Graham and Dilcher 1998; Muller 1981) and Boraginaceae s.str. (upper Miocene of Spain: Van Campo 1976). Dated phylogenies generally place the origin of the stem group of Boraginales (Boraginaceae s.l.) into the late Cretaceous (81 Ma: Wikström et al. 2004; 77.47 Ma: Magallón and Castillo 2009) or mid-Cretaceous (Bell et al. 2010). Gottschling et al. (2004) postulated an early diversification of Boraginales predating the late Cretaceous. Their data support the divergence of several modern genera in Ehretiaceae and Heliotropiaceae before 80–100 Ma, which would imply that the divergence of families of the order would have taken place no later than the mid-Cretaceous.

Different from the above circumscription of relationships (Weigend et al. 2014), Lennoaceae, clearly nested in Ehretiaceae s.l., are here treated as a separate family for practical reasons. Also, doubtfully monophyletic Hydrophyllaceae are retained in their classical circumscription until their relationships are better resolved based on a wider sampling.

Asterids II (Campanulids)

Conspectus of the Families of Aquifoliales

1. Inflorescences epiphyllous 2
– Inflorescences not epiphyllous 3
2. Flowers unisexual and plants dioecious. 1/4, E Asia **Helwingiaceae**
– Flowers bisexual. 1/4, New World tropics **Phyllonomaceae**
3. Gynoecium pluriloculate. 1/>600, cosmopolitan **Aquifoliaceae**
– Gynoecium uniloculate (probably pseudomonomerous) 4
4. Petals fused at least at base, stamens epipetalous (not *Citronella*). 5/ca. 45, Old World tropics and subtropics, rarely America (*Citronella*) **Cardiopteridaceae**
– Petals free, stamens not epipetalous. 12/ca. 90, mainly Old World tropics, rarely America (*Discophora*) **Stemonuraceae**

Aquifoliales, sister to all other Campanulids (Tank and Donoghue 2010), comprise Cardiopteridaceae, Stemonuraceae, Aquifoliaceae, Helwingiaceae and Phyllonomaceae. The order Aquifoliales in this circumscription was first resolved by Soltis et al. (2000) and has been recovered since in several other studies (Bremer et al. 2002; Winkworth et al. 2008; Tank and Donoghue 2010; Soltis et al. 2011). Before Soltis et al. (2000), Aquifoliales were considered to consist of Aquifoliaceae, Helwingiaceae and Phyllonomaceae only (APG 1998). No morphological synapomorphy of this clade is known. Aquifoliales consist of two subclades. The first contains Cardiopteridaceae and Stemonuraceae, of which all genera except *Cardiopteris* had previously been included in Icacinaceae. Detailed reallocation of genera of Icacinaceae to Cardiopteridaceae and the newly erected Stemonuraceae was first suggested by Kårehed (2001). Stevens (2001 onwards) lists entire leaf margins, adaxial carpel alone fertile and integument vascularized as possible synapomorphies of the two families, but toothed or lobed leaves can be found in Cardiopteridaceae, and gynoecium structure is well known only for *Cardiopteris* (Kong et al. 2014) and *Gonocaryum* (Fagerlind 1945). Tobe (2012) suggested that a pseudomonomerous gynoecium may represent a synapomorphy for the two families, but no detailed data on gynoecium structure appear to be available for Stemonuraceae. The second subclade contains Aquifoliaceae, Helwingiaceae and Phyllonomaceae, of which the latter two are sister to each other in more recent analyses (Tank and Donoghue 2010; Soltis et al. 2011). Following Stevens (2001 onwards), nodes 1:1, the presence of stipules, the absence of a style, and one ovule/carpel may be synapomorphies of this clade, but there are more than one ovule/carpel in Phyllonomaceae, and short styles have been described for most members of these three families. The sister group relationship between Helwingiaceae and Phyllonomaceae is well supported by the shared possession of epiphyllous inflorescences. The stem group age of the order has been estimated at ca. 100 Ma (Magallón and Castillo 2009).

Conspectus of the Families of Escalloniales

1. Glandular annual herbs. 1/1, western Australia **Eremosynaceae**
- Plants woody at least at base 2
2. Leaves opposite to subopposite, flowers 4-merous. 1/80, India, New Caledonia, Australia **Polyosmaceae**
- Leaves alternate, flowers 5-merous 3
3. Creeping subshrubs, entirely glabrous, leaf apices tridentate. 1/1, Southern Andes **Tribelaceae**
- Mostly shrubs or small trees, variously pubescent at least in flowers, leaf apices not tridentate. 4/45, S America, Australia and Tasmania, Réunion **Escalloniaceae**

Escalloniales, a clade first identified by Lundberg (2001) and confirmed by, for example, Winkworth et al. (2008; as *Escallonia* clade), Tank and Donoghue (2010; as Escalloniaceae) and Soltis et al. (2011; as Escalloniaceae), are here treated to comprise Eremosynaceae, Escalloniaceae, Polyosmaceae and Tribelaceae. These four largely south hemispherical families were all included in a broader Escalloniaceae in APG III (2009). Although the order is clearly monophyletic, relationships within it are still essentially unresolved (Lundberg 2001; Bremer et al. 2002; Winkworth et al. 2008; Tank and Donoghue 2010), so that treatment as four separate families does not contradict the molecular data available. Escalloniales are herbaceous to more often woody plants with mostly alternate and exstipulate leaves. Flowers are bisexual and actinomorphic with free petals. The mostly inferior ovary often consists of two carpels, ovules are mostly tenuinucellate, and fruits often are capsules, more rarely drupes. No clear synapomorphy uniting these four families has been identified. In historical systems, and before the advent of molecular analyses, they had been associated with Saxifragaceae in Engler's (1928) circumscription. The stem group age of the order (Escalloniaceae and Polyosmaceae) has been estimated at ca. 97 Ma (Magallón and Castillo 2009). Escalloniales are sister to all other Campanulids except Aquifoliales and Asterales (Tank and Donoghue 2010).

Conspectus of the Families of Bruniales

1. Ericoid shrubs or small trees, leaves alternate, stipulate, petals free. 6/81, S Africa **Bruniaceae**
- Trees or shrubs, leaves opposite, exstipulate, petals fused. 2/5 C to S America **Columelliaceae**

Bruniaceae, with six genera and 81 species mainly in the western and eastern Cape of South Africa, and Columelliaceae, with two genera and five species in the Andes of South America, are reasonably well supported as sister families (Winkworth et al. 2008; Tank and Donoghue 2010; Soltis et al. 2011) and have been treated as **Bruniales** by APG III (2009). The order is sister to Apiales, Paracryphiales and Dipsacales (Tank and Donoghue 2010). Apart from being woody and having bisexual flowers with a mostly half-inferior ovary of mostly two and rarely five (*Desfontainia*) fused carpels, there is little to connect the two families. Whereas Bruniaceae are ericoid shrubs without iridoids, have alternate and stipulate or exstipulate leaves, free petals, weakly crassinucellate ovaries and 1-seeded dry fruits, Columelliaceae, with *Columellia* and *Desfontainia* as rather divergent elements, are trees or shrubs with iridoids, have exstipulate opposite leaves, fused petals, tenuinucellate ovules and capsules or berries. Earlier analyses of 109 non-DNA characters and of a combined matrix of the non-DNA and *rbc*L sequence data both indicated Columelliaceae to be part of Dipsacales. Bruniaceae were sister to such an expanded order in the combined analysis and very closely related to the order in the other analysis (Backlund and Donoghue 1996). Stem group ages of Columelliaceae and Bruniaceae have been estimated at ca. 97 and 94 Ma respectively (Magallón and Castillo 2009). The sister group relationship of the South African Bruniaceae and the South American Columelliaceae may represent an ancient vicariance event.

Conspectus of the families of Dipsacales

1. Corolla usually actinomorphic; nectary of various types (on the ovary roof, adaxially on the petal bases, urceolate in the inflorescences, or a nectarine stigma) or absent; stylodia 3–5, short or absent. 5/175-210, mainly N hemisphere, centred in E Asia, also S America, Africa and SE Australia **Adoxaceae**
– Corolla usually zygomorphic; trichome nectary basally within the flower tube; style simple, elongated 2
2. Mainly herbs, sometimes subshrubs. Calyx variously modified; corolla with strong, mostly single transpetal veins 3
– Shrubs or lianas, rarely perennial herbs (*Triosteum*). Calyx not modified; corolla without transpetalar veins or more than one 4
3. Inflorescences forming dense involucrate heads, rarely thyrsoidal (*Triplostegia*); epicalyx present, simple or double (*Triplostegia*). 14/300 (+1/2 *Triplostegia*), mainly extratropical Old World, especially E Mediterranean and Irano-Turanian region, rare in C and S Africa and C and E Asia **Dipsacaceae** (incl. ***Triplostegia***)
– Inflorescences not forming involucrate heads, epicalyx absent. 5/400, mainly in the N hemisphere and mountainous regions of S America (especially Andes) **Valerianaceae**
4. Flowers in "verticillasters", involucel (epicalyx) formed by four fused bracteoles with spiny margin present, enclosing the achene. 2/12, Balkans to China, centred in E Asia **Morinaceae**
– Inflorescences different, epicalyx with spiny margin absent, but sometimes supernumerary bracts at the base of the inferior ovary present. 13/230, (+ 1/6 *Zabelia*), mainly temperate to boreal N hemisphere **Caprifoliaceae** (+ ***Zabelia*** inc. sed.)

Dipsacales, with the Adoxaceae s.l. (incl. *Sambucus* and *Viburnum*), Caprifoliaceae, Dipsacaceae, Morinaceae and Valerianaceae, were recognized as a "natural" group as early as the 19th century. The assignment of certain genera (e.g. *Heptacodium*, *Triplostegia*, *Viburnum*, *Zabelia*) to families and the circumscription of families varied through the years, and in some cases discussion continues until today. Dipsacales are ancestrally woody (mainly shrubs) although herbs are common, the opposite leaves lack true stipules, nodal anatomy is mainly 3:3 (also in the closely related Paracryphiales and Apiales) but also often 5:5 in the Morinaceae-Dipsacaceae-Valerianaceae clade (Neubauer 1978); flowers are arranged in mostly terminal, cymose or paniculate inflorescences which often are corymbiform, monotelic or frequently lack terminal flowers, corollas are tubular at the base with early sympetaly (excl. *Adoxa* with late sympetaly; Erbar 1994), there are (1–)3–5 (the unithecal split stamens of *Adoxa* should not be counted as two, see below) stamens which are adnate to the corolla, and ovaries are (semi)inferior with unitegmic, mostly tenuinucellate ovules. Route 1 secoiridoids are commonly present.

The present volume treats Dipsacales as containing the following families: Adoxaceae s.l. (incl. Sambucaceae and Viburnaceae), Caprifoliaceae (incl. Diervillaceae and Linnaeaceae), Dipsacaceae (incl. *Triplostegia*), Morinaceae and Valerianaceae. According to various recent studies, however, the phylogeny of the Dipsacales is Adoxaceae [Diervillaceae [Caprifoliaceae [Linnaeaceae [Morinaceae [Dipsacaceae + Valerianaceae]]]]]. For a monophyletic Caprifoliaceae s.l., Caprifoliaceae s.str. (incl. Linnaeaceae), Morinaceae, Dipsacaceae and Valerianaceae must be combined into one family as already suggested by Judd et al. (1994). Alternatively, Caprifolieae, Diervilleae and Linnaeeae could each be raised to family rank (with Adoxaceae, Morinaceae, Dipsacaceae and Valerianaceae as remaining families of the order) in order to have monophyletic taxa (Backlund and Pyck 1998), as recently done in Flora of China (2011). Judd et al. (1994) and later APG II (2003) and III (2009) argued in favour of the first option (combining several families in Caprifoliaceae) in order to escape problems with the monophyly of Caprifoliaceae in a more restricted circumscription and to keep the number of families in APG classifications at bay (Stevens 1998). Indeed, the position of *Zabelia* is still unclear (Landrein et al. 2012) and conceivably an additional family will be needed for this genus. Adoxaceae were not included in the expanded Caprifoliaceae by Landrein et al. (2012) although *Sambucus* and *Viburnum*, now members of Adoxaceae s.l., were included in Caprifoliaceae

s. str. in most earlier classifications because of their similarity to that family.

Whatever the best solution for the classification of Dipsacales, the order has 100% bootstrap support in Soltis et al. (2011), and infraordinal relationships are well understood for most parts. Stevens (2001 onwards) suggested pollen grains trinucleate when shed and a vascularised testa as possible synapomorphies of the order. Bud scales (reduced in herbaceous taxa), the typically multicellular club-shaped glandular hairs (Weberling 1977), and the position of the ovule (basically apotropous but turned lateral in the mature flower; Lagerberg 1909) may be other candidates for possible synapomorphies. Paracryphiales (Sphenostemonaceae + Paracryphiaceae) were resolved as a sister group of Dipsacales in the phylogenies of Tank and Donoghue (2010, strongly supported) and Soltis et al. (2011, no bootstrap support). Stevens (2001 onwards) suggested that the presence of true tracheids, serrate laminas and terminal inflorescences may represent synapomorphies of Dipsacales + Paracryphiales.

Of *Adoxa*, *Viburnum* and *Sambucus* as major genera of Adoxaceae s.l., the latter two were mostly included in Caprifoliaceae in the past (e.g. Fritsch 1892; Höck 1897; Bessey 1915; Wagenitz 1964; Fukuoka 1972; Takhtajan 1980; Cronquist 1981; Goldberg 1986). Among recent authors, only Thorne (1983) included both *Sambucus* and *Viburnum* in Adoxaceae. Earlier in the 19th century, however, a close relationship between *Adoxa*, *Sambucus* and *Viburnum* was frequently accepted: Röper (1834, cited in Meissner's Plantarum vascularium genera, pt. 2: p. 111, 1838), Koch (1837: 323) and Meissner (1838) favoured an inclusion of *Adoxa* in Caprifoliaceae close to or together with *Sambucus* and *Viburnum*. Also Payer (1857: p. 413) suggested a close relationship of *Sambucus* and *Adoxa*, and Röper (1860: p. 4) stated: "... one is almost inclined to name *Adoxa*, *Sambucus Moschatellina*" ["... man fast dazu neigen möchte, *Adoxa* als *Sambucus Moschatellina* zu bezeichnen"]. Ascherson (1864), following the system of Alexander Braun (1864), included *Adoxa*, *Sambucus* and *Viburnum* in Caprifoliaceae subfamily Sambucoideae (Adoxaceae s.l. of today) with actinomorphic rotate and campanulate flowers, distinct stylodia or sessile stigmas and one ovule per locule; also Hooker (1873) included these three genera in Caprifoliaceae tribe Sambuceae, and Eichler (1875) concluded that *Adoxa* belonged to Caprifoliaceae close to *Sambucus*.

While a close relationship of *Adoxa* to *Sambucus* and *Viburnum* (and of these three genera to other Dipsacales) was frequently accepted in the 19th century, this was only exceptionally the case up to the 1980s. This is somewhat surprising considering Lagerberg's (1909) detailed embryological studies in support of a close relationship of *Adoxa* and *Sambucus*. He explicitly rejected monogeneric Adoxaceae. The embryological studies by Moissl (1941) of Caprifoliaceae s.l. emphasized the similarities between Sambucus and Viburnum, especially the shared presence of a glandular tapetum and of strongly reduced sterile ovules. Otherwise, the distinctness of *Sambucus* and *Viburnum* from other Caprifoliaceae (in their traditional circumscription) was mostly acknowledged, but the Adoxaceae s.str. were rarely considered. This changed in the last decades of the 20th century when various new characters were used. Donoghue (1983), in a formal cladistic analysis using morphological characters, obtained evidence supporting Ascherson's (1864) classification. Wagenitz and Laing (1984) used the morphology of floral nectaries as argument in support of the view that *Viburnum* and *Sambucus* differ from Caprifoliaceae, and for the monophyly of Caprifoliaceae s.l. sensu APG III (= Dipsacales excl. Adoxaceae s.l.): the flowers of all genera investigated have a nectary in the floral tube composed of unicellular epidermal hairs and an underlying nectary tissue. In contrast, Adoxaceae s.l. have various types of nectaries: *Adoxa* has multicellular hairs on the corolla, *Viburnum* has a disk nectary on the roof of the inferior ovary with nectary slits, and in *Sambucus*, nuptial nectaries are normally absent (only *S. javanica* has urceolate nectaries in the inflorescence, probably modified from extrafloral nectaries common in the genus, and *S. ebulus* has a nectarine stigma; Vogel 1997).

When listing characters separating Adoxaceae s.l. from other Dipsacales, Wagenitz and Laing (1984) included the glandular tapetum as typical for the family (incorrectly scored as amoeboid by Judd et al. 1994), as well as a superficial cork cambium (*Sambucus* and *Viburnum*; Solereder 1899) and the armpalisade cells of the mesophyll. Corolla vascularization (Gustafsson 1995; Backlund and Donoghue 1996) and karyological data (Jianquan et al. 1999; Benko-Iseppon and Morawetz 2000) further supported the close relationship of *Adoxa* to *Viburnum* and *Sambucus*. Donoghue et al. (1992) and Judd et al. (1994), based on morphological cladistic analyses, Backlund and Bremer (1997), based on a taxonomically more restricted *rbc*L analysis, and Eriksson and Donoghue (1997), based on ITS sequences and morphology, found relationships in Adoxaceae as basically still accepted today: (*Viburnum* (*Adoxa*/*Sambucus*)). Interestingly, while evidence for removing *Viburnum* and *Sambucus* from Caprifoliaceae and linking them to *Adoxa* instead (as variously accepted earlier, see above) kept accumulating, it was considered unclear whether the *Adoxa*/*Sambucus*/*Viburnum* group and the Caprifoliaceae (plus Dipsacaceae, Morinaceae and Valerianaceae) were closely related at all (cf. Donoghue 1983; Wagenitz and Laing 1984; Wagenitz 1992).

The basic phylogeny of the Dipsacales (Adoxaceae [Diervillaceae [Caprifoliaceae [Linnaeaceae [Morinaceae [Dipsacaceae + Valerianaceae]]]]]) as recently found by Soltis et al. (2011) was first correctly recovered by Donoghue et al. (2003) based on a combined ITS and chloroplast DNA sequence dataset. In principle, this made possible the identification of synapomorphic characters of various subclades, but the relationships of certain problematic genera (especially *Heptacodium* and *Zabelia*; see Jacobs et al. 2011) and different options of character optimization influence the interpretation of various supposed synapomorphies. Judd et al. (1994) suggested reduced calyx lobes with one vascular trace, rotate corollas, semi-tectate-reticulate pollen grains (Reitsma and Reuves 1975; Böhnke-Gütlein and Weberling 1981), short stylodia and drupaceous fruits as possible synapomorphies for Adoxaceae.

For the Caprifoliaceae s.l. (sensu APG III), on the other hand, Judd et al. (1994) suggested long zygomorphic corolla tubes (only slightly zygomorphic in the basal Diervilleae and part of Caprifolieae), nectaries with unicellular hairs, large pollen grains, fully inferior ovaries with capitate stigmas on long styles, reduced carpel vasculature (Wilkinson 1949), and small chromosomes (Sax and Kribs 1930) as possibly apomorphic. The study of corolla vascularisation provided another character (Gustaffsson 1995; Backlund and Donoghue 1996): in nearly all Caprifoliaceae s.l. sensu APG, the main petal veins show lateral connections ("transpetalar veins") which are lacking in Adoxacaeae s.l. and all studied outgroups (Paracryphiales not included).

Above the basal Diervilleae/Diervillaceae, Caprifoliaceae s.l. sensu APG, i.e. Caprifoliaceae s.s. + Linnaeeae/Linnaeaceae + Morinaceae + Dipsacaceae + Valerianaceae, have no clear synapomorphies (Stevens 2001 onwards). As shown by Jacobs et al. (2011), three related fruit characters (fruit an achene, two sterile carpels, pericarp as outermost mechanical layer) either are synapomorphic for this clade with a reversal within Lonicereae or evolved in parallel in *Heptacodium* and at the base of the Linnaeaceae-Morinaceae-Dipsacaceae-Valerianaceae clade (see below). *Heptacodium* has also been suspected, however, of being of hybrid origin (Zhang et al. 2002; Jacobs et al. 2011).

Linnaeeae/Linnaeaceae+Morinaceae+Dipsacaceae+Valerianaceae (clade "Linnina" of Donoghue et al. 2001) typically have two of the three carpels aborted and achenes (cypselas; Bell et al. 2011). Supernumerary inflorescence bracts are possibly synapomorphic for this clade (Donoghue et al. 2001), or alternatively for the less inclusive Morinaceae/Dipsacaceae/Valerianaceae clade (Zhang et al. 2003). Four didynamous stamens were added by Stevens (2001 onwards). Morinaceae/Dipsacaceae/Valerianaceae (clade "Valerina" of Donoghue et al. 2001) share various morphological characters—e.g. herbaceous habit (and bud scales absent) with basal rosette and a taproot, vascular flank-bridge between lateral bundles in stem (see Neubauer 1979), nodes 5:5, epicalyx (modified or absent in Valerianaceae), enlarged embryo, and the occurrence of monoterpenoids and alkaloids (Donoghue et al. 2001; Stevens 2001 onwards; Zhang et al. 2003; Jacobs et al. 2011). In Dipsacaceae/Valerianaceae young shoots have an endodermis (Solereder 1899), but data for Morinaceae seem to be lacking.

Inflorescences are basically cymose, mostly thyrses, often corymbiform, but also complex panicles, sometimes pseudo-umbellate in Adoxaceae s.l., and condensed into capitula (e.g. *Adoxa*, *Heptacodium*, Dipsacaceae), among other modifications (Troll and Weberling 1966; Landrein and Prenner 2013). In Adoxaceae s.l., terminal flowers are normally present, in Caprifoliaceae s. l. sensu APG typically absent (truncated monotelic to polytelic), with the exception of the basal Diervilleae/Diervillaceae (Wagenitz and Laing 1984; Landrein and Prenner 2013). In the *Adoxa* clade, elongated thyrses (*Sinadoxa*) or raceme-like inflorescences (*Tetradoxa*) are probably plesiomorphic (Donoghue et al. 2001). The peculiar cube-like inflorescence of *Adoxa*, a condensed depauperate thyrse (Weberling 1966 cited in Wagenitz and Laing 1984), is derived from these. Concerning the capituliform inflorescences of Dipsacaceae, however, Hofmann and Göttmann (1990) pointed out that their internal structure could easily be described as racemose with the lateral flowers lacking prophylls. Only the phylogenetic position of the family would favour the interpretation of an evolutionary reduction to their terminal flowers of the dichasial paracladia of an original thyrse. Supernumerary bracts below the flowers (*Abelia* spp., *Dipelta*, *Linnaea*, *Patrinia*; in some spp. forming a 3-lobed wing in fruit, as in some *Valeriana* spp.) and the homologous development of an epicalyx (*Morina*, Dipsacaceae incl. *Triplostegia*) are conspicuous features in many genera of Caprifoliaceae s.l. sensu APG (Hofmann and Göttmann 1990).

The evolution of reproductive characters has been analyzed based on the mostly well-resolved and stable phylogeny of the order (Donoghue et al. 2002; Jacobs et al. 2010a, 2010b). Numbers of stamens and petals were found to have evolved independently in the usually zygomorphic flowers of Caprifoliaceae s.l. sensu APG clade. They mostly covary in the sister clade Adoxaceae with typically actinomorphic flowers. In Caprifoliaceae s.l. the number of stamens varies from one to five and the number of petals from three and five. With very few exceptions (cf. *Valeriana* p.p.: former genera *Belonanthus* and *Phyllactis*) flowers with 1–4 stamens have five petals (e.g. *Fedia*, *Centranthus*, *Morina*, *Triplostegia*, *Valerianella*). In Adoxaceae, however, the 3–5 corolla lobes correlate with 3–5 stamens, although the stamens are split more or less deeply into two halves in *Adoxa*, *Tetradoxa* and *Sinadoxa*, resulting in apparently twice the number of unithecal stamens (e.g. Yang et al. 2011). The ancestral condition of the gynoecium in the order probably is 3-carpellate as present in *Viburnum*, *Sambucus* p.p. and part of the Caprifoliaceae s.l. sensu APG. In the Adoxaceae an increase of carpel number to up to five (in *Adoxa*, *Tetradoxa* and *Sambucus* p. p.) as well as a reduction (*Sinadoxa*: bicarpellate, unilocular and uniovular) can be observed. Of the three locules in *Viburnum*, only one has a single ovule and the other locules are sterile, while in *Sambucus* all 3–5 locules are fertile. In the Caprifoliaceae s.l. sensu APG, a bicarpellate ovary with many ovules is present in the basal Diervilleae (Diervillaceae) clade (an apomorphic condition). In the following Caprifolieae clade, some genera lack compressed sterile locules (*Leycesteria* has five or more and *Lonicera* 2–5 locules), while the others (*Heptacodium*, *Symphoricarpos*, *Triosteum*) may have one or two sterile locules with aborted ovules. Two sterile locules are also typical for the clade Linnaeaceae+Morinaceae+Dipsacaceae+Valerianaceae, and in Dipsacaceae the ovary is pseudomonomerous. Thus, there is a tendency in Dipsacales to a reduction in number of fertile locules, which took place in parallel in *Viburnum* and one or more times in the Caprifoliaceae sensu APG. While in the Caprifoliaceae s.l. sensu APG the progressive reduction of ovules is well visible in the extant genera, in the ovary of Adoxaceae s.l. only vestigial archesporial tissue as a remnant of aborted ovules above the fertile ovule can be found (see Moissl 1941; Wilkinson 1949; Erbar 1994). Drupes with one or more pyrenes are typical for Adoxaceae s.l., and the Caprifoliaceae-Valerianaceae clade is more variable. The basal Diervilleae (Diervillaceae) have septicidal capsules with wind-dispersed seeds, which is unique in Dipsacales. Drupes with pyrenes occur only in *Heptacodium*, *Symphoricarpos* and *Triosteum* in tribe Caprifolieae. The other two genera of this tribe, *Leycesteria* and *Lonicera*, have baccate fruits with three to several seeds, probably a derived condition. In the clade comprising the rest of Caprifoliaceae s.l. sensu APG (Linnaeeae, Morinaceae, Dipsacaceae and Valerianaceae), the fruits are achenes with 1(2) seeds. This is correlated with the reduction of two of the locules (see above). Jacobs et al. (2010a, 2011) stated that evolution took place from small embryos and copious endosperm in the basal clades of Caprifoliaceae s.l. sensu APG to larger embryos in Morinaceae, Dipsacaceae and basal Valerianaceae (*Patrinia*, *Nardostachys*). The endpoint of this evolution is the condition seen in most

Valerianaceae where endosperm is basically absent and the embryos fill the entire seed. Furthermore, the embryo in the basal clades is protected by the sclerified seed coat (Diervilleae) or endocarp (Caprifolieae), while in the derived clades the protection is transferred to the pericarp or the epicalyx (Dipsacaceae).

The stem group age of the order is probably Mid Cretaceous, and the basal split into two clades occurred in the Tertiary, but estimates vary considerably (see Stevens 2001 onwards). Bremer et al. (2004) estimated the stem group age of Dipsacales as 111 Ma, and the crown group age as 101 Ma. Bell and Donoghue (2005a) suggested that major lineages with considerable morphological differences only appeared in the Eocene and more recently in the Miocene (especially in the herbaceous clades Valerianaceae and Dipsaceae), and that further diversification in these families took place even more recently in the Mediterranean region in Dipsacaceae and in alpine habitats of western S America in Valerianaceae. Disjunctions between the Old and New Worlds are common in the order. According to the study of Adoxaceae and Valerianaceae by Moore and Donoghue (2007), species diversification correlated less with morphological innovations but rather with shifts to mountainous habitats. The evolution from shrubs to herbs occurred three or more times in the order: in the clade of Morinaceae/Dipsacaceae/Valerianaceae, in *Adoxa* and related genera, and in *Sambucus* where several species are rhizomatous perennial herbs, a condition which evolved possibly more than once in this genus (cf. phylogeny in Eriksson and Donoghue 1997). Jacobs et al. (2010b) speculated that ovary position, seed characters and fruit colour suggest paedomorphosis in the evolution from *Sambucus*-like ancestors to *Adoxa*.

Backlund (1996), using Bremer's ancestral area and DIVA analyses, concluded that the origin of the Apiales-Dipsacales clade is located in SE Asia. Donoghue et al. (2003) confirmed a probable origin in eastern Asia, with the plants originally inhabiting the understorey of temperate forests (but see Clement and Donoghue 2011). From there, the order spread into both hemispheres of the Old and New World. Herbaceous lineages occupy alpine habitats with considerable diversification especially in western South America (Bell and Donoghue 2005a, 2005b). The sister group of Dipsacales, the Paracryphiales, is centred in the tropical western Pacific region (New Guinea, Queensland and New Caledonia). The past distribution of members of this order is disputed due to ambiguous fossils from Sweden possibly close to *Quintinia* (see Stevens 2001 onwards); Friis et al. (2013) even question that *Quintinia* belongs into Paracryphiales in spite of the data of Soltis et al. (2011).

Conspectus of the families of Paracryphiales

1. Fruit a berry with 1 or 2 seeds, filaments very short to absent. 1/9, New Caledonia, N Queensland, New Guinea, Moluccas, Celebes **Sphenostemonaceae**
– Fruit a capsule with several seeds, filaments distinct, sometimes short. 2/25, New Caledonia, tropical Australia, New Guinea, Philippines, Vanuatu, New Zealand **Paracryphiaceae**

Paracryphiaceae with two genera and ca. 25 species in New Guinea, the Philippines, Vanuatu, New Caledonia, tropical Australia and New Zealand, and Sphenostemonaceae with one genus of nine species in New Guinea, the Moluccas, Celebes, New Caledonia and northern Queensland, were recovered as a well-supported clade by Tank and Donoghue (2010) and Soltis et al. (2011) and are regarded as the only family (Paracryphiaceae) of the **Paracryphiales** by APG III (2009). Such treatment is justified by the fact that *Paracryphia* seems to be more closely related to *Sphenostemon* than to *Quintinia* as second genus of Paracryphiaceae as treated here. Both families contain woody plants with alternate and exstipulate leaves and, as far as presently known, free petals. Apart from that, few similarities and no obvious synapomorphies have been identified. However, in both families both bi- and unisexual flowers with or without petals can be found, and an almost verticillate leaf arrangement has been described for some species of both. Considering their essentially identical distribution range and the characters discussed, their close relationships to each other appears plausible although it had never been suggested before. The age of both families has been estimated at ca. 97 Ma (Magallón and Castillo 2009), and the order is sister to Dipsacales (Tank and Donoghue 2010).

Bibliography

Albach, D.C., Soltis, P.S., Soltis, D.E., Olmstead, R.G. 2001. Phylogenetic analysis of Asterids based on sequences of four genes. Ann. Missouri Bot. Gard. 88: 163–212. 2001.

APG 1998. An ordinal classification for the families of flowering plants. Ann. Missouri Bot. Gard. 85: 531–553.

APG II. 2003. An update of the Angiosperm Phylogeny Group classification for the orders and families of flowering plants: APG II. Bot. J. Linn. Soc. 141: 399–436.

APG III. 2009. An update of the Angiosperm Phylogeny Group classification for the orders and families of flowering plants: APG III. Bot. J. Linn. Soc. 161: 105–121.

Ascherson, P. 1864. Flora der Provinz Brandenburg, der Altmark und des Herzogthums Magdeburg. Zum Gebrauche in Schulen und auf Excursionen. 1. Abteilung. Berlin: August Hirschwald, pp. 265–267.

Backlund, A. 1996. Phylogeny of the Dipsacales. Acta Universitatis Upsaliensis, Uppsala.

Backlund, A., Bremer, B. 1997. Phylogeny of the Asteridae s. str. based on rbcL sequences, with particular reference to the Dipsacales. Plant Syst. Evol. 207: 225–254.

Backlund, A., Donoghue, M.J. 1996. Morphology and phylogeny of the order Dipsacales. In: Backlund, A., Phylogeny of the Dipsacales. Acta Universitatis Upsaliensis, Uppsala, 55 pp + 18 pp apps.

Backlund, A., Pyck, N. 1998. Diervillaceae and Linnaeaceae, two new families of caprifolioids. Taxon 47: 657–661.

Baillon, H.E. 1877. Organogénie florale des *Garrya*. Compt. Rend. Assoc. Franç. Avancem. Sci. 1877: 561–566. Reprinted as Adansonia 12: 262–269. 1879.

Baillon, H.E. 1880. Rubiacées-Dipsacacées. In: Baillon, H., Histoire des plantes 7. Paris: Hachette, pp. 257–546.

Baillon, H. 1891: Histoire des plantes. Paris: Hachette. 476 pp.

Bartling, F.G. 1830. Ordines naturales plantarum eorumque characteres et affinitates; adjecta generum enumeratione. Göttingen: Dieterich.

Bell, C.D., Donoghue, M.J. 2005a. Dating the Dipsacales: comparing models, genes, and evolutionary implications. Amer. J. Bot. 92: 284–296.

—. 2005b. Phylogeny and biogeography of Valerianaceae (Dipsacales) with special reference to the South American valerians. Organisms Diversity Evolution 5: 147–159.

Bell, C.D., Edwards, E.J., Kim, S.-T., Donoghue, M.J. 2001. Dipsacales phylogeny based on chloroplast DNA sequences. Harvard Pap. Bot. 6: 481–499.

Bell, C.D., Soltis, D.E., Soltis, P.S. 2010. The age and diversification of the angiosperms re-revisited. Amer. J. Bot. 97: 1296–1303.

Benko-Iseppon, A.M., Morawetz, W. 2000. Viburnales: cytological features and a new circumscription. Taxon 49: 5–16.

Bentham, G., Hooker, J.D. 1883. Genera plantarum 2/2. London: A. Black. 1279 pp.

Berchtold, F., Presl, J.S. 1820. Rad 32. Chrastawcowité (Dypsacaceae Jussieu). In: Přirozenosti Rostlin. Prague: Jos. Krause, p. 255.

Bessey, C.E. 1915. The phylogenetic taxonomy of flowering plants. Ann. Missouri Bot. Gard. 2: 108–164.

Böhnke-Gütlein, E, Weberling, F. 1981. Palynologische Untersuchungen an Caprifoliaceae I. Sambuceae, Viburneae und Diervilleae. Acad. Wiss. Abh. Math.-Naturwiss. Kl. [Trop. Subtrop. Pflanzenwelt] 34: 131–189.

Bremer, K., Backlund, A., Sennblad, B., Swenson, U., Andreasen, K., Hjertson, M., Lundberg, J., Backlund, M., Bremer, B. 2001. A phylogenetic analaysis of 100 + genera and 50+ families of euasterids based on morphological and molecular data with notes on possible higher level morphological synapomorphies. Plant Syst. Evol. 229: 137–169.

Bremer, B., Bremer, K., Heidari, N., Erixon, P., Olmstead, R.G., Anderberg, A.A., Källersjö, M., Barkhordarian, E. 2002. Phylogenetics of asterids based on 3 coding and 3 non-coding chloroplast DNA markers and the utility of non-coding DNA at higher taxonomic levels. Mol. Phylogenet. Evol. 24: 274–301.

Bremer, K., Friis, E.M., Bremer, B. 2004. Molecular phylogenetic dating of asterid flowering plants shows Early Cretaceous diversification. Syst. Biol. 53: 496–505.

Brummitt, R. 2011. Valid publication of the family name Pteleocarpaceae. Kew Bull. 66: 1–3.

Chandler, M.E.J. 1964. The lower tertiary floras of Southern England 4. London: British Museum. 150 pp.

Chase, M.W., Soltis, D.E., Olmstead, R.G., Morgan, D., Les, D.H., Mishler, B.D., Duvall, M.R., Price, R.A., Hills, H.G., Qiu, Y.-L., Kron, K.A., Rettig, J.H., Conti, E., Palmer, J.D., Manhart, J.R., Sytsma, KJ., Michaels, H.J., Kress, W.J., Karol, K.G., Clark, W.D., Hedrén, M., Gaut, B.S., Jansen, R.K., Kim, K.-J., Wimpee, C.F., Smith, J.F., Furnier, G.R., Strauss, S. H., Xiang, Q.-Y., Plunkett, G.M., Soltis, P.S., Swensen, S.M., Williams, S.E., Gadek, P.A., Quinn, C.J., Eguiarte, L.E., Golenberg, E., Learn, G.H. Jr., Graham, S.W., Barrett, S.C.H., Dayanandan, S., Albert, V.A. 1993. Phylogenetics of seed plants: An analysis of nucleotide sequences from the plastid gene *rbc*L. Ann. Missouri Bot. Gard. 80: 528–580.

Clement, W.L., Donoghue, M.J. 2011. Dissolution of *Viburnum* section *Megalotinus* (Adoxaceae) of southeast Asia and its implications for morphological evolution and biogeography. Int. J. Plant Sci. 172: 559–573.

Cosner, M.E., Jansen, R.K., Lammers, T.G. 1994. Phylogenetic relationships in the Campanulales based on *rbcL* sequences. Plant Syst. Evol. 190: 79–95.

Cronquist, A. 1981. An integrated system of classification of flowering plants. New York: Columbia University Press.

Dahlgren, R. 1983. General aspects of angiosperm evolution and macrosystematics. Nord. J. Bot. 3: 119–149.

Dahlgren, R., Jensen, S.R., Nielsen, B.J. 1977. Seedling morphology and iridoid occurrence in *Montinia caryophyllacea* (Montiniaceae). Bot. Not. 130: 329–332.

Diane, N., Hilger, H.H., Gottschling, M. 2002. Transfercells in the seeds of Boraginales. J. Bot. Linn. Soc. 140: 155–164.

Di Fulvio de Basso, T.E. 1990. Endospermogenesis y taxonomía de la familia Hydrophyllaceae y su relación con las demas Gamopetalas. Acad. Nac. Cienc. Exact. Físic. Natur. Buenos Aires 5: 73–82.

Donoghue, M.J. 1983. The phylogenetic relationships of *Viburnum*. In: Platnick, N.I., Funk, V.A. (eds.) Advances in Cladistics 2. New York: Columbia Univ. Press, pp. 143–211.

Donoghue, M.J., Eriksson, T., Reeves, P.A., Olmstead, R. G. 2001. Phylogeny and phylogenetic taxonomy of Dipsacales, with special reference to *Sinadoxa* and *Tetradoxa* (Adoxaceae). Harvard Pap. Bot. 6: 459–479.

Donoghue, M.J., Bell, C.D., Winkworth, R.C. 2003. The evolution of reproductive characters in Dipsacales. Int. J. Plant Sci. 164 (5 Suppl): S453–S464.

Eich, E. 2008. Solanaceae and Convolvulaceae: Secondary metabolites. Berlin: Springer.

Eichinger, A. 1907. Vergleichende Entwickelungsgeschichte von *Adoxa* und *Chrysosplenium*. Mitt. bayr. bot. Ges., Bd. II, Nr. 5: 65–74, Nr. 6: 81–93.

Eichler, A.W. 1875. Blüthendiagramme. Erster Theil. Leipzig: W. Engelmann.

Engler, A. 1897. Nachtrag zu den Adoxaceae. In: Nat. Pflanzenfam. ed.1, Vol. 4, 4. Leipzig: W. Engelmann, p. 190.

Engler, A. 1903. Syllabus der Pflanzenfamilien, ed. 3. Berlin: Gebrüder Borntrager.

Engler, A. 1928. Saxifragaceae. In: Natürl. Pflanzenfam. ed. 2, 18a: 74–226. Leipzig: W. Engelmann.

Erbar, C. 1995. On the floral development of *Sphenoclea zeylanica* (Sphenocleaceae, Campanulales) - SEM investigations on herbarium material. Bot. Jahrb. Syst. 117: 469–493.

Eriksson, T., Donoghue, M.J. 1997. Phylogenetic relationships of *Sambucus* and *Adoxa* (Adoxoideae, Adoxaceae) based on nuclear ribosomal ITS sequences and preliminary morphological data. Syst. Bot. 22: 555–573.

Ettingshausen, C. von. 1888. Contributions to the Tertiary flora of Australia. Mem. Geol. Surv. New South Wales 2: 1–189.

Fagerlind, F. 1945. Bau des Gynöceums, der Samenanlage und des Embryosackes bei einigen Repräsentanten der Familie Icacinaceae. Sven. Bot. Tidskr. 39: 346–364.

Ferguson, D.M. 1999. Phylogenetic analysis and relationships in Hydrophyllaceae based on *ndhF* sequence data. Syst. Bot. 23: 253–268.

Friis, E.M., Pedersen, K.R., Endress, P.K. 2013. Floral structure of extant *Quintinia* (Paracryphiales, Campanulids) compared with the Late Cretaceous *Silvianthemum* and *Bertilanthus*. Int. J. Plant Sci. 174: 647–664.

Fritsch, K.1892. Caprifoliaceae. In: Engler, A., Prantl, K. (eds.) Die natürlichen Pflanzenfamilien, ed. 1, Vol. 4, 4. Leipzig: W. Engelmann, pp. 156–169.

Fu, P.P., Yang, Y.C., Xia, Q., Chou, M.C., Cui, Y.Y., Lin, G. 2002. Pyrrolizidine alkaloids-tumorigenic components in Chinese herbal medicines and dietary supplements. J. Food Drug Analysis 10: 198–211.

Fukuoka, N. 1972. Taxonomic study of the Caprifoliaceae. Mem. Fac. Sci. Kyoto Univ., Ser. Biol. 6: 15–58.

Goldberg, A. 1986. Classification, evolution and phylogeny of the families of dicotyledons. Smithsonian Contr. Bot. 58: 1–314.

Gottschling, M., Mai, D.H., Hilger, H.H. 2002. The systematic position of *Ehretia* fossils (Ehretiaceae, Boraginales) from the European Tertiary and implications for character evolution. Rev Palaeobot. Palynol. 121: 149–156.

Gottschling, M., Diane, N., Hilger, H.H., Weigend, M. 2004. Testing hypotheses on disjunctions present in the Primarily Woody Boraginales: Ehretiaceae, Cordiaceae, and Heliotropiaceae, inferred from ITS1 sequence data. Int. J. Plant Sci. 165: 123–135.

Graham, A., Dilcher, D.L. 1998. Studies in neotropical paleobotany. XII. A palynoflora from the Pliocene Rio Banano Formation of Costa Rica and the Neogene vegetation of Mesoamerica. Amer. J. Bot. 85: 1426–1438.

Graham, A., Jarzen, D.M. 1969. Studies in neotropical paleobotany. I. The Oligocene communities of Puerto Rico. Ann. Missouri Bot. Gard. 56: 308–357.

Greilhuber, J. 1979. C-Band distribution, DNA-content and base composition in *Adoxa moschatellina* (Adoxaceae), a plant with cold-sensitive chromosome segments. Plant Syst. Evol. 13: 243–259.

Gürke, M. 1893. Borraginaceae. In: Natürl. Pflanzenfam. IV, 3a. Leipzig: Engelmann, pp. 71–131

Gustafsson, M.H.G. 1995. Petal venation in the Asterales and related orders. Bot. J. Linnean Soc. 118: 1–18.

Hallier, H. 1912. L'origine et le système phylétique des angiospermes exposées à l'aide de leur arbre généalogique. Arch. Neêrlandaise, 2nd ser., B. (Sci. nat.) 1: 146–234.

Harris, E.M. 1999. Capitula in the Asteridae: A widespread and varied phenomenon. Bot. Rev. 65: 348–369.

Hegnauer, R. 1966. Chemotaxonomie der Pflanzen: Vol. 4: Dicotyledoneae: Daphniphyllaceae bis Lythraceae. Basel: Birkhäuser. 556 pp.

Hirai, N., Sakashita, S., Sano, T., Inoue, T., Ohigashi, H., Premasthira, C., Asakawa, Y., Harada, J., Fujii, Y. 2000. Allelochemicals of the tropical weed *Sphenoclea zeylanica*. Phytochemistry 55: 131–140.

Höck, F. 1897. Dipsacaceae. In: Engler, A., Prantl, K. (eds.) Die natürlichen Pflanzenfamilien, ed. 1, IV, 4–5. Leipzig: Wilhelm Engelmann. pp. 183–189.

Hofmann, U., Göttmann, J. 1990. *Morina* L. und *Triplostegia* Wall. ex DC. im Vergleich mit Valerianaceae und Dipsacaceae. Bot. Jahrb. Syst. 111: 499–553.

Hooker, J.D. 1873. Caprifoliaceae. In: Bentham and Hooker, Genera Plantarum, Vol. 2,1. London: Lovell Reeve, Williams & Norgate, pp. 1–7.

Howarth, D.G., Donoghue, M.J. 2009. Duplications and expression of *DIVARICATA*-like genes in Dipsacales. Mol. Biol. Evol. 26: 1245–1258.

Howarth, D.G., Martins, T., Chimney, L., Donoghue, M.J. 2011. Diversification of *CYCLOIDEA* expression in the evolution of bilateral flower symmetry in Caprifoliaceae and *Lonicera* (Dipsacales). Ann. Bot. 107: 1521–1532.

Hufford, L. 1992. Rosidae and their relationships to other nonmagnoliid dicotyledons: A phylogenetic analysis using morphological and chemical data. Ann. Missouri Bot. Gard. 79: 218–248.

Jacobs, B., Donoghue, M.J., Bouman, F., Huysmans, S., Smets, E. 2008. Evolution and phylogenetic importance of endocarp and seed characters in *Viburnum* (Adoxaceae). Int. J. Plant Sci. 169: 409–431.

Jacobs, B., Bell, C., Smets, E. 2010a. Fruits and seeds of the *Valeriana* clade (Dipsacales): Diversity and evolution. Int. J. Plant Sci. 171: 421–434.

Jacobs, B., Huysmans, S., Smets, E. 2010b. Evolution and systematic value of fruit and seed characters in Adoxaceae (Dipsacales).Taxon 59: 850–866.

Jacobs, B., Pyck, N., Smets, E. 2010c. Phylogeny of the Linnaea clade: Are *Abelia* and *Zabelia* closely related? Mol. Phylogenet. Evol. 57: 741–752.

Jacobs, B., Geuten, K., Pyck, N., Huysmans, S., Jansen, S., Smets, E. 2011. Unraveling the phylogeny of *Heptacodium* and *Zabelia* (Caprifoliaceae): An interdisciplinary approach. Syst. Bot. 36: 231–252.

Janssens, S.B., Knox, E.B., Huysmans, S., Smets, E.F., Merckx, V.S.F.T. 2009. Rapid radiation of *Impatiens* (Balsaminaceae) during Pliocene and Pleistocene: Results of a global climate change. Mol. Phylogenet. Evol. 52: 806–824.

Jianquan, L., Tingnong, H., Guoying, Z., Anmin, L. 1999. Karyomorphology of *Sinadoxa* (Adoxaceae)and its systematic significance. Caryologia 52 (3–4): 159–164.

Judd, W.S., Olmstead, R.G. 2004. A survey of tricolpate (eudicot) phylogenetic relationships. Amer. J. Bot. 91: 1627–1644.

Judd, W.S., Sanders, R.W., Donoghue, M.J. 1994. Angiosperm family pairs: preliminary phylogenetic analyses. Harvard Pap. Bot. 5: 1–51.

Judd, W.S., Campbell, C.S., Kellogg, E.A., Stevens, P.F., Donoghue, M.J. 2008. Plant systematics: A phylogenetic approach, ed. 3. Sunderland, Mass.: Sinauer.

Jussieu, A.-L. de. 1789. Genera plantarum. Paris: Herissant & Barrois.

Kårehed, J. 2001. Multiple origin of the tropical forest tree family Icacinaceae. Amer. J. Bot. 88: 2259–2274.

Kong, D.-R., Schori, M., Lu, S.-G., Li, L., Peng, H. 2014. Floral development of *Cardiopteris*, with emphasis on gynoecial structure and ovular morphology. J. Syst. Evol. 52: 629–642.

Lagerberg, T. 1909. Studien über die Entwicklungsgeschichte und systematische Stellung von Adoxa Moschatellina L. Kongl. Svenska Vetenskapsacad. Handl. ser. 2, 44(4): 1–86.

Landrein, S., Prenner, G. 2013. Unequal twins? Inflorescence evolution in the twinflower tribe Linnaeeae (Caprifoliaceae s.l.). Int. J. Plant Sci. 174: 200–233.

Landrein, S., Prenner, G., Chase, M.W., Clarkson, J.J. 2012. *Abelia* and relatives: Phylogenetics of Linnaeeae (Dipsacales-Caprifoliaceae s.l.) and a new interpretation of their inflorescence morphology. Bot. J. Linn. Soc. 169: 692–713.

Luebert, F., Brokamp, G., Wen, J., Weigend, M., Hilger, H. H. 2011a. Phylogenetic relationships and morphological diversity in Neotropical *Heliotropium* (Heliotropiaceae). Taxon 60(3): 663–680.

Luebert, F., Hilger, H.H., Weigend, M. 2011b. Diversification in the Andes: Age and origins of South American *Heliotropium* lineages (Heliotropiaceae, Boraginales). Mol. Phylogenet. Evol. 61(1): 90–102.

Lundberg, J. 2001. Phylogenetic studies in the Euasterids II with particular reference to Asterales and Escalloniaceae. Doctoral Dissertation. Uppsala, Sweden: Uppsala University.

Magallón, S., Castillo, A. 2009. Angiosperm diversification through time. Amer. J. Bot. 96: 349–365.

Meissner, C.F. 1836–1843. Plantarum vascularium genera, 2 parts. Leipzig: Weidmann.

Moissl, E. 1941. Vergleichende embryologische Studien über die Familie der Caprifoliaceae. Österr. Bot. Zeitschr. 90: 153–212.

Moore, M.J., Hassan, N., Gitzendanner, M.A., Bruenn, R. A., Croley, M., Vandeventer, A., Horn, J.W., Dhingra, A., Brockington, S.F., Latvis, M., Ramdial, J., Alexandre, R., Piedrahita, A., Xi, Z., Davis, C.C., Soltis, P. S., Soltis, D.E. 2011. Phylogenetic analysis of the plastid inverted repeat for 244 species: Insights into deeper-level angiosperm relationships from a long, slowly evolving sequence region. Int. J. Plant Sci. 172: 541–558.

Muller, J. 1981. Fossil pollen records of extant angiosperms. Bot. Rev. 47: 1–142.

Neubauer, H.F. 1978 (publ. 1979). On nodal anatomy and petiole vascularization of some Valerianaceae and Dipsacaceae. Phytomorphology 28: 431–436.

Novák, T. 1904. Über den Blütenbau von *Adoxa moschatellina*. Österr. Bot. Zeitschr. 54: 1–7.

Olmstead, R. 2013. Phylogeny and biogeography of Solanaceae, Verbenaceae, and Bignoniaceae: A comparison of continental and intercontinental diversification patterns. Bot. J. Linnean Soc. 171: 80–102.

Olmstead, R.G., Jansen, R.K., Kim, K.-J., Wagstaff, S.J. 2000. The phylogeny of the Asteridae s.l. based on chloroplast ndhF sequences. Mol. Phylogenet. Evol. 16: 96–112.

Payer, J.-B. 1857. Traité d'organogénie comparée de la fleur, 2 vols. Paris: Victor Masson.

Prosperi, C.H., Cocucci, A.E. 1979. Importancia taxonómica de la calosa de los tubos polínicos en Tubiflorae. Kurtziana 12–13: 75–81.

Qiu, Y.-L., Li, L., Wang, B., Xue, J.-Y., Hendry, T.A., Li, R.-Q., Brown, J.W., Liu, Y., Hudson, G.T., Chen, Z.-D. 2010. Angiosperm phylogeny inferred from sequences of four mitochondrial genes. J. Syst. Evol. 48: 391–425.

Refulio-Rodríguez, N.F., Olmstead, R. 2014. Phylogeny of Lamiidae. Amer. J. Bot. 101: 287–299.

Reitsma, T.J., Reuvers, A.A.M.L. 1975. Adoxaceae. Rev. Palaeobot. Palynol. 19: NEPF 4: 71–74.

Riedl, H., 1997. Boraginaceae. In: Kalkman, C., Kirkup, D. W., Nooteboom, H.P., Stevens, P.F., de Wilde, W.J.J. O. (eds.) Flora Malesiana, Series I - Seed Plants. Leiden: Rijksherbarium / Hortus Botanicus, pp. 43–144.

Roeper, J.A.C. 1860. Vorgefasste botanische Meinungen. Rostock: Stiller'sche Hofbuchhandlung.

Sax, K, Kribs, D.A. 1930. Chromosomes and phylogeny in Caprifoliaceae. J. Arnold Arbor. 11: 147–153.

Solereder, H. 1899. Systematische Anatomie der Dicotyledonen. Stuttgart: Ferdinand Enke.

Soltis, D.E., Soltis, P.S., Chase, M.W., Mort, M.E., Albach, D.C., Zanis, M., Savolainen, V., Hahn, W.H., Hoot, S. B., Fay, M.F., Axtell, M., Swensen, S.M., Prince, L.M., Kress, W.J., Nixon, K.C., Farris, J.S. 2010. Angiosperm phylogeny inferred from 18S rDNA, *rbcL*, and *atpB* sequences. Bot. J. Linn. Soc. 133: 381–461.

Soltis, D.E., Smith, S.A., Cellinese, N., Wurdack, K.J., Tank, D.C., Brockington, S.F., Refulio-Rodriguez, N. F., Walker, J.B., Moore, M.J., Carlsward, B.S., Bell, C. D., Latvis, M., Crawley, S., Black, C., Diouf, D., Xi, Z., Rushworth, C.A., Gitzendanner, M.A., Sytsma, K.J., Qiu, Y.L., Hilu, K.W., Davis, C.C., Sanderson, M.J., Beaman, R.S., Olmstead, R.G., Judd, W.S., Donoghue, M.J., Soltis, P.S. 2011. Angiosperm phylogeny: 17 genes, 640 taxa. Amer. J. Bot. 98: 704–730.

Stevens, P.F. 1998. What kind of classification should the practicing taxonomist use to be saved? In: Dransfield, J., Coode, M.J.E., Simpson, D.A. (eds.) Plant Diversity in Malesia III. Royal Botanic Gardens, Kew, pp. 295–319.

Stevens, P.F. 2001 onwards. Angiosperm phylogeny web site, version 12, July 2012 (and more or less continuously updated since). http://www.mobot.org/MOBOT/research/APweb (accessed April 2013).

Suri, O.P., Jamwal, R.S., Suri, K.A., Atal, C.K. 1980. Ehretinine, a novel pyrrolizidine alkaloid from *Ehretia aspera*. Phytochemistry 19: 1273–1274.

Takhtajan, A. 1980. Outline of the classification of flowering plants (Magnoliophyta). Bot. Rev. (Lancaster) 46: 225–359.

Takhtajan, A. 2009. Flowering plants, 2nd edn. Springer Science and Business Media.

Tang, Y.-C., Li, L.-Q. 1994. The phytogeography of Caprifoliaceae s.str. with its implications for understanding eastern Asiatic flora. Acta Phytotax. Sinica 32: 197–218.

Tank, D.C., Donoghue, M. J. 2010. Phylogeny and phylogenetic nomenclature of the Campanulidae based on an expanded sample of genes and taxa. Syst. Bot. 35: 425–441.

Thomasson, J.R. 1977. Late cenozoic grasses, borages and hackberries from southwestern Nebraska. Contributions to Geology, University of Wyoming 16: 39–43.

Thomasson, J.R. 1979. Late Cenozoic grasses and other angiosperms from Kansas, Nebraska, and Colorado: Biostratigraphy and relationships to living taxa. Kansas Geol. Surv. Bull. 128: 1–68.

Thorne, R.F. 1983. Proposed new realignments in the Angiosperms. Nordic J. Bot. 3: 85–117.

Thulin, M., Johansson, A.N.B. 1996. Taxonomy and biogeography of the anomalous genus *Wellstedia*. In: Van der Maesen, L.J.G., van der Burgt, X.M., van Medenbach de Rooy, J.M. (eds.) The Biodiversity of African Plants - Proc. XIVth AETFAT Congr. Dordrecht, Boston, London: Kluwer, pp. 73–86.

Tobe, H. 2012. Floral structure of *Cardiopteris* (Cardiopteridaceae) with special emphasis on the gynoecium: systematic and evolutionary implications. J. Plant Res. 125: 361–369.

Troll, W., Weberling, F. 1966. Die Infloreszenzen der Caprifoliaceen und ihre systematische Bedeutung. Abh. Akad. Wiss. Lit. Mainz. Math. Naturw. Kl. 4: 455–605.

Van Campo, E. 1976. La flore sporopollénique du gisement Miocène terminal de Venta del Moro (Espagne). Ph.D. Thesis, Univ. Montpellier.

Vogel, S. 1997. Remarkable nectaries: structure, ecology, organophyletic perspectives. I. Substitutive nectaries. Flora 192: 305–333.

Wagenitz, G. 1964. 10. Reihe Dipsacales (Rubiales p. p.). In: Melchior, H. (ed.) A. Engler's Syllabus der Pflanzenfamilien, ed. 12, 2. Berlin: Wallich, pp. 472–478.

Wagenitz, G. 1992. The Asteridae: evolution of a concept and its present status. Ann. Missouri Bot. Gard. 79: 209–217.

Wagenitz, G., Laing, D. 1984. Die Nektarien der Dipsacales und ihre systematische Bedeutung. Bot. Jahrb. Syst. 104: 483–507.

Wagstaff, S.J., Martinsson, K., Swenson, U. 2000. Divergence estimates of *Tetrachondra hamiltonii* and *T. patagonica* (Tetrachondraceae) and their implications for austral biogeography. New Zeal. J. Bot 38: 587–596.

Weberling, F. 1957. Morphologische Untersuchungen zur Systematik der Caprifoliaceen. Abh. Akad. Wiss. Lit. Mainz. Math. Naturw. Kl. 1957: 1–50.

Weberling, F. 1977. Vergleichende und entwicklungsgeschichtliche Untersuchungen über die Haarformen der Dipsacales. Beitr. Biol. Pflanzen 53(1): 61–89.

Weberling, F., Hildenbrand, M. 1982. Zur Tapetumentwicklung bei *Triosteum* L., *Leycesteria* Wall. und *Kolkwitzia* Graebn. (*Caprifoliaceae*). Beitr. Biol. Pflanzen 57: 481–486.

Weberling, F., Hildenbrand, M. 1986. Weitere Untersuchungen zur Tapetumentwicklung der *Caprifoliaceae*. Beitr. Biol. Pflanzen 61: 3–20.

Weigend, M., Hilger, H.H. 2010. Codonaceae - a newly required family name in Boraginales. Phytotaxa 10: 26–30.

Weigend, M., Luebert, F., Selvi, F., Brokamp, G., Hilger, H. H. 2013. Multiple origins for Hounds tongues (*Cynoglossum* L.) and Navel seeds (*Omphalodes* Mill.) - the phylogeny of the borage family (Boraginaceae s.str.). Mol. Phylogenet. Evol. 68: 604–618.

Weigend, M., Luebert, F., Gottschling, M., Couvreur, T.L. P., Hilger, H.H., Miller, J. 2014. From capsules to nutlets—phylogenetic relationships in the Boraginales. Cladistics 30(5): 508–518.

Wikström, N., Savolainen, V., Chase, M.W. 2004. Angiosperm divergence times: Congruence and incongruence between fossils and sequence divergence estimates. In: Donoghue, P.C.J., Smith, M.P. (eds.) Telling the evolutionary time: molecular clocks and the fossil record. Boca Raton: CRC Press, pp. 142–165.

Wilkinson, A.M. 1949. Floral anatomy and morphology of *Triosteum* and of the Caprifoliaceae in general. Amer. J. Bot. 36: 481–489.

Winkworth, R.C., Bell, C.D., Donoghue, M.J. 2008. Mitochondrial sequence data and Dipsacales phylogeny: Mixed models, partitioned Bayesian analyses, and model selection. Mol. Phylogenet. Evol. 46: 830–843.

Wydler, H. 1850. Über *Adoxa moschatellina*. Flora 7: 433–437.

Xu, L., Lu, L., Li, D.-Z., Wang, H. 2011. Evolution of pollen in Dipsacales. Plant Divers. Resources 33: 249–259.

Yang, Q., Hong, D.Y., Malécot, V., Boufford, D.E. 2011. Adoxaceae. In: C.Y. Wu et al. (eds.) Flora of China 19: 570–614.

Zhang, W.H., Chen, Z.D., Li, J.-H., Chen, H.B., Tang, Y.C. 2002. Phylogeny of the Dipsacales s.l. based on chloroplast *trn*L-F and *ndh*F sequences. Mol. Phylogenet. Evol. 26: 176–189.

Adoxaceae

Adoxaceae E. Mey., Preuss. Pfl.-Gatt. 198 (1839) (Adoxeae), nom. cons.
Viburnaceae Raf. (1820) (*Viburnidia*).
Sambucaceae Batsch ex Borkh. (1797) (*Sambuci*).

A. Backlund and V. Bittrich

Perennial herbs, small to moderately sized shrubs, or small trees, buds sometimes perulate; plants glabrous or with indumentum of unicellular or multicellular stellate (rarely lepidote) hairs; extrafloral nectaries sometimes present in various parts of the plants. Leaves opposite (very rarely 3-verticillate), pinnately veined and petiolate, simple, ternate, biternate, pinnatisect or pinnate, rarely bipinnate; leaf margin entire or, more often, serrate; pseudostipules occasionally present. Flowers in sometimes complex monotelic inflorescences, forming corymbs (sometimes some flowers sterile and showy), pseudo-umbels, heads, racemes, or few-flowered cymes in interrupted spikes. Flowers (with very few exceptions) bisexual, ± actinomorphic, semi-epigynous to epigynous; sepals 2–5, ± connate , lobes small, 1-traced, persistent; petals 3–5(6), connate, with corolla tube sometimes very short, lobes imbricate or valvate; stamens antesepalous, adnate to the corolla and isomerous with the petals, though sometimes split to the filament base or nearly so, giving the appearance of twice the number of unithecal stamens; anthers basifixed or dorsifixed, dithecal, tetrasporangiate; ovary syncarpous with 2–5 carpels, uni- or plurilocular, with some locules occasionally sterile; stylodia short or absent; stigma capitate or lobed, then lobes isomerous with locule number; one fertile ovule per locule, sometimes additionally some vestigial ovules present, anatropous, bitegmic, tenui- or weakly crassinucellate; nectary of different types or absent. Fruit a drupe with one single pyrene in *Viburnum* or with few separable pyrenes in Adoxeae and *Sambucus*. Endosperm oily, copious, cellular.

Adoxaceae comprise five genera and approximately 175 to 210 species, distributed mainly across the northern hemisphere, extending southward to South America (*Sambucus* and *Viburnum*), Africa and Australia.

Vegetative Morphology. Adoxaceae vary in size from less than 20 cm high perennial herbs in *Adoxa* to trees of *Viburnum* reaching 20 m. Many of the herbaceous species have ± branched subterranean rhizomes. Winter buds are perulate in *Sambucus*, perulate (those of the outer pair sometimes fused) or eperulate in *Viburnum*. Accessory buds occur in at least some species of *Sambucus* (*S. nigra*, *S. racemosa*). Extrafloral nectaries are sometimes present in *Sambucus* in various parts of the plant—e.g. branchlets, petioles, and branches of the inflorescence. Pseudostipules of the leaves or leaflets can be glands (e.g. *S. javanica*, *S. adnata*), and in *Viburnum* spp. glands are found on the petiole apex (e.g. *V. opulus*) or midvein base (e.g. *V. inopinatum*). Leaves are deciduous or evergreen, nearly always opposite (but mostly 3-verticillate in, for example, *Viburnum triphyllum*) and often with petioles basally fusing; in Adoxeae, basal leaves are rosulate. Venation pinnate, also in the compound leaves of subfamily Adoxoideae. Leaf margin entire or, more often, serrate or dentate. Pseudostipules occasionally present in *Sambucus* and *Viburnum* (Weberling 1957).

Vegetative Anatomy and Ultrastructure. Rhizome, when present (also in some shrubby *Sambucus* spp.), with a dorsal and a ventral flattened collateral vascular bundle separated by pith. Lenticels prominent in most species. Cork cambium, when present, initially superficial (Solereder 1899). Most but not all species of *Sambucus* have tanniniferous tubes (coenocytes) in the

J.W. Kadereit and V. Bittrich (eds.), *Flowering Plants. Eudicots*, The Families and Genera of Vascular Plants 14,
DOI 10.1007/978-3-319-28534-4_2, © Springer International Publishing Switzerland 2016

internodes (cf. Solereder 1899), which can be more than 30 cm long (Zobel 1986). Detailed studies of wood anatomy have been presented by Carlquist (1992). Wood partially storied, parenchyma scanty, paratracheal. Perforation plates are simple in Sambuceae, scalariform in Viburneae and Adoxeae. Both helical thickenings and vesturing are absent in the vessels, but all studied species have sieve element plastids of the S-type. Xylem with libriform fibres. A starch sheath has been reported from the parenchymatous tissue. Nodes uni-, tri- or penta-lacunar. In *Sambucus nigra*, besides 3 gaps corresponding to 3 normal leaf traces for each leaf, on each side of the node additionally one gap produces one strand bifurcating shortly after leaving the stem vascular ring. Each of the resulting strands produces a larger strand entering the leaf and a weaker strand entering the pseudostipule (Neubauer 1977). Secondary thickening, when present, takes place with a conventional cambial ring. Calcium oxalate crystals are absent in Adoxeae, but abundant in Sambuceae and Viburneae as crystal sand or druses. Leaf mesophyll includes cubical and arm-palisade cells and 4–5 additional layers of spongy mesophyll, but lacks mucilage cells. Plants glabrous in Adoxeae, with ± dense indumentum of unicellular, seriate or multicellular stellate, rarely lepidote hairs in Sambuceae and Viburneae respectively. Glandular hairs with uniseriate stalks and pluricellular heads have been found in *Sambucus* and *Viburnum* (Solereder 1899). Stomata are anomocytic and present mainly on the abaxial side of the leaves. Hydathodes are present in Adoxeae (e.g. Sturm 1910). Domatia were observed in the axils of the nerves on the lower leaf surface of *Viburnum clemensae* (erroneously described as "glandular pitted" by Kern and van Steenis 1951, cf. Puff et al. 1997), and different types of leaf domatia were described from leaves of *Viburnum* spp. from Korea (O'Dowd and Pemberton 1994).

Inflorescence. Inflorescences are monotelic, sometimes condensed and ± pseudanthial, or frequently partly pseudo-umbellate, corymbose or elongated panicles or thyrses, mostly large and repeatedly branching (Troll and Weberling 1966); flowers rarely in racemes or interrupted spikes of few-flowered cymes. The peripheral

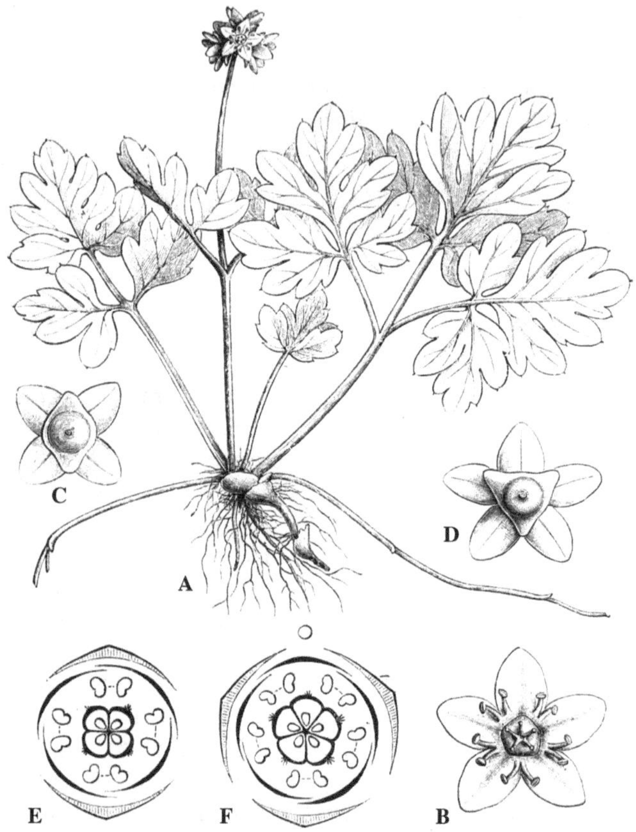

Fig. 1. Adoxaceae. *Adoxa moschatellina*. **A** Flowering plant. **B** Open lateral flower. **C** Terminal flower from below. **D** Lateral flower from below. **E** Floral diagram of a terminal flower. **F** Floral diagram of a lateral flower (from Fritsch 1891, Fig. 58 pr. p., modified)

flowers in the corymbose inflorescences of some *Viburnum* species (e.g. *V. spicatum*) are sterile with an enlarged showy corolla.

Flower Structure and Floral Anatomy (Figs. 1 and 2). Flowers are usually bisexual. *Viburnum clemensae*, endemic to Borneo and basalmost branch in the phylogeny of *Viburnum* (Clement and Donoghue (2011), is dioecious; the staminodes of the female flowers lack anthers, the sterile ovary of the male flowers have the roof nectary typical for the genus (Puff et al. 1997). *Sambucus australis* has dimorphic flowers and is probably functionally dioecious (cf. Bolli 1994). Floral anatomy of *Sambucus* and *Viburnum* was studied in detail by Wilkinson (1948a, 1948b) and anatomy and ontogeny of *Adoxa* by Erbar (1994). The lateral flowers in *Adoxa moschatellina* and those of *Sambucus ebulus* are unusual in that,

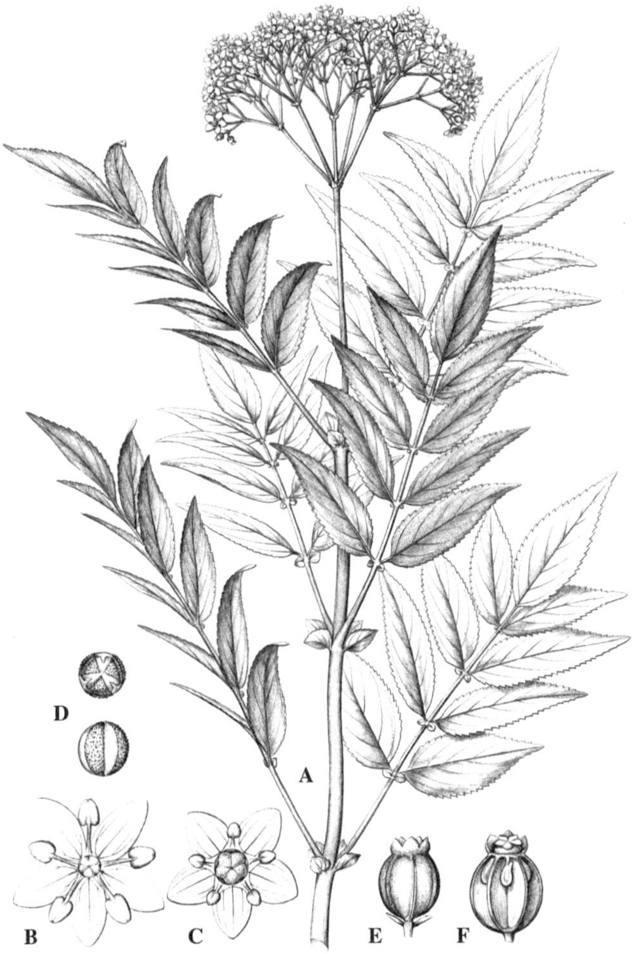

Fig. 2. Adoxaceae. *Sambucus australis*. **A** Flowering branch. **B** Male flower. **C** Female flower. **D** Pollen grains. **E** Ovary with calyx. **F** Ovary with opened locules (from Müller 1885, t. 99 pr. p., modified)

different from the vast majority of higher eudicots, the second sepal appears in median abaxial position and the odd petal in median adaxial position ("lobelioid position", Payer 1857; Eichler 1875; Erbar 1994). The sepals of *Adoxa* (normally 2 in the terminal, and 3 in the lateral flowers) were considered by Eichler (1875) and Novák (1904) to be involucral bracts. The ontogenetic studies by Erbar (1994) showed, however, that two additional sepals are regularly initiated and normally aborted. This supports the interpretation of Payer (1857) and others about the presence of a calyx and suggests that the flowers of *Adoxa* are isomerous throughout (Erbar 1994). *Viburnum* and *Sambucus* show early sympetaly as typical for Dipsacales in general (Erbar 1991; Reidt and Leins 1994). *Adoxa*, however, lacks an early ring primordium and the slow lateral extension of the petals bases sometimes does not even lead to a complete ring structure outside the stamens; the border between the corolla tube and petal-stamen tube is obscure (Erbar 1994; Roels and Smets 1994). Nuptial nectaries are present in Adoxeae and *Viburnum*, but very rare in *Sambucus*. In Adoxeae the petal lobes are each furnished with a nectary at the base of the upper side, composed of cushion-like groups of multicellular sessile or short-stalked clavate trichome glands. These were first described by Novák (1904) and interpreted by him (and recently again by Deyuan and Boufford 2011) as of staminodial nature, but by Lagerberg (Lagerberg 1904 in Eichinger 1907) as emergences as they lack any vascularization. In *Viburnum* an ovary roof disk nectary with nectary slits occurs. In *Sambucus javanica* yellow urceolate nectaries occur in the corymbose inflorescences. These were generally regarded as transformed flowers (e.g. Schwerin 1909), but a detailed study by Vogel (1997) showed them to be homologous to the morphologically and anatomically similar extrafloral nectaries which are very common in *Sambucus* (see above). *Sambucus ebulus* has a nectarine stigma concealed between the stamens, which attracts pollinating bees and wasps (Vogel 1997). In *Adoxa*, *Sinadoxa* and *Tetradoxa*, each stamen is divided either to the base (*Adoxa*) or only in the upper part (*Tetradoxa*), resulting in two monothecous anthers. Anthers are dorsifixed or basifixed to ventrifixed in *Sambucus*, non-versatile, opening with longitudinal slits, extrorse in Adoxoideae with the exception of *Sambucus javanica*, which has latrorse anthers, and generally introrse in *Viburnum*. Gynoecium syncarpous, of (2)3–5 carpels, (1)3–5-locular, semi-inferior to inferior, with one pendant fertile ovule per locule. *Viburnum* and *Sambucus* differ from the Caprifoliaceae by the presence of distinct lateral and dorsal carpellary traces (Wilkinson 1949). Placentation is axile or axile-apical in *Adoxa* and *Sambucus*. The ovary of *Viburnum* is pseudomonomerous, for the most part 1-locular and contains one fertile ovule. According to Wilkinson (1948b), in *Viburnum* placentation of the one fertile ovule is parietal and the chamber in which it develops belongs to one of the sterile carpels. Vestigial ovules or archaesporial tissue are present in the two aborted carpels; their number varies according to the species and the degree of atrophism may vary within the same ovary (Wilkinson 1948b). In *Sambucus*, Wilkinson (1948a) found vestigial archaesporial tissue at the base of the three-lobed style canal in parietal position, reminding of the situation in *Viburnum*. Similarly, vestigial archaesporial tissue was observed in *Adoxa* in the same position as described by Wilkinson (1948a), protruding into the basal lumen of the style base (Lagerberg 1909; Erbar 1994). The vestigial archaespore is always found above the fertile ovules. Styles are separate in Adoxeae, short and mostly connate in *Sambucus* spp. and *Viburnum* or absent in *Sambucus* spp. Stigma capitate or 3–5-lobed, dry or rarely moist and papillate.

Embryology. Endothecium with fibrous thickenings, microsporogenesis simultaneous and the initial microspore tetrads tetrahedral. Anther wall with one initial layer and a binucleate, glandular (secretory) tapetum (Lagerberg 1909; Moissl 1941) of the common dicot-type, which is reorganized into a false periplasmodium. Pollen grains 3-nucleate. Embryo sac development follows the Adoxa type in Adoxoideae and Polygonum or Allium type in *Viburnum*. Endosperm formation cellular, endosperm copious and oily with a smooth surface. Embryogenesis is of the Asterad type, resulting in a small but well-developed embryo with two cotyledons without chlorophyll detected. Ovules are anatropous, pendent, basically apotropous (e.g. Eichinger 1907), but micropyle and raphe become orientated laterally at anthesis (*Adoxa*, *Sambucus*, Lagerberg 1909; Fritsch 1891, Fig. 51E: *Sambucus australis*), unitegmic, tenuinucellate or in *Viburnum* weakly crassinucellate (Moissl 1941). The presence of vestigial archaespores, probably last remnants of sterile ovules, in addition to the fertile ovules appears to be a common feature of all studied members of the family (Moissl 1941; Erbar 1994). The ovules have an integumentary tapetum (Eichinger 1907).

Pollen Morphology. Adoxaceae pollen is highly uniform. The pollen grains are prolate or widely ellipsoidal in equatorial view and obtusely triangular in polar view, and size ranges approximately from 20–35 μm along the polar axis and 15–25 μm along the equatorial axis, giving an R/E-ratio of ≈ 1.2. The exine is thin, isonexinous, with distinct columellae and a reticulate, semitectate surface. Apertures are tricolporate or trizono-colporate, and arranged along the obtuse angles. Ectoapertures are comparably long and broad, often deeply sunken and occasionally with an equatorial bridge. Colpus membranes are granular with a fastigium in Adoxeae, but psilate and lacking a fastigium in Sambuceae and Viburneae (Erdtman 1952; Reitsma and Reuvers 1975; Donoghue 1985).

Karyology. The base chromosome number probably is $x = 9$, but great variation occurs (cf. generic descriptions). A detailed cytological study of *Viburnum* by Egolf (1962) reported considerable infrageneric variation in chromosome numbers with $2n = 16$ and $2n = 18$ being most common. A few species showed infraspecific variation—e.g. *V. carlesii* with $2n = 18$, 20 and 22, and in other species different ploidy levels were found. In a karyological analysis of Dipsacales, several characteristic features of Adoxaceae were described. These include coarsely reticulate interphase nuclei, continuous condensing behaviour and cold induced regions (CIRs) of the chromosomes, all of which clearly set Adoxaceae apart from the clade of the other Dipsacales families, i. e. Caprifoliaceae s.s., Diervillaceae, Linnaeaceae, Morinaceae, Dipsacaceae and Valerianaceae (Benko-Iseppon and Morawetz 1993, 2000).

Fruits and Seeds. Important contributions to the knowledge of fruit and seed characters in the family have been provided by Jacobs et al. (2008, 2010). All members of the family have drupaceous fruits with one (*Viburnum*) to several pyrenes (*Adoxa*, *Sambucus*). The fruits are black or coloured in *Viburnum* and *Sambucus*, but green in *Adoxa*. Pyrene size, outer form, and shape and grooving in cross-section can be useful to distinguish between species. The endocarp has an inner layer of fibres and an outer layer of sclereids. An antiraphe is present in the seeds of *Adoxa* and *Viburnum*, but not in *Sambucus*. The mature seed coat of *Adoxa* and *Sambucus* shows a single or few layers of flattened cells. Most *Viburnum* species have a parenchymatous exotesta with columnar or rectangular cells while flattened cells are rare. In some species one or two additional layers of probably meso- and endotestal origin are present. The endosperm of *Viburnum* differs from that of *Adoxa* and *Viburnum* in that it is ruminated and the peripheral cells are anatomically different from the others. The embryo of *Sambucus* fills most of the seed length, while in *Adoxa* and *Viburnum* only about one quarter. A possible autapomorphy of *Viburnum* is the presence of small amorphous crystals in the endosperm. The weakly sclerified endocarp sclereids of *Adoxa* and the small seeds were considered by Jacobs et al. (2010) as paedomorphic tendencies in the evolution of these small herbs.

Pollination, Reproductive Systems and Dispersal. Adoxaceae are generally entomophilous and are pollinated mainly by generalists of Diptera, moths or Coleoptera; some taxa are apparently wind-pollinated (Charlebois et al. 2010). Sterile flowers in the inflorescences of some *Viburnum* spp. were shown to enhance the reproductive success (Jin et al. 2010). *Adoxa moschatellina* showed low seed set in Britain (Holmes 2005), and vegetative propagation occurs by the rhizomes. Peruzzi and Passalacqua (2006) report propagation by bulbils in sterile *A. moschatellina* plants from triploid populations. The flowers of Adoxaceae are generally bisexual. Exceptions are some functionally unisexual individuals of *Sambucus nigra* var. *peruviana*, and *Sambucus australis* from South America and *Viburnum clemensae*, endemic to Borneo, which are both dioecious (Bolli 1994; Puff et al. 1997). Dispersal of *Sambucus* and *Viburnum* is generally by birds. For *Adoxa moschatellina* Müller-Schneider (1967 reported that during fruit development the peduncle curves down to the soil and that the fruits are eaten by snails. Seeds from snail faeces showed a 75 % germination success.

Phytochemistry. The following compound classes have been reported: secoiridoids, iridoids from groups 6 (all taxa studied), 10 (*Adoxa* and *Viburnum*), 3, 5 and 7 (*Viburnum*), with groups assigned following Jensen et al. (1975), tannins and cyanogenic glucosides (especially sambunigrin) derived from phenylalanin (*Sambucus*, but not in all species, Bolli 1994), amines, amenthoflavones (*Viburnum*), several alkaloids (Willaman 1970), and a number of flavonols, mainly quercetin, kaempferol, polyphenols, and acidic compounds including caffeic, p-coumaric, sinapic, urosolic, and ferulic acids (Bate-Smith 1961; Harborne and Williams 1971). Arbutin, proanthocyanins, cyanidin, leuco-anthocyanins and trihydroxy constituents such as delphinidin, leucodelphinidin, ellagic acid and myricetin have been reported as absent.

Distribution and Habitats. Adoxaceae occur mainly in temperate, less often subtropical climates, typically in forests, with centres of distribution in eastern Asia and the mountains of Latin America. For *Viburnum*, Clement and Donoghue (2011) suggested that multiple shifts between temperate and subtropical regions occurred in Asia, and that it is possible that the genus originated and initially diversified in subtropical mountains of SE Asia.

Subdivision of the Family. Parts of Adoxaceae as circumscribed here have been included in various families including Caprifoliaceae s.lat., Araliaceae, Cornaceae, Hydrangeaceae and Saxifragaceae in earlier classifications. Based on a number of phylogenetic analyses (e.g. Shi-you 1987; Donoghue et al. 1992, 2001; Backlund 1996; Backlund and Bremer 1997; APG 1998), Adoxaceae can be divided into subfamily Adoxoideae, comprising tribes Adoxeae with *Adoxa*, *Sinadoxa* and *Tetradoxa* and Sambuceae with the single genus *Sambucus*, and subfamily Opuloideae with the monogeneric tribe Viburneae containing *Viburnum*.

Affinities. Earlier suggestions regarding the classification of members of Adoxaceae followed two lines of reasoning. On the one hand, *Adoxa* has often been regarded as closely related to Saxifragaceae (e.g. Hutchinson 1973) or Araliaceae (Candolle 1830; Hallier 1905), and to Caprifoliaceae ‘through’ *Viburnum* and *Sambucus* (e.g. Baillon 1880; Hara 1983; Takhtajan 1987; Dahlgren 1991). On the other hand, *Viburnum*, and often *Sambucus*, have been regarded as related to Hydrangeaceae or Cornaceae (e.g. Jussieu 1789; Candolle 1830), Caprifoliaceae (e.g. Agardh 1858; Hooker 1873; Baillon 1880; Hutchinson 1973; Takhtajan 1997), Araliaceae, to all of these families (Dahlgren 1983), or even to the Rubiaceae (Lamarck 1778). In recent years a wider circumscription of Adoxaceae, including *Adoxa*,

Sambucus, Sinadoxa, Tetradoxa and *Viburnum*, has been suggested by, for example, Donoghue et al. (1992). The monophyly of so widely circumscribed Adoxaceae is now strongly supported. Former hypotheses of a relationship to Hydrangeaceae and Cornaceae, however, have been rejected on both morphological, palynological, molecular and karyo-morphological grounds (e.g. Olmstead et al. 1993; Gustafsson et al. 1996; Benko-Iseppon and Morawetz 2000). The extended discussion of the anatomical interpretation of the flowers of Adoxeae had resulted in suggesting affinities ranging from *Chrysosplenium* (Saxifragaceae s.str.) to *Panax* (Araliaceae) and Cornaceae. These suggestions have been firmly refuted (e.g. Eichinger 1907; Ronse DeCraene et al. 1998). Information obtained from anatomical, phytochemical and even serological studies had been suggestive of a position closer to Apiaceae (e.g. Metcalfe and Chalk 1950; Hegnauer 1969; Hillebrand and Fairbrothers 1970). The position of Adoxaceae as sister group to the rest of the Dipsacales is now well established.

PARASITES. There are records of four well-documented parasitic fungi from *Adoxa*: *Melanotaenium adoxae* and *Puccinia argentata, P. albescens* and *P. adoxae*. The smut *Melanotaenium adoxae* belongs to a genus occurring in approximately 10 families of eudicotyledons, most of which are essentially herbaceous. The rust *Puccinia argentata* shows only the aecial stage in *Adoxa* while other stages are found on species of *Impatiens*, and the two last species, *Puccinia albescens* and *P. adoxae*, are considered to be closely related to *Puccinia argentata* (Savile 1979). The *Viburnum* leaf beetle (*Pyrrhalta viburni*, larvae and adults) can cause severe defoliation in cultivated viburnums.

PALAEOBOTANY. A large number of palaeobotanical records are known for the family. The oldest fossil known has been assigned to *Sambucus*, and was found in England in strata from the lower Tertiary, estimated to be 23–65 Ma old (Chandler 1961). Additional collections found in strata from the Tertiary include *Viburnum weberi, Sambucus celtifolia, S. princeps* and *S. pulchella*, all from the Rhineland (Weyland 1938; Kilpper 1969). *Viburnum hungaricum* from Hungary (Hably and Szakály 1989) and *Viburnum protofurcatum* from Japan (Uemura 1988) were dated to the Miocene. These generic assignments indicate that the two subfamilies Adoxoideae and Opuloideae possibly were differentiated already in the lower Tertiary.

ECONOMIC IMPORTANCE AND CONSERVATION. Many members of *Viburnum* and *Sambucus* are important ornamental plants—e.g. varieties of *V. carlesii, V. opulus* and *S. nigra* and *S. racemosa*. The flowers of *S. nigra* are used for flavouring soft drinks (Elderflower cordial), liqueurs and pancakes, the fruits of some *Sambucus* spp. are edible (e.g. *S. nigra*). Several *Sambucus* species have been used medicinally.

In particular in tribe Adoxeae, there are several comparatively short-lived taxa with very limited distribution areas which could be highly sensitive to habitat disturbance.

KEY TO THE GENERA

1. Herbs less than 30 cm high. Stamens 3–5, each divided more or less deeply into monothecate half-stamens 2
– Shrubs or small trees, rarely robust herbs >50 cm high. Stamens 5, undivided 4
2. Inflorescence forming small heads of 5–9 flowers (temperate Eurasia) **1. *Adoxa***
– Inflorescence elongated (China) 3
3. Stamens cleft to base. Styles absent, ovary 1-locular **3. *Sinadoxa***
– Stamens cleft apically, T-shaped. Styles present, ovary 4-locular **2. *Tetradoxa***
4. Leaves compound. Ovary with 3–5 locules **4. *Sambucus***
– Leaves simple. Ovary 1-locular **5. *Viburnum***

I. SUBFAM. ADOXOIDEAE Syme (1865) ('Adoxeae').

1. TRIBE ADOXEAE Dumort. (1827).

1. ***Adoxa*** L. Fig. 1

Adoxa L., Sp. Pl. (1753); Deyuan & Boufford, Flora of China 19: 613–614 (2011). Peruzzi & Passalacqua, Nord. J. Bot. 24: 249–256 (2006), karyol., infrasp. tax.

Delicate perennial herbs <20 cm high, musk-scented; subterranean rhizome with scales. Basal rosulate leaves ternate or biternate, cauline leaves opposite or solitary, entire, 3-partite or ternate, leaf margins coarsely serrate. Inflorescences terminal, compact, cube-like, 3–9-flowered. Corolla rotate, greenish-yellow, terminal flowers usually with 2 sepals and 4 corolla lobes, lateral flowers usually with 3(4) sepals and 5(6) corolla lobes; stamens and petals forming a basal tube; stamens isomerous with corolla lobes, but filaments split to the base giving the appearance of a double number of stamens, anthers extrorse; gynoecium usually isomerous with petals, stylodia distinct, short, stigma capitate. Fruits drupaceous with persistent calyx and containing (1–)4–5 pyrenes with coriaceous endocarp and endospermous seeds. Chromosome counts available suggest $x = 9$ (8) as base number, but with complicated series for *A. moschatellina* with $2n = 18, 36$, and 54 in Europe, North America and Siberia, 45, 54, and 56 in Japan, and $2n = 108$ for *Adoxa orientalis* Nepomn. The penta- and hexaploid plants scarcely produce any fruits, but show vegetative propagation through bulbils (cf. Peruzzi and Passalacqua 2006).

Two species: *A. moschatellina* L. (incl. *A. inodora* (Falconer ex C. B. Clarke) Nepomn. and *A. orientalis* Nepomn.), circumpolar in temperate regions of the northern hemisphere. The variability in the number of sepals and petals in this species was already described by Wydler (1850). The other species, *A. xizangensis* G. Yao, is endemic to Sichuan, Xizang and Yunnan (Deyuan and Boufford 2011).

2. ***Tetradoxa*** C.Y. Wu

Tetradoxa C.Y. Wu, Acta Bot. Yunnanica 3: 383–388 (1981).

Stems up to 20 cm tall. Basal leaves 1–3, cauline leaves 2, opposite, 3-cleft. Inflorescences raceme-like, 4–5-flowered. Terminal or axillar flowers ebracteolate, greenish, calyx 4-lobed; corolla rotate, lobes 4 (lateral flowers) or 5 (terminal flowers), internally minutely papillate with a glandular base; petals and stamens forming a basal tube; stamens 4, T-shaped, split half-way, anthers extrorse, reniform; ovary tetralocular, stylodia distinct, attenuate at apices, stigmas small and entire. Fruits not known.

One species, *Tetradoxa omeiensis* (H. Hara) C.Y. Wu, Mount Omei and Ya-an, Sichuan (Szechuan) province, China, at 2300 m (Hara 1983). *Tetradoxa* is synonymized in the Flora of China (Deyuan and Boufford 2011) under *Adoxa*. Both genera form sister-taxa in phylogenetic analyses (e.g. Jacobs et al. 2010).

3. ***Sinadoxa*** C.Y. Wu, Z.L. Wu & R.F. Huang

Sinadoxa C.Y. Wu, Z.L. Wu & R.F. Huang, Acta Phytotax. Sinica 19: 203–210 (1981).

Plants up to 60 cm high, plants caespitose, arising from an erect rhizome. Basal leaves rosulate, 1–2 pinnate-ternate, ovate-lanceolate, petiolate; cauline leaves two, opposite, ovate-triangular. Inflorescences up to 8 cm long, indefinite, flowers in groups of 3–5 in an interrupted spike, inconspicuous, greenish-yellow. Calyx cup-shaped, fleshy, three-lobed, connate, open in bud; corolla 2–3 mm in diameter, yellow-brownish, with 3 or 4 oblong-ovate saccate and winged lobes, the tips of which are punctuated with glands on the inside; bases of petal lobes with digitate nectaries; stamens 3 or 4, isomerous, antesepalous, with filaments split almost to the base, there fused with the petal tube; anthers extrorse, yellow, roundish; ovary bicarpellate, semi-inferior, unilocular, with one pendant ovule; style reduced, simple, stigma 1, small and entire. Fruits not known.

One species, *Sinadoxa corydalifolia* C.Y. Wu, Z.L. Wu & R.F. Huang, found 4000 m above sea level in damp gravelly locations near the glacial zone at Yushu and Nangqen, in the southern Quinghai Province of China.

2. Tribe Sambuceae A. Rich. ex Duby (1828) ('Sambucineae').

4. ***Sambucus*** L. Fig. 2

Sambucus L., Sp. Pl. 269 (1753); Bolli, Dissert. Bot. 223: 1–227 (1994), monogr.; Eriksson and Donoghue, Syst. Bot. (1997), phylog., evol.

Ebulus Fabr. (1763).

Tripetalus Lindl. (1839).

Perennial herbs, shrubs or small trees up to ca. 7 m high, deciduous or semi-evergreen; young stems and branches with a prominent pith with tannin ducts, and in woody taxa with conspicuous lenticels; winter buds perulate. Leaves pinnate, decussate, variable in leaflet size and shape, marginal teeth sometimes glandular; pseudostipules common; extrafloral nectaries may be present in various parts of the plant, such as nodes, pseudostipules, branchlets, petioles, and branches of the inflorescence. Inflorescence monotelic, usually a corymb, more rarely a corymbose thyrse. Flowers hermaphroditic except in the probably dioecious *Sambucus australis* (dimorphic flowers), and *S. nigra* ssp. *peruviana* (functionally unisexual individuals recorded). Flowers actinomorphic, pentamerous, occasionally tetramerous (*S. gaudichaudiana*) or even trimerous (*S. australasica* and *S. gaudichaudiana*); corolla lobe aestivation imbricate or, in Old World species, valvate; stamens with extrorse anthers (latrorse in *S. javanica*); ovary 3(–5)-carpellate; nectary absent or rarely with yellow urn-shaped nectaries in the inflorescence (*S. javanica*) or stigma nectarine (*S. ebulus*). Drupe with 3–5 pyrenes, calyx remains present, colour various. Base chromosome number probably $x = 9$ (8), but variable with complicated series giving $2n = 36$ in most taxa, 37 in *S. australis*, 38 in *S. nigra* ssp. *caerulea* and *S. sieboldiana*, 40 in *S. australasica*, $\approx$72 in *S. nigra* ssp. *peruviana*, probably a natural tetraploid (Bolli 1994).

About 10 spp. in temperate to subtropical regions and on tropical mountains, mainly of the northern hemisphere, two spp. in E Australia, one sp. in SE South America, *S. ebulus* ssp. *africana* (Engl.) Bolli in eastern Africa. Bolli (1994) reduced several species to subspecies, with six alone in *S. nigra* L., of which at least *S. caerulea* Raf. must be accepted as species (cf. Eriksson and Donoghue 1997). *Sambucus javanica* Blume may grow as a perennial herb, shrub or treelet, depending on environmental conditions. Some species are ornamental and widely cultivated.

II. Subfam. Opuloideae Raf. (1820) (*Opulidia*).

3. Tribe Viburneae O. Berg (1866).

5. ***Viburnum*** L.

Viburnum L., Sp. Pl. 267 (1753); Egolf, J. Arnold Arb. 43: 132–172 (1962), karyol.; Yang & Malécot in Flora of China 19: 570–611 (2011), reg. rev.; Winkworth & Donoghue, Amer. J. Bot. 92(4): 653–666 (2005), phylog. and infrageneric groups; Clement & Donoghue, Int. J. Plant Sci. 172 (4): 559–573 (2011), phylog., evol., biogeogr.
Opulus Mill. (1754).

Shrubs or small trees (rarely up to 20 m high), very rarely dioecious (*V. clemensae*), perulae present or absent; branchlets glabrous or pubescent with simple, fascicled or stellate, rarely lepidote or glandular hairs; winter buds perulate or naked. Leaves opposite or rarely 3-verticillate, simple, entire or 3–5-lobed, margin entire or dentate; pseudostipules present or absent. Inflorescence usually many-flowered, often corymbose or paniculate-cymose, sometimes peripherally with sterile showy flowers. Flowers sometimes strongly fragrant, calyx small, 5-toothed; corolla form various; stamens 5; anthers introrse; ovary semi-inferior or inferior, 3-carpellar, 1 locule with a single fertile ovule, 2 sterile; style simple, short, stigmas subcapitate to 3-lobed. Fruit a usually red or blackish drupe, rarely yellow, with 1 pyrene; endosperm ruminated. $2n = 16$ and 18, occasionally complex series including $2n = 18$, 20, 22 (*V. carlesii*), $2n = 32$ in several species, all from Himalayas or SE Asia, $2n = 36$, 40, and 72 also reported.

Estimated number of species varies from ca. 160 (Clement and Donoghue 2011) to 200 (Yang et al. 2011). Distributed in temperate and subtropical parts of the northern hemisphere, with centres of diversity in East Asia and south-eastern North America to western South America.

Several different subdivisions of *Viburnum* have been suggested, of which the most widely used is the division into nine sections suggested by Wilkinson (1948b). More recently, Winkworth and Donoghue (2005 and Clement and Donoghue (2011) have proposed various informal infrageneric clade names and species groups. *Viburnum clemensae* Kern from Borneo was found to be sister to the rest of the genus.

Selected Bibliography

Agardh, J.G. 1858. Theoria systematis plantarum; accedit familiarum phanerogamarum. Lund: C.W.K. Gleerup.

APG (Angiosperm Phylogeny Group) 1998. An ordinal classification of flowering plants. Ann. Missouri Bot. Gard. 85: 531–553.

APG III. 2009. An update of the angiosperm phylogeny group classification for the orders and families of flowering plants: APG III. Bot. J. Linn. Soc. 161: 105–121.

Backlund, A. 1996. Phylogeny of the Dipsacales. Comprehensive Summaries of Uppsala Dissertations from the Faculty of Science and Technology 243. Doctoral thesis, Department of Systematic Botany, Uppsala University.

Backlund, A., Bremer, B. 1997. Phylogeny of the Asteridae s.str based on rbcL sequences, with particular reference to the Dipsacales. Plant Syst. Evol. 207: 225–254.

Bate-Smith, E.C. 1961. The phenolic constituents of plants and their taxonomic significance 1. Dicotyledons. J. Linn. Soc. 58: 95–173.

Benko-Iseppon, A.M., Morawetz, W. 1993. Cold-induced chromosome regions and karyosystematics in *Sambucus* and *Viburnum*. Bot. Acta 106: 183–191.

Benko-Iseppon, A.M., Morawetz, W. 2000. Viburnales: cytological features and a new circumscription. Taxon 49: 5–16.

Bentham, G., Hooker, J.D. 1873. Genera Plantarum. London: Lovell Reeve & Co.

Bolli, R. 1994. Revision of the Genus *Sambucus*. Dissertationes Botanicae 223: 1–227.

Bremer, K., Backlund, A., Sennblad, B., Swenson, U., Andreasen, K., Hjertson, M., Lundberg, J., Backlund, M., Bremer, B. 2001. A phylogenetic analysis of 100+ genera and 50+ families of euasterids based on morphological and molecular data. Plant Syst. Evol. 229: 137–169.

Candolle, A.P. de 1830. Prodromus Systematis Naturalis Regni Vegetabilis. Paris: Treuttel & Würtz.

Carlquist, S. 1992. Wood anatomy of sympetalous dicotyledon families: A summary, with comments on systematic relationships and evolution of the woody habit. Ann. Missouri Bot. Gard. 79: 303–332.

Chandler, M.E.J. 1961. The Lower Tertiary Floras of Southern England. London: The Trustees of the British Museum.

Charlebois, D., Byers, P., Finn, C.E., Thomas, A.L. 2010. Elderberry: botany, horticulture, potential. Horticult. Rev. 37: 213–280.

Clement, W.L., Donoghue, M.J. 2011. Dissolution of *Viburnum* section *Megalotinus* (Adoxaceae) of southeast Asia and its implications for morphological evolution and biogeography. Int. J. Plant Sci. 172: 559–573.

Dahlgren, R. 1983. General aspects of angiosperm evolution and macrosystematics. Nord. J. Bot. 3: 119–149.

Dahlgren, G. 1991. Steps toward a natural system of the dicotyledons: Embryological characters. Aliso 13: 107–166.

Donoghue, M.J. 1981. Growth patterns in woody plants with examples from the genus *Viburnum*. Arnoldia 41: 2–23.

Donoghue, M.J. 1983. A preliminary analysis of phylogenetic relationships in *Viburnum* (Caprifoliaceae s.l.). Syst. Bot. 8: 45–58.

Donoghue, M.J. 1985. Pollen diversity and exine evolution in *Viburnum* and the Caprifoliaceae sensu lato. J. Arnold Arboretum 66: 421–469.

Donoghue, M.J., Olmstead, R.G., Smith, J.F., Palmer, J.D. 1992. Phylogenetic relationships of Dipsacales based on rbcL sequences. Ann. Missouri Bot. Gard. 79: 333–345.

Donoghue, M.J., Eriksson, T., Reeves, P.A., Olmstead, R.G. 2001. Phylogeny and phylogenetic taxonomy of Dipsacales, with special reference to *Sinadoxa* and *Tetradoxa* (Adoxaceae). Harvard Pap. Bot. 6: 459–479.

Drude, O. 1884. Über die verwandtschaftlichen Beziehungen von *Adoxa* zu *Chrysosplenium* und *Panax*. Bot. Jahrb. 5: 441–447.

Eames, A.J. 1961. Morphology of the Angiosperms. New York: McGraw-Hill.

Egolf, D.R. 1962. A cytological study of the genus *Viburnum*. J. Arnold Arboretum 43: 132–172.

Eichinger, A. 1907. Vergleichende Entwicklungsgeschichte von *Adoxa* und *Chrysosplenium*. Mitt. bayr. bot. Ges., Bd. II, Nr. 5: 65–74, Nr. 6: 81–93.

Erbar, C. 1994. Contributions to the affinities of *Adoxa* from the viewpoint of floral development. Bot. Jahrb. Syst. 116(2): 259–282.

Erdtman, G. 1952. Pollen Morphology and plant taxonomy. Stockholm: Almqvist & Wiksell.

Eriksson, T., Donoghue, M.J. 1997. Phylogenetic analyses of *Sambucus* and *Adoxa* (Adoxoideae, Adoxaceae) based on nuclear ribosomal ITS sequences and preliminary morphological data. Syst. Bot. 22: 555–573.

Fritsch, K. 1891. Adoxaceae. In: Engler, A., Prantl, K. (eds.) Die natürlichen Pflanzenfam. IV, 4. Leipzig: W. Engelmann, pp. 170–171.

Gustafsson, M.G.H., Backlund, A., Bremer, B. 1996. Phylogeny of the Asterales sensu lato based on rbcL sequences with particular reference to the Goodeniaceae. Plant Syst. Evol. 199: 217–242.

Hably, L., Szakály, M. 1989. The Catalogue of Leaf-Fossil types preserved in Hungary. Budapest: Akadémiai Kiadó.

Hallier, H. 1905. Provisional scheme of the natural (phylogenetic) system of flowering plants. New Phytol. 4: 151–162.

Hara, H. 1983. A revision of Caprifoliaceae of Japan with reference to allied plants in other districts and the Adoxaceae. Tokyo: Academia Scientific Book.

Harborne, J.B., Williams, C.A. 1971. 6-Hydroxyluteolin & scutellarein as phyletic markers in higher plants. Phytochemistry 10: 369–378.

Hegnauer, R. 1969. Chemical evidence for the classification of some plant taxa. In: Harborn J.B., Swain, T. (eds.) Perspectives in phytochemistry. London, New York: Academic Press.

Hillebrand, G.R., Fairbrothers, D.E. 1970. Serological investigation of the systematic position of the Caprifoliaceae. I. Correspondence with selected Rubiaceae and Cornaceae. Amer. J. Bot. 57: 810–815.

Holmes, D.S. 2005. Sexual reproduction in British populations of *Adoxa moschatellina* L. Watsonia 25: 265–273.

Hutchinson, J. 1973. The families of flowering plants. Oxford: Clarendon Press.

Jacobs, B., Donoghue, M.J., Bouman, F., Huysmans, S., Smets, E. 2008. Evolution and phylogenetic importance of endocarp and seed characters in *Viburnum* (Adoxaceae). Int. J. Plant Sci. 169: 409–431.

Jacobs, B., Huysmans, S., Smets, E. 2010. Evolution and systematic value of fruit and seed characters in Adoxaceae (Dipsacales). Taxon 59: 850–866.

Jensen, S.R., Nielsen, B.J., Dahlgren, R. 1975. Iridoid compounds, their occurrence and systematic importance in the angiosperms. Bot. Not. 128: 148–180.

Jin, B., Wang, L., Wang, J., Teng, N.J., He, X.D., Mu, X.J., Wang, Y.L. 2010. The structure and roles of sterile flowers in *Viburnum macrocephalum* f. *keteleeri* (Adoxaceae). Plant Biol. 12: 853–862.

Jussieu, A.-L. de 1789. Genera Plantarum Secundum Ordines Naturales Disposita, Juxta Methodum in Horto Regio Parisiensi Exaratam, Anno 1774. Paris: Vidaum Herissant.

Kamelina, O.P. 1980. Sravnitelnaja embriologia semejstv Dipsacaceae i Morinaceae. Leningrad (St. Petersburg): Nauka.

Kern, J.H., van Steenis, C.G.G.J. 1951. Caprifoliaceae. In: van Steenis, C.G.G.J. (ed.) Flora Males. I, 4. Batavia: Noordhoff-Kolff, pp. 175–194.

Kilpper, K. 1969. Verzeichnis der im mittleren und unteren Rheinland gefundenen Großreste von Tertiärpflanzen (von 1821 bis 1968). Essen: Ruhrland- und Heimatsmuseum der Stadt Essen.

Lagerberg, T. 1904. Organografiska Studier öfver *Adoxa Moschatellina* L. Arkiv Botanik vol. 3.

Lagerberg, T. 1909. Studien über die Entwicklungsgeschichte und systematische Stellung von *Adoxa moschatellina* L. Kongl. Svenska Vetenskapsakad. Handl., 2nd ser. 44: 1–86.

Metcalfe, C.R., Chalk, L. 1950. Anatomy of the dicotyledons. Oxford: Clarendon Press.

Moissl, E. 1941. Vergleichende embryologische Studien über die Familie der Caprifoliaceae. Österr. Bot Z. 90: 153–212.

Müller, C.A. 1885. Caprifoliaceae. In: Martius, C.F. et al. (eds.) Flora brasiliensis, vol. 6(4): 333–338, t. 99. Leipzig: F. Fleischer.

Müller-Schneider, P. 1967. Zur Verbreitungsbiologie des Moschuskrautes (*Adoxa moschatellina*). Vegetatio Acta Bot. 15: 27–32.

Neubauer, H.F. 1977. Morphologische Beobachtungen an Sämlingen von *Sambucus nigra*. Phyton (Austria) 18: 57–69.

Novák, T. 1904. Über den Blütenbau von *Adoxa moschatellina*. Österr. Bot. Zeitschr. 54: 1–7.

O'Dowd, D.J., Pemberton, R. 1994. Leaf domatia in Korean plants: floristics, frequency, and biogeography. Vegetatio 114: 137–148.

Olmstead, R.G., Bremer, B., Scott, K.M., Palmer, J.D. 1993. A parsimony analysis of the Asteridae sensu lato based on rbcL sequences. Ann. Missouri Bot. Gard. 80: 700–722.

Payer, J.-B. 1857. Traité d'organogénie comparé de la fleur. Paris: Masson.

Peruzzi, L., Passalacqua, N.G. 2006. On a new subspecies of *Adoxa moschatellina* (Adoxaceae), apoendemic in Calabria (S Italy). Nord. J. Bot. 24: 249–256.

Plouvier, V. 1992 [1993]. Chemotaxonomy of the Caprifoliaceae and relationship with some allied families. Adansonia 14: 461–472.

Puff, C., Buchner, R., Sugau, J.B. 1997. *Viburnum clemensae* (Caprifoliaceae s.l.), an unusual dioecious species from the high mountains of Sabah (Borneo). Sandakania 9: 35–54.

Reidt, G., Leins, P. 1994. Das Initialstadium der sympetalen Krone bei *Sambucus racemosa* L. und *Viburnum farreri* Stearn. Bot. Jahrb. Syst. 116: 1–9.

Reitsma, T.J., Reuvers, A.A.M.C. 1975. Adoxaceae. Rev. Palaebot. Palynol. 19: 71–74.

Roels, P., Smets, E. 1994. A comparative floral ontogenetical study between *Adoxa moschatellina* and *Sambucus ebulus*. Belgian J. Bot. 127: 235–254.

Ronse DeCraene, L.-P., Roels, P., Smets, E., Backlund, A. 1998. The floral development and floral anatomy of *Chrysosplenium alternifolium*, an unusual member of the Saxifragaceae. J. Plant Res. 111: 573–580.

Sax, K., Kribs, D.A. 1930. Chromosomes and phylogeny in Caprifoliaceae. J. Arnold Arbor. 11: 147–153.

Savile, D.B.O. 1979. Fungi as aids in higher plant classification. Bot. Rev. 45: 377–503.

Savolainen, V., Fay, M.F., Albach, D.C., Backlund, A., van der Bank, M., Cameron, K.M., Johnson, S.A., Lledó, M.D., Pintaud, J.-C., Powell, M., Sheahan, M.C., Soltis, D.E., Soltis, P.S., Weston, P., Whitten, W.M., Wurdack, K.J., Chase, M.W. 2000. Phylogeny of the eudicots: a nearly complete familial analysis based on rbcL gene sequences. Kew Bull. 55: 257–309.

Schwerin, F.G. von 1909. Monographie der Gattung *Sambucus*. Mitt. Deutsch Dendrol. Ges. 18: 1–56.

Shi-you, L., Zhu-hua, N. 1987. Studies on the family Adoxaceae. Bull. Bot. Res. 7: 93–109.

Solereder, H. 1899. Systematische Anatomie der Dicotyledonen. Stuttgart: Ferdinand Enke.

Sprague, T.A. 1927. The morphology and taxonomic position of the Adoxaceae. J. Linn. Soc. Bot. 47: 471–487.

Sturm, C. 1910. Monographische Studien über *Adoxa Moschatellina* L. (Mitteilungen aus dem Botanischen Museum der Universität Zürich 51). Vierteljahresschr. Naturf. Ges. Zürich 54: 391–462.

Takhtajan, A. 1987. Systema Magnoliophytorum. St. Petersburg (Leningrad): Officina Editoria NAUKA, Sectio Leninopolitana.

Takhtajan, A. 1997. Diversity and classification of flowering plants. New York: Columbia University Press.

Troll, W., Weberling, F. 1966. Die Infloreszenzen der Caprifoliaceen und ihre systematische Bedeutung. Abh. Akad. Wiss. Lit. Mainz. Math. Naturw. Kl. 4: 455–605.

Uemura, K. 1988. Late Miocene Floras in Northeast Honshu, Japan. Tokyo: National Science Museum.

Vogel, S. 1997. Remarkable nectaries: structure, ecology, organophyletic perspectives. I. Substitutive nectaries. Flora 192: 305–333.

Weberling, F. 1957. Morphologische Untersuchungen zur Systematik der Caprifoliaceen. Abh. Math.-Naturwissensch. Kl. Akad. Wissensch. Lit. Mainz 1957: 1–50.

Weyland in Wuppertal-Elberfeld, H. 1938. III. Zweite Ergänzungen und Berichtigungen zur Flora der Blätterkohle und des Polierschiefers von Rott im Siebengebirge. Köln: Geologisch-Mineralogischen Institut der Universität Köln.

Wilkinson, A.M. 1948a. Floral anatomy and morphology of some species of the tribes Linnaeeae and Sambuceae of the Caprifoliaceae. Amer. J. Bot. 35: 365–371.

Wilkinson, A.M. 1948b. Floral anatomy and morphology of the genus *Viburnum*. Amer. J. Bot. 35: 455–465.

Wilkinson, A.M. 1949. Floral anatomy and morphology of *Triosteum* and of the Caprifoliaceae in general. Amer. J. Bot. 36: 481–489.

Willaman, J.J., Li, H.-L. 1970. Alkaloid-bearing plants and their contained alkaloids. Lloydia 33: 1–286.

Winkworth, R.C., Donoghue, M.J. 2005. *Viburnum* phylogeny based on combined molecular data: implications for taxonomy and biogeography. Amer. J. Bot. 92: 653–666.

Wu, C.-Y., Zhang, T.B. 1981. Another new genus of Adoxaceae, with special references on the infrafamiliar evolution and the systematic position of the family. Acta Bot. Yunnanica 3: 383–388.

Wu, C.-Y., Wu, Z.-L., Huang, R.-F. 1981. *Sinadoxa* C.Y. Wu, Z.L. Wu, et R.F. Huang, genus novum familiae Adoxacearum. Acta Phytotax. Sinica 19: 203–210.

Wydler, H. 1850. Über *Adoxa moschatellina*. Flora N.F. 7: 433–437.

Yang, Q., Hong, D.Y., Malécot, V., Boufford, D.E. 2011. Adoxaceae. In: Wu, C.Y. et al. (eds.) Flora of China 19: 570–614.

Zobel, A.M. 1986. Ontogenesis of tannin-containing coenocytes in *Sambucus racemosa* L. III. The mature coenocyte. Ann. Bot. 58: 849–858.

Aquifoliaceae

Aquifoliaceae Bercht. & J. Presl, Přir. Rostlin 2: 440 (1825), nom cons.

P.-A. Loizeau, V. Savolainen, S. Andrews, G. Barriera, and R. Spichiger

Trees or shrubs, rarely climbers; base of trunk straight, rarely with small buttresses; stem without latex or resin. Leaves usually evergreen, rarely deciduous, spirally arranged, rarely opposite or subopposite, simple, coriaceous to chartaceous, stipulate; plants dioecious. Flowers small, hypogynous, regular, unisexual by abortion, without disk, often 4–6-merous or up to 23-merous, in axillary cymose inflorescences, sometimes reduced to one flower. Sepals small, more or less connate at the base or rarely distinct; petals imbricate, mostly connate at the base, rarely free; stamens usually isomerous, alternating with the petals, and connate at the base. Carpels isomerous, united to form a superior ovary, with as many locules as carpels, each generally containing one ovule. Fruit a drupe, red, black, yellow, white, orange or purple, rarely green but mostly variously red, containing 1–6(−23) pyrenes. Seed small with abundant oily and proteinaceous endosperm, without starch.

A family of irregular cosmopolitan distribution; comprising one genus with more than 600 species.

Vegetative Morphology and Anatomy. The stem is straight and up to 30 m high, and rarely has small buttresses. The leaves are simple, evergreen or rarely deciduous, spirally arranged (2/5 for the leaves, bracts and cymes), rarely opposite or subopposite, orbiculate, oblanceolate, elliptic, ovate or lanceolate. Their length is approximately 2–15 cm, with a maximum of 35 cm in *I. megaphylla* from Sarawak and a minimum of 4 mm in *I. microphylla*. The base of the lamina is acute to obtuse, the apex truncate to acute, often acuminate, and the margin can be serrate, crenate, or entire, rarely spinose. Entire or spiny leaves can be present in the same species (e.g. *I. dipyrena* and *I. dimorphophylla*), or on the same plant (e.g. *I. aquifolium*). Stipules present, often caducous, small. Colleters (stipular on the node at the base of the leaves, standard and lachrymiform found on the leaf teeth or crenations, and sessile found on the margins of the floral bracts) have been identified in all the nine south American species studied (Gonzalez and Tarragó 2009). Stipular colleters replace stipules in these species. The leaves are coriaceous, rarely chartaceous, rarely pubescent, and punctate or not. Hairs simple and unicellular. The cortex is grey or ochre; when slashing the trunk, a thin green layer appears under the bark; the wood is yellow and its sap oxidizes rapidly, becoming black. This provides important field characters. Wood anatomy has been extensively studied by Baas (1973[1974]). Based on the observation of 81 species of *Ilex*, he concluded that the wood is characterized by conspicuous growth rings, and numerous narrow but relatively short vessel elements with few bars per perforation plate. Thickenings on vessel and fibre walls are spiral, and the fibre-tracheids frequently have numerous conspicuously bordered tangential wall pits (Baas 1973[1974]). *Ilex* has been cited as an example of a genus with trilacunar and unilacunar nodes (Baas 1975; Gonzalez and Tarragó 2009).

Inflorescence Structure. The flowering unit is the cyme, which can be reduced to one flower, or increased to up to 15–31 flowers. They are then solitary dichasia, thyrses or thyrsoids. Most of the species have thyrses, which are borne axillary, rarely terminally. Thyrses can be either non-proliferating, or proliferating when the post-flowering growth is vegetative. In case of the former, the axis of the thyrse can be contracted, so that it appears like a fascicle. Solitary dichasia are

J.W. Kadereit and V. Bittrich (eds.), *Flowering Plants. Eudicots*, The Families and Genera of Vascular Plants 14,
DOI 10.1007/978-3-319-28534-4_3, © Springer International Publishing Switzerland 2016

formed when the bracts of proliferating thyrses develop as leaves. Some species have simple thyrsoids whereas others have a "fascicle of thyrsoids" with alternate or decussate arrangements (Loizeau and Spichiger 1992).

The different classifications of *Ilex* have been based on the structure of the inflorescences. Loesener (1901, 1908) describes his subgenera and lower rank taxa on the basis of inflorescence type (e.g. solitary cymes, fascicles, racemes, or panicles). Hu (1949–1950) in her work on *Ilex* of China uses the cyme as the flowering unit, while Loizeau and Spichiger (1992) propose a phylogenetic classification.

Flower Morphology (Fig. 3). The flowers are hypogynous, regular, unisexual by abortion, often 4–6- or up to 23-merous in all whorls. The first flower of a cyme or the terminal flower of the thyrsoid can sometimes have one extra sepal and petal. The calyx is glabrous or pubescent, valvate, with small lobes and a more or less connate base. The calyx is persistent, rarely caducous (*I. mucronata*), semipersistent in *I. collina*. The corolla is often white or cream, rarely green, yellow, pink, purple, red or chocolate-coloured, with imbricate petals, and 1–10 mm in length. The petals are mostly connate at the base to up to half of their length. Some species have distinct petals (*I. collina*, *I. mucronata* and *I. vismiifolia*). In staminate flowers, the stamens alternate with the petals, are usually glabrous, and have a white filament. The anthers are basifixed and introrse. The sterile ovary is conical. In pistillate flowers, staminodes with sterile anthers are present; these are glabrous, rarely pubescent. The syncarpous ovary is globose, with as many locules as carpels. The style is terminal, very short or absent, and has a distinct stigma, with lobes poorly developed. One ovule is found in each locule, but additional aborted ovules can occur. The ovules are generally apical-axile, pendant, anatropous, apotropous, unitegmic, tenuinucellar, and with a large funiculus (Corner 1976). Because of the presence of staminodes in the pistillate flower and a sterile ovary in the staminate flower, some authors have considered *Ilex* to have bisexual flowers.

Embryology. Ives (1923) observed that the embryo is in a very immature condition when the drupe falls off the plant. It continues growing, very slowly, for a period of 8–12 months.

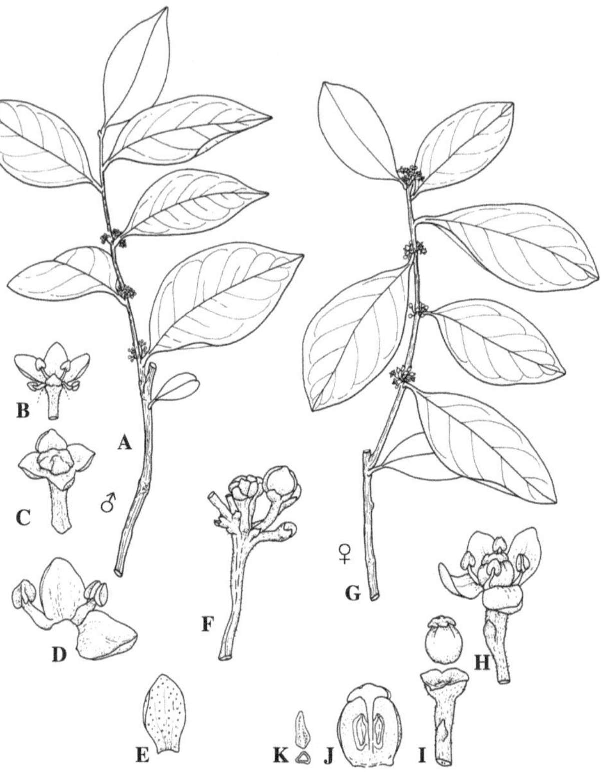

Fig. 3. Aquifoliaceae. **A–K** *Ilex vismiifolia*. **A** Branch with male inflorescences. **B** Male flower (one petal removed). **C** Male flower showing the calyx and the sterile ovary. **D** Male flower showing petals and stamens. **E** Petal. **F** Male inflorescence. **G** Branch with female inflorescences. **H** Female flower. **I** Female flower showing the ovary and the calyx. **J** Ovary, vertical section. **K** Pyrene. (Loizeau 1994, with permission from Boissiera)

Pollen. Pollen morphology has been extensively studied by Lobreau (1969). Pollen is tricolporate, isopolar, clavate, the exo-apertures are furrows with granular membranes, and the endo-apertures are oval and small. Individual pollen grains are between 17–47 μm × 18–48 μm, being larger in species of higher latitude or altitude (Lobreau-Callen 1977).

Pollination. Pollination usually is accomplished by insects (ants, bees, flies, wasps, butterflies and beetles) attracted by nectar produced at the base of the petals (Tsang and Corlett 2005).

Fruit and Seed Dispersal. Drupe with (1–)4–6 (–23) pyrenes, red, brown or black, usually globose; exocarp membranous or papery; mesocarp fleshy. Pyrenes with endocarp smooth, leathery, woody, or stony, striate, striate-sulcate, or rugose, and/or pitted (Chen and al. 2008). The fruits are generally dispersed by birds. Germination in nature is delayed and requires from one to three years, with a germination of more or less half of the seeds (Ives 1923; Arrieta and Suárez 2004).

KARYOLOGY. Chromosome numbers of about 30 species of *Ilex*, mostly Asian, are known (Goldblatt 1981, 1984). Four different base numbers, $x = 17$, 18, 19, 20, were found. The majority of the species (21) have been reported to be diploid and to have $2n = 40$ chromosomes. Polyploidy occurs in some species: *I. pedunculosa* ($2n = 120$) is hexaploid, and *I. anomala* ($2n = 80$) and *I. verticillata* ($2n = 72$) are tetraploid. Most *Ilex* species have $2n = 40$ chromosomes.

PHYTOCHEMISTRY. Aquifoliaceae are sometimes saponiferous or tanniniferous, accumulate proanthocyanins and rarely cyanogenic compounds, but lack ellagic acid or iridoid compounds. Alikaridis (1987) studied the natural constituents of approximately 50 species worldwide; their stimulating and vomitive effects are related to the presence of very high quantities of purine bases (caffeine, theobromine, theophylline or adenine). Moreover, they contain vitamins (C, B1, B2), nicotinic acid, and carotene (Alikaridis 1987). According to Lewis et al. (1991), *I. guayusa* has the highest concentration of caffeine and methylxanthine known.

AFFINITIES. Various hypotheses on the affinities of Aquifoliaceae have been published. Cronquist (1981, 1988) included Aquifoliaceae in Celastrales, near Celastraceae, and included *Phelline* (now Phellinaceae, Asterales) and *Sphenostemon* (now Paracryphiaceae, Paracryphiales) in the family. The inclusion in Celastrales was based on the morphology of the ovule with a dorsal raphe (Bentham and Hooker 1862) and the thick funiculus of Aquifoliaceae which was considered by Loesener (1908) to be homologous to the aril in the Celastraceae. Hallier (1903) placed Aquifoliaceae in his "Umbellifloren" (Umbelliferae/Apiaceae and their relatives) because of the small embryo and their common unisexuality. Baillon (1891) placed Aquifoliaceae between Ericaceae (because of the descendent placenta) and Ebenaceae (because of the drupaceous fruit). Tieghem (1898) placed Aquifoliaceae near Solanaceae because of the unitegmic ovule and the isomerous stamens alternating with the petals. Based on pollen, Lobreau (1969) regarded Aquifoliaceae as related to Gentianales and Campanulales. Finally, Dahlgren (1983) placed the family in Cornales because of the unitegmic ovules, whereas Thorne (1992) placed Aquifoliaceae in Theales, as did Baas (1973[1974], 1975) on the basis of wood anatomy. Based on several molecular phylogenetic studies, Aquifoliaceae are now well placed within the order Aquifoliales together with *Helwingia* and *Phyllonoma*, two genera with epiphyllous inflorescences, at the base of the euasterids (Savolainen et al. 2000a, 2000b; Soltis et al. 2000; Bremer et al. 2001). This placement of *Ilex* in a clade containing mostly gamopetalous plants is supported by the petals which are often connate at the base of the tube, and by the unitegmic ovules. Putative synapomorphies for the family may include the unisexual flowers and fleshy fruits (Bremer et al. 2001). Since *Nemopanthus* has been shown to be included in *Ilex* (Powell et al. 2000), Aquifoliaceae, Helwingiaceae and Phyllonomaceae are three monotypic families which may be better united in a single family (Powell et al. 2000; Savolainen et al. 2000a, 2000b; Soltis et al. 2000; Bremer et al. 2001). APG II (2003) and APG III (2009) proposed to keep Helwingiaceae and Aquifoliaceae as separate families.

ECOLOGY AND DISTRIBUTION. *Ilex* occurs in all tropical areas of the world and extends into temperate regions to up to 63° N (America, Eurasia) and 35° S (America, Africa) (Baas 1974) from lowland to montane forests and up to 4000 m altitude in the Andes (Loizeau 1994). However, its distribution is very irregular and most species occur in South America and Asia. In China 204 species have been recorded (Chen et al. 2008). Andrews (2002) assumed that there are some 120 species in Southeast Asia, and Loizeau (1994) estimated that up to 300 species are found in America. There exist only two to three species in Europe, *I. aquifolium* (also in N Africa), *I. colchica* and *I. perado* subsp. *iberica*, one in Australia, *I. arnhemensis*, and one in Africa south of the Sahara, *I. mitis*. Twenty-six species are native to eastern North America (Kartesz 1994).

PALAEOBOTANY. Pollen of a type attributed to *Ilex* is known from Turonian deposits in Australia and Coniacian deposits in Africa. The pollen became cosmopolitan in the Palaeocene (Muller 1970) and pollen fossils have been found in Alaska, Iceland, western North America, southern South America, Siberia, New Zealand and southern Australia (Loizeau et al. 2005).

Economic Importance and Uses. Many species of *Ilex* are of horticultural importance and are widely grown in parks and gardens throughout the world for their foliage and decorative berries. These include (Andrews 1982, 1983a, 1983b, 1984a, 1984b, 1985, 1989, 1991, 1992) *I. aquifolium*, the common holly and its many cultivars, *I.* x *altaclerensis* (*I. aquifolium* x *I. perado*), *I. opaca*, the American holly (Eisenbeiss and Dudley 1973), *I. cornuta*, the horned holly, and *I. crenata*, the Japanese holly (Dudley and Eisenbeiss 1992). In China *I. purpurea* (*I. chinensis* is a misapplied name) is commonly planted as an ornamental tree, and the fruiting branches are used to decorate temple courts and halls from December to February. Large quantities of glossy-berried branches are sold at the Chinese New Year (Hu 1949–1950). The leaves of over 60 species of *Ilex* are used for beverages. The most widely known is a tea ("tereré") made from *I. paraguariensis*, the "Yerba Maté" or Paraguay Tea (Loizeau 1988). This species is native to Argentina, Uruguay, Paraguay and Brazil and is drunk throughout South America. The Yaupon or *I. vomitoria* from south-eastern North America and Mexico has been used by certain native North Americans who brewed a tea called the Black Drink or "Cassena". This was said to restore lost appetite, guarantee health and give courage and agility in war (Hudson 1979; Alikaridis 1987). Several species are known for their medicinal properties and the leaves of *I. aquifolium* were used as an infusion for smallpox, catarrh and pleurisy. The drupes acted as an emetic and purgative and have been used for dropsy and to prevent bleeding (Coles 1657). *Ilex guayusa* (see Phytochemistry) has been used for centuries as a medicine, an emetic and as a stimulant (Lewis et al. 1991). The roots of the widespread Southeast Asian *I. cymosa* are said to have medicinal properties, especially in the treatment of fevers (Andrews 2002). *Ilex mitis*, the only species to occur naturally south of the Sahara, has been used as an enema for colic in children, as a purgative and is used in witchcraft (Andrews 1994). The wood of *I. aquifolium* is hard and white and has been much prized for ornamental use. The strong, straight shoots when stripped of their bark are made into whip handles and walking sticks (Dallimore 1908). The leaves were used as winter food crop for livestock in England, Wales and northwest France from the Middle Ages onwards (Spray 1981). The wood of *I. cissoidea* from Sumatra, Sulawesi and Borneo has been used as firewood and possibly for making matches. The dried leaves were pounded into a powder and used as a soap (Andrews, ined.), as were the leaves of *I. mitis*, while the wood is used for furniture, ceilings, brake blocks, railway sleepers or for firewood (Andrews 1994).

One genus:

Ilex L. Fig. 3

Ilex L., Sp. Pl.: 125 (1753); Loesener, Nova Acta Acad. Caes. Leop.-Carol. German. Nat. Cur. 78 (1901) and 89 (1908); Loizeau, Boissiera 48 (1994), rev. of Peruvian spp; Chen et al., Fl. China 11: 359–438 (2008), rev. of Chinese spp.; Manen et al., Pl. Syst. Evol. 235: 79–98 (2002), evol.; Manen et al., Mol. Phyl. Evol. 57: 961–977 (2010), evol. *Nemopanthus* Raf. (1819), nom. cons.

Molecular studies have shown hybridization and introgression even between distantly related lineages of the genus (Manen et al. 2002, 2010).

Selected Bibliography

Alikaridis, F. 1987. Natural constituents of *Ilex* species. J. Ethnopharmacol. 20: 121–144.

Andrews, S. 1982. Hollies in Ireland. Moorea 1: 5–15.

Andrews, S. 1983a. Notes on some *Ilex* x *altaclerensis* clones. Plantsman 5: 65–81.

Andrews, S. 1983b. Eight motley hollies. Holly Soc. J. 1(2): 11–12.

Andrews, S. 1984a. An account of *Ilex* in and around Paris with a catalogue of herbarium and living specimens. Holly Soc. Amer. Bull. 22: 1–146.

Andrews, S. 1984b. More notes on clones of *Ilex* x *altaclerensis*. Plantsman 6: 157–166.

Andrews, S. 1985. Holly berries of a varied hue. Garden (London 1975+) 110: 518–522.

Andrews, S. 1989. Hollies. In: Brickell, C. (ed.) The Royal Horticultural Society gardeners' encylopedia of plants and flowers: 70–71, 490–491. London: Dorling Kindersley.

Andrews, S. 1991. Hollies - underutilised trees and shrubs for cultivation in southern Africa. S. African Nurseryman 10(2): 18–20.

Andrews, S. 1992. *Ilex, Nemopanthus*. In: Huxley, A., Griffiths, M., Levy, M. (eds.) The New Royal Horticultural Society dictionary of gardening 2: 644–648, 3: 300. London: The Macmillan Press.
Andrews, S. 1994. FSA contributions 1: Aquifoliaceae. Bothalia 24: 163–166.
Andrews, S. 2002. Aquifoliaceae. In: Soepadmo, E., Saw, L. G., Chung, R.C.K. (eds.) Tree Flora of Sabah and Sarawak 4: 1–27. Kuala Lumpur: Forest Research Institute Malaysia.
Angiosperm Phylogeny Group 2003. An update of the Angiosperm Phylogeny Group classification for the orders and families of flowering plants: APG II. Bot. J. Linn. Soc. 141: 399–436.
Angiosperm Phylogeny Group 2009. An update of the Angiosperm Phylogeny Group classification for the orders and families of flowering plants: APG III. Bot. J. Linn. Soc. 161: 105–121.
Arrieta, S., Suárez, F. 2004. Germination and seed bank depletion of holly (*Ilex aquifolium* L.) in four microhabitat types. Seed Sci. Res. 14: 305–313.
Baas, P. 1973 [1974]. The wood anatomical range in *Ilex* (Aquifoliaceae) and its ecological and phylogenetic significance. Blumea 21: 193–258.
Baas, P. 1975. Vegetative anatomy and affinities of Aquifoliaceae, *Sphenostemon*, *Phelline* and *Oncotheca*. Blumea 22: 311–407.
Baillon, H. 1891. Ilicacées. Histoire des Plantes 11: 211–220. Paris: Librairie Hachette.
Bentham, G., Hooker, J.D. 1862. Ilicineae. Genera plantarum 1: 355–357. London: Reeve, Williams and Norgate.
Bremer, K., Backlund, A., Sennblad, B., Swenson, U., Andreasen, K., Hjertson, M., Lundberg, J., Backlund, M., Bremer, B. 2001. A phylogenetic analysis of 100+ genera and 50+ families of euasterids based on morphological and molecular data with notes on possible higher level morphological synapomorphies. Plant Syst. Evol. 229: 137–169.
Chen, S.-K., Ma, H.-Y., Feng, Y.-X., Barriera, G., Loizeau, P.-A. 2008. Aquifoliaceae. In: Wu, Z.-G., Raven, P.H., Hong, D.-Y. (eds.) Flora of China 11: 359–438. Beijing: Science Press. 622 pp.
Coles, W. 1657. Adam in Eden or Natures Paradise. The history of plants, fruits, herbs and flowers, etc. London: 385–386.
Corner, E.J.H. 1976. The seeds of dicotyledons. Cambridge: Cambridge University Press.
Cronquist, A. 1981. An integrated system of classification of flowering plants. New York: Columbia University Press.
Cronquist, A. 1988. The evolution and classification of flowering plants, ed. 2. New York: The New York Botanical Garden.
Dahlgren, R. 1983. General aspects of angiosperm evolution and macrosystematics. Nordic J. Bot. 3: 119–149.
Dallimore, W. 1908. Holly, Yew and Box. London: John Lane.
Dudley, T.R., Eisenbeiss, G.K. 1992. International checklist of cultivated *Ilex*. Part 2, *Ilex crenata* Thunberg ex J. A. Murray. Natl. Arbor. Contr. 6.
Eisenbeiss, G.K., Dudley, T.R. 1973. International checklist of cultivated *Ilex*. Part 1, *Ilex opaca* Natl. Arbor. Contr. 3.
Goldblatt, P. 1981. Index to plant chromosome numbers 1975–1978. Monogr. Syst. Bot. Missouri Bot. Gard. 5.
Goldblatt, P. 1984. Index to plant chromosome numbers 1979–1981. Monogr. Syst. Bot. Missouri Bot. Gard. 8.
Gonzalez, A.M., Tarragó, J.R. 2009. Anatomical structure and secretion compounds of colleters in nine *Ilex* species (Aquifoliaceae) from southern South America. Bot. J. Linn. Soc. 160: 197–210.
Hallé, F., Oldeman, R.A.A., Tomlinson, P.B. 1978. Tropical trees and forests. An architectural analysis. Berlin: Springer.
Hallier, H. 1903. Vorlaüfiger Entwurf des natürlichen (phylogenetischen) Systems des Blüthenpflanzen. Bull. Herb. Boissier II, 3: 306–317.
Hu, S.-Y. 1949–1950. The genus *Ilex* in China [I]-V. J. Arnold Arbor. 30: 233–344, 348–387; 31: 39–80, 214–240, 241–263.
Hu, S.-Y. 1967. The evolution and distribution of the species of Aquifoliaceae in the Pacific area (1)-(2). J. Jap. Bot. 42: 13–27, 49–58.
Hudson, C.M. (ed.) 1979. Black drink: a native American tea. Athens: The University of Georgia Press.
Ives, S.A. 1923. Maturation and germination of seeds of *Ilex opaca*. Bot. Gaz. (Crawfordsville) 76: 60–77.
Kartesz, J.T. 1994. Aquifoliaceae. A synonymized checklist of the vascular flora of the United States, Canada, and Greenland, ed. 2, 1: 50–51. Portland: Timber Press.
Lewis, W.H., Kennelly, E.F., Bass, G.N., Wedner, H.J., Elvin-Lewis, M.P., Fast, W.D. 1991. Ritualistic use of the holly *Ilex guayusa* by Amazonian Jívaro Indians. J. Ethnopharmacol. 33: 25–30.
Lobreau, D. 1969. Les limites de "l'ordre" des célastrales d'après le pollen. Pollen Spores 11: 499–555.
Lobreau-Callen, D. 1977. Les variations dimensionnelles du pollen du genre *Ilex* (Aquifoliaceae) et leurs rapports avec le climat. Bull. Soc. Bot. France 122 [Morphol. Pollinique]: 179–199.
Loesener, T. 1901. Monographia Aquifoliacerum. Pars I. Nova Acta Acad. Caes. Leop.-Carol. German. Nat. Cur. 78.
Loesener, T. 1908. Monographia Aquifoliacerum. Pars II. Nova Acta Acad. Caes. Leop.-Carol. German. Nat. Cur. 89.
Loesener, T. 1942. Aquifoliaceae. In: Engler, A., Prantl, K. (eds.) Die Natürlichen Pflanzenfamilien, ed. 2, 20b: 36–86. Leipzig: Wilhelm Engelmann.
Loizeau, P.-A. 1988. L'or vert du Paraguay. Mus. Genève II, 287: 16–20.
Loizeau, P.-A. 1994. Les Aquifoliaceae péruviennes. Boissiera 48.
Loizeau, P.-A., Spichiger, R. 1992. Proposition d'une classification des inflorescences d'*Ilex* L. (Aquifoliaceae). Candollea 47: 97–112.
Loizeau, P.-A., Barriera, G., Manen, J.-F., Broennimann, O. 2005. Towards an understanding of the distribution of *Ilex* L. (Aquifoliaceae) on a World-wide scale. In: Friis, I., Balslev, H. (eds.) Plant diversity and complexity patterns. Local, regional and global dimensions. Biol. Skrifter 55: 501–520.
Manen, J.-F., Boulter, M.C., Naciri-Graven, Y. 2002. The complex history of the genus *Ilex* L. (Aquifoliaceae): evidence from the comparison of plastid and nuclear

DNA sequences and from fossil data. Plant Syst. Evol. 235: 79–98.

Manen, J.-F., Barriera, G., Loizeau, P.-A., Naciri, Y. 2010. The history of extant *Ilex* species (Aquifoliaceae): evidence of hybridization within a Miocene radiation. Mol. Phylogen. Evol. 57: 961–977.

Muller, J. 1970. Palynological evidence on early differentiation of angiosperms. Biol. Rev. Cambridge Philos. Soc. 45: 417–450.

Powell, M., Savolainen, V., Cuenoud, P., Manen, J.-F., Andrews, S. 2000. The mountain holly (*Nemopanthus mucronatus*) revisited with molecular data. Kew Bull. 55: 341–347.

Savolainen, V., Chase, M.W., Morton, C.M., Hoot, S.B., Soltis, D.E., Bayer, C., Fay, M.F., de Bruijn, A., Sullivan, S., Qiu, Y.-L. 2000a. Phylogenetics of flowering plants based upon a combined analysis of plastid atpB and rbcL gene sequences. Syst. Biol. 49: 306–362.

Savolainen, V., Fay, M.F., Albach, D.C., Backlund, A., van der Bank, M., Cameron, K.M., Johnson, S.A., Lledó, M.D., Pintaud, J.-C., Powell, M., Sheahan, M.C., Soltis, D.E., Soltis, P.S., Weston, P., Whitten, W.M., Wurdack, K.J., Chase, M.W. 2000b. Phylogeny of the eudicots: a nearly complete familial analysis based on *rbcL* gene sequences. Kew Bull. 55: 257–309.

Soltis, D.E., Soltis, P.S., Chase, M.W., Mort, M.E., Albach, D.C., Zanis, M., Savolainen, V., Hahn, W.H., Hoot, S. B., Fay, M.F., Axtell, M., Swensen, S.M., Nixon, K.C., Farris, J.S. 2000. Angiosperm phylogeny inferred from a combined data set of 18S rDNA, *rbcL*, and *atpB* sequences. Bot. J. Linn. Soc.133: 381–461.

Spray, M. 1981. Holly as a fodder crop in England. Agric. Hist. Rev. 29(2): 97–110.

Thorne, R.F. 1992. Classification and geography of the flowering plants. Bot. Rev. (Lancaster) 58: 225–348.

Tieghem, P. van 1898. Structure de quelques ovules et parti qu'on en peut tirer pour améliorer la classification. J. Bot. (Morot) 12: 197–220.

Tsang, A.C.W, Corlett, R.T. 2005. Reproductive biology of the *Ilex* species (Aquifoliaceae) in Hong Kong, China. Can. J. Bot. 83: 1645–1654.

Aucubaceae

Aucubaceae Bercht. & J. Presl, Přir. Rostlin 2: 91 (1825) (Aucubeae).

Q.-Y. Xiang

Small evergreen trees or shrubs; branchlets opposite, with conspicuous leaf scars, often pubescent when young, sometimes pendulous with aerial roots; plants dioecious. Leaves simple, pinnately veined, opposite, without stipules, stoutly petiolate; blade thick leathery to thick chartaceous, dark green and shiny above, light green below, sometimes variegated, pubescent or glabrous; margin serrate or glandular serrate, rarely entire. Flowers small, unisexual, epigynous, actinomorphic. Both male and female flowers arranged in terminal thyrsoid compound inflorescences; female inflorescences usually smaller than male ones. Male flowers subtended by one small, lanceolate and pubescent bracteole; calyx minute, 4-dentate; petals 4, appearing free, valvate in bud, oblong or ovate, 3–4 mm long, purple red to yellow or greenish, floral disk fleshy, slightly 4-lobed; stamens 4, alternating with petals; filaments conical, short; anthers yellow, dorsifixed, bithecate , rarely unithecate, dehiscing with longitudinal slits; female flowers similar to male ones but pedicel articulate at the base of the ovary and subtended by one or two lanceolate bracteoles; style simple, short and thick; stigma capitate, straight or bent to one side, slightly 2–4-lobed; ovary unicarpellate, cylindric or ovoid, inferior, 1-locular; ovule 1, pendant, anatropous with dorsal raphe, unitegmic, crassinucellate. Fruit a fleshy 1-seeded drupe, cylindric or ovate, red when mature, black when dry, crowned by persistent calyx teeth, style, and stigma. Endosperm copious; embryo minute.

A monogeneric family consisting of only *Aucuba* with 8–11 species distributed in eastern Asia (Japan, China, Vietnam, to eastern Himalayas).

Vegetative Morphology. The shoot of *Aucuba japonica* shows a typical dichasial branching system and a decussate phyllotaxis that is maintained from the vegetative to the reproductive phase (Hara 1980). The sequence of initiation of successive foliar appendages is very precise in its order. A majority of the shoot apices changes into the reproductive phase to form a terminal inflorescence in the mature plants. Some shoot meristems continue their vegetative growth when they are not transformed into the reproductive phase. Aerial roots may develop on branchlets (Konta 1994).

Vegetative Anatomy. Stomata are of the anomocytic type and mainly confined to the lower leaf surface. The wood is diffuse-porous with a tendency to being semi-ring-porous, showing distinct (e.g., *A. japonica*) or indistinct (*A. chinensis*) growth rings (Noshiro and Baas 1998). Vessels are mostly solitary or rarely in multiples or clusters of 2 or 3; perforation scalariform with 27–100 bars. In *A. chinensis* bar number is less variable, ranging from 27 to 35. Intervessel wall pitting uncommon; if present, it is often opposite or rarely scalariform, and without vestures. Average length of vessel elements ranges from 850–1085 μm, and average vessel diameter from 35–39 μm. Helical thickenings are present in both vessels and vascular tracheids. Fibers have minutely bordered pits in their radial walls. Axial parenchyma is present and mostly scanty paratracheal, occasionally diffuse. Rays consist of a mixture of procumbent, square, and upright cells (Adams 1949; Li and Chao 1954; Noshiro and Baas 1998).

Inflorescence and Flower Morphology. Both male and female inflorescences are basically a thyrsoid type of monotelic synflorescence in which the paracladia are dichasia (Hara 1980). Larger numbers of flowers are produced in male

J.W. Kadereit and V. Bittrich (eds.), *Flowering Plants. Eudicots*, The Families and Genera of Vascular Plants 14,
DOI 10.1007/978-3-319-28534-4_4, © Springer International Publishing Switzerland 2016

inflorescences (Hara 1980). Reidt and Leins (1994) observed the corolla originating from a low ring-wall, the resulting sympetalous part being so small, however, that it is easily overlooked in adult flowers.

Embryology. Embryo sac development usually is of the Polygonum type with some exceptions. In some ovules, an eight-nucleate mature embryo sac is formed by the fusion of two four-nucleate embryo sacs (Sato 1971). Such ovules have 4–5 parietal layers, differing from those containing the typical Polygonum type embryo sac with 6–7 parietal layers (Sato 1971). This unusual type of embryo sac development occurs at a much higher rate in *A. japonica* var. *borealis* than in *A. japonica* var. *japonica* (Sato 1976). Typically, each ovule contains one mature embryo sac.

Pollen Morphology. Pollen grains are spheroidal or subprolate in shape and 36–64 (polar) × 29–55 (equatorial) μm in size, 3-colporate, and radially symmetrical. Endoapertures are simple-diffuse and without lamellation of the endexine, indistinct, more or less circular or oblong. Exine 2–3.5 μm thick. Sexine thicker than nexine, intectate, psilate. Psila composed of columella and expanded head. There is little variation in pollen morphology among the three species examined (*A. chinensis*, *A. japonica*, and *A. himalaica*; Ferguson 1977; Ferguson and Hideux 1978).

Karyology. Chromosome numbers were reported for three species: *A. japonica*: $2n = 16, 32$; *A. chinensis* and *A. himalaica*: $2n = 16$. Plants of *A. japonica* from north and central Japan (var. *japonica* and var. *borealis*) are tetraploids, and those from southern Japan are diploids (var. *australis*). Tetraploid *A. japonica* is more vigorous than the diploid material (Viinikka 1970; Kurosawa 1971; see Hara 1972).

Reproductive Biology. The flowers are visited by a variety of insects, including small bees, flies, and beetles. Effects of floral display size and flowering phenology on fruit set was investigated in *A. japonica* by Abe (2001). The results indicated that display size did not affect fruit set; the proportion of flowers setting fruit remained nearly constant independent of the total number of flowers per inflorescence, the total number of inflorescences per plant, and the total number of flowers per plant. However, flowering phenology did affect fruit set. Fruit set was most abundant when male and female flowering was most abundant. Female flowers remained receptive until they received pollen, and fruit set was generally not pollen-limited unless male flowers were scarce (Abe 2001). Asexual reproduction via pendulous branchlets bearing aerial roots was observed in *A. japonica* (Konta 1994). New plantlets can grow from these branchlets when they have contact with soil.

Phytochemistry. Fruits of *A. japonica* contain 2.75 % lipids including 0.45 % fatty acids of their dry weight. Of the fatty acids, 26.37 % are C16, 71.02 % C18, and 2.62 % C20. Among the C18 fatty acids, petroselinic acid is dominant (33.78 %). Other C18 fatty acids include stearic acid (2.38 %), oleic acid (26.99 %), linoleic acid (17.98 %), and linolenic acid (24.09 %; Breuer et al. 1987). Iridoids (aucubin) and several flavonols and flavone glycosides were found in *Aucuba* (Bate-Smith et al. 1975; Bernini et al. 1984; Iwashina et al. 1997). Two anthocyanins were isolated from the fruits of *A. japonica* (Ishikura 1971). Tannins, procyanidins, and cyanogenic compounds were not detected in the genus (Bate-Smith et al. 1975).

Affinities. The affinities of *Aucuba* have long been controversial. The genus has commonly been treated either as a member of Cornaceae (e.g., Harms 1898; Wangerin 1910; Cronquist 1981, 1988), or as a monotypic family (e.g., Airy-Shaw 1973; Takhtajan 1980, 1987) allied to Cornaceae. Molecular sequence data suggested that *Aucuba* is clearly not a close relative of *Cornus*, but sister of the western American *Garrya*, and both are closely related to Eucommiaceae (Xiang et al. 1993; Xiang and Soltis 1998; Soltis et al. 2000). *Aucuba* and *Garrya* are also similar in phytochemistry (Bate-Smith et al. 1975; Kleiman and Spencer 1982; see Eyde 1988). Both genera contain the decarboxylated iridoid aucubin, petroselinic acid in their fruit oils, and no tannins. *Aucuba* and *Garrya* were grouped together as Garryaceae in the APG III classification. The family was placed in Garryales that also include Eucommiaceae (APG III 2009).

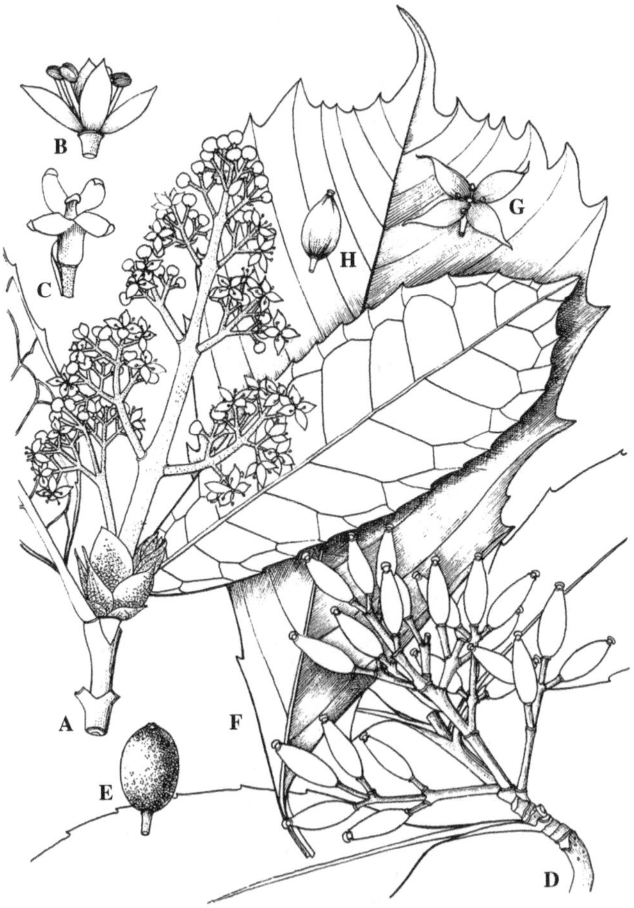

Fig. 4. Aucubaceae. **A–E** *Aucuba chinensis* var. *chinensis*. **A** Male flowering branch. **B** Male flower. **C** Female flower. **D** Fruiting branch. **E** Fruit. **F–H** *A. obcordata*. **F** Leaf. **G** Male flower. **H** Fruit. (from Xiang and Boufford 2006, with permission from Missouri Bot. Garden Press, St. Louis and Science Press, Beijing)

Distribution and Habitats. *Aucuba*, as the only genus of the family, contains 8–11 species distributed in eastern Asia, from Japan to the Himalayas. All species occur in areas south of the Yellow River in China. Most species are confined to China. Only *A. japonica* extends its distribution to Japan and Korea, and *A. himalaica* to the Himalayan region. Plants of *Aucuba* grow in subalpine coniferous forests below 2800 m and in broad-leaved evergreen or broad-leaved evergreen-deciduous mixed forests, favoring habitats with rich soil and abundant moisture (Hu and Soong 1990).

Economic Importance. Some species of *Aucuba* are used in folk medicine; wood of a few species is used for carving, cigarette pipes, walking canes, etc. *Aucuba japonica* is commonly cultivated as an ornamental because of its bright red showy fruits in the winter season, and shiny coriaceous and yellow variegated evergreen leaves (Hara 1972; Hu and Soong 1990).

Only one genus:

Aucuba Thunb. Fig. 4

Aucuba Thunb., Nov. Gen. 3: 61 (1783); Fl. Jap. 4, t. 12 & 13 (1784); Xiang, Boufford in Wu and Raven, Flora of China 14: 222–226 (2005).

Description as for family.

Aucuba chinensis Benth., *A. himalaica* Hook. f. & Thomson, and *A. japonica* Thunb. are three long-recognized species. Several new species and varieties were added to the genus from the flora of China mainly in the last two decades.

Flowering Jan.–Feb., fruit maturing in the following Feb.

Selected Bibliography

Abe, T. 2001. Flowering phenology, display size, and fruit set in an understory dioecious shrub, *Aucuba japonica* (Cornaceae). Amer. J. Bot. 88: 455–461.

Adams, J.E. 1949. Studies in the comparative anatomy of the Cornaceae. J. Elisha Mitchell Sci. Soc. 65: 218–244.

Airy-Shaw, H.K. 1973. A dictionary of the flowering plants and ferns, 8th ed. Cambridge: Cambridge Univ. Press.

APG III (Angiosperm Phylogeny Group) 2009. An update of the Angiosperm Phylogeny Group classification for the orders and families of flowering plants: APG III. Bot. J. Linn. Soc. 161: 105–121.

Bate-Smith, E.C., Ferguson, I.K., Hutson, K., Nielsen, B.J., Swain, T. 1975. Phytochemical interrelationships in the Cornaceae. Biochem. Syst. Ecol. 3: 79–89.

Bernini, R.C., Iavarone, C., Trogolo, C. 1984. 1-*O*-*B*-D-Glucopyranosyleucommiol, an iridoid glucoside from *Aucuba japonica*. Phytochemistry 23: 1431–1433.

Breuer, B., Stuhlfauth, T., Fock, H., Huber, H. 1987. Fatty acids of some Cornaceae, Hydrangeaceae, Aquifoliaceae, Hamamelidaceae and Styracaceae. Phytochemistry 26: 1441–1445.

Cronquist, A. 1981. An integrated system of classification of flowering plants. New York: Columbia University Press.

Cronquist, A. 1988. The evolution and classification of flowering plants, 2nd ed. New York: New York Botanical Garden.

Eyde, R.H. 1964. Inferior ovary and generic affinities of *Garrya*. Amer. J. Bot. 51: 1083–1092.

Eyde, R.H. 1988. Comprehending *Cornus*: puzzles and progress in the systematics of the dogwoods. Bot. Rev. 54: 233–351.

Ferguson, I.K. 1977. World pollen and spore flora 6: Cornaceae Dum. Stockholm: Almquist and Wiksell.

Ferguson, I.K., Hideux, M.J. 1978. Some aspects of the pollen morphology and its taxonomic significance in Cornaceae sens. lat. Proc. IV Int. Palynol. Conf., Lucknow 1: 240–249 (printed in 1980).

Hara, H. 1972. Corresponding taxa in North America, Japan and the Himalayas. In: Valentine, D.H. (ed.) Taxonomy, phytogeography and evolution. London, New York: Academic Press, pp. 61–72.

Hara, N. 1980. Shoot development of *Aucuba japonica* I. Morphological study. Bot. Mag. Tokyo 93: 101–116.

Harms, H. 1898. Cornaceae. In: Nat. Pflanzenfam. III, 8. Leipzig: Engelmann, pp. 250–270.

Hu, W. K., Soong, T.P. 1990. Cornaceae. In: Fang, W.P., Hu, W.K. (eds.) Flora reipublicae popularis sinicae 56: 3–108. Beijing: Science Press.

Ishikura, N. 1971. Pelargonidin glycosides in fruits. Experientia 27: 1006.

Iwashina, T.K., Kamenosono, K., Hatta, H. 1997. Flavonoid glycosides from leaves of *Aucuba japonica* and *Helwingia japonica* (Cornaceae): Phytochemical relationship with the genus *Cornus*. J. Japan. Bot. 72: 337–346.

Kleiman, R., Spencer, G.F. 1982. Search for new industrial oils: XVI. Umbelliflorae - Seed oils rich in petroselinic acid. J. Amer. Oil Chemist's Soc. 59: 29–38.

Konta, F. 1994. Development of aerial roots and asexual propagation in *Orixa japonica* (Rutaceae) and *Aucuba japonica* (Cornaceae). Bull. Natn. Sci. Mus. (Tokyo) Ser. B 20(4): 157–161.

Kurosawa, S. 1971. Cytotaxonomical studies on the genus *Aucuba*. J. Jap. Bot. 46: 231–238.

Li, H.-L., Chao, C.-Y. 1954. Comparative anatomy of the woods of the Cornaceae and allies. Quat. J. Taiwan Mus. 7: 119–136.

Noshiro, S., Baas, P. 1998. Systematic wood anatomy of Cornaceae and allies. IAWA J. 19: 43–97.

Reidt, G., Leins, P. 1994. Das Initialstadium der sympetalen Krone bei *Sambucus racemosa* L. und *Viburnum farreri* Stearn. Bot. Jahrb. Syst. 116: 1–9.

Sato, Y. 1971. Embryological study in *Aucuba japonica* Thunb., with special reference to the unusual development of the embryo sac. Sci. Rep. Tohoku Univ. Ser. IV (Biol.), 35: 201–206.

Sato., Y. 1976. Embryological studies of some cornaceous plants. Sci. Rep. Tohoku Univ. Ser. IV (Biol.) 37: 117–130.

Soltis, D.E., Soltis, P.S., Chase, M.W., Mort, M.E., Albach, D.C., Zanis, M., Savolainen, V., Hahn, W.H., Hoot, S. B., Fay, M.F., Axtell, M., Swensen, S.M., Price, L.M., Kress, W.H., Nixon, K.C., Farris, J.S. 2000. Angiosperm phylogeny inferred from 18S rDNA, *rbc*L, and *atp*B sequences. Bot. J. Linn. Soc. 133: 381–461.

Takhtajan, A.L. 1980. Outline of the classification of flowering plants (Magnoliophyta). Bot. Rev. 46: 225–359.

Takhtajan, A.L. 1987. Sistema magnoliofitov. Leningrad: NAUKA.

Viinikka, Y. 1970. A comparative study of mitotic and meiotic chromosomes of *Aucuba japonica* Thunb. Ann. Bot. Fennici 7: 203–211.

Wangerin, W. 1910. Cornaceae. In: Engler, A. (ed.) Das Pflanzenreich IV, 229, Heft 4: 1–101. Leipzig: W. Engelmann.

Xiang, Q.-Y., Boufford, D.E. 2005 (Text), 2006 (Illustrations). Aucubaceae. In: Wu, Z.Y., Raven, P.H., Hong, D.Y. (eds.) Flora of China, vol. 14. St. Louis: Missouri Bot. Garden Press, Beijing: Science Press, pp. 222–226, fig. 410.

Xiang, Q.-Y., Soltis, D.E. 1998. *Rbc*L sequence data define a cornaceous clade and clarify relationships of Cornaceae sensu lato. In: Boufford, D. E., Ohba, H. (eds.) Sino-Japanese flora - its characteristics and diversification. Univ. Mus. Bull. 37: 123–137.

Xiang, Q.-Y., Soltis, D.E., Morgan, D.R., Soltis, P.S. 1993. Phylogenetic relationships of *Cornus* L. sensu lato and putative relatives inferred from *rbc*L sequence data. Ann. Missouri Bot. Gard. 80: 723–734.

Boraginaceae

Boraginaceae Juss., Gen. Pl.: 128 (1789) ('Borragineae'), nom. cons.

M. Weigend, F. Selvi, D.C. Thomas, and H.H. Hilger

Annual, biennial or perennial, rarely monocarpic herbs, subshrubs and rarely erect or lianescent shrubs, not aromatic; primary root usually persistent, often developed as strong taproot, rarely primary and secondary roots thickened and spindle-shaped or primary root splitting from above and forming root pleiocorm; stems terete, rarely angular, then often winged from decurrent leaf bases, erect, rarely ascending, sprawling, decumbent or procumbent, prostrate, sometimes forming rhizomes or stolon tubers; indumentum usually present on whole plant, often strongly developed, scabrid to hispid or strigose, sometimes uncinate, very rarely with stiffly deflexed trichomes for climbing or lanate or consisting of stellate trichome complexes or largely absent (trichomes reduced to basal cells), trichomes often scabrid, often mineralized with Si or Ca, often inserted on a multicellular basal tubercle or cystolith cell, rarely unicellular, simple and smooth or uniseriate and gland-tipped. Leaves alternate, rarely opposite, simple, entire, exstipulate, cauline leaves usually sessile to semiamplexicaulous, lamina linear, narrowly ovate to subcircular, usually decurrent on petiole, more rarely base distinctly rounded or cordate and distinctly petiolate, basal leaves usually forming an ephemeral or persistent rosette, sessile or tapering into a petiole. Inflorescences terminal or axillary, frondose, bracteose or ebracteose, paraclades monochasial or dichasial, lax or very dense, usually scorpioid and contracted into boragoids, these paraclades present as simple terminal inflorescence or combined into complex thyrsoids, sometimes with extensive accessory paraclades and metatopia. Inflorescences sometimes congested into terminal "heads" or strongly reduced to axillary or terminal single flowers. Flowers pentamerous, hypogynous, bisexual, often proterandrous; perianth biseriate, sepals united at base or nearly to apex, usually radially symmetrical, sometimes slightly or distinctly unequal, calyx tubular to rotate, membranaceous, usually accrescent in fruit, spreading or closing; petals usually united to form distinct tube, rarely united only basally, corolla actinomorphic, rarely zygomorphic or irregular, rotate, hypocrateriform, infundibular, campanulate or tubular, rarely curved or geniculate, corolla tube internally usually appendaged near base with ring-shaped intrusion or 5 or 10 free scales (basal scales = annulus) and near throat with 5 crescent- or scale-shaped intrusions (= faucal scales, fornices), these closing the tube and/or appressed to anthers and/or protruding to form a ring around opening of tube, erect, recurved or incurved, papillose to pubescent, usually contrastingly coloured (often yellow or white vs. blue or white corolla lobes); corolla lobes reflexed to porrect, narrowly triangular to (usually) subcircular; aestivation usually quincuncial, rarely contorted; stamens epipetalous and antesepalous, rarely unequal or some abortive, filaments free from each other, usually short and anthers included in tube or partially or completely exserted, inserted at the same or different heights in tube, filament lengths and/or point of insertion differing between morphs in taxa with stigma height polymorphism or heterostyly; anthers usually free from each other, dorsifixed, dithecous, tetrasporangiate, opening by longitudinal slits, sometimes anthers coherent, connective sometimes appendaged; gynoecium 2-carpellate, syncarpous, usually with basal nectary disc; ovary bilocular, but secondarily subdivided into four chambers by false septa, 4-lobed in flower; style overtopped by the nutlets (gynobasic), usually simple, rarely 2–4 stylodia, included or exserted,

J.W. Kadereit and V. Bittrich (eds.), *Flowering Plants. Eudicots*, The Families and Genera of Vascular Plants 14,
DOI 10.1007/978-3-319-28534-4_5, © Springer International Publishing Switzerland 2016

stigmas capitate to oblong; ovules anatropous-epitropous, usually basal or sub-basal, erect to pendulous, unitegmic, tenuinucellate. Fruit a dry schizocarp separating into four 1-seeded, rarely two 2-seeded, indehiscent mericarpids, sometimes fewer by abortion or fusion; mericarpids ("nutlets", "eremocarpids") usually equal, rarely heteromorphic (in individual fruit, in fruits on different parts of the plant or in different morphs of a population), nutlet shape and size highly variable, ovoid to subcircular, tetrahedral, lenticellate, bowl-shaped or disc-shaped, often with differentiated, lobed/glochidiate or membranaceous ("winged") margin, surface smooth and shiny to papillose and dull, pubescent, verrucose or with complex glochidia, sometimes with base strongly modified as ring and/or with basal elaiosome. Mature seeds with scanty endosperm, cotyledons fleshy, embryo usually straight.

The family in the narrow definition here adopted contains ca. 85 genera and 1600–1700 species; cosmopolitan, but largely absent from the wet tropics.

Vegetative Morphology. Boraginaceae are predominantly herbaceous, relatively few species are suffruticose or fruticose. Many species are ephemeral, winter/spring flowering annuals with life spans of a few weeks, mostly in temperate, desert and high alpine environments (e.g. some *Myosotis*, *Cryptantha*, *Microcaryum*). Winter annuals are common in warm-temperate and Mediterranean climates (Mediterranean *Cynoglossum*, some *Lithospermum*, many *Amsinckia*, *Cryptantha*, Schaal and Leverich 1982; Parsons and Cuthbertson 2001), but rare in temperate climates (*Omphalodes scorpioides*). Longer-lived, diffusely branched annuals [*Asperugo*, *Bothriospermum*, *Hackelia* p.p. ("*Austrocynoglossum*")] and biennials (many *Cynoglossum*, *Echium*) are also found. Some of the annuals are likely amongst the smallest angiosperms (*Microcaryum*: ca. 10 mm). Perennials are usually polycarpic, while monocarpic (hapaxanthic) perennials are rare (some *Echium*). Amongst perennial herbs a wide range of growth forms is present, including tiny, high-alpine cushion-forming plants, usually with soft, shaggy pubescence (*Chionocharis*, some *Eritrichium*, *Myosotis*), and prostrate or diffusely decumbent herbs with rooting stems and distichous phyllotaxy (e.g. some *Myosotis*, *Plagiobothrys*, *Trigonotis*). Coarse, large-leaved, often stoloniferous taxa, sometimes with gradual transitions to rhizomatous herbs and geophytes, are found in Boragineae (e.g. *Symphytum*, *Moritzia*, *Trachystemon*, *Pulmonaria*), but also Cynoglosseae (*Brachybotrys*, *Cynoglossum*) and Lithospermeae (*Ancistrocarya*). Thick, sympodial rhizomes are rare, but found in *Brachybotrys* (Cynoglosseae) and *Paramoltkia* (Lithospermeae). Distinct stolon tubers are very rare (e.g. *Symphytum tuberosum*, *S. bulbosum*). Pleiocorm herbs (i.e. perennial underground shoots permanently coherent via persistent primary root) are common in the family and are typical for, amongst others, *Mertensia*, *Rindera*, *Trichodesma* and most *Lithospermum*. These pleiocorm herbs usually have large, basal leaf rosettes, but these are absent in some taxa (many *Lithospermum*). Root pleiocorms are sometimes formed by the lengthwise splitting of the primary root from top to bottom in, for example, *Borago pygmaea*, some *Symphytum* and *Mertensia maritima*. Subshrubs, dwarf and small shrubs (<1.5 m) are comparatively frequent in Lithospermeae (e.g. *Lithodora*, *Lobostemon*, some *Lithospermum*) and Echiochileae (*Echiochilon*, some *Antiphytum*). They usually have narrow leaves with revolute margins and are more less distinctly ericoid. Some shrubby species are strongly stoloniferous, forming large colonies (*Glandora oleifolia*). Larger shrubs are even rarer, but ca. 25 *Echium* species endemic to Macaronesia show clear "insular woodiness". *Selkirkia berteroi* from the Juan Fernandez Islands is a shrub with leaves crowded at the shoot apices and reaches some 2 m in height. The probably largest species is *Trichodesma scottii*, a shrub up to 4 m high from Socotra. Leaves are generally alternate, sometimes crowded at the shoot apices (*Selkirkia*) or at the base of annual shoots, rarely verticillate (*Brachybotrys*). Opposite leaves are found in *Thaumatocaryon* (*Th. tetraquetrum*), *Antiphytum* and most *Trichodesma* and, at least in the basal part of the axis, are also characteristic of *Pectocarya* p.p. ("*Harpagonella*"), *Suchtelenia* and many *Plagiobothrys*. The leaves lack pulvini and stipules; the lamina is always simple and entire. *Amsinckia* and some *Cryptantha* have deeply bifid cotyledons which are the only divided leaves in the family. Leaves usually have a flat, membranaceous, (widely) ovate lamina, ranging from a few millimetres (some *Myosotis*) to 1.5 m

(*Myosotidium*) in length. Narrow, sometimes even needle-shaped leaves with revolute margins are found in some shrubby species of genera such as *Antiphytum*, *Echiochilon*, *Lithodora* and *Glandora*. Coloured leaf patterning, very rare in the temperate flora, is known only from Boragineae—e.g. in *Pulmonaria* (esp. *P. officinalis* group: white and green) and *Symphytum* (e.g. *S. ibericum*: white, purple and green). Leaf venation is mostly brochidodromous or (semi-)craspedodromous, rarely hyphodromous. Scale-like cataphylls are largely restricted to the rhizomes, stolons and stolon tubers of the few taxa with these modifications. Adaptations to periodical exposure to fire exist in several species of *Trichodesma*, *Lithodora*, *Echiostachys* and *Lobostemon*. Loss of the aerial parts through fire leads to vigorous vegetative re-growth from the pleiocorm and subsequent mass flowering and fruiting in *Lobostemon* and *Trichodesma*. Recovery may be very rapid, with reestablishment of the aerial parts in *Lithodora hispidula* taking as little as 4 months (Türkmen and Düzenli 2005).

Vegetative Anatomy. Vegetative anatomy of Boraginaceae s.str. is poorly understood, with few recent studies available (Solereder 1899; Jodin 1903; Metcalfe and Chalk 1950). The indumentum is usually very well developed and characteristically rough (Fig. 5). A characteristic feature is the presence of usually stalkless cystoliths in the basal cells of the trichomes, sometimes also the sheathing cells. Trichomes are often incrusted with minerals based on x-ray analysis (EDX) silica and calcium are the principal components; especially the short, stiff trichomes are strongly silicified (Hilger et al. 1993). Some species of *Cerinthe*, *Omphalodes* and *Mertensia* have a poorly developed indumentum or are virtually glabrous, leaves are then usually glaucous from epicuticular waxes. Trichome characters have been reported by several researchers and have been used for grouping taxa such as Macaronesian *Echium* species, but the results do not correspond well with DNA phylogenies. Selvi and Bigazzi (2001) investigated the leaf anatomy of more than 50 specific and subspecific taxa belonging to 14 genera of tribe Boragineae. They described seven trichome types, likely covering the majority of trichomes found in Boraginaceae: 1. uniseriate, eglandular, usually erecto-patent, sharply pointed, base surrounded by a conspicuous basal tubercle of 20–30 cells; 2. as 1. but smaller and with only 10–20 basal cells; 3. as 1. but trichome sitting on an inconspicuous ring of 6–8 cells; 4. very short trichome with pointed tip and densely micropapillose surface; 5. hooked (uncinate) trichome without a distinct base; 6. minute trichome tipped with 3-celled glands; 7. long, uniseriate, gland-tipped trichome with elongate head cell. Secretory trichomes functioning as salt glands have been reported from the halophilic *Mertensia maritima*. Straight but stiffly deflexed trichomes are found in subscandent *Hackelia* p.p. ("*Austrocynoglossum*") and *Asperugo* and possibly elsewhere. Stem anatomy has been studied in only few species (e.g. Park 1982) and shows no striking peculiarities. In the primary stem, open collateral vascular bundles are arranged in a cylinder. Cortical and medullary bundles as well as an internal phloem are absent. Secondary thickening develops from a cambial ring. Cork is usually initiated from a superficial or deep-seated cork cambium. Further studies focussed mainly on the woody Macaronesian *Echium* species (Carlquist 1970; Aldridge 1981) which are pachycaulous with a broad pith. Plants with long-lived shoots have distinct growth-rings, corresponding to annual growth and likely reflecting the seasonality of moisture availability. Xylem rays are present or absent; if present, they are uniseriate or multiseriate with usually erect or isodiametric cells, procumbent ray cells are rare. Axial parenchyma is scanty paratracheal and composed of two to five strands. Vessels are relatively short (between 100 and 300 μm) and narrow

Fig. 5. Boraginaceae. Indumentum of *Onosma echioides*. Bar = 500 μm

(40–65 μm) in both herbaceous and woody members. Carlquist (1970) found a negative correlation between vessel length and the degree of xeromorphy in Macaronesian *Echium*. Pits are bordered, alternate and circular. Libriform fibres (fibre-tracheids) are short (300–400 μm) and with vestigial borders. Storied wood structure probably is characteristic for all *Echium* species (Carlquist 1970). Oil bodies are absent (Lersten et al. 2006), but Carlquist (1970) observed resin-like compounds in nearly all *Echium* woods. Nodes are unilacunar and the leaf trace is a complex vascular strand leaving a rather wide lacuna (Neubauer 1977). A persistent leaf basal meristem is lacking. Roots are apparently mostly diarch (Jodin 1903) or tetrarch. Periderm develops from the pericycle. Growth rings may be present in the secondary root xylem of *Symphytum officinale* (Dietz and Ullmann 1997) and *Anchusa officinalis* (von Arx and Dietz 2006), thus allowing for a calculation of age on the basis of annual growth rings in the (persistent) primary root. Root hairs on seedling roots of Boraginaceae develop on an epidermis derived from the root cap and the trichoblasts are arranged in a radial pattern (Clowes 2000). The seedling root of *Echium plantagineum* shows a unique pattern with files of long root hairs producing cells separated by one or two files of cells bearing short hairs (Tsai et al. 2003). The leaf anatomy of Boraginaceae is not well studied, but apparently leaves are mostly bifacial with 2–3 layers of palisade parenchyma on the adaxial and spongiose parenchyma on the abaxial side, but isobilateral leaves are also common. The mesophyll commonly contains calcium crystals (raphides, druses or solitary-prismatic crystals). Stomata distribution is hypo- to amphistomatic and the stomata are mostly anomocytic or anisocytic, but other types of stomata have also been reported (Dasti et al. 2003). Stomata distribution may vary among leaves of a single shoot (Schittengruber 1953). In Boragineae, Selvi and Bigazzi (2001), investigating 14 genera and 49 species, found no correlation of leaf anatomy with taxonomy, but with habitat. Outer cell walls in the adaxial epidermal cells are thicker in more xerophilic or xerotolerant taxa than in mesophilic taxa. Mesophilic taxa have exclusively hypostomatic leaves, while xerophytic (incl. psammophytic) species show increased stomatal density on the adaxial surface. Transfer cells are present in the minor leaf veins in at least 17 genera. They usually have ingrowths distributed around the periphery ("type A") of specialized phloem parenchyma (Pate and Gunning 1969). Gunning et al. (1970) also found transfer cells in the shoot axes, associated with the xylem. Boraginaceae have S-type sieve-element plastids (Behnke 1981). Nuclear crystalloids are a common component of sieve elements in the Boraginaceae (Fisher et al. 1989). They are distinct from the P-protein of sieve-element protoplasts. Furthermore, nuclei of phloem parenchyma cells may also occur, but in much fewer species. Fisher et al. (1989) presented a list of species with the distribution of both types of nuclear inclusions, but there is no obvious correlation with the systematic position of the taxa. Chloroplast and nuclear crystalloids differ even within individual genera, only *Amsinckia* apparently shows a unique loose structure of nuclear crystalloids.

Inflorescence Structure. This paragraph is partly based on Buys and Hilger (2003). Detailed investigations of inflorescence architecture are rare. Boraginaceae have exclusively "closed" (monotelic) inflorescences and paraclades are usually cymose. Schwarzer (2007) found six different synflorescence types investigating only three South American genera (*Cryptantha*, *Pectocarya*, *Plagiobothrys*), including both homoeocladic and heterocladic thyrsoids, anthocladia and creeping special thyrsoids, indicating a high degree of diversification in this character complex. Overall inflorescences are usually frondose, frondose-bracteose, bracteose or ebracteose. Bract morphology often changes abruptly within the inflorescence and the pherophylls of paraclades (mostly frondose) may be morphologically widely different from those subtending dichasial ramifications (mostly bracteose) and these, in turn, may differ widely from the bracts in monochasial portions (distinctly bracteose or absent). Extensive recaulescence may occur with bracts moving far up lateral branches (esp. *Symphytum*). Bracts are usually green, but may be conspicuously coloured and take over the role of pollinator attraction from the obscured corolla in *Cerinthe*. The basic unit of Boraginaceae inflorescences is usually a scorpioid (= coiled) cincinnus, the so-called boragoid: the mostly shortly pedicellate and often ebracteose flowers are separated by initially very short internodes and arranged in two alternating rows. Flowers usually all face in the same direction. Loss of alpha-prophylls and re- and concaulescence of bracts may contribute to the overall architecture of the boragoid (esp. in Boragineae and *Cynoglossum*). Boragoids are often paired and arise from a first dichasial division (from the leaf axils of a pair of opposite bracts preceding an often well-developed terminal flower). These "paired boragoids" or "double boragoids" are a common basic unit of Boraginaceae inflorescences. An individual boragoid or double boragoid may make up the entire terminal inflorescence, in which case the inflorescence can be classified as a mono- or dichasial cymoid. Terminal cymoids may be strongly condensed and few-flowered or even reduced to one-flowered cymoids, with peduncle and internodes overtopped and obscured by foliage leaves, so that there are terminal "heads" (some *Lithodora*) or apparently single, terminal flowers (flowers open sequentially and buds are not immediately visible: cushion-forming *Eritrichium* and *Myosotis*). Commonly, the boragoids and more often the double boragoids are found as paraclades of more or less complex thyrsoids, rarely di- or pleiothyrsoids. Typically, paraclades are well-spaced and have elongated internodes at least basally, immediately revealing the thyrsoidal nature of the inflorescence. However, internodes in the paraclades and between the paraclades may be very short, leading to a basically cylindrical inflorescence resembling a spike (many *Echium*, "miner's candle" = *Cryptantha* spp.). In some cases the lateral paraclades are additionally reduced to one or two flowers, the overall inflorescence then superficially resembling a spike (*Lobostemon* sect. *Argentei*, Levyns 1934). A superficially capitate inflorescence is formed (some *Cryptantha*, *Plagiobothrys*) when the internodes separating the paraclades are completely reduced. Elongation of internodes, especially during fruit development, is also found in a wide range of taxa. In these cases the bracts are usually frondose. Bract and flower position clearly reveal the monochasial (cincinnal) nature of these inflorescences. These monochasial anthocladia are common in some mostly annual taxa such as *Asperugo*, *Bothriospermum* and some *Anchusa*.

Flower Morphology and Anatomy. Flowers are usually erect, sometimes held horizontally, nodding (*Pulmonaria*, *Symphytum*) or pendulous (*Onosma*, *Cerinthe*). Sepals, petals and stamens originate centripetally as five separate primordia each in quincuncial order and in alternate and aequidistant position, the first sepal arising is median adaxial. Carpel primordia are in a median position. The petal ring fuses ("late sympetaly", Erbar and Leins 1996) on the abaxial side of the stamen primordia, which are subsequently lifted upwards to form the stamen-corolla tube (Hilger 1985). The calyx is at least basally united and the united portion is very short (<10 % of overall sepal length) or very long with very short free lobes. Calyx lobes are mostly narrowly ovate-acuminate, sometimes linear (e.g. *Rochelia*) or broadly ovate-acuminate with a cordate base (*Trichodesma*). In *Asperugo* the calyx is folded and bilabiate, forming two roughly hemispherical, dentate lobes. Calyx aestivation is apert to valvate. Supernumerary sepals are present in *Plagiobothrys* spp. and *Rochelia* (up to nine in *R. cancellata*). Sepals have a more or less pronounced scabrid to lanate or uncinate indumentum. The corolla is usually conspicuous, the largest flowers of the family are found in American *Lithospermum* p.p. (>90 mm, "*Macromeria*"). Very small flowers with corollas barely reaching 1 mm in length and obscured by the calyx are found in *Cryptantha* and *Microcaryum*. Radially symmetrical, infundibuliform to hypocrateriform corollas with faucal scales protruding from the throat are the most common type in all tribes. (Sub-)Rotate corollas in usually deflexed flowers are found in *Borago*, *Trachystemon*, *Brachybotrys*, *Cystostemon* and *Trichodesma*. More or less campanulate or tubular corollas with porrect or reflexed lobes are characteristic for some groups (*Symphytum*, *Onosma*, *Mertensia*). Distinctly zygomorphic flowers are uncommon, but found scattered in Lithospermeae (slightly curved and widened: *Echium*, *Lobostemon*), Boragineae (strongly geniculate and narrowed: *Lycopsis*) and Echiochileae (*Echiochilon*). Corolla lobes are usually well developed, rectangular, oblong, triangular or suborbicular, spreading, erect or reflexed. The margin of the corolla lobes is usually entire, but fimbriate in *Arnebia fimbriopetala* and incised in *Lithospermum incisum*. Corolla lobes may be connected by smaller additional lobes (*Alkanna*) or trichome tufts (*Arnebia*). Aestivation of the corolla lobes is usually quincuncial, rarely contort (*Myosotis*). The stamen-corolla tubes rarely persist in fruit (*Anchusa samothracica*). Corollas are commonly blue, purple, pink or white. Orange corollas are largely restricted to Lithospermeae, yellow corollas are common in Lithospermeae and Boragineae but rare elsewhere. Corollas are, apart from the faucal scales, usually uniformly coloured, rarely bicolorous (*Onosma*, *Podonosma*). Colour changes are widespread—e.g. in *Symphytum* and *Pulmonaria* (pink to blue or dark purple)—and often also affect the colour of the faucal scales. Corolla colour polymorphism (usually blue and white) is occasionally found and has been the source of several cultivars. Inner basal scales sometimes cover the nectary disc and/or ovary. They may be annular or scale-shaped, hairy or glabrous; sometimes they are reduced to a barely differentiated fold. The five scales at the filament bases in *Lobostemon* are likely homologous to the basal scales in other genera. In the vast majority of taxa the throat of the corolla tube has characteristic invaginations, the so-called faucal scales or fornices. They often (partially or completely) close the entrance to the corolla tube and are usually well visible from outside by contrasting colours, commonly white or yellow. These scales originate as adaxial folds of the petal, then grow inwards as invaginations. Faucal scales are absent in Trichodesmeae and many Lithospermeae and a few other genera, most of them with tubular flowers. Faucal scales are usually exserted from the tube (hidden in *Alkanna*), but, apart from few exceptions (*Solenanthus albiflorus*, Czukavina and Meling 1982), always inserted above the filament bases. Arrangement is commonly in one series (but two series in *Trachystemon*). The scales are variable in shape, ranging from oblong, trapeziform or rectangular to emarginate or bifid and form either a closed ring around the throat (e.g. *Myosotis*) or are distinct and prominent (e.g. *Anchusa*, *Cynoglossum*). In *Symphytum circinnale* the long (8 mm) and narrowly triangular scales sheath the style. In some taxa the faucal scales are reduced to folds or hairy ridges or papillose tufts above or between the anthers.

In most taxa the stamens are inserted near the middle of the tube, in some taxa nearly at the base (e.g. *Neatostema*, *Anchusa* subg. *Buglossoides*, *Melanortocarya*) or at different levels (e.g. *Glandora*). Androecial zygomorphy—e.g. arrangement of anthers at different levels in the tube—occurs in several species of *Alkanna* and *Anchusa* and in taxa with zygomorphic flowers (*Echium*). In zygomorphic flowers the three abaxial stamens are usually inserted at a higher level than the two adaxial stamens. Stamens are usually all fertile except in *Anchusella*, where only two have anthers and three are staminodial. Anthers are usually included, but distinctly exserted from the tube in, for example, *Solenanthus* or *Rindera*. Extreme heteranthery with one very long and exserted stamen occurs in *Nonea* ("*Elizaldia*") *heterostemon* and *Caccinia monandra*. Anthers are of variable size, often more or less ovoid, rarely sagittate (*Onosma*, *Cerinthe*), basifixed to dorsifixed, often connivent to coherent and occasionally appendaged. Appendages may be basal (*Borago*) or apical (*Cystostemon*) and are long and spirally twisted in *Trichodesma*. *Lobostemon* has hairy staminal scales at or below the filament insertion within the tube. The pollen sacs open with longitudinal slits.

The gynoecium is bicarpellate, syncarpous, superior, each carpel giving rise to two uniovulate globular, triangular or erect to curved lobes. The following description is based on Hilger (1985). In flower primordia, the gynoecium bulges out in the centre of the flower. The two carpels are in a median position. After an initially conical growth of the entire ovary, the four ovular lobes begin to overtop the initially terminal stylar primordium, leading to a gynobasic style. Outgrowths—often referred to as "true" - corresponding to carpel boundary and "false" septum - from the ovary roof (apical septa) and from the base (basal septa)—divide and separate the primary two ovarial chambers into four. At maturity, nutlets are attached (apparently) directly to the receptacle or to the more or less conical gynobase. The corresponding scars left after detachment are termed "areole" on the gynobase and "cicatrix" on the nutlet. Deviations from the common pattern of gynoecium structure and merosity are described under "Fruit and Seed". The style is simple in most species, rarely forked (e.g. *Echium*) to twice forked with four stigmas (*Arnebia*). The style is mostly included in taxa with hypocrateriform corollas, but may be long exserted in tubular corollas. The style usually persists, but plays no role in fruit dispersal. Stigmas are entire or shallowly bilobed, capitate, conical or cylindrical. Subterminal stigmas are sometimes formed by the extension of a sterile style elongation beyond the stigma (*Buglossoides*, some *Lithospermum*). The stigmatic surface is dry (Heslop-Harrison 1981) and usually has unicellular receptive papillae of various, mostly lageniform (flask-like) shapes.

Embryology. The last comprehensive work on the embryology of the family was published by Wunderlich (1991). Anthers are tetrasporangiate with four-layered walls and development follows the Dicotyledonous type (Khaleel 1977). The cells of the glandular tapetum are either mononucleate or two- to multinucleate (up to 10 in *Pentaglottis sempervirens*, Mascré 1922). There is no differentiation into an inner and an outer tapetum. Pollen formation is simultaneous. Pollen grains are predominantly shed in the three-nucleate stage. Two-nucleate grains have been reported for *Mertensia platyphylla* (Khanna 1964) and *Trichodesma* (Khaleel 1977), but apparently the number of nuclei may vary even within species (*Trichodesma stocksii*, Wunderlich 1991). The ovules are epitropic, unitegmic and tenuinucellate. Their position varies from upright in Lithospermeae and Boragineae to horizontal or pendant in Cynoglosseae. Integuments may consist of up to 20 cell layers (*Cynoglossum denticulatum*). Integumentary rostra serving as obturators are present in *Myosotis* (Svensson 1925), *Lepechiniella* p.p. ("*Microparacaryum*", Briechle and Hilger 1988), *Cynoglossum* and *Trichodesma* (Khanna 1964). An integumentary tapetum is not developed. There is usually one archesporial cell (rarely 2 to 3) developing directly into the megaspore mother cell. Usually, the embryo sac develops according to the *Polygonum* type, the bisporic type was reported for *Anchusa officinalis*. Aposporic embryo sac formation has been reported either from the integument or from the chalaza (*Cynoglossum denticulatum*). The mature embryo sac is usually spherical. The Boragineae are apparently unique in the formation of a "diverticle", i.e. a lateral outgrowth directed towards the funicle (Svensson 1925; Pitot 1937). The synergids are well developed and in some cases larger than the egg cell. Polar nuclei are adjacent to the egg apparatus and usually fuse simultaneously with fertilization. Fertilization is porogamous. The endosperm is either cellular (reported only from *Myosotis arvensis*), intermediate (sensu Wunderlich 1959, modifications from cellular endosperm) or nuclear (the "*Borago*" type of Svensson 1925). Mature seeds are virtually exendospermous with one or two cell layers of endosperm, the remaining space being take up by the large embryo. Embryogeny is variable (Souèges 1948; Crété 1963; Briechle and Hilger 1988), corresponding to the first period and megarchetypes II, III, V of Souèges (1948) or the *Aster* and *Chenopodium* type of Johansen (1950). Embryo development may vary in individual genera or even species. The suspensor, if present, is short, usually consisting of only one, rarely two cell layers. The radicle is very short, the cotyledons contain lipids and proteins as storage reserves (McGill et al. 2002). Cotyledons are broadly ovate to nearly subcircular and shortly acuminate cotyledons with fine pubescence appear to be the most common type. In species with glabrous and glaucous adult leaves the cotyledons are also glabrous and glaucous (e.g. *Mertensia*, some *Omphalodes*). A few genera have deeply bifid cotyledons (*Amsinckia*, *Cryptantha* p.p.).

Pollen Morphology (by Oliver Mohr). Studies comparing pollen across floras or the family show that Boraginaceae are remarkably eurypalynous (Fig. 6). Differences are found in shape, size, exine sculpturing and number, nature and position of apertures and pseudoapertures, colpus margin differentiation and presence or absence of mesocolpial or apocolpial fields and/or an endocingulum (Fig. 6). The smallest pollen grains of the family (and in Angiosperms) are apparently found in Cynoglosseae, namely in *Cryptantha* (ca. 4 × 2 μm in *C. clevelandii*, Hargrove and Simpson 2003), *Myosotis* and *Trigonotis* (ca. 5 × 2 μm, Grau and Leins 1968). The largest pollen grains reach up to 55 × 40 μm (*Anchusa*, Boragineae, Clarke 1977) and large grains may in general be more typical of Boragineae and Lithospermeae. Pollen shape is commonly prolate or dumbbell-shaped. Exine sculpturing is variable and ranges from psilate (*Buglossoides*, *Cynoglossum*), perforate-foveolate (*Echium*, *Nonea*), microechinate (*Onosma*, *Cerinthe*), microreticulate (*Moltkia*) and verrucate (*Mertensia*, *Symphytum*) to granulate (*Borago*, *Symphytum*—following the terminology of

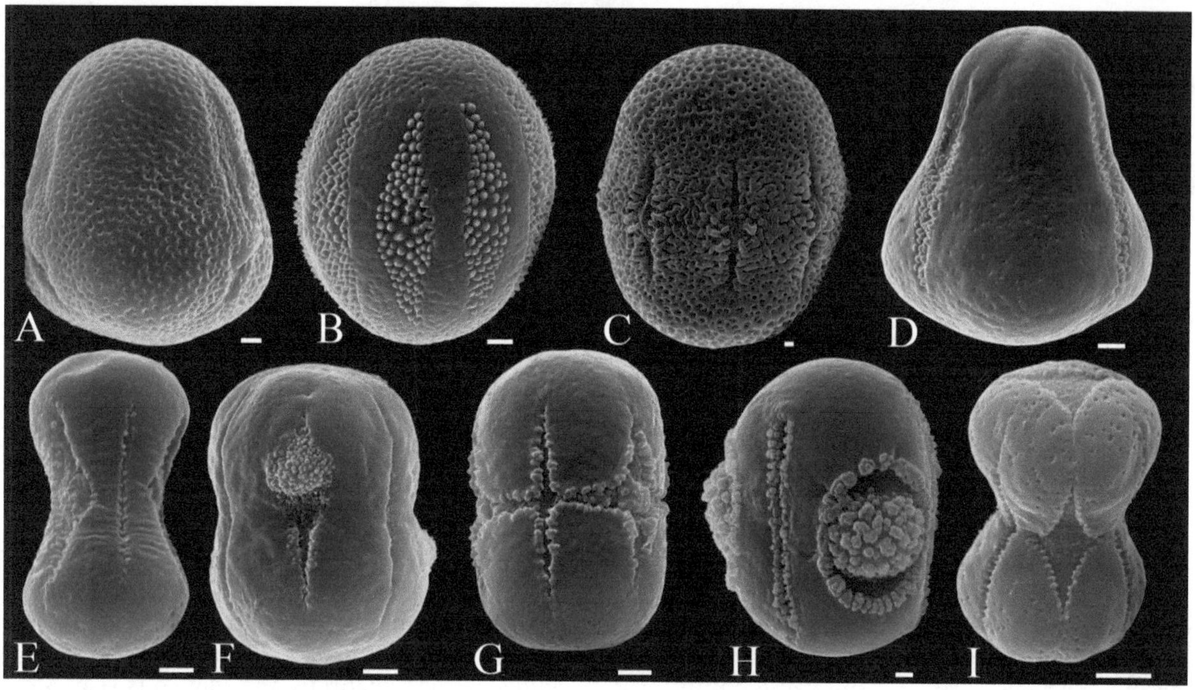

Fig. 6. Boraginaceae pollen grains. **A** *Echium vulgare*, ovate, 3-colporate. **B** *Moltkia petraea*, rounded, 8-colporate, echinate ornamentations of apertural surface. **C** *Nonea vesicaria*, 8-colporate, reticulate, with change of exine ornamentation towards the equatorial level. Transition between ornamentation of apocolpium/mesocolpium and region of endocingulum. **D** *Alkanna hirsutissima*, pear-shaped, 3-colporate. **E** *Trigonotis rockii*, 6 apertures (simple colpi and colporate apertures alternating), perforate-foveolate exine surface, pores less dense in region of mesocolpium. **F** *Eritrichium nanum*, 3 colporate alternating with 3 colpate apertures, the pores characteristically arranged (2 on one, the third on the other sphere of the pollen grain), in the centre of the image colporate aperture with porus on the right side. In the lower left, upper part of porus on the back side, third one hidden. **G** *Cynoglossum creticum*, 6 apertures (as in *T. rockii*, 3 simple colpi, 3 colporate apertures). **H** *Solenanthus watieri* (≡ *Pardoglossum watieri*), typical ("*Pardoglossum*") apertural peculiarity, 2 types of apertures: simple ones (long, slit-like, parallel sided) and compound apertures (more or less circular, densely granulate). **I** *Myosotis azorica*, top right the apocolpial field on polar area, apertures compare *T. rockii*. Bars = 1 μm

Halbritter et al. 2008, examples of genera not necessarily exclusively with one type). Ornamentation can be homogeneous over the entire grain, but it can also be differentiated between apo- and mesocolpia: in *Nonea vesicaria* the apocolpia are perforate-foveolate, mesocolpia are fossulate and reticulate, in *Trigonotis rockii* the density of perforate-foveolate sculpturing decreases towards the equatorial constriction, in some *Lithospermum* the mesocolpium is psilate, but the apocolpium is echinate. Apertures are usually colporate, but porate pollen is also known (some Lithospermeae). The vast majority of taxa has 3–4-aperturate grains. In Echiochileae the pollen is often 2–3-colporate and usually more or less oblong to rectangular in equatorial view (Lönn 1999). Higher aperture numbers are found mainly in Lithospermeae and Boragineae with 6–8 in *Cerinthe* (Sahay 1979) and 8-colporate in many (American) *Lithospermum*, 10–12(−15)-colporate (zonoaperturate) in *Borago officinalis*, some *Moltkia* and 15-colporate in *Nonea heterostemon* (Selvi et al. 2006). Apertures are mostly median, but apertures close to one pole are common in Lithospermeae. Some species of *Arnebia* (Lithospermeae) and possibly also some *Lappula* (Cynoglosseae) have dumbbell-shaped grains with bi-aperturate colpi (Khatamsaz 2001). Syncolpate pollen grains with neighbouring colpi anastomosing near the poles (but leaving a small island of exine at the pole itself) are known from, for example, some Lithospermeae (Avetisjan 1956). Syncolpate pollen grains with colpi united across the poles are found in *Heterocaryum* (Cynoglosseae, Avetisjan 1956). The *Lappula* group (Cynoglosseae) shows a very peculiar aperture position with two pores located closer to one pole and the third one closer to the opposite pole (Avetisjan 1956). Endoapertures are mostly lalongate or circular. The ornamentation of the ectoapertural margins ranges from straight and smooth (*Cerinthe major*), frayed (*Echium vulgare*), thickened with rounded or elongated ornamentation (*Mertensia ciliata*, *Solenanthus apenninus*) to spinulose (*Anchusella*). The colpus membrane is smooth or granulose with spherical or conical ornamentation (e.g. *Gastrocotyle*). The mean exine thickness ranges from 0.11 μm (*Cryptantha*), 0.5 μm (*Cynoglossum*, *Lappula*) to 3 μm (*Borago*). Heterocolpate pollen is characteristic of the vast majority of Cynoglosseae (but normally tricolporate; e.g. in *Trichodesma*, *Craniospermum*, *Caccinia*, Avetisjan 1956). They usually have 3–4 colporate apertures alternating with 3–4 slit-like pseudocolpi (pseudo-, "simple and compound" apertures) not functional for pollen tube growth. This is not proven by pollen germination studies but corroborated by TEM investigations (Hargrove and Simpson 2003). *Cynoglossum* p.p. ("*Pardoglossum*") has circular, complex and granular apertures alternating with narrow slit-like pseudocolpi.

Endocingular modifications of the exine along the equator, leading either to a granular reduction of the exine or a slit-like incision forming a belt around the grain are frequent in Boragineae, where they are present as a granular equatorial belt, often diffusely delimited from the remainder of the exine (Bigazzi and Selvi 1998). Conversely, in Cynoglosseae there is often a sharp incision (without or with rugose ornamentation) uniting the colpi and pseudocolpi in the equatorial plain (Avetisjan 1956; Hargrove and Simpson 2003). Both the width of the endocingulum and the ornamentation of its surface show considerable variation. Triangular or circular areas in the apocolpium are known from some species of *Myosotis*, *Trigonotis* and *Cryptantha* (Hargrove and Simpson 2003), but appear not to be present in all species of these genera. In *Myosotis* and *Trigonotis* they are delimited by the margins of the anastomosing colpi, while in *Cryptantha* they are not connected to the colpi and have a distinctly modified margin. Large, rugose, circular apocolpial fields, usually at only one pole, are occasionally found in *Thaumatocaryon* (Boragineae). In this genus there are always mesocolpial fields of exactly the same shape and sculpture. Rugose, more or less circular mesocolpial fields are found in *Moritzia* and *Thaumatocaryon*, but also in some *Anchusa* s.l. (*Phyllocara aucheri*, Bigazzi and Selvi 1998) and are best considered as derived from the rugose endocingulum that is common in Boragineae. All four subgroups of Boraginaceae here recognized thus have some unique pollen characters which are typical of a large part of the tribe (e.g. only 2 apertures in Echiochileae, pear-shaped, heteropolar in Lithospermeae, a rugose endocingulum or mesocolpial fields in Boragineae, heterocolporate in Cynoglosseae). However, none of these modifications is present in all representatives of the respective groups. Prolate, 3-colporate pollen grains are found in at least some representatives of all of them. Some smaller subgroups can, however, be clearly delimited by pollen morphology.

Karyology. The first karyological surveys of Strey (1931), Smith (1932) and Britton (1951) already revealed considerable variation in base and somatic chromosome numbers, ploidy levels and karyotype morphology. Approximately 630 specific and subspecific taxa, representing ca. 35 percent of the Boraginaceae, have been surveyed at least for chromosome number. The taxonomical and geographical distribution of sampling is highly uneven, with numerous data from Europe and North America and very few data from Asia, Africa and South America. Metaphasic A-chromosomes range from 1–2 to ca. 10 μm (some *Anchusa*); the smallest appear to be more typical of Cynoglosseae (e.g. *Rochelia*, *Lappula*, *Asperugo*), the larger of Lithospermeae and especially Boragineae. Metacentric, submetacentric and subtelocentric chromosomes form the complement of most taxa. Bimodal karyotypes are rare (*Onosma*). Supernumerary B-chromosomes have been observed in only a few genera (see below).

Somatic complements vary between $2n = 8$ (*Amsinckia lunaris*) and $2n = 144$ (*Symphytum tuberosum*). Base numbers do not seem to match the tribal subdivision of the family, although some numbers occur with higher frequency in given taxonomic groups. Echiochileae are karyologically unknown. Boragineae are the most comprehensively studied group. Primary base numbers are $x = 6, 7, 8, 9$ and 10, with $x = 8$ being particularly common; of possible secondary origin are $x = 11$, 12, 13 and 15. In *Anchusa* s.l. (*Anchusa*, *Anchusella*, *Lycopsis*, *Hormuzakia*, *Phyllocara* and some *Gastrocotyle*) diploidy with $2n = 16$ is prevalent, in addition there are some tetraploids ($2n = 4x = 32$ in *Anchusa* spp.) and hexaploids (*Lycopsis arvensis*, $2n = 6x = 48$). *Symphytum* is probably the karyologically most variable genus of the family, showing aneuploidy, dysploidy ($x = 7$, 8, 10, 11, 12 and 15), auto- and allopolyploidy associated with hybridisation (up to dodecaploidy). *Nonea* ($x = 7$, 8, 9, 10, 15, up to decaploid) and *Pulmonaria* ($x = 7$, 8, 9, 11, $x = 8$, 9, 10, 11, 12, 13 in diploid or polyploid cytotypes) are also karyologically diverse. Supernumery B-chromosomes have been observed in both *Nonea* and *Pulmonaria*. In those Boragineae investigated, 23% are polyploids, while the rest are exclusively diploid or at least include diploid cytotypes. Lithospermeae have fewer polyploids (16%, with ca. 38% of the species of the tribe surveyed). Primary base numbers are $x = 6$, 7, 8, 9, 10 and (possibly secondarily) 11, 12, 13 and 15, with $x = 7$ and 8 being most common and $x =$ 6 and 12 rare. The lowest base number, $x = 6$, is known from *Pontechium*, *Huynhia* and *Moltkiopsis* ($2n = 12/24$) and some diploid species of *Onosma* ($2n = 12$). Diploids predominate in *Echium* ($2n = 16$, rarely 10, some tetraploids with $2n = 32$), *Lobostemon* and *Echiostachys* ($2n = 14$, some tetraploids $2n = 28$), *Arnebia* ($2n =$ 14, 16, 22, 24), *Moltkia* ($2n = 14$, 16) and *Cerinthe* ($2n = 16$ or 18). *Onosma*, the largest genus of Boraginaceae (>150 spp.), is karyologically particularly diverse. The most common base numbers are $x = 6$ and 7, polyploidy is based on $x = 6$, 7, 8, 10 and 11. Several species show high intraspecific polymorphism and unstable karyotypes. Bimodal karyotypes and other phenomena, such as permanent anorthoploidy, have also been reported. *Lithodora*, *Lithospermum* and *Buglossoides* mostly have the base numbers $x = 7$ and 8, sometimes $x = 9$, 11, 12, 13; $2n = 20$ seems unique to the Italian endemic *B. calabra*. Polyploidy, mainly based on $x = 7$, is widespread in *Lithospermum* s.l. and in annual species complexes of *Buglossoides*, while the perennial species of *Buglossoides* are mostly diploid. *Lithodora* is rarely diploid ($n = 13$, $2n = 16$), more often tetraploid based on $x = 7$, 8, 9 and 10. An extreme case of high-level polyploidy is known for the monotypic Balkan genus *Paramoltkia* ($2n = 96$, 114, assuming $x = 8$: 12-ploid or 14-ploid). Cynoglosseae are karyologically relatively homogeneous, with diploids with $2n = 24$ being the most common type. Descending aneuploidy is present in several taxa with $2n = 22$ and $2n =$ 20. In *Myosotis* aneuploidy is common ($x = 11$, 10, 9, 8, 7). Polyploidy or aneuploidy are present in over 50% of the species of *Myosotis* studied and may have played an important role in the worldwide diversification of the genus. Polyploidy is less frequent. The karyologically most diverse group in Cynoglosseae appear to be the closely allied genera *Amsinckia* and *Cryptantha*. In *Amsinckia* primary numbers range from $x = 4$ to $x = 8$, plus several odd base numbers of likely secondary origin such as 13, 15, 17 and 19. Intraspecific aneuploidy is documented in *A. intermedia* and *A. menziesii*. In *Cryptantha* the base numbers $x = 6, 7, 8, 9$ and 10 have been reported, $2n = 24$ is present in only few species and higher levels of polyploidy are reported from *C. haplostachya* and *C. calycotricha* ($2n = 8x = 64$).

Pollination and Reproductive System (by Markus Ackermann). Zoophilous allogamy is the most common breeding system in Boraginaceae, but selfing is also common across the family and cleistogamy is reported for several genera (see below). Abiotic pollination is unknown. Selfing is typical of small-flowered annual species but also of at least some perennials (e.g. *Anchusa crispa*, Quilichini et al. 2001). In many species fruit set is close to 100% in the complete absence of pollinators. Cleistogamous flowers have been reported from *Cryptantha* (Grau 1983), *Lithospermum* (Lord 1981) and *Neatostema*. Chasmogamous and cleistogamous flowers are generally found on the same plant, with cleistogamous flowers produced near the ground or in the ground in *Cryptantha* p.p. ("*Geocarya*", Higgins 1971), on the lower, very short branches of the inflorescence (*Lithospermum*) or in poorly developed individuals and towards the end of the growing season (*Neatostema*). Conversely, there are many species which are obligate outcrossers and have developed special mechanisms to avoid selfing: both self-incompatibility and dichogamy are common, protogyny is widespread and proterandry has been reported from, for example,

Trichodesma and *Alkanna*. Herkogamy in the form of a stigma-height dimorphism is frequently observed in *Anchusa* and *Lithodora* (Al-Shehbaz 1991), heterostyly in, for example, *Amsinckia*, *Arnebia*, *Cryptantha*, *Glandora*, *Lithodora* and *Paracaryum*, where it is often associated with pollen heteromorphy (Ganders 1979). True dioecy has not been reported from Boraginaceae, but gynodioecy is known from *Echium* and *Myosotis*. Pollinators are attracted mostly by flower colour and less by scent produced from osmophores (Maurer 1961). Scent is usually not very strong, but may be pronounced especially in some white-flowered species.

Flowers are usually conspicuously coloured, often bicolorous and the corolla limb and faucal scales then show contrasting coloration. White is probably the most common colour in Boraginaceae, followed by blue, purple, yellow and pink. True blue flowers, rare in most other plant families, are particularly common. Yellow corollas are particularly common in Lithospermeae and some Cynoglosseae (*Amsinckia*), red, pink or brownish corollas are found across the family. Colour changes of the corolla during anthesis are relatively common (Weiss 1995). Either the whole corolla, the centre of the flower (androecial parts or faucal scales) or only the nectar guides (e.g. *Amsinckia*, *Huynhia*) change colour. Colour change is probably a visual signal for the pollinators, indicating changes in reward availability. Characteristic UV-reflection and absorption patterns have been documented on the corolla limbs (Rosen and Barthlott 1991). Faucal scales and/or trichome patches largely limit access to the nectar at the base of the tube to animals with sufficiently long and narrow mouth parts and/or foretarsi. In addition, they may serve as pollen exchangers: pollen released by the anthers is trapped by the papillose or pubescent scale surface; pollen is then transferred to the hairy head or body of visiting insects brushing against the scales while trying to reach the nectar at the base of the ovary.

Nectar is the primary floral reward. It is produced by a nectar disc as basal part of the gynoecium and secreted through modified stomata. Own observations show that within the insect-pollinated taxa nectar production (standing crop) is between 0.03–10.1 μl per flower and sugar concentrations range from 16 to 59% (Ackermann, Mittelbach, Weigend, unpubl. data, ca. 100 spp.). Nectar is usually rich in glucose, fructose and sucrose; smaller amounts of maltose and arabinose have been found in some taxa (e.g. *Anchusa*, Nepi et al. 2010). Flower visits and pollination follow common patterns. In cultivation in Central Europe the vast majority of Boraginaceae across floral types are visited primarily by long-tongued bumblebees (*Bombus*) and these are also the most important pollinators in some taxa in their natural habitat. However, in nature many species are pollinated by apparently quite specialized, often oligolectic bees, especially *Anthophora* and *Osmia*. A particular case of narrow specialization has been reported for *Cynoglottis barrelieri*, pollinated by *Colletes anchusae* (Apoideae, Müller and Kuhlmann 2003). Species with narrow corolla tubes are primarily pollinated by Diptera such as beeflies and tachinid flies, while butterflies are reported to visit flowers especially with large limbs and long and narrow corolla tubes. Pendulous or deflexed flowers with connivent or coherent anthers of, for example, *Borago*, *Caccinia*, *Cerinthe*, *Onosma*, *Symphytum* and *Trichodesma* are buzz-pollinated by various bees (Teppner 2011). Though here pollen is the primary reward, flowers may still be nectariferous (Gilbert et al. 2001). Ornithophily by hummingbirds is known for some North American species of *Lithospermum* with corolla tubes up to 9 cm in length. Red-flowered *Lobostemon* in South Africa is sunbird-pollinated. In Tenerife (Canary Islands), native birds have been observed to visit the red-flowered *Echium wildpretii* and the nectar properties of this species correspond to ornithophily (Dupont et al. 2004). Small lizards (*Gallotia* spp.) have also been observed entering the flowers of *Echium wildpretii* and drinking the dilute nectar, but it has not been verified whether they pollinate the flowers (Olesen and Valido 2004).

Fruit. The fruits of Boraginaceae provide the single most important set of characters for their classification and taxonomy. During fruit maturation the four parts of the ovary usually develop into four separate, highly characteristic nutlets (= mericarpids, Fig. 7), the calyx is typically accrescent and pedicel and internodes of the

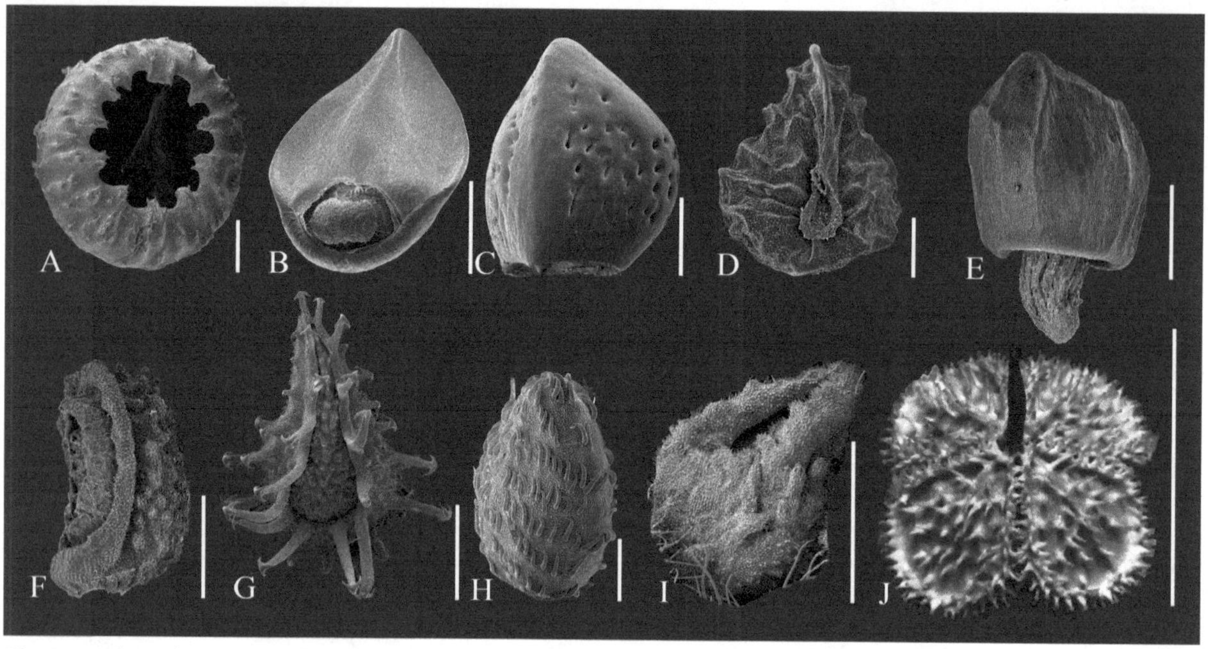

Fig. 7. Nutlets of Boraginaceae. **A** *Omphalodes linifolia*. **B** *Myosotis involucrata* (syn. *Trigonocaryum involucratum*). **C** *Lithospermum cinerascens*. **D** *Plagiobothrys congestus*. **E** *Borago pygmaea*. **F** *Antiotrema dunnianum*. **G** *Lappula squarrosa*. **H** *Lasiocaryum densiflorum*. **I** *Microula pseudotrichocarpa*. **J** *Cynoglossum officinale*, fruit. A, B, D, E, F, G, I bars = 1 mm, C, H bars = 250 µm, J bar = 1 cm. (G from Hilger 1985, with permission from Borntraeger-Cramer)

inflorescence elongate. There are several exceptions from this basic pattern. Two two-seeded twin-nutlets develop in *Cerinthe*. Reduction in the number of nutlets occurs frequently, there are only two nutlets in *Rochelia*, a single nutlet is common in, for example, *Buglossoides*, *Trichodesma*, *Caccinia*, *Moritzia* and *Thaumatocaryon*. Mericarpid multiplication (up to 10) is only known from *Trigonotis* p.p. ("*Zoelleria*"), but its ontogeny has not been investigated. Nutlets are usually dispersed individually and separately from the calyx. Taxa with glochidiate, epizoochorous fruits usually have small and spreading to reflexed sepals (e.g. *Cynoglossum*, *Lappula*, *Hackelia*). In some taxa the calyx is accrescent, but the base becomes wider and the calyx lobes become more or less flaring, providing a wide opening for the nutlets to fall out (e.g. *Symphytum*). In other taxa the mature nutlets remain enclosed in the calyx, with the calyx then usually having uncinate (hooked) trichomes which facilitate epizoochory (some *Myosotis* spp., *Asperugo*, *Pectocarya*, *Moritzia*). In *Moritzia* and *Thaumatocaryon* the single fully developed nutlet does not become detached from the gynobase, but remains firmly attached to it during dispersal in the closed calyx. In *Asperugo* the calyx folds transversally and closes over the nutlets with two dentate lobes and, in *Pectocarya* p.p. ("*Harpagonella*"), two fused sepals arch over and enclose one of the two nutlets. *Cryptantha circumscissa* has circumscissile calyx dehiscence. In *Suchtelenia*, the broad, wing-like, persistent calyx ensures wind-dispersal. Three basically different types of heterocarpy are found in Cynoglosseae, with fruits differing between individuals, on different parts of a plant (amphicarpy, known only from a few species of *Cryptantha* in Chile) or nutlets differing in the individual fruit (heteromericarpidy). Heteromericarpidy again follows three different patterns and is relatively frequent in widely different groups of the Cynoglosseae. In *Suchtelenia* there is a 2+2 pattern: two nutlets detach and are dispersed separately, the other two nutlets remain firmly attached to the gynobase and the strongly enlarged calyx, but all nutlets are morphologically similar. In *Heterocaryum* a 2+(1+1)-pattern is found: two opposite mericarpids are identical in shape and ornamentation, the other pair differs from the first pair and the corresponding two nutlets additionally differ from each other. The most widespread form of heteromericarpidy is a 3+1 pattern, with three nutlets becoming detached and a fourth one remaining firmly attached to the gynobase (and sometimes the plant). This is found in several distantly related groups of Cynoglosseae (e.g. *Lappula*, *Lepechiniella*, *Amsinckia*, *Plagiobothrys*, *Cryptantha*). The fourth nutlet is here dispersed either with the calyx or with fragments of the mother plant. The three nutlets which are dispersed separately are usually larger, have a broader rim/wing and/or are distinctly glochidiate, the fourth is smaller and the surface less strongly ornamented and/or unwinged (Hilger et al. 1985). In almost all Boragineae and in some *Lithodora* and *Myosotis*, an often peg-shaped plug of gynobase tissue breaks off with the nutlet. This structure is, at least in Boragineae and *Myosotis*, an elaiosome serving ant dispersal. Nutlet orientation is intimately linked to both the shape of the gynobase and the shape of the nutlet. In Boragineae, Lithospermeae, Echiochileae and some taxa of Cynoglosseae (e.g. *Myosotis*, *Trigonotis*, *Bothriospermum*), the erect nutlets sit on a flat gynobase. The nutlets of *Trichodesma*, *Caccinia* and *Suchtelenia* are sunken into the convex gynobase. Nutlets in Cynoglosseae are mostly elevated on their adaxial side, i.e. sit in an oblique position due to the pyramidal shape of the gynobase. They always have a distinct adaxial and abaxial surface and are often dorsiventrally compressed and differentiated. The "nutlet margin" thus formed may be noticeable as a more or less distinct ridge or may be spectacularly elaborated in the form of a ring-shaped, comb-like or wing-like, sometimes glochidiate, usually simple but sometimes double structure. The ventral side of the nutlet is often convex or more or less sharply keeled, but concave in the mature fruits of *Mertensia*. The dorsal side is usually flat or concave, rarely convex. The nutlets are usually broadest at the base, but sometimes indistinctly (some *Lithospermum*, *Plagiobothrys*, *Alkanna*, *Podonosma*, *Stenosolenium*) or distinctly stipitate (*Thaumatocaryon*, *Echiochilon*). Nutlets, especially in Boragineae and Lithospermeae, are often slightly incurved (e.g. *Borago*, *Echium*) to nearly horizontally incurved (e.g. *Anchusa*, *Alkanna*). Ornamentation of the nutlet surface is widely variable across the family. Entirely smooth and shiny nutlets occur in relatively few taxa. The common name of *Lithospermum* ("stone seed", "marble seed") reflects the white, shiny, porcelain-like appearance of the fruit.

The fruit wall itself has an arrangement of layers characteristic for each tribe (F1 exocarp = outer epidermis, F2 sclerenchyma, F3 parenchyma, F4 endocarp = inner epidermis, Seibert 1978). Boragineae, Lithospermeae and Echiochileae share a common fruit wall type and differ from Cynoglosseae (Seibert 1978; Hilger 1985). The pericarp is also sometimes incrusted with silica and/or calcium carbonate, especially in Lithospermeae, to a lesser degree in Echiochileae. Outer exocarp cell walls often have a distinct secondary sculpturing varying from micropapillose to verrucose and shortly spinulose. Very rarely the nutlet surface is pubescent (*Lasiospermum*, some *Trigonotis*). Commonly the nutlets are ornamented with irregular crests, wrinkles or glochidia (most Cynoglosseae). These glochidia (Figs. 8, 9) are usually complex, multicellular, one- to many-hooked structures, often including subepidermal tissues. Their stiffness depends on the degree of silica incrustation in the cell walls. Nutlet margins are widely variable in Cynoglosseae. In *Cynoglossum* there is often a row of free glochidia along the margin, whereas the bases of these glochidia are more or less united, forming a

Fig. 8. Boraginaceae. Fruit of *Cynoglossum furcatum* with four densely glochidiate nutlets and the gynobasic style arising from the centre. (phot. H.J. Ensikat)

Fig. 9. Boraginaceae. Glochid of *Cynoglossum columnae*. Bar = 100 μm

small flat rim or wing along the margin. As a further elaboration these glochidia can fuse completely and loose their glochidiate nature entirely, forming a flat and narrow wing in *Mattiastrum* or a broad one in *Rindera*. In *Omphalodes*, *Lepechiniella* p.p. and *Paracaryum* the wing is thinly membranaceous, incurved and partly closes over the dorsal side of the nutlet, leaving a more or less wide "aperture". The lateral wing can be displaced into a different position by asymmetrical growth of the nutlet body relative to the margin. In *Bothriospermum* and *Antiotrema* this leads to an inverse morphology with the originally "dorsal" (abaxial) marginal wing facing the style, i.e. displaced into an adaxial position. The "lateral wing" is here turned into an adaxial, umbilicus-like structure ("pseudo-aperture", often mistaken for a cicatrix).

Seeds, enclosed by a testa consisting only of outer epidermal cells, are exalbuminous. Some taxa show more or less pronounced dormancy. Cold and moist stratification overcomes this dormancy or at least dramatically increases germination rates in many temperate and Mediterranean Boraginaceae such as *Amsinckia*, *Arnebia*, *Buglossoides*, *Pulmonaria*, *Mertensia* and *Lithospermum*. Increased germination after exposure to smoke has been reported for several species of *Cryptantha* (Keeley and Fotheringham 1998) and is likely present in other groups such as *Echium* and South African *Lobostemon* and *Echiostachys*. Seed longevity is relatively high under dry storage even at relatively high temperatures (several years in some *Lithospermum*, Baskin and Baskin 1991).

Dispersal. Zoochory is likely the most important dispersal mechanism in Boraginaceae, followed by anemochory and hydrochory. Epizoochory is possibly the single most important mechanism. In the majority of Cynoglosseae the nutlets have glochidia permitting the attachment of the fruits to fur and/or feathers (e.g. *Cynoglossum*, *Lindelofia*, *Solenanthus*, *Hackelia*,, *Lappula*). Some Cynoglosseae are very widespread, indicating the efficiency of this dispersal mechanism. At least one endemic Malagasy species of *Cynoglossum* has secondarily lost its glochidia for dispersal (*C. tsaratananse*, Miller 2005). In individual taxa throughout the family, the calyx encloses mature nutlets (some or all) and has uncinate trichomes or glochidia for animal dispersal (e.g. *Moritzia*, *Myosotis*, *Amsinckia*). Exozoochorous dispersal is reported for glochidiate nutlets, but even the only "rough" (verrucose) nutlets of *Echium plantagineum* appear to attach readily to fur and are transported in this manner (Parsons and Cuthbertson 2001). In several taxa the fruiting calyx remains firmly attached to the gynobase, and the entire fruit remains firmly attached to the plant, leading to "whole plant dispersal" or to gradual plant disintegration and fruit release. In many species of *Pectocarya* the inflorescence axis disarticulates at fruit maturity and the fruits are dispersed with parts of the inflorescence axis. In one species of *Pectocarya* ("*Harpagonella*") the mature nutlets are additionally enclosed by two united, accrescent and incurved sepals with glochidia. The enlarged calyx in *Asperugo* also encloses the mature nutlets and is densely covered with uncinate trichomes. However, it remains firmly attached to the plant and the entire plant is covered with uncinate trichomes, so that here large parts of the plant are dispersed on the fur of larger mammals. A similar case may be present in *Rochelia*, where the nutlets remain attached to the gynobase and are additionally enclosed by incurved trichomatose sepals. Granivory, i.e. active collection of fruits for food by animals (and subsequent loss or hoarding), may play a very important role in the dispersal of a large number of Boraginaceae. The superficially very different fruits of *Lappula*, *Myosotis*, *Cryptantha* and *Lithospermum* are commonly found in rodent faeces and/or middens or around ant mounds. Birds are likely important for the dispersal for the stone-like seeds of Lithospermeae, esp. *Lithospermum* (Best and Smartt 1986), but these are also accidentally consumed by grazing animals and pass into the faeces undigested and viable (e.g. *Echium plantagineum*, Parsons and Cuthbertson 2001). Ants are attracted by the elaiosomes on the nutlet (Bresinsky 1963). Nutlets with large elaiosomes are characteristic for Boragineae, but also *Myosotis* p.p. (incl. "*Trigonocaryum*") and possibly some *Lithodora* (Thomas et al. 2008). Anemochory is largely restricted to Cynoglosseae, namely to taxa with distinctly winged nutlets (*Lepechiniella*, *Mattiastrum*, *Omphalodes*, *Paracaryum*, *Rindera* and *Suchtelenia*), where some nutlets remain attached to the dramatically enlarged calyx. Hydrochory has been reported as dispersal by sea water for the coastal *Mertensia maritima* (Ridley 1930; Skarpaas and Stabbetorp 2001) and *Anchusa crispa* (Quilichini and Debussche 2001). In both species, fruits tolerate exposure to sea water and float for a considerable time (>9 and 1 week respectively). Rain splash dispersal has been reported for *Trigonotis brevipes*: the nutlets are splashed out of the cup-shaped calyces by raindrops (Nakanishi 2002), and it is likely present in a number of *Trigonotis* species with similar fruits. Dispersal by flash-floods has been suggested for *Alkanna orientalis* (Wolff et al. 1997). Whole-plant dispersal by wind or flash-floods is the likely dispersal mechanism for *Lappula* p.p. ("*Sclerocaryopsis*"), where all nutlets remain attached to the mother plant, but the mother plant becomes strongly lignified when fruiting and breaks off near the base.

Phytochemistry (by Dietrich Ober). Phytochemistry of Boraginaceae has been studied mostly because of their (former or present) use for medicinal preparations and for dyes. The bulk of the studies concentrated on the most common representatives of Cynoglosseae (*Cynoglossum*), Boragineae (*Anchusa*, *Borago*, *Symphytum*) and Lithospermeae (*Lithospermum*), whereas many species-rich groups remain entirely unstudied. No reports have been published on tribe Echiochileae. On the basis of the data available, the most characteristic secondary metabolites for Boraginaceae appear to be fructanes, allantoin, γ-linolenic acid as transport and storage compounds and naphthoquinones of the alkannin type and pyrrolizidine alkaloids as typical secondary metabolites. The characteristic storage carbohydrates are fructanes, a group of β-D-fructofuranosides with up to 40 fructose residues attached to a sucrose moiety and belonging to the inulins. They are found predominantly in the underground organs of perennial taxa (up to 5.15% of fresh weight of *Symphytum officinale* roots). They form a demulcent mucilage and several species were used in folk medicine due to this property. Recently, fructo-oligosaccharide cinnamoyl ester derivatives have been identified in *Lindelofia stylosa* with sucrose, 1-kestose and nystose esterified at both ends to caffeic acid derivatives (Choudhary et al. 2006). The cyclit D-bornesitol seems to be widespread within Boraginaceae (Hegnauer 1964), but absent from the other families of the Boraginales (Hegnauer 1997).

The seeds of Boraginaceae have a high oil content, with main component and often also the corresponding tetraenic acid, octadecatetraenoic acid (ODA, = stearidonic acid, C18:4 $\Delta^{6,9,12,15}$, 5–21 %,). These two fatty acids are characteristic of Boraginaceae s.str. separating this family from all other Boraginales which lack these compounds (Hegnauer 1989). In humans GLA and ODA act as metabolic precursors of eicosanoids such as prostaglandins and leukotrienes. Due to their ability to lower blood cholesterol levels, oils rich in GLA are of high commercial value for use in dietary supplements (Cook et al. 2002). The oils contain a range of different tocopherols at varying levels (Velasco and Goffman 1999).

Nitrogen storage and transport is mainly mediated by S-Allantoin in Boraginaceae (Hegnauer 1989). This storage compound contains equal amounts of nitrogen and carbon and its nitrogen mobilization is achieved via enzymatic degradation (Kim et al. 2007). Conversely, the nitrogen containing pyrrolizidine alkaloids (PAs) do not serve as storage compounds and show neither turnover nor enzymatic breakdown in the living plant. They are produced by the plant as chemical defence against herbivores and are poisonous and strong feeding deterrents. They are accumulated in their polar, salt-like *N*-oxide form primarily in young leaves and in the reproductive organs of the plant (reviewed in Hartmann and Ober 2000). Most of the PAs found in species of the Boraginaceae are 1,2-dehydropyrrolizidines that are known to be responsible for acute and chronic poisoning of humans and livestock by causing particularly irreversible liver damage and photosensitivity in cattle and other domestic animals including poultry (Stegelmeier et al. 1999; Authority ANZF 2001; Stewart and Steenkamp 2001). Several species belonging to Boraginaceae are traditionally used as medicinal plants or remedies, but the presence of PAs in many of them spoils their value for medicinal purposes (Roeder 1995, 2000). PAs are esters of a necine base, a bicyclic amino alcohol that is esterified to one or more necic acids. Within the Boraginaceae (as well as in the Heliotropiaceae), PA-structures that belong to the lycopsamine type dominate. These lycopsamine type PAs are characterized by a unique C_7-necic acid that is synthesized via an intermediate of valine biosynthesis (Hartmann and Witte 1995; Weber et al. 1999). For the biosynthesis of the necine base moiety, homospermidine is the first specific intermediate that is synthesized by homospermidine synthase (HSS) from putrescine and spermidine as substrates (Ober and Hartmann 1999). Based on sequence data of HSS it was shown that this alkaloid-specific enzyme was recruited early in the evolution of the Boraginales by duplication of the gene encoding deoxyhypusine synthase. In spite of this common origin of PA biosynthesis in members of the Boraginales, the site of PA biosynthesis varies significantly between *Symphytum officinale* (biosynthesis in shoots and roots), *Cynoglossum officinale* (exclusively in the roots) and *Heliotropium indicum* (Heliotropiaceae, exclusively in the shoots; Frölich et al. 2007). These observations were supported recently by immunolocalization experiments, showing that in these three species HSS expression is restricted to completely different cell types (Niemüller et al. unpublished).

Naphthoquinones are widespread in Boraginaceae and the liposoluble alkannin and its stereoisomer shikonin are major components of the deep red pigments found especially in the taproot of many species of the Boraginaceae (esp. Lithospermeae, Hegnauer 1989), but occasionally also in the aerial parts (*Plagiobothrys*, Hegnauer 1997). These red pigments are quite common in Lithospermeae with the two economic sources *Lithospermum erythrorhizon* from E Asia and *Alkanna tinctoria* from W Asia and have also been reported from Cynoglosseae (*Plagiobothrys myosotoides* and *P. arizonicus*). Herbarium sheets of species containing this dye can easily be identified by the intense colouration of the paper. The presence of the red colorant may be species-specific or be present only in individuals or populations of otherwise uncoloured species (e.g. *Buglossoides arvensis*). The use of alkannin, shikonin and their derivatives as pigments for food-colouring and cosmetics and for several medical applications can be traced back many centuries and was reviewed recently (Papageorgiou et al. 1999). These compounds have been shown to possess strong wound healing, antitumor, antimicrobial, anti-inflammatory and antithrombotic properties (Papageorgiou et al. 2006). Alkannin and shikonin are prenylated naphthoquinones that are synthesized from two key precursors, 4-hydroxybenzoate and geranyldiphosphate. With respect to the stereospecificity of many biosynthetic pathways it is remarkable that plants are able to produce both enantiomers simultaneously. Nevertheless, the composition of this mixture varies: 3-alkannin, the S-enantiomer, dominates mainly in European plants (for example more than 95% in *Alkanna tinctoria*), while shikonin, the R-enantiomer, is the dominant compound (84–93%) in *Lithospermum erythrorhizon* from E Asia (Hegnauer 1997; Papageorgiou et al. 2006).

Rosmarinic acid, an ester of caffeic acid with α-hydroxydihydrocaffeic acid, is found in several plant families, but is most characteristic for members of the Boraginaceae and subfamily Nepetoideae of Lamiaceae (Petersen and Simmonds 2003; Harley et al. 2004). Rosmarinic acid shows various biological activities of which the main activities are adstringent, antioxidative, anti-inflammatory, antimutagen, antibacterial and antiviral (Petersen and Simmonds 2003). Further caffeic acid derivatives that are characteristic for members of the Boraginaceae are lithospermic acid, an ester of three hydroxycinnamic acid moieties and lithospermic acid B, a dimer of rosmarinic acid (Petersen and Simmonds 2003). This polyphenol is found in *Anchusa, Echium, Symphytum* and the eponymous *Lithospermum* species (Hegnauer 1989) and was discussed as a plant-derived antifertility agent due to its gonadotropine inhibitory effects (Farnsworth et al. 1975). Other secondary metabolites have only been reported sporadically. Traces of volatile oils (i.a. thymol and carvacrol) have been reported from *Echium* and *Onosma*. Cyanogenic compounds have been found in only very few species (e.g. (2S)-dhurrin from *Borago officinalis*, Van Valen 1979; Hegnauer 1989). Triterpenoids and triterpenoid derivatives such as phytosterols, triterpen alcohols and saponins (glycosides of nonpolar triterpenoids with surface active properties) have been reported from several distantly related species in Boragineae and Cynoglosseae (*Anchusa officinalis, Caccinia glauca, Trichodesma africanum, Symphytum officinale*, Hegnauer 1964, 1989, 1997). The flavonoid glycoside rutin is apparently relatively widespread in the family (Hegnauer 1989). In the dark blue bee pollen of *Echium plantagineum* the anthocyanin petunidin-3-O-rutinoside was identified as the main pigment, while in the violet-blue flowers of *Lobostemon* delphinidin-3,5-diglucoside was identified as the most common pigment (Van Wyk et al. 1997). Cyanidin, delphinidin and malvidin are the dominant chromophores of Boraginaceae flower anthocyanins (Hegnauer 1997). Flavone aglycones were identified in the exudates on the plant surface of *Nonea* species (Wollenweber et al. 2002).

Subdivisions and Relationships. Boraginaceae lend their name to the order Boraginales in the Lamiid clade (APG III, Angiosperm Phylogeny Group 2009). Molecular, morphological and phytochemical data clearly indicate that Boraginaceae in their narrow circumscription are monophyletic (Gilg 1907; Nowicke and Miller 1989; Wang and Harley 2004; APG III 2009). *Wellstedia*, traditionally included in Boraginaceae and at one time assigned to a distinct subfamily Wellstedioideae Pilg. (Pilger 1912), is likely the sister group of Boraginaceae and should be treated as Wellstediaceae Novák (Weigend et al. 2014). Traditionally, Boraginaceae have been treated as a family with the four subfamilies Boraginoideae, Heliotropioideae, Cordioideae and Ehretioideae (Gürke 1893). However, capsule-fruited Hydrophyllaceae appear to be nested in Boraginaceae in this classical circumscription (Ferguson 1996). The former subfamilies of Boraginaceae can be redefined as Ehretiaceae (incl. Cordiaceae, Lennoaceae and Hoplestigmataceae), Boraginaceae (s.str.) and Heliotropiaceae, all of which are morphologically well-defined. Hydrophyllaceae represent a more complex problem, since they are evidently not monophyletic and their fruit morphology is so different from that of the other groups of Boraginales that none of its subgroups/genera can be readily incorporated into one of the families of the Boraginales without compromising their morphological integrity. A group of genera of Hydrophyllaceae appears to be monophyletic and can be redefined as Hydrophyllaceae s.str., but *Codon* appears to be sister to Boraginaceae s.str. plus Wellstediaceae. Formerly included in Boraginaceae s.str. as subfam. Codonoideae (Retief and van Wyk 2005), it is, however, so different in fruit and flower morphology that we prefer segregation as the monogeneric Codonaceae (Weigend and Hilger 2010; Weigend et al. 2013, 2014). *Hydrolea* has been segregated from Hydrophyllaceae into Solanales as Hydroleaceae.

The infrafamilial classification of Boraginaceae is in urgent need of revision. Up to 13 tribes and eight subtribes have been recognized (Popov 1953), with many monospecific genera and subtribes. Recent attempts at improving this system (Ovczinnikova 2009) caused additional confusion. These and other infrafamilial classifications concentrated on moving individual "aberrant" genera into species-poor tribes or subtribes, but leaving the bulk of genera in the established tribes, leading to an equally complex and confusing infrafamilial classification. The definitions of these small (sub-)tribes are mostly based on evidently apomorphic characters such as twin-nutlets (Cerintheae), zygomorphic flowers (Echieae), increased nutlet number (Zoellerieae), "inverted" nutlets (Bothriosperminae), reduction to two nutlets (Rochelieae) or united sepals curving over the fruit (Harpagonelleae). Some of these tribes become obsolete by the synonymization of their type (and only) genera under larger genera with different tribal placement (*Zoelleria* in *Trigonotis*; *Harpagonella* in *Pectocarya*, *Trigonocaryum* in *Myosotis*).

The most recent listing of tribes was provided by Riedl (1997), recognizing the six tribes

Boragineae, Cynoglosseae DC., Eritrichieae Benth. & Hook., Lithospermeae Dumort., Myosotideae Reichenb. and Trigonotideae Riedl, but molecular studies (e.g. Weigend et al. 2010, 2013) retrieved Eritrichieae, Myosotideae and Trigonotideae as nested within Cynoglosseae sensu lato. Weigend et al. (2013) retrieved a total of four strongly supported lineages, corresponding to the Boragineae, Echiochileae (Riedl) Långström & M.W. Chase, Lithospermeae and Cynoglosseae. Additional tribes (e.g. Eritrichieae, Myosotideae, Trichodesmeae Zak., Craniospermeae DC. ex Meisn., Asperugeae Zak. ex Ovczinnikova) could be recognized, but these are either species-poor or even monospecific, or would have to be extensively redefined. Relationships within three of the major tribes are relatively well understood and have been studied with a combination of molecular and morphological characters: Lithospermeae (Seibert 1978; Thomas et al. 2008; Cecchi and Selvi 2009; Weigend et al. 2009), Boragineae (Guşuleac 1923, 1928, 1931; Hilger et al. 2004; Weigend et al. 2010) and Echiochileae (Lönn 1999; Långström and Chase 2002). These studies led to a consolidated view of generic limits and to the inclusion of several small or monospecific genera into larger genera (e.g. *Elizaldia* into *Nonea*, *Paraskevia* into *Anchusa*, *Lasiarrhenum*, *Onosmodium*, *Macromeria*, *Nomosa* into *Lithospermum*), while also recognizing or confirming some segregate genera which had previously been overlooked or not generally accepted (e.g. *Buglossoides*, *Glandora*, *Melanortocarya*, *Pontechium*). Generic limits and approximate relationships between genera are thus largely resolved for these three tribes.

The largest and taxonomically and morphologically most complex tribe, Cynoglosseae, has been largely resolved by Weigend et al. (2013). It comprises more than half of the species of the family, many of them in several large and/or widespread and/or heterogeneous genera such as *Cryptantha*, *Cynoglossum*, *Microula*, *Lappula*, *Hackelia* and *Omphalodes*. Additionally, numerous small, often monospecific genera have been described, such as *Amblynotus*, *Austrocynoglossum*, *Gyrocaryum*, *Mimophytum*, *Omphalolappula* and *Sinojohnstonia*. These are generally poorly defined and were uncritically segregated from larger genera without a study of the "parent genera" across their range. Comprehensive phylogenetic data have so far not been published but preliminary molecular data indicate that, for example, the majority of generic segregates of *Cynoglossum* (*Cynoglossopsis*, *Ivanjohnstonia*, *Lindelofia*, *Paracynoglossum*, *Pardoglossum*) are nested in *Cynoglossum*. Also, *Hackelia*, *Lappula*, *Microula* and *Eritrichium* are paraphyletic in relation to several segregate genera (such as *Amblynotus*, *Austrocynoglossum*, *Omphalolappula*, *Setulocarya*). Weigend et al. (2013) retrieved the following clades: Trichodesmeae (incl. *Lasiocaryum*, *Microcaryum*, *Trichodesma* and *Caccinia*), *Mertensia* group (e.g. *Asperugo*, *Mertensia*), *Omphalodes* group (e.g. *Omphalodes*, *Myosotidium*), *Lappula* group (= Eritrichieae s. str., e.g. *Suchtelenia*, *Heterocaryum*, *Eritrichium*, *Hackelia*, *Rochelia*), *Myosotis* group (= Myosotideae s.str., *Myosotis*, *Decalepidanthus*, *Trigonotis*), and Core-Cynoglosseae (e.g. *Amsinckia*, *Bothriospermum*, *Cryptantha*, *Cynoglossum*, *Microula*). Large-scale generic realignments will be required in Cynoglosseae, with nearly all larger genera polyphyletic and/or paraphyletic (e.g. *Cynoglossum*, *Omphalodes*, *Lappula*, *Eritrichium*, *Hackelia*; Weigend et al. 2013). The present classification and the genera here recognized therefore have to be considered as highly provisional.

In the present treatment we recognize the four well-supported tribes Boragineae, Cynoglosseae, Echiochileae and Lithospermeae. Within Cynoglosseae, genera are further arranged into a total of 10 informal groups, based on morphological and/or molecular data. Genera are recognized when they are well-characterized on the basis of morphological and/or molecular data, but even morphologically distinctive genera are synonymised when they are known to be derived from within other genera (e.g. *Actinocarya* in *Microula*, *Ivanjohnstonia* in *Cynoglossum*). Two genera, while clearly referable to Cynoglosseae, remain unplaced to genus group, because data on their affinities are currently equivocal. Trichodesmeae (composed of the *Trichodesma* and the *Lasiocaryum* groups, sister to each other) are sister to all remaining genera and genus groups of Cynoglosseae. The *Trichodesma* group consists of two genera (*Caccinia*, *Trichodesma*) of robust herbs or shrubs ranging from Africa to Central and South Asia and Australia. They have very short corolla tubes and mostly spreading corolla lobes with long exserted anthers. Unlike Cynoglosseae, they have tricolporate pollen (versus heterocolpate). The *Lasiocaryum* group consists of three small genera (*Lasiocaryum*, *Microcaryum*, *Chionocharis*) of small, annual to perennial herbs from E Asia. They share the presence of turbinate, mostly pubescent nutlets not found elsewhere in the family. The *Omphalodes* group consists of only three genera, with *Omphalodes* polyphyletic (Weigend et al. 2013) and Core-*Omphalodes* likely paraphyletic to the two other genera of the group, i.e. *Myosotidium* and *Selkirkia*. All three are softly pubescent to glabrescent, annual or perennial herbs or shrublets with nutlets with a distinct but variable dorso-marginal wing and flower generally long-pedicellate. *Omphalodes* is here already widely defined and includes the segregates *Mimophytum* (N America) and *Omphalotrigonotis* and *Sinojohnstonia* (E Asia). The genus is in urgent need of a critical revision across its range and the European species appear to be paraphyletic in relation to the segregate genera proposed. *Myosotidium*, from the Chatham Islands, is morphologically quite aberrant in the family, but molecular data clearly retrieve it as nested in *Omphalodes*. The *Myosotis* group comprises the four genera *Trigonotis*, *Brachybotrys*, *Decalepidanthus* and *Myosotis*, of which all but *Myosotis* are restricted to C and E Asia. They are small-flowered herbs with soft, often appressed pubescence and small, lentil-shaped to dorsally keeled or obliquely tetrahedral, smooth and dark brown to black nutlets. The *Mertensia* group (two closely allied north temperate genera, *Mertensia* and *Anoplocaryum*) consists of perennial, often large-leaved, often more or less glabrous or glabrescent herbs, mostly with large, blue, campanulate corollas. The nutlets of these two genera are differentiated from all other genera by shape: They are irregularly ovoid, sometimes indistinctly winged and are attached to the shortly pyramidal gynobase with a short, submedial stipe arising from the flat or concave adaxial side. Aberrant *Asperugo* is retrieved as sister to these two genera. The *Lappula* group (Eritrichieae s.str.) consists primarily of the three large genera *Eritrichium*, *Hackelia* and *Lappula* (Eurasia and Americas), none of which is monophyletic as currently circumscribed. Central Asian *Heterocaryum* and *Suchtelenia* were retrieved as sister to this clade. Several small genera such as *Austrocynoglossum* are here treated as synonyms, but some well-characterized, albeit doubtfully monophyletic genera such as *Rochelia* are maintained as separate. Genera of the *Lappula* group are mainly short-lived, often annual, herbaceous and the flowers have a usually narrowly pyramidal gynobase with glochidiate, often marginally winged nutlets. The *Bothriospermum* group consists of the three E Asian genera *Bothriospermum*, *Antiotrema* and *Thyrocarpus*. The nutlets have a two-layered wing, which is unique in the family. In two of the genera the "wings" are displaced into an adaxial position, which is also unique in the family. The *Cryptantha* group is largely American and consists of small-flowered taxa with dense, much-branched inflorescences and variable nutlet morphology. The nutlets usually have a submedial or medial attachment and a distinct dorsal keel or carina, are mostly small and, apart from *Pectocarya*, unwinged. The subdivision into four genera, followed in floristic treatments, has been shown to be unnatural. Hasenstab-Lehman and Simpson (2012) have found *Cryptantha* to be paraphyletic to *Amsinckia* and *Plagiobothrys*, and *Plagiobothrys* to be polyphyletic in *Crypthantha*. They therefore proposed a subdivision of *Cryptantha* into several segregate genera. The group is very diverse and its largest genus is *Cryptantha* with ca. 190 spp. The relationships between the *Cryptantha* group and the *Cynoglossum* group are unclear and *Microula* probably intercalates between the two. Both genus and species limits are highly controversial. The taxonomically most confusing taxon is *Cynoglossum* s.l., technically characterized by mostly large, glochidiate, sometimes conspicuously winged fruits. It is subcosmopolitan in distribution and highly paraphyletic and polyphyletic (Weigend et al. 2013). The bulk of the monospecific segregates (e.g. *Ivanjohnstonia*, *Cynoglossopsis*) are here treated as synonyms of *Cynoglossum*. Other, often small and always geographically restricted genera have been segregated, such as the Mediterranean/W Asian *Lindelofia*, *Mattiastrum*, *Paracaryum*, *Rindera*, *Trachelanthus* and *Solenanthus*, American *Dasynotus* and African *Afrotysonia*. They are defined by evidently derived types of either fruit (variously winged) or flower (corolla lobes porrect, anthers exserted) morphology. These genera are here provisionally maintained as separate, but may have to be either synonymised in the future, saved by the segregation of additional entities from *Cynoglossum* or by the transfer of species between genera. All of them are likely to be nested in *Cynoglossum* as currently defined (Weigend et al. 2013).

Distribution and Habitats. All four tribes have their centre of diversity in Eurasia. Over 70 of the 85 genera here recognized are found in and the majority of them are restricted to Eurasia. The centre of diversity for Boragineae, Lithospermeae and Echiochileae is in W Asia and the Mediterranean basin, with all narrowly endemic genera in Boragineae and Echiochileae and most narrowly endemic genera of Lithospermeae found there. The centre of diversity of Cynoglosseae is E and C Asia. Australia has only representatives of Cynoglosseae and no endemic genera apart from the Chatham Islands (New Zealand) endemic *Myosotidium*. Africa and America have representatives of all four tribes, but Africa has only three endemic genera (*Afrotysonia*, *Echiostachys*, *Lobostemon*) and the Americas seven (*Amsinckia*, *Antiphytum*, *Cryptantha*, *Dasynotus*, *Moritzia*, *Pectocarya*, *Thaumatocaryon*). The only largely American subgroup of Cynoglosseae is the *Cryptantha* group in W North and W South America (*Cryptantha*, *Plagiobothrys*, *Amsinckia*, *Pectocarya*, ca. 300 spp.), but a handful of species of *Plagiobothrys* are found in NW Asia and Australia. The other subgroups are either endemic to E Asia (*Lasiocaryum* group, *Bothriospermum* group, *Craniospermum*) or have their centre of diversity in C and E Asia (*Lappula* group, *Cynoglossum* group, *Myosotis* group, *Trichodesma* group). Several genera are very widespread and present on several continents—e.g. *Trichodesma* (Australia, Asia, Africa), *Lithospermum* and *Lappula* (Eurasia, Africa, S and N America) and *Eritrichium* (Eurasia, North America). *Hackelia*, *Lappula*, *Cynoglossum* and *Myosotis* are present on all continents with a centre of diversity for *Hackelia* in W North America and for *Myosotis* in both the Mediterranean basin and New Zealand. Boraginaceae are a predominantly (warm-) temperate and Mediterranean climate family and in the tropics largely restricted to higher elevations. The *Trichodesma* group, Echiochileae and *Cystostemon* (Lithospermeae) are the only groups with a largely (sub-)tropical range and are found in subtropical and tropical semi-desert and desert habitats. Another exception are some species of *Trigonotis* (Cynoglosseae), which are found in SE Asian rainforests and are the only group of Boraginaceae s.str. in this habitat. Boraginaceae grow in nearly all habitats, but the majority prefers open, sunny places. A very large number of species in genera such as *Cryptantha*, *Pectocarya*, *Microcaryum*, *Lasiocaryum*, *Myosotis*, *Cynoglossum*, *Amsinckia* and *Anchusa* grow in open, disturbed ground without closed vegetation cover. There are several coastal species in *Mertensia*, *Lithospermum*, *Anchusa*, *Omphalodes*, numerous species of steppe habitats in *Cynoglossum*, *Lindelofia*, *Myosotis*, *Rindera* and *Lithospermum* and some species of extremely high elevations (>3500 m, *Hackelia* and *Plagiobothrys* in South America, *Eritrichium* in Eurasia, *Chionocharis* in Asia). Mesophytic forest herbs are less numerous, but are common in Boragineae (*Brunnera*, *Symphytum*, *Pulmonaria*, *Trachystemon*) and Cynoglosseae (*Omphalodes*, *Myosotis*, *Mertensia*, *Hackelia*, *Brachybotrys*), but rare in Lithospermeae (*Ancistrocarya*). True hydro- and helophytes are apparently lacking in the family, but some species of *Myosotis* are common in at least seasonally inundated habitats. The majority of taxa appears to prefer neutral or basic soils, often calcareous ground, but there are a few species which prefer distinctly acidic soils in, for example, *Mertensia* and *Decalepidanthus*. Serpentine soils have been colonized by several taxa of Lithospermeae (*Halacsya*, *Paramoltkia*, *Onosma*, Cecchi and Selvi 2009), Boragineae (*Anchusa*) and Cynoglosseae (*Rindera*, *Cryptantha*, *Plagiobothrys*) as true serpentinophytes.

Islands are generally poor in species of Boraginaceae, with the only exception of radiations of *Echium* in Macaronesia (Canary Islands, Madeira, Cape Verde Islands; Böhle et al. 1996) and *Myosotis* in New Zealand (Winkworth et al. 2002). In both cases a dramatic diversification of (esp. vegetative) morphological characters took place. In *Echium* woody ornithophilic taxa arose from herbaceous mainland ancestors, in *Myosotis* a wide range of flower morphologies and colours and growth forms arose from the rather uniform north temperate ancestors. A recent revision of

Cynoglossum in Madagascar and the Comoro Islands recognized six species, five of which are endemic and one of which shows a striking loss of dispersability (glabrous nutlets). Three island endemic (and monotypic) genera of Boraginaceae have been described from the Pacific, namely the shrubby *Selkirkia* (Juan Fernandez Islands, Chile), the spectacular, large-leaved *Myosotidium* (Chatham Islands, New Zealand) and the shrubby *Nesocaryum*, (Isla San Ambrosio, Chile). However, *Nesocaryum* is here included in *Cryptantha* and is probably closely allied to shrubby Chilean species of that genus. Boraginaceae also include some very widespread agricultural weeds; see "Economic Importance".

Palaeobotany. The fossil record of Boraginaceae is poor and restricted to records of pollen and nutlets. Unambiguous boraginaceous fruits have been found in the Late Miocene (late Clarendonian) Ogallala Series from Texas to South Dakota in North America. Preserved are complete nutlets or parts of them. Extinct genera include *Biorbia* (Elias 1932) and *Prolithospermum* (Elias 1942), *Eliasiana* (Thomasson 1987) and *Prolappula* (Thomasson 1979) with well-preserved glochids. Fossils assigned to extant genera belong to *Cryptantha* (Elias 1932) and *Lithospermum* (Gabel 1987). The anatomy of the F2a pericarp layer in *L. dakotense* is identical to that of extant *Lithospermum* species. Nutlets of *Prolithospermum* have been suggested to resemble those of *Pentaglottis sempervirens* (Thomasson 1979), a clade which has only two early divergent genera (*Moritzia* and *Thaumatocaryon*, Weigend et al. 2010) in the New World. The North American fossils apparently include taxa falling into all three of the four tribes of Boraginaceae here recognized. Fossil pollen supposedly of the *Symphytum* type was found by Van Campo (1976) (cited by Muller 1981) from the upper Miocene of Spain.

Economic Importance. Boraginaceae have little economic importance as food plants due to their content of toxic pyrrolizidine alkaloids and their indumentum. A few species are used as pot herbs, especially *Borago officinalis* and to a lesser extent *Mertensia maritima* are traditionally used for that purpose in Europe. The roots of *Mertensia maritima* are occasionally consumed in Alaska and native Americans formerly consumed the fruits of several species of *Plagiobothrys*, *Lithospermum* and *Amsinckia* raw and ground and their leaves as a green vegetable (Moerman 1998). The red colour of the roots of several species served to make important dyes in both Eurasia and North America. Mediterranean *Alkanna tinctoria* was used to obtain "alkanna" or "alkanet" since ancient times and was employed ("dyers' bugloss") by, for example, women as make-up, hence the German name "Schminkwurz" (make-up plant) for the species. Alkanet is still used to some extent for dying cosmetics and textiles, but is no longer prepared by extraction in camel urine, the method formerly considered to give the best results (Radcliffe 1927). A tincture of *Alkanna* is still used in microscopy for detection of oils and fats. The use of various species of *Arnebia* for dying is still widespread in India (Kaul 1997). *Lithospermum erythrorhizon* is used in Japan to obtain a purple dye and employed in large quantities in, for example, the production of lipstick. Similarly, red roots of various species of *Plagiobothrys* and *Lithospermum* were used by native North American tribes for obtaining a red dye to paint skin and leather (Moerman 1998). Since ancient times species of Boraginaceae were employed for medicinal purposes. One of the oldest records of medicinal use of plants in Europe is for *Lithospermum officinale*. An evidently medicinal preparation of the fruits was found in remains of the Mierzanowice culture (from about 1750–1600 B. C.) in southern Poland (Baczyńska and Lityńska-Zając 2005). A wide range of species used to be employed in pharmaceutical preparations, but the vast majority of these are obsolete due to the high PA-concentrations and concomitant health hazards. Several medicinal Asian species of *Arnebia* and *Lithospermum* (Duke and Ayensu 1985) have recently been removed from the European market for that reason (Roeder 2000). The European "medicinal" herbs *Anchusa officinalis*, *Alkanna tinctoria*, *Cynoglossum officinale*, *Borago officinalis*, *Lithospermum officinale*, *Myosotis* spp. are no longer used (Roeder 1995). Comfrey (*Symphytum officinale* and *S. x uplandicum*), however, is still used extensively in commercially available ointments for blunt injuries, and a low-PA cultivar has been registered for that purpose.

Also, homoeopathic preparations of *Lithospermum* p.p. ("*Onosmodium*" *virginianum*) are widely marketed. The most important current market for Boraginaceae products are seed oils of Boraginaceae, which are very rich in highly unsaturated fatty acids and tocopherols and very low in PAs. Until recently mainly Borage oil was on the market (*Borago officinalis*), but there is now also a considerable market for Comfrey oil (*Symphytum officinale* and *S. x uplandicum*) and *Echium* oil (*E. plantagineum*, *E. vulgare*). The latter is the best dietary source of stearidonic acid known so far. A number of species of Boraginaceae are commonly grown as ornamentals, with biennial cultivars of *Myosotis* (mainly of *M. alpestris* and *M. latifolia*) probably commercially the most important and most widely used. Additionally, some annual species of *Anchusa* (*A. capensis*), *Echium* (*E. plantagineum*) and a range of perennial herbs such as *Omphalodes verna* and *O. cappadocica*, *Brunnera macrophylla*, several species of *Symphytum*, *Pulmonaria*, *Lithodora* are also widely grown in the temperate zone. The economic importance of Boraginaceae is possibly highest in its negative aspects. Weedy species of, for example, *Echium*, *Cynoglossum*, *Buglossoides* and *Amsinckia* cause enormous economic damage due to the invasion of native vegetation, degradation of pastures, as agricultural weeds and by livestock poisoning especially in Australia and North America (e.g. Parsons and Cuthbertson 2001).

Conspectus of Boraginaceae

I. Tribe Echiochileae (Riedl) Långström & M. W. Chase (2002)
   Genera 1–4

II. Tribe Lithospermeae Dumort. (1827)
   Genera 5–28

III. Tribe Boragineae Bercht. & J. Presl (1820)
   Genera 29–45

IV. Tribe Cynoglosseae Bercht. & J. Presl (1823)
   1. *Trichodesma* group
      Genera 46–47
   2. *Lasiocaryum* group
      Genera 48–50
   3. *Omphalodes* group
      Genera 51–53
   4. *Myosotis* group
      Genera 54–57
   5. *Mertensia* group
      Genera 58–60
   6. *Lappula* group
      Genera 61–66
   7. *Bothriospermum* group
      Genera 67–69
   8. *Cryptantha* group
      Genera 70–74
   9. *Cynoglossum* group
      Genera 75–84
   10. unplaced
      Genus 85

Key to the Genera of Borginaceae

1. Gynobase flat; nutlets erect or incurved, round in transverse section or laterally compressed (rarely dorsiventrally compressed, then smooth, black or dark brown, < 5 mm, lentil-shaped or oblique tetrahedral, usually ventrally keeled, sometimes also dorsally, neither winged nor pubescent nor glochidiate); cicatrix basal or suprabasal, usually subcircular 2

– Gynobase shortly pyramidal to subulate; rarely flat (then nutlets distinctly winged and/or pubescent and/or with narrowly triangular cicatrix); nutlets straight, not or dorsiventrally compressed, sometimes triangular in cross-section, marginally keeled or winged and/or pubescent and/or glochidiate; cicatrix usually suprabasal to apical, rarely subcircular, usually triangular to narrowly triangular (**Cynoglosseae excl. *Myosotis* group**) 51

2. Shrublets, rarely annuals; leaves opposite throughout or only below, basal leaves rarely present (then calyx > corolla, corolla < 3 mm long: *Ogastemma*); corolla small (<10 mm), zygomorphic or radially symmetrical, tube narrow, curved in zygomorphic taxa, faucal scales absent but throat with ciliate or papillate appendages; nutlets slightly laterally compressed, smooth, rugose or verrucose, never pubescent, glochidiate or dark and shiny, ventrally keeled and often also dorsally keeled; cicatrix in proximal half of nutlet, at lower end of ventral keel or terminating in a downward-projecting stipitate prolongation (**Echiochileae**) 3

– Perennials, sometimes annuals or shrubs, leaves alternate throughout (if opposite, then large rosette leaves present: *Thaumatocaryon*), basal leaves usually present; corolla often larger (>10 mm long), radially symmetrical, rarely zygomorphic with wide tube, faucal scales usually present in taxa with hypocrateriform flowers; nutlets variable, sometimes dorsiventrally compressed, broadly sessile or with narrowed base, rarely smooth, rugose or verrucose, sometimes winged or angular, ventrally keeled or grooved; cicatrix subapical to basal, nutlets rarely with basal stipitate prolongation 6

3. Annual to 25 cm high; calyx longer than corolla; stigma terminal **3. *Ogastemma***

- Perennials or shrublets, rarely annuals, then much larger; corolla always longer than calyx; stigma subterminal and overtopped by sterile tip, rarely terminal 4

4. Corolla hypocrateriform, radially symmetrical, tube cylindrical and limb spreading, with 5 invaginations at throat **1. *Antiphytum***
- Corolla, infundibuliform, zygomorphic or not, tube conical or cylindrical, limb half erect, without invaginations at throat but densely pubescent 5

5. Nutlets ovoid with cordate base, adaxially with longitudinal narrow sulcus, cicatrix basal to subbasal, subcircular to triangular or boomerang-shaped with downcurved ends; corolla mostly zygomorphic, tube conical, often constricted **2. *Echiochilon***
- Nutlets erect, ovoid, smooth and shiny or ornamented, adaxially indistinctly keeled, shortly stipitate with basal subcircular cicatrix; corolla radially symmetrical, tube short, cylindrical **4. *Sericostoma***

6. Nutlets small, 1–3(5) mm, obliquely tetrahedral, lentil-shaped or with flat abaxial side and ventrally keeled, dark brown to black **(Cynoglosseae – *Myosotis* group)** 7
- Nutlets mostly larger (>5 mm), variously shaped and coloured, but never tetrahedral or lentil-shaped **(Lithospermeae & Boragineae)** 10

7. Rhizomatous perennial (rhizome ca. 1 cm thick) with large leaves crowded near end of stem; inflorescence a terminal boragoid of nodding flowers; corolla widely campanulate with porrect limb, violet-blue; anthers coherent around style (NE Asia) **57. *Brachybotrys***
- Leaves (at least initially) in basal rosette, often with above-ground runners or below-ground stolons, sometimes with creeping shoots and erect inflorescences, inflorescences variable; flowers never widely campanulate with porrect limb; anthers never coherent around style 8

8. Perennials with underground stolons, leaves in dense rosettes, narrowed at base but epetiolate; inflorescences mono- or dichasial, with 1–3 (rarely more) leaves, elevated above rosette, flowers campanulate, infundibuliform to hypocrateriform, usually blue or pink (Himalayas) **56. *Decalepidanthus***
- Annuals or perennials, often with creeping shoots, rarely cushion-forming, without underground stolons, basal leaf rosette usually absent in perennial taxa, if present then leaves long petiolate; inflorescence shoots usually repeatedly branched, rarely dichasial, then flowers small, rotate to hypocrateriform, white, or densely cushion-forming herb with flowers apparently single directly from rosette 9

9. Corolla with contort aestivation; nutlets lentil-shaped, never stipitate, but sometimes with elaiosome **54. *Myosotis***
- Corolla with valvate aestivation; nutlets (obliquely) tetrahedral, often stipitate, never with elaiosome **55. *Trigonotis***

10. Nutlets without well-developed, plug-shaped elaiosome at base (if present evergreen shrubs), pericarp mineralized, hard, white, grey, rarely brown, smooth and shiny or perforate to verrucose; annuals or perennials, subshrubs or shrubs, sometimes ericoid **(Lithospermeae)** 11
- Nutlets with well-developed, plug-shaped elaiosome and ± thickened basal ring around cicatrix, rarely single nutlet firmly enclosed in calyx, then elaiosome absent; pericarp not calcified, dark brown or black, tessellate, verrucose and/or papillose; plants mostly hispid, coarse-leaved perennials with basal leaf rosettes, annuals or perennials, never shrubs or subshrubs, never ericoid **(Boragineae)** 35

11. Anthers coherent, connective protracted into long terminal appendage 12
- Anthers not coherent, connectives not protracted into conspicuous terminal appendages 14

12. Corolla with lobes as long as or longer than corolla tube, spreading or reflexed (Africa, SW Asia) **13. *Cystostemon***
- Corolla tubular, infundibuliform or hypocrateriform; corolla lobes much shorter than tube, erect, spreading or reflexed 13

13. Corolla cylindrical, calyx lobes narrow, elongate, more or less parallel, separated by a very narrow, usually closed sinus (Europe, Asia) ***11. Onosma***
- Corolla subspherical, calyx lobes more or less triangular, separated by an open triangular sinus (E Asia) **12. *Maharanga***

14. Nutlets narrow (>7 x as long as wide), with hook-shaped tip, smooth and white; inflorescence ebracteose above (E Asia) **28. *Ancistrocarya***
- Nutlets wider (<3 x as long as wide), apex rounded or acute, but never protracted into hook, smooth or rough, grey, brown, mottled or white; inflorescence usually bracteose or frondose-bracteose 15

15. Leaves subglabrous and glaucous; nutlets united in pairs with each nutlet 2-locular and 2-seeded **10. *Cerinthe***
- Leaves hispid or pubescent, never glaucous; 4 separate nutlets present or 2–3 by abortion (i.e. nutlets unilocular and not united in pairs) 16

16. Nutlets shortly stipitate at base, densely tuberculate; annual herb; corolla mauve **9. *Stenosolenium***
- Nutlets broadly sessile at base, smooth to densely tuberculate; annual or perennial herbs or subshrubs, corolla rarely pink, usually yellow orange, white or blue 17

17. Nutlet body strongly incurved (90–130°, when slightly incurved compare *Echium* and *Halacsya*); plants sometimes glandular above 18
- Nutlets straight (or rarely slightly bent in *Echium* and *Halacsya*); plants eglandular 19

18. Corolla infundibuliform, lobes rounded, half-spreading to spreading; throat usually glandular,

faucal appendages borne low in the corolla throat, deeply included; stamens inserted at different levels **5. *Alkanna***

– Corolla tubular with lobes porrect or reflexed, not spreading; throat glabrous (eglandular) and lacking faucal appendages; all stamens inserted at one level 20

19. Corolla tubular with very short, triangular, reflexed corolla lobes, bicolorous (pink/blue tube with yellow/orange apex); anther apices exserted **6. *Podonosma***

– Corolla tubular with usually long, triangular, porrect corolla lobes, uniformly coloured, mostly blue, rarely yellow, anthers included or long exserted **20. *Moltkia***

20. Thick sympodial rhizome present; leaves all cauline (basal leaves absent); lobes of fruiting calyx disarticulating at base and falling off individually; corolla lobes adaxially finely papillose (Balkans) **21. *Paramoltkia***

– Rhizome absent; basal and cauline leaves present (though often not simultaneously); lobes of fruiting calyx not disarticulating at base and not falling off individually; corolla lobes adaxially not papillose 21

21. Corolla usually zygomorphic and/or with five conspicuous and densely pubescent scales borne below filament attachment 22

– Corolla radially symmetrical, rarely slightly zygomorphic (then long and greenish-yellow on shrubs with ovate, coarsely veined leaves); annulus absent or present, if present not situated at filament base but near base of tube 26

22. Corolla bright yellow, deeply lobed (ca. 1/2 of length), annulus absent; inflorescences on long, annual shoots with only 1–3 elongating monochasia; thecae with apiculate hairs along margins (Balkans) **18. *Halacsya***

– Corolla white, pink, red or blue, less deeply lobed (<1/3 of length), annulus present; inflorescences terminal, dense, often complex, dense terminal thyrsoids or monochasia reduced and few-flowered; thecae without hairs 23

23. Annulus 1 mm or less from base of corolla tube, composed of minute collar or ring of 5–10 minute, sparingly hairy lobules 24

– Annulus 1.5–6 mm from base of corolla tube; five conspicuous densely pubescent bulges or scales borne below filament attachment (S Africa) 25

24. Stigma apically bifid **14. *Echium***

– Stigma entire **15. *Pontechium***

25. Perennials with basal leaf rosette; annulus present as five conspicuous, densely pubescent bulges at point of filament attachment **17. *Echiostachys***

– Subshrubs or shrubs without basal leaf rosette; annulus present as five pubescent scales at point of filament attachment **16. *Lobostemon***

26. Style deeply 2- or 4-divided, each branch terminating in a single stigma, corolla predominantly yellow to orange-red with tube at least 2.5 times longer than limb **7. *Arnebia***

– Style undivided, stigma single, rarely shallowly bilobed or apparently two (*Glandora*), then corolla blue or purple (never yellow) with tube less than 2.5 times longer than limb 27

27. Corolla tube abaxially glabrous, corolla throat without glandular hairs 28

– Corolla tube abaxially pubescent and/or corolla throat glandular 29

28. Annual; corolla tubular with very short lobes, < 7 mm long, annulus present as villose ring; calyx at anthesis united for 1/3 of its length, basal part elongating postflorally, then united for ca. 1/2 to 2/3 of its length (NW Africa and Canary Islands) **19. *Mairetis***

– Subshrubs or dwarf shrubs; corolla hypocrateriform, > 10 mm long, annulus absent; calyx free nearly to base (W and SE Europe, N Africa and SW Asia) **24. *Lithodora***

29. Annual, filaments inserted nearly at base of corolla tube; corolla yellow, small (<7 mm), with lobes ad- and abaxially glandular **23. *Neatostema***

– Annuals, perennial herbs or subshrubs, filaments inserted higher up in corolla tube and/or corolla distinctly larger (>11 mm), corolla lobes eglandular or glandular only abaxially at base 30

30. Filaments attached to the corolla tube at distinctly different levels 31

– Filaments attached at the same level (or rarely at different levels in *Glandora prostrata*, but then corolla blue to purple and calyx not circumscissile) 32

31. Perennial; corolla sulphur-yellow with spreading limb and blackish spots at sinuses, 18–25 mm across, tube without basal annulus; calyx not circumscissile at base; nutlets symmetrical with straight ventral keel (Caucasus) **8. *Huynhia***

– Small shrub; corolla blue or purple, subtubular, 4–5 mm across; calyx circumscissile; tube with hairy annulus; nutlets slightly asymmetrical with oblique ventral keel **22. *Moltkiopsis***

32. Receptacle with areoles cupulate, rarely oblique with ventral depression; nutlets with distinct peg-like appendage, rarely appendage minute; annulus and corolla invaginations absent **27. *Glandora***

– Receptacle with areoles flat to slightly concave; nutlet appendage absent or minute; corolla with annulus and/or faucal scales or vertical pleats 33

33. Corolla blue or purple, 15–20 mm long, throat and tube with 5 distinct vertical, pubescent and/or glandular pleats and 5 congregations of glandular hairs beneath filament attachment **25. *Buglossoides* (sect. *Margarospermum*)**

– Corolla yellow or orange, whitish or, when blue or purple, corolla much shorter (<10 mm) or corolla invaginations absent; faucal scales absent or present or when weakly developed vertical invaginations present, then plants annual to biennial, corolla at most 10 mm long and congregations of glandular hairs directly beneath the attachment of the filaments absent 34

34. Annuals; corolla with 5 vertical lines of simple or glandular hairs inside; nutlets tuberculate and tumulose **25. *Buglossoides*** (sect. ***Buglossoides***)
– Perennials, rarely subshrubs, shrubs, biennials or annuals; corolla usually with faucal scales; nutlets usually smooth, white and shiny, sometimes with scattered perforations, rarely brown, tuberculate, rugose or tumulose **26. *Lithospermum***

Boragineae (from 10*)

35. Fruit with a single nutlet remaining firmly enclosed in accrescent and closed calyx at maturity, distinctly stalked, elaiosome absent (S & C America) 36
– Usually more than one nutlet developing, calyx accrescent, but with spreading lobes at maturity and nutlets dispersed separately from calyx, elaiosome present (Old World) 37
36. Calyx externally with uncinate trichomes **29. *Moritzia***
– Calyx externally without uncinate trichomes **30. *Thaumatocaryon***
37. Corolla distinctly zygomorphic with oblique, half-spreading limb 38
– Corolla actinomorphic with regular and spreading limb 39
38. Inflorescence simple; corolla tube slightly curved; fertile stamens 2, 3 staminodial **32. *Anchusella***
– Inflorescence branched; corolla tube sharply incurved at middle; all 5 stamens fertile **35. *Lycopsis***
39. Corolla lobes revolute and rolled outwards, abaxially pubescent; faucal scales in two series; filaments pubescent **45. *Trachystemon***
– Corolla lobes neither revolute nor rolled, abaxially glabrous; faucal scales in one series; filaments glabrous 40
40. Flowers on long pedicels (up to 30 mm); corolla tube very short or nearly absent; faucal scales glabrous; filaments with acute or rounded appendage at base **38. *Borago***
– Flowers sessile or on short pedicels (10 mm); corolla tube always present and well-developed; faucal scales pubescent or papillose at least along margins; filaments without basal appendage 41
41. Corolla narrowly clavate to subcylindrical; style exserted; scales narrowly triangular to lanceolate, papillose along margins **44. *Symphytum***
– Corolla hypocrateriform, infundibuliform or rotate; style included; scales never triangular-lanceolate, ± rounded and shorter, pubescent or ciliate-papillose on entire surface 42
42. Nutlets with an excentric, stalked attachment at base **43. *Pentaglottis***
– Nutlets without excentric stalked attachment at base 43
43. Corolla rotate to subrotate, with short tube (ca. 2 mm) and ± flat limb; scales shortly ciliate-papillose 44
– Corolla hypocrateriform or infundibuliform with longer tube (5 mm); scales ± densely hairy or penicillate 46
44. Plants rhizomatous; cymes ebracteate **39. *Brunnera***
– Plants not rhizomatous; cymes bracteate 45
45. Perennials; corolla 6–8 mm in diam., well exceeding calyx; nutlets erect-oblong, without lateral beak **33. *Cynoglottis***
– Annuals; corolla 3–4.5 mm in diam., not or hardly exceeding calyx; nutlets transversely ovoid with lateral beak **34. *Gastrocotyle***
46. Nutlets erect-ovoid, with smooth and ± pubescent surface, constricted at base above distinct collar-like ring 47
– Nutlets transversely ovoid with a lateral beak or oblong-erect and parallel-sided, with a ribbed, tuberculate or variously sculptured surface, usually glabrous or glabrescent, with ± thickened basal annulus but not constricted at base 48
47. Annual; corolla hypocrateriform with 5 hairy scales prolonged downward into the tube along veins; stamens inserted near base of tube **40. *Melanortocarya***
– Perennials; corolla infundibuliform with a ring of hairs at throat; stamens inserted at or above middle of tube **42. *Pulmonaria***
48. Inflorescence usually subcapitate and with sessile flowers, bracts frondose, erecto-patent, much longer than flowers 49
– Inflorescence with ovate-triangular bracts not much longer than flowers, ± lax at least in fruit; flowers with distinct pedicels 50
49. Corolla hypocrateriform, blue; scales densely white-hairy; nutlets subhemispherical, helmet-shaped, with a lateral cleft and a plicate-dentate basal ring **36. *Hormuzakia***
– Corolla infundibuliform, pinkish to purple; scales shortly hairy-papillose, yellowish; nutlets transversely reniform, with a thin basal ring **37. *Phyllocara***
50. Plants usually without glandular hairs; corolla usually blue, rarely white or yellow, faucal scales triangular-oblong to linear, at least as long as wide, hairy or penicillate; fruiting calyx not or slightly accrescent and mostly cylindrical to urceolate **31. *Anchusa***
– Plants usually with glandular hairs; corolla usually not blue, white, yellow or orange-pink, purple to dark purple, dark brown, rarely dark-violet, scales replaced by tufts of hairs, shorter than wide or absent; fruiting calyx strongly accrescent and spherical-sacciform **41. *Nonea***

Cynoglosseae (from 1*)

51. Calyx strongly two-lipped in fruit, completely enclosing the nutlets; nutlets dorsiventrally flattened with cicatrix asymmetrically placed to one side of flat adaxial side of nutlet **60. *Asperugo***
- Calyx not strongly two-lipped; nutlets variable, but cicatrix always symmetrically in centre of adaxial side 52
52. Anthers exserted from corolla tube, either all equal and with connective appendages spirally twisted into a cone or very unequal with at least one anther much longer than the rest 53
- Anthers included or exserted, but always equal and never with appendages twisted into cone 54
53. Glaucous, coarsely hispid herbs with obovate leaves; anthers very unequal with at least one much longer than the rest **47. *Caccinia***
- Densely pubescent, often sericeous herbs or shrubs with oblong to narrowly ovate, acuminate leaves; anthers, all equal and with connective appendages spirally twisted into a cone **46. *Trichodesma***
54. Nutlets turbinate to ovate, not winged, adaxially with ovate suprabasal cicatrix (reaching to middle of nutlet), appressedly pubescent; dwarf annuals or perennials 55
- Nutlets not turbinate (usually dorsiventrally compressed, sometimes acuminate from ovoid base), cicatrix variable, often triangular, nutlets smooth, glochidiate or verrucose, very rarely pubescent, then distinctly winged and pubescence spreading or erect; often perennials 56
55. Dense, cushion-forming perennial; flowers arising directly from leaf rosettes; leaves fan-shaped from linear base **50. *Chionocharis***
- Annuals with basal rosette, not cushion-forming; inflorescences with distinct peduncle; leaves ovate to elliptical **48. *Lasiocaryum***
56. Corolla throat without faucal scales; stamens exserted; nutlets with cupular emergence abaxially **85. *Craniospermum***
- Corolla throat with faucal scales; stamens included, rarely exserted; nutlets variable, sometimes with cupular emergence abaxially or adaxially 57
57. Nutlets with double, cupular or incurved wing (pseudo-cicatrix) ventrally or apically (only E Asia) **(Cynoglosseae - *Bothriospermum* group)** 58
- Nutlets unwinged or with simple, usually dorsomarginal, rarely dorsal wing, sometimes wing flat, cupular or incurved 60

Cynoglosseae - *Bothriospermum* group

58. Anthers exserted from corolla; biennials or perennials **68. *Antiotrema***
- Anthers included in corolla; annuals or biennials 59
59. Nutlets with apical, cupular and incurved double-wing, the outer dentate/laciniate, the inner entire **69. *Thyrocarpus***
- Nutlets with adaxial, cupular and incurved double-wing, both wings entire **67. *Bothriospermum***
60. Nutlets slightly dorsiventrally compressed, rugose or verrucose, never glochidiate or pubescent, sometimes with flattened margin, cicatrix in proximal half of nutlet, stalked on flat or concave adaxial side of nutlet **(Cynoglosseae - *Mertensia* group)** 61
- Nutlets dorsiventrally compressed or not, smooth, rugose, verrucose, glochidiate and/or pubescent, sometimes with flattened margin and/or membranaceous wing; cicatrix variable, if stalked then adaxial side convex and attachment suprabasal or basal 62

Mertensia group

61. Perennials, mostly robust; inflorescence bracteose to ebracteose; corolla with long cylindrical, rarely conical tube and then abruptly widened, lobes porrect to half-spreading, faucal scales absent or, if present, not visible from the side **58. *Mertensia***
- Small annuals, erect to ascending; inflorescence frondose; corolla hypocrateriform with cylindrical tube and spreading lobes, conspicuous yellow faucal scales exposed (E & C Asia) **59. *Anoplocaryum***
62. Ovary two-parted, 1–2 nutlets in ripe fruit, these acuminate, oblique-ovoid in lateral view, overtopped by the incurved linear (rarely cordate-acuminate) calyx lobes **64. *Rochelia***
- Ovary four-parted, typically four nutlets in ripe fruit, these often dorsiventrally compressed, circular to ovoid in dorsal view, often with marginal wing, never overtopped by linear incurved calyx lobes 63
63. Rhizomatous perennials with large leaves (>30 cm long), adaxially glabrous or subglabrous (but not glaucous) 64
- Plants not rhizomatous or not perennial, leaves usually smaller, if glabrous then glaucous and plants annual 65
64. Leaves glabrous, dark green and glossy adaxially, abaxially sericeous, nutlets with coriaceous spreading wing, glabrous; flower hypocrateriform (Chatham Islands) **52. *Myosotidium***
- Leaves not glossy adaxially, subglabrous on both sides, nutlets not winged, only with slightly elevated rim abaxially, densely glochidiate; corolla cylindrical with erect to slightly spreading limb (Asia) **84. *Trachelanthus***
65. Shrub to 2 m tall, with leaves only at the end of branches; flowers in widely and diffusely branched pleiothyrses (Juan Fernandez Islands, Chile) **53. *Selkirkia***
- Plants herbaceous or suffruticose, much smaller; inflorescences variable, but often compact or sparsely branched **66**

66. Annuals, glabrous and glaucous; fruits always winged, wing often incurved
 51. ***Omphalodes*** (Mediterranean, annual group)
- Annuals, perennials or subshrubs, usually pubescent to hispid, if glabrous then fruits dorsally distinctly convex and not winged 67

67. Calyx spreading and enlarged in fruit, saucer-shaped; nutlets dorsally convex, 1–4, if 3–4 then 1–2 much larger and firmly attached to gynobase and partly sunken in it, remainder small, abortive, easily detached; annuals, pubescent or glabrous and glaucous (C Asia) 62. ***Suchtelenia***
- Calyx not spreading and enlarged in fruit, rarely enlarged, but then fruits dorsally concave; fruits all equal in size; annuals or perennials and subshrubs, never glabrous and glaucous 68

68. All four nutlets firmly attached to gynobase at maturity; gynobase columnar or pyramidal and fruits accordingly parallel or basally divergent 69
- At least 1–2 nutlets detached from gynobase at maturity; not parallel, not differentiated into two opposite pairs and/or not attached to the gynobase along their entire adaxial surface 70

69. Gynobase pyramidal to narrowly pyramidal, nutlets at least basally divergent; nutlets triangular-ovate, all four equal, rarely one differing from the other three, surface glochidiate, tuberculate or with horn-shaped processes 63. ***Lappula***
- Gynobase columnar and nutlets parallel, attached to the columnar gynobase along their entire adaxial surface; nutlets narrowly ovate to oblong, glochidiate, differentiated into two opposite pairs, nutlets of one pair distinctly wider than those of the other
 61. ***Heterocaryum***

70. Nutlets 2, one enclosed by upper calyx lobes, the second free; calyx irregular, 2 upper lobes fused, accrescent and incurved over nutlets, at maturity with 5–10 strong basal processes, 3 lower calyx lobes free nearly to base
 73. ***Pectocarya*** **p.p. (*Harpagonella*)**
- Nutlets 4, rarely fewer; calyx regular, with 5 lobes 71

71. Nutlets erect and more or less parallel on flat to narrowly pyramidal gynobase, triangular-ovate (rarely turbinate or flattened), not winged, at most laterally slightly differentiated into a flattened ridge, sometimes with ovate to subcircular depression dorsally (depression usually much narrower than nutlet) or apically, then nutlets less than 3(5) mm in diam. 72
- Nutlets basally divergent on widely pyramidal gynobase, subcircular, ovate or oblong in dorsal view; distinctly winged laterally or dorsally, wing membranaceous, incurved to spreading or with clearly differentiated margin differentiated from nutlet body, sometimes consisting of connected bases of flattened glochidia, nutlets > 5 mm in diam. 77

72. Diminutive annual (<5 cm), leaves very closely spaced; inflorescence superficially umbellate with individual flowers long pedicellate, pedicels longer than leaves; nutlets < 1 mm 49. ***Microcaryum***
- Plants larger, pedicels shorter than leaves and inflorescences, usually with elongated, scorpioid cymes at least in fruit, never umbellate; nutlets mostly > 1 mm 73

73. Length of deflexed pedicel much exceeding fruit diameter; nutlets long glochidiate (glochidia about as long as or longer than nutlet diameter); dorsomarginal wing absent or poorly differentiated, then consisting of the indistinctly confluent flattened bases of the glochidia 66. ***Hackelia***
- Fruits shortly pedicellate (pedicel < fruit diameter); nutlets not glochidiate or glochidia shorter than nutlet diameter; dorsomarginal wing present or absent 74

74. Gynobase shortly pyramidal to nearly flat; nutlets with narrowly triangular cicatrix along their entire adaxial side, usually dorsally with oblong to subcircular depression, rarely dorsally keeled or cicatrix basal and stipitate, then nutlet flattened and spinose or glochidiate 74. ***Microula***
- Gynobase pyramidal to subulate; nutlets with short cicatrix submedially or suprabasally, not stipitate, without dorsal depression 75

75. Corolla without faucal scales, throat open, orange or yellow; (winter) annuals 71. ***Amsinckia***
- Corolla with faucal scales, throat closed, corolla usually white with yellow faucal scales, rarely yellow, then plants perennial 76

76. Nutlets ventrally keeled, keel with longitudinal groove; leaves always alternate; ephemeral to perennial herbs, sometimes shrublets (Americas)
 70. ***Cryptantha***
- Nutlets ventrally keeled, keel not grooved; leaves opposite or alternate; ephemeral to perennial herbs, never shrublets (SW USA and SW South America, Australia, NE Asia (Kamchatka)) 72. ***Plagiobothrys***

77. Nutlets oblong with concave back, spreading to recurved and attached with small cicatrix near apex (and nutlet body extending outside the calyx), arranged in two opposite pairs in form of a St. Andrews cross (one angle < 90°, the other angle > 90°); margin coarsely dentate from half-spreading to spreading flattened glochidia; ephemeral herbs with linear leaves, flowers minute, very shortly pedicellate
 73. ***Pectocarya***
- Nutlets variable, broadly attached or attached in the middle or near the base, rarely attached near apex, but always arranged regularly (same angle between all nutlets); nutlet margin, flower and habit variable 78

78. Corolla hypocrateriform, tube 1–1.5 cm long, white; nutlets without glochidia, widely turbinate from hemispherical base, cicatrix small, situated centrally on adaxial side, sides, base and dorsomarginal rim smooth and shiny, pale brown, concave dorsal depression roughly pubescent (NW USA)
 77. ***Dasynotus***

– Corolla cylindrical or infundibuliform to hypocrateriform, tube shorter than 1 cm and/or differently coloured; nutlets often glochidiate, with concave or convex back, but never turbinate from hemispherical base, cicatrix usually larger; dorsomarginal rim usually glochidiate or protracted into membranaceous wing 79

79. Nutlets at apex free from gynobase, overtopping apex of gynobase, nutlets usually < 6 mm and/or lacking glochidia; nutlets always with differentiated dorsomarginal rim or wing 80

– Nutlets attached to apex of gynobase or along their entire length, not overtopping apex of gynobase, nutlets usually > 6 mm and/or densely glochidiate; nutlets with or without differentiated dorsomarginal rim or wing; rarely nutlets attached only in the middle of the adaxial side, then nutlets depressedly globose and glochidiate **(Cynoglosseae s.str.)** 82

80. Nutlets spreading on nearly flat gynobase, strongly compressed, dorsolateral margins all in one plane; leaves obovate to cordate-acuminate, basal rosette usually present, if absent then leaves long petiolate with rounded to cordate base, plants pilose or pubescent, but never densely so (not canescent, sericeous or hispid) **51. *Omphalodes*** (perennial species, American species)

– Nutlets spreading to oblique on widely pyramidal to subulate gynobase, dorsolateral margins not in one plane; leaves linear, oblong, ovate or obovate, base narrowed or decurrent, never distinctly petiolate, plants pubescent, sometimes canescent, hispid or sericeous 81

81. Gynobase widely pyramidal, nutlets attached in the middle of the adaxial side; perennials, often cushion-forming and vegetative rosettes present when in flower, often canescent **65. *Eritrichium***

– Gynobase narrowly pyramidal, nutlets attached nearly along their entire length, rarely attached only below apex; mostly annuals or biennials, never cushion-forming and vegetative rosettes absent when in flower, not canescent **63. *Lappula***

Cynoglosseae s.str.

82. Nutlets with concave back, 2–4 mm long; annuals with narrowly obovate leaves and without basal leaf rosette at flowering; nutlets more or less parallel on narrowly pyramidal gynobase (C and W Asia, NE Africa) **75. *Lepechiniella***

– Nutlets > 5 mm long and/or with convex back; biennials or perennials, rarely subshrubs, with ovate or obovate leaves, rarely robust annuals with basal leaf rosette at flowering; nutlets divergent on pyramidal gynobase 83

83. Nutlets distinctly winged, wing clearly demarcated from disc, without glochidia on adaxial side 84

– Nutlets not winged, sometimes with indistinct, flattened, wing-like margin, with glochidia on entire surface 86

84. Corolla tubular, with small acute or obtuse teeth and weakly developed faucal scales **82. *Rindera***

– Corolla funnel-shaped to hypocrateriform, with well-developed limb and faucal scales 85

85. Marginal wing of nutlet spreading **80. *Mattiastrum***

– Marginal wing of nutlet incurved **81. *Paracaryum***

86. Corolla cylindrical to narrowly funnel-shaped, corolla lobes narrow (oblong to narrowly obovate), porrect or half-spreading; faucal scales small, not closing the corolla throat 87

– Corolla hypocrateriform, corolla lobes widely ovate to subcircular; faucal scales large, more or less closing the corolla throat 88

87. Plants densely pubescent to sericeous, inflorescence compact, internodes shorter than individual boragoids in flower **83. *Solenanthus***

– Plants subglabrous, inflorescence lax, internodes longer than individual boragoids in flower **84. *Trachelanthus***

88. Corolla limb spreading, with faucal scales elevated and protracted at mouth of tube forming an extension, filaments and anthers long exserted; pedicels much longer than flowers **78. *Afrotysonia***

– Corolla limb spreading to half erect, faucal scales not elevated and not protracted at mouth of tube, filaments always included in tube, but anthers sometimes half-exserted; pedicels roughly as long as flowers or shorter 89

89. Anthers (but not filaments) half-exserted from corolla tube; corolla tube longer than limb **79. *Lindelofia***

– Anthers included in corolla tube, corolla tube shorter or equal to limb **76. *Cynoglossum***

Genera of Boraginaceae

I. Tribe Echiochileae (Riedl) Långström & M.W. Chase (2002).

Lithospermeae Dumort. subtr. Echiochilinae Riedl (1967).

Erect shrublets, rarely herbs, mostly perennial, sometimes annual, old branches with exfoliating bark; well-developed persistent primary root present. Leaves opposite basally or throughout, small, narrowly ovate or obovate or linear, sessile to subsessile, indumentum dense, often sericeous and appressed, sometimes hispid and/or glandular. Inflorescences mostly with long, straight monochasia, rarely much and shortly branched.

Flowers subsessile to shortly pedicellate; calyx divided nearly to base, sometimes asymmetrical and tetramerous, lobes usually unequal; corolla radially symmetrical or zygomorphic, tube narrow and curved in zygomorphic taxa, faucal scales absent but throat with ciliate or papillate appendages or indistinct invaginations; pollen bi- or sometimes tricolporate; gynobase flat or shortly pyramidal, style with sterile tip and two subterminal stigmas, rarely with terminal stigma (*Ogastemma*). Nutlets smooth, rugose or verrucose, not or laterally compressed, often with narrowed base, ventrally and sometimes dorsally keeled, cicatrix at lower end of ventral keel or terminating a downward-projecting stipitate prolongation at base, erect or incurved.

Four genera, three in W Eurasia, one in the Americas. The group consists of predominantly shrubby species, mostly with small flowers and narrow leaves with a sericeous, only rarely hispid indument, the corollas always lack typical faucal scales.

1. ***Antiphytum*** DC. ex Meisn.

Antiphytum DC. ex Meisn., Pl. Vasc. Gen. I: 280, II: 188 (1840); Johnston, Contr. Gray Herb. 68: 48–52 (1923), rev.

Perennial herbs or shrublets; branched from base, erect to procumbent; plants densely pubescent with appressed, scabrid trichomes, often glandular above. Leaves cauline, bases often shortly connate and sheathing, sometimes basal leaf rosette present at flowering time. Inflorescences straight, frondose cymoids, sometimes strongly branched (annual taxa) or reduced. Corolla infundibuliform, limb spreading to half-erect, with indistinct invaginations at throat, white or blue; stamens included; style filiform, forked. Nutlets erect, ovoid, nearly smooth to ornamented, ventral side distinctly keeled, keel terminating in short stipe with basal subcircular cicatrix.

10 to 15 species, disjunct between Mexico/SW USA and SW South America.

2. ***Echiochilon*** Desf.

Echiochilon Desf., Fl. Atlant. 1: 166, t. 47 (1798); Johnston, J. Arnold Arbor. 38: 255–294 (1957), rev.; Lönn, Bot. J. Linn. Soc. 130: 185–259 (1999), rev., morph., phylog.; Långström & Chase, Pl. Syst. Evol. 234: 137–153 (2002), phylog.

Perennial (rarely annual) herbs or shrublets; branched from base, procumbent to erect; subglabrous to densely pubescent with erect, hooked and/or appressed, scabrid trichomes, sometimes with glandular trichomes. Leaves cauline, narrowly ovate to narrowly oblong. Inflorescence terminal, straight, frondose-bracteose cymoids. Flowers erect but corolla often curved (arcuate) to spreading corolla narrowly infundibuliform, actinomorphic to zygomorphic, then bilabiate with 2-lobed upper and 3-lobed lower lip, tube longer than calyx, faucal scales absent but throat densely pubescent; stamens included to exserted, filaments often inserted at different levels in tube; style filiform with 2 subterminal stigmas. Nutlets ovoid with cordate base, smooth to verrucose, adaxially with narrow longitudinal sulcus, cicatrix basal to subbasal, subcircular to triangular or boomerang-shaped with down-curved ends.

14 species, N Africa to SW Asia, Macaronesia.

3. ***Ogastemma*** Brummitt

Ogastemma Brummitt, Kew Bull. 36: 679 (1982); Lönn, Bot. J. Linn. Soc. 130: 185–259 (1999), rev.
Megastoma (Benth. & Hook. f.) Coss. & Durieu ex Bonnet & Barratte (1895), nom. illeg.

Annual herb; branched from base, stems erect. Inflorescences terminal, straight, frondose-bracteose cymoids. Flowers erect, subsessile; calyx lobed to base, lobes unequal, elongating and closing over the fruit; corolla narrowly infundibuliform, short (equalling longest calyx lobes), tube long (>> lobes), white, faucal scales absent but throat densely pubescent; stamens included; style filiform, short with 2 subterminal stigmas. Nutlets erect, ovoid, apex conical to rostrate, verrucose, ventral side with narrow sulcus; cicatrix submedian, narrowly triangular.

One species, *O. pusillum* (Coss. & Durieu ex Bonnet & Barratte) Brummitt, Macaronesia, N Africa, to Arabian Peninsula.

4. ***Sericostoma*** Stocks

Sericostoma Stocks in Wight, Icon. Pl. Ind. Orient. (Wight) 4: 14, t. 1377 (1848); Lönn, Bot. J. Linn. Soc. 130: 185–259 (1999), rev.

Perennial shrublet; branched from base, branches erect to procumbent, densely pubescent with erect and/or appressed, scabrid trichomes. Inflorescences frondose to frondose-bracteose, dense, little branched thyrsoids, clearly delimited from vegetative shoots by 1–3 bracts per flower. Calyx lobed to base, lobes equal, slightly accrescent in fruit; corolla infundibuliform, tube short, limb half-erect, white, faucal scales absent but throat villous; stamens exserted; style filiform, stigmas 2, subterminal, horizontal. Nutlets erect, ovoid, smooth and shiny or ornamented, indistinctly keeled, shortly stipitate with basal subcircular cicatrix.

One species, *S. pauciflorum* Stocks, Pakistan and W India.

II. Tribe Lithospermeae Dumort. (1827).

Tribe Cerintheae Dumort. (1829).
Tribe Echieae Dumort. (1829).

Annual to perennial herbs, subshrubs or shrubs, sometimes rhizomatous or stoloniferous; well-developed, sometimes massive and red-staining primary root present, rarely absent (rhizomatous and stoloniferous taxa). Leaves alternate throughout, small to very large, linear to widely ovate or elliptical, sessile to shortly petiolate, indumentum usually dense, often hispid, rarely sericeous and appressed or glandular, glabrous in *Cerinthe*. Inflorescences with typical boragoids, rarely reduced and few-flowered, often in complex thyrsoids. Flower subsessile to pedicellate; calyx divided nearly to base to largely united with only free teeth, lobes usually equal; corolla radially symmetrical or zygomorphic, then tube wide, faucal scales often present and well-developed, but sometimes missing and replaced by pubescent or glandular patches; pollen with three or (mostly) more apertures and (sometimes) pseudoapertures; gynobase flat, style with or without sterile tip and two, rarely four terminal or subterminal stigmata. Nutlets hard and walls incrusted with calcium carbonate, often smooth and shiny, sometimes perforate, foveate, rugose or verrucose, often white or beige, rarely brownish or mottled, sometimes incurved or laterally compressed, often with narrowed base, ventrally and sometimes dorsally keeled, widely ovoid to subspherical.

23 genera, 21 in Eurasia, especially W Asia and the Mediterranean, two endemic to Africa, only one genus also in the Americas (*Lithospermum*). Lithospermeae are characterized by a flat gynobase in combination with ovoid or incurved, smooth or rugose, always very hard (mineralized) nutlets, which are typically keeled both ventrally and dorsally.

5. ***Alkanna*** Tausch

Alkanna Tausch, Flora 7: 234 (1824), nom. cons., non Adans. (1763); Rechinger, Ann. Naturhist. Mus. Wien 68: 191–220 (1965), reg. rev.

Perennial, rarely annual herbs or subshrubs, indumentum often glandular; thick, sometimes woody taproot present, often dark red. Leaves cauline, usually also basal, oblong to obovate, sessile. Inflorescences frondose-bracteose cymoids or thyrsoids. Flowers subsessile, erect to pendulous; calyx lobed almost to base, often basally accrescent; corolla infundibuliform, regular, rarely slightly zygomorphic, faucal scales inconspicuous, inserted at constricted tube below throat, sometimes three inserted higher; basal scales usually annular; stamens included, filaments mostly inserted at different levels in the tube; style included, stigma capitate to nearly bilobed. Nutlets 1–2(4), obliquely horizontal, more or less incurved, adaxially keeled, usually rugose to verrucose, base distinctly stipitate. $2n$ = 14, 20, 22, 28, 30, 37.

Ca. 40 species, Mediterranean and SW Asia.

6. ***Podonosma*** Boiss.

Podonosma Boiss., Diagn. Pl. Orient. II, 11: 113 (1849); Johnston, J. Arnold Arbor. 35: 1–81 (1954), rev.; Riedl, Anzeiger Österr. Akad. Wiss. Math.-Nat. Kl. (Wien) 101: 354–362 (1964), rev.

Perennial herbs; indumentum hispid and often glandular. Leaves ovate to oblong, sessile, sometimes semiamplexicaulous. Inflorescences frondose-bracteose. Flowers pendulous, pedicellate; calyx lobed almost to base, accrescent in fruit; corolla cylindrical with lobes triangular, reflexed, blue to pink with yellow to orange apex, faucal scales absent, annulus 10-lobed, pubescent; stamens partially exserted, connective protracted into an acute apical appendage; style exserted, stigma minutely bilobed. Nutlets strongly incurved with apical beak, ventrally keeled, base shortly stipitate.

Three species, NE Africa, E Mediterranean and SW Asia.

7. ***Arnebia*** Forssk.

Arnebia Forssk., Fl. Aegypt.-Arab. 62 (1775); Johnston, J. Arnold Arbor. 35: 1–81 & 158–166 (1954), rev.; Riedl, Österr. Bot. Z. 109: 45–80 (1962), part. rev.; Riedl, Österr. Bot. Z. 111: 149–153 (1964), part. rev.; Riedl, Ann. Naturhist. Mus. Wien 75: 209–222 (1971), part. rev.; Sadat, Mitt. Bot. Staatss. München 28: 1–210 (1989), reg. rev.

Annual or perennial herbs, sometimes with well-developed pleiocorm; thick, sometimes woody taproot present, often dark red. Leaves all cauline or basal and cauline, narrowly oblong, lanceolate or elliptic. Inflorescences (often dense) frondose-bracteose cymoids (perennials) or thyrsoids (annuals). Flowers erect to pendulous, sessile or shortly pedicellate, distylous; calyx lobed almost to base, accrescent, often hardening and tightly enclosing nutlets in fruit; corolla narrowly infundibuliform or hypocrateriform, regular or obscurely zygomorphic, yellow or cream, rarely blue or purplish, corolla lobes ovate or subcircular, sometimes erose or lacerate, faucal scales or folds absent, annulus absent (perennials) or present (annuals), then usually pubescent; stamens inserted at one level in tube, included or sometimes partially exserted (depending on stylar morph); style included or slightly exserted (depending on stylar morph), usually shortly once or twice forked, each branch ending in stigma. Nutlets 1–4, straight or slightly incurved at the apex, ovoid or subglobose, ventrally keeled or ventral keel obscure, dorsally rounded, sometimes keeled in the distal part, ornamented or rarely almost smooth. $2n$ = 14, 16, 22, 24.

About 30 species, mainly SW and C Asia, Himalaya, also NE Africa and SE Europe.

8. ***Huynhia*** Greuter

Huynhia Greuter, Willdenowia 11: 37 (1981).

Perennial herb from stout pleiocorm; indumentum densely patent-pilose. Leaves basal and cauline, basal ones oblong, cauline ones narrowly ovate. Inflorescence a dense, bracteose cymoid. Flowers distylous; calyx divided nearly to base with lobes obtuse, not hardening and without thickened nerves or angular projections in fruit; corolla hypocrateriform, tube narrow, puberulent outside, without faucal scales or annulus, limb spreading; stamens inserted at two different levels below the throat (3+2); stigma capitate-bilobed. Nutlets erect, ovoid-subglobose, apically acute and shortly beaked, ventrally keeled, finely tuberculate-scrobiculate.

One species, *H. pulchra* (Willd. ex Roemer & Schultes) Greuter & Burdet, E Anatolia, Caucasus and NW Iran.

9. ***Stenosolenium*** Turcz.

Stenosolenium Turcz., Bull. Soc. Imp. Naturalistes Moscou 13: 253 (1840); Johnston, J. Arnold Arbor. 35: 1–81 & 158–166 (1954), rev.

Annual herbs with erect to ascending stems; indumentum hispid; thin, purplish-brown taproot present. Leaves basal and cauline, obovate to linear. Inflorescences frondose-bracteose cymoids. Flowers erect, shortly pedicellate; calyx lobed to base, accrescent in fruit; corolla narrowly infundibuliform, tube yellowish, limb usually purple, faucal scales absent, annulus hairy; stamens included, filaments inserted spirally; style short, included, forked shortly below the apex, with two terminal stigmas. Nutlets straight, oblique ovoid, ventrally keeled, tuberculate, cicatrix on a short, stout, basal-ventral stipe.

One species, *S. saxatile* (Pall.) Turcz., NE Asia. Doubtfully distinct from *Arnebia*.

10. ***Cerinthe*** L. Fig. 10

Cerinthe L., Sp. Pl. 136 (1753); Johnston, J. Arnold Arbor. 35: 1–81 & 158–166 (1954), rev.; Selvi, Cecchi & Coppi, Taxon 58: 1307–1325 (2009), rev., morph., phylog.

(Winter-)Annual, biennial or perennial herbs, glabrous and glaucous. Leaves often irregularly blotched, trichomes reduced to tubercles; leaves basal and cauline or all cauline, obovate to oblong, cauline leaves sessile to amplexicaulous. Inflorescences frondose to frondose-bracteose cymoids of 1–2 scorpioid monochasia. Flowers pendulous; calyx lobed to middle or nearly to the base, lobes very unequal, narrowly ovate or oblong, accrescent in fruit; corolla cylindrical with short lobes, yellow, often tinged with violet, red or blue, faucal scales absent, annulus a glabrous ring or collar; stamens included or partially exserted, anthers appendaged apically and basally, apically

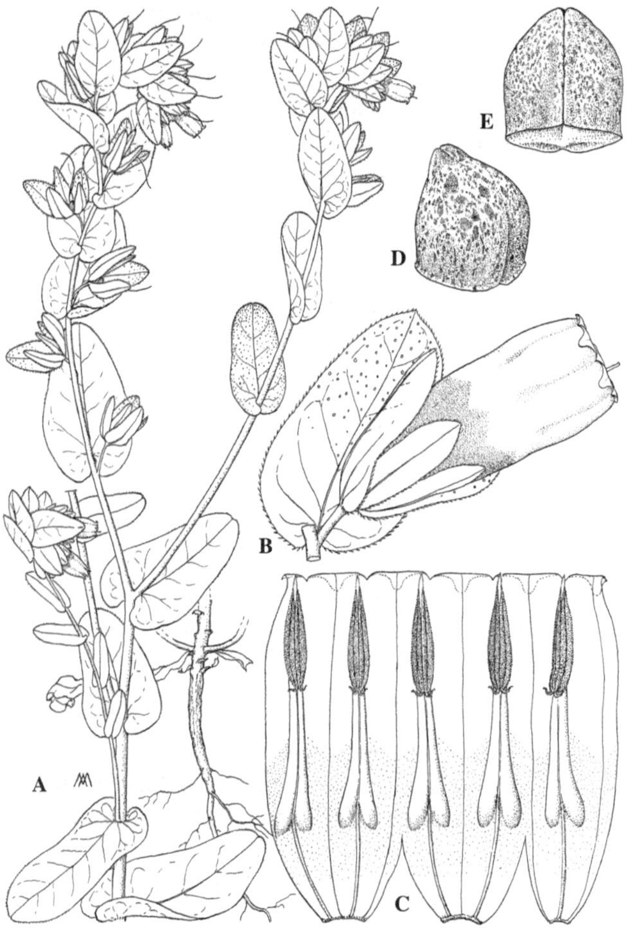

Fig. 10. Boraginaceae. Tribe Lithospermeae: *Cerinthe major* subsp. *major*. **A** Habit. **B** Flower with bract. **C** Corolla, longitudinal section. **D, E** Nutlets. (Selvi et al. 2009, drawn by A. Maury)

coherent; style long, exserted, stigmas 1 or 2. Nutlets pair-wise united into 2 two-seeded carpids; erect, broadly ovoid to subglobose, smooth. $2n = 16, 18, 24, 36$.

Seven to ten species from Europe and NW Africa to W Asia.

11. ***Onosma*** L.

Onosma L., Sp. Pl. ed. 2: 196 (1762); Johnston, J. Arnold Arbor. 32: 201–225 (1951), reg. rev.
Chorianthа Riedl (1961).

Biennial or perennial herbs, dwarf shrubs or subshrubs; indumentum dense and hispid, often with stellate trichome complexes, rarely sericeous; strong taproot usually present, sometimes with well-differentiated pleiocorm or rhizomatous. Leaves basal and cauline, narrowly obovate to linear. Inflorescences bracteose or frondose-bracteose cymoids. Flowers pendulous at anthesis, rarely erect, pedicellate with pedicel elongating in fruit; calyx lobed nearly to base, often accrescent and connivent in fruit; corolla tubular to narrowly infundibuliform, lobes short, yellow, whitish or cream-coloured, rarely blue, often changing colour during anthesis, faucal scales absent, annulus present, often irregularly lobed; stamens included or partially to completely exserted, anthers connivent, forming a tube, connective conspicuously protracted into a short to elongate, attenuate to oblong appendage; style long, usually exserted, stigmas 1 or 2. Nutlets erect, ovoid, usually smooth, ventrally keeled, dorsally usually convex. $2n = 12, 14, 16, 18, 20, 21, 22, 24, 26, 27, 28, 30, 32, 38, 44, 50, 51$.

Circa 150 species, NW Africa, Europe and Asia.

12. ***Maharanga*** DC.

Maharanga DC., Prodr. (DC.) 10: 71 (1846); Johnston, J. Arnold Arbor. 35: 1–81 (1954), rev.

Biennial or perennial herbs or subshrubs; indumentum hispid, often with stellate trichome complexes; strong taproot usually present, sometimes with well-differentiated pleiocorm. Leaves basal and cauline, narrowly (ob-)ovate. Inflorescences cymoids, frondose-bracteose. Flowers pendulous to erect, pedicellate, pedicels elongating in fruit; calyx lobed to the middle or slightly below, weakly accrescent and incurved in fruit; corolla urceolate to subspherical, contracted at both ends, usually widest near the middle, usually with inflexed longitudinal folds below the sinuses of the lobes and below the lobes inflated with lamellae protruding between the calyx lobes, corolla lobes recurved, much shorter than the tube, faucal scales absent, annulus of 5 basal scales, hairy; stamens included or apex slightly exserted; style long, included or slightly exserted, stigmas 1 or 2. Nutlets slightly incurved, ovoid, ventrally keeled, dorsally rounded, ornamented. $2n = 14$.

Nine species in C & E Asia. Closely related to *Onosma* and sometimes synonymised with it.

13. ***Cystostemon*** Balf. f.

Cystostemon Balf. f., Proc. Roy. Soc. Edinb. 12: 82 (1883); Miller & Riedl, Roy. Bot. Gard. Edinb. 40: 11–21 (1982), rev.
Vaupelia Brand (1914).

Perennial herbs, subshrubs or shrubs, branched from base; strong lignescent taproot present, usually crowned with lignescent pleiocorm; leaves all cauline, sessile to shortly petiolate, lamina narrowly ovate to elliptical. Inflorescences frondose-bracteose, simple to many-branched thyrsoids, bracteoles sometimes present. Flowers long pedicellate; calyx lobed nearly to base, sometimes accrescent in fruit; corolla hypocrateriform with tube $<<$ limb, limb spreading, faucal scales absent; stamens long exserted, filaments very short and flattened, sometimes inflated or with adaxial triangular to oblong, basal appendage, anthers oblong to narrowly oblong, connective conspicuously protracted into linear appendage, all connective appendages together forming a narrow, straight cone above anthers; style exserted, stigma capitate. Nutlets oblique ovoid with acuminate apex, more or less angular, keeled at least ventrally, densely rugose.

About 15 species in SW Arabia, tropical Africa.

14. ***Echium*** L.

Echium L., Sp. Pl. 139 (1753); Bramwell, Lagascalia 2: 37–115 (1972), reg. rev.
Megacaryon Boiss. (1875).

Annual, biennial or poly- or monocarpic perennial herbs, rarely shrubs with many erect, ascending or decumbent stems (Macaronesian species); indumentum dense and hispid, sometimes sericeous. Leaves basal and cauline or cauline only (shrubs), usually narrowly obovate to broadly ovate. Inflorescence mostly bracteose, densely branched thyrsoids with many short boragoids as branches. Flowers usually erect to horizontal, shortly pedicellate; calyx divided almost to base, often accrescent in fruit; corolla zygomorphic, narrowly to broadly infundibuliform, mostly blue, rarely pink, red, yellow or white, faucal scales absent, annulus of 5–10 lobes or swellings present; stamens with long filaments, usually at least 2 exserted; style long, exserted, stigmas 2. Nutlets 4 or 1–2 by abortion, straight to slightly incurved, more or less tuberculate and rugose. $2n = 8, 10, 12, 14, 16, 24, 28, 32, 64$.

About 60 species, primarily in Mediterranean Europe and Macaronesia, N Africa, C Europe and W Asia. A few species introduced to the Americas and Australia. *Echium plantagineum* L. is considered a noxious weed.

15. ***Pontechium*** U.-R. Böhle & Hilger

Pontechium U.-R. Böhle & Hilger, Taxon 49: 743 (2000).

Biennial herb, stems erect, simple; indumentum densely hispid; taproots present. Leaves basal and cauline, narrowly obovate or narrowly elliptical, sessile. Inflorescences bracteose thyrsoids. Flowers horizontal to erect, dark red, shortly pedicellate; calyx lobed almost to base, lobes linear, weakly accrescent in fruit; corolla narrowly infundibuliform, zygomorphic, red, abaxially pubescent, adaxially glabrous, lobes ovate to subsemiorbicular, ascending, faucal scales absent, annulus of 5–10 lobes or united as ring, glabrous; stamens long exserted, anthers versatile; style long, exserted, pubescent, stigma bicapitate. Nutlets slightly incurved, ovoid, ventrally keeled, dorsally rounded, keeled in the distal part, tuberculate. $2n = 24$.

One species, *P. maculatum* (L.) U.-R. Böhle & Hilger, E Europe to W Asia.

16. ***Lobostemon*** Lehm.

Lobostemon Lehm., Linnaea 5: 378, t. 5. (1830); Levyns, Bot. J. Linn. Soc. 49: 393–451 (1934), rev.; Buys, S. Afr. J. Bot. 72: 383–390 (2006), morphol.

Shrubs, dwarf shrubs, rarely subshrubs; indumentum densely hispid. Leaves all cauline, linear or narrowly ovate, rarely oblong or narrowly obovate, sessile. Inflorescences frondose-bracteose thyrsoids, paraclades sometimes reduced to 1–2 flowers. Flowers erect to pendulous, sessile or shortly pedicellate; calyx lobed almost to base, often zygomorphic, slightly accrescent in fruit; corolla infundibuliform to campanulate, usually zygomorphic, blue, pink, white or red, lobes ascending to horizontally spreading, faucal scales and annulus absent but pubescent staminal scales present; stamens exserted or some or all included; style included or exserted, stigma bilobed. Nutlets usually 1–2, straight or slightly incurved, ovoid, ventrally keeled, dorsally convex but often keeled distally, usually ornamented. $2n = 14, 28, 42$.

About 30 species, endemic to South Africa.

17. ***Echiostachys*** Levyns

Echiostachys Levyns, J. Linn. Soc., Bot. 49: 445 (1934).

Perennial herbs from persistent pleiocorm; indumentum spreading or tomentose. Leaves basal and cauline, linear, narrowly obovate or narrowly elliptic. Inflorescences very dense, spike-like bracteose thyrsoids. Flowers horizontal to erect, shortly pedicellate; calyx lobed almost to base, lobes linear, accrescent in fruit; corolla narrowly infundibuliform, slightly zygomorphic, white, blue or pink, lobes ovate, porrect to half-spreading, faucal scales and annulus absent but with tufts of trichomes at point of filament insertion or with trichomes along the filaments and sometimes on the tube below; stamens exserted, filaments pubescent at base; style often long, exserted, usually pubescent, minutely forked at apex, stigmas 2. Nutlets 4 or 1–2 by abortion, straight, ovoid, ventrally keeled, dorsally convex but apically keeled, almost smooth or sculptured with minute white trichomes. $2n = 14, 28$.

Three species, endemic to South Africa.

18. ***Halacsya*** Dörfl.

Halacsya Dörfl., Herb. Norm. Sched. Cent. 44: 103 (1902); Johnston, J. Arnold Arbor. 35: 158–166 (1954); Cecchi & Selvi, Taxon 58: 700–714 (2009), phylog.

Subshrub with erect or ascending stems from stout rhizome; indumentum of appressed scabrid trichomes. Leaves basal and cauline, basal ones narrowly obovate with narrowed base, cauline narrowly oblong or narrowly ovate, sessile. Inflorescences cymoids of 1–2 frondose-bracteose, scorpioid monochasia. Flowers suberect, shortly pedicellate; calyx lobed almost to base, accrescent in fruit; corolla infundibuliform, slightly zygomorphic, glabrous, lobes ascending, widely ovate, yellow, faucal scales and annulus absent; stamens included or partially exserted, connivent, connective protracted into a short, acute appendage; style exserted, stigma entire or obscurely bilobed. Nutlets usually 1–2, slightly incurved in upper half, ovoid, ornamented, ventrally keeled. $2n = 22$.

One species, *H. sendtneri* (Boiss.) Dörfl., Balkans (C Serbia, N Albania and NW Greece), serpentinophyte.

19. ***Mairetis*** I.M. Johnst.

Mairetis I.M. Johnst., J. Arnold Arbor. 34: 4 (1953).

Annual herb with ascending to erect branches; primary root present. Leaves all cauline, sessile, narrowly obovate to narrowly oblong. Inflorescences cymoids of 1–2 bracteose, scorpioid monochasia. Flowers erect to pendulous, subsessile; calyx divided to middle, strongly accrescent in fruit; corolla cylindrical, tube yellow, throat and lobes blue, glabrous, faucal scales absent, annulus present as ring of trichomes; stamens included; style short, included, stigma capitate. Nutlets 4, slightly incurved, pyriform, verrucose, ventrally keeled.

One species, *M. microsperma* (Boiss.) I.M. Johnst., NW Africa (W Morocco and Canary Islands).

20. ***Moltkia*** Lehm.

Moltkia Lehm., Neue Schriften Naturf. Ges. Halle 3(2): 3 (1817).

Subshrubs or perennial herbs from pleiocorm, indumentum of appressed scabrid trichomes, sometimes hispid; massive taproot present. Leaves basal and cauline or all cauline, basal leaves narrowly obovate or linear, sometimes shortly petiolate, cauline leaves linear to narrowly obovate, sessile. Inflorescences frondose-bracteose or bracteose cymoids of 1–2 monochasia or with additional paraclades distally. Flowers deflexed, pendulous or ascending, shortly pedicellate; calyx lobed almost to base, lobes linear, weakly accrescent in fruit; corolla narrowly infundibuliform, lobes porrect, ovate, blue or rarely yellow, faucal scales and annulus absent; stamens exserted, rarely included; style exserted, stigma entire or emarginate. Nutlets usually 1–2, incurved, ovoid, smooth or ornamented, ventrally keeled, dorsally convex. $2n = 16, 18, 108, 112$.

Six species in S Europe and SW Asia.

21. ***Paramoltkia*** Greuter

Paramoltkia Greuter, Willdenowia 11: 38 (1981); Cecchi & Selvi, Taxon 58: 700–714 (2009), phylog.

Perennial herb from thick, sympodial rhizome, indumentum sparse, shortly but roughly pubescent. Leaves all cauline, sessile, narrowly ovate-acuminate, median cauline larger than the basal ones. Inflorescences frondose-bracteose cymoids with 1–2 scorpioid monochasia. Flowers pendulous to erect, subsessile or shortly pedicellate; calyx lobed almost to base, lobes linear, accrescent in fruit; corolla narrowly infundibuliform, glabrous, lobes broadly ovate to subsemicircular, porrect, violet-purple, faucal scales and annulus absent; stamens included, filaments shorter than anthers; style long, exserted, stigma entire or obscurely bilobed. Nutlets 1–2, ovoid, slightly incurved in upper half, ventrally keeled, dorsally convex, smooth. $2n = 108, 112$.

One species, *P. doerfleri* (Wettst.) Greuter & Burdet, endemic to the SW Balkan peninsula (NE Albania and Kosovo/SW Serbia), serpentinophyte.

22. ***Moltkiopsis*** I.M. Johnst.

Moltkiopsis I.M. Johnst., J. Arnold Arbor. 34: 2 (1953).

Subshrub, old bark white; indumentum densely ciliate-hispid. Leaves cauline, ovate–acuminate, sessile. Inflorescences frondose-bracteose cymoids of 1–2 scorpioid monochasia. Flowers erect, sessile or subsessile, calyx lobed almost to base, corolla subtubular, abaxially pubescent, lobes blue to purple, faucal scales absent but with 5 patches of trichomes in throat, annulus indistinctly 10-lobed, often pubescent; stamens included, attached at distinctly different heights, with 3 filaments longer; style usually included, stigma 1, terminal; calyx circumscissile above the base at maturity. Nutlets usually 2–3, almost straight, ovoid, slightly asymmetric, smooth or with few tubercles, ventrally keeled. $2n = 12, 24$.

One species, *M. ciliata* (Forssk.) I.M. Johnst., NE Africa, SW Asia.

23. ***Neatostema*** I.M. Johnst.

Neatostema I.M. Johnst., J. Arnold Arbor. 34: 5 (1953).

Erect annual herb. Leaves basal and cauline, narrowly obovate to linear. Inflorescences frondose-bracteose thyrsoids or reduced to cymoids. Flowers erect, subsessile, cleistogamous flowers sometimes present; calyx lobed almost to base, lobes linear to narrowly triangular, accrescent and connivent in fruit; corolla narrowly infundibuliform, lobes orange-yellow, pubescent, partly glandular, faucal scales absent but throat with 5 patches or a ring of trichomes, annulus 10-lobed, pubescent; stamens included; style very short, included, stigma capitate to obscurely 2-lobed. Nutlets ovoid, incurved, verrucose, ventrally keeled, dorsally convex, with weakly prominent keel in distal part. $2n = 28$.

One species, *N. apulum* (L.) I.M. Johnst., Mediterranean and Macaronesia.

24. ***Lithodora*** Griseb.

Lithodora Griseb., Spicil. Fl. Rumel. 2: 85 (1844); Browicz, Ann. Mus. Goulandris 7: 39–48 (1986); reg. rev.; Thomas et al., Taxon 57: 79–97 (2008), phylog., genus delimitation.

Dwarf shrubs or shrubs, with sericeous or hispid pubescence; strong, lignescent taproot present. Leaves cauline, sessile, narrowly oblong. Inflorescences frondose-bracteose to bracteose cymoids of 1–2 monochasia, sometimes subcapitate, rarely reduced to single flower. Flowers deflexed to erect, (sub-)sessile, with stigma–height dimorphism; calyx lobed almost to base, lobes linear or narrowly triangular, accrescent in fruit; corolla narrowly infundibuliform, generally glabrous, lobes ovate to subsemicircular, blue, purple, rarely white, faucal scales and annulus absent; stamens included or partially exserted (depending on stylar morphs); style long or short and included (depending on stylar morph), stigmas 2, terminal, juxtaposed. Nutlets 1–2, straight to strongly incurved, ovoid, ornamented, ventrally keeled, dorsal side convex, cicatrix basally with a pyramidal to asymmetrical appendage (elaiosome?), areoles deeply cupulate. $2n = 26, 28, 35, 38, 40$.

Three to five species in W and SE Europe, N Africa and SW Asia.

25. *Buglossoides* Moench[1]

Buglossoides Moench, Meth. 418 (1794); Clermont et al., Feddes Repert. 114: 58–70 (2003), reg. rev., morph.; Johnston, J. Arnold Arbor. 35: 1–81 & 158–166 (1954), genus delimitation.
Aegonychon Gray (1821).
Margarospermum (Rchb.) Opiz (1839).

Annual or perennial herbs or subshrubs, with erect, ascending or decumbent stems, some with arching stems rooting at the apex; indumentum of appressed or spreading trichomes. Inflorescences many-flowered, frondose-bracteose cymoids. Flowers erect, (sub-)sessile; calyx lobed nearly to base, accrescent in fruit, sometimes becoming oblique or asymmetric through asymmetrical swelling of pedicel; corolla narrowly infundibuliform, lobes subcircular to oblong, half-spreading, metallic blue to pink or white, sometimes with elevated, white, central crease, faucal scales absent but with 5 distinct trichome bands on faucal invaginations; annulus ring-shaped or 10-lobed, sometimes missing; stamens included; style short, included, stigmas 2, subterminal or terminal. Nutlets 1(2) (perennials) or 4 (annuals), erect to slightly incurved, ovoid, white and smooth (perennials) or brown and verrucose (annuals), dorsally convex. $2n$ = 14, 16, 20, 24, 28, 36, 42.

About 10 species, Mediterranean and SW Asia, a single species, *B. zollingeri* (A. DC.) I.M. Johnst., in E Asia., the genus may not be monophyletic and falls into two very distinct groups, sect. *Buglossoides* with annual, small- and mostly pale-flowered species with verrucose nutlets and sect. *Margarospermum* (Rchb.) I.M. Johnst. with perennial, large- and metallic blue-flowered species with smooth nutlets. *B. arvensis* (L.) I.M. Johnst. is a cosmopolitan weed.

26. *Lithospermum* L.

Lithospermum L., Sp. Pl. 1: 132 (1753); Johnston, J. Arn. Arbor. 35: 1–81 (1952), rev. of *Lithospermum*; Johnston, J. Arn. Arbor. 35: 1–81 (1954a), rev. Lithospermeae; Johnston, J. Arn. Arbor. 35: 158–166 (1954b), rev. Lithospermeae; Turner, Phytologia 77: 38–44 (1994), rev. *Lasiarrhenum*; Turner, Phytologia 78: 39–60 (1995a), rev. *Onosmodium*; Turner, Phytologia 77: 393–407 (1995b), rev. *Macromeria*; Weigend et al., Molec. Phylog. Evol. 52: 755–768 (2009), genus delimitation; Cohen & Davis, Brittonia 61: 101–111 (2009), genus delimitation.
Lasiarrhenum I.M. Johnst. (1924).
Macromeria D. Don (1832).
Nomosa I.M. Johnst. (1954).
Onosmodium Michx. (1808).
Perittostema I.M. Johnst. (1954).
Psilolaemus I.M. Johnst. (1954).
Ulugbekia Zakirov (1961).

Perennial, rarely annual herbs, subshrubs or shrublets, indumentum hispid or sericeous, strong primary root always developed, sometimes red to purple. Leaves cauline and sometimes some basal, subsessile, narrowly to widely ovate or obovate, rarely elliptical or nearly linear, base rounded or attenuate, rarely decurrent. Inflorescences frondose-bracteose to bracteose, monochasial or dichasial cymoids, rarely thyrsoids. Flowers mostly erect, rarely pendulous or deflexed, shortly pedicellate, chasmogamous, rarely cleistogamous, sometimes distylous; calyx divided nearly to base, slightly accrescent in fruit; corolla usually symmetrical, tubular, infundibuliform or hypocrateriform, rarely slightly zygomorphic, often papillose and/or glandular in throat, lobes reflexed to porrect, sometimes with erose to ciliate margin, white, yellow or orange, faucal scales and basal scales present or absent; stamens included to long exerted; style filiform, included, sometimes exserted, stigma small, two-lobed, surpassed by sterile, two-lobed tip. Nutlets erect, rarely slightly incurved, ovoid to pyriform, sometimes slightly dorsiventrally compressed or ventrally keeled, dorsally convex, fruit wall smooth to irregularly grooved, rarely verrucose. $2n$ = 14, 24, 28.

About 80 species in Eurasia, Africa and both Americas, centre of diversity in North America. *Ulugbekia* and the American segregate genera have recently been synonymised with *Lithospermum* based on molecular and morphological data (Cohen and Davis 2009; Weigend et al. 2009).

[1] Cecchi et al. (2014) transferred two additional species to *Glandora* and segregated the perennial taxa of *Buglossoides* (sect. *Margarospermum*) into the re-instated genus *Aegonychon* Gray.

27. *Glandora* D.C. Thomas, Weigend & Hilger[1]

Glandora D.C. Thomas, Weigend & Hilger, Taxon 57: 92 (2008).

Much-branched dwarf shrubs, often with lignescent underground stolons, sometimes mat-forming; indumentum mostly sericeous. Leaves cauline, sessile, narrowly oblong, narrowly elliptic or narrowly obovate. Inflorescences frondose-bracteose cymoids of 1–2 weakly scorpioid monochasia. Flowers erect, (sub-)sessile to shortly pedicellate; distylous; calyx lobed almost to base, lobes linear or narrowly triangular, accrescent in fruit; corolla infundibuliform, abaxially pubescent, adaxially usually with glandular trichomes in the throat, otherwise glabrous or with 5 pubescent patches on base of lobes or in the tube, lobes half-spreading, blue, purple or rarely white, faucal scales and annulus absent; stamens included or partially exserted (depending on stylar morph); style long or short and included (depending on stylar morph), stigmas 2, juxtaposed at the minutely forked or notched style apex. Nutlets 1–3, straight or rarely distally slightly incurved, ovoid, smooth, rarely ornamented, ventrally keeled, dorsally convex, cicatrix often with a peg-like appendage (elaiosome?), areoles cupulate. $2n = 16, 26, 28, 32, 40, 50$.

Eight species in S and SW Europe and N Africa.

28. *Ancistrocarya* Maxim.

Ancistrocarya Maxim., Bull. Acad. Imp. Sci. Saint Péters bourg III, 17: 443 (1872); Johnston, J. Arnold Arbor. 35: 1–81 & 158–166 (1954), genus delimitation.

Perennial herb with erect simple stems; thick, shortly branched rhizome present; indumentum of appressed scabrid trichomes. Leaves cauline, crowded in upper third of shoot, sessile, obovate-acuminate with long cuneate base. Inflorescences loose thyrsoids with up to 6 monochasia, more often reduced to a cymoid, with 1–2 bracts at the base of each monochasium, otherwise ebracteose. Flowers erect to half-spreading, shortly pedicellate; calyx lobed almost to base, lobes linear to narrowly ovate, slightly accrescent in fruit; corolla narrowly infundibuliform, lobes ascending to spreading, pale blue or white, faucal scales absent, annulus of 10 tufts of trichomes; stamens included; style short, included, stigmas 2, terminally juxtaposed. Nutlets 1–2, narrowly ovoid with apex tapering and uncinate, smooth, ventrally keeled, dorsally convex.

One species, *A. japonica* Maxim., in C and S Japan, Korea. An easily recognizable taxon, but possibly nested in *Lithospermum*.

III. Tribe Boragineae Rchb. (1831).

Anchuseae W.D.J. Koch (1837).
Symphyteae D. Don (1832).

Perennial, rarely annual herbs, sometimes rhizomatous or sometimes with stolon tubers; plants hispid, rarely sericeous, erect, rarely ascending or appressed, uncinate trichomes sometimes present. Leaves often cauline and basal (basal leaf rosette usually present), mostly large and widely ovate-acuminate, more rarely narrowly ovate or obovate, oblong or cordate, distinctly petiolate or with decurrent leaf bases. Inflorescence thyrsoids or cymoids, frondose or frondose-bracteose, rarely bracteose. Calyx united for most of its length or deeply divided, mostly tubular–cylindrical at anthesis, often accrescent in fruit; corolla radially symmetrical or zygomorphic, infundibular to hypocrateriform, rarely rotate, faucal scales large, conspicuous, often exserted from throat, in one or rarely two series, pubescent or papillose, rarely glabrous (*Borago*); stamens 5 (rarely 2 by abortion, *Anchusella*), pollen mostly isocolpate, mostly 4-colporate (rarely up to 15 zonocolporate), often with mesocolpial fields; gynobase flat, style with truncate or capitate stigma, rarely bilobed or shallowly bifid, with flask-shaped papillae, these with or without apical plate-like cap. Nutlets erect or incurved, more or less laterally compressed, usually tessellate and papillose, ventrally keeled, base narrowed into thin stalk (*Pentaglottis*, *Thaumatocaryon*, *Moritzia*) or more commonly widened into broad, collar-like ring and with large, white elaiosome (from gynobase tissue), areola basal. Nutlets dispersed individually or only one nutlet present and firmly enclosed in accrescent calyx (only *Moritzia*, *Thaumatocaryon*).

17 genera, 15 in W Eurasia and N Africa, two in South America. Boragineae are mostly mesophilic herbs, often with large leaves and a dense, hispid indumentum. The nutlets are typically

keeled both ventrally and dorsally and have a papillose surface. With the exception of *Moritzia* and *Thaumatocaryon*, the fruits are provided with an elaiosome.

29. ***Moritzia*** DC. ex Meisn.

Moritzia DC. ex Meisn., Pl. Vasc. Gen. [Meisner] 280 (1840); Johnston, Contr. Gray Herb. Harvard Univ. 78: 1–118 (1927), rev.; Fernandes-Soares, Iheringia 17: 28–33 (1973), rev.; Weigend et al., Syst. Bot. 35: 409–419 (2010), phylog., morph.

Perennial herbs, with several erect flowering stems from short, erect rhizome and with arching, frondose runners, rooting at apex; indumentum sericeous to hispid, sometimes reduced. Leaves (ob-)ovate, in dense basal rosette and some much smaller cauline leaves. Inflorescences ebracteose thyrsoids with narrowly spaced paraclades, these with peduncles much shorter than monochasia, initially very dense and nearly capitate, then elongating. Flowers erect, (sub)sessile; calyx divided about 1/2 of its length, abaxially with very short scabrid and much longer uncinate trichomes, basally asymmetrically accrescent and enclosing the nutlet; corolla hypocrateriform, tube short, lobes erect to spreading, white or blue, faucal scales widely subcircular to ovate; stamens included or rarely exserted; style filiform, stigma capitate. Nutlet 1 by abortion, erect, straight, oblique-ovoid, laterally compressed with distinct dorsal and ventral keel, base shortly stipitate ending in small, circular cicatrix.

Three species in NW and SW South America. Allied to *Thaumatocaryon*, but clearly distinct.

30. ***Thaumatocaryon*** Baill.

Thaumatocaryon Baill., Bull. Soc. Linn. Paris, 2: 839 (1890); Johnston, Contr. Gray Herb. Harvard Univ. 78: 1–118 (1927), rev.; Fernandes-Soares, Iheringia 17: 28–33 (1973), rev.; Weigend et al., Syst. Bot. 35: 409–419 (2010), phylog., morph.

Perennial herbs, with several erect flowering stems from short, erect rhizome and arching, frondose or scale-leafed runners, rooting at apex. Leaves in loose basal rosette and some much smaller cauline leaves alternate or opposite, basal ones sometimes attenuate into long petiole. Inflorescences bracteose thyrsoids with widely spaced paraclades (peduncles equalling monochasia), ascending, initially dense, later with elongating internodes. Flowers erect, (sub-)sessile; calyx lobed to middle, lobes triangular-ovate, abaxially with very short scabrid trichomes; corolla hypocrateriform, tube short, lobes subcircular to oblong, erect to spreading, white or blue, faucal scales subcircular, not clearly delimited, densely papillose to pubescent; stamens included; style filiform, long, stigma capitate. Nutlet 1 by abortion, erect, straight, oblique-ovoid, laterally compressed with distinct dorsal and ventral keel, base shortly stipitate ending in small, circular cicatrix.

Three species in SW South America. Allied to *Moritzia*, but clearly distinct.

31. ***Anchusa*** L.

Anchusa L., Sp. Pl.: 133 (1753); Guşuleac, Bul. Fac. Şti. Cernauti 1: 73–123 & 235–325 (1927), reg. rev.; Hilger et al., Ann. Bot. 94: 201–212 (2004), phylog.; Selvi & Bigazzi, Bot. J. Linn. Soc. 142: 431–454 (2003), reg. rev.

Annual, biennial or perennial herbs; indumentum hispid-setose to coarsely tuberculate-strigose. Leaves basal and cauline, linear, lanceolate, ovate or oblanceolate, entire or crispate-undulate. Inflorescences bracteose thyrsoids, ± lax and elongating in fruit. Calyx divided to 1/3 or to base, often slightly accrescent in fruit; corolla hypocrateriform to broadly infundibuliform, tube long, rarely limb reduced, faucal scales triangular-oblong, pubescent or penicillate; filaments inserted near or above middle of tube, stamens mostly included; style included to slightly exserted (stylar polymorphism in some species), stigma capitate-ovoid, bilobed. Nutlets ovoid, horizontally inflexed, adaxially shortly beaked or oblong-erect and rounded at apex, rugose-tuberculate; cicatrix with ± incrassate basal ring. $2n = 16, 22, 32$.

About 35 species in Europe, Africa and W Asia, centre of diversity in the Mediterranean region.

32. ***Anchusella*** Bigazzi, Nardi & Selvi Fig. 11

Anchusella Bigazzi, Nardi & Selvi, Pl. Syst. Evol. 205: 253 (1997).

Annual herbs; indumentum hispid-strigose. Leaves basal and cauline, sessile, narrowly ovate,

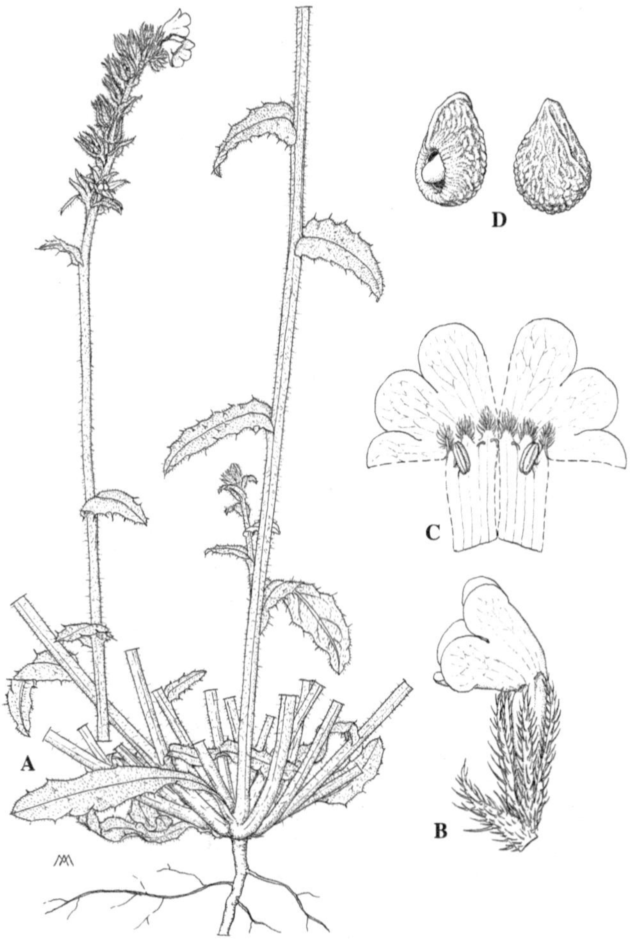

Fig. 11. Boraginaceae. Tribe Boragineae: *Anchusella cretica*. **A** Habit. **B** Flower. **C** Open corolla. **D** Nutlets. (Bigazzi et al. 1997, original drawn by A. Maury)

margins repand-dentate. Inflorescences monochasial, bracteose cymoids. Calyx lobes acute, spreading in fruit; corolla obliquely infundibuliform, zygomorphic, tube slightly curved, limb oblique, lobes unequal, rounded, blue with white lines or white with dark-purple spots. Filaments inserted in upper part of tube, two stamens functional, three staminodial; style included, weakly curved, stigma shallowly bifid with corniculate lobes. Nutlets widely ovoid to orbicular, horizontally inflexed (long axis parallel to hypanthium), nearly smooth to ornamented, cicatrix with scarcely thickened ring. $2n = 16$.

Two species in the central to E Mediterranean.

33. ***Cynoglottis*** (Guşul.) Vural & K. Tan

Cynoglottis (Guşul.) Vural & K. Tan, Notes Roy. Bot. Gard. Edinb. 41: 71 (1983); Selvi & Bigazzi, Pl. Biosystems 132: 113–142 (1998), rev. reg.; Bigazzi & Selvi, Bot. J. Linn. Soc. 136: 365–378 (2001), cytogeography; Selvi et al., Pl. Syst. Evol. 246: 195–209 (2004), phylog.

Perennial herbs; indumentum shortly hispid-setose. Leaves oblong-spathulate to almost linear, margins entire or slightly undulate-crenate. Inflorescences bracteose thyrsoids. Flowers shortly pedicellate; corolla subrotate, tube very short (equalling or shorter than calyx), lobes ovate-orbicular, spreading, blue, faucal scales finely papillose; filaments inserted slightly below scales; style short, included, stigma truncate, slightly bilobed. Nutlets erect, oblong-ovoid, ornamented; cicatrix with thin ring. $2n = 18, 36$.

Two species, one in Europe and one in Asia Minor.

34. ***Gastrocotyle*** Bunge

Gastrocotyle Bunge, Ann. Sci. Nat., Bot. 3, 12: 363 (1949); Bigazzi, Selvi, Fiorini, Edinb. J. Bot. 56: 239–251 (1999), genus delimitation; Guşuleac, Bul. Fac. Şti. Cernauti 2: 394–461 (1928), genus delimitation.

Annual herbs; indumentum coarsely setose-hispid. Leaves narrowly obovate with entire or obscurely erose-dentate margins. Inflorescences dichotomously branched frondose cymoids (= flowers apparently axillary); corolla infundibuliform, subrotate to subcampanulate, pink or blue, tube short, equalling calyx or slightly longer, faucal scales widely oblong, emarginate, papillose; stamens included; style included, stigma cylindrical with spherical and trichoid papillae. Nutlets widely ovoid, horizontally inflexed (long axis parallel to hypanthium), adaxially shortly beaked, ornamented, cicatrix with thickened ring. $2n = 16, 18, 20$.

Two species, one in the Saharo-Sindic region and one in the southern Balkans.

35. ***Lycopsis*** L.

Lycopsis L., Sp. Pl.: 139 (1753); Bigazzi et al., Giorn. Bot. Ital. 129: 121 (1995), morph.; Guşuleac, Bul. Fac. Şti. Cernauti 1: 73–123, 235–325 (1927), rev.; Selvi & Bigazzi, Pl. Biosystems 132: 113–142 (1998), rev. reg.

Annual herbs; indumentum hispid-strigose. Inflorescences initially dense bracteose cymoids. Flowers subsessile; calyx accrescent or spreading in fruit; corolla zygomorphic, blue; tube sharply bent at middle, equalling calyx, lobes unequal, subcircular to oblong, faucal scales white-papillose; filaments inserted in the middle of tube, stamens included; style included, stigma bilobed, capitate-ovoid. Nutlets widely ovoid, horizontally inflexed (long axis parallel to hypanthium), adaxially shortly beaked, ornamented, cicatrix with thickened basal ring. $2n = 16, 48$.

Two species in Europe and W Asia.

36. ***Hormuzakia*** Guşul.

Hormuzakia Guşul., Publ. Soc. Nat. Bucur. 6: 83 (1923); Bigazzi et al., Edinb. J. Bot. 56: 239–251 (1999), genus delimitation; Guşuleac, Bul. Fac. Şti. Cernauti 2: 394–461 (1928), genus delimitation; Selvi & Bigazzi, Biosystems 132: 113–142 (1998), reg. rev.

Annual or perennial herbs; indumentum white, coarsely hispid; massive, succulent primary root present. Leaves cauline and basal, basal ones linear to narrowly obovate, margins obscurely erose-dentate; cauline leaves linear, sessile. Inflorescences frondose, compact cymoids with densely aggregated, sessile flowers, rarely internodes elongating; calyx divided to base, accrescent and spherical in fruit; corolla hypocrateriform, tube slightly longer than calyx, lobes acute, blue, faucal scales prominent, oblong-erect, densely white-pilose; annulus present as pubescent ring; filaments inserted below or at the same level as faucal scales; style included, stigma capitate. Nutlets 2–4, hemispherical and helmet-shaped with a lateral cleft, papillose; cicatrix with a plicate-dentate ring. $2n = 16$.

Two species in the SE Mediterranean.

37. ***Phyllocara*** Guşul.

Phyllocara Guşul., Bul. Fac. Şti. Cernauti 1: 119 (1927); Bigazzi et al., Edinb. J. Bot. 56: 239–251 (1999), genus delimitation.

Annual herb with decumbent to ascending shoots, these simple or dichotomously branched above; indumentum hispid-strigose. Inflorescences frondose, dense and compact cymoids. Flowers subsessile; calyx slightly accrescent and deflexed in fruit; corolla pinkish to purple, narrowly infundibuliform, tube longer than calyx, lobes half-spreading; faucal scales oblong, erect, yellowish, papillose to shortly pubescent; stamens included; style included, stigma capitate-bilobed. Nutlets reniform, horizontally inflexed (long axis parallel to hypanthium), finely rugose-tuberculate; cicatrix with thin basal ring. $2n = 16$.

One species, *P. aucheri* (DC.) Guşul., Irano-Turanian region.

38. ***Borago*** L.

Borago L., Sp. Pl. 137 (1753); Guşuleac, M. Bul. Fac. Şti. Cernauti 2: 394–461 (1928), rev.; Selvi et al., Ann. Bot. 98: 857–868 (2006), phylog.

Annual or perennial herbs; indumentum hispid. Leaves basal and cauline or cauline only, basal leaves ovate to oblong-obovate, petiolate, cauline leaves sessile or semiamplexicaulous. Inflorescences lax, bracteose cymoids. Flowers long pedicellate, pedicels deflexed in fruit; calyx divided to base, accrescent in fruit; corolla subrotate with very short tube and (half-)spreading to reflexed, ovate-acuminate lobes (subg. *Borago*) or narrowly campanulate with suberect lobes (subg. *Buglossites*), blue or whitish; faucal scales glabrous, trapezoid; filaments inserted near base of tube, with an acute or rounded appendage at base, anthers exserted or included, connivent, mucronate; style included; stigma capitate. Nutlets oblong-ovoid, erect, cicatrix with a thin basal ring. $2n = 12, 16, 18, 30, 32, 48$.

Five species, mainly W Mediterranean, one species (*B. officinalis* L.) widely cultivated as culinary herb and often naturalized.

39. ***Brunnera*** Steven

Brunnera Steven, Bull. Soc. Imp. Naturalistes Moscou 24 (1): 582 (1851); Bigazzi & Selvi, Bot. J. Linn. Soc. 136: 365–378 (2001), cytogeogr.; Guşuleac, M. Bul. Fac. Şti. Cernauti 2: 394–461 (1928), rev.; Selvi et al., Pl. Syst. Evol. 246: 195–209 (2004), phylog.

Perennial herbs, shortly bristly-setose; rhizomes present, robust; stems erect, usually branched only above. Basal leaves long petiolate, ovate-cordate or ovate-lanceolate. Inflorescence of ebracteose lax cymes. Flowers short pedicellate, pedicels elongate in fruit; calyx divided to 3/4, slightly accrescent in fruit; corolla blue, subrotate with short tube, lobes oblong; faucal scales trapezoid, ciliate, closing throat, white; stamens inserted near top of tube, included; style included, stigma capitate, gynobase flat. Nutlets obliquely oblong-ovoid, sculptured, cicatrix surrounded by slightly thickened basal ring. $2n = 12$, 14, 24, 36, 72.

Three species in W Asia and SE Mediterranean.

40. ***Melanortocarya*** Selvi, Bigazzi, Hilger & Papini

Melanortocarya Selvi, Bigazzi, Hilger & Papini, Taxon 55: 915 (2006).

Annual herb; indumentum hispid. Leaves basal and cauline, ovate-acuminate. Inflorescences often simple frondose cymoids with few subsessile flowers, pedicels deflexed in fruit; calyx divided to base, accrescent; corolla hypocrateriform, tube slightly longer than calyx, lobes spreading, trapezoid; corolla sky blue, faucal scales shaggy pubescent, pubescence extending downward into tube; filaments inserted near base of tube, stamens included; style very short; stigma bilobed. Nutlets erect, oblong-ovoid, black, smooth and glossy, but minutely puberulent. $2n = 20$.

Only *M. obtusifolia* (Willd.) Selvi, Bigazzi, Hilger & Papini, SE Mediterranean.

41. ***Nonea*** Medik. Fig. 12

Nonea Medik., Philos. Bot. 1: 31 (1789); Sadat, Mitt. Bot. Staatss. München 28: 1–210 (1989), reg. rev.; Selvi et al., Taxon 51: 719–730 (2001), morphol.; Selvi et al., Taxon 55: 907–918 (2006), phylog.
Elizaldia Willk. (1852).

Annual or perennial herbs; indumentum hispid and usually glandular. Inflorescences frondose to frondose-bracteose cymoids, ± elongated in fruit. Flowers shortly pedicellate; calyx accrescent, spherical-sacciform and deflexed in fruit; corolla radially symmetrical or weakly zygomorphic, narrowly or broadly infundibuliform, white, yellow, blue or purple to almost black, faucal scales absent but throat with 5 tufts or a ring of trichomes, rarely glabrous; filaments inserted near middle of tube or in distal half, anthers included or partially exserted at corolla mouth, rarely one stamen completely exserted (*N. heterostemon*); style included, stigma capitate-bilobed. Nutlets broadly ovoid to reniform, horizontally inflexed (long axis parallel to hypanthium), incurved or erect, reticulate, cicatrix with incrassate, rarely dentate ring. $2n = 14$, 16, 18, 20, 22, 28, 30, 32, 40, 44, 59, 60.

About 35 species in Europe, N Africa and W Asia.

Fig. 12. Boraginaceae. Tribe Boragineae: *Nonea persica*. **A** Habit. **B** Flower. **C** Corolla tube. **D** Open corolla. **E** Nutlet. (Selvi and Bigazzi 2001, drawn by A. Maury)

42. ***Pulmonaria*** L.

Pulmonaria L., Sp. Pl. 135 (1753); Bolliger, Phaner. Monogr. 8: 1–215 (1982), reg. rev.; Kerner, Monographia Pulmonarum, Oeniponte: 1–52, tab. I-XIII (1878), rev.; Selvi et al., Taxon 55: 907–918 (2006), phylog.
Paraskevia W. Sauer & G. Sauer (1980).

Perennial herbs with short rhizomes; indumentum pubescent and usually glandular. Leaves basal and cauline, basal rosette with large, ovate leaves, usually developing after flowering, rarely before (*P. cesatiana*), cauline leaves smaller, semiamplexicaulous. Inflorescences bracteose cymoids, rarely weakly branched thyrsoids. Flowers shortly pedicellate, usually distylous (except *P. cesatiana*); calyx divided to 1/3, cylindrical at anthesis, calyx lobes porrect, triangular-ovate, calyx ventricose-campanulate in fruit; corolla infundibuliform, purple, pink or blue, faucal scales absent but throat with a ring of trichomes; filaments inserted at or below throat; style included or slightly exserted (depending on stylar morph), filiform, stigma capitate-bilobed. Nutlets erect, ovoid, smooth to pubescent, constricted at base above a distinct collar-like ring. $2n$ =14, 16, 22, 18, 20, 21, 22, 26, 28, 30, commonly 14 or 28.

About 17 species, Europe, one species extending to E Asia.

43. ***Pentaglottis*** Tausch

Pentaglottis Tausch, Flora 12: 643 (1829).
Caryolopha Fisch. & Trautv. (1837).

Perennial herb with thick rhizome; indumentum hispid. Leaves basal and cauline, basal leaves large, ovate-acuminate, long petiolate, cauline leaves much smaller, sessile. Inflorescences bracteose-frondose thyrsoids. Flowers shortly pedicellate; calyx divided to base, accrescent in fruit; corolla hypocrateriform, bright blue, tube about as long as calyx, lobes spreading, oblong to trapezoid, faucal scales shortly trapezoid, white, shortly pubescent; filaments inserted in distal half of tube, included; style included, stigma capitate. Nutlets small, excentrically stalked, obliquely ovoid, beaked, surface finely scabrid, cicatrix with a slightly thickened annulus. $2n = 22$.

One species, *P. sempervirens* (L.) Tausch ex L. H. Bailey, SW Europe.

44. ***Symphytum*** L.

Symphytum L., Sp. Pl. 136 (1753); Kurtto, Acta Bot. Fenn. 19: 177–192 (1982), rev.; Sandbrink et al., Proc. Kon. Ned. Akad. v. Wetensch. 93: 295–334 (1990), phylog.
Procopiania Guşul. (1928).

Perennial herbs with persistent, later splitting primary root (root-pleiocorm) or stoloniferous with stolon tubers or rhizomatous; indumentum hispid. Leaves basal and cauline, widely ovate-acuminate with cuneate, rounded to subcordate base, basal leaves long petiolate, cauline leaves subsessile or decurrent. Inflorescences bracteose thyrsoids. Flowers long pedicellate (pedicel to 8 mm); calyx cylindrical to campanulate, divided to 1/3 or nearly to base, accrescent in fruit, calyx lobes narrowly triangular-ovate; corolla cylindrical to campanulate, sometimes hypocrateriform with spreading to reflexed lobes and very short tube ("*Procopiania*"), faucal scales linear or subulate, papillose along margins; stamens included, rarely exserted, connectives sometimes with apical appendages; style filiform, exserted, stigma small, entire. Nutlets ovoid, erect or oblique, verruculose and often rugose, obliquely keeled, cicatrix surrounded by collar-like thickened toothed ring, teeth clasping receptacle. $2n$ = 18, 20, 22, 24, 28, 30, 32, 34, 36, 40, 42, 44, 48, 56, 60, 64, 72, 84, 96, 104, 120, 128, 144.

About 35 species in Europe and W Asia, apparently including several interspecific hybrids.

45. ***Trachystemon*** D. Don

Trachystemon D. Don, Edinburgh New Philos. J. 13: 239 (1832).

Perennial herb from thick, branched rhizome; indumentum shortly hispid. Leaves cauline and basal, basal leaves large, ovate-acuminate from cordate base, long petiolate, developing at flowering time, cauline leaves much smaller, sessile. Inflorescences bracteose thyrsoids. Flowers long pedicellate, deflexed; calyx cup-shaped, lobes ovate, accrescent in fruit; corolla with short, conical tube and very long lobes (> tube), lobes nearly linear, reflexed and spirally twisted, violet-blue, faucal scales in two series of 5, white, the lower villous, the upper thick and papillose, protruding from throat; filaments inserted in distal half of tube, pink, pubescent at base, anthers long-exserted and connivent; style long-exserted, violet-purple, stigma punctiform. Nutlets ovoid, horizontally inflexed (long axis parallel to hypanthium), adaxially shortly beaked, beak keeled, papillose-tuberculate, cicatrix with thin ring. $2n = 56$.

One species, *T. orientalis* (L.) D. Don, Black Sea region.

IV. Tribe Cynoglosseae W.D.J. Koch (1837).

Asperugeae Zakirov ex Ovczinnikova (2007).
Cryptantheae Brand (1925), nom. illegit.

Eritrichieae Gürke (1893).
Myosotideae Rchb.f. (1858).
Trigonocaryeae Kerimov (2005).

Annual or perennial herbs, sometimes minute and ephemeral, rarely subshrubs or shrubs, sometimes rhizomatous or stoloniferous, basal leaf rosette usually present (at least initially), leaves (ob-)ovate-acuminate, (ob-)ovate, elliptical or linear, usually with cuneate or decurrent base, rarely cordate and/or distinctly petiolate; plants hispid, sericeous, pubescent or villous, trichomes erect, rarely ascending or appressed, uncinate trichomes sometimes present. Inflorescence frondose to ebracteose thyrsoids or cymoids. Calyx variously united, often only in lower third, rotate to campanulate, sometimes accrescent in fruit; corolla radially symmetrical, mostly hypocrateriform or infundibuliform, faucal scales large, conspicuous, often exserted from throat, pubescent or papillose, rarely glabrous or absent; stamens equal or rarely unequal (*Caccinia*), anthers sometimes with long connective appendages; pollen eurypalynous, but mostly heterocolporate with 3–4 apertures and 3–4 pseudoapertures, rarely tricolporate (*Caccinia, Trichodesma*); gynobase widely to narrowly pyramidal to subulate, rarely almost flat, style with truncate or capitate stigma, rarely bilobed. Nutlets 1–4(10), straight or spreading, rarely recurved (never incurved), often oblique with apex close to style base and nutlet base widely divergent, dorsiventrally compressed, sometimes ventrally keeled, rarely ovoid to subcylindrical or lenticellate, then usually with ventral keel, variously papillose or glochidiate, rarely smooth, often with distinct dorsomarginal wing, wing spreading or erect to incurved, nutlets usually with ovate to triangular cicatrix, often in median or subapical position, rarely basal, elaiosome usually absent (present in some *Myosotis*). Nutlets dispersed singly or remaining enclosed in accrescent calyx or attached to the calyx or the plant.

Subcosmopolitan, 40 genera, centre of diversity in E Asia. Cynoglosseae, as the largest subgroup of Boraginaceae, are highly heterogeneous in both generative and vegetative characters, but the presence of dorsiventrally flattened, often marginally keeled or winged fruits is common to the majority of representatives. The gynobase is usually pyramidal to subulate and nutlet attachment then ventral rather than basal.

IV.1 Trichodesma Group

Robust perennial or biennial herbs, often (sometimes tall) shrubs; plants hispid to sericeous, sometimes leaves additionally glaucous. Leaves elliptical to obovate, subsessile. Calyx divided to base or to the middle; corolla radially symmetrical or slightly zygomorphic; pollen tricolporate. Nutlets often 1 or up to 4, oblique to horizontal, dorsiventrally compressed and usually winged.

Two genera, Africa, Asia, Australia.

46. ***Trichodesma*** R. Br. Fig. 13

Trichodesma R. Br., Prodr.: 496 (1810), nom. cons.; Brummitt, Kew Bull. 37: 429–450 (1982), reg. rev.; Sadat, Mitt. Bot. Staatss. München 28: 167–191 (1989), reg. rev.

Robust annual herbs or shrubs to small trees, often with lignescent to ligneous pleiocorm; indumentum hispid to sericeous, rarely subglabrous. Leaves all cauline, mostly sessile and elliptical, alternate or frequently opposite almost throughout; inflorescence bracteose to frondose cymoids or few-branched thyrsoids. Flowers long pedicellate, deflexed; calyx divided almost to base, sepals ribbed or cordate-winged near base, accrescent in fruit; corolla campanulate to hypocrateriform, tube shorter than calyx; faucal scales absent, rarely trapezoid and deeply lobed and apparently 10; stamens 5–6, with short filaments and long anthers, these usually widely exserted, connective with a long terminal appendage, all appendages together forming a narrow, spirally twisted cone above anthers; style filiform, stigma capitate, gynobase shortly pyramidal. Nutlets large, 1–4, oblique or horizontal, ovoid to subcircular in outline, often with a narrow, erect or spreading, dentate wing margin, smooth or rugose to glochidiate; cicatrix large, medial to subapical. $2n$ = 14, 22, 24, 44.

A genus of about 40 to 50 species from S Africa to S and SE Asia and Australia.

47. ***Caccinia*** Savi

Caccinia Savi, Cose Botaniche 1, 7, t. 1 (1832); Papava, Not. Syst. Geogr. Inst. Bot. Tbiliss. 12: 39–45 (1944), reg. rev.
Heliocarya Bunge (1871).

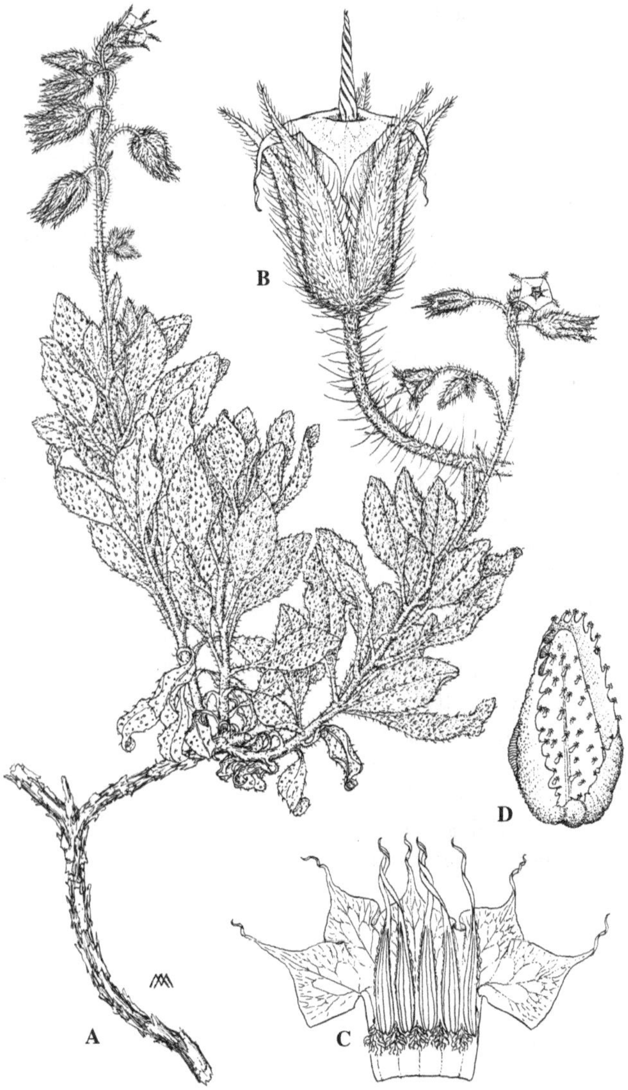

Fig. 13. Boraginaceae. Tribe Cynoglosseae, *Trichodesma* group: *Trichodesma cinereum*. **A** Habit. **B** Flower. **C** Open corolla. **D** Nutlet. (Mosti and Selvi 2007, original drawn by A. Maury)

Robust biennial or perennial herbs, sometimes extensively rhizomatous; indumentum sparsely hispid, leaves often glaucous. Shoots ascending or erect. Leaves all cauline or some basal, obovate to narrowly oblong, ciliate from hispid trichomes. Inflorescences many-branched thyrsoids, bracteose-frondose to bracteose. Flowers spreading, pedicellate; calyx lobed for 1/2 to 3/4, strongly accrescent in fruit; corolla hypocrateriform, narrow tube much longer than calyx, lobes asymmetrically spreading, narrowly ovate-triangular; faucal scales oblong to rectangular; stamens inserted near apex of tube, filaments very unequal, at least one very long exserted, others short or reduced; gynobase shortly pyramidal, style filiform, stigma capitate. Nutlets 1(2), large, horizontal, straight, ovoid to subspherical, completely covering the receptacle, dorsally often keeled, rugose and with glochidiate wing; cicatrix large, subapical. $2n = 24$.

About six species from Iran to southern C Asia.

IV.2 Lasiocaryum Group

Small to minute herbs. Indumentum sericeous to villous. Leaves obovate to spathulate, subsessile. Calyx divided to base; corolla radially symmetrical, hypocrateriform. Nutlets 4, erect, ovoid to subcylindrical, pubescent at least dorsally or rugose, cicatrix triangular, triangular-ovate to elliptical, suprabasal on adaxial side.

Three genera, essentially Himalayan, sister to *Trichodesma* + *Caccinia* (Weigend et al. 2013).

48. ***Lasiocaryum*** I.M. Johnst.

Lasiocaryum I.M. Johnst., Contr. Gray Herb. 75: 45 (1925); Banerjee & Bhattacharjee, Bull. Botan. Soc. Bengal 39: 33–39 (1985), reg. rev.
Oreogenia I.M. Johnst. (1924) non *Orogenia* Watson (1871).
Setulocarya R.R. Mill & D.G. Long (1996)

Small annual herbs from basal rosette. Indumentum pubescent to villous. Leaves basal and cauline, obovate to elliptic. Inflorescences bracteate cymoids, initially dense, later elongated; corolla hypocrateriform, blue or whitish, tube equalling calyx; faucal scales small, broadly rectangular, basal scales present; filaments inserted at middle of tube, anthers included; style short, stigma capitate, included; gynobase columnar. Nutlets erect, turbinate with longitudinal keel on abaxial side, pubescent on abaxial side; cicatrix broadly elliptic, reaching from base to upper half of nutlet.

Four species from N India to China.

49. ***Microcaryum*** I.M. Johnst.

Microcaryum I.M. Johnst., Contr. Gray Herb. 73: 63 (1924); Banerjee & Bhattacharjee, Bull. Botan. Soc. Bengal 39: 33–39 (1985), reg. rev.

Minute (<5 cm) ephemeral herb, stiffly erect, unbranched or sparsely branched; indumentum scabrid to villous. Leaves sessile, narrowly obovate. Inflorescences bracteose cymoids, all pedicels long and overtopping main axis, appearing umbellate. Pedicels strongly elongated in fruit; calyx divided nearly to base, lobes acute; corolla equalling calyx, blue to white, faucal scales ovate to crescent-shaped, basal scales five; stamens included; style included. Nutlets oblong-ovoid, 1 mm long, rugose, glabrous, adaxially with basal triangular attachment scar continuing into a broad, raised sulcus, abaxially rounded to slightly keeled.

One species, *M. pygmaeum* (C.B. Clarke) I.M. Johnst., from NE India to China. Clearly allied to and distinct from *Lasiocaryum*.

50. ***Chionocharis*** I.M. Johnst.

Chionocharis I.M. Johnst., Contr. Gray Herb. 73: 65–66 (1924).

Small cushion-forming perennial herb, simple or dichotomously branched, individual stems clavate in outline by crowded, densely overlapping leaves. Leaves all cauline and crowded towards the shoot apex, flabellate, basally sparsely pubescent, distally long and densely villous with soft, silky trichomes. Flowers apparently solitary, scarcely protruding from cushions; calyx villous; corolla hypocrateriform, tube short (< lobes), faucal scales broadly rectangular, white or yellow; filament insertion unknown, anthers included; style short, stigma capitate, included; gynobase shortly subulate. Nutlets erect, ovate to cylindrical, black, basally appressed pubescent, apically long pubescent; cicatrix elliptic, reaching from base to upper half of nutlet.

One species, *C. hookeri* (C.B. Clarke) I.M. Johnst., high alpine (3500–5800 m), NE India to China.

IV.3 Omphalodes Group

Medium-sized to large annual or perennial herbs, often rhizomatous. Indumentum pubescent or leaves at least adaxially glabrous. Leaves ovate to oblong, lower leaves (if present) petiolate. Corolla hypocrateriform; stamens included; gynobase widely pyramidal, style linear, short, stigma capitate, included. Nutlets 4, rarely 1, oblique or horizontal on gynobase, ovoid to subcircular, glabrous or pubescent, sometimes sparsely glochidiate, dorso-marginally differentiated, cicatrix in median position, ovate to elliptical.

Three genera, one widespread but local in northern hemisphere, one only on Juan Fernandez Islands, one only on Chatham Islands.

51. ***Omphalodes*** Mill.[2]

Omphalodes Mill., Gard. Dict. Abr. ed. 4. (1754); Nesom, Sida 13: 25–30 (1988), reg. rev.
Gyrocaryum Valdés (1983).
Mimophytum Greenm. (1905).
Omphalotrigonotis W.T. Wang (1984).
Sinojohnstonia Hu (1936).

(Winter-)Annual, biennial or perennial herbs, often rhizomatous or stoloniferous, aerial stems stiffly erect or diffuse; indumentum soft pubescent, often appressed, sometimes glabrous and glaucous. Leaves initially basal and cauline, basal ones (long-) petiolate, sometimes later absent, lamina broadly ovate-acuminate from cordate base or ovate to oblong, cauline leaves sometimes amplexicaulous. Inflorescences cymoids, rarely thyrsoids, ebracteose or frondose below and ebracteose above, rarely frondose-bracteose, sometimes rooting near apex and forming a new leaf rosette. Flowers erect, pedicellate, pedicels sometimes elongating and curved in fruit; calyx lobed nearly to base, rotate, rarely campanulate, not or strongly accrescent in fruit; corolla hypocrateriform, rarely infundibuliform, white or bright blue; lobes > tube, faucal scales present, broadly rectangular, often emarginate. Nutlets oblique-obconical in lateral view, ovoid to circular in dorsal view, glabrous, rarely pubescent, dorsomarginal wing spreading, erect or incurved, membranaceous or coriaceous, entire or serrate, sometimes glochidiate. $2n = 19, 20, 22, 24, 28, 36, 42, 48$.

[2] Two lineages of *Omphalodes* recognized as not belonging to the core-group of the genus have been removed to the novel genera *Memoremea* A. Otero, Jim.-Mejías, Valcárcel & P. Vargas and *Nihon* A. Otero, Jim.-Mejías, Valcárcel & P. Vargas. *Memoremea* is the first-branching lineage of the *Mertensia*-group and *Nihon* falls into the *Bothriospermum*-group (Weigend et al. 2013; Otero et al. 2014).

About 20–30 species, Eurasia and N America. A widespread and highly heterogeneous taxon. The segregate genera proposed may be monophyletic, but the remainder of the genus is then likely to be paraphyletic. *Omphalodes* is therefore here defined broadly until a critical study of the genus across its range is undertaken. Poorly known *Gyrocaryum* from Spain is provisionally included here. Both *Myosotidium* and *Selkirkia* are probably nested in *Omphalodes* (Weigend et al. 2013).

52. ***Myosotidium*** Hook.

Myosotidium Hook. Bot. Mag. 85: t. 5137 (1859).

Robust rosulate perennial herb from thick rhizome; indumentum sparse, soft sericeous on abaxial leaf surface, glabrous and dark green on adaxial leaf surface. Leaves coriaceous when dry, basal ones petiolate, up to 1.5 m long, broadly ovate to reniform, cauline leaves cordate to elliptical. Inflorescence a much-branched thyrsoid. Flowers pedicellate; calyx divided to base with oblong, marginally ciliate lobes; corolla hypocrateriform, blue or white; tube shorter than lobes, lobes subcircular, faucal scales depressedly rectangular, carnose. Nutlets large (10–15 mm), flat, with a spreading, unevenly serrated wing, black. $2n = 40, 42$.

A single species, *M. hortensia* (Decne.) Baill., endemic to Chatham Islands (New Zealand), probably nested in *Omphalodes*.

53. ***Selkirkia*** Hemsl.[3]

Selkirkia Hemsl., Bot. Challenger III, 1: 47 (1884).

Sparsely branched shrub; indumentum sparse, appressed, leaves adaxially glabrous. Leaves crowded at branch apices, petiolate, lamina ovate to elliptical, apex abruptly acuminate, chartaceous. Inflorescences widely branched ("corymbose") ebracteose thyrsoids. Flowers pedicellate; calyx slightly accrescent in fruit; corolla white, hypocrateriform, tube short and wide, lobes widely subcircular, half erect to spreading, usually pubescent at least abaxially, faucal scales present, semicircular, annulus present; gynobase shortly pyramidal, much shorter than nutlets in fruit. Nutlets (1–)4, erect, straight, triangular-ovoid in dorsal view, densely glochidiate but with thick glochidia lacking reflexed tips, ventrally keeled, indistinctly winged; cicatrix large, medial, subcircular.

A single species, *S. berteri* Hemsl., Juan Fernandez Islands (Chile). A close affinity to *Hackelia* has been suggested, but *Selkirkia* is probably nested in *Omphalodes*.

IV.4 Myosotis Group

Tiny to medium-sized annual or perennial herbs, sometimes ephemeral, often rhizomatous or stoloniferous. Indumentum pubescent or puberulent, rarely sericeous or villous. Leaves elliptical, (ob-)ovate-acuminate or oblong, lower leaves (if present) petiolate. Gynobase nearly flat to shallowly pyramidal; style filiform, stigma sub-capitate, sometimes obscurely bilobed. Nutlets 4, rarely fewer or up to 10, small (< calyx), erect, lenticellate to tetrahedral, usually with lateral, often also with ventral keels, dark brown to black, rarely triangular, glabrous, smooth or papillose, cicatrix circular to elliptical, suprabasal, rarely cicatrix distinctly stalked.

Four genera, three in E Asia, one subcosmopolitan (*Myosotis*).

54. ***Myosotis*** L.

Myosotis L., Sp. Pl.: 131 (1753); Grau & Leins, Ber. Deutsch. Bot. Ges. 81: 107–115 (1968), palyn., infragen.; Schuster, Feddes Repert. 74: 39–98 (1967), partial rev.; Winkworth et al., Mol. Phyl. Evol. 24: 180–193 (2002), phylog.
Exarrhena (A. DC.) O.D. Nikiforova (2000).
Gymnomyosotis (A. DC.) O.D. Nikiforova (2000).
Strophiostoma Turcz. (1840).
Trigonocaryum Trautv. (1875).

Annual, biennial or perennial herbs, rarely decumbent or stoloniferous or cushion-forming; indumentum pubescent, rarely glabrescent or villous, uncinate trichomes often present. Leaves basal (at least initially) and cauline. Inflorescences cymoids, often monochasial, ebracteose at least above.

[3] New molecular data retrieve *Selkirkia* and *Myosotidium* (the "Pacific *Omphalodes*") together, deeply nested in *Ompalodes*, in a clade together with native South American *Cynglossum*. Generic realignments are in progress in order to accommodate these results.

Flowers pedicellate, pedicel often elongating and curved in fruit; calyx lobed 1/3 to 1/2, often with uncinate trichomes and accrescent in fruit, enclosing nutlets; corolla hypocrateriform to infundibuliform, tube short (= lobes), blue to pink, white or yellow, throat yellow or white, lobes oblong to subcircular, contort in bud; faucal scales semicircular to narrowly rectangular and largely reduced; anthers included, rarely exserted; style included, rarely exserted. Nutlets (1–)4, erect, usually ovoid, dorsiventrally compressed with more or less distinct marginal keel and ventrally keeled, usually smooth and shiny, black; cicatrix basal, subcircular, sometimes with extruding white elaiosome ("*Strophiostoma*", "*Trigonocaryum*"). $2n$ = 12, 14, 16, 18, 20, 22, 24, 28, 30, 32, 36, 40 44, 46, 48, 52, 54, 64, 66, 72, 84, 86, 88.

About 80–100 species, subcosmopolitan with centres of diversity in the Mediterranean and New Zealand.

55. ***Trigonotis*** Steven

Trigonotis Steven, Bull. Soc. Imp. Naturalistes Moscou 24 (1): 603 (1851); Starchenko, Izv. Sib. Otd. Akad. Nauk SSSR, Ser. Biol, Nauk 36–44 (1979), reg. rev.
Zoelleria Warb. (1892).

Annual, biennial or perennial herbs, rosette-forming and rhizomatous or with decumbent, rooting shoots with distichous phyllotaxy; indumentum densely pubescent to subglabrous. Leaves cauline, sometimes also basal, basal leaves long petiolate, cauline leaves sessile, ovate-acuminate to elliptical-acuminate. Inflorescences cymoids, often elongated scorpioid monochasia or boragoids, ebracteose at least above. Flowers subsessile to long pedicellate; calyx divided 1/3 to 1/2, lobes triangular-ovate, often spreading in fruit; corolla hypocrateriform, white or rarely blue, tube short (< calyx), lobes subcircular, imbricate in bud; faucal scales small, semicircular to depressedly rectangular, often emarginate, white to yellow or orange-red; stamens inserted in middle of corolla tube, anthers included; style usually included. Nutlets (1–)4(10), erect to half-erect, usually tetrahedral, sessile or stipitate, adaxially keeled, usually smooth, glabrous or minutely pubescent to tuberculate, rarely narrowly winged ("*Stephanocaryum*"), cicatrix very small, at basal end of adaxial keel. $2n$ = 36, 48.

About 60 species, mainly E & SE Asia, west to SE Russia.

56. ***Decalepidanthus*** Riedl

Decalepidanthus Riedl, Österr. Bot. Z. 110: 608–612 (1963); Dickoré & Hilger, Phytotaxa 226: 131–143 (2015).
Oreocharis (Decne.) Lindl. (1846), nom. rej. non Benth. (1876).
Pseudomertensia Riedl (1967).
Scapicephalus Ovcz. & Czukav. (1974).

Perennial herbs with dense, basal leaf rosettes, rhizomatous and stoloniferous; indumentum pubescent to sericeous. Leaves mostly basal, long petiolate, lamina ovate-acuminate to obovate. Inflorescences initially dense, postflorally elongate ebracteose cymoids or weakly developed thyrsoids. Flowers pendulous or spreading, shortly pedicellate; calyx divided 1/2 to nearly to base, shorter than corolla tube; corolla campanulate, infundibuliform to hypocrateriform, blue or pink, lobes often short, porrect to spreading, faucal scales triangular to narrowly oblong, rarely absent, annulus of 10 basal scales; stamens included to long exserted; style filiform, included or exserted. Nutlets erect, straight, ovoid, dorsiventrally compressed, adaxially sharply keeled, smooth.

About 10 species, Himalayas.

57. ***Brachybotrys*** Maxim. ex Oliv.

Brachybotrys Maxim. ex Oliv., Hooker's Icon. Pl. 13: 43, t. 1254 (1878).

Perennial herb from slender, branched rhizomes, aerial stems unbranched, erect, basally with scale leaves, foliage leaves crowded in upper half; indumentum sparsely tomentose pubescent to appressed pubescent above, subglabrous below. Inflorescences ebracteose, dichasial cymoids. Flowers pendulous, long pedicellate; calyx divided to base, lobes linear, slightly accrescent in fruit and apparently enclosing the fruit; corolla campanulate, purple, tube short (< lobes), lobes porrect, triangular-ovate, faucal scales triangular to depressedly rectangular. Anthers exserted from tube; style filiform, long-exserted. Nutlets tetrahedral, smooth to pubescent, black. $2n$ = 24.

A single species, *B. paridiformis* Maxim. ex Oliv., NE Asia (Russia, China, Korea).

IV.5 Mertensia Group[2]

Small to large perennial herbs, often with pleiocorm; indumentum sparse, pubescent or puberulent, often glabrous and glaucous. Leaves elliptical or ovate-acuminate, lower leaves (if present) petiolate. Gynobase shallowly pyramidal; style exserted or included, stigma capitate. Nutlets 4, oblique on pyramidal gynobase, ovoid to subcircular in dorsal view, weakly dorsiventrally flattened, dorsally more or less convex, ventrally flat or concave, brown to black, smooth or irregularly verrucose or rugose, sometimes very indistinctly winged along margins, cicatrix medial, on short ± distinctive stalk.

Two genera, one Asia, one widespread in the northern hemisphere (*Mertensia*).

58. ***Mertensia*** Roth

Mertensia Roth, Catal. Bot. 1: 34 (1797), nom. cons.; Williams, Ann. Missouri Bot. Gard. 24: 17–159 (1937), reg. rev.; Popov, Bot. Mater. Notul. Syst. 15: 248–266 (1953), reg. rev.; Matthews, Proc. Utah Acad. Sci. Arts Lett. 45: 590–602 (1968), reg. rev.; Starchenko, Bot. Zhurn. 64: 1666–1669 (1979), reg. rev.

Perennial herbs with pleiocorm or root-pleiocorm; indumentum puberulent to pubescent, often subglabrous to glabrous and glaucous; massive taproot usually present, sometimes also root tuber. Leaves basal and cauline, basal leaves long petiolate or with decurrent margins, lamina ovate to elliptical, rarely flabellate, membranaceous to thinly succulent. Inflorescences cymoids or thyrsoids, monochasia ebracteose, but each subtended by one large bract. Flowers long pedicellate, usually pendulous; calyx divided 1/2 to nearly to base, short (< corolla tube), not or slightly accrescent in fruit; corolla narrowly infundibuliform to campanulate, blue or pink to white, lobes ovate to semicircular, obtuse; faucal scales usually small, triangular; filaments inserted between faucal scales or slightly below. Nutlets oblique, with flat, rounded, rarely inflated back, margin sometimes narrowly winged, apex sometimes distinctly acuminate, smooth or rugulose to verrucose, ventrally sharply keeled above and more or less convex below; cicatrix submedial, small, triangular-ovate, elevated onto small stipe. $2n = 24, 48, 72$.

About 40 species in E Asia, N America, one reaching NW Europe.

59. ***Anoplocaryum*** Ledeb.

Anoplocaryum Ledeb., Fl. Ross. 3: 154 (1847).

Perennial herbs with stems branching from base, decumbent to ascending; indumentum pubescent; persistent taproot present. Leaves obovate to oblong. Inflorescence short bracteose cymoids, strongly elongating in fruit. Flowers pedicellate; calyx divided nearly to base, spreading in fruit; corolla hypocrateriform, blue, tube equalling limb; faucal scales depressedly rectangular, papillose, white to yellow; anthers included. Nutlets small, ventrally keeled above and more or less convex below; cicatrix submedial, small, triangular-ovate, elevated onto small stipe; black, finely tuberculate; cicatrix submedial, small, narrowly triangular.

Five species in C Asia.

60. ***Asperugo*** L.

Asperugo L., Sp. Pl. 1: 138 (1753).

Lax to scrambling annual herb; indumentum on stem retrorsely uncinate, otherwise scabrid. Leaves basal and cauline, distinctive rosette absent, ovate to elliptical, lower leaves opposite to pseudoverticillate, long petiolate, upper leaves sessile, alternate. Inflorescences frondose cymoids (mostly monochasial), later very lax. Flowers subsessile, pedicels elongating and recurved in fruit; calyx bilobed almost to base, strongly accrescent, 2-lipped in fruit with two broadly triangular lobes, one lip 7-, the other 8-dentate, enclosing the nutlets; corolla infundibuliform to hypocrateriform, very small, blue to white, faucal scales small, crescent-shaped; stamens included; style included, stigma capitate, gynobase subulate. Nutlets laterally flattened, finely tuberculate, cicatrix suprabasal and asymmetrically placed to one side of flattened nutlet. $2n = 24$.

A single species, *A. procumbens* L., in Europe and Asia. Morphologically highly divergent, but retrieved with *Mertensia* (Weigend et al. 2013).

IV.6 Lappula Group

Small to medium-sized annual or perennial herbs, sometimes ephemeral or cushion-forming. Indumentum mostly hispid, sometimes pubescent, sericeous or villous. Leaves elliptical, (ob-)

ovate-acuminate or oblong, rarely linear, lower leaves (if present) sessile or petiolate. Calyx usually divided nearly to base; corolla hypocrateriform to infundibuliform; anthers included; gynobase (narrowly) pyramidal to subulate, rarely widely pyramidal; style filiform, included, stigma capitate. Nutlets 1–4, oblique on pyramidal gynobase or parallel to subulate gynobase (perpendicular to receptacle), ovate to subcircular in dorsal view, glochidiate, rarely only verrucose or nearly smooth, cicatrix triangular or triangular-ovate, medial.

Eight genera recognized here, but generic limits highly problematical, essentially an Old World group with centre of diversity in C Asia. *Heterocaryum* and *Suchtelenia* are apparently more closely allied to each other and sister to genera 61–64. *Selkirkia*, albeit morphologically superficially similar to *Hackelia*, may indeed be closer to Chilean *Cynoglossum*.

61. ***Heterocaryum*** A. DC.

Heterocaryum A. DC., Prodr. (DC.) 10: 144 (1846).

Annual herbs; taproot present. Leaves all cauline (in flower), linear, sessile. Inflorescences bracteose cymoids, elongating in fruit. Flowers shortly pedicellate, pedicels usually elongating in fruit; calyx accrescent in fruit; corolla cylindrical to narrowly infundibuliform, tube short (< calyx), faucal scales minute, depressedly rectangular; gynobase columnar, subulate in fruit, winged between areoles, nutlets oblong, heteromorphic (3+1 or 2+2), either with dentate-glochidiate wing or not winged, permanently attached to gynobase along entire adaxial surface. $2n = 24, 48$.

About six species in W Asia. Closely related to and sometimes included in *Lappula*.

62. ***Suchtelenia*** Karel. ex Meisn.

Suchtelenia Karel. ex Meisn., Pl. Vasc. Gen. [Meisner] 1: 279 (1840); Kerimov & Askerova, Bot. Zh. 90: 264–267 (2005), syst.

Annual herb; indumentum hispid, more often subglabrous and glaucous. Leaves subopposite below, alternate above, slightly succulent. Inflorescences few-flowered, ebracteose cymoids. Flowers pedicellate; calyx divided 1/2 to 3/4, strongly accrescent and patelliform in fruit; corolla minute, infundibuliform, blue, faucal scales depressedly rectangular; gynobase widely pyramidal, ultimately hollow. Nutlets (1–)4, ovoid with convex back, smooth or spinulose, adaxial side sunken in gynobase, heteromorphic, either persisting or leaving cavities or frames when abscising.

A single species, *S. calycina* (C.A. Mey.) A. DC., in C Asia, infraspecific taxa are variously recognized or not.

63. ***Lappula*** Moench[4]

Lappula Moench, Meth. (Moench): 416 (1794); Goloskokov, Fl. Rast. Res. Kaz. 1975: 69–80 (1975), reg. rev.; Ovczinnikova, Pjak & Ebel, Turczaninowia 7: 5–13 (2004), reg. rev.; Voytenko & Oparina, Bot. Zhurn. 70: 865–875 (1985), morph.; Sadat, Mitt. Bot. Staatss. München 28: 18–51 (1989), reg. rev.; Zhu et al., Flora of China 16: 403–413 (1995), reg. rev.
Omphalolappula Brand (1931).
Sclerocaryopsis Brand (1931).

Annual or rarely biennial or perennial herbs, stems stiffly erect or decumbent, rarely trailing; indumentum hispid or pubescent, rarely uncinate; well-developed taproot usually present. Leaves basal and cauline, narrowly (ob-)ovate to linear, sessile or shortly petiolate. Inflorescences erect, frondose, frondose-bracteose, bracteose, distally sometimes ebracteose thyrsoids, rarely cymoids, elongating in fruit. Flowers (sub-)sessile to pedicellate, pedicel sometimes elongating and recurved in fruit; calyx accrescent in fruit; corolla cylindrical or infundibuliform, sometimes hypocrateriform, blue to whitish, faucal scales present, often saccate; gynobase pyramidal to subulate. Nutlets oblique to perpendicular to receptacle, often heterocarpic or heteromericarpic, nutlets triangular-ovate to ovoid in dorsal view, dorsally concave, rarely convex, usually with dorsomarginal wing of either confluent bases of glochidia or distinctly membranaceous, often hyaline, wing spreading to incurved, sometimes dentate, rarely nutlets glabrous and firmly attached to gynobase

[4] Synonymy and placement are highly tentative: these three genera are very poorly understood and – depending on generic type species - *Lepechiniella* may turn out to be a member of the *Lappula*-group, and possibly synonymous with *Lappula*, while *Microparacaryum* undoubtedly belongs here.

("*Sclerocaryopsis*"), cicatrix triangular to circular, medial. $2n = 22, 24, 46, 48$.

A difficult, probably paraphyletic genus of about 50–60 species in Eurasia, Australia, N Africa, W North America; some species introduced worldwide, centre of diversity in E Asia (>40 species in China). The Australian segregate *Omphalolappula* represents a specialized lineage within *Lappula* and is not recognized at generic rank here. Delimitation from the other genera of the group here recognized (esp. *Heterocaryum*) is also doubtful, as are the limits between *Eritrichium* and *Lappula*.

64. ***Rochelia*** Rchb.

Rochelia Rchb., Flora 7(1): 243 (1824), nom. cons.

Annual herbs; indumentum hispid, rarely pubescent. Leaves linear, 1-veined. Inflorescences bracteose cymoids. Flowers pedicellate; calyx lobes linear to lanceolate, rarely widely ovate, accrescent and mostly incurved in fruit; corolla minute, infundibuliform, light blue; tube straight or slightly curved; faucal scales sometimes present; gynobase subulate, columnar in fruit, ovary 2-locular. Nutlets 1–2, obliquely subpyriform, laterally compressed, glochidiate or smooth, permanently attached to the gynobase. $2n = 20$.

About 15 species, W Europe to C, SC & SW Asia. Nested in *Lappula* according to Weigend et al. (2013).

65. ***Eritrichium*** Schrad. ex Gaudin

Eritrichium Schrad. ex Gaudin, Fl. Helvetica, 2: 4, 57 (1828); Sadat, Mitt. Bot. Staatss. München 28: 11–17 (1989), reg. rev.; Zhu et al., Flora of China 16: 378–390 (1995), reg. rev.
Amblynotus I.M. Johnst. (1924).
Sauria M.S. Bajtenov (1996).
Tianschaniella B. Fedtsch. (1951).

Annual, biennial or perennial herbs, often cushion-forming, caespitose; indumentum sericeous, strigose or villous. Leaves basal and cauline, basal ones petiolate, cauline ones sessile, oblanceolate to spathulate. Inflorescences frondose-bracteose cymoids or ebracteose distally. Flowers pedicellate; calyx often rotate in fruit; corolla hypocrateriform, rarely infundibuliform, blue or white, tube equalling calyx, faucal scales depressedly rectangular; filaments inserted on middle of corolla tube; gynobase shortly pyramidal. Nutlets 4, turbinate to ovoid, mostly dorsiventrally compressed, abaxial side usually flat or very slightly concave, with narrow wing and triangular marginal teeth or glochidiate appendages forming an ultimately incurving wing-like margin; cicatrix supramedial. $2n = 20, 22, 24, 28, 36, 46, 48$.

Some 50 species, Asia to Europe and N America. The alpine cushion-forming species from Eurasia and North America apparently are monophyletic, but the remainder may represent several unrelated lineages, some of them close to *Lappula*. *Sauria*, a very imperfectly known genus, is only tentatively included here.

66. ***Hackelia*** Opiz

Hackelia Opiz in Bercht. & Opiz, Ökon.-techn. Fl. Böhmens 2(2): 146 (1839); Gentry & Carr, Mem. New York Bot. Gard. 26: 121–227 (1976), reg. rev.
Austrocynoglossum Popov ex R.R. Mill (1989).
Embadium J.M. Black (1931).

Perennial, rarely annual herbs, often with pleiocorm or rhizomatous; indumentum hispid, pubescent or sericeous; well-developed taproot often present. Leaves basal and cauline, ovate-acuminate to oblong, basal ones long petiolate, cauline ones shortly petiolate to subsessile. Inflorescences ebracteose or basally bracteose thyrsoids, rarely cymoids. Flowers pedicellate, pedicel elongating and reflexed in fruit; corolla hypocrateriform, tube short (< lobes), faucal scales semicircular to triangular; gynobase shortly pyramidal. Nutlets oblique to nearly perpendicular to receptacle, triangular-ovoid in dorsal view, usually glochidiate on disc, margin winged with crest of separate or basally confluent glochidia, sometimes incurved, ventrally keeled; cicatrix medial or submedial, rarely subapical, ovate. $2n = 24, 48$.

About 45 species, mostly W North America, S America, Asia, Australia, Europe.

IV.7 Bothriospermum Group[2]

Small to medium-sized annual or perennial herbs, lax and trailing or compact with distinct basal leaf rosette. Indumentum pubescent, sometimes hispid or villous. Leaves basal and cauline, obovate-acuminate or oblong from cuneate base. Calyx divided nearly to base; corolla hypocrateriform to infundibuliform; gynobase flat, widely pyramidal; style filiform, included, stigma usually capitate. Nutlets typically 4, erect or incurved on pyramidal gynobase, ovate to circular in dorsal view, usually glochidiate, rarely only verrucose, "dorso-"marginal wing often double, erect or incurved, often much smaller than nutlet diameter ("pseudocicatrix"), displaced into a dorsal-apical or even ventral position, cicatrix triangular or triangular-ovate, medial.

Three genera, all E Asia, evidently closely related and sharing the unique character of a distinct "double wing". Apparently sister to the *Cynoglossum* group s.str. (excl. American species) and *Lepechiniella*.

67. ***Bothriospermum*** Bunge

Bothriospermum Bunge, Enum. Pl. Chin. Bor. 47 (1833); Zhu et al., Flora of China 16: 418–420 (1995), reg. rev.

Annual or biennial herbs, sometimes trailing; indumentum densely scabrid to hispid; roots fibrous. Leaves alternate, all cauline or some in basal rosette, (ob-)ovate-acuminate, base cuneate to decurrent. Inflorescences frondose to frondose-bracteose cymoids, sometimes anthocladal. Flowers erect, pedicellate, pedicels often elongating in fruit and incurved; calyx deeply lobed, lobes narrowly ovate-acuminate, slightly accrescent in fruit; corolla hypocrateriform to infundibuliform, tube short, lobes subcircular to oblong, blue or white, faucal scales semicircular to depressedly rectangular; gynobase flat or very shallowly pyramidal, style filiform, included, stigma capitate. Nutlets erect or slightly incurved, ovoid, back densely tuberculate, convex, ventral (style facing) side concave, with a "pseudoaperture" formed by 2 concentric structures, outer one a thickened, elevated margin, inner one a membranaceous, incurved, sometimes dentate wing; cicatrix usually basal, triangular-ovate to subcircular. $2n = 24$.

Five species in C & E Asia.

68. ***Antiotrema*** Hand.-Mazz.

Antiotrema Hand.-Mazz., Anz. Akad. Wiss. Wien. Math.-Nat. Kl. 57: 239 (1920).

Perennial herb with perennial basal rosette; indumentum dense, scabrid to hispid. Leaves basal and cauline, narrowly (ob-)ovate. Inflorescences on long, erect scapes, sparsely branched above, ebracteose, sometimes frondose-bracteose at base. Flowers pendulous, shortly pedicellate, pedicels elongating and recurved in fruit; corolla infundibuliform, tube short, lobes subcircular, blue or purple, faucal scales oblong, papillose, white; stamens exserted; stigma terete. Nutlets erect, ovoid, ventrally (adaxially) with double, incurved cupular wing, outer wing dentate, inner wing membranaceous.

One species, *A. dunnianum* (Diels) Hand. Mazz., W China.

69. ***Thyrocarpus*** Hance

Thyrocarpus Hance, Ann. Sci. Nat. Bot. IV, 18: 225 (1862); Zhu et al., Flora of China 16: 430–431 (1995), reg. rev.

Annual herbs, with 1(3) stems, these stiffly erect and branched only above, densely pubescent with scabrid and hispid trichomes. Leaves cauline and (at least initially) basal, (ob-)ovate-acuminate to elliptical, margin entire or indistinctly serrate. Inflorescences lax, frondose to frondose-bracteose cymoids. Flowers erect, pedicellate, pedicel elongating in fruit; calyx with narrowly ovate-acuminate lobes, slightly accrescent in fruit; corolla lobes subcircular to oblong, blue or white, faucal scales oblong to nearly linear, sometimes emarginate, white; stamens included; stigma terete. Nutlets held horizontally, slightly incurved, ovoid to subcircular, abaxially tuberculate, "ventral" (upper) side parallel to receptacle with a "pseudoaperture" with 2 concentric wings, outer one often erect, dentate, inner one more or less involute and, hiding abaxial nutlet surface, cicatrix apical, ovate to subcircular. $2n = 24$.

Three species, China, Vietnam.

IV.8 Cryptantha Group[5]

Small to medium-sized, ephemeral, annual or perennial herbs, usually compact with distinct basal leaf rosette, rarely shrublets (some

[5] *Dasynotus* is now firmly placed at the base of the New World *Cryptantha*-group, together with several North American species of *Cynoglossum* (Weigend et al. 2013). Resolution in this clade is very poor and the description of two additional genera, *Adelinia* and *Andersonglossum*, to accommodate the representatives of *Cynoglossum* in this group, appears premature. *Oncaglossum* from Mexico likely also belongs here (Cohen 2015).

Cryptantha) or creeping and rooting at nodes (some *Plagiobothrys*); indumentum pubescent, sometimes hispid, villous or sericeous, very rarely glaucous and hispid. Leaves basal and cauline, rarely cauline only, linear to narrowly oblong or elliptical, more rarely obovate-acuminate, sessile with narrowed base. Corolla hypocrateriform to infundibuliform; gynobase widely or narrowly pyramidal to subulate. Style filiform, included or exserted, stigma capitate. Nutlets typically 4, erect, spreading or recurved, ovate-acuminate to oblong in dorsal view, verrucose, rugulose or glochidiate, dorsomarginal wing usually absent or present as distinct glochidiate teeth with united bases (only *Pectocarya*) or as indistinct, flattened nutlet margin (some *Cryptantha*), nutlets ventrally sharply keeled, keel sometimes with distinct groove, cicatrix triangular or triangular-ovate, suprabasal, rarely apical (*Pectocarya*).

Five genera, American, *Plagiobothrys* also Australia and NE Asia, *Microula* Asian. A further subdivision of *Cryptantha* into several genera based on molecular data has been proposed by Hasenstab-Lehman and Simpson (2012), but the genus is here adopted in its traditional circumscription. This group is retrieved with several species of *Cynoglossum* nested in it (Weigend et al. 2013) and generic realignments will be required.

70. ***Cryptantha*** Lehm. ex G. Don

Cryptantha Lehm. ex G. Don, Gen. Hist. 4(1): 373 (1837); Grau, Mitt. Bot. München 18: 379–400 (1982), reg. rev.; Higgins, Brigham Young Univ. Sci. Bull., Biol. Series 13: 1-63 (1971), part. rev.; Johnston, Contr. Gray Herb. 74: 1–114 (1925), reg. rev.; Pérez-Moreau, Darwiniana 20: 155–188. (1976), reg. rev.
Nesocaryum I.M. Johnst. (1927).

Annual, biennial or perennial herbs, rarely shrublets, stems stiffly erect, one main stem from the base or with several long basal branches, sometimes cushion-forming; indumentum densely scabrid to uncinate, rarely sericeous or villous. Leaves basal (at least initially) and cauline, lamina narrowly obovate to linear. Inflorescences ebracteose or basally bracteose complex thyrsoids, rarely reduced to cymoids, sometimes very dense and subcapitate. Flowers erect, subsessile, sometimes distylous, geoflorous and cleistogamous flowers sometimes present; calyx deeply lobed, sometimes circumscissile in fruit; corolla narrowly infundibuliform or salverform, shorter or longer than calyx, usually white with yellow throat, rarely yellow, faucal scales usually absent, rarely present, basal scales 5–10 (conspicuous in perennials); stamens included or exserted (depending on floral morph); style included or exserted (depending on floral morph). Nutlets straight, triangular-ovoid, dorsally convex, smooth to ornamented, rarely indistinctly winged, ventrally keeled and keel with groove, keel often forked towards base with small, triangular to triangular-ovate cicatrix; nutlets often heteromericarpous (3+1), those in geocarpic flowers often widely differing. $2n = 12, 14, 20, 24, 36, 62, 64, 120, 124$.

About 160 species, SW North America and W South America. Several groups have been recognized at genus level in the past, including the Isla San Ambrosio (Chile) endemic *Nesocaryum*. We here treat the genus in a wide sense. It is clearly closely allied to both *Plagiobothrys* and *Amsinckia* and generic limits as currently defined are almost certainly artificial.

71. ***Amsinckia*** Lehm.

Amsinckia Lehm., Sem. Hort. Bot. Hamburg 1831: 7 (1831), nom. cons.; Macbride, Contr. Gray Herb. 49: 1–16 (1917), reg. rev.; Ray & Chisaki, Amer. J. Bot. 44: 529–536 (1957), reg. rev., morph.

(Winter-)annual, stiffly erect herbs; indumentum densely hispid, scabrid to uncinate, rarely setose and glaucous. Leaves initially in dense basal rosette, cauline leaves (sub-)sessile, narrowly obovate to linear. Inflorescences ebracteose or basally bracteose cymoids or thyrsoids. Flowers erect, subsessile, sometimes distylous; calyx lobes equal or unequal; corolla narrowly infundibuliform or hypocrateriform, orange to orange-yellow, often with 5 darker spots on lobes, faucal scales usually absent; stamens included or exserted (depending on stylar morph); style included or exserted (depending on stylar morph). Nutlets (1–)4, usually heteromericarpous (3+1), erect, straight, triangular-ovoid, dorsally convex or keeled, smooth or ornamented, ventrally sharply keeled, cicatrix basal at lower end of keel, small, triangular to triangular-ovate; cotyledons deeply bifid. $2n = 8, 10, 12, 14, 24, 26, 30, 32, 34, 38$.

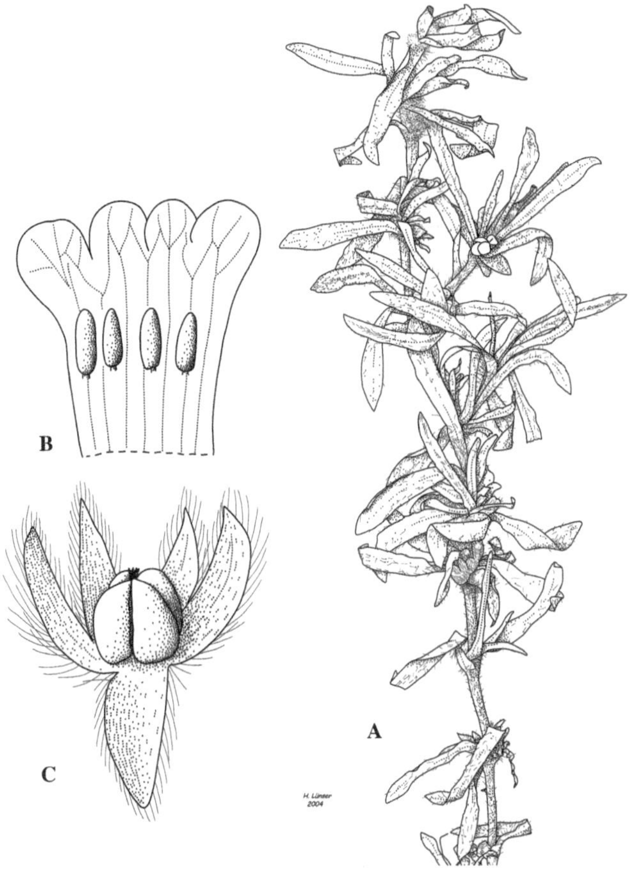

Fig. 14. Boraginaceae. Tribe Cynoglosseae, *Cryptantha* group: *Plagiobothrys humilis*. **A** Habit. **B** Open corolla. **C** Fruit. (orig., drawn by H. Lünser)

About 15 species, SW North America and SW South America, two species widely naturalized nearly worldwide.

72. ***Plagiobothrys*** Fisch. & C.A. Mey. Fig. 14

Plagiobothrys Fisch. & C.A. Mey., Index Sem. Hort. Bot. Petrop. 2: 46 (1836); Johnston, Contr. Gray Herb. 68: 57–80 (1923), synopsis; Johnston, J. Arnold Arbor. 3: 3–82 (1932), reg. rev.
Allocarya Greene (1887).

Annual to perennial herbs, stiffly erect or with creeping, rooting shoots; indumentum sparse and pubescent or scabrid to uncinate. Leaves (at least initially) basal and cauline, some or most opposite, (sub-)sessile, (ob-)ovate, oblong or linear. Inflorescence ebracteose or irregularly bracteose cymoids, rarely thyrsoids, sometimes strongly congested into capitula. Flowers erect, subsessile to shortly pedicellate; corolla infundibuliform, white, lobes erect to spreading, usually pubescent at least abaxially, faucal scales present, usually yellow, annulus present, indistinct; stamens included; style included. Nutlets (1–)4, erect, straight, (triangular-)ovoid, dorsally convex, more or less ornamented, ventrally sharply keeled, keel not grooved; cicatrix basal to medial, at lower end of the strong ventral keel or rarely terminating on a stipitate prolongation. $2n = 24$, 34, 36, 48, 58, 72.

A difficult, possibly paraphyletic genus of ca. 70 species, most in SW North America and Andean South America, five species in Australia, one in NE Asia (Kamchatka).

73. ***Pectocarya*** DC. ex Meisn.

Pectocarya DC. ex Meisn., Pl. Vasc. Gen. [Meisner] 1: 279; 2: 188 (1840).
Harpagonella A. Gray (1876).

Ephemeral herbs with horizontally spreading branches; indumentum scabrid to hispid. Leaves initially basal, later only cauline. Inflorescences frondose to frondose-bracteose, irregularly branched thyrsoids or cymoids. Flowers erect, (sub-)sessile; calyx spreading at maturity or highly asymmetrical ("*Harpagonella*": 2 upper lobes fused, strongly accrescent in fruit, at maturity with 5–10 large basal processes with uncinate trichomes, 3 lower lobes free nearly to base); corolla narrowly infundibuliform, white, throat often yellow, faucal scales present, minute, semicircular, yellow, annulus indistinct; stamens included; gynobase very shortly pyramidal, style very short, included. Nutlets 4, horizontally spreading to reflexed, narrowly obovate, with more or less distinct marginal wing of basally confluent glochidia or rarely half-erect, obovoid, pubescent, one enclosed by upper calyx lobes, the second free, pedicel reflexed in fruit, axis disarticulating ("*Harpagonella*"). $2n = 24$, 72.

15 species, mainly SW North America and SW South America. A highly distinctive genus, apart from aberrant monotypic *Harpagonella*, which is here included following Veno (1979).

74. ***Microula*** Benth.

Microula Benth. in Benth. & Hook., Gen. Pl. 2: 853 (1876); Banerjee & Banerjee, Bull. Bot. Surv. India 15: 71–75

(1973), reg. rev.; Wang, Acta Phytotaxon. Sin. 18: 266–283 (1980), rev.; Zhu et al., Flora of China 16: 391–401 (1995).
Actinocarya Benth. (1876).
Glochidocaryum W.T. Wang (1957).
Metaeritrichium W.T. Wang (1980).
Schistocaryum Franch. (1891).
Tretocarya Maxim. (1881).

Annual, biennial or perennial herbs, acaulescent or with erect, ascending or prostrate stem; indumentum sparse and setose, pubescent, villous or sericeous. Leaves usually basal and cauline, sessile to petiolate, elliptical, oblong or obovate. Inflorescences initially dense, frondose-bracteose cymoids, later elongating and spike-like. Flowers pedicellate, pedicels often elongating in fruit. Calyx divided nearly to base; corolla infundibuliform to hypocrateriform, blue or whitish, faucal scales crescent-shaped; stamens included; style included, stigma subglobose; gynobase nearly flat. Nutlets erect to spreading, ovoid, sometimes strongly flattened ("*Setulocarya*", "*Actinocarya*"), rarely turbinate, usually tuberculate, shortly pubescent, rarely glochidiate or setose, usually with an abaxial triangular to elliptic rimmed pseudoaperture/foveola and adaxial keels, cicatrix basal, small, circular to elongate. $2n = 24$.

A genus of 30 species from N India to China, centre of diversity is China with 29 species. Close to the *Cryptantha* group, but likely an early derived group, sister to an African clade of *Cynoglossum* (Weigend et al. 2013). Proposed generic segregates are clearly linked to the core of the genus via morphological intermediates.

IV.9 Cynoglossum Group

Medium-sized to mostly robust perennial, rarely annual or ephemeral herbs, often with distinct basal leaf rosette; indumentum pubescent, sometimes hispid, villous or sericeous. Leaves basal and cauline, rarely cauline only, narrowly oblong or elliptical to widely ovate, usually with cuneate, sometimes with cordate base. Calyx divided nearly to base, erect to spreading; corolla hypocrateriform to infundibuliform, rarely cylindrical; gynobase widely pyramidal; style filiform, included or exserted, stigma capitate. Nutlets typically 4, large, oblique on gynobase, ovate-acuminate to subcircular in dorsal view, usually glochidiate all around, rarely glabrous, dorsomarginal rim often differentiated and delimitating a convex back or differentiated into a distinct, flat or incurved wing, cicatrix triangular or triangular-ovate, subapical, rarely central.

Nine genera, subcosmopolitan. Genus limits extremely doubtful and *Cynoglossum* in its current circumscription apparently paraphyletic in relation to the remaining genera (Weigend et al. 2013).

75. *Lepechiniella* Popov[4]

Lepechiniella Popov, Fl. USSR 19: 713 (1953); Shmida, Pl. Syst. Evol. 129: 323–326 (1978), reg. rev.
Brandella R.R. Mill (1986).
Microparacaryum (Popov ex Riedl) Hilger & Podlech (1985).

Annual herbs with slender taproot and stems branched from base. Leaves basal (rosette) and cauline or only cauline, linear-lanceolate. Inflorescence cymoids of scorpioid monochasia, ebracteose. Flowers small, sessile or short-pedicellate, pedicels sometimes elongating and recurved in fruit; calyx divided to base, spreading in fruit; corolla infundibuliform, tube equalling lobes, faucal scales present. Nutlets with spreading or incurved wing, wing usually deeply dentate, teeth often with glochidiate apices, often heterocarpous with individuals differing in nutlet morphology or heteromericarpous, then often 1 of the 4 nutlets unwinged and firmly attached to gynobase; cicatrix large, ovate.

About six species in NE Africa to SW Asia. Systematics and affinities of this genus are unclear, but it appears to be closer to the *Cynoglossum* group than to the *Lappula* group, in spite of the superficial similarity to the latter (small, annual plants with small nutlets).

76. *Cynoglossum* L. Figs. 8, 9, 15

Cynoglossum L., Sp. Pl.: 134 (1753); Brand in Engl., Pflanzenreich, IV, 252 (Heft 78): 115–153 (1921), rev.; Mill & Miller, Notes Roy. Bot. Gard. Edinb. 41: 473–482. 1984), reg. rev.; Miller, Adansonia III, 27: 113–127 (2005), reg. rev.
Adelocaryum Brand (1915).
Cynoglossopsis Brand (1931).
Ivanjohnstonia Kazmi (1975).
Oncaglossum Sutorý (2010).[5]
Paracaryopsis (Riedl) R.R. Mill (1991).
Paracynoglossum Popov (1953).
Pardoglossum Barbier & Mathez (1973).

Fig. 15. Boraginaceae. Tribe Cynoglosseae, *Cynoglossum* group: *Cynoglossum columnae*. **A, B** Habit. **C** Open corolla. **D** Fruit. **E** Glochidium from nutlet surface. (orig., drawn by A. Maury)

Robust, biennial, perennial, rarely annual herbs; massive primary root and sometimes pleiocorm present. Leaves basal and cauline, basal ones often long petiolate. Inflorescences ebracteose, rarely basally bracteose thyrsoids, considerably elongating in fruit. Flowers pedicellate; calyx accrescent in fruit; corolla blue, dark purplish red, purple or yellow-green, rarely white, hypocrateriform, campanulate, rarely cylindrical or infundibuliform, tube usually short (< calyx), lobes subcircular, faucal scales depressedly rectangular, oblong or crescent-shaped; stamens included. Nutlets usually large, ovate to subcircular in outline, glochidiate, back convex or shallowly concave with distinct, glochidiate margin, cicatrix large, apical to central, nutlets often awned. $2n = 24, 48$.

A difficult and undoubtedly para- and polyphyletic taxon with 80–100 species, subcosmopolitan. Several segregate genera have been proposed, but no comprehensive study of the genus has been carried out yet. Recent data indicate that the genus is both polyphyletic and paraphyletic (Weigend et al. 2013). Even recently segregates have been proposed without any study of infrageneric phylogeny. Some species (*C. officinale* L., *C. amabile* Stapf & J.R. Drumm., *C. glochidiatum* Wall. ex Benth.) are widely introduced weeds.

77. ***Dasynotus*** I.M. Johnst.[5]

Dasynotus I.M. Johnst., J. Arnold Arbor. 29: 233 (1948).

Perennial, rhizomatous herb; indumentum pubescent to scabrid. Leaves basal and cauline, basal ones long petiolate. Inflorescences ebracteose or with a few frondose bracts at base, cymoids or thyrsoids. Flowers erect, long pedicellate, pedicels recurved in fruit; calyx spreading in fruit; corolla large, hypocrateriform, white, tube short; lobes subcircular; faucal scales large, narrowly oblong, ciliate to papillose, apex bifid and recurved, usually yellow, annulus distinct; stamens included (rarely exserted); style included. Nutlets (1–)4, large, shortly cylindrical, subcircular in dorsal view, adaxially keeled, dorsally with low, cup-shaped rim, initially dorsally setose, later shiny glabrescent; cicatrix central or supramedial, subcircular.

One species, *D. daubenmirei* I.M. Johnst., endemic to NW N America (Idaho). Closely allied to American species of *Cynoglossum* and the *Cryptantha* group.

78. ***Afrotysonia*** Rauschert

Afrotysonia Rauschert, Taxon 31: 558 (1982); Mill, Notes Roy. Bot. Gard. Edinb. 43: 467–475 (1986), rev.
Tysonia Bolus (1890), nom. illegit. non Fontaine (1889) nec F. Muell. (1896).

Tall, perennial herbs; pleiocorm with persistent leaf bases present. Leaves basal and cauline, basal ones long petiolate, ovate to lanceolate; indumentum pubescent to scabrid. Inflorescence bracteose-frondose to bracteose thyrsoids, multiply branched and corymbose. Flowers pendulous or deflexed, pedicels elongating in fruit; calyx accrescent in fruit; corolla campanulate or shortly hypocrateriform, pink, white to dull yellow, faucal scales depressedly rectangular to crescent-shaped, annulus of 10 emarginate scales; anthers widely exserted; style exserted. Nutlets large, either dorsiventrally or laterally compressed, densely glochidiate or largely glabrous, winged with undulate, cartilaginous wing or unwinged, cicatrix apical, triangular to lanceolate.

Three species in southern and E Africa. Doubtfully monophyletic and doubtfully distinct from *Cynoglossum*.

79. ***Lindelofia*** Lehm.

Lindelofia Lehm., Neue Allg. Deutsche Garten-Blumenzeitung 6: 351 (1850); Sadat, Mitt. Bot. Staatss. München 28: 100–117 (1989), reg. rev.

Perennial herbs from massive pleiocorm; indumentum pubescent to scabrid. Leaves basal and cauline, basal leaves long petiolate, ovate-acuminate to elliptical. Inflorescences ebracteose or basally bracteose cymoids or thyrsoids. Flowers nodding to pendulous; calyx slightly accrescent and spreading in fruit; corolla infundibuliform, blue, tube usually longer than calyx; lobes orbicular-obtuse, faucal scales oblong, rarely reduced; anthers exserted; style widely exserted, thickened and persistent in fruit. Nutlets ovate, with distinct disc, back flattened to concave, with multiseriate glochidia along margin, glochidia often fused into narrow wing; ventral part of nutlet spinulose or tuberculate; cicatrix central, extending to apex, large, triangular-ovate, nutlets awned. $2n = 24$.

About 10–20 species, Asia. Nested in an Asian clade of *Cynoglossum* (Weigend et al. 2013) and with unclear species limits.

80. ***Mattiastrum*** (Boiss.) Brand

Mattiastrum (Boiss.) Brand, Repert. Spec. Nov. Regni Veg. 14: 150 (1915); Sadat, Mitt. Bot. Staatss. München 28: 60–99 (1989), reg. rev.

Biennial or perennial herbs often from well-developed pleiocorm; indumentum hispid, scabrid or villous. Leaves basal and cauline, narrowly (ob-)lanceolate. Inflorescences ebracteose or basally bracteose thyrsoids or cymoids. Flowers pedicellate, pedicels often reflexed in fruit; calyx reflexed in fruit; corolla infundibuliform to subcampanulate, blue to brownish-purple, rarely yellow, faucal scales linear to depressedly rectangular; anthers included or only tips exserted; style included. Nutlets subcircular to ovate or narrowly ovate in outline, often dorsally keeled and glochidiate, with a broad, flat wing; cicatrix central, extending to apex, large, ovate, nutlet awned. $2n = 24$.

About 50 species, Asia. An imperfectly known genus, often included in *Paracaryum*, nested in *Cynoglossum* s.str. (Weigend et al. 2013)

81. ***Paracaryum*** **Boiss.**

Paracaryum Boiss., Diagn. Pl. Orient. I, 11: 128 (1849); Hilger et al., Pl. Syst. Evol. 148: 291–312 (1985), genus delimitation.

Biennial or perennial herbs often from well-developed pleiocorm; indumentum hispid, scabrid or villous. Leaves basal and cauline, narrowly (ob-)lanceolate. Inflorescences ebracteose or basally bracteose thyrsoids or cymoids. Flowers small, pedicellate, pedicels often reflexed in fruit; calyx spreading to reflexed in fruit; corolla infundibuliform to subcampanulate, blue to brownish-purple, faucal scales crescent-shaped; anthers included or only tips exserted; style included. Nutlets ovate in outline, tuberculate or nearly smooth, surrounded by a scarious, strongly inflexed, mostly denticulate-glochidiate wing, leaving only a small aperture, often dorsally keeled and glochidiate, with a broad, flat wing; cicatrix central extending to apex, large, ovate, nutlet awned. $2n = 24$.

About 10–15 species, SW Asia, mainly in the Irano-Turanian region, nested in *Cynoglossum* s. str. (Weigend et al. 2013).

82. ***Rindera*** Pall.

Rindera Pall., Reise Russ. Reich 1: 486 (1771).
Bilegnum Brand (1915).

Perennial herbs with well-developed pleiocorm; indumentum sericeous or villous. Flowers in ebracteose cymoids, these often corymbose or paniculate in outline. Flowers long pedicellate; calyx reflexed in fruit; corolla tubular-campanulate, usually purple, faucal scales crescent-shaped to ligulate, rarely absent, sometimes a second whorl of appendages near middle or base of tube; anthers exserted or reaching mouth of corolla tube; style usually exserted. Nutlets large, circular to ovate in outline, dorsally concave, glabrous, with wide, flat, spreading wings, ventrally keeled, with very few glochidia or surface tuberculate or even smooth and shiny; glochidia rarely along margins or 1 row of large, flat glochidia along keel; cicatrix central, extending to apex, large, ovate, nutlet awned. $2n = 22$.

About 20–25 species, E Europe to C Asia. Close to and doubtfully distinct from *Paracaryum* and *Mattiastrum*, nested in *Cynoglossum* s.str. (Weigend et al. 2013).

83. ***Solenanthus*** Ledeb.

Solenanthus Ledeb., Ic. Fl. Ross. 8, t. 26 (1829); Fl. Alt. 1: 193 (1829).

Biennial or perennial herbs with well-developed pleiocorm; indumentum densely appressed pubescent. Leaves basal and cauline, basal leaves large, long petiolate. Inflorescence mostly large, many-branched, ebracteose thyrsoids, paraclades dense, often coiled in bud, becoming open thyrsoidal and elongated in fruit; calyx usually enlarging in fruit; corolla tubular-infundibuliform to subcampanulate, mostly reddish-purple, faucal scales inserted near middle of corolla tube; anthers exserted; style exserted. Nutlets ovate to subcircular in outline, glochidiate, sometimes with distinctly elevated margin, cicatrix central, extending to apex, large, ovate, nutlet awned. $2n = 24, 48$.

About 10 species, S Europe to Asia, doubtfully monophyletic and nested in *Cynoglossum* s. str. (Weigend et al. 2013).

84. ***Trachelanthus*** Kunze

Trachelanthus Kunze, Bot. Zeitung (Berlin) 8: 665 (1850).

Perennial herbs with well-developed pleiocorm; indumentum sparse, pubescent to glabrescent. Leaves basal and cauline, basal leaves very large, long petiolate. Inflorescences lax thyrsoids. Flowers pedicellate, erect or nodding; corolla cylindrical, lobes erect, subequalling tube, yellow, faucal scales crescent-shaped, sometimes small; anthers included but nearly equalling lobes; style exserted. Nutlets large, widely ovate in outline, ventrally densely spiny, glochidiate or disc smooth, not winged, cicatrix central, extending to apex, large, ovate, nutlet awned.

Four species, W Asia. Nested in *Cynoglossum* s.str. (Weigend et al. 2013).

IV.10 Craniospermum Group

85. ***Craniospermum*** Lehm.

Craniospermum Lehm., Pl. Asperif. Nucif. 3: 336 (1818); Ovczinnikova, Bot. Zhurn. 85: 77–87 (2000), rev.

Biennial or perennial herbs, often with pleiocorm; indumentum hirsute to floccose. Leaves basal and cauline, narrowly ovate. Inflorescences ebracteose or basally bracteose cymoids or thyrsoids. Flowers erect, sessile or short pedicellate; calyx lobes narrowly ovate to linear, elongating in and enclosing the fruit; corolla cylindrical, lobes erect to half-spreading, (triangular-)ovate, faucal scales absent, sometimes scale-like emergences between corolla lobes present; stamens inserted at middle of corolla tube, anthers long exserted; gynobase shortly pyramidal, style exserted, persistent, stigma depressedly capitate to discoid. Nutlets erect, ovoid, transversely rugose, dorsally or dorso-apically with pseudoaperture with narrow, coriaceous, unequally dentate wing almost closing over dorsal disc, cicatrix basal.

Four or five species in C and E Asia. The genus represents an isolated lineage firmly placed in Cynoglosseae.

Selected Bibliography

Aldridge, A.E. 1981. Anatomy and evolution in Macaronesian *Echium* (Boraginaceae). Plant Syst. Evol. 138: 9–22.

Al-Shehbaz, I.A. 1991. The genera of Boraginaceae in the Southeastern United States. J. Arnold Arbor., Suppl. Ser. 1: 1–169.

APG (Angiosperm Phylogeny Group) III, Bremer, B., Bremer, K., Chase, M.W. et al. 2009. An update of the Angiosperm Phylogeny Group classification for the orders and families of flowering plants: APG III. Bot. J. Linn. Soc. 161: 105–121.

Authority ANZF 2001. Pyrrolizidine alkaloids in food. A toxicological review and risk assessment. Technical Report Series No. 2. Canberra and Wellington: ANZFA, 16 pp.

Avetisjan, E.M. 1956. Morphology of the microspores in Boraginaceae. Trudy Bot. Inst. Akad. Nauk Armen. SSR 10: 1–66.

Baczyńska, B., Lityńska-Zając, M. 2005. Application of *Lithospermum officinale* L. in early Bronze Age medicine. Veget. Hist. Archaebot. 14: 77–80.

Baskin, J.M., Baskin, C.C. 1991. An eight-year greenhouse germination study of the cedar glade endemic *Onosmodium molle* subsp. *molle*. Nat. Areas J. 11: 190–192.

Behnke, H.-D. 1981. Sieve-element characters. Nord. J. Bot. 1: 381–400.

Best, T.L., Smartt, R.A. 1986. Feeding ecology of mourning doves (*Zenaida macroura*) in southeastern New Mexico. Southwest. Nat. 31: 33–38.

Bigazzi, M., Nardi, E., Selvi, F. 1997. Anchusella, a new genus of Boraginaceae from the Central-Eastern Mediterranean. Pl. Syst. Evol. 205: 241–264.

Bigazzi, M., Selvi, F. 1998. Pollen morphology in the Boragineae (Boraginaceae) in relation to the taxonomy of the tribe. Plant Syst. Evol. 213: 121–151.

Böhle, U.-R., Hilger, H.H., Martin, W.F. 1996. Island colonization and evolution of the insular woody habit in *Echium* L. (Boraginaceae). Proc. Natl. Acad. Sci. USA 93: 11740–11745.

Bresinsky, A. 1963. Bau, Entwicklungsgeschichte und Inhaltsstoffe der Elaiosomen. Studien zur myrmekochoren Verbreitung von Samen und Früchten. Bibl. Bot. 126: 1–54.

Briechle, M., Hilger, H.H. 1988. Die Embryogenese von *Microparacaryum intermedium* und die embryologische Klassifizierung der Boraginaceae nach Souèges. Flora 181: 45–59.

Britton, D.M. 1951. Cytogenetic studies on the Boraginaceae. Brittonia 7: 233–266.

Buys, M.H, Hilger, H.H. 2003. Boraginaceae cymes are exclusively scorpioid and not helicoid. Taxon 52: 719–724.

Carlquist, S. 1970. Wood anatomy of *Echium* (Boraginaceae). Aliso 7: 183–199.

Cecchi, L., Selvi, F. 2009. Phylogenetic relationships of the monotypic genera *Halacsya* and *Paramoltkia* and the origins of serpentine adaptation in circum-Mediterranean Lithospermeae (Boraginaceae): insights from ITS and matK DNA sequences. Taxon: 700–714.

Cecchi, L., Coppi, A., Hilger, H. H., Selvi F. 2014. Non-monophyly of Buglossoides (Boraginaceae: Lithospermeae): Phylogenetic and morphological evidence for the expansion of *Glandora* and reappraisal of *Aegonychon*. Taxon 63: 1065–1078.

Choudhary, M.I., Begum, A., Abbaskhan, A., Rehman, S.-u., Rahman, A.-U. 2006. Cinnamate derivatives of fructo-oligosaccharides from *Lindelofia stylosa*. Carbohydr. Res. 341: 2398–2405.

Clarke, G.C.S. 1977. Boraginaceae. Rev. Palynol. Palaeobot. 24: A59–A101.

Clowes, F.A.L. 2000. Pattern in root meristem development in angiosperms. New Phytol. 146: 83–94.

Cohen, J.I. 2015. *Adelinia* and *Andersonglossum* (Boraginaceae), two new genera from New World species of *Cynoglossum*. Syst. Bot. 40: 611–619.

Cohen, J. I., DAVIS, J. I. 2009. Nomenclatural changes in Lithospermum (Boraginaceae) and related taxa following a reassessment of phylogenetic relationships. Brittonia 61: 101–111.

Cook, D., Grierson, D., Jones, C., Wallace, A., West, G., Tucker, G. 2002. Modification of fatty acid composition in tomato (*Lycopersicon esculentum*) by expression of a borage delta 6-desaturase. Mol. Biotechnol. 21: 123–128.

Crété, P. 1963. Embryo. In: Maheshwari, P. (ed.) Recent advances in the embryology of angiosperms. Delhi: International Society of Plant Morphologists, pp. 171–220.

Czukavina, A., Meling, E. 1982. Generis *Solenanthus* Ledeb. (Boraginaceae) species nova e Tadzhikistania. Novosti Sist. Vyssh. Rast. 19: 161–164.

Dasti, A.A., Bokhari, T.Z., Malik, A.S., Akhtar, R. 2003. Epidermal morphology in some members of family Boraginaceae in Baluchistan. Asian J. Plant Sci. 2: 42–47.

Dietz, H., Ullmann, I. 1997. Age-determination of dicotyledonous herbaceous perennials by means of annual rings: exception or rule? Ann. Bot. 80: 377–379.

Duke, J.A., Ayensu, E.S. 1985. Medicinal Plants of China. Medicinal Plants of the World. Algonac, MI: Reference Publications, 705 pp.

Dupont, Y.L., Hansen, D.M., Rasmussen, J.T., Olesen, J.M. 2004. Evolutionary changes in nectar sugar composition associated with switches between bird and insect pollination: the Canarian bird-flower element revisited. Funct. Ecol. 18: 670–676.

Elias, M.K. 1932. Grasses and other plants from the tertiary rocks of Kansas and Colorado. Univ. Kansas Sci. Bull. 20: 333–367.

Elias, M.K. 1942. Tertiary prairie grasses and other herbs from the high plains. Geol. Soc. Amer. Special Papers 41: 1–176.

Erbar, C., Leins, P. 1996. Distribution of the character states "early sympetaly" and "late sympetaly" within the "sympetalae tetracyclicae" and presumably allied groups. Bot. Acta 109: 427–440.

Erbar, C., Porembski, S., Leins, P. 2005. Contributions to the systematic position of *Hydrolea* (Hydroleaceae) based on floral development. Plant Syst. Evol. 252: 71–83.

Farnsworth, N.R., Bingel, A.S., Cordell, G.A., Crane, F.A., Fong, H.H. 1975. Potential value of plants as sources of new antifertility agents I. J. Pharm. Sci. 64: 535–598.

Ferguson, D.M. 1996. Phylogenetic analysis of Hydrophyllaceae and a preliminary study of biogeographic patterns and life history evolution. Amer. J. Bot. 83: 155–156.

Fisher, D.D., Thorsch, J., Esau, K. 1989. Inclusions in nuclei and plastids of Boraginaceae and their possible taxonomic significance. Can. J. Bot. 67: 3608–3617.

Frölich, C., Ober, D., Hartmann, T. 2007. Tissue distribution, core biosynthesis and diversification of pyrrolizidine alkaloids of the lycopsamine type in three Boraginaceae species. Phytochemistry 68: 1026–1037.

Gabel, M.L. 1987. A fossil *Lithospermum* (Boraginaceae) from the tertiary of South Dakota. Amer. J. Bot. 74: 1690–1693.

Ganders, F.R. 1979. The biology of heterostyly. New Zeal. J. Bot. 17: 607–635.

Gilbert, F., Azmeh, S., Barnard, C., Behnke, J., Collins, S. A., Hurst, J., Shuker, D. 2001. Individually recognizable scent marks on flowers made by a solitary bee. Anim. Behav. 61: 217–229.

Gilg, E. 1907. Die systematische Stellung der Gattung *Hoplestigma* und einiger anderer zweifelhafter Gattungen. Bot. Jahrb. Syst. 40, Beiblatt Nr. 93: 76–84.

Grau, J. 1983. Life form, reproductive biology and distribution of the Californian/Chilean genus *Cryptantha*. In: Kubitzki K. (ed.) Dispersal and distribution. Sonderbd. Naturwiss. Ver. Hamburg, pp. 231–240.

Grau, J., Leins, P. 1968. Pollenkorntypen und Sektionsgliederung der Gattung *Myosotis*. Ber. Deutsch. Bot. Ges. 81: 107–115.

Gunning, B.E.S., Pate, J.S., Green, L.W. 1970. Transfer cells in the vascular system of stems: Taxonomy, association with nodes, and structure. Protoplasma 71: 147–171.

Gürke, M. 1893, 1897. Borraginaceae (Asperifoliaceae). In: Engler, A., Prantl, K. (eds.) Die natürlichen Pflanzenfamilien, IV, 3a, 3b. Leipzig: Engelmann, pp. 71–131 (1893), 377 (1897).

Guşuleac, M. 1923. Beiträge zur Systematik der Anchuseae. Publ. Soc. Nat. Bucur. 6: 79–92.

Guşuleac, M. 1928. Die monotypischen und artenarmen Gattungen der Anchuseae (*Caryolopha, Brunnera, Hormuzakia, Gastrocotyle, Phyllocara, Trachystemon, Procopiania* und *Borago*). Bul. Fac. Ştiinte Cernăuti 2: 394–461.

Guşuleac, M. 1931. Die monotypischen und artenarmen Gattungen der Anchuseae (*Caryolopha, Brunnera, Hormuzakia, Gastrocotyle, Phyllocara, Trachystemon, Procopiania* und *Borago*). Feddes Repert. 29:113–125.

Halbritter, H., Weber, M., Zetter, R., Frosch-Radivo, A., Buchner, R., Hesse M. 2008. PalDat - Illustrated handbook on pollen terminology. Wien: http://www.paldat.org/Paldat_Terminology_small.pdf.

Hargrove, L., Simpson, M.G. 2003. Ultrastructure of heterocolpate pollen in *Cryptantha* (Boraginaceae). Int. J. Plant Sci. 164: 137–151.

Harley, R.M., Atkins, S., Budantsev, A.L., Cantino, P.D., Conn, B.J., Grayer, R., Harley, M.M., De Kok, R., Krestovskaja, T., Morales, R., Paton, A.J., Ryding, O., Upson, T. 2004. Labiatae. In: Kubitzki, K. (ed.) The Families and Genera of Vascular Plants, vol. VII. Heidelberg: Springer, pp. 167–275.

Hartmann, T., Ober, D. 2000. Biosynthesis and metabolism of pyrrolizidine alkaloids in plants and specialized insect herbivores. In: Leeper, F.J., Vederas, J.C. (eds.) Topics in Current 5 Chemistry. Berlin: Springer, pp. 207–244.

Hartmann, T., Witte, L. 1995. Chemistry, biology and chemoecology of the pyrrolizidine alkaloids. In: Pelletier, S.W. (ed.) Alkaloids: Chemical and biological perspectives. Oxford: Pergamon Press, pp. 155–233.

Hasenstab-Lehman, K.E., Simpson, M.G., 2012. Cat's eyes and popcorn flowers: Phylogenetic systematics of the genus *Cryptantha* s.l. (Boraginaceae). Syst. Bot. 37: 738–757.

Hegnauer, R. 1964. Chemotaxonomie der Pflanzen. Bd. 3. Dicotyledoneae: Acanthaceae–Cyrillaceae. Basel, Stuttgart: Birkhäuser, 743 pp.

Hegnauer, R. 1989. Chemotaxonomie der Pflanzen. Bd. 8. Nachträge zu Band 3 und Band 4 (Acanthaceae bis Lythraceae). Basel, Boston, Berlin: Birkhäuser, 718 pp.

Hegnauer, R. 1997. Phytochemistry and chemotaxonomy. In: Kalkman, C., Kirkup, D.W., Nooteboom, H.P., Stevens, P.F., de Wilde, W.J.J.O. (eds.) Flora Malesiana Series I Seed Plants, vol. 13. Leiden: Publications Department Rijksherbarium, pp. 52–58.

Heslop-Harrison, Y. 1981. Stigma characteristics and angiosperm taxonomy. Nord. J. Bot. 1: 401–420.

Higgins, L.C. 1971. A revision of *Cryptantha* subgenus *Oreocarya*. Brigham Young Univ. Sci. Bull., Biol. Series 13: 1–63.

Hilger, H.H. 1985. Ontogenie, Morphologie und systematische Bedeutung geflügelter und glochidientragender Cynoglosseae- und Eritrichieae-Früchte (Boraginaceae). Bot. Jahrb. Syst. 105: 323–378.

Hilger, H.H., Balzer, M., Frey, W., Podlech, D. 1985. Heteromerikarpie und Fruchtpolymorphismus bei *Microparacaryum*, gen. nov. (Boraginaceae). Plant Syst. Evol. 148: 291–312.

Hilger, H.H., Hoppe, J.R., Hofmann, M. 1993. Energiedispersive Röntgenmikroanalyse (EDX) von Boraginaceae subfam. Boraginoideae - Klausenoberflächen. Sind Silicium- und Calcium - Einlagerungen in die Fruchtwand systematisch verwertbare Merkmale? Flora 188: 387–398.

Hilger, H.H., Selvi, F., Papini, A., Bigazzi, M. 2004. Molecular systematics of Boraginaceae tribe Boragineae based on ITS1 and trnL sequences, with special reference to *Anchusa* s.l. Ann. Bot. 94: 201–212.

Jodin, H. 1903. Recherches anatomiques sur les Borraginées. Ann. Sci. Nat., 8e sér., Bot. 17: 263–346.

Johansen, D.A. 1950. Plant embryology. Waltham MA: Chronica Botanica, 305 pp.

Johnston, I.M. 1927. Studies in the Boraginaceae. VI. A revision of South American Boraginoideae. Contr. Gray Herb. 78: 3–118.

Johnston, I.M. 1953. Studies in the Boraginaceae. XXIV, A. Three genera segregated from *Lithospermum* 1–7; B. Supplementary notes on *Lithospermum*. J. Arnold Arbor. 34: 7–16.

Johnston, I.M. 1953. Studies in the Boraginaceae. XXV. A revaluation of some genera of the Lithospermeae. J. Arnold Arbor. 34: 258–299.

Jonová, M. 1926. Anatomie a morfologie trichomu u Borraginacei s ohledem na systematiku této celedi (L'anatomie et la morphologie des trichomes des Borraginées à l'égard du système de cette famille). Vestnik Král. Ceské Spolecn. Nauk, Tr. Mat.-Prir. 2: 1–66.

Kaul, M.K. 1997. Medicinal plants of Kashmir and Ladakh: Temperate and cold arid Himalaya. Delhi: Indus Publishing Company, 173 pp.

Keeley, J.E., Fotheringham, C.J. 1998. Smoke-induced germination in California chaparral. Ecology 79: 2320–2336.

Khaleel, T.F. 1977. Embryology of *Trichodesma* (Boraginaceae). Bot. Notiser 130: 441–452.

Khanna, P. 1964. Embryology of *Mertensia*. J. Indian Bot. Soc. 43: 192–202.

Khatamsaz, M. 2001. Pollen morphology of Iranian Boraginaceae family and its taxonomic significance. Iranian J. Bot. 9: 27–40.

Kim, K., Park, J., Rhee, S. 2007. Structural and functional basis for (S)-allantoin formation in the ureide pathway. J. Biol. Chem. 282: 23457–23464.

Knight, A.P., Walter, R.G. (eds.) 2001. A guide to plant poisoning of animals in North America. Jackson, WY: Teton New Media, 367 pp.

Långström, E., Chase, M.W. 2002. Tribes of Boraginoideae (Boraginaceae) and placement of *Antiphytum*, *Echiochilon*, *Ogastemma* and *Sericostoma*: A phylogenetic analysis based on atpB plastid DNA sequence data. Plant Syst. Evol. 234: 137–153.
Lersten, N.R., Czlapinski, A.R., Curtis, J.D., Freckmann, R., Horner, H.T. 2006. Oil bodies in leaf mesophyll cells of angiosperms: overview and a selected survey. Amer. J. Bot. 93: 1731–1739.
Levyns, M.R. 1934. A revision of *Lobostemon* Lehm., and a discussion of the species problem. Bot. J. Linn. Soc. 49: 393–451.
Lönn, E. 1999. Revision of the three Boraginaceae genera *Echiochilon*, *Ogastemma* and *Sericostoma*. Bot. J. Linn. Soc. 130: 185–259.
Lord, E.M. 1981. Cleistogamy: a tool for the study of floral morphogenesis, function and evolution. Bot. Rev. 47: 421–449.
Mascré, M. 1922. Sur l'étamine des Borraginées. Compt. Rend. Hebd., Séances Acad. Sci. Paris 175: 987–989.
Maurer, E.S. 1961. The scent of flowers and leaves; a search for fragrance among the minor natural orders. 4. The borage family (Boraginaceae). Manufact. Chem. 32: 176–178.
McGill, C.R., McIntosh, J.C., Outred, H.A., Fountain, D.W. 2002. Seed storage and seed storage reserves in Chatham Island forget-me-not (*Myosotidium hortensia*, Boraginaceae). New Zeal. J. Bot. 40: 337–346.
Metcalfe, C.R., Chalk, L. 1950. Anatomy of the dicotyledons. Vol. 2. Boraginaceae. Oxford: Clarendon Press, pp. 945–954.
Miller, J.S. 2005. A synopsis of the genus *Cynoglossum* L. (Boraginaceae) in Madagascar and the Comoro Islands. Adansonia III, 27: 113–127.
Moerman, D.E. 1998. Native American ethnobotany. Portland, OR: Timberland Press, 927 pp.
Mosti, S., Selvi, F. 2007. *Trichodesma cinereum* Mosti & Selvi (Boraginaceae), a new species from Oman. Candollea 62: 205–210.
Muller, J. 1981. Fossil pollen records of extant angiosperms. Bot. Rev. 47: 1–142.
Müller, A., Kuhlmann, M. 2003. Narrow flower specialization in two European bee species of the genus *Colletes* (Hymenoptera: Apoidea: Colletidae). Eur. J. Entomol. 100: 631–635.
Nakanishi, H. 2002. Splash seed dispersal by raindrops. Ecol. Res. 17: 663–671.
Nepi, M., Selvi, F., Pacini, E. 2010. Variation in nectarsugar profile of *Anchusa* and allied genera (Boraginaceae). Bot. J. Linn. Soc. 162: 616–627.
Nesom, G.L. 1988. Synopsis of the species of *Omphalodes* (Boraginaceae) native to the New World. Sida 13: 25–30.
Neubauer, H.F. 1977. Über Knotenbau und Blattgrund von *Cordia myxa* L., *Anchusa officinalis* L. und *Borago officinalis* L. (Boraginaceae). Bot. Jahrb. Syst. 98: 362–371.
Niemüller, D., Reimann, A., Ober, D. [unpublished]. Tissue specific expression of homospermidine synthase in species of the Boraginales.
Nowicke, J.W., Miller, J.S. 1989. Pollen morphology and the relationships of Hoplestigmataceae. Taxon 38: 12–16.
Ober, D., Hartmann, T. 1999. Homospermidine synthase, the first pathway-specific enzyme of pyrrolizidine alkaloid biosynthesis, evolved from deoxyhypusine synthase. Proc. Natl. Acad. Sci. USA 96: 14777–14782.
Olesen, J.M. Valido, A. 2004. Lizards and birds as generalized pollinators and seed dispersers of island plants. In: Fernandez-Palacios, J.M., Morici, C. (eds.) Ecologia insular/Island ecology. Asociacion Espanola de Ecologia Terrestre (AEET). La Palma: Cabilda Insular de La Palma, pp. 229–249.
Otero, A., Jiménez-Mejía, P., Valcárcel, V. & Vargas, P. 2014. Molecular phylogenetics and morphology support two new genera (*Memoremea* and *Nihon*) of Boraginaceae s.s. Phytotaxa 173: 241–277.
Ovczinnikova, S. 2009. On the position of the tribe Eritrichieae in the Boraginaceae system. Botanica Serbica 33: 141–146.
Pakeman, R.J., Engelen, J., Attwood, J.P. 1999. Rabbit endozoochory and seedbank build-up in an acidic grassland. Vegetatio 145: 83–90.
Papageorgiou, V.P., Assimopoulou, A.N., Couladouros, E. A., Hepworth, D., Nicolaou, K.C. 1999. The chemistry and biology of alkannin, shikonin and related naphthazarin natural products. Angew. Chem. Int. Edn. 38: 270–301.
Papageorgiou, V.P., Assimopoulou, A.N., Samanidou, V. F., Papadoyannis, I.N. 2006. Recent advances in chemistry, biology and biotechnology of alkannins and shikonins. Curr. Org. Chem. 10: 2123–2142.
Park, C.U. 1982. Study of the secondary xylem in herbaceous dicotyledons I. Boraginaceae, Cruciferae, and Euphorbiaceae. Trans. Missouri Acad. Sci 16: 25–36.
Parsons, W.T., Cuthbertson, E.G. 2001. Noxious weeds of Australia, ed. 2. Collingwood: CSIRO Publishing, 712 pp.
Pate, J.S., Gunning, B.E.S. 1969. Vascular transfer cells in angiosperm leaves. A taxonomic and morphological survey. Protoplasma 68: 135–156.
Petersen, M., Simmonds, M.S. 2003. Rosmarinic acid. Phytochemistry 62: 121–125.
Pilger, R. 1912. Die Gattung *Wellstedia* in Südwestafrika. Bot. Jahrb. Syst. 46: 558–561.
Pitot, A. 1937. L'ovaire du *Symphytum orientale* L. Bull. Soc. bot. France 84: 393–400.
Popov, M.G. 1953. Boraginaceae. In: Shishkin, B., Bobrov, E. (eds.) Vol. 19. Tubiflorae. Moskva, Leningrad: Izdatel'stvo Akademii Nauk SSSR, pp. 97–691, 703–718.
Quilichini, A., Debussche, M. 2001. Seed dispersal and germination patterns in a rare Mediterranean island endemic (*Anchusa crispa* Viv., Boraginaceae). Acta Oecol. 21: 303–313.
Quilichini, A., Debussche, M., Thompson, J.D. 2001. Evidence for local outbreeding depression in the Mediterranean island endemic *Anchusa crispa* Viv. (Boraginaceae). Heredity 87: 190–197.
Radcliffe, C.E. 1927. The Stockholm Papyrus. J. Chem. Educ. 4: 979–1002.
Retief, E., van Wyk, A.E. 2005. Boraginaceae: Codonoideae, a new subfamily based on *Codon*. Bothalia 35: 78–80.

Ridley, H.N. 1930. The dispersal of plants throughout the world. Ashford: Reeve, 744 pp.
Riedl, H. 1997. Boraginaceae. In: Kalkman, C., Kirkup, D. W., Nooteboom, H.P., Stevens, P.F., de Wilde, W.J.J. O. (eds.) Flora Malesiana Series I Seed Plants 13. Leiden: Publications Department Rijksherbarium, pp. 43–144.
Roeder, E. 1995. Medicinal plants in Europe containing pyrrolizidine alkaloids. Pharmazie 50: 83–98.
Roeder, E. 2000. Medicinal plants in China containing pyrrolizidine alkaloids. Pharmazie 55: 711–726.
Rosen, D., Barthlott, W. 1991. Ökologische Aspekte der Ultraviolett-Reflexion von Blumen in Mitteleuropa, besonders in der Eifel. Decheniana 144: 72–112.
Sahay, S.K. 1979. Palynotaxonomy of Boraginaceae and some other families of Tubiflorae. Biol. Mem. 4: 117–205.
Schaal, B.A., Leverich, W.J. 1982. Survivorship patterns in an annual plant community. Oecologia 54: 149–151.
Schittengruber, B. 1953. Die Stomaverteilung an Blättern von *Pulmonaria* - Arten. Phyton 5: 128–132.
Schwarzer, C. 2007. Systematische Untersuchungen an den peruanischen Vertretern der Gattungen *Pectocarya* DC. ex Meisn., *Amsinckia* Lehm., *Plagiobothrys* Fisch. & C.A. Mey. und *Cryptantha* Lehm. ex G. Don (Boraginaceae). Diploma thesis, Systematic Botany, FU Berlin.
Seibert, J. 1978. Fruchtanatomische Untersuchungen an Lithospermeae (Boraginaceae). Bibl. Bot. 44. Vaduz: Cramer.
Selvi, F., Bigazzi, M. 2001. Leaf surface and anatomy in Boraginaceae tribe Boragineae with respect to ecology and taxonomy. Flora 196: 269–285.
Selvi, F., Bigazzi, M., Hilger, H.H., Papini, A. 2006. Molecular phylogeny, morphology and taxonomic re-circumscription of the generic complex *Nonea/Elizaldia/Pulmonaria/Paraskevia* (Boraginaceae-Boragineae). Taxon 55: 907–918.
Selvi, F., Cecchi, L. Coppi, A. 2009. Phylogeny, karyotype evolution and taxonomy of Cerinthe L. (Boraginaceae). Taxon 58: 1307–1325.
Shevchenko, N.K., Baimuradov, T., Iaraeva, R.M. 1973. *Trichodesma* poisoning in poultry. Veterinariia 49: 99–100.
Skarpaas, O., Stabbetorp, E. 2001. Diaspore ecology of *Mertensia maritima*: effects of physical treatments and their relative timing on dispersal and germination. Oikos 95: 374–382.
Smith, S.G. 1932. Cytology of *Anchusa* and its relation to the taxonomy of the genus. Bot. Gaz. 94: 394–403.
Solereder, H. 1899. Systematische Anatomie der Dicotyledonen. Stuttgart: Ferdinand Enke, 984 pp.
Souèges, R. 1948. Embryogénie et classification, 3e fascicule: Essai d'un système embryogénique (Partie spéciale: Première période du système). Actualités scientif. industr. 1060: 1–107.
Stegelmeier, B.L., Edgar, J.A., Colegate, S.M., Gardner, D. R., Schoch, T.K., Coulombe, R.A., Molyneux, R.J. 1999. Pyrrolizidine alkaloid plants, metabolism and toxicity. J. Nat. Toxins 8: 95–116.
Stewart, M.J., Steenkamp, V. 2001. Pyrrolizidine poisoning: A neglected area in human toxicology. Therap. Drug Monit. 23: 698–708.
Strey, M. 1931. Karyologische Studien an Borraginoideae. Planta 14: 682–730.
Svensson, H.G. 1925. Zur Embryologie der Hydrophyllaceen, Borraginaceen und Heliotropiaceen mit besonderer Rücksicht auf die Endospermbildung. Uppsala Univ. Årsskr. 2: 3–175.
Teppner, H. 2011. Flowers of Boraginaceae (*Symphytum*, *Onosma*, *Cerinthe*) and *Andrena symphyti* (Hymenoptera-Andrenidae). Morphology, pollen portioning, vibratory pollen collection, nectar robbing. Phyton 50: 145–180.
Thomas, D.C., Weigend, M., Hilger, H.H. 2008. Phylogeny and systematics of *Lithodora* (Boraginaceae - Lithospermeae) and its affinities to the monotypic genera *Mairetis*, *Halacsya* and *Paramoltkia* based on ITS1 and trnLUAA-sequence data and morphology. Taxon 57: 79–97.
Thomasson, J.R. 1979. Late cenozoic grasses and other angiosperms from Kansas, Nebraska, and Colorado: Biostratigraphy and relationships to living taxa. Kansas Geol. Surv. Bull. 128: 1–68.
Thomasson, J.R. 1987. Late Miocene plants from Northeastern Nebraska (USA). J. Paleontol. 61: 1065–1079.
Tsai, S.L., Harris, P.J., Lovell, P.H. 2003. The root epidermis of *Echium plantagineum* L.: a novel type of pattern based on the distribution of short and long root hairs. Planta 217: 238–244.
Türkmen, N., Düzenli, A. 2005. Changes in floristic composition of *Quercus coccifera* macchia after fire in the Çukurova region (Turkey). Ann. Bot. Fennici 42: 453–460.
Van Campo, E. 1976. La flore sporopollénique du gisement Miocène terminal de Venta del Moro (Espagne). PhD Thesis, Univ. Montpellier.
Van Valen, F. 1979. Contribution to the knowledge of cyanogenesis in Angiosperms. 12. Cyanogenesis in Boraginaceae. Proc. K. Ned. Akad. Wet., Ser. C 82: 171–176.
Van Wyk, B.E, Winter, P.J.D., Buys, M.H. 1997. The major flower anthocyanins of *Lobostemon* (Boraginaceae). Biochem. Syst. Ecol. 25: 39–42.
Velasco, L., Goffman, F.D. 1999. Chemotaxonomic significance of fatty acids and tocopherols in Boraginaceae. Phytochemistry 52: 423–426.
Veno, B.A. 1979. A revision of the genus *Pectocarya* (Boraginaceae) including reduction to synonymy of the genus *Harpagonella* (Boraginaceae). Ph.D. dissertation, University of California, Los Angeles, pp. 1-201.
von Arx, G., Dietz, H. 2006. Growth rings in the roots of temperate forbs are robust annual markers. Plant Biol. 8: 224–233.
Wang, W.-M., Harley, M.M. 2004. The Miocene genus *Fupingopollenites*: comparisons with ultrastructure and pseudocolpi in modern pollen. Rev. Palaeobot. Palynol. 131: 117–145.
Weber, S., Eisenreich, W., Bacher, A., Hartmann, T. 1999. Pyrrolizidine alkaloids of the lycopsamine type: Biosynthesis of trachelanthic acid. Phytochemistry 50: 1005–1014.
Weigend, M., Hilger, H.H. 2010. Codonaceae - a newly required family name in Boraginales. Phytotaxa 10: 26–30.

Weigend, M., Selvi, F., Gottschling, M., Hilger, H.H., 2009. Marbleseeds are Gromwells - systematics and evolution of *Lithospermum* L. s.l. (Boraginaceae tribe Lithospermeae) based on molecular and morphological data. Mol. Phylogenet. Evol. 52: 755–768.

Weigend, M., Gottschling, M., Selvi, F., Hilger, H.H. 2010. Fossil and extant Western Hemisphere Boragineae, and the polyphyly of "Trigonotideae" Riedl (Boraginaceae: Boraginoideae). Syst. Bot. 35: 409–419.

Weigend, M., Luebert, F., Selvi, F., Brokamp, G., Hilger, H. H., 2013. Multiple origins for Hounds tongues (*Cynoglossum* L.) and Navel seeds (*Omphalodes* Mill.) - the phylogeny of the borage family (Boraginaceae s.str.). Mol. Phylogenet. Evol. 68: 604–618.

Weigend, M., Luebert, F., Gottschling, M., Couvreur, T.L. P., Hilger, H.H., Miller, J. 2014. From capsules to nutlets—phylogenetic relationships in the Boraginales. Cladistics 30(5): 508–518.

Weiss, M.R. 1995. Floral color change: A widespread functional convergence. Amer. J. Bot. 82: 167–185.

Winkworth, R.C., Grau, J., Robertson, A.W., Lockhart, P.J. 2002. The origins and evolution of the genus *Myosotis* L. (Boraginaceae). Mol. Phylogenet. Evol. 24: 180–193.

Wolff, K., El Akkad, S., Abbott, R.J. 1997. Population substructure in *Alkanna orientalis* (Boraginaceae) in the Sinai desert, in relation to its pollinator behaviour. Mol. Ecol. 6: 365–372.

Wollenweber, E., Wehde, R., Dörr, M., Stevens, J.F. 2002. On the occurrence of exudate flavonoids in the borage family (Boraginaceae). Z. Naturforsch., C: Biosci. 57: 445–448.

Wunderlich, R. 1959. Zur Frage der Phylogenie der Endospermtypen bei den Angiospermen. Österr. Bot. Z. 106: 203–293.

Wunderlich, R. 1991. Zur Frage nach der systematischen Stellung der Limnanthaceae. Stapfia 25: 1–59.

Zhu, G., Riedl, H., Kamelin, R.V. 1995. Boraginaceae. In: Wu, Z.Y., Raven, P.H. (eds.) Flora of China, vol. 16: Gentianaceae through Boraginaceae, pp. 329–431. Beijing: Science Press and St. Louis: Missouri Botanical Garden Press.

Bruniaceae

Bruniaceae R. Br. ex DC., Prodr. 2: 43 (1825), nom. cons.

R. Claßen-Bockhoff

Decumbent to erect shrubs or small trees. Leaves alternate, densely set, sessile or shortly petiolate, small, simple, linear to lanceolate, subterete, mostly sclerophyllous, persistent, with brownish mucro-like tip; stipules minute when present; trichomes uniseriate, non-glandular. Inflorescences spike- or head-like when present, terminal or lateral, solitary or in compound clusters. Flowers in inflorescences or solitary on bracteate short-shoots or leafy shoots, mostly small, white, pinkish or yellowish, bisexual, actinomorphic, mainly pentamerous, tetracyclic, mostly (half-)epigynous; sepals (4)5, imbricate, calyx tube often adnate to ovary; petals (4)5, free, imbricate, distinct, sometimes clawed, with conspicuous swellings; androecium haplostemonous, stamens (4)5, filaments free or sometimes adnate to the petals, equal or rarely unequal; anthers dithecal, tetrasporangiate, dorsifixed, versatile in tribe Brunieae, introrse, dehiscing longitudinally; carpels (1)2(–5), stylodia usually 2, simple or bifid, stigmas usually minute; gynoecium coenocarpous, ovaries usually (half-)inferior, (1)2(–5)-locular; gynoecial nectaries of the mesenchymatous type, ovules (1)2–4(–12) in each locule, weakly crassinucellate, unitegmic, anatropous. Fruits mainly dry and 1-seeded, dehiscent or indehiscent. Seeds small, occasionally arillate; endosperm often fleshy; embryo straight, very small.

The family is native to South Africa and is distributed in the Western and Eastern Cape provinces, with a single species in KwaZulu-Natal. It contains six genera with 81 species.

Vegetative Morphology. All Bruniaceae are woody, mostly shrubs of medium size, but ranging from rock-crevice species (e.g. *Thamnea* spp., *Linconia cuspidata*) to small trees (e.g. *Berzelia* spp). All species are sympodially branched. Dependent on the rhythm of branching, the habit is characterized either by regular sympodial branches flowering terminally (Figs. 16B, L, 17C) or, in case of monopodial inserts, by branches which grow on for several years and flower exclusively laterally (Figs. 16A, D, 17A) or laterally and terminally (Fig. 17B). Many Bruniaceae have vegetative branches among the flowering ones. Total life span depends on whether the plant is a reseeder or resprouter. Fire determines the age of the reseeders. It destroys the plants but favours germination (see Fig. 16C). After 2–15 years plants start to flower. At least 21 species are resprouters with a lignotuber (Carlquist 1978). de Lange and Boucher (1993b) determined the age of a lignotuber of *Audouinia capitata* to at least 90 years. The successful regeneration from lignotubers reduces the dependence on sexual reproduction. The branches are usually densely covered with spirally arranged leaves persisting for about three or four years (Fig. 16). The leaves are flat or, more often, subterete. They are predominantly small (about 5 mm long) and acicular or scale-like (Fig. 18A–C). The midvein in most species is much larger than the lateral veins, resulting in triangular or triquetrous leaf shapes in cross section. All leaves have a relatively long distinctive brownish to black tip ('apiculum', Figs. 16E, 18C) with phellogen activity (Niedenzu and Harms 1930; Carlquist 1991). Minute stipules are present in all genera but have not been found in all species. Some *Berzelia* species (e.g. *B. abrotanoides*, *B. alopecuroides*, *B. lanuginosa*, *B. galpinii*) have very conspicuous (green to) red swellings at the bases of the flowering branches. These effigurations are massive, anatomically homogenous and often subdivided into two parts, indicating that they possibly originate from stipules. Their function is unknown.

J.W. Kadereit and V. Bittrich (eds.), *Flowering Plants. Eudicots*, The Families and Genera of Vascular Plants 14,
DOI 10.1007/978-3-319-28534-4_6,

Fig. 16. Growth form, inflorescences and fruits of Bruniaceae. **A** *Brunia albiflora*, sympodial unit with monopodial inserts and lateral inflorescences. **B** *Staavia radiata*, regular sympodial branching pattern. **C** *Berzelia lanuginosa*, adult plant (left) and seedlings after a controlled fire (right). **D** *Audouinia capitata*, spike-like clusters of one-flowered short shoots. **E** *Staavia dodii*, straw-blossom with black tips on involucral bracts. **F** *Berzelia abrotanoides*, inflorescences and berry-like infructescences. **G** *Berzelia abrotanoides*, terminal flower of the inflorescence flowering first. **H** *Brunia fragarioides*, compound inflorescences. **I** *Staavia dodii*, open capsule with mature seed. **J** *Thamnea teres*, growth form of a rock-crevice species. **K** *Brunia bullata*, minute, solitary flowers on leafy shoots. **L** *Brunia paleacea*, sympodial shoot system. n, n-1, n-2 seasonal growth units

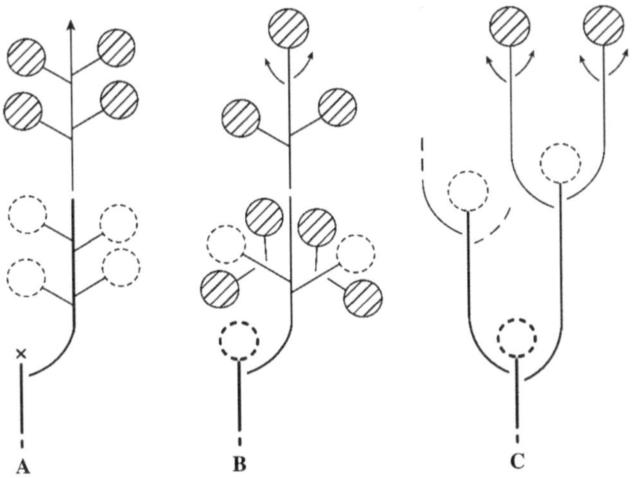

Fig. 17. Basic branching patterns and inflorescence position, schematically. All Bruniaceae grow as sympodia, but differ in the rhythm of branching. **A, B** Main axis grows on for several seasons with exclusively lateral (A) or lateral and terminal (B) inflorescences. **C** Seasonal sympodial branching with terminal inflorescences. Dotted circles: inflorescences of past season; hatched circles: inflorescences of present season; arrows: renewal shoots; discontinuous lines indicate start of a new season (after Claßen-Bockhoff 2000)

Vegetative Anatomy Including Ultrastructure. Wood and leaf anatomy have been studied by Carlquist (1978, 1991). According to him, wood of Bruniaceae is very 'primitive'. It is characterized by the presence of long vessels (average vessel-element length exceeds 1000 μm in some species) with scalariform perforation plates having numerous bars, fully or vestigially bordered. Tracheids are imperforate and thick-walled with bordered pits. Axial parenchyma is largely diffuse. Rays are heterocellular, with most or all cells upright, uniseriate and multiseriate. Growth rings vary from absent to conspicuous. Crystals, rhomboidal or smaller aggregates (druses), are present in rays of most species. Deposits of amorphous dark-staining substances (tannins?) are present in the ray cells. The wood of Bruniaceae can be considered xeromorphic, and Carlquist (1978) associated the unusually small diameter of the tracheary elements and the high density of vessels with resistance to water stress. The leaves of Bruniaceae also show distinctive xeromorphic adaptations. They are only rarely bifacial (e.g. *Brunia cordata*), most often they are transitional between an isolateral structure and an 'inverse' bifacial condition which is correlated with the tendency for leaves to be appressed to the stems. The very thick cuticle often shows a relief of longitudinal ridges. Scale-, rod- or flake-like waxes are often present all over the leaf surface or in distinct strands (see also Fehrenbach and Barthlott 1988). Epidermal cells are often markedly papillate or dome-shaped. Stomata are present all over the leaf surface or they are restricted to the abaxial or adaxial side. In case of a thick midvein, they are restricted to lateral parts of the leaf (Fig. 18C). In some species of *Brunia* subgen. *Brunia* and in the showy involucral leaves of *Staavia glutinosa*, the stomata on the abaxial side are restricted to the basal part of the leaves. The stomata are anomocytic, but often surrounded by 4–7 relatively small cells. The cuticular domes overlying the guard cells are simple or, in *Linconia*, *Audouinia* and *Thamnea teres*, have a rim around the outer surface (Fig. 18D). Trichomes are often found on the leaf margin (Fig. 18B) or on the surfaces of leaves (Fig. 16K) and may be lost as leaves mature. Trichomes are either straight or more rarely helically coiled and often have longitudinal cuticular striations. Tannins and other dark-staining materials are very characteristic of mesophyll cells. According to Carlquist (1991), *Audouinia*, *Berzelia*, *Brunia* and *Thamnea* have a large strand of fibres in the midvein and rhomboidal crystals in bundle sheath cells. The remaining two genera, *Linconia* and *Staavia*, have few or no fibres in veins and have druses in mesophyll cells. The characteristic mucro-like apicula of the leaves are thin-walled and contain massive deposits of resin-like substances. A volatile component may be released when cell walls break and leaves of Bruniaceae have a characteristic scent. At the base of the apiculum lies an active meristem, much like that of a cork wart, which produces a few layers of cells externally (Carlquist 1991). The sieve-element plastids of the Bruniaceae are of the S-type (Behnke 1981).

Inflorescences. Bruniaceae have solitary flowers (Linconieae, Audouinieae) or inflorescences (Brunieae) which can be secondarily reduced to single flowers (*Brunia bullata*, *B. callunoides*). According to Claßen-Bockhoff (2000), solitary flowers are arranged at the top of leafy shoots (*Thamnea*, *Brunia bullata*, *B. callunoides*; Fig. 16K) or at the end of lateral bracteate short-

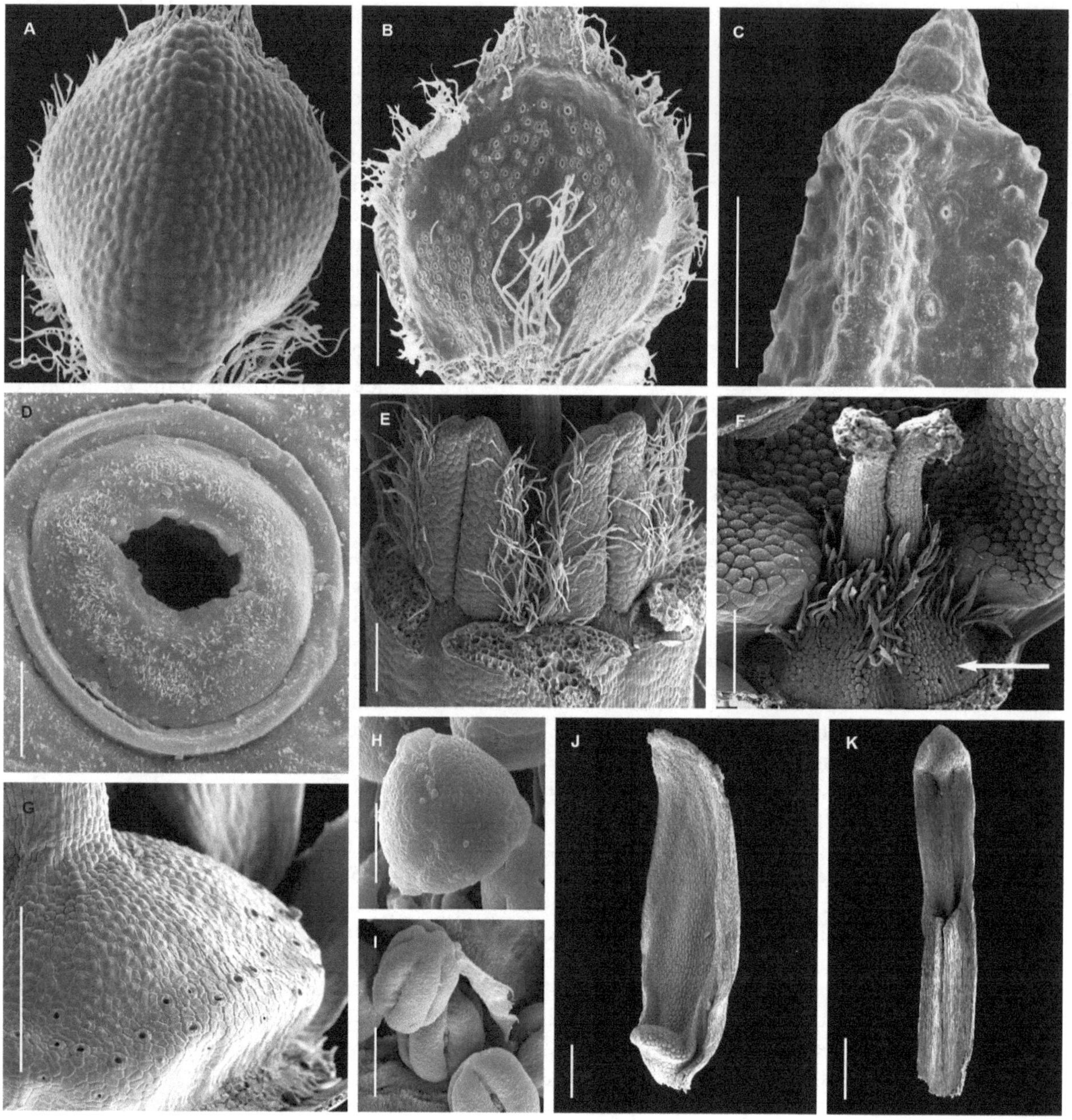

Fig. 18. Leaf and flower characteristics. **A, B** *Brunia microphylla*, scale-like leaves from abaxial (A, without stomata) and adaxial side (B, with stomata). **C** *Staavia glutinosa*, upper part of involucral bract from abaxial side; note the massive midrib, the lateral arrangement of stomata and the 'brunioid' leaf tip. **D** *Linconia alopecuroides*, stomatal aperture with cuticular rim. **E** *Brunia monogyna*, view of the petal scars of a late bud stage illustrating choripetaly. **F** *Staavia verticillata*, section of flower (stamens removed) showing the two stylodia, the basal petal bulges and the nectarostomata on the flanks of the half-inferior ovary (arrow). **G** *Thamnea thesioides*, inferior ovary with nectarostomata. **H** *Thamnea matroosbergensis*, tricolporate pollen grain. **I** *Brunia compacta*, pentacolporate pollen grains. **J** *Staavia dodii*, petal with basal bulge. **K** *Berzelia abrotanoides*, petal with two narrowly associated ridges on lower half of petal. Bars: 10 μm H, I; 20 μm D; 200 μm E, F, G; 250 μm A, B, C, J; 500 μm K

shoots. Each flower is then surrounded by (4)8 (–16) pale bracts which differ markedly from and abruptly follow the foliaceous branch-subtending leaves. Except *Thamnea teres* (Fig. 16J), the short-shoots are aggregated in loose (*Audouinia* p.p.) or dense (*Audouinia* p.p., *Linconia*) spike-like clusters (Fig. 16D). Inflorescences are closed or open spikes or heads in lateral, terminal or lateral and terminal positions (Fig. 17A–C). Foliage is found in the floral part of the branches providing the flower-subtending bracts and 1–3 bracteoles below each lateral flower. In *Brunia fragarioides* (Fig. 16H), *B. powrieae* and *Staavia zeyheri*, the heads are densely aggregated and form 'compound heads', but of different morphologies with respect to the relative position of the innovation zone (Claßen-Bockhoff 2000). Dependent on the length of the filaments and the appearance of attractive involucral bracts, inflorescences look like pincushions (*Berzelia*, *Brunia* subgen. *Brunia*, *Brunia dregeana*; Fig. 16G), strawflowers (*Staavia* p.p., *B. paleacea*; Fig. 16E, L) or rather inconspicuous 'knobs'. Within a single inflorescence, terminal flowers tend to open first (Fig. 16G). The flowering sequence of the lateral flowers is usually acropetalous or synchronous, rarely divergent; only in *Staavia* is it strictly basipetal.

Flower Structure and Floral Anatomy. Flowers are small to medium-sized, bowl-shaped, brush-like or tubular; they are white, cream, yellow, pink, deep rose or magenta. The perianth is adnate to the ovary in most species. The calyx is (4)5-lobed, often ribbed and adnate to the ovary. The calyx-lobes are free almost to the base, well-developed, often covered with unicellular hairs on the outer side, and generally have a brown tip. In some species they are persistent and crown the fruits (e.g. *Audouinia* except *A. capitata*). The petals are free, even in the supposedly sympetalous flowers of *Brunia* subgen. *Lonchostoma*, where the tubes result from the postgenital fusion of petals and stamens (Leinfellner 1964; Fig. 18E). The petals are corolline, glabrous, sessile or unguiculate (*Linconia*, *Audouinia*, *Thamnea*) and have diverse adaxial swellings (Fig. 18J, K). They have been quoted as an example of diplophylly (Leinfellner 1965) and were supposed to produce nectar (Dahlgren and Van Wyk 1988), but both interpretations were rejected by recent morphological studies (Quint and Claßen-Bockhoff 2006b). The petal swellings are formed rather late and are not vascularised. According to their position, size and shape, six types are distinguishable which in part are phylogenetically informative. Possible functions could be to protect the ovary and/or nectar, to subdivide the flower to guide the pollinator's mouthparts, or to fix the stamens in a certain position. The stamens are incurved in bud and generally caducous. They are included in (Fig. 16D) or exserted beyond the petals (*Berzelia*, *Brunia dregeana*, *Brunia* subgen. *Brunia*; Fig. 16G) and free or adnate to the base of the petals. In *Brunia* subgen. *Lonchostoma* they are almost completely secondarily fused with the petals, placing the anthers between the corolla lobes. The filaments are linear, equal or, in some species of *Brunia* subgen. *Brunia*, unequal. The anthers have a very narrow connective. They are distinctly sagittate and have a conspicuous sterile tip in Linconieae (*Linconia*, Fig. 19A), linear (to linear-oblong) with thecae connate with the connective along their entire length in Audouinieae (*Audouinia*, *Thamnea*; Fig. 19B), and versatile in Brunieae (*Berzelia*, *Brunia*, *Staavia*; Fig. 19C). According to Endress and Stumpf (1991), the endothecium is one-layered in *Berzelia lanuginosa* and *Brunia noduliflora*. The ovary is (half-) inferior in most species (Fig. 18G), but it varies from inferior (*Staavia*) to almost superior (*Brunia monogyna*). It usually consists of two carpels (3–5 in *Audouinia capitata*, apparently one in *Berzelia*, *Brunia* subgen. *Mniothamnea* p.p.) forming a coenocarpous ovary with as many locules as carpels. Usually there are two stylodia (1–5), each with a small, slightly lateral stigma. They are free or united (Fig. 18F, G) and then often appear as one style. The stylar canal is hollow in *Audouinia capitata* (de Lange et al. 1993d). Nectary tissue is present on the upper flanks of the ovary (Quint and Claßen-Bockhoff 2006b). The nectaries are of the mesenchymatous type, and they discharge their nectar via sap-holes (nectarostomata; Fig. 18F: arrow, G).

Embryology. Saxton (1910) studied the ovules of *Audouinia capitata*, *Berzelia lanuginosa*, *Brunia noduliflora* and *Staavia glutinosa* and found great similarities among these species. The ovules are unitegmic with a massive integument and a long micropylar canal. The nucellus is small and weakly crassinucellate. The embryo sac develops according to the Polygonum type. The developing embryo sac very soon absorbs the cells above it and reaches the apex of the nucellus before the second nuclear division. Starch grains are found in the embryo sac of *Berzelia* and *Brunia* p.p. An

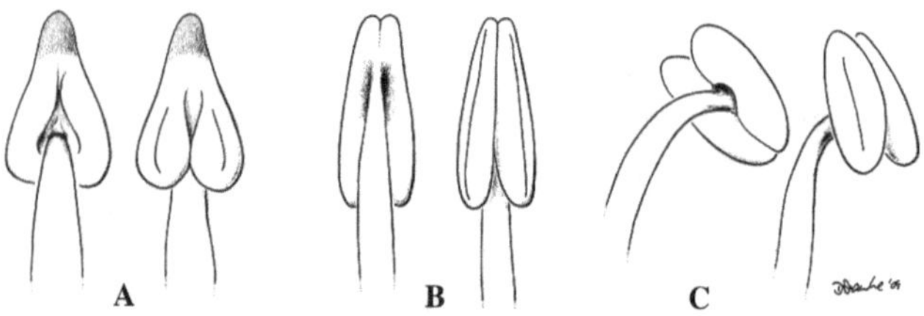

Fig. 19. Anther morphology showing the diagnostic characters for the three subfamilies. **A** Linconieae, anthers with conspicuous sterile tip. **B** Audouinieae, anthers lacking sterile tip; thecae connate with the connective along their entire length. **C** Brunieae, anthers versatile, sterile tip absent (from Quint and Claßen-Bockhoff 2006a)

endothelium is formed and a hypostase is present and particularly prominent in *Brunia* p.p. (Dahlgren and Van Wyk 1988). de Lange et al. (1993b, c, d) described the embryology of *Audouinia capitata* in detail and confirmed the data of Saxton (1910).

Pollen Morphology. There is a high degree of palynological diversity. Following Hall (1988), Linconieae, Audouinieae (except *Thamnea diosmoides* and *T. hirtella*) and *Berzelia* are characterized by consistently tricolporate pollen with a foveolate or psilate tectum. In Brunieae, colpus number varies from 3–10 (maximum in *Brunia* subgen. *Lonchostoma*) and tectum ornamentation from psilate to granular, verrucose or coarsely reticulate. Pollen diameter (polar axis) ranges between 11 μm (*Brunia villosa*) and 50 μm (*Thamnea massoniana*). A quarter of the species have 20% or more (up to 96%) of their pollen grains with irregular appendages, distorted in shape or reduced in size. In *Berzelia intermedia* pollen size (12–19 μm mean width) and frequency of distortion (45–96%) vary from one locality to another.

Karyology. Goldblatt (1981) examined 23 species of four genera (*Audouinia, Berzelia, Brunia, Staavia*) karyologically. He found similar and uniformly small chromosomes in all species, ranging in size at mitotic metaphase from 1.5–2.5 μm. The basic chromosome number for Bruniaceae has been suggested to be $n = 11$. The diploid number of $2n = 22$ has been recorded only in *Audouinia capitata*, while the other species investigated are at least tetraploid on the base of $n = 10$–12. Exact counts were not obtained in all species. *Staavia radiata, Berzelia abrotanoides, B. ecklonii* and *B. intermedia* were found to have $n = 20$–23. They are usually tetraploid and only in case of *B. intermedia* diploid (ca. 42), tetraploid (ca. 80–86) and hexaploid (ca. 120).

Pollination. Flowers are usually unspecialized and present pollen and nectar. Pollen grains are dispersed in monads. Comprehensive field studies in *Audouinia capitata* (Wright et al. 1991) revealed that small pollen-feeding beetles (Nitidulidae) appear to be the most common pollen vectors. Pollen dispersal by these beetles is limited to short distances (less than 4 m) which leads to limited gene flow within and between *A. capitata* populations. Schardt (2010) recently confirmed promiscuous pollination in several species of Bruniaceae with ants, beetles and flies being the most frequent pollinators, and bees and even butterflies being rarer. Specialisation to longer-tongued insects (flies, e.g. Bombyliidae) was only found in the narrow-tubular flowers of *Brunia* subgen. *Lonchostoma* which mechanically exclude other insects from nectar.

Reproductive Systems. Comprehensive field studies in *Audouinia capitata* (de Lange et al. 1993a, b, c, d) revealed that the reproductive success of plants is limited by inefficient pollination, low pollen viability, a high degree of self-incompatibility, a high incidence of aberrant endosperm development and embryo abortion, poor seed set and high seedling mortality. The general picture is that many Bruniaceae lack reproductive success and only rarely reproduce successfully by seed (see also Hall 1988). According to Schardt (2010), *Brunia africana* is autogamous; *Staavia radiata, Brunia monogyna* and *Brunia africana* are protandrous and *Brunia paleacea, B. alopecuroides, Berzelia abrotanoides* and *B. lanuginosa* protogynous. The two types of dichogamy are most likely associated with two different types of stigmatic surfaces (Schardt 2010). Flowering sequence is often synchronous within the whole plant. Anthesis of an individual plant lasts from about three weeks in the mass flowering *Berzelia lanuginosa* to more than three months in the sparsely flowering *Staavia radiata*.

Fruit and Seed. Fruits are mainly 1-seeded, dry, nutlets (*Audouinia, Thamnea, Berzelia, Brunia* p.p.) or capsules (*Linconia*, *Staavia*, Fig. 16I, *Brunia* p.p.). Very rarely they are 2–3-seeded (e.g. *Brunia* subgen. *Brunia* p.p., *Brunia* subgen. *Lonchostoma*). In *Audouinia capitata* the thick, strongly sclerified pericarp has thinner regions where the fruit fractures after fire (de Lange and Boucher 1993a). In *Berzelia alopecuroides* the sepals become red and fleshy and the whole inflorescence forms a berry-like infructescence (Fig. 16F). However, the sepals dry up very soon and the minute fruits are released separately as in all other Bruniaceae with indehiscent fruits. In *Berzelia albiflora* seeds are retained on the plant in fire-protected cones (serotinous infructescences) and are only released after fire. According to Nel (1986), *Brunia noduliflora* retains seeds for three years, *Brunia paleacea* (Fig. 16L) for two years and *Berzelia lanuginosa* for one year. The wetter the habitat the shorter the seeds seem to be kept. The mature seeds have a thick cuticle, copious endosperm, and a very small, straight embryo. They are arillate in some genera (e.g. *Staavia*; Fig. 16I). Dark-stained cells with amorphous material are abundant in the seed coat. Under natural conditions seeds only germinate after fire. In *Audouinia* seed germination is initiated by chemical compounds contained in smoke (de Lange and Boucher 1990, 1993b, 1993c). *Staavia dodii* can be grown from seed after erosion of the testa by concentrated sulphuric acid and subsequent treatment with gibberellic acid (Wolff 1990). Fruits and seeds are auto-, (anemo- ?) or myrmecochorous but in general do not germinate far from the source plant.

Phytochemistry. A study of leaf flavonoids based on six species of Bruniaceae by Jay (1968) revealed the presence of leucocyanidin, leucodelphinidin, generally also quercetin and traces of myricetin, but not ellagic acid or kaempferol. Methylated flavonols (isorhamnetin and methyl-3-quercetin) were detected and contribute to the distinctness of the Bruniaceae. Iridoids were not detected in the samples studied by Jensen et al. (1975). Scott (1999) examined leaf flavonoid patterns in 55 species of Bruniaceae and some putative allies. Proanthocyanidins were recorded as present in all genera of Bruniaceae except *Linconia*. Only procyanidins were found in *Audouinia*, *Thamnea* and *Brunia* subgen. *Brunia* p.p.; the remaining genera have both prodelphinidins and procyanidins. The methylated flavonol isorhamnetin was detected in all genera and 70% of the species examined. Myricetin is lacking in those genera of Bruniaceae which contain only procyanidin. In *Brunia* p.p., 5-deoxy flavonol fisetin appears to be present; flavonoids lacking a 5-OH function were recorded in *Staavia*, *Berzelia* and *Brunia* p.p. Quercetin-3-methyl ether, recorded as present in *Staavia* and *Brunia* p.p. by Jay (1968), was not detected in any other species examined. Common flavones were not detected in Bruniaceae except in *Brunia* p.p. which contains luteolin and/or chrysoeriol.

Subdivision and Relationships Within the Family. The taxonomic history of the Bruniaceae dates back to Brown (1818). Niedenzu (1891) subdivided the family into Audouinieae and Brunieae on the basis of differences in anther morphology. Based on Brongniart (1826), Sonder (1862), Niedenzu (1891), Dümmer (1912) and Niedenzu and Harms (1930), Pillans (1947) wrote the last revision of the family. He distinguished twelve genera using number of stylodia, number of locules and ovules, fusion of floral parts, inflorescence structure, anther shape and stamen length as the main discriminating characters. Since that time new data have been accumulated about wood anatomy (Carlquist 1978), chromosome numbers (Goldblatt 1981), pollen morphology (Hall 1988), leaf anatomy (Carlquist 1991), autecology in *Audouinia capitata* (de Lange and Boucher 1990, 1993a, b, c; de Lange et al. 1993a, b, c, d; Wright et al. 1991), flower morphology (Leinfellner 1964, 1965; Quint and Claßen-Bockhoff 2006b), phytochemistry (Scott 1999), growth form and inflorescence morphology (Claßen-Bockhoff 2000), pollination ecology (Schardt 2010) and taxonomy including the description of five new species (Strid 1968; Powrie 1969a, 1969b; Oliver and Oliver 1999; Hall et al. 2010). As a general conclusion *Audouinia* was postulated to be the most basal genus with the only palaeodiploid chromosome number and the only trilocular ovary in the family.

However, molecular data identified *Linconia* as sister to all remaining Bruniaceae (Quint and Claßen-Bockhoff 2006a). These authors confirmed the obvious misinterpretation of *Pseudobaeckea teres* and some *Brunia* species and clearly showed that *Raspalia* is polyphyletic with morphologically similar species in four different subclades. In a new classification of the family (Claßen-Bockhoff et al. 2011), three tribes with six genera were accepted, i.e. Linconieae with only *Linconia*, Audouinieae with *Audouinia* (including *Tittmannia*) and *Thamnea* (including *Pseudobaeckea teres*), and Brunieae with *Staavia* (including *Raspalia staavioides*), *Berzelia* (including *Brunia* p.p.) and an expanded *Brunia* (including *Nebelia*, *Raspalia* p.p., *Pseudobaeckea* p.p., *Mniothamnea* and *Lonchostoma* in five subgenera). The three tribes are morphologically supported by anther morphology (Fig. 19A–C), thus largely confirming the early subdivision of the family by Niedenzu (1891). Results of a molecular clock analysis revealed that Bruniaceae are a family of ancient origin (Quint and Claßen-Bockhoff 2008). The major diversification of the family may have happened in relatively recent times (18–3 Ma) simultaneously with the establishment of the present Mediterranean climate in their distribution range.

Affinities. Since Bruniaceae were established, the family has been placed in the widely circumscribed Rosales (e.g. in Hallier 1912; Cronquist 1981), Saxifragales (Takhtajan 1980) or Hamamelidales (Hutchinson 1969). Dahlgren and Van Wyk (1988) followed a concept first used by Van Tieghem (1897) and placed Bruniaceae close to the monogeneric Grubbiaceae both either in Ericales (after Cronquist 1988) or in a separate order, Bruniales, near Ericales. In contrast to them, Thorne (1976, 1983) placed the family in Pittosporales suborder Bruniineae together with the four monogeneric families Geissolomaceae, Myrothamnaceae, Roridulaceae and Hydrostachyaceae, all native to South Africa.

Molecular data (e.g. Bremer et al. 2001, 2002; Winkworth et al. 2008) clearly indicate that Bruniaceae are sister to the Apiales-Paracryphiales-Dipsacales-clade. The Angiosperm Phylogeny Group (APG III 2009) combines Bruniaceae and Columelliaceae (western South America) in a separate order Bruniales.

Distribution and Habitats. Bruniaceae are a characteristic element of the Mediterranean-type fynbos biome. With only one species in southern KwaZulu-Natal/Eastern Cape (*Brunia trigyna*), they are near-endemic to the Cape Floristic Region (CFR) and represent one of the 33 'Cape floral clades' (Linder 2003). As an adaptation to periodic fires, many species have lignotubers from which they are able to resprout. Fire stimulates flowering in *Audouinia capitata* (de Lange and Boucher 1993b) and germination in many species. A mykorrhizal association has been found in *Staavia dodii* (Moll and Gubb 1981). Species are generally confined to the geological formation of Table Mountain Sandstone with soils characteristically low in nutrients. *Berzelia cordifolia* occurs in the Bredasdorp area (Agulhas Plain) and appears to be a limestone endemic (see Quint and Claßen-Bockhoff 2008). Bruniaceae tend to occur in mesic microhabitats (e.g. cloud-covered south-eastern mountain slopes, marshes, stream banks) where they generally grow in isolated patches of a few individuals. Only about a quarter of the species are locally frequent, the remainder being rare and often declining. Hall (1987, 1988) postulated a great age and a formerly widespread distribution for bruniaceous ancestors, only remnants of which have survived in the climatically separated southern tip of Africa. There is no indication of where the ancestral stock of the family might have originated, but the remarkably high diversity and the limited geographical range suggest long-term isolation. With respect to the assumed closest relatives of Bruniaceae in South America, the clade is of biogeographic interest.

Palaeobotany. Fossil pollen grains similar to those of Bruniaceae have been recorded from Hondeklip Bay in Namaqualand (de Villiers and Cadman 1997), but the age of these deposits is not clear (Linder 2003).

Economic Importance. Cut flowers are exported as 'Cape greens'. Besides the 'stompies' (*Brunia* subgen. *Brunia*) and 'kolkols' (*Berzelia*), there are other plants of great horticultural potential (*Audouinia capitata*, *Staavia dodii*). However, they are difficult to propagate.

Conservation. Twenty-three species of Bruniaceae are listed as critically rare, threatened or recently extinct (Hilton-Taylor 1996). Incorrect burning practices, overpicking, low pollinator efficiency, sterile pollen grains and inadequate sexual fertility account most for the rarity of species.

KEY TO GENERA OF BRUNIACEAE

1. Anthers with conspicuous sterile tip (Linconieae) **1. *Linconia***
- Anthers without sterile tip 2
2. Thecae connate with the connective along their entire length; anthers not versatile; stomata often surrounded by cuticular rim (Audouinieae) 3
- Thecae free at base; anthers versatile; stomata never surrounded by cuticular rim (Brunieae) 4
3. Flowers arranged in spike-like clusters, each flower on a bracteate short-shoot; pedicels present; stomata surrounded by cuticular rim **2. *Audouinia***
- Flowers solitary, each on a leafy shoot or a bracteate short shoot (only *Thamnea teres*); pedicels absent (except *T. hirtella*); stomata rarely surrounded by cuticular rim (only *T. teres*) **3. *Thamnea***
4. Stylodia 2, connate; inflorescences with basipetal flowering sequence **4. *Staavia***
- Styles 1 or 2, distinct (except *B. monogyna*); flowering sequence not as above 5
5. Pollen 3-colporate; stamens exserted; inflorescences lateral **5. *Berzelia***
- Pollen 5- to 10-colporate; stamens shorter than corolla (except *Brunia* subgen. *Brunia*); terminal inflorescence usually present **6. *Brunia***

GENERA OF BRUNIACEAE

1. **Tribe Linconieae** Quint & Class.-Bockh. (2011).

Anthers with conspicuous sterile tip; stomata surrounded by a cuticular rim (Figs. 18D, 19A).

1. ***Linconia*** L.

Linconia L., Mant. Pl. Altera: 216 (1771); Pillans, J. S. African Bot. 13: 130 (1947) rev.; Claßen-Bockhoff et al., Taxon 60: 1144 (2011), classif.

Erect undershrubs or rock crevice shrublets, lignotubers present. Leaves imbricate (*L. alopecuroidea*) or spreading, small and ericoid (*L. cuspidata, L. ericoides*), shortly petiolate; stipules minute; hypostomatic. Flowers pedicellate, terminal on bracteate short-shoots, arranged in spike- or head-like clusters at top of vegetative branches that continue to grow after flowering, usually with four bracts per flower. Petals bright pink (*L. alopecuroidea, L. ericoides*) or creamy white (*L. cuspidata*), with two ventral thick ridges converging towards base of petal; stamens included; anthers sagittate with pale, sterile tip, blue (white in *L. cuspidata*); pollen 3-colporate; ovary half to two-thirds inferior, bilocular (rarely trilocular in *L. alopecuroidea*) with two ovules per locule; stylodia two, free. Fruit 1- or 2-seeded, dry, dehiscent.

Three species, south-western region of Western Cape extending to Heidelberg (Humansdorp *vide* Pillans 1947). Relatively dry rock crevices or in moist fynbos (fine-leaved shrub vegetation) in partial shade of vegetation on southern slopes, at medium (500–600 m) or high altitudes (900–1500 m). Flowering September to November.

2. **Tribe Audouinieae** Nied. (1891).

Anthers without sterile tip, thecae connate with the connective (Fig. 19B); stomata often surrounded by a cuticular rim.

2. ***Audouinia*** Brongn. Fig. 16D

Audouinia Brongn., Ann. Sci. Nat. (Paris) 8: 384, t. 38, fig. 1A–G (1826); Pillans, J. S. African Bot. 13: 126 (1947), rev. p.p.; Claßen-Bockhoff et al., Taxon 60: 1145 (2011), classif.
Tittmannia Brongn. (1826), nom. cons.

Erect ericoid shrubs, densely branched, lignotubers present in *A. capitata*. Leaves sessile, imbricate, appressed, linear; stipules ± present; amphistomatic (*A. hispida* hypostomatic), stomata surrounded by cuticular rim. Flowers shortly pedicellate, terminal on bracteate short-shoots, arranged in spike-like clusters at top of vegetative branches that continue to grow after flowering; flowers small and white to pinkish with 6–8 subtending bracts or, in *A. capitata*, larger and vivid red with 6–16 bracts. Petals unguiculate or ovate, with minute bulges (*A. capitata*) or two thick converging ridges on the adaxial side; stamens included or shortly exserted; anthers linear to linear-oblong, thecae parallel, free only in basal part; pollen tricolporate; ovary in *A. capitata* half-inferior and mostly 3-locular (also 4- and 5-locular), in remaining species almost completely inferior and bilocular (rarely 3-locular in *A. laevis*), with two ovules per locule; stylodia two or three (*A. capitata*), united. Fruit usually 1-seeded, dry, indehiscent, crowned by persistent floral leaves (except *A. capitata*). $2n = 22$ (*A. capitata*).

Five species, rare, scattered in fynbos vegetation, in SW Cape. Relatively well-drained soils, thriving in partial shade of vegetation at low to medium altitudes (up to 340 m; *A. capitata* (L.) Brong.) or between rocks at higher altitudes of > 700 m. Flowering May, June (*A. esterhuyseniae* (Powrie) A.V. Hall), July, August (*A. capitata*) or September, October.

3. ***Thamnea*** Sol. ex Brongn. Fig. 16J

Thamnea Sol. ex Brongn., Ann. Sci. Nat. (Paris) 8: 386, t. 38 (1826); Pillans, J. S. African Bot. 13: 143 (1947), rev.; Claßen-Bockhoff et al., Taxon 60: 1145 (2011), classif.
Pseudobaeckea Nied. (1891) p.p.

Ericoid dwarf shrubs, prostrate, often in rock crevices; lignotubers present. Leaves ± imbricate, sessile, slightly appressed; stipules minute (not seen in *T. teres*); hypostomatic (epistomatic in *T. teres*), stomata not surrounded by cuticular rim (except *T. teres*). Flowers solitary, sessile (except *T. hirtella*), on bracteate short-shoots with 8–12 bracts (*T. teres*) or at top of leafy shoots subtended by uppermost involucre-like leaves. Petals free, unguiculate, recurved, white to pink, with two separate, rather narrow, parallel ridges; stamens included or shortly exserted; anthers linear or linear-oblong, thecae parallel, free only in basal part; pollen tricolporate, except *T. ustulata* and *T. hirtella* with 4 or 5 pollen colpi, tectum granular; ovary inferior (up to half-inferior in *T. teres*), uni- or bilocular with 2–8 ovules per locule, topped by ring-shaped nectary in some species; styles one or two, ± shortly exserted, united (free in *T. teres*). Fruit 1-seeded, dry, indehiscent.

Nine species, restricted to mountain slopes in the south-western part of the Cape. On well-drained soils at high altitudes (>1000 m; except *T. thesioides* Dümmer, ca. 700 m). Flowering December–February (*T. massoniana* Dümmer), April–July (*T. hirtella* Oliv.), October–November (*T. teres* (Dümmer) Class.-Bockh. & E.G.H. Oliv.) or September–November.

3. **Tribe Brunieae** Quint & Class.-Bockh. (2011).

Anthers without sterile tip, versatile (Fig. 19C); stomata not surrounded by cuticular rim.

4. ***Staavia*** Dahl Fig. 16B, E, I

Staavia Dahl, Observ. Bot. 15 (1787); Pillans, J. S. African Bot. 13: 169 (1947), rev.; Claßen-Bockhoff et al., Taxon 60: 1146 (2011), classif.
Raspalia Brongn. (1826) p.p.

Erect, ericoid undershrubs with clear sympodial branching, lignotubers present or absent. Leaves erect-spreading, imbricate, petiolate, amphistomatic; stipules present. Inflorescences closed heads with basipetal flowering sequence; usually terminal on main axis of sympodial shoot, often enriched by lateral inflorescences, in *S. zeyheri* densely aggregated in compound cluster; uppermost leaves forming an involucre, straw-like and very showy (white) in some species. Flowers cup-shaped with copious nectar; each of the lateral flowers subtended by a bract and two bracteoles; petals white, with swollen homogeneous petal bulge; stamens shorter than petals; anthers ovate to oblong, thecae parallel, versatile; pollen tri- to hexacolporate, foveolate; ovary (half-)inferior, bilocular, uniovulate; stylodia two, fused, included. Fruit 1- or 2-seeded, dry, dehiscent; seed nutlike, with cup-shaped aril. 2n = ca. 44, 46 (up to 88 in *S. radiata*).

Eleven species, concentrated in south-western region of Western Cape, extending eastwards to Riversdale and northwards to Nieuwoudtville in Northern Cape. In dry, marshy or mountainous areas with moisture-bearing clouds, sea level to 1200 m. Flowering all year round (*S. radiata* (L.) Dahl), May–August (*S. brownii* Dümmer, *S. zeyheri* Sond.) or September–November.

5. ***Berzelia*** Brongn.[1] Fig. 16C, F, G

Berzelia Brongn., Ann. Sci. Nat. (Paris), 8: 370, t. 35 (1826). Pillans, J. S. African Bot. 13: 186 (1947), rev. p.p.; Claßen-Bockhoff et al., Taxon 60: 1147 (2011), classif.
Brunia Lam. (1785) p.p.

Ericoid shrubs or small trees; lignotuber absent or present. Leaves petiolate, closely set or imbricate, spreading or erect; amphistomatic (except *B. alopecuroides* (Thunb.) Sond.); stipules present. Inflorescences pincushion-like, closed heads with

[1] A proposal to conserve the name *Berzelia* Brongn. (1826, Bruniaceae) over *Berzelia* Mart. (1825, Amaranthaceae) will be submitted to the ICBN committee by E.G.H. Oliver.

acropetal or divergent flowering sequence, lateral on vegetative branches that continue to grow after flowering, often arranged in large racemose clusters with corymb-like shape; conspicuous red swellings of unknown function often present at base of the uppermost flowering branches. Flowers tetra- or pentamerous, small, lateral ones subtended by a bract and two bracteoles; petals pale white to yellowish (red in *B. stokoei*), with two narrowly associated ridges present on basal half of petal; stamens clearly exserted; thecae parallel, anthers versatile; pollen tricolporate; ovary half-inferior to almost entirely inferior, unilocular, uniovulate; style 1, exserted. Fruit 1-seeded, dry, indehiscent; *B. alopecuroides* with red, berry-like infructescences caused by fleshy calyx lobes.

Sixteen species, common to critically rare, concentrated in south-western region of Western Cape extending eastwards to Grahamstown in Eastern Cape. Some species are exported as 'Cape Greens'. Dry to moist soils, predominantly sandstone substrates but also on limestone (*B. cordifolia* Schltdl.), 0 to 2000 m. Flowering September–December.

6. ***Brunia*** Lam. Fig. 16A, H, K, L

Brunia Lam., Encycl. 1(2): 474 (1785), nom. cons.; Pillans, J. S. African Bot. 13: 178 (1947), rev. p.p.; Claßen-Bockhoff et al., Taxon 60: 1147 (2011), classif.
Nebelia Neck. ex Sweet (1830).
Raspalia Brongn. (1826), p.p.
Mniothamnea Nied. (1891).
Lonchostoma Wikstr. (1818).
Pseudobaeckea Nied. (1891), p.p.

Ericoid shrubs or undershrubs, erect or prostrate, rarely small trees; lignotubers absent or present. Leaves small, appressed or spreading, linear to ovate, often 3-angular in cross section; sessile or petiolate, amphi-, epi- or hypostomatic; stipules present or absent. Inflorescences oblong to globose heads, with or without terminal flower, rarely compound heads (*B. fragarioides*, *B. powrieae*) or solitary on top of leafy shoots (*B. bullata*, *B. callunoides*) in terminal and/or lateral position, often arranged in racemose or corymbose clusters, main axis growing on in some species; involucral bracts present in subgen. *Brunia* and showy in *B. paleacea*. Flowers small to minute (except subgen. *Lonchostoma*), lateral flowers subtended by a bract and two bracteoles; petals free or postgenitally fused with stamens into a tube (subgen. *Lonchostoma*); white to yellowish, rarely reddish; petal bulges usually forming a thick cushion with a vertical subdivision, in subgen. *Brunia* petals with two closely aligned longitudinal ridges in basal half, in subgen. *Lonchostoma* bulges reduced to a low ridge or absent; stamens usually 5, included or exserted and occasionally unequal in length (*B. dregeana* and subgen. *Brunia*); anthers rotund, oblong or subglobose, yellow, rarely red to black; thecae parallel, versatile; pollen polycolporate, mainly foveolate; ovary half-inferior to almost inferior, rarely superior (subgen. *Lonchostoma*); uni- or bilocular, usually uniovulate, but also biovulate and rarely up to 10 ovules per locule (subgen. *Lonchostoma*); styles 1 or 2, distinct. Fruit 1- or 2-seeded, dry, indehiscent or dehiscent.

37 species in five subgenera, concentrated in the south-western region of Western Cape, extending eastwards to the Uitenhage district in Eastern Cape; fynbos plants, often restricted to mountain slopes, some critically rare. *Brunia trigyna* (Schltr.) Class.-Bockh. & E.G.H. Oliv. is only known from a few sites in southern KwaZulu-Natal and Eastern Cape. Usually on moist soils, along riversides or even in swamps (*B. fragarioides* Willd.), if growing in drier areas then receiving humidity from moisture-bearing clouds at high altitudes; 0–2200 m. Flowering (September) October–December(–February), May–August (only *B. squalida* E. Mey. ex Sond.).

Acknowledgement I thank Gillan Scott, Constantia, for her assistance in writing the chapter 'Phytochemistry', and the editors of the International Journal of Plant Sciences and Opera Botanica Belgica for permission to reuse figures already published in these journals.

Selected Bibliography

Angiosperm Phylogeny Group (APG III). 2009. An update of the Angiosperm Phylogeny Group classification for the orders and families of flowering plants: APG III. Bot. J. Linn. Soc. 161: 105–121.
Behnke, H.D. 1981. Sieve-element characters. Nord. J. Bot. 1: 381–400.
Bremer, K., Backlund, A., Sennblad, B., Swenson, U., Andreasen, K., Hjertson, M., Lundberg, J., Backlund, M., Bremer, B. 2001. A phylogenetic analysis of 100+ genera and 50+ families of euasterids based on morphological and molecular data with notes on possible

higher level morphological synapomorphies. Plant Syst. Evol. 229: 137–169.
Bremer, B., Bremer, K., Heidari, N., Erixon, P., Olmstead, R.G., Anderberg, A.A., Källersjö, M., Barkhordarian, E. 2002. Phylogenetics of asterids based on 3 coding and 3 non-coding chloroplast DNA markers and the utility of non-coding DNA at higher taxonomic levels. Mol. Phylogenet. Evol. 24: 274–301.
Brongniart, A. 1826. Mémoire sur la famille des Bruniacées. Ann. Sc. Nat. 8: 357–389.
Brown, R. 1818. Appendix B. Containing characters of three new species of plants. In: Abel, C., Narrative of a journey in the interior of China. London: Longman.
Carlquist, S. 1978. Wood anatomy of Bruniaceae: correlations with ecology, phylogeny and organography. Aliso 9: 323–364.
Carlquist, S. 1991. Leaf anatomy of Bruniaceae: ecological, systematic and phylogenetic aspects. Bot. J. Linn. Soc. 107: 1–34.
Claßen-Bockhoff, R. 2000. Inflorescences in Bruniaceae. With general comments on inflorescences in woody plants. Opera Bot. Belgica12: 5–310.
Claßen-Bockhoff, R., Oliver, E.G.H., Hall, A.V., Quint, M. 2011. New classification of the Bruniaceae based on molecular and morphological data. Taxon 60: 1138–1155.
Cronquist, A. 1981. An integrated system of classification of flowering plants. New York: Columbia Univ. Press.
Cronquist, A. 1988. The evolution and classification of flowering plants, ed. 2. New York: New York Botanical Garden.
Dahlgren, R., Van Wyk, A.E. 1988. Structures and relationships of families endemic to or centered in South Africa. 5. Bruniaceae A.P. de Candolle. In: Goldblatt, P. Lowry, P.P. (eds.) Modern Systematic Studies in African Botany. Monogr. Syst. Bot. Miss. Bot. Gard. 25: 1–94.
Dümmer, R.A. 1912. An enumeration of the Bruniaceae. J. Bot. 50, Suppl. 2: 1–37.
Endress, P.K., Stumpf, S. 1991. The diversity of stamen structures in 'Lower' Rosidae (Rosales, Fabales, Proteales, Sapindales). Bot. J. Linn. Soc. 107: 217–293.
Fehrenbach, S., Barthlott, W. 1988. Mikromorphologie der Epicuticular-Wachse der Rosales s. l. und deren systematische Gliederung. Bot. Jahrb. Syst. 109: 407–428.
Goldblatt, P. 1981. Chromosome cytology of Bruniaceae. Ann. Miss. Bot. Gard. 68: 546–550.
Hall, A.V. 1987. Evidence of a Cretaceous alliance for the Bruniaceae. S. African J. Sci. 83: 58–59.
Hall, A.V. 1988. Systematic palynology of the Bruniaceae. Bot. J. Linn. Soc. 96: 285–296.
Hall, A.V., Oliver, E.G.H., Claßen-Bockhoff, R. 2010. Bruniaceae: new species of *Thamnea* and *Brunia* from western Cape, South Africa. Bothalia 40: 96–101.
Hallier, H. 1912. L'origine et le système phylétique des angiospermes exposés á l'aide de leur arbre généalogique. Arch. Néerl. Sci. Exact. Nat., sér. 3, 1: 146–234.
Hilton-Taylor, C. 1996. Red data list of South African plants. Strelitzia 4.
Hutchinson, J. 1969. Evolution and phylogeny of flowering plants. Dicotyledons: Facts and theory. London: Academic Press.
Jay, M. 1968. Distribution des flavonoides chez les Bruniacées. Taxon 17: 484–488.
Jensen, S.R., Nielsen, B.J., Dahlgren, R. 1975. Iridoid compounds, their occurrence and systematic importance in the angiosperms. Bot. Not. 128: 148–180.
Lange, J.H. de, Boucher, C. 1990. Autecological studies on *Audouinia capitata* (Bruniaceae). I. Plant-derived smoke as a seed germination cue. S. African J. Bot. 56: 700–703.
Lange, J.H. de, Boucher, C. 1993a. Autecological studies on *Audouinia capitata* (Bruniaceae). 4. Seed production, seed banks and seedling recruitment. S. African J. Bot. 59: 145–155.
Lange, J.H. de, Boucher, C. 1993b. Autecological studies on *Audouinia capitata* (Bruniaceae). 8. Role of fire in regeneration. S. African J. Bot. 59: 188–202.
Lange, J.H. de, Boucher, C. 1993c. Stimulation of in vitro pollen germination in *Audouinia capitata* by the simultaneous use of several carbohydrates. S. African J. Bot. 59: 270–272.
Lange, J.H. de, Boucher, C., Van der Walt, J.J.A. 1993a. Autecological studies on *Audouinia capitata* (Bruniaceae). 3. Pollination biology. S. African J. Bot. 59: 135–144.
Lange, J.H. de, Van der Walt, J.J.A., Boucher, C. 1993b. Autecological studies on *Audouinia capitata* (Bruniaceae). 5. Seed development, abortion and pre-emergent reproductive success. S. African J. Bot. 59: 156–167.
Lange, J.H. de, Van der Walt, J.J.A., Boucher, C. 1993c. Autecological studies on *Audouinia capitata* (Bruniaceae). 6. Nutritional aspects of the developing ovule. S. African J. Bot. 59: 168–177.
Lange, J.H. de, Thompson, G.R., Swart, C.R. 1993d. Autecological studies on *Audouinia capitata* (Bruniaceae). 7. Nutritional aspects in the mature seed. S. African J. Bot. 59: 178–187.
Leinfellner, W. 1964. Über die falsche Sympetalie bei *Lonchostoma* und anderen Gattungen der Bruniaceen. Österr. Bot. Z. 111: 345–353.
Leinfellner, W. 1965. Sind die Kronblätter der Bruniaceen peltat gebaut? Österr. Bot. Z. 111: 500–526.
Linder, H.P. 2003. The radiation of the Cape flora, southern Africa. Biol. Rev. 78: 597–638.
Moll, E.J., Gubb, A.A. 1981. Aspects of the ecology of *Staavia dodii* in the south western Cape of South Africa. In: Synge, H. (ed.) The biological aspects of rare plant conservation, pp. 331–342. New York: Wiley & Sons.
Nel, J.P. 1986. Ontekologiese studie van *Berzelia lanuginosa*, *Brunia nodiflora* en *Nebelia paleacea* (Bruniaceae). B.Sc. (Hons) (unpublished), University of Stellenbosch.
Niedenzu, F. 1891. Bruniaceae. In: Engler, A., Prantl, K. (eds.) Die natürlichen Pflanzenfamilien III, 2a: 131–136. Leipzig: Engelmann.
Niedenzu, F., Harms, H. 1930. Bruniaceae. In: Engler, A., Prantl, K. (eds.) Die natürlichen Pflanzenfamilien, ed. 2, 18a: 288–303. Leipzig: Engelmann.
Oliver, E.G.H., Oliver, I.M. 1999. Bruniaceae. A new species of *Linconia* from Western Cape. Bothalia 29: 256–258.
Pillans, N.S. 1947. A revision of Bruniaceae. J. S. African Bot. 13: 121–206.

Powrie, E. 1969a. Types of Bruniaceae in the Thunberg Herbarium. J. S. African Bot. 35: 327–339.
Powrie, E. 1969b: A new species of *Tittmannia* (Bruniaceae). J. S. African Bot. 35: 363–366.
Quint, M. 2004. Evolution of Bruniaceae: evidence from molecular and morphological studies. Dissertation, University of Mainz, Germany (Cuvillier, Göttingen).
Quint, M., Classen-Bockhoff, R. 2006a. Phylogeny of Bruniaceae based on *mat*K and ITS sequence data. Int. J. Plant Sci. 167: 135–146.
Quint, M., Claßen-Bockhoff, R. 2006b. Floral ontogeny, petal diversity and nectary uniformity in Bruniaceae. Bot. J. Linn. Soc. 150: 459–477.
Quint, M., Claßen-Bockhoff, R. 2008. Ancient or recent? Insights into the temporal evolution of the Bruniaceae. Org. Divers. Evol. 8: 293–304.
Saxton, W.T. 1910. The ovule of the Bruniaceae. Trans. Roy. Soc. S. Africa 2: 27–31.
Schardt, L. 2010. Evolution of South African Bruniaceae: evidence from ecological and geographical investigations. Dissertation, University of Mainz, Germany.
Scott, G. 1999. A chemosystematic and cladistic study of the Southern African endemic family Bruniaceae. Dissertation, University of Cape Town, South Africa.
Sonder, O.W. 1862. Bruniaceae. In: Harvey, W.H., Sonder, O.W. (eds.) Flora Capensis 2: 309–324. London: Reeve.
Strid, A. 1968. A new species of *Lonchostoma* (Bruniaceae). Bot. Not. 121: 312–316.
Takhtajan, A. 1980. Outline of the classification of flowering plants (Magnoliophyta). Bot. Rev. (Lancaster) 46: 225–359.
Thorne, R.F. 1976. A phylogenetic classification of the Angiospermae. Evol. Biol. 9: 36–106.
Thorne, R.F. 1983. Proposed new realignments in the Angiosperma. Nordic J. Bot. 3: 85–117.
Van Tieghem, T. 1897. Sur les caractères et les affinités des Grubbiacées. J. Bot. (Morot) 11: 127–138.
Villiers, S.E. de, Cadman, A. 1997. The palynology of tertiary sediments from a palaeochannel in Namaqualand, South Africa. Palaeontol. Afr. 34: 69–99.
Winkworth, R.C., Lundberg, J., Donoghue, M.J. 2008. Towards a resolution of campanulid phylogeny, with special reference to the placement of Dipsacales. Taxon 57: 53–65.
Wolff, M.G. 1990. Success with *Staavia dodii*. Veld & Flora 76: 48–49.
Wright, M.G., Visser, D., Lange, J.H. de 1991. Autecological studies on *Audouinia capitata* (Bruniaceae). 2. Insects as pollen vectors. S. African J. Bot. 57: 260–263.

Caprifoliaceae (with *Zabelia* incert. sed.)

Caprifoliaceae Juss., Gen. Pl. 210 (1789), nom. cons.
Diervillaceae (Raf.) Pyck, Taxon 47: 658 (1998).
Linnaeaceae (Raf.) Backlund, Taxon 47: 658 (1998).

U. Hofmann and V. Bittrich

Tall to medium-sized shrubs or lianas (species of *Lonicera*), rarely dwarf shrubs (*Linnaea*) or perennial herbs (*Triosteum*), winter buds with several scales. Leaves mostly deciduous, opposite, rarely whorled, blades sometimes sinuous to deeply lobed, margin entire or rarely serrate or dentate; normal stipules absent, rarely the bases of opposite leaves form intrapetiolar stipule-like structures of often varying form on subsequent nodes (*Leycesteria* spp., *Lonicera* spp.). Inflorescence a thyrse, mostly without terminal flower, or simplified to a spike, raceme or head; all or only the lower cymes often in the axils of foliage leaves. Flowers showy, actinomorphic to ± zygomorphic, haplostemonous, sometimes with supernumerary bracts at the base of the ovary; calyx normally 5-merous, regular or rarely bilabiate, with a short tube and 5 equal or rarely unequal short to long, persistent or deciduous teeth; corolla normally 5-merous, gamopetalous, funnel-shaped or campanulate, regular or normally more or less bilabiate, tube normally longer than the 5 imbricate lobes, sometimes basally gibbous or with a short abaxial spur; nectary usually 1 in the corolla tube at the base of the abaxial petal, or rarely 3–5; stamens 5 of ± equal length or 4, didynamous; anthers introrse, tetrasporangiate, inserted in the corolla tube; style simple, elongate, slender, stigma capitate, discoid or lobed; ovary inferior, of 2–5, very rarely 7–8 (–10) carpels, all or only 1–3 fertile (sterile locules often with abortive ovules), more or less constricted beneath the calyx, placentation usually axile, rarely few ovules inserted on free septa in the distal part of the ovary. Fruit a many-seeded septicidal capsule or berry, a drupe with 2–4 pyrenes or a 1–2-seeded achene, sometimes with accrescent bracts; endosperm copious.

Thirteen genera (excl. *Zabelia*) with ca. 230 species, distributed mostly in north-temperate zones, occasionally in the tropics.

Vegetative Morphology and Anatomy. Most Caprifoliaceae are shrubs, some up to 5 m high, or woody vines (species of *Lonicera*); only few are dwarf shrubs (*Linnaea*) or perennial herbs (*Triosteum*). Vegetative propagation is rare: by suckers, e.g. in *Symphoricarpos*, *Lonicera korolkovii*, *Diervilla*, or by runners, e.g. in *Lonicera japonica*. Ascending serial accessory buds occur in several genera. Annual shoots mostly are terminated by vegetative buds, terminal flowers occur only in Diervilleae, mostly in *Diervilla* and occasionally in *Weigela* (often deformed). The terminal vegetative buds of non-flowering shoots can decay, resulting in branching of the main branches. Old branches can be replaced by basal innovation branches with upright growth, e. g. in *Symphoricarpos*. The bark of older shoots consists of several layers and detaches in stripes. The first periderm arises inside the pericyclic and wide-celled sclerenchyma (directly outside in *Heptacodium*). Alternating layers of thin-walled and thick-walled lignified cells are present in the phellem. In *Weigela* further periderms are initiated in the phloem. Calcium oxalate druses are present in the phloem. The wood is mostly diffuse-porous with inconspicuous growth rings, but distinct growth rings and ring-porous wood occur in *Symphoricarpos* and *Lonicera involucrata* (Jacobs et al. 2011). Vessel element length is 325–700 μm, exceptionally up to 1000 μm (*Leycesteria* spp., *Weigela coraeensis*), the diameter is small (less than 40 μm in diameter). Vessel are normally solitary or rarely in groups of two. Scalariform perforations (with 10 to 20 bars, *Weigela*

Hofmann Ursula (deceased).

J.W. Kadereit and V. Bittrich (eds.), *Flowering Plants. Eudicots*, The Families and Genera of Vascular Plants 14,
DOI 10.1007/978-3-319-28534-4_7, © Springer International Publishing Switzerland 2016

with up to 40 bars) predominate but, in *Symphoricarpos*, *Lonicera* and *Heptacodium*, perforations are mostly simple, in *Lonicera* they are mixed with perforation plates with only two bars. The main part of the wood consists of thick-walled fibre-like elements with several bordered pits. The pit cavity is conspicuous; the apertures are crossed slits of the same length as the cavity diameter. In some genera, tracheary members and fibres have helical thickenings (e.g. *Abelia*, *Dipelta ventricosa*, *Kolkwitzia*, *Leycesteria*, *Lonicera* and *Symphoricarpos*), but these are absent in *Diervilla* and *Weigela* (Jacobs et al. 2011). Axial parenchyma is lacking (*Leycesteria formosa*, *Lonicera* spp.) or sparsely diffuse in, for example, *Heptacodium*, *Abelia graebneriana*, *Diervilla*, *Lonicera thibetica* and *Weigela*. The rays consist of large uniseriate wings of erect cells with one or few small groups of square to slightly procumbent cells in two to three rows in the centre. The uniseriate parts with erect cells predominate. The pith may persist or perish. The resulting cavity is rather wide in most species of *Leycesteria*.

The normally petiolate and mostly deciduous leaves are opposite (rarely in whorls of three) and decussate, but on lateral branches arranged in two rows by torsion of the internodes. Nodal anatomy is reported as 3:3, but it is unclear which genera have been studied. In *Triosteum* and some species of *Leycesteria*, *Symphoricarpos* and *Lonicera* subg. *Caprifolium*, the blade of the sessile paired leaves are basally connate, forming a perfoliate disc. Normal stipules are lacking. Interpetiolar stipules are mentioned frequently in the older literature, however, which has led to problems in delimitation of the family from Rubiaceae.

Cases of apparent stipules were studied in detail by Weberling (1957) in species of *Leycesteria* and *Lonicera* and some individuals of *Symphoricarpos*. He discovered that the fused bases ("Unterblatt" regions) of opposite leaves can produce small leaf-like outgrowths. Different from normal stipules, however, these pseudostipules show neither an early appearance nor prolepsis. In *Triosteum erythrocarpum*, the bases of opposite leaves are broadly fused and produce lobes in the interpetiolar region, but these can hardly be confused with stipules. The blades are simple and entire. Only in species of Caprifolieae can the lower leaves on fast growing twigs be lobed or even pinnatifid (e.g. *Triosteum*). The leaf margin is entire or finely dentate. The venation is camptodromous and brochidodromous. Free vein ends occur.

The leaves are strictly bifacial with one layer of palisade parenchyma in deciduous and up to three layers in evergreen species. The palisade cells are short to isodiametric. The spongy tissue consists of star cells with ± long arms parallel to the surface. Calcium oxalate druses are mostly found in the spongy tissue, never in the epidermis. The anomocytic stomata are confined to the abaxial leaf surface. Trichomes can be present, more often on the lower leaf surface. The simple and often lignified hairs are all one-celled and not separated by a transversal wall from the epidermal cell. Glandular trichomes consist of a short, few-celled stalk and a head of some storeys of glandular cells. Leaf epicuticular wax crystals were studied by Theisen and Barthlott (1996). Wax platelets are typical for *Abelia* and *Linnaea* of tribe Linnaeeae, *Weigela* of Diervilleae, and a few species of *Lonicera* of Caprifolieae. No wax crystals were found on leaves of *Diervilla* (Diervilleae), *Dipelta* and *Kolkwitzia* (Linnaeeae), and *Heptacodium* and *Triosteum* of tribe Caprifolieae, while wax tubules characterized all species studied of *Leycesteria* and *Symphoricarpos*, and most species studied of *Lonicera*, also of tribe Caprifolieae.

INFLORESCENCE STRUCTURE (Figs. 20, 21). Basically the inflorescences are thyrses or can be derived from these by reduction. Only *Diervilla* mostly has a terminal flower (occasionally present in *Weigela*). The inflorescences of *Leycesteria* and *Heptacodium* end in a bud which never develops further; in the other genera the abortive shoot end is inconspicuous. The thyrses consist of ca. five (more in *Symphoricarpos*) to two pairs of cymes, in a few species (*Linnaea*, *Heptacodium*) often of only one. The cymes often have three flowers; they have one basal dichasial ramification, further ramification often is irregularly monochasial. Only few species have more than three orders of flowers. In several genera the cymes are reduced to the terminal flower of first order (e.g. *Symphoricarpos*, *Linnaea*, species of *Weigela* and *Abelia*). In *Lonicera* subg. *Lonicera* the terminal flower of the cymes is missing (Fig. 21A). Here the thyrses consist of several paired flowers

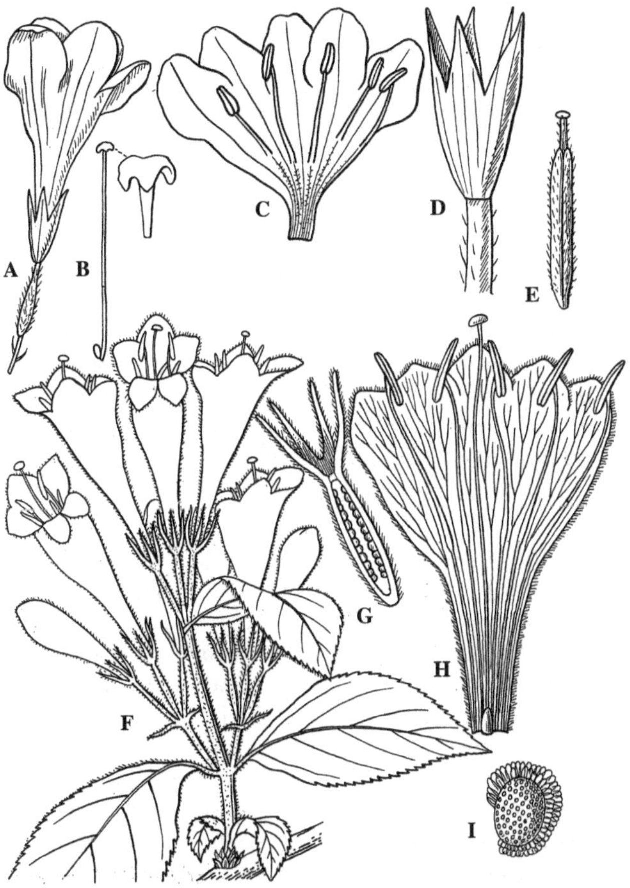

Fig. 20. Caprifoliaceae. *Weigela florida* (**A–E**). **A** Flower in lateral view. **B** Style, stigma and basal nectary. **C** Corolla cut open with stamens. **D** Calyx. **E** Fruit. *Weigela floribunda* (**F–I**). **F** Flowering branch. **G** Calyx and ovary in vertical section. **H** Corolla cut open with stamens and basal nectary. **I** Seed. (From Schneider 1912, fig. 469, modified)

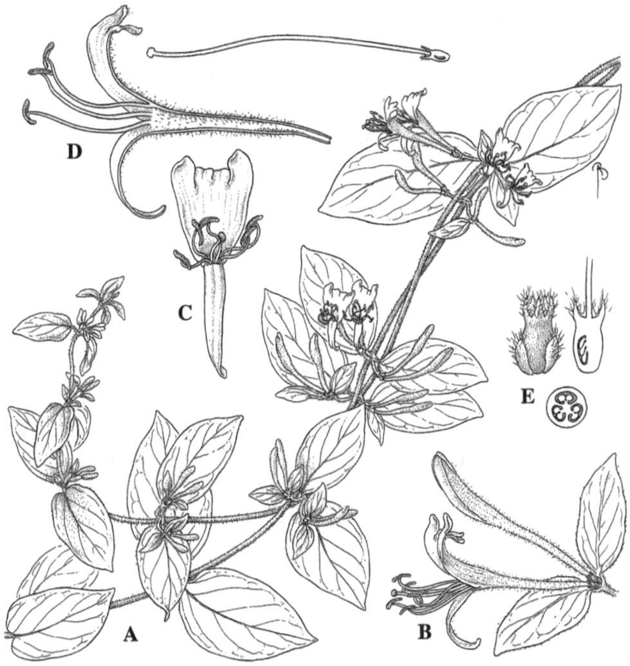

Fig. 21. Caprifoliaceae. *Lonicera japonica*. **A** Flowering branch. **B** Detail of the inflorescence with a bud and an open flower. **C** Open flower in apical view. **D** Corolla in vertical section and gynoecium. **E** Entire fruit in lateral view, in vertical and in transverse section. (Acevedo-Rodríguez 2005; with kind permission of artist Bobbi Angell)

of second order. The internode between bracteoles and flower is never elongated (Troll and Weberling 1966). When flowers appear to be stalked above the bracteoles, there are two pairs of additional bracteoles at the base of the flowers (e.g. in *Dipelta*). Such additional bracteoles were only found in all genera of Linnaeeae. The additional bracteoles are sterile except in *Abelia aschersoniana* and *Kolkwitzia*, where one flower is formed in one axil of the uppermost pair of bracteoles. In *Kolkwitzia* the additional bracteoles are not always strictly paired; the upper two are always in contact with the base of the ovary or mostly connate with it. The bracts at least of the lower cymes are often foliose. Lateral thyrses can be formed in leaf axils below the terminal thyrse. On the background of a new phylogenetic analysis, the inflorescence evolution in Linnaeeae was recently investigated in detail by Landrein and Prenner (2013), leading to some different interpretations compared to Troll and Weberling (1966).

Flower Morphology and Anatomy. The normally pentamerous flowers are mostly large and showy, they are rarely actinomorphic (*Leycesteria*, few species of *Lonicera* and *Symphoricarpos*), more often slightly (corolla shape, nectary) or distinctly zygomorphic as in the Linnaeeae with an androecium of only four instead of five stamens. The green calyx is inconspicuous, actinomorphic or bilabiate, with a basal tube and free sepal teeth which rarely are shorter than the tube (e.g. *Symphoricarpos* species). The white, pinkish or red, rarely yellow corolla is campanulate or funnel-shaped and often constricted beneath into a narrow tube. The five imbricate corolla lobes are shorter than or equalling the tube. The partition into lips (two lobes forming the upper lip in Diervilleae and Linnaeeae) is usually rather weak, but evident in most species of *Lonicera*, where four lobes form the upper lip. In *Triosteum*, the partition into four upper and one lower lobe is less clear. The zygomorphy of the campanulate, weakly bilabiate flowers of Diervilleae and some Linnaeeae is reinforced by a sap mark (mostly yellow or orange, often with reticulate venation) on the abaxial part of the corolla tube. The tube is basally gibbous in some species of, for example, *Triosteum* and *Lonicera*, or rarely has a short abaxial spur (*Lonicera*) containing the generally single nectary of the flower. The strongly developed hump-like or club-like nectary of *Diervilla* and *Weigela* apparently is situated on the ovary (Fig. 20B) and thus was confused with a gynoecial gland (Behrens 1879; Yang et al. 2011), but in longitudinal sections the connection of the nectary to the corolla tube is clearly visible (Wagenitz and Laing 1984). Five nectaries are found only in clearly actinomorphic flowers (*Leycesteria*) and, in the Mexican genus *Vesalea*, the nectary is located in 3 or 4 grooves of the corolla tube. In the nectaries of Caprifoliaceae the inner epidermis of the corolla tube consists of papillose cells secreting nectar through the cuticle (Wagenitz and Laing 1984). Beneath the nectary papillae, the corolla tube is thickened by division of the mesophyll cells, or a cushion protruding into the corolla tube is formed—e.g. in *Weigela* and *Abelia*. There often are additional phloem strands or phloem and xylem strands (connected with the corolla bundles) supplying the nectary tissue when large quantities of nectar are produced. The stamens alternate with the corolla lobes, and the filaments are inserted in the lower part of the corolla tube. There are no anastomoses between the vascular bundles of the filaments and the five corolla bundles. The anthers are introrse, dithecous with four pollen sacs, and open by two slits. The androecium consists of 5 ± similar stamens in Caprifolieae and Diervilleae and 4 (by complete loss of the adaxial stamen, no staminodes have been reported) didynamous stamens in Linnaeeae.

The gynoecium is composed of 2 to 5 carpels, rarely (in *Leycesteria* spp.) 7 or 8 (or even up to ten). The style is long and slender, the small stigma is capitate, discoid or lobed. The ovary is completely inferior and more or less constricted beneath the calyx. Anatomy of the ovary was studied in detail by Wilkinson (1948a, 1948b, 1949). In Diervilleae and *Leycesteria* and *Lonicera* of Caprifolieae, all carpels are fertile. In *Triosteum*, only 3 of 4, and in *Dipelta* 2 of 4 carpels are fertile. In the other genera only 1 of 3 carpels is fertile (see Fig. 22B). In ovaries with only fertile carpels, many seeds develop; the ovules are inserted in two axile rows mainly in the symplicate region but also in the upper synascidiate region of the ovary. In the upper part the septa may remain separated with some ovules inserted in a parietal position (Wilkinson 1948a). When there are fertile and sterile carpels in the ovary (*Abelia, Diabelia, Dipelta, Heptacodium, Kolkwitzia, Linnaea, Symphoricarpos, Triosteum, Vesalea*), the fertile carpels are always uniovulate with a hanging, axile ovule at the top of the synascidiate region. In the sterile carpels, there are mostly some sterile axile ovules in two rows. In these ovaries, the septa also mostly become free distally, but no sterile ovules are attached to them. Apically, a sterile neck is present in the ovary of several genera, probably due to the abortion of ovules in this region and the related closure of the locules (Wilkinson 1948a). The genera with ovaries containing fertile and sterile locules do not form a monophyletic group, and genera with all

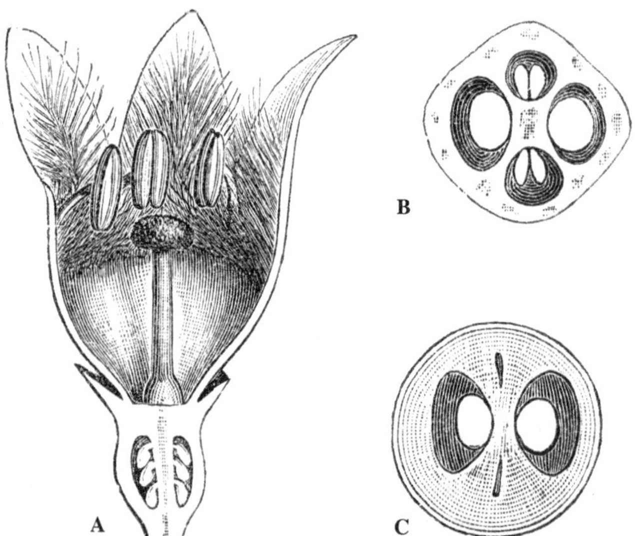

Fig. 22. Caprifoliaceae. *Symphoricarpos racemosus*. **A** Flower in vertical section. **B** Ovary in transverse section, the median locules with several sterile ovules, the lateral ones uni-ovulate and fertile. **C** Fruit in transverse section. (From Fritsch 1892, fig. 53)

locules fertile, like *Leycesteria* or *Lonicera*, do not appear in a basal position in the phylogeny of the family (e.g. Jacobs et al. 2010). The wall of the inferior ovary normally has only corolla bundles and bundles leading to the calyx and stamens, the latter being fused in the ovary walls; there are no separate dorsal carpel bundles, but lateral carpellary supply occurs in *Triosteum* (Wilkinson 1949). A central bundle is frequently present.

Embryology. The anthers are tetrasporangiate and introrse, normally with a 4-layered wall when young. In *Leycesteria formosa* and *Weigela*, the cells of the epidermis disconnect by enlargement of the fibrous endothecium during maturation of the anther. In other genera this character is only weakly expressed. The endothecium is often not strictly one-layered. In *Weigela* there are 2 to more layers of fibrous cells. An amoeboid tapetum has been reported for *Leycesteria*, *Lonicera*, *Triosteum*, *Symphoricarpos* and *Kolkwitzia*, all members of tribe Caprifolieae (Weberling and Hildebrand 1982). Young tapetum cells can have two nuclei. The ovules are anatropous, unitegmic and tenuinucellar. The archesporial cell develops directly into the megaspore mother cell. The chalazal megaspore from a linear tetrad develops into a Polygonum-type embryo sac. The embryo is small and linear, endosperm formation is cellular (Davis 1966; Johri et al. 1992).

Pollen Morphology. Pollen is ± spheroidal to oblate, medium to large. The pollen of *Diervilla* and *Weigela* was described as atectate tripororate (Böhnke-Gütlein and Weberling 1981), while Donoghue (1985) only speaks of an absence of columellae. Mostly the pollen is tectate with columellae of various sizes, tricolporate, more rarely tricolpate with short colpi (e.g. *Leycesteria formosa*, *Lonicera myrtillus*, *L. obovata*; e.g. Perveen and Qaiser 2007; Xu et al. 2011). Other aperture numbers (2–5) are occasionally found. A cavity below the ectoaperture (fastigium) was observed in the pollen of *Diervilla*, *Weigela*, *Leycesteria*, *Lonicera*, *Symphoricarpos*, *Abelia* and *Triosteum* (Jacobs at al. 2011). The pollen grains are rarely smooth and then sometimes fossulate (*Lonicera myrtillus*, *L. thibetica*, *Symphoricarpos albus*, *Triosteum aurantiacum*, Donoghue 1985; Perveen and Qaiser 2007), or granulate (*Heptacodium* sp.), or most frequently echinate. Tricolporate grains with an echinate exine were found in ten genera investigated by SEM by Chia-Chi and Chao-Xing (1988). A scabrate exine was observed only in *Lonicera* sect. *Isoxylosteum*, while species of three other sections of *Lonicera* showed the typical echinate exine. Sometimes the surface shows a combination of spinules and small perforations (e.g. *Leycesteria formosa*, *Triosteum himalayanum*) or a combination of spinules, gemmae and/or granules in Diervilleae (Jacobs et al. 2011). As regards the common echinate pollen surface, the spinules (and gemmae and granules) of *Diervilla* and *Weigela*, different from other genera, would not be supratectate structures if their pollen is atectate as described by Böhnke-Gütlein and Weberling (1981). Pollen grains are three-celled at anthesis.

Karyology. Chromosome counts of about a quarter of caprifoliaceous species are available (IPCN 2010). The base number is $x = 9$ in Caprifolieae and Diervilleae, and $x = 8$ in most Linnaeeae. *Heptacodium* is unusual with $2n = 28$. Most counts were made in *Lonicera* but no correlation between chromosome number and infrageneric classification was found. Diploid chromosome numbers dominate in *Lonicera* and other Caprifolieae. All Diervilleae are tetraploid or octoploid except *Weigela florida* with both ploidy levels. In Linnaeeae only few species have been examined, and all seem to be tetraploid. Benko-Iseppon and Morawetz (2000) made detailed cytological studies of the Dipsacales and found various differences between Adoxaceae s.l. (incl. *Sambucus* and *Viburnum*) on the one hand and Caprifoliaceae, Morinaceae and Valerianaceae on the other. The latter families lack cold induced chromosome regions (CIRs), have smaller chromosomes and show a condensing behaviour which is different from that in Adoxaceae.

Pollination. All Caprifoliaceae are insect-pollinated except for few hummingbird-pollinated red-flowered honeysuckles (e.g. *Lonicera sempervirens*, *L. ciliosa*) in the USA. In *Lonicera*, some white-flowered species—e.g. the European *L. caprifolium* and *L. periclymenum*, or the Asian *L. hildebrandiana* and *L. japonica*, with a long, narrow corolla tube and spreading lips, and a strong odour when open in the evening—are pollinated by hawkmoths. Here, stamens and style are exserted from the corolla tube. The white flowers turn yellow after pollination. Mostly bilabiate species of *Lonicera* with smaller flowers and shorter tubes and the few species with actinomorphic, hanging flowers may be visited by bees and bumblebees. The large flowers of Diervilleae, *Dipelta*, *Kolkwitzia* and *Abelia* of Linnaeeae, often with yellow to orange sap marks, are probably pollinated by large bumblebees. The inconspicuous flowers of some species of *Symphoricarpos* are frequently visited by honeybees in European gardens.

Fruit, Seeds and Dispersal. The Diervilleae are unique in the Dipsacales in having long, slender septicidal capsules with two valves, and numerous very small and light seeds adapted to wind dispersal. Seed anatomy was studied in detail by Jacobs et al. (2009). The one integument has usually a more or less strongly sclerified exotesta, a parenchymatous mesotesta, which becomes more or less compressed and contains a vascular bundle or raphe, and one or two layers of endotesta, the inner usually weakly sclerified. In Diervilleae the exotesta lacks sclerified outer tangential cell walls, the sclerification of the radial and inner tangential walls results in a U-shaped pattern. In *Leycesteria* and in most *Lonicera* spp. a similar U-shaped pattern was observed, but the outer tangential wall is also moderately sclerified. Crystals, especially druses, occur in the seed coat of most species of Caprifolieae studied so far. They are lacking, for example, in *Leycesteria*, *Triosteum hirsutum* and *Lonicera involucrata*. In *Weigela* they mostly have narrow wings. Caprifolieae, except *Heptacodium*, have red, blue or dark berries or pink, blue or white drupes and probably are dispersed by birds. Donoghue et al. (2003) noted parallel evolution from red to bright white fruits in *Triosteum* and *Symphoricarpos*. Interestingly, the seeds of these drupes also have sclerified exotestal walls, although the sclerified endocarp acts as a protective layer (Jacobs et al. 2009). The seeds of berries are smaller than the pyrenes of the drupes. Seeds are smooth or slightly granulate with fleshy albumen and a small embryo. *Heptacodium* (probably basal in Caprifolieae) and the Linnaeeae have 1-seeded achenes (2-seeded in *Dipelta*). Similar to Morinaceae, Dipsacaceae and Valerianaceae with similar fruits, where the pericarp serves as a protective layer, the seeds have compressed, parenchymatous seed coats (Jacobs et al. 2009). In *Dipelta*, the upper additional bracteoles enlarge after flowering and form wings around the achene for wind dispersal. The bristly fruits of *Kolkwitzia* decay on the plant during European winters. Dispersal of the 1-seeded achenes of other Linnaeeae and *Heptacodium* is unknown. Jacobs et al. (2011) noted a surprising similarity in fruit and seed morphology of *Heptacodium* (basal in Caprifolieae), the Linnaeeae and *Zabelia* (incert. sed.). Seeds of the family contain no starch, but protein and oils.

Phytochemistry. Most data are summarized in Hegnauer (1989). There exist no family-specific chemical compounds. Accumulation of phenolic heterosides (especially β-glycosides) and non-phenolic heterosides (e.g. saponins) has been recorded; saponins might cause the toxicity of *Lonicera* fruits, in immature fruits also condensed tannins and alkaloids were found. Iridoids and secoiridoids are common in the family. Alkaloids were detected in *Lonicera* and *Symphoricarpos*, coumarins in *Diervilla*, *Weigela*, *Symphoricarpos* and *Lonicera*. Flavones and flavonols were found in genera of Caprifolieae and Linnaeeae.

Affinities, Subdivision and Relationships Within the Family. Caprifoliaceae traditionally have been assigned to Dipsacales. There exists morphological and anatomical (cf. Wilkinson 1948b, 1949) and molecular evidence (e.g. Jacobs et al. 2010) that *Sambucus* and *Viburnum*, generally placed in the family, should be placed elsewhere (see Adoxaceae). The same applies to the genera of Alseuosmiaceae (Asterales) and Carlemanniaceae (Lamiales), earlier often included in Caprifoliaceae s.l. The remaining Caprifoliaceae had already been divided into Caprifolieae (as Lonicereae) and Linnaeeae by Fritsch (1892), who grouped *Triosteum* with *Viburnum*. Rehder (1940, 1949) finally established the subdivision of the family into Caprifolieae (as Lonicereae), Diervilleae and Linnaeeae but disregarded *Triosteum* and *Heptacodium*. Both authors placed *Symphoricarpos* in Linnaeeae, probably because of the structure of its ovary, but it is better considered a member of Caprifolieae (Judd et al.

1994; Jacobs et al. 2010). The morphological subdivision of Caprifoliaceae into three tribes as accepted here is basically supported by the molecular findings of Backlund and Pyck (1998), Zhang et al. (2003), Theis et al. (2008) and Winkworth et al. (2008). The relationship between these tribes and the position of *Heptacodium* are still not fully understood, although it now seems probable that *Heptacodium* belongs into Caprifolieae (Jacobs et al. 2009, 2010). Some morphological characters, however, would rather suggest a position of *Heptacodium* closer to the Linnaeeae, Morinaceae, Dipsacaceae and Valerianaceae (e.g. Jacobs et al. 2011). There also exists evidence for the hypothesis that *Heptacodium* is the result of a hybridization of a member of Caprifolieae and a member of Linnaeeae (see Landrein et al. 2012). According to various recent studies, the phylogeny of the Dipsacales basically is Adoxaceae [Diervilleae [Caprifolieae [Linnaeeae [Morinaceae [Dipsacaceae + Valerianaceae]]]]]. This implies, however, that Caprifoliaceae might better be combined with Morinaceae, Dipsacaceae and Valerianaceae into one family as already suggested by Judd et al. (1994) and adopted by APG III (2009). Alternatively, Caprifolieae, Diervilleae and Linnaeeae could be raised to family rank to have monophyletic taxa (Backlund and Pyck 1998) as was recently done in the Flora of China. A broader taxon sampling for phylogenetic studies would be useful before a final decision is made; especially the relationships of *Heptacodium* and *Zabelia* still need to be better resolved.

Distribution and Habitats. Caprifoliaceae have a mainly northern temperate distribution with centres of diversity in E Asia and North America. Using parametric methods for a historical biogeographical reconstruction of tribe Caprifolieae, Smith (2009) analyzed divergence times and possible dispersal events. The results indicated an origin of the tribe in Asia, diversification between 36 and 51 million years ago, and several migrations throughout the Cainozoic into Europe and especially North America through Beringia. Similar results were reported by Bell and Donoghue (2005) using a different method of biogeographical analysis.

Typical habitats for members of the family are forests and forest margins, but also scrub and mountain slopes.

Economic Importance. On account of their hardiness, showy flowers and sometimes fruits (*Symphoricarpos*), several species are important ornamentals. The flowering season is rather long, starting with *Lonicera standishii* or *L. fragrantissima* in February and March and ending with *Heptacodium* in late August. The twining species of *Lonicera* species are used for bowers. Some ornamentals are naturalized in N America (e.g. the weedy *Lonicera japonica*) and/or Europe (e.g. *Leycesteria formosa* in W Europe).

Key to the Genera

1. Corolla lobes and stamens of the same number, normally 5, occasionally 4 — 2
– Corolla lobes 5, stamens 4 (**Linnaeeae** and ***Zabelia***) — 8
2. Fruit a berry, drupe or achene; ovary of 5 (rarely more), 4 or 3 (rarely 2) carpels, subglobose or ovoid at flowering time (**Caprifolieae**) — 3
– Fruit a woody capsule; ovary of two fertile carpels, thin, terete and long at flowering time (**Diervilleae**) — 7
3. Ovary of 5 (rarely more) or 3 (rarely 2) equal, fertile, many-seeded carpels; fruit a many-seeded berry — 4
– Ovary of 4 or 3 carpels, one or two sterile, the sterile ones compressed, with several abortive, the fertile ones with 1 fertile ovule; fruit a drupe or 1-seeded achene — 5
4. Flowers in 3 (rarely more)-flowered or 1-flowered cymes in the axils of conspicuous bracts and bracteoles, forming an erect or hanging spike with several whorls; flowers actinomorphic with 5 nectaries inside at the base of the corolla tube, ovary of 5 carpels (rarely more) — **5. *Leycesteria***
– Flowers in 3-flowered cymes in the axils of mostly inconspicuous bracts, forming short erect spikes or heads, or mostly in 2-flowered cymes in the axil of foliage leaves, rarely of bracts; flowers zygomorphic or nearly actinomorphic with only one abaxial nectary, or actinomorphic with 5 nectaries, but then in 2-flowered cymes — **6. *Lonicera***
5. Perennial herbs with a terminal spike of several 6- or 2-flowered whorls; fruit a drupe with (2–)4 pyrenes — **4. *Triosteum***
– Shrubs; flowers in spikes or racemes on lateral branches or in several heads with involucrum; fruit a drupe with 2 pyrenes or an achene — 6
6. Shrubs with stolons, leaves mostly <5 cm; flowers in spikes or racemes, fruit a spongy drupe with 2 pyrenes — **7. *Symphoricarpos***
– Shrubs without stolons, leaves 8–15 cm with two conspicuous lateral veins; flowers in several heads of one or two 6-flowered whorls with an involucre; fruit a 1-seeded achene — **3. *Heptacodium***

7. Small shrubs with subterranean stolons; flowers inconspicuous, $\leq$2 cm, bilabiate with long lobes, yellow **1. *Diervilla***
- Shrubs >1 m without stolons, flowers >2 cm, weakly zygomorphic, corolla lobes broader than long, shorter than the tube, white or red, rarely pale yellow **2. *Weigela***
8. Subshrubs with small evergreen leaves and thin creeping shoots; flowers campanulate, nodding, mostly in pairs on upright, bifurcate flowering shoots; calyx deciduous **8. *Linnaea***
- Shrubs; inflorescence various; calyx persistent 9
9. Ovary and calyx becoming hidden by big, reniform, accrescent bracts; 2 of the locules with fertile ovule, achene 2-seeded **10. *Dipelta***
- Ovary always visible; 1–2 of the locules with fertile ovule; achene 1–2-seeded 10
10. Several flowers fused in pairs in the ovary region, bracts and ovary very bristly **9. *Kolkwitzia***
- All flowers free; bracts and ovary smooth or pilose 11
11. Floral tube not saccate at base, nectary trichomes in 3 or 4 separate grooves (Mexico) **13. *Vesalea***
- Floral tube saccate at base above the homogeneous nectary (Asia) 12
12. Petioles dilatated at base, enclosing axillary buds; corolla hypocrateriform; stigma green and mucilaginous ***Zabelia***
- Petioles not dilated, axillary buds exposed; corolla campanulate to funnel-shaped or bilabiate, stigma whitish, not mucilaginous 13
13. Flowers in terminal pairs, opening simultaneously **11. *Diabelia***
- Flowers terminal or axillary, single or when in pairs opening consecutively **12. *Abelia***

Genera of Caprifoliaceae

1. ***Diervilla*** Miller

Diervilla Miller, Gard. Dict. Abr. ed. 4 (1754).

Shrubs, 0.5–2 m tall. Leaves deciduous, acuminate, serrate, base rounded. Petiole short or ± absent. Thyrse terminal, often with terminal flower, foliose in the lower part; cymes with up to 5 orders of flowers. Calyx actinomorphic, segments 5, linear, 2–5 mm long; corolla weakly zygomorphic, dilated upwards, saccate at base above the abaxial conspicuous hump-like nectary, 1–1.5 cm long, lobes longer than wide, about as long as the tube, light yellow, turning red, middle lobe of the lower lip with darker yellow sap mark; stamens 5, exserted; stigma capitate; ovary 2-locular, slender, ovules numerous; basal parts thickening to form an elongate capsule, 0.8–1.5 cm long, splitting into two apically joined valves, abruptly contracted into a slender beak. Seeds numerous, light brown, ca. 1 mm long. $2n = 36$.

Three not clearly differentiated species, eastern N America.

2. ***Weigela*** Thunb. Fig. 20

Weigela Thunb., Kongl. Vetensk. Acad. Nya Handl. 1: 137, t. 5 (1780); Hara, Ginkgoana 5: 1–336 (1983), rev. Jap. spp.
Weigelastrum Nakai (1936).
Macrodiervilla Nakai (1936).

Mostly tall, deciduous shrubs, 1.5–5 m. Leaves shortly petiolate, acuminate, serrate. Thyrse sometimes with end flower; cymes 1–3-flowered in the axils of foliage leaves. Calyx tube narrowly cylindric, limb divided to middle or nearly to base into 5 lobes, actinomorphic or bilabiate; corolla almost actinomorphic or slightly bilabiate, funnel-shaped or tubular-campanulate, 2–4 cm long, showy, tube distinctly inflated in the upper part, abaxially at base with a conspicuous hump-like nectary; margin 5-lobed, lobes broad, much shorter than the tube, pale yellow, white turning rose or red to dark red, some species with yellow or orange sap mark; stamens 5, free or connected by hairs in a row below the stigma on the dorsal side of the flower, anthers weakly exserted; ovary, 2-locular, slender, petiole-like, with sterile apex, ovules numerous; style slender, often exserted; stigma discoid or 2-lobed. Capsule narrowly cylindrical or fusiform, thick-walled, splitting into 2 valves mostly from the top, leaving column and beak. Seeds numerous, small, often winged. $2n = 36, 72$.

About ten species in E Asia.

3. ***Heptacodium*** Rehder

Heptacodium Rehder in Sargent, Pl. Wilson. 2: 617 (1916); Zhang, Taxon 51(3): 499–505 (2002), karyol.

Deciduous shrubs up to 7 m tall, leaves coarse, abruptly attenuate, blade with three main veins, margin smooth or undulate. Sessile flowers in a reduced thyrse with only one or two pairs of 3-flowered cymes forming a head around a terminal bud, each head within an involucre of 10–12 subtending bracts and bracteoles in two rows; many such heads form a conspicuous, thyrsoid inflorescence of higher order. Calyx with 5 long, bristly teeth; corolla white, funnel-shaped, tube slightly curved, bilabiate, lobes ± equal; Stamens 5; ovary 3-locular, with only one fertile, uniovulate carpel, the other locules with numerous sterile ovules. Fruit a leathery achene topped by the accrescent calyx. $2n = 28$.

One species, *Heptacodium miconioides* Rehder, W China.

4. *Triosteum* L.

Triosteum L., Sp. Pl. 176 (1753); Gould & Donoghue, Harv. Papers Bot. 5: 157–166 (2000), phylog., biogeogr.

Small caespitose to rather tall, coarse herbs with simple shoots from a persistent rhizome. Leaves medium to large, simple or the lower ones deeply lobed, sessile, some auriculate, the lower ones sometimes united basally into a disk-like perfoliate sheath. Flowers in short axillary or terminal spikes of several whorls of 3-flowered cymes, rarely only 2 flowers per whorl. Calyx actinomorphic or bilabiate, with short or long lobes; corolla bilabiate, greenish to purplish outside, upper lip 4-lobed, tube funnel-shaped, saccate above the abaxial nectary gland; stamens 5; ovary with 3–5 locules, the adaxial one sterile, the others fertile and uniovulate. Fruit a large, dry, white, red or orange drupe with rather dry mesocarp and 2–4 pyrenes. $2n = 18, 36$.

About six species, E Asia (3 spp.), eastern N America (3 spp.).

5. *Leycesteria* Wall.

Leycesteria Wall. in Roxb., Fl. Indica 2: 181 (1824).

Deciduous shrubs up to 2 m high with erect, hollow, rather short-lived shoots, rarely evergreen and shoots with narrow, persistent pith. Leaves large, simple, long acuminate, entire or the lower leaves sinuate with some lobes; leaves of some species joined by small interpetiolar stipule-like structures. Flowers in upright or hanging spike-like thyrses in the axils of ± conspicuous bracts, with several whorls of mostly 6 or only 2 sessile flowers. Flowers small, radiate, weakly zygomorphic, pentamerous throughout; sepals small, unequal, tube very short; corolla white, rarely yellow, with 5 nectary glands at the base of the corolla tube; ovary of 5(7–10) fertile, many-ovuled carpels. Fruit an ovoid berry, like the persistent calyx with glandular hairs. $2n = 18$.

Six species, W Himalaya to S China.

6. **Lonicera** L. Fig. 21

Lonicera L., Sp. Pl. 173 (1753); Rehder (1903), Annual Rep. Missouri Bot. Gard. 14: 27–232 (1903), rev.; Akhter, Flora of Pakistan 174 (1986).
Caprifolium Miller (1754).
Chamaecerasus Medicus (1789).

Upright or twining, rarely creeping shrubs. Leaves deciduous, rarely evergreen, petiolate or sessile, sometimes connate at base, interpetiolar stipule-like structures sometimes present. Flowers in thyrses with several lateral 2–3-flowered cymes, often in the axils of foliage leaves, rarely of bracts. Individual flowers sessile, subtended by bracts and bracteoles, pentamerous, zygomorphic, rarely actinomorphic; pairs of flowers may be united basally or along most of the inferior ovary; calyx short, tube almost truncate or with 5 teeth, limb persistent or deciduous; corolla white or yellow, purple or scarlet, tubular to funnel-shaped to campanulate, with regular or subregular 5-lobed limb or mostly bilabiate with a 4-lobed upper lip; tube short or elongated and slender, rarely with 5–3, normally with 1 abaxial nectary (rarely 5) and gibbous at base, rarely calcarate; stamens 5, inserted in the corolla-tube, usually exerted; ovary normally 3-locular, occasionally 2- or rarely 4–5-locular; all locules equal, with (2)3–8 ovules; style elongated, stigma capitate. Fruit a 1- or few-seeded fleshy berry, red, white, black or blackish-blue. $2n = 18, 36, 54$.

Ca. 180 species in the temperate and subtropical regions of the northern hemisphere, extending northward to the arctic circle and south to the Malayan Archipelago, southern Asia, N Africa, Madeira and Mexico.

The genus can be divided into two well-defined monophyletic subgenera (Theis et al. 2008): Subg. *Caprifolium*: Leaves free, flowers in whorled 3-flowered cymes. Subg. *Lonicera*: Upper leaves and bracts often broadly connate. Flowers in 2-flowered cymes.

7. ***Symphoricarpos*** Duhamel Fig. 22

Symphoricarpos Duhamel, Traité Arbr. Arbust. 2: 295 (1755).

Deciduous, 1–2 m tall shrubs, some with stolons. Leaves simple, lower leaves sometimes lobed, sometimes interpetiolar stipule-like structures present. Flowers small or medium, with inconspicuous bracteoles, clustered in opposite spikes, or flower solitary, axillary. Flowers ± radiate, weakly zygomorphic, normally calyx, corolla and androecium 5-merous, rarely 4-merous, nectary glands 5 or 1; calyx short, campanulate, teeth as long as tube; corolla cup-shaped or tubular, pink or whitish; ovary with four locules, the median ones with several sterile ovules, the lateral ones fertile and uniovulate. Fruit a white, red or dark blue berry-like drupe with rather dry and mealy mesocarp and 2 small pyrenes. $2n = 18, 54, 72$.

About 15 species in N America and Mexico (*S. microphyllus* Kunth), one species endemic in China.

8. ***Linnaea*** L. Fig. 23

Linnaea L., Sp. Pl.: 631 (1753).

Evergreen dwarf subshrub, with long creeping thin stems, leaves persistent, very small, ovate, with 1–3 pairs of obtuse teeth. Flowers mostly 2 on erect lateral branches, nodding, peduncles and pedicels elongated, with a pair of small mostly sterile bracteoles and 2 pairs of additional thickish bracteoles, covered with glandular hairs. Calyx deeply 5-toothed, with glandular hairs; teeth deciduous; corolla funnel-shaped, slightly zygomorphic, 5-lobed in the upper part, white to pink, inside pubescent, with a small nectary inside the base of corolla tube; stamens 4, didynamous; ovary small, ovoid, 3-locular, consisting of 1 fertile uniovulate locule and 2 locules with sterile ovules; style filiform, stigma capitate. Fruit a small ovoid 1-seeded achene, enclosed by the 2 pairs of glandular additional bracteoles. $2n = 32$.

Fig. 23. Caprifoliaceae. *Linnaea borealis*. **A** Part of a flowering plant. **B** Leaf forms. **C** Corolla cut open with stamens. **D** Gynoecium and calyx in vertical section. (From Schneider 1912, fig. 435)

One species, *Linnaea borealis* L., widely distributed in subarctic, subalpine and cold temperate regions of the Northern Hemisphere.

9. ***Kolkwitzia*** Graebner

Kolkwitzia Graebner, Bot. Jahrb. Syst. 29: 593 (1900).

Deciduous shrub, up to 3 m high, winter buds perulate. Uppermost cymes of thyrse reduced to one flower, the others mostly with 2 or 3 flowers; base of the flowers with 2–4 additional bracts, the 2 uppermost fused with the ovary; only one of these bracts may support an additional sessile flower with only two bracteoles. Flowers of different age to some extent fused at their bases; sepals 5, narrow, pilose, spreading, persistent; corolla ca. 1. 5 cm long, campanulate, slightly zygomorphic, 5-lobed, pink, yellow in throat, puberulous; stamens 4, didynamous, included in corolla tube; ovary ellipsoid, narrowing into a long beak, hispid, 3-4-locular, 2 locules with several sterile ovules, 1–2 locules fertile, uniovulate. Fruit an ovoid, 1(2)-seeded achene, covered by the accrescent bristly bracts. $2n = 32$.

One species, *Kolkwitzia amabilis* Graebner, in C and N China.

10. ***Dipelta*** Maximowicz

Dipelta Maximowicz, Bull. Acad. Imp. Sci. St. Pétersbourg 24: 50 (1877).

Deciduous shrubs up to 6 m tall, bark peeling in layers or stripes. Leaves rather large, shortly petiolate. Flowers large, up to 4 cm long, mostly in a few-flowered raceme, normally without end flower, lower flowers in the axil of foliage leaves; bracteoles normally sterile, only the uppermost whorl consisting rarely of 3-flowered cymes. Flower base with 4 additional bracteoles in two pairs, those of the median one unequal and small, those of the upper, transversal one large, transversely oval to reniform, auriculate or peltate, enclosing the ovary at flowering time; calyx with 5 lanceolate or subulate teeth or with a wide, funnel-shaped tube and short obtuse teeth, persistent; corolla actinomorphic or weakly bilabiate, tubular to campanulate, base of the tube wide or narrow, pale or deep rose outside, white with yellow or orange sap mark inside; stamens 4, didynamous, included; ovary 4-locular, lateral locules fertile with one ovule, median ones with several sterile ovules. Fruit a 2-seeded achene with wings formed by the accrescent additional bracteoles. $2n = 18$.

Three species in China.

11. ***Diabelia*** Landrein

Diabelia Landrein, Phytotaxa 3: 35 (2010).
Linnaea subg. *Abelia* sect. *Bilaciniatae* ser. *Serratae* Graebn. (1901).

Deciduous shrubs; leaves opposite, margin entire or serrate. Flowers in pairs at the end of short shoots, opening simultaneously, each with 3 pairs of small bracts directly below the ovary, with additional flowers sometimes present in the axil of these bracts. Sepals 2–5, accrescent in fruit; corolla bilabiate, 5-lobed, white, yellow, pink or red, basally saccate above the nectary gland; stamens didynamous, included or exserted; ovary 3-locular, 2 locules with sterile ovules, 1 locule with a single fertile ovule; style filiform, stigma capitate. Fruit a leathery 1-seeded achene.

Three species, forests of China and Japan.

12. ***Abelia*** R. Br.

Abelia R. Br. in Abel, Narr. J. China, App. B.: 376 (1818).

Small and partly prostrate to medium-sized deciduous, rarely evergreen shrubs. Leaves often rather small, shortly petiolate, opposite or rarely whorled. Flowers in thyrses with few 3-flowered cymes, these often reduced to 1 flower, some species only with the uppermost pair of reduced cymes with recaulescent bracts; some species with 4 additional, normally sterile inconspicuous bracts directly below the ovary (*A. aschersoniana* has additional flowers as described for *Kolkwitzia*). Calyx persistent, accrescent, with 5 or 4 narrow lobes or only 2 broad lobes; corolla 5-lobed, weakly zygomorphic, campanulate to funnel-shaped, white, pink to purplish, some with orange sap mark, basally saccate above the nectary; stamens 4, didynamous; ovary 3-locular, 2 locules with sterile ovules, 1 fertile uniovulate locule. Fruit a 1-seeded, basally ovoid achene with a sterile beak. $n = 16$, $2n = 32$.

About 40 species in S and E Asia. The genus *Zabelia* (Rehder) Makino (see below), formerly often included in *Abelia*, probably forms an independent branch within Dipsacales (Landrein et al. 2012). Morphologically, *Abelia* differs from Zabelia by the basally not dilated petioles, exposed axillary buds, corolla form, and the white and not mucilaginous stigma (Yang and Landrein 2011).

13. ***Vesalea*** M. Martens & Galeotti

Vesalea M. Martens & Galeotti, Bull. Acad. Roy. Sci. Bruxelles 11(1): 241 (1844); J.A. Villarreal, Caprifoliaceae, Flora de Veracruz, fasc. 126 (2002) (sub *Abelia*).

Shrub to 1.5 m tall, trichomes simple and glandular. Leaf margin sometimes with glandular teeth. Flowers axillary at the end of short shoots, paired, pendant, red or scarlet, with 4 bracts at the base of the ovary; calyx lobes 5; corolla tube not saccate over the nectaries, nectar glands located in three or four grooves, corolla limb more or less bilabiate, 5-lobed; stamens 4, didynamous; ovary 3-locular, 2 locules with sterile ovules, one fertile, uniovulate; stigma subcapitate. Achene rostrate with persistent accrescent calyx.

Two or three species, Mexico. The phylogenetic analysis of Jacobs et al. (2010) suggests that *Vesalea* has to be separated from *Abelia* to avoid paraphyly of the latter. In the phylogeny, *Vesalea* is basal in the *Linnaea* clade s.s. (*Vesalea, Abelia, Dipelta, Linnaea, Kolkwitzia*).

Genus incertae sedis:

Zabelia (Rehder) Makino

Zabelia (Rehder) Makino, Makinoa 9: 175. 1948; Jacobs et al., Syst. Bot. 26: 231–252 (2011), phyl., morph., palyn.; Landrein et al., Bot. J. Linn. Soc. 169: 692–713 (2012), phyl., morph., evol.
Abelia sect. *Zabelia* Rehder (1911).

Deciduous shrubs, wood with distinct growth rings; branches usually with 6 longitudinal fissures corresponding to furrows in the xylem, often with stiff retrorse hairs when young, nodes thickened. Leaves opposite, margin entire or dentate, rarely lobed; petioles short, dilated, connate at base, enclosing axillary buds; interpetiolar remains of the fusion of petiole bases may appear stipule-like. Thyrse terminal, congested, composed of sessile, 1–3-flowered cymes; free bracts sometimes present at the base of the flowers. Calyx actinomorphic, lobes 4 or 5, persistent, spreading; corolla hypocrateriform, zygomorphic, 4- or 5-lobed; corolla tube cylindrical, abaxially gibbous at base or not, nectary inconspicuous; stamens included, didynamous; anthers introrse, pollen grains prolate and 3-zonocolporate, about 79–88 × 70–84 μm, tectum psilate, endocingulum present at the inner nexine wall layer. Ovary 3-locular, 2 locules with sterile ovules, the other with 1 fertile ovule; stigma capitate, mucilaginous. Fruit a leathery, more or less strongly curved achene, topped by persistent, slightly accrescent calyx lobes. Endosperm fleshy, embryo small. $n = 9, 54$; $2n = 36, 108$.

Six species, Afghanistan to China and Japan.

The relationships of *Zabelia* are currently unclear. It is here annexed to the Caprifoliaceae treatment with hesitation; originally it was described as a section of *Abelia* (see above). Jacobs et al. (2010, 2011) suggested a position outside Caprifoliaceae s.s. and as a basal branch of Morinaceae. Besides DNA sequence data, such a position was mainly supported by palynological characters, i.e. a psilate exine and an endocingulum at the inner nexine wall layer. The analyses of Landrein et al. (2012), however, based on an improved sampling and additional markers, rather supported a position of *Zabelia* as sister to the clade formed by Morinaceae, Dipsacaceae and Valerianaceae.

Selected Bibliography

Acevedo-Rodríguez, P. 2005. Vines and climbing plants of Puerto Rico and the Virgin Islands. Smithsonian Institution. Contrib. U. S. National Herbarium 51: 1–483.

APG III 2009. An update of the Angiosperm Phylogeny Group classification for the orders and families of flowering plants: APG III. Bot. J. Linn. Soc. 161: 105–121.

Backlund, A., Pyck, N. 1998. Diervillaceae and Linnaeaceae, two new families of caprifolioids. Taxon 47: 657–661.

Behrens, W.J. 1879. Die Nectarien der Blühten. Anatomisch-physiologische Untersuchungen. Flora 62: 113–123.

Bell, C.D., Donoghue, M.J. 2005. Dating the Dipsacales: comparing models, genes, and evolutionary implications. Amer. J. Bot. 92: 284–296.

Benko-Iseppon, A.M., Morawetz, W. 2000. Viburnales: cytological features and a new circumscription. Taxon 49: 5–16.

Böhnke-Gütlein, E., Weberling, F. 1981. Palynologische Untersuchungen an Caprifoliaceae I. Sambuceae, Viburneae und Diervilleae. Acad. Wiss. Abh. Math.-Naturwiss. Kl. [Trop. Subtrop. Pflanzenwelt] 34: 131–189.

Carlquist, S. 1988. Comparative wood anatomy. New York: Springer.

Chia-Chi, H., Chao-Xing, H. 1988. Pollen morphology of Caprifoliaceae from China and its taxonomic significance. Acta Phytotax. Sinica 26: 343–352.

Davis, G.L. 1966. Systematic embryology of the Angiosperms. New York: Wiley.

Donoghue, M.J. 1985. Pollen diversity and exine evolution in *Viburnum* and the Caprifoliaceae sensu lato. J. Arnold Arbor. 66: 421–469.

Donoghue, M.J., Bell, C.D., Winkworth, R.C. 2003. The evolution of reproductive characters in Dipsacales. Int. J. Plant Sci. 164(5 suppl): S453–S464.

Fritsch, K. 1892. Caprifoliaceae. In: Engler & Prantl, Die natürlichen Pflanzenfamilien IV, 4: 159–169. Leipzig: W. Engelmann.

Gould, K.R., Donoghue, M.J. 2000. Phylogeny and biogeography of *Triosteum* (Caprifoliaceae). Harvard Pap. Bot. 5: 157–166.

Graebner, P. 1901. Die Gattung *Linnaea* (einschließlich *Abelia*). Bot. Jahrb. Syst. 29: 120–145.

Hara, H. 1983. A revision of Caprifoliaceae of Japan with reference to allied plants in other districts and the Adoxaceae. Ginkgoana 5: 1–336.

Hegnauer, R. 1989. Chemotaxonomie der Pflanzen. Band 8, Nachträge zu Band 3 und Band 4 (Acanthaceae bis Lythraceae). Basel: Birkhäuser.

Index to plant chromosome numbers (IPCN) 1979–. P. Goldblatt & D.E. Johnson (eds.) Missouri Botanical Garden, St. Louis. http://www.tropicos.org/Project/IPCN (accessed May 2010).

Jacobs, B., Lens, F., Smets, E. 2009. Evolution of fruit and seed characters in the *Diervilla* and *Lonicera* clades (Caprifoliaceae, Dipsacales). Ann. Bot. 104: 253–276.

Jacobs, B., Pyck, N., Smets, E. 2010. Phylogeny of the *Linnaea* clade: Are *Abelia* and *Zabelia* closely related? Mol. Phylogenet. Evol. 57: 741–752.

Jacobs, B., Geuten, K., Pyck, N., Huysmans, S., Jansen, S., Smets, E. 2011. Unraveling the phylogeny of *Heptacodium* and *Zabelia* (Caprifoliaceae): An interdisciplinary approach. Syst. Bot. 26: 231–252.

Johri, B.M., Ambegaokar, K.B., Srivastava, P.S. 1992. Comparative embryology of Angiosperms. New York: Springer.

Judd, W.S., Sanders, R.W., Donoghue, M.J. 1994. Angiosperm family pairs: preliminary phylogenetic analyses. Harvard Pap. Bot. 5: 1–51.

Krüssmann, G. 196. Handbuch der Laubgehölze. 2 Vols. Berlin: Paul Parey.

Landrein, S., Prenner, G. 2013. Unequal twins? Inflorescence evolution in the twinflower tribe Linnaeeae (Caprifoliaceae s.l.). Int. J. Plant Sci. 174: 200–233.

Landrein, S., Prenner, G., Chase, M.W., Clarkson, J.J. 2012. *Abelia* and relatives: Phylogenetics of Linnaeeae (Dipsacales–Caprifoliaceae s.l.) and a new interpretation of their inflorescence morphology. Bot. J. Linn. Soc. 169: 692–713.

Perveen, A., Qaiser, M. 2007. Pollen Flora of Pakistan – LV. Caprifoliaceae. Pakistan J. Bot. 39(5): 1393–1401.

Rehder, A. 1903. Synopsis of the genus *Lonicera*. St. Louis: Missouri Botanical Garden Press.

Rehder, A. 1912. *Lonicera*. In: Schneider, C.K. 1912. Illustriertes Handbuch der Laubholzkunde. Jena: Gustav Fischer, pp. 632–753.

Rehder, A. 1940. Manual of cultivated trees and shrubs hardy in North America. Ed. 2, pp. 827–876. New York: Macmillan.

Rehder, A. 1949. Bibliography of Cultivated Trees and Shrubs, hardy in the cooler temperate regions of the northern hemisphere. Jamaica plain, Mass.: Arnold Arboretum of Harvard University, pp. 598–630.

Schneider, C.K. 1912. Illustriertes Handbuch der Laubholzkunde: Charakteristik der in Mitteleuropa heimischen und im freien angepflanzten angiospermen Gehölz-Arten und Formen mit Ausschluss der Bambuseen und Kakteen, Bd. 2. Jena: G. Fischer.

Smith, S.A. 2009. Taking into account phylogenetic and divergence-time uncertainty in a parametric biogeographical analysis of a Northern Hemisphere plant clade Caprifolieae. J. Biogeogr. 36: 2324–2337.

Theis, N., Donoghue' M.J., Li, J. 2008. Phylogenetics of the Caprifolieae and *Lonicera* (Dipsacales) based on nuclear and chloroplast DNA sequences. Syst. Bot. 33(4): 776–783.

Theisen, I., Barthlott, W. 1996. Die Mikromorphologie epicuticularer Wachse bei den Caprifoliaceae. Feddes Repert. 107: 31–35

Troll, W., Weberling, F. 1966. Die Infloreszenzen der Caprifoliaceen und ihre systematische Bedeutung. Akad. Wiss. Lit. (Mainz), Abh. Math.-Nat. Kl. Nr. 4: 459–605.

Wagenitz, G., Laing, B. 1984. Die Nektarien der Dipsacales und ihre systematische Bedeutung. Bot. Jahrb. Syst. 104: 483–507.

Weberling, F. 1957. Morphologische Untersuchungen zur Systematik der Caprifoliaceen. Akad. Wiss. Lit. (Mainz), Abh. Math.-Nat. Kl. 1: 3–50.

Weberling, F. 1966. Zur systematischen Stellung der Gattung *Heptacodium* Rehd. Bot. Jahrb. Syst. 85: 253–258.

Weberling, F., Hildenbrand, M. 1982. Weitere Untersuchungen zur Tapetumentwicklung der Caprifoliaceae. Beitr. Biol. Pflanzen 61: 3–20.

Wilkinson, A.M. 1948a. Floral anatomy and morphology of some species of the tribe Lonicereae of the Caprifoliaceae. Amer. J. Bot. 35: 261–271.

Wilkinson, A.M. 1948b. Floral anatomy and morphology of some species of the tribes Linnaeeae and Sambuceae of the Caprifoliaceae. Amer. J. Bot. 35: 365–371.

Wilkinson, A.M. 1949. Floral anatomy and morphology of *Triosteum* and of the Caprifoliaceae in general. Amer. J. Bot. 36: 481–489.

Winkworth, R.C., Bell, C.D., Donoghue, M.J. 2008. Mitochondrial sequence data and Dipsacales phylogeny: Mixed models, partitioned Bayesian analyses, and model selection. Mol. Phylogenet. Evol. 46: 830–843.

Wittrock, V.B. 1907. *Linnaea borealis* L. species polymorpha et polychroma. Acta Horti Bergiani 4(7): 5–187.

Xu, L., Lu, L., Li, D.-Z., Wang, H. 2011. Evolution of pollen in Dipsacales. Plant Divers. Resources 33: 249–259.

Yang, Q., Landrein, S. 2011. Linnaeaceae. In: C.Y. Wu et al. (eds.) Flora of China 19: 642–648.

Yang, Q., Barrie, F.R., Bell, C.D. 2011. Diervillaceae. In: C. Y. Wu et al. (eds.) Flora of China 19: 615.

Zhang, W.-H., Chen, Z.-D., Li, J.-H., Chen, H.-B., Tanga, Y.-C. 2003. Phylogeny of the Dipsacales s.l. based on chloroplast trnL-F and ndhF sequences. Mol. Phylogenet. Evol. 26: 176–189.

Cardiopteridaceae

Cardiopteridaceae Blume, Rumphia 3: 205 (1847), nom. cons.
Peripterygiaceae King (1895).

M. SCHORI

Twining lianas (*Cardiopteris*) or evergreen trees or shrubs; laticifers present (*Cardiopteris*) or absent; buttresses absent or present (*Citronella*); indument of simple to malpighiaceous hairs. Leaves simple, alternate, sometimes appearing distichous, entire to toothed (*Citronella*) or palmately lobed (*Cardiopteris*), penninerved or palmatinerved (*Cardiopteris*), exstipulate. Inflorescences axillary, occasionally ramiflorous (cauliflorous in *Pseudobotrys*), flowers arranged in spikes, fascicles, cincinnate panicles or thyrses. Flowers bisexual or unisexual, plants then andromonoecious or dioecious, actinomorphic, usually 5-merous; calyx persistent, articulated at the base, small, generally cupular, sepals 4–5, bases connate, lobes free; petals 4–5, minute to conspicuous, fused at least at the base (free in *Citronella*), often inflexed at apex, lobes imbricate (*Cardiopteris*) to valvate, midrib usually thickened ventrally; stamens isomerous, epipetalous (except *Citronella*), filaments filiform or broad and flattened, glabrous, anthers bisporangiate, basifixed to medifixed, introrse, longitudinally dehiscent; staminodes in pistillate flowers present, resembling stamens or much smaller but pollen sacs sterile; disk absent or present (staminate *Cardiopteris*, some *Gonocaryum*); ovary superior, pseudomonomerous, carpels evidently 3, unilocular (or with a partial pseudoseptum), ovules 2, apical, pendent, anatropous (orthotropous in *Cardiopteris*), unitegmic (ategmic in *Cardiopteris*); styles 0–2, deciduous to persistent, stigma capitate and often minute, less commonly large and sessile (*Gonocaryum*); ovary rudiment in staminate flowers small, stigma absent or much reduced. Fruit a drupe, or a samara with accrescent terminal appendage (*Cardiopteris*). Seed 1, testa vascularised. Embryo small, endosperm generally abundant.

Five genera with about 45 species, distributed mainly in the tropics of the Old World, rapidly decreasing in number towards the subtropics.

VEGETATIVE MORPHOLOGY. Most members of the family are small to medium-sized trees or shrubs; *Cardiopteris* has been reported as an herbaceous vine but forms wood (Lens et al. 2008) and may develop stems to 12 cm diameter. Branchlets and petioles are puberulous or glabrous. Leaves simple, petiolate, alternate, sometimes appearing distichous. Leaves mostly oblong-elliptic, sometimes lanceolate. Leaf bases typically obtuse. Leaf margin entire to toothed or lobed, usually flat. Texture ranges from thinly chartaceous to thickly coriaceous. Leaves penninerved with the secondary veins 2- to 10-paired, rarely triplinerved. Tertiary veins typically obscure. Domatia absent or present (*Citronella*) (Kårehed 2001).

VEGETATIVE ANATOMY. Laticifers present in *Cardiopteris* (Sleumer 1971). Stomata paracytic (*Citronella*, *Pseudobotrys*), intermediate between cyclocytic and anomocytic (*Gonocaryum*) or anomocytic (*Pseudobotrys*). Polar T-pieces absent (*Citronella*) to distinct in *Gonocaryum* and *Pseudobotrys*. Stomatal guard cells with inner and/or outer stomatal ledges (Van Staveren and Baas 1973). Several types of hairs are found in Cardiopteridaceae and include, amongst others, normal hairs (hairs with an elongate or tapering body), papillar hairs (*Gonocaryum*) and malpighiaceous hairs (*Citronella*) (Van Staveren and Baas 1973). Thin-base hairs (long, tapering hairs with a thick wall and restricted lumen in the body and a very thin-walled base) are present in *Citronella* and *Gonocaryum*. Druses are present in all genera except *Pseudobotrys*. Rhombic crystals occur in *Gonocaryum* and *Citronella*, and crystal sand in

J.W. Kadereit and V. Bittrich (eds.), *Flowering Plants. Eudicots*, The Families and Genera of Vascular Plants 14,
DOI 10.1007/978-3-319-28534-4_8, © Springer International Publishing Switzerland 2016

Gonocaryum and *Leptaulus* (Heintzelman and Howard 1948). Lens et al. (2008) reported prismatic and styloid-like crystals in the rays of all genera except *Leptaulus* (no prismatics) and *Cardiopteris*. Growth ring boundaries absent or obscure, somewhat evident in *Leptaulus* and *Pseudobotrys*. Wood diffuse-porous. Vessels mostly solitary, with simple (*Gonocaryum*, *Leptaulus*) to scalariform (exclusively so in *Citronella*, *Pseudobotrys*) perforation plates. Intervessel pits 5–10 μm in horizontal diameter, nonvestured, opposite to alternate. Rays uniseriate to multiseriate. Wall thickenings and tyloses absent. Fibres non-septate, thick- to thin-walled, (1,175–)2,190–4,240(–4,900) μm long, with distinctly bordered pits, borders 5–9 μm in horizontal diameter (Lens et al. 2008). *Cardiopteris* forms abnormal wood characterised by non-lignified parenchyma with embedded fibres and tracheids. Other details of wood anatomy of Cardiopteridaceae given by Lens et al. (2008) do not necessarily apply to Cardiopteridaceae in the present circumscription because the authors included *Metteniusa* and *Dendrobangia*, which do not belong to the family.

Inflorescence Structure. Inflorescences of Cardiopteridaceae may be paniculate, fasciculate or thyrsoid. *Cardiopteris* has cincinnate thyrses, which have been erroneously described as cincinnate panicles in the literature. Inflorescences may be axillary, terminal, ramiflorous, or cauliflorous (*Pseudobotrys*). Solitary flowers are occasionally found (*Leptaulus*). Bracts are often present. Flowers (sub)sessile or on short pedicels.

Flower Structure (Fig. 24A–G). Flowers are small but are often clustered. The calyx is usually cupular, 4–5-lobed, imbricate to valvate, radially symmetrical. The corolla is radially symmetrical, with 4–5 free or connate petals, valvate to imbricate. In most genera, the petals have a thickened midrib and an inflexed tip. The inner surface may be papillate (some *Gonocaryum*) or pubescent (*Leptaulus*). Stamen number is equal to petal number. The stamens are usually included (except *Citronella*) and are glabrous. The filaments vary from extremely short (*Cardiopteris*, some *Gonocaryum*) to long and filiform (*Citronella*, *Leptaulus*, *Pseudobotrys*) and are epipetalous (except *Citronella*). The anthers are basifixed

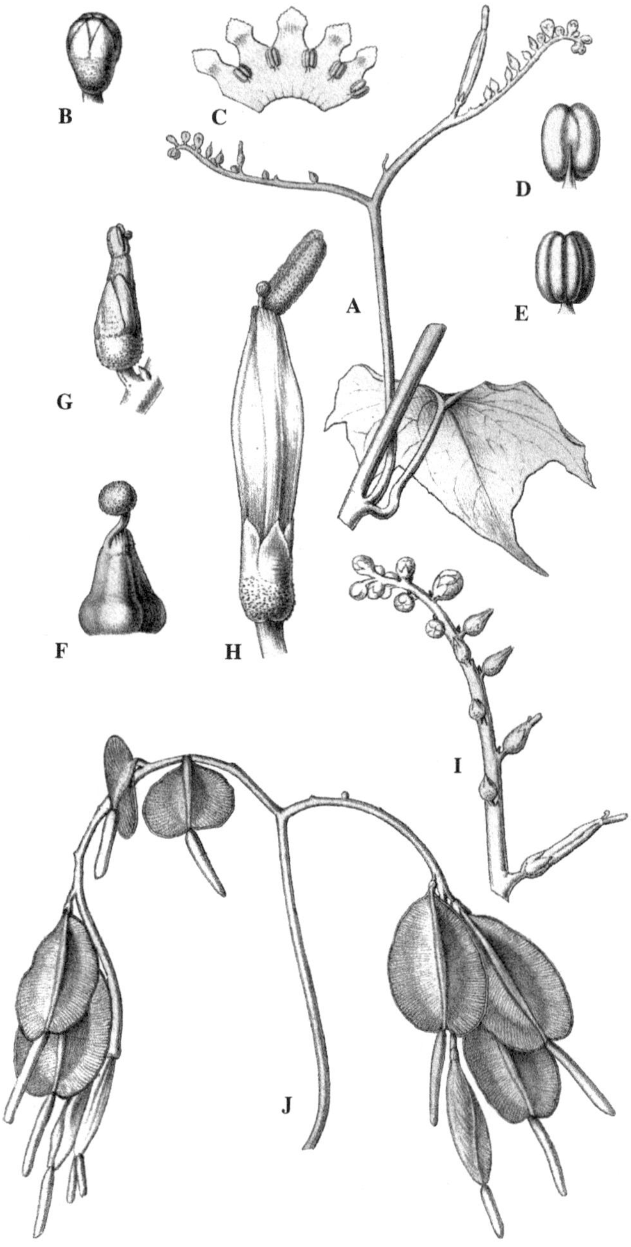

Fig. 24. Cardiopteridaceae. *Cardiopteris lobata*. **A** Inflorescence and leaf. **B** Bud. **C** Corolla with epipetalous stamens. **D** Anther, dorsal view. **E** Anther, ventral view. **F** Ovary and capitate stigma. **G** Flower shortly after anthesis. **H** Young fruit, showing persistent style and accrescent carpel apex. **I** Inflorescence detail. **J** Mature winged fruits with fleshy appendage. From Beccari (1877, t. 8), modified

to medifixed, longitudinally dehiscent, and introrse, smooth or echinate (some *Leptaulus*). Staminodes in pistillate flowers resemble the stamens but are often reduced. Some species of

Gonocaryum and flowers of *Cardiopteris* have a disk. The ovary is superior, 3-carpellate and unilocular, with two pendent apical ovules, only one of which develops. In *Cardiopteris*, two carpels form a bifid ovary apex, while the third carpel gives rise to the style (Kong et al. 2014). In *Citronella* and *Pseudobotrys* there is a partial pseudoseptum in the locule. The stigma is sessile (*Gonocaryum*) or on a short to long style, and is capitate to lobed. Two styles may be present in *Citronella*. The ovary rudiment in staminate flowers is minute and much shorter than the stamens or petals, conical to cylindrical, with a very reduced or absent stigma.

Embryology. Embryology has been investigated in *Leptaulus*, *Citronella* (Mauritzon 1936) and *Gonocaryum* (Fagerlind 1945) but both authors described or compared these taxa within the context of Icacinaceae s.l., so that a reinvestigation of embryology is warranted. The two pendent ovules are anatropous, unitegmic, and crassinucellate to tenuinucellate (Mauritzon 1936). Kong et al. (2002) reported that *Cardiopteris* has ategmic, orthotropous ovules which are tenuinucellate with the egg-cell at the chalazal end, and additional observations (Kong et al., unpublished data) have shown that a haustorium develops and the nucellus thickens asymmetrically after fertilisation to form a vascularised bundle encircling the seed. In *Leptaulus*, *Citronella*, and *Gonocaryum*, the vascular bundle in the chalaza extends up to the top of the single integument. In the linear tetrad of megaspores, the chalazal megaspore develops into a Polygonum-type embryo sac while the other megaspores disintegrate. The synergids are pyriform and hooked, with the filiform apparatus. The polar nuclei fuse at the time of fertilisation and the 3 antipodal cells are ephemeral. Endosperm development is cellular in *Cardiopteris*, but otherwise nuclear. Embryogeny is of Asterad or Solanad type in *Cardiopteris*.

Pollen Morphology. The pollen of Cardiopteridaceae is quite diverse and varies from tricolporate (*Pseudobotrys*, *Citronella*, *Cardiopteris*) to 3–4-porate (*Leptaulus*) or cryptoaperturate (*Gonocaryum*), and from spheroidal to prolate or oblate. The exine ranges from reticulate (*Citronella*) to (micro)perforate (*Leptaulus*, *Pseudobotrys*), while *Gonocaryum* and *Cardiopteris* have very distinctive pollen with a more complex exine. *Gonocaryum* has a semi-tectate reticulate exine with ornamented muri. *Cardiopteris* has a smooth to rugose or ridged mesocolpial rim around each reticulate face of the triangular pollen grain, with smooth to rugose heteropolar regions (Lobreau-Callen 1982; Schori and Furness 2014).

Karyology. A chromosome number of $n = 14$ was reported for *Leptaulus* (Stevens 2012).

Pollination and Reproductive Systems. The flowers of Cardiopteridaceae are relatively small, although *Pseudobotrys* and *Leptaulus* may have larger flowers, while the flowers of *Cardiopteris* are minute. Colour ranges from white, green, and yellow to purple. *Leptaulus* has tubular flowers but the other genera have relatively open flowers. Buds of *Cardiopteris* often contain insect larvae; these buds are relatively large, with thickened petals and no functional stamens or ovary. Pollinators are not documented for any species. Genera within Cardiopteridaceae exhibit three different reproductive systems. *Gonocaryum* is functionally dioecious, *Leptaulus* and *Pseudobotrys* have bisexual flowers, and *Citronella* and *Cardiopteris* are polygamous. *Citronella* may be andro- or gynomonoecious, while *Cardiopteris* might be androdioecious. Tobe (2012) reported functionally pistillate flowers with no corolla, but the corolla is caducous (Kong et al. 2014).

Fruit and Seed. The fruits of Cardiopteridaceae are predominantly indehiscent drupes, although *Cardiopteris* has a symmetrically winged samara with a fleshy appendage formed by the bifid ovary apex (Fig. 24J). Parthenocarpy has been observed in *Leptaulus*, *Gonocaryum* and *Pseudobotrys*. The drupes of *Citronella* and *Pseudobotrys* have an invaginated fruit wall which protrudes into the locule as a partial pseudoseptum; the seed curves around this protrusion. Seeds have a vascularised testa, although in *Cardiopteris* it is formed from nucellar tissue. The embryo is generally small, while endosperm is abundant. Ruminate endosperm is present in *Gonocaryum* and *Leptaulus*.

Dispersal. The winged fruits of *Cardiopteris* are presumably wind dispersed, but all other genera in the family have a fleshy drupe which is dispersed by birds or mammals. *Cardiopteris* fruits have a terminal fleshy appendage which may attract animals, although dispersers have not been observed and the appendage may be deciduous.

Phytochemistry. Secoiridoids have been reported from *Gonocaryum calleryanum* (Kaneko et al. 1995; Chan et al. 1998).

Subdivisions and Relationships Within the Family. No well-sampled molecular phylogeny for Cardiopteridaceae has yet been published. The family is heterogeneous in terms of its morphology, although epipetalous stamens are shared by four genera and three genera have a bilobed ovary apex with a lateral style (evident in early stages of floral development in *Gonocaryum*; Kong, unpublished data). All genera except *Pseudobotrys* were placed in the family on the basis of molecular data (Kårehed 2001).

Affinities. All genera, excluding *Cardiopteris*, were formerly included in Icacinaceae, but as early as 1941 had emerged as a distinct group on evidence of wood anatomy (Bailey and Howard 1941). Recent morphological and molecular analyses have supported the distinctness of the group and have found it be widely separated from Icacinaceae sensu stricto. For further discussion, see Kårehed (2002). Molecular data place Cardiopteridaceae in Aquifoliales as sister to Stemonuraceae (Kårehed 2001; Winkworth et al. 2008). Morphologically, the family is quite variable and difficult to characterise. *Cardiopteris* in particular, as a liana with latex and a samara, is quite different from the other genera except in its ovary structure.

Metteniusa was considered to be part of Cardiopteridaceae, but González et al. (2007) have shown that Metteniusaceae belong near Garryales. *Dendrobangia* was also included in Cardiopteridaceae, but molecular data (Byng et al. 2014) show that it is more closely related to *Metteniusa* and does not belong in Aquifoliales.

Distribution and Habitats. The family occurs in all major tropical wet forest regions. *Citronella* (21 spp.) is amphipacific; *Leptaulus* (6–10 spp.) is found in Guineo-Congolan Africa and Madagascar; *Gonocaryum* (9–12 spp.) grows from India to China and the Solomons; *Cardiopteris* (3 spp.) occurs from India and China south to Papua New Guinea; *Pseudobotrys* (2 spp.) is endemic to Papua New Guinea. Evergreen tropical rainforest is the typical habitat, although some species grow in coastal or swamp forests.

Palaeobotany. Cardiopteridaceae as currently circumscribed do not have a clear fossil record. Palaeobotanical literature which mentions *Cardiopteris* refers to a fossil fern taxon.

Economic Importance. *Citronella gongonha* is used as a substitute for *Ilex paraguariensis* in maté. *Citronella moorei* and *C. smythii* are used as timber species in Australia, primarily for cabinet work. *Gonocaryum* is used on a local scale for furniture, cabinetry and carved utensils in the Philippines (Nasution 1998). Oil from the seed of *Gonocaryum* is used for soap and as a lubricant (Hua and Howard 2008).

Key to the Genera

1. Liana with latex; leaves cordate, palmatinerved; fruit a samara — 1. ***Cardiopteris***
– Trees or shrubs without latex; leaves penninerved; fruit a drupe — 2
2. Flowers unisexual; plants dioecious — 3. ***Gonocaryum***
– Flowers unisexual or bisexual; plants bisexual or polygamous — 3
3. Flowers tubular; petals free at the apex only; Africa and Madagascar — 4. ***Leptaulus***
– Flowers not tubular; petals entirely to mostly free; Asia, Pacific, and South America — 4
4. Petals free, stamens not epipetalous; inflorescences not cauliflorous — 2. ***Citronella***
– Petals fused at base; stamens epipetalous; inflorescences cauliflorous — 5. ***Pseudobotrys***

Genera of Cardiopteridaceae

1. *Cardiopteris* Wall. ex Royle — Fig. 24.

Cardiopteris Wall. ex Royle, Ill. Bot. Himal. Mts. 136 (1834).

Peripterygium Hassk. (1843).
Sioja Buch.-Ham. ex Lindl. (1836), nom. nud.

Liana with whitish latex. Leaves simple or lobed, cordate, palmatinerved. Inflorescences axillary panicles, cincinnate. Flowers bisexual or staminate, (4)5-merous; calyx cupular, lobed; petals fused below, tips free, imbricate; stamens with filaments free at apex only, anthers medifixed; disk present, below base of pistillode in staminate flowers, on base of ovary in bisexual flowers; ovary angular, apex bilobed, lateral style short with a capitate stigma; ovary apex developing into a fleshy accrescent appendage in fruit. Samara winged on either side of ovary.

Three spp., India, Vietnam, Myanmar, Thailand, China, Philippines, Malaysia, Indonesia, New Guinea.

2. ***Citronella*** D. Don

Citronella D. Don, Edinb. New Phil. J. 13: 243 (1832); R.A. Howard, Contrib. Gray Herb. 142: 60–89, t. 4–6 (1942), rev.
Briquetina J.F. Macbr. (1931).
Chariessa Miq. (1856).
Pleuropetalon Bl. (1850), non Hook. f. (1845).
Sarcanthidion Baill. (1874).
Villaresia Ruiz & Pavon (1803), non Ruiz & Pavon (1793).
Villaresiopsis Sleumer (1940).

Trees or shrubs; buttresses absent or occasionally present, thin. Leaves elliptic to ovate, generally coriaceous, domatia present. Inflorescences terminal and/or (supra-)axillary, paniculate or thyrsoid, cymes of secondary branches scorpioid, forked or not. Flowers bisexual or polygamous (monoecious), 5-merous, sessile, subtended by a minute bract, fragrant or not; sepals fused at base, tips imbricate; petals free, valvate or subimbricate, apex inflexed, midrib keeled within; stamens in bisexual and staminate flowers free, filaments fleshy, subulate; anthers introrse, basifixed; staminodes in pistillate flowers similar but much reduced; ovary in perfect flowers with bilobed apex and subapical style; styles 1(2); stigma small, capitate, bilobed. Ovary in pistillate flowers without bilobed apex or style; stigma sessile, capitate; ovary in staminate flowers reduced, lacking both style and stigma; disk absent. Drupe fleshy, endocarp woody. Seed 1, hippocrepiform around a partial pseudoseptum.

About 21 spp.: three in Philippines, Indonesia, to New Guinea; six in Australia, Solomon Islands, New Caledonia, Loyalty Island, New Hebrides, Fiji, Samoa and Tonga; 12 in Central and South America.

3. ***Gonocaryum*** Miq.

Gonocaryum Miq., Fl. Ind. Bat. Suppl. 1: 343 (1861).
Phlebocalymna Griff. ex Benth. & Hook. f. (1862).

Shrubs or trees. Leaves oblong-elliptic, coriaceous, venation prominent below. Inflorescences thyrsoid, axillary, supra-axillary or cauliflorous. Flowers unisexual, 5-merous; calyx cupular, lobes free; petals valvate, connate at base, slightly keeled within, inner surface sometimes papillose; staminate flowers: stamens adnate to petals at base, filaments narrow to broad, fleshy, anthers introrse, dorsifixed; ovary rudiment small, pubescent, disk absent or present; style and stigma absent; pistillate flowers: staminodes rudimentary, ovary sessile, disk absent or present, style conical, stigma capitate. Drupe longitudinally ribbed. Seed with a thin testa, endosperm ruminate.

Nine to 12 spp., Myanmar to China, throughout Malesia to the Bismarck Islands.

4. ***Leptaulus*** Benth.

Leptaulus Benth. in Benth. & Hook. f., Gen. 1: 351, 996 (1862).
Acrocoelium Baill. (1892).

Trees. Leaves ovate-elliptic to obovate, coriaceous. Inflorescences of clustered flowers in thyrses, axillary or terminal, flowers subtended by bracts. Flowers bisexual, 5-merous; calyx cupular, lobed, imbricate; petals free at apex, fused below and forming a tube, with a tuft of hairs above or between each anther; stamens epipetalous, filaments free near apex only, anthers basifixed; ovary asymmetrical with bifid apex, single style attached laterally near the apex; style base persistent in fruit; stigma minute; disk absent. Drupe ellipsoid to globose. Endosperm ruminate. $n = 14$.

Six to 10 spp., Africa and Madagascar.

5. ***Pseudobotrys*** Moeser

Pseudobotrys Moeser, Repert. Spec. Nov. Regni Veg. 10: 310 (1912).

Shrub to small tree. Leaves ovate-elliptic to obovate, coriaceous. Inflorescences cauliflorous, paniculate or fasciculate. Flowers bisexual, 5-merous, pedicel articulated with the calyx, subtended by 2 small bracteoles; sepals connate at base, lobes imbricate; petals linear, reflexed, valvate, apex inflexed, bases fused; stamens epipetalous, filaments filiform and free for most of their length, anthers introrse, basifixed; ovary ovoid, style filiform, stigma small, subcapitate; disk absent. Drupe with thin fleshy exocarp, woody endocarp, verrucose, partial pseudoseptum in the locule.

Two spp., New Guinea.

Selected Bibliography

Bailey, I.W., Howard, R.A. 1941. The comparative morphology of the Icacinaceae. II. Vessels. J. Arnold Arbor. 22: 171–187.

Beccari, O. 1877. Sulla *Cardiopteris lobata*. Nuovo Giorn. Bot. Ital. 9: 100–108, t. 8.

Byng, J.W., Bernardini, B., Joseph, J.A., Chase, M.W., Utteridge, T.M.A. 2014. Phylogenetic relationships of Icacinaceae focusing on the vining genera. Bot. J. Linn. Soc. 176: 277–294.

Chan, Y.Y., Leu, Y.L., Lin, F.W., Li, C.Y., Wu, Y.C., Shi, L. S., Liou, M.J., Wu, T.S. 1998. A secoiridoid and other constituents of *Gonocaryum calleryanum*. Phytochemistry 47: 1073–1077.

Fagerlind, F. 1945. Bau des Gynöceums, der Samenlage und des Embryosackes bei einigen Repräsentanten der Familie Icacinaceae. Svensk Bot. Tidskr. 39: 346–364.

Gonzáles, F., Betancur, J., Maurin, O., Freudenstein, J.V., Chase, M.W. 2007. Metteniusaceae, an early-diverging family in the lamiid clade. Taxon 56: 795–800.

Heintzelman, C.E., Howard, R.A. 1948. The comparative morphology of the Icacinaceae. V. The pubescence and the crystals. Amer. J. Bot. 35: 42–52.

Hua, P., Howard, R.A. 2008. Icacinaceae. In: Wu, Z., Raven, P.H. (eds.) Flora of China 11. St. Louis: Missouri Botanical Garden Press, pp. 505–514.

Kaneko, T., Sakamoto, M., Ohtani, K., Ito, A., Kasai, R., Yamasaki, K., Padorina, W.G. 1995. Secoiridoid and flavonoid glycosides from *Gonocaryum calleryanum*. Phytochemistry 39: 115–120.

Kårehed, J. 2001. Multiple origin of the tropical forest tree family Icacinaceae. Amer. J. Bot. 88: 2259–2274.

Kårehed, J. 2002. Not just hollies – the expansion of Aquifoliales. In: Evolutionary Studies in Asterids emphasising Euasterids II. Uppsala, Sweden, Acta Univ. Uppsala, thesis 761.

Kong, D.-R., Peng, H., Liang, H.-X. 2002. A new type of embryo sac in *Cardiopteris* and its systematic implication. Acta Bot. Sinica 44: 496–498.

Kong, D.-R., Schori, M., Lu, S.-G., Li, L., Peng, H. 2014. Floral development of *Cardiopteris*, with emphasis on gynoecial structure and ovular morphology. J. Syst. Evol. 52: 629–642.

Lens, F., Kårehed, J., Baas, P., Jansen, S., Rabaey, D., Huysmans, S., Hamann, T., Smets, E. 2008. The wood anatomy of the polyphyletic Icacinaceae s.l., and their relationships within asterids. Taxon 57: 525–552.

Lobreau-Callen, D. 1982. Structures et affinités polliniques des Cardiopterygaceae, Dipentodontaceae, Erythropalaceae et Octoknemataceae. Bot. Jahrb. Syst. 103: 371–412.

Mauritzon, J. 1936. Embryologische Angaben über Stackhousiaceae, Hippocrateaceae und Icacinaceae. Svensk Bot. Tidskr. 30: 541–550.

Nasution, R.E. 1998. *Gonocaryum* Miq. In: Sosef, M.S.M., Hong, L.T., Prawirohatmodjo, S. (eds.) Plant Resources of South-East Asia no. 5(3). Timber trees: Lesser-known timbers. Leiden: Backhuys, pp. 262–263.

Schori, M., Furness, C.A. 2014. Pollen diversity in Aquifoliales. Bot. J. Linn. Soc. 175: 169–190.

Sleumer, H.O. 1971. Cardiopteridaceae. Flora Malesiana 7: 93–96.

Stevens, P.F. (2001 onwards). Angiosperm Phylogeny Website. Version 12, July 2012 [and more or less continuously updated since]. http://www.mobot.org/MOBOT/research/APweb/.

Tobe, H. 2012. Floral structure of *Cardiopteris* (Cardiopteridaceae) with special emphasis on the gynoecium: systematic and evolutionary implications. J. Plant Res. 125(3): 361–369.

Van Staveren, M.G.C., Baas, P. 1973. Epidermal leaf characters of the Malesian Icacinaceae. Acta Bot. Neerl. 22: 329–359.

Winkworth, R.C., Lundberg, J., Donoghue, M.J. 2008. Toward a resolution of Campanulid phylogeny, with special reference to the placement of Dipsacales. Taxon 57: 53–65.

Codonaceae

Codonaceae (Retief & A.E. van Wyk) Weigend & Hilger, Phytotaxa 10: 27 (2010).
Boraginaceae subfam. Codonoideae Retief & A.E. van Wyk, Bothalia 35: 79 (2005).

M. Weigend and H.H. Hilger

Annual to perennial herbs or shrublets, densely branched, spiny; strong taproot present; stems terete, erect; indumentum of white spines, stiff, unicellular trichomes with 3–12 cystolithic foot cells, and 2–5-celled, uniseriate, gland-tipped trichomes. Leaves alternate, exstipulate, petiolate, pedicels often bent towards shoot apex and lamina held ± parallel to shoot axis, lamina (narrowly) ovate with acuminate apex and rounded base, margin entire to coarsely serrate, lamina adaxially and along the margins densely spiny, abaxially with spines only on the very prominent midvein, venation pinnate with lateral veins ascending. Inflorescence frondose-bracteose, initially scorpioid, later straight monochasia, rarely reduced to a single terminal flower. Flowers erect, with 10–20 perianth elements, actinomorphic, bisexual, hypogynous; calyx divided nearly to base, calyx lobes unequal in width with narrow and wide calyx lobes alternating, densely spinose and pubescent; corolla white or yellow, sympetalous, campanulate to saucer-shaped, lobes half-erect to spreading, oblong with rounded apex; stamens as many as petals, filaments basally hairy, epipetalous, inserted near base of the tube, basally incurved and forming tight fascicle in the middle of the flower, anthers dorsifixed, exserted; ovary bicarpellate, nearly two-locular by deeply intruding placentae, style terminal, divided into two long lobes for ca. 1/2 of its length, elongating after pollen is shed; stigmas minute. Fruit a bivalved capsule, loculicidal, seeds numerous, ellipsoid to ovoid. Testa with deeply reticulate epidermis, endosperm copious.

A single genus with two species in SW Africa.

Vegetative Morphology. Both species of *Codon* are essentially perennial, aseasonal plants which grow and flower as long as sufficient moisture is available. In cultivation they reach the fruiting stage in ca. 4 months and may then die off if moisture is not sufficient in nature (then behaving as annuals). The densely spinose aspect of the plants together with their peculiar incurved leaves are good diagnostic features. Seedlings are phanerocotylar, with two subcircular, densely pubescent cotyledons.

Vegetative Anatomy. Very little is known about the vegetative anatomy of the group. The leaves of *C. schenckii* are bifacial, amphistomatic and the stomata are anomocytic. The indumentum consists of three different trichome types: stiff, unicellular trichomes with 3–12 cystolithic foot cells, 2–5-celled, uniseriate, gland-tipped trichomes and stiff, white spines, up to 1 cm long. The latter consist of a many-celled foot with numerous, linear foot cells making up the bulk of the length of the spine, and a single unicellular trichome at the tip. Wood structure of *C. royenii* was described by Carlquist and Eckhardt (1984). In the primary stem with a voluminous pith, small open collateral vascular bundles are arranged in a closed cylinder. Cortical and medullary bundles as well as an internal phloem are absent. The scarce secondary thickening develops from a cambial ring. Perforation plates are simple. Diffuse axial parenchyma is scarce. Ray cells are erect. Cortex is thin, with few rows of peripheral collenchyma. Nodes are unilacunar and the leaf trace is a complex vascular front. Anatomy of roots is unknown.

Floral Structure. The flowers of *Codon* are quite atypical for Boraginales and have been described in some detail by Brand (1913) and Gess (1999). The perianth is irregularly (10–)12 (–20)-merous, and calyx lobes are unequal with

J.W. Kadereit and V. Bittrich (eds.), *Flowering Plants. Eudicots*, The Families and Genera of Vascular Plants 14,
DOI 10.1007/978-3-319-28534-4_9, © Springer International Publishing Switzerland 2016

Fig. 25. Codonaceae. *Codon royenii*. Flower. Photo H.H. Hilger

narrow lobes alternating with wider lobes (Fig. 25). The calyx is densely covered with spines, as is the rest of the plant, but not the inner floral organs. The corolla is sympetalous with oblong lobes, and is whitish and campanulate in *C. royenii* but yellow and saucer-shaped in *C. schenckii*. At the base of the corolla tube, nectar is secreted into chambers which are formed by the filament bases protruding from the walls of the corolla tube. These chambers between the filament bases are distally closed by tufts of hairs inserting on the filaments themselves. The location of nectar secretion is unclear. Both the apex of the corolla lobes and the ovary wall are usually dotted with reddish-brown, lentil-shaped hollow protuberances which are likely to be oil or resin glands. The filaments arise from the corolla tube and are basally inflexed so as to form a tight fascicle in the centre of the flower. Hairy filament bases thus close access to the nectary chambers. This may have a function in the exclusion of illegitimate flower visitors and/or protect the nectar against evaporation. Both anthers and stigmata are exserted from the corolla. The superior ovary is ovoid to nearly spherical and crowned by a deeply bifid style. The T-shaped placentae are intrusive and meet in the centre of the ovary and thus nearly divide it into two cells.

Pollen Morphology. Pollen of *Codon* was studied by Constance and Chuang (1982). Pollen grains are tricolpate to tricolporate without pseudocolpi, subprolate to prolate, isopolar and with a microreticulate tectum. Grains are 15–25 μm long and ca. 10 μm wide. Doughnut-shaped tapetal orbicules similar to those of *Wellstedia* and some Hydrophyllaceae have been reported (Retief and van Wyk 2005).

Karyology. A haploid chromosome number of $n = 17$ has been reported for *Codon royenii* (Constance 1963).

Phytochemistry. Virtually nothing is known about the phytochemistry of *Codon*, but alkaloids appear to be absent (Raffauf 1996).

Fruit and Seed. The mature fruit is an ovoid capsule with apical-loculicidal dehiscence. Seeds are black to dark brown, irregularly ellipsoid to subspherical, or more or less conical and ca. 1.5–2 × 1–1.5 mm large. The testa is deeply and irregularly reticulate: the outer periclinal cell walls of the testa epidermis are collapsed in the mature seed, and anticlinal walls are very irregularly developed, some very low, others very high. Anticlinal walls of adjacent cells protrude over neighbouring cell walls, giving the entire seed surface an extremely irregular appearance. The seeds contain copious endosperm and a small, straight embryo.

Dispersal. The fruits remain erect and firmly attached to the plants, and the seeds are gradually dispersed by wind. In cultivation seedlings emerge irregularly over a long period of time (ca. 4 months), in nature they likely emerge after occasional rain falls in winter (June–July).

Affinities. *Codon* has been universally accepted as a member of Boraginales and has been traditionally considered a member of Hydrophyllaceae. It is quite isolated in Hydrophyllaceae on the basis of its polymerous flowers, spinose aspect (with unique spine anatomy) and geographical distribution. Retief and van Wyk (2005) provided an overview of its taxonomic history. Brand (1913) included *Codon* in Hydrophyllaceae tribe Phacelieae, but Constance (1963) considered it a quite isolated genus in that family, and suggested that it either belongs into a different family or should be moved into a family of its own. Pollen morphology has been used to argue for a close affinity between *Codon* and certain groups in Hydrophyllaceae and *Wellstedia* (Constance and Chuang 1982; Retief and van Wyk 2005; Retief et al. 2005). However, pollen of *Codon* is very unspecialized and represents one of the most common types in angiosperms (prolate, tricolporate, microreticulate). Similar pollen is found in both Hydrophyllaceae and some members of Boraginaceae s.str., but also elsewhere in Boraginales and Asterids. A molecular phylogenetic analysis by Ferguson (1999) retrieved *C. schenckii* as sister to Boraginaceae s. str. (with *Wellstedia* not included in the study), contradicting the widely held view of a close affinity with Hydrophyllaceae. The capsular fruit, polymerous flowers and unique spines are equally at odds with a classification in Boraginaceae s.str. and with a classification in Hydrophyllaceae. Retief and van Wyk (2005) consequently decided to move *Codon* into a distinct subfamily of Boraginaceae which they considered to comprise several more subfamilies treated as separate families by other authors (e.g. Cordiaceae, Hydrophyllaceae, Heliotropiaceae, Gottschling et al. 2001; Luebert and Wen 2008). We prefer assigning family rank to each of these groups since they comprise large, morphologically divergent and readily characterized lineages. The segregate families can be easily distinguished morphologically. Recent molecular data (Luebert and Wen 2008) support the placement of *Codon* close to Boraginaceae s.str. (and separate from Hydrophyllaceae s.str.) as found by Ferguson (1999). Weigend et al. (2013, 2014) clearly retrieve *Codon* and *Wellstedia* (= Wellstediaceae, another isolated African taxon of Boraginales) as consecutive sister taxa of Boraginaceae s.str. The same relationships were also found by Refulio-Rodríguez and Olmstead (2014). An inclusion of these two taxa in Boraginaceae s.str. would thus be possible without creating a paraphyletic taxon. Due to their disparate vegetative and generative morphology, however, an inclusion of Wellstediaceae and Codonaceae in Boraginaceae s.str. would make this morphologically homogeneous family unnecessarily heterogeneous and difficult to characterize. Both Wellstediaceae and Codonaceae have two-valved capsular fruits (versus four nutlets in Boraginaceae), but share a pollen type found in several lineages of Boraginaceae s.str.

Distribution and Habitats. *Codon* is restricted to southwestern Africa, ranging from Namibia to the northwest of South Africa, primarily in the Succulent Karoo. It is found in semi-arid savannas, but also in sandy washes (e.g., with *Augea*, *Sisyndite*, *Kissenia*) in arid regions.

Plant–Animal Interactions. The floral biology of *Codon* is relatively well understood. Gess (1999) and Gess and Gess (2004) reported large carpenter bees [e.g. *Xylocopa lugubris*] as pollinators of *C. royenii*. Flowers provide copious amounts of nectar, well hidden at the base of the corolla and accessible only to sufficiently large and strong visitors. Carpenter bees are polylectic and are likely not specialized pollinators of *Codon*. Conversely, at least three species of pollen wasps (Vespidae: Masarinae) are apparently largely dependent on *Codon* (*Jugurtia codoni*, two species of *Quartinia*), but without being effective pollinators.

Uses. *Codon* has no major uses, but the flowers with their copious nectar are considered a delicacy by the Nama people (Gess 1999), who call the genus “honey bush”.

Only one genus:

Codon L. Fig. 25

Codon L., Syst. Nat., ed. 12. 2: 292 (1767).

Description as for family. Two species, *Codon schenckii* Schinz and *C. royenii* L. in SW Africa.

Selected Bibliography

Brand, A. 1913. Hydrophyllaceae. In: Engler, A. (ed.) Das Pflanzenreich IV. 251. Leipzig: Engelmann, pp. 1–210.

Carlquist, S., Eckhart, V.C. 1984. Wood anatomy of Hydrophyllaceae. II. Genera other than *Eriodictyon*, with comments on parenchyma bands containing vessels with large pits. Aliso 10: 527–546.

Constance, L. 1963. Chromosome number and classification in Hydrophyllaceae. Brittonia 15: 273–285.

Constance, L., Chuang, T.I. 1982. SEM survey of pollen morphology and classification in Hydrophyllaceae (waterleaf family). Amer. J. Bot. 69: 40–53.

Ferguson, D.M. 1999. Phylogenetic analysis and relationships in Hydrophyllaceae based on *ndhF* sequence data. Syst. Bot. 23: 253–268.

Gess, S. 1999. *Codon* flowers - chalices of hidden nectar. Veld & Flora March 1999: 30–31.

Gess, S.K., Gess, F.W. 2004. Distributions of flower associations of pollen wasps (Vespidae: Masarinae) in southern Africa. J. Arid Environ. 57: 17–44.

Gottschling, M., Hilger, H.H., Wolf, M., Diane, N. 2001. Secondary structure of the ITS1-transcript and its application in a reconstruction of the phylogeny of Boraginales. Plant Biol. 3: 629–636.

Gürke, M. 1897. Borraginaceae (Asperifoliaceae). In: Engler, A., Prantl, K. (eds.) Die natürlichen Pflanzenfamilien, IV, 3a. Leipzig: W. Engelmann, pp. 71–131.

Luebert, F., Wen, J. 2008. Phylogenetic analysis and evolutionary diversification of *Heliotropium* Sect. *Cochranea* (Heliotropiaceae) in the Atacama Desert. Syst. Bot. 33: 390–402.

Peter, A. 1897. Hydrophyllaceae. In: Engler, A., Prantl, K. (eds.) Die natürlichen Pflanzenfamilien, IV, 3a. Leipzig: W. Engelmann, pp. 54–71.

Raffauf, R.F. 1996. Plant alkaloids - a guide to their discovery and distribution. Binghampton, NY: Haworth Press.

Refulio-Rodríguez, N.F., Olmstead, R.G. 2014. Phylogeny of Lamiidae. Amer. J. Bot. 101: 287–299.

Retief, E., van Wyk, A.E. 2005. Codonoideae, a new subfamily based on *Codon*. Bothalia 35: 78–80.

Retief, E., van Wyk, A.E., Condy, G. 2005. *Codon royenii*. Flowering Plants of Africa 59: 114–121.

Weigend, M., Hilger, H.H. 2010. Codonaceae - a newly required family name in Boraginales. Phytotaxa 10: 26–30.

Weigend, M., Luebert, F., Selvi, F., Brokamp, G., Hilger, H. H. 2013. Multiple origins for Hounds tongues (*Cynoglossum* L.) and Navel seeds (*Omphalodes* Mill.) - the phylogeny of the borage family (Boraginaceae s.str.). Mol. Phylogen. Evol. 68: 604–618.

Weigend, M., Luebert, F., Gottschling, M., Couvreur, T.L. P., Hilger, H.H., Miller, J. 2014. From capsules to nutlets—phylogenetic relationships in the Boraginales. Cladistics 30(5): 508–518.

Columelliaceae

Columelliaceae D. Don, Edinb. New Philos. J. 6: 46, 49 (1828), nom. cons., including Desfontainiaceae Endl. ex Pfeiff. (1873).

A. Backlund

Shrubs or treelets. Leaves opposite, simple, pinnately veined and petiolate, slightly connate; leaves with spinulose margins (*Desfontainia*) or finely serrate (*Columellia*), coriaceous, with mucronate glands at tips of teeth; stipules absent. Flowers in terminal or axillary cymes, bisexual, ±actinomorphic, perianth hypogynous (*Desfontainia*) or semi-epigynous (*Columellia*); sepals 5, fused, valvate or slightly imbricate in bud, persistent; petals 5, fused, imbricate in bud; stamens 2 (3) (*Columellia*) or 5 (*Desfontainia*), oppositisepalous, adnate to corolla, anthers basifixed, with prominent connective, dithecal and tetrasporangiate, introrse in *Desfontainia*, undeterminable in *Columellia* due to plicate thecae; ovary syncarpous of 2 carpels and bilocular (*Columellia*) or of 5 carpels and (imperfectly) pentalocular (*Desfontainia*); style short, simple, stigma lobed, isomerous with locule number, fertile ovules numerous, placentation parietal and multiseriate on intrusive placentae which are centrally fused, in *Desfontainia* occasionally only fused in the basal part of the ovary; ovules anatropous, unitegmic, tenuinucellate. Fruit in *Columellia* a loculi- and septicidal capsule opening with four valves, in *Desfontainia* a thin-walled berry. Seeds numerous, minute, oblong, endosperm copious, embryo minute, straight and without chlorophyll.

Columelliaceae comprise two genera and five species, distributed in montane rain or cloud forests from Costa Rica to Cape Horn, up to 3600 meters.

Vegetative Morphology, Anatomy and Ultrastructure. Young branches, leaves, and exterior parts of the sepals and petals in *Columellia* usually sericeous or pubescent, in *Desfontainia* glabrous. Leaf teeth, apices, and calyx lobes mucronate. Stomata anomocytic, on abaxial side of leaves. Midveins with a sclerenchymatous sheath. Hydathodes present in *Columellia*. Nodes unilacunar in both genera. Vessel element perforation plates scalariform, pits bordered. Endodermis in stem, cork cambium with stone-cells, pericyclic in *Columellia*, but reported as superficial in *Desfontainia*. Calcium oxalate druses in both genera, in *Desfontainia* prominently in nodes (Stern and Brizicky (1969) for *Columellia*; Mennega (1980) for *Desfontainia*).

Inflorescence. Inflorescences lax, cymose, few-flowered, apical on branches in *Columellia*, flowers apical and solitary in *Desfontainia*.

Flower Structure and Floral Anatomy. Corolla sympetalous, broadly campanulate in *Columellia*, long and tubular in *Desfontainia* (Fig. 26), with lateral connections between petals of the type present in all members of Dipsacales studied (Stern and Brizicky 1969; Backlund and Donoghue 1996). The two stamens of *Columellia* alternate with the lateral and posterior corolla lobes. Flowers with three stamens, or sessile to subsessile stamens have been described, and in these cases the position of the third stamen varies. Van Tieghem (1903) suggested a basic number of five stamens and subsequent abortion of three. This can be corroborated from anatomical sections of the corolla of *Columellia*, which is furnished with five vascular strands, two of which lead to the two stamens and three of which terminate at the level where the two stamens are inserted (Backlund, unpubl. data). The anthers of the two developing stamens are strongly plicate and highly conspicuous. Style simple, short and with bilobate (*Columellia*) or sometimes obscurely capitate (*Desfontainia*) stigma. Floral nectary disk present on top of the ovary in *Columellia*. Gynoecium

J.W. Kadereit and V. Bittrich (eds.), *Flowering Plants. Eudicots*, The Families and Genera of Vascular Plants 14, DOI 10.1007/978-3-319-28534-4_10, © Springer International Publishing Switzerland 2016

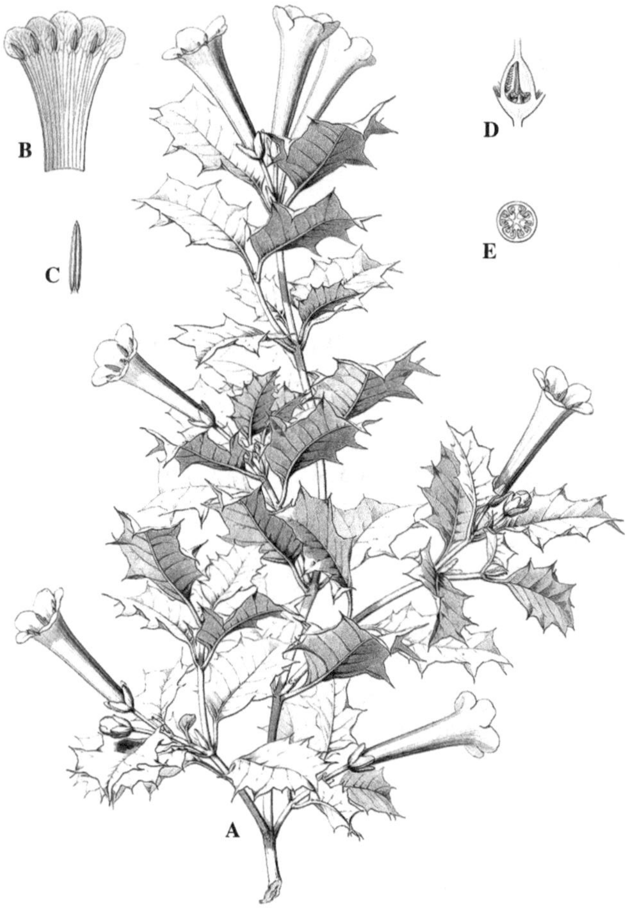

Fig. 26. Columelliaceae. *Desfontainia spinosa*. **A** Branch with flowers and imbricate buds. **B** Corolla, longitudinally opened. **C** Anther. **D** Fruit, longitudinal section, with incompletely separated locules and numerous seeds. **E** Gynoecium, transverse section, with intrusive placentae. (From Gay 1854)

syncarpous, in *Columellia* of two carpels, bilocular. In *Desfontainia* of (3–)5 carpels, pentalocular, occasionally reported as 5-locular at base but imperfectly septate opening into one locule at the top of the ovary. Placentation parietal with intrusive placentae.

Embryology. Endothecium with fibrous thickenings, microsporogenesis simultaneous, initial microspore tetrads tetrahedral. Anther wall with multi-nucleate, glandular (in *Desfontainia*) or amoeboid (in *Columellia*) tapetum of the common dicot type, which is reorganized into a false periplasmodium (Maldonade de Magnano 1986). Ovules numerous, anatropous, unitegmic, tenuinucellate, without hypostases. Internal epidermis of integument differentiates into endothelium with fibrous thickenings. Embryo sac development follows the Polygonum type. Embryogeny of Solanad type in *Desfontainia*, not studied in *Columellia*. Endosperm formation cellular. Micropylar and chalazal haustoria as well as an episperm of sclerids present. Food reserves as protein bodies remain in the fully developed seed. The embryo is small but well developed, without chlorophyll (Hakki 1980).

Pollen Morphology. Pollen tricolporate, oblate to suboblate, crassisexinous with a finely reticulated exine. Size 22.5–25 × 29–34 μm. Tectum with microperforations in *Desfontainia*. Columellae layer prominent in both genera. Pollen shed at three-celled stage in *Columellia*, not studied in *Desfontainia*.

Karyology. Base chromosome number $x=7$ for *Desfontainia* (Hakki 1980), no data available for *Columellia*.

Pollination. Pollination studies have not been reported for either *Columellia* or *Desfontainia*.

Fruit and Seed. The fruit is a loculi- and septicidal capsule opening with four valves (*Columellia*) or a berry with a thin, leathery pericarp (*Desfontainia*), in both cases with numerous small seeds.

Phytochemistry. *Desfontainia* contains iridoids of the seco- and ordinary type, including secoxyloganin, dimethylsecologanin, loganin, loganic acid, sweroside (Houghton and Lian 1986a) and triterpenoids (Houghton and Lian 1986b). Presence of alkaloids noted by Bisset (1980) but refuted by Houghton and Lian (1985). No phytochemical studies of *Columellia* have yet been published. Both genera have, however, been reported as medicinal, and *Desfontainia* even as psychoactive (Schultes 1977, 1981).

Affinities and Subdivisions Within the Family. Traditionally *Columellia* and *Desfontainia* have been treated as (often considered somewhat aberrant) members of a number of larger families, viz. Aquifoliaceae, Ebenaceae, Escalloniaceae, Gentianaceae, Gesneriaceae, Loganiaceae, Oleaceae, Rubiaceae, Saxifragaceae, Scrophulariaceae and Solanaceae. It has been stated that “It is clear, though, that the variety of families proposed as relatives of *Columellia* (Columelliaceae) could not be much more diverse" in a thorough review of the affinities of *Columellia* (Stern and Brizicky 1969, p. 39). For a review of the affinities of *Desfontainia* see, e.g. Leeuwenberg (1969). Similarities between *Columellia* and *Desfontainia* have also occasionally been emphasised—e.g. by Hallier (1910). Palynology (Mennega 1980) and similarities in wood anatomy (e.g. Carlquist 1992) are among the strongest morphological arguments for treating *Desfontainia* and *Columellia* as sister genera. The two genera have also been treated as monogeneric families, i.e. Columelliaceae and Desfontainiaceae (e.g. Takhtajan 1997). On the basis of phylogenetic analyses of molecular, embryological, anatomical and morphological information, a wider circumscription of Columelliaceae, including the former Desfontainiaceae, has been suggested (e.g. Backlund 1996; Backlund and Donoghue 1996; Backlund and Bremer 1997). The monophyly of this more widely circumscribed Columelliaceae appears to be strongly corroborated, but little is still known about the relationships of the family. Some features are indicative of a position not far from *Escallonia* and *Forgesia* of the Escalloniaceae. Affinities to euasterids II sensu APG (1998) appear, however, to be unambiguous. This is also supported in more recent molecular analyses (e.g. Soltis et al. 2011) where Columelliaceae form a strongly supported group, which in turn is loosely associated with Bruniaceae. This placement of this assemblage, tentatively referred to as Bruniales, is also poorly supported in analyses based on 17 genes (Soltis et al. 2011).

Distribution and Habitats. Both genera are distributed from Costa Rica (*Desfontainia*) through the mountains of Colombia (*Columellia*) southwards, with the range of *Desfontainia* extending along the Andes to southern Chile. Their usual habitat is wet forest on the western side of the mountains.

Economic Importance and Conservation. Both genera have horticultural value but are rarely seen in cultivation. The wood of *Columellia* is very hard and used for making utensils and handles. An infusion or decoction of *Columellia* leaves is reputed to be "wonderfully effective for tertian fevers" (MacBride 1961) and used also as a stomachic in Peru. This practice appears not to have spread to Ecuador where the only reported use of *Columellia* is as fire wood (Ruiz and Pavón 1794, in MacBride 1961, and Brizicky 1961). *Desfontainia* leaves have been reported to be used as a narcotic and stomachic in areas east of Pasto in Colombia. They are also being used as a source of yellow dye in textiles by Mapuche Indians in Chile.

Both genera are found in the threatened Andean cloud forest, but their distribution ranges are comparatively wide and both are generally locally abundant.

Key to the Genera of Columelliaceae

1. Leaf margin finely serrate; flowers yellow, stamens two, rarely three **1. *Columellia***
– Leaf margin spinulate; flowers orange, stamens five **2. *Desfontainia***

1. ***Columellia*** Ruiz & Pav.

Columellia Ruiz & Pav., Fl. Peruv. Prodr. 3, pl. 1 (1794), nom. cons.
Uluxia Juss. (1818).
Columellia Ruiz et Pavón, 1794, is conserved against the earlier homonym *Columella* Loureiro, Fl. Cochinch. 85 (1790), nom. rejic. vs. *Cayratia* Jussieu (1818), nom. cons. For further references see Bullock (1960), in Taxon 9: 221.

Inflorescences terminal, lax, with up to ten bright yellow flowers. Stamens two, rarely three, with conspicuously plicate anthers.

The genus contains four species possibly better regarded as one species, *Columellia oblonga* Ruiz & Pav. with two subspecies (Stern and Brizicky 1969). It is distributed at altitudes above 1600 m in the South American Andes from Colombia to Bolivia.

2. ***Desfontainia*** Ruiz & Pav. Fig. 26

Desfontainia Ruiz & Pav., Fl. Peruv. Prodr. 29, pl. 5 (1794).
Linkia Pers., Syn. 1: 219 (1805).

Shrubs, treelets or trees always with opposite, entire, and spinulose leaves.

One variable species, *D. spinosa* Ruiz & Pav., distributed in wet forests from Costa Rica to Cape

Horn. In a study by Weigend (2001) it has been proposed that this species should be divided in three morphologically less variable species.

Selected Bibliography

Angiosperm Phylogeny Group (APG). Bremer, K., Chase, M.W., Stevens, P.F. (compilers, in alphabetical order). Anderberg, A.A., Backlund, A., Bremer, B., Briggs, B.G., Endress, P.K., Fay, M.F., Goldblatt, P., Gustafsson, M.H.G., Hoot, S.B., Judd, W.S., Källersjö, M., Kellogg, E.A., Kron, K.A., Les, D.H., Morton, C. M., Nickrent, D.L., Olmstead, R.G., Price, R.A., Quinn, C.J., Rodman, J.E., Rudall, P.J., Savolainen, V., Soltis, D.E., Soltis, P.S., Sytsma, K.J., Thulin, M. (contributors, in alphabetical order). 1998. An ordinal classification of flowering plants. Ann. Missouri Bot. Gard. 85: 531–553.

Backlund, A. 1996. Phylogeny of the Dipsacales. Doctoral thesis, Systematic Botany, Acta Univ. Ups. 243. Uppsala.

Backlund, A., Bremer, B. 1997. Phylogeny of the Asteridae s.str. based on *rbc*L sequences, with particular reference to the Dipsacales. Plant Syst. Evol. 207: 225–254.

Backlund, A., Donoghue, M.J. 1996. Phylogeny of the Asteridae s.str. based on *rbc*L sequences, with particular reference to the Dipsacales. In: Backlund, A. Phylogeny of the Dipsacales, paper IV. Doctoral thesis, Department of Systematic Botany, Uppsala University.

Bisset, N.G. 1980. Phytochemistry. In: Leeuwenberg, A.J. M. (ed.) Loganiaceae, in Engler and Prantl, Die natürlichen Pflanzenfamilien, 2nd edn, vol. 28b(1). Berlin: Duncker & Humblot, pp. 211–237.

Brizicky, G.K. 1961. A synopsis of the genus *Columellia* (Columelliaceae). J. Arnold Arbor. 42: 363–372.

Carlquist, S. 1992. Wood anatomy of sympetalous dicotyledon families: A summary, with comments on systematic relationships and evolution of the woody habit. Ann. Missouri Bot. Gard. 79: 303–332.

Fritsch, K. 1894. Columelliaceae. In: Engler, A., Prantl, K., Die natürlichen Pflanzenfamilien IV, 3b. Leipzig: Wilhelm Engelmann, pp. 186–188.

Gay, C. 1854. Atlas de la historia fisica y politica de Chile, vol. 1: t. 56. Paris: E. Thunot.

Hakki, M.I. 1980. Embryology. In: Leeuwenberg, A.J.M. (ed.) in Engler und Prantl, Die natürlichen Pflanzenfamilien, 2nd edn, Fam. Loganiaceae, vol. 28b(1). Berlin: Duncker & Humblot, pp. 8–96.

Hallier, H. 1910. Über Phanerogamen von unsicherer oder unrichtiger Stellung. Med. Herb. Leiden 1.

Houghton, P.J., Lian, L.M. 1985. Iridoids from *Desfontainia spinosa*. Phytochemistry 24: 1841–1842.

Houghton, P.J., Lian, L.M. 1986a. Iridoids, Iridoid-triterpenoid congeners and lignans from *Desfontainia spinosa*. Phytochemistry 25: 1907–1912.

Houghton, P.J., Lian, L.M. 1986b. Triterpenoids from *Desfontainia spinosa*. Phytochemistry 25: 1939–1944.

Leeuwenberg, A.J.M. 1969. Notes on American Loganiaceae IV, Revision of *Desfontainia* Ruiz et Pav. Acta Bot. Neerl. 18: 669–679.

MacBride, J.F. 1961. Flora of Peru. Chicago: Field Museum of Natural History.

Maldonado de Magnano, S. 1986. Estudios embriologicos en *Desfontainia spinosa* (Desfontainiaceae). Darwiniana 27: 207–224.

Mennega, A.M.W. 1980. Anatomy of the secondary xylem. In: Loganiaceae, ed. by A.J.M. Leeuwenberg, in Engler and Prantl, Die nat. Pflanzenfamilien, 2nd edn, vol. 28b(1). Berlin: Duncker & Humblot.

Ruiz H., Pavón J. 1794, Florae Peruvianae et Chilensis prodromus. Madrid: Sancha.

Schultes, R.E. 1977. De Plantis Toxicariis e Mundo Novo Tropicale commentationes XV. Botanical Museum Leaflets, Harvard University 25: 99–104.

Schultes, R.E. 1981. Iconography of New World plant hallucinogens. Arnoldia 41: 80–125.

Soltis, D.E., Smith, S.A., Cellinese, N., Wurdack, K.J., Tank, D.C., Brockington, S.F., Refulio-Rodriguez, N. F., Walker, J.B., Moore, M.J., Carlsward, B.S., Bell, C. D., Latvis, M., Crawley, S., Black, C., Diouf, D., Xi, Z., Rushworth, C.A., Gitzendanner, M.A., Sytsma, K.J., Qiu, Y.L., Hilu, K.W., Davis, C.C., Sanderson, M.J., Beaman, R.S., Olmstead, R.G., Judd, W.S., Donoghue, M.J., Soltis, P.S. 2011. Angiosperm phylogeny: 17 genes, 640 taxa. Amer. J. Bot. 98: 704–730.

Stern, W.L., Brizicky, G.K. 1969. Comparative anatomy and relationships of Columelliaceae. J. Arnold Arbor. 50: 36–72.

Takhtajan, A. 1997. Diversity and Classification of Flowering Plants. New York: Columbia University Press.

Tieghem, P. v. 1903. Sur les Columelliacées. Ann. Sc. nat. 17–18: 155–164.

Weigend, M. 2001. *Desfontainia* Ruiz & Pav. (Desfontainiaceae) revisited: A first step back towards alphadiversity. Bot. Jahrb. Syst. 123: 281–310.

Dipsacaceae (inclusive *Triplostegia*)

Dipsacaceae Jussieu, Gen. Pl.: 194 (1789), nom. cons.

V. Mayer

Annual, biennial or perennial herbs, rarely woody semi-shrubs or shrubs. Leaves opposite, sometimes whorled, often in a basal rosette; entire or toothed to deeply pinnatifid or pinnately dissected, stipules 0; indumentum of uni- or multicellular, sometimes glandular hairs. Flowers in dense involucrate capitula; receptacle almost flat, hemispherical or cylindrical, mostly with scaly bracts, when scales absent with hairs or naked. Flowers bisexual, marginal flowers often female, 4- or 5-merous, actinomorphic or zygomorphic, with an epicalyx of 4 fused and strongly modified bracts surrounding the ovary; calyx small, cupuliform or rarely divided into 4–5 lobes or more often modified into 4 or 5 spiny setae, in some genera setae pappus-like and multiplied to up to 25, persistent or non-persistent; stamens 4 (in one species 2), filaments attached to corolla tube; gynoecium inferior with one fertile carpel, unilocular, ovule anatropous, unitegmic; style slender, stigma 1-, 2- or 3-lobed or capitate; flowers proterandrous. Fruit a dry, single-seeded nutlet (cypsela) enclosed in the quadrangular or cylindrical, ± lignified epicalyx, in *Triplostegia* a double epicalyx; the latter variously differentiated, often surmounted by a persistent calyx; *Knautia* with a fleshy appendage (elaiosome). Seeds with a large spathulate embryo and fleshy, oily endosperm. $x = 5, 7, 8, 9, 10$.

Fourteen genera in the core group of Dipsacaceae with about 300 species, *Triplostegia* as debated fifteenth genus. This genus differs from core Dipsacaceae by its thyrsoidal inflorescence instead of a capitulum and a double epicalyx covered with glandular hairs. Due to the polymorphism of *Knautia* and *Scabiosa* higher species numbers (up to 350) are often reported. Dipsacaceae is an extratropical family of the Old World, mainly distributed in the E Mediterranean and Irano-Turanian region, extending sparsely into C and S Africa and C and E Asia.

Vegetative Morphology. Most species of Dipsacaceae as well as *Triplostegia* are perennial herbs. They are hemicryptophytes with a basal rosette (Fig. 27). In *Lomelosia*, *Pterocephalus*, *Pseudoscabiosa* and *Scabiosa* chamaephytes with lignified stems and dwarf shrubs are common, the only species which is somewhat tree-like is the chamaephytic *Pterothamnus centennii*, reaching 2 m height. Biennial herbs are found in *Dipsacus* and a few species of *Scabiosa*. Annuals are rare and occur most often in *Lomelosia*, but also in *Cephalaria*, *Knautia*, *Pterocephalidium*, *Pterocephalus*, *Scabiosa* and *Sixalix*. The primary root can develop into a long, stout tap-root (e.g. *Scabiosa*). Adventitious root formation is found in *Succisa* and *Succisella*, in *Succisa pratensis* the root appears as if bitten off ("devil's bit-scabious"). *Succisella* propagates vegetatively through runners; *Succisa* and *Knautia* through elongated rhizomes. Stems are always erect and often sympodially branched. They have cauline leaves except in species of *Bassecoia*. Leaf arrangement is opposite, the basal leaves often form a rosette. The leaves are estipulate, sessile or shortly petiolate. Leaf form is simple with crenate or lyrate margins in all species of *Dipsacus*, *Pseudoscabiosa*, *Succisa* and *Succisella*. In all other genera they often vary within a single plant from simple to 1- to 2-pinnatifid (in *Lomelosia hymettia*, even up to 9-pinnatifid). Leaves are prickly in some *Dipsacus* species and have stellate hairs in *Pseudoscabiosa* and a few *Cephalaria* species. Connate cauline leaves are sometimes found in *Dipsacus* and in *Cephalaria syriaca*, connate involucral bracts exist in *Pycnocomon*.

J.W. Kadereit and V. Bittrich (eds.), *Flowering Plants. Eudicots*, The Families and Genera of Vascular Plants 14,
DOI 10.1007/978-3-319-28534-4_11, © Springer International Publishing Switzerland 2016

Fig. 27. Dipsacaceae. *Scabiosa ochroleuca*. **A** Basal leaf rosette. **B** Flowering and fruiting shoot. **C** Capitulum with receptacular bracts. **D** Flowers. **E** Diaspore. **F** Fruit s.str. (without epicalyx) and calyx setae. (Modified from Reichenbach 1850)

Vegetative Anatomy. Wood anatomy of all species examined by Carlquist (1982) and Solereder (1899) shows some primitive features: all species have tracheids rather than the expected libriform fibres. In the secondary xylem the perforation plates are predominantly simple. Scalariform perforation plates occur rarely in the primary xylem. According to Carlquist (1982: 445), "wood structure of Dipsacaceae appears to be basically like that of a shrubby family", thus indicating that woodiness may be primary in the family. The vascular bundles of the axis are at first separate, but soon become interconnected into a continuous ring by phloem and prosenchymatous elements. Vessels are usually in radial rows and many have spiral thickenings. The sieve-tube plastids of Dipsacaceae belong to the S-type (Behnke 1994). Cork in the stem originates subepidermally in *Knautia sylvestris*, in *Cephalaria*, *Dipsacus* and some *Knautia* species inside the pericycle (Solereder 1899). An endodermis with thickened walls has been reported for *Dipsacus*. Medullar rays are only 1–2 cells wide, multiseriate rays occur in *Scabiosa africana*. The rays consist of relatively thin-walled cells which are lignified and in some cells bordered pits occur. Perforated ray cells were observed in *Pterocephalus dumetorum*. The pith is usually wide and consists of thin-walled or collenchymatous cells, and stems become frequently hollow when mature. Crystals of calcium oxalate are often present in cells of the parenchymatous tissues (cortex, pith) and are usually arranged in clusters. In *Scabiosa africana* rhomboidal crystals are present in virtually every ray cell (Carlquist 1982).

Leaves are dorsiventral (*Dipsacus pilosus*, *Knautia arvensis*, *K. dipsacifolia*, *Scabiosa columbaria* and presumably most other species) or centric (*Dipsacus azureus*, *Lomelosia caucasica*); isobilateral leaves in the upper part of the stem occur in *Dipsacus laciniatus* and *D. silvestris* (Metcalfe and Chalk 1950). Stomata are generally present on both sides of the leaves; stomata development is anomocytic and rarely anisocytic. Petiolar vascularization in Dipsacaceae is trilacunar, the lateral traces bifurcate. In contrast to Valerianaceae, the median trace does not ramify (Neubauer 1978). Hairs are unicellular and usually lignified in *Dipsacus*, *Knautia* and *Succisa*, sometimes emerging from multicellular emergences. In many species of *Cephalaria*, *Dipsacus*, *Knautia*, *Lomelosia*, *Scabiosa* and *Sixalix*, glandular hairs with a short stalk and a head composed of a few cells occur. Additionally, in *Knautia* glands with a long uniseriate stalk and a multicellular head are found. The prickles on stem and leaf margins in *Dipsacus* are of epidermal origin, as are the glandular leaf teeth which are found on the leaf margin of *Knautia* and *Cephalaria*. Teeth and plumose hairs on the calyx setae originate as outgrowths of the epidermal cells. In *Triplostegia* the branches and epicalyces are covered with capitate-glandular hairs (Fig. 28). Vascular bundles of the leaf veins are not accompanied by sclerenchyma. In the mesophyll, clusters of calcium oxalate crystals are often found. Crystalliferous idioblasts with thick walls have been recorded in bracts of *Cephalaria* (Metcalfe and Chalk 1950).

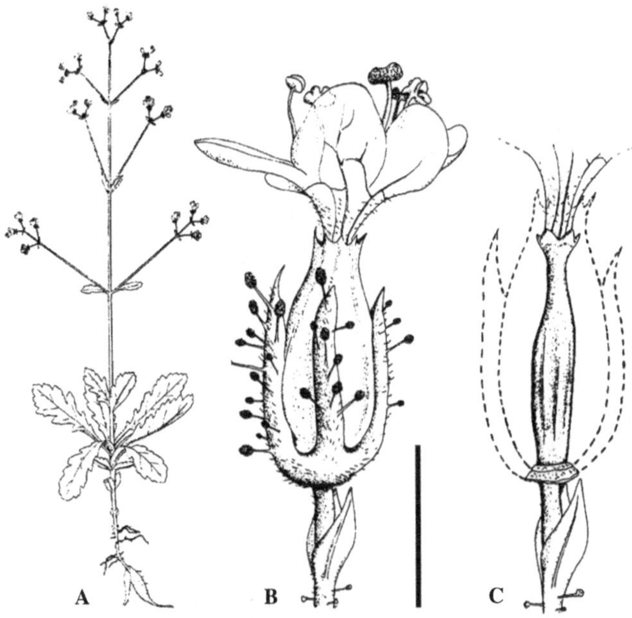

Fig. 28. Dipsacaceae. *Triplostegia glandulifera*. **A** Flowering plant. **B** Flower with epicalyces. **C** Epicalyces opened. Scale bar 2 mm. (From Backlund and Nilsson 1997)

Inflorescence Structure. The flowers of Dipsacaceae are grouped in dense, involucrate heads which are either cylindrical (*Dipsacus*), globose (*Cephalaria*, rarely *Dipsacus*) or radiant (*Lomelosia*, *Scabiosa*, a.o.). The heads contain 5–10 (*Knautia orientalis*, *Cephalaria syriaca*) or up to 115 (*Sixalix atropurpurea*) usually hermaphroditic flowers (Günthart 1904). In *Cephalaria*, *Knautia*, *Pterocephalus*, *Scabiosa* and *Sixalix*, the capitulum often has enlarged marginal flowers which are usually female. In the few-flowered capitula of *Knautia orientalis*, the lower corolla lobes are lanceolate and oblong, such that the capitulum resembles a flower of Caryophyllaceae. In *Triplostegia* the small flowers are aggregated in a more or less few-flowered terminal cymose inflorescence (Weberling 1961) (Fig. 28). The receptacular scales are linear-lanceolate in most genera (Fig. 27C), in *Cephalaria* they are scarious, in *Dipsacus* ± spine-tipped and large. In *Knautia* and *Pterocephalus* the receptacle lacks scales and is hairy instead.

The phylogenetic origin of the capitulum of Dipsacaceae is still unclear. Van Tieghem (1909) and others interpreted the capitulum as a racemose, botrytic structure. Some authors (e.g. Čelakovsky 1893) interpreted it as a Valerianaceae-like thyrsoid structure with an indeterminate main shoot lacking a terminal flower. According to this hypothesis, the axillary side-branches are shortened and form a capitulum. Szabó (1930) considered the capitulum as homologous to a dichasium with contracted side-branches lacking a terminal flower. The pattern of primordia emergence, however, does not indicate that the capitulum is a compound inflorescence (Mayer, unpubl.). A very peculiar character is the opening sequence of the flowers. Contrary to the expected acropetal sequence, a ring in the middle sector of the capitulum opens first, while those in the centre and outside of the ring remain closed (Fig. 29A). Different but unsatisfactory explanations were given for this observation, such as differences in nutrient supply (Goebel 1924) or pressure of neighbouring flowers (Günthart 1904).

Flower Structure and Floral Anatomy. The single flower usually appears in the axil of a bract, the inferior ovary closely surrounded by an epicalyx (= involucel, outer calyx; Fig. 29B). The epicalyx originates from two median and two transverse bracts which are fused and surround the inferior ovary like a tube. The calyx is persistent in most genera and caducous only in *Cephalaria*, *Dipsacus*, *Knautia* and *Pterocephalidium*. The sepals are variously modified in the family. Four stiff and rough persistent calyx setae ± densely covered with small teeth occur in *Pseudoscabiosa saxatilis* and *P. limonifolia*, 4–5 stiff setae in *Succisa* or consistently five in *Lomelosia*, *Pycnocomon rutifolium*, *Scabiosa* (except *S. triandra*) and *Sixalix*. *Lomelosia brachiata* has 8–10 relatively broad, fringed setae. 12–16 toothed setae are found in *Bassecoia siamensis* and *B. bretschneideri*. In *Pseudoscabiosa grosii* the calyx comprises 4–6, in *Pycnocomon intermedium* 5 (-8) relatively long and plumose setae and, in *Pterocephalus*, *Pterocephalidium* and *Pterothamnus*, the calyx is consistently modified to 12–24 pappus-like plumose setae. A reduction of calyx setae into short lobes is found in *Succisella* and *Scabiosa triandra*. In *Cephalaria*, *Dipsacus* and *Knautia* it is cupuliform and often pubescent, sometimes with 8 apical teeth in *Cephalaria* or frequently with (6–)8–16(–24) apical awns or teeth in *Knautia*. The corolla is 4-merous (*Bassecoia*, *Cephalaria*, *Dipsacus*, *Knautia*, *Pseudoscabiosa*, *Pterocephalidium*, *Pterothamnus*, *Succisa*

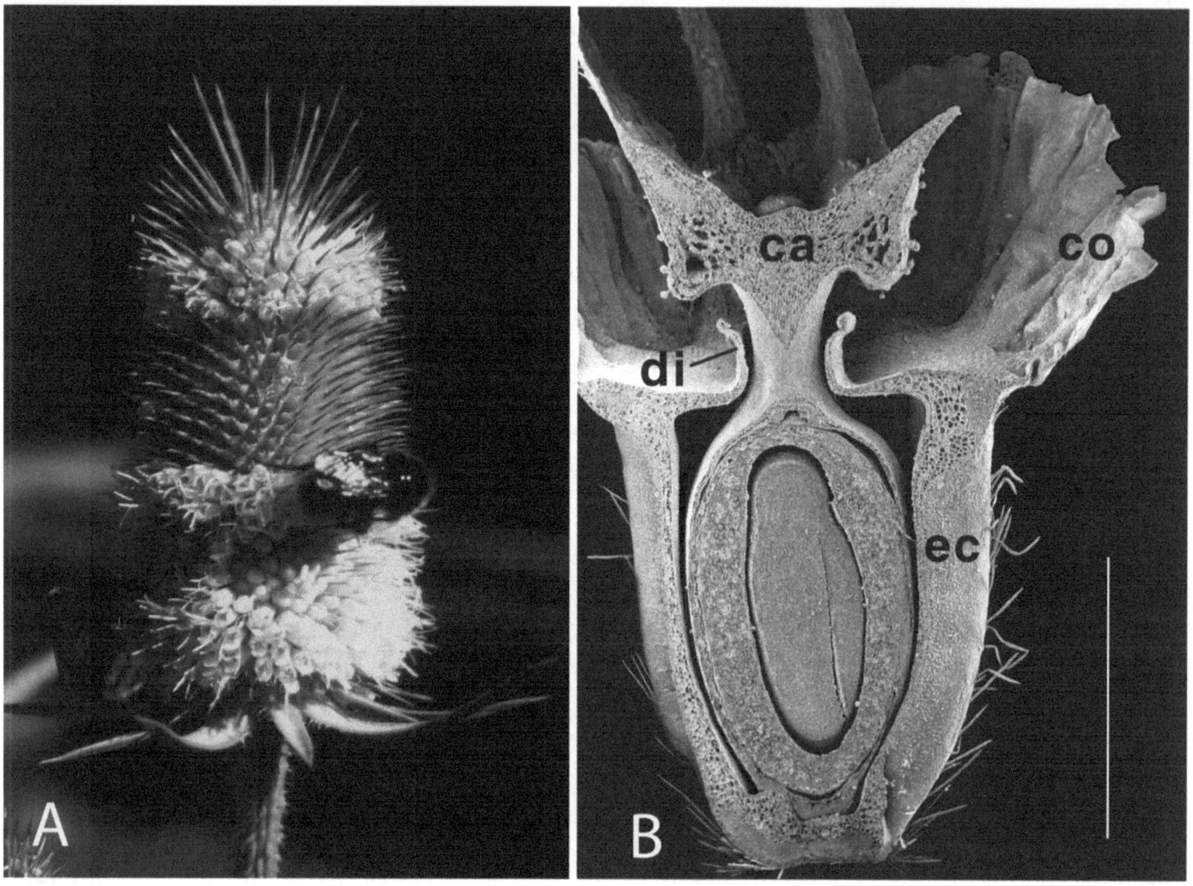

Fig. 29 Dipsacaceae. **A** *Dipsacus fullonum*. Flowering sequence of the inflorescence. **B** *Scabiosa columbaria*. Longitudinally sectioned diaspore. The fruit s.str. is enclosed by a tubular epicalyx (ec). At the apical end the tube is closed by the diaphragm (di) which surrounds the calyx (ca) stalk. In some Dipsacaceae a membranous corona (co) is present. Bar: 2 mm. (Orig.)

and *Succisella*) or 5-merous (*Lomelosia*, *Pycnocomon*, *Pterocephalus*, *Scabiosa*, except *S. suaveolens* with 4-merous flowers and *Sixalix*) and actinomorphic or zygomorphic. In *Pterocephalus afghanicus*, *P. khorassanicus* and *P. frutescens* 4- and 5-merous flowers occur in the same capitulum. The question whether flowers are originally 4-merous, and the 5-merous condition results from the incision of the [median] lobe, or whether 4-merosity is due to loss of a primordium is still not solved. Only four vascular strands are found in 4-merous flowers (Van Tieghem 1909; Szabó 1923), whereas in 5-merous flowers the strand of the fifth lobe results from a ± deep bipartition of another strand (Molliard 1895; Alvarado 1927). A nectary consisting of unicellular trichomes is present at the base of the corolla, forming a ring or protuberance (Wagenitz and Laing 1984). In all species the androecium consists of four stamens alternating with the corolla lobes, with only two stamens in *Pterocephalidium diandrum*. The androecium of Dipsacaceae can be regarded as tetramerous since no trace of a fifth stamen can be found (Van Tieghem 1909; Szabó 1923). The filaments are glabrous and attached to the corolla tube. During ripening the bent filaments become straight and the two posterior stamens rise earlier than the anterior ones. The filaments are attached to the anthers by a thin joint on the dorsal side; thus the anthers become versatile. Anthers are dorsifixed, sagittate, tetrasporangiate and dithecal, opening by longitudinal slits. The gynoecium is basically formed of two carpels with one of them reduced, the ovary therefore is unilocular and pseudomonomerous. Stigmata are capitate with oblong or plain stigma heads (*Bassecoia*, *Lomelosia*, *Pseudoscabiosa*, *Pterocephalus*, *Pterocephalidium*, *Pterothamnus*, *Pycnocomon*, *Scabiosa*, *Succisa*, *Succisella*) or 2–3-lobed in *Knautia* species. *Cephalaria* and *Dipsacus* possess very long, single-lobed, laterally decurrent stigmata. This is also the case in the annuals *Pterocephalus papposus* and *Scabiosa parviflora*. The stigmata are non-papillate and dry. *Triplostegia* has a minute, dentate calyx, 4(5)-merous corolla, four stamens and a pseudomonomerous inferior gynoecium with a capitate stigma. According to Peng et al. (1995, p. 506), the style of *Triplostegia* "has three distinct xylems (. . .), suggesting that the gynoecium is composed of three carpels." The most peculiar feature of *Triplostegia* is the double epicalyx that subtends each flower (Fig. 28B, C).

Embryology. The anther wall is 4-layered, comprising the epidermis, a fibrous endothecium, an ephemeral middle layer and a secretory tapetum of large multinucleate vacuolate cells (Kamelena 1980). The tapetal cells extend pseudo-amoeboidally into the locule. According to Kamelena (1980), there is no evidence of periplasmodium formation in *Knautia arvensis*, *Scabiosa lucida* and *S. bipinnata* whereas in *Cephalaria ambrosioides*, *Dipsacus leschenaultii* and *Lomelosia caucasica* a periplasmodium has been reported (Johri et al. 1992). The tetrads are tetrahedral and isobilateral. The pollen grains are trinucleate. The ovary is inferior and unilocular with a single anatropous unitegmic tenuinucellate ovule. Placentation is apical. The integument is massive with a very long and narrow micropyle. A prominent thick-walled hypostase and an endothelium differentiate. The development of the embryo has been studied mainly in *Scabiosa* (Frey 1976), where the chalazal megaspore lies very close to the chalazal region and develops into a Polygonum-type embryo sac. In the embryo sac the polar nuclei fuse before fertilization, the three antipodal cells are 2–4-nucleate (Kamelena 1980). The nuclear division is followed by an oblique vertical wall and the development corresponds to the Piperad type. The albuminous seeds have a large, straight, spathulate, chlorophyllous embryo and a fleshy, oily endosperm. In *Triplostegia* and all genera of core Dipsacaceae the endosperm is copious (see Mayer and Ehrendorfer (2013), their Figs. 3A–E, 6D–F, 7B, C), as in the basal lineages of Valerianacae (Jacobs et al. 2011). This contrasts with the widespread report of only poorly developed endosperm.

Pollen Morphology. The pollen grains are tricolpate, subequiaxial to prolate (= longiaxial) or triporate, oblate (= breviaxial) to subequiaxial (Verlaque 1981; Clarke and Jones 1981). Measuring 80–180 μm at the longest axis, they are comparably large (Verlaque 1981). The exine ornamentation consists of few echinae (over 1 μm in height) and relatively dense microechinae (less than 1 μm in height). Exceptions are pollen grains of *Bassecoia* with monomorphic echinae rather loosely scattered over the surface. The structure of the aperture is complex with a thickened margin (margo) surrounding the aperture on the inside. A wide halo is present in *Dipsacus* and *Cephalaria*. Especially in *Dipsacus* the echinae on the aperture membrane are very long and branched, forming a loosely organised operculum. The pollen grains have a tectum of complex structure, the columellae are often branched. In *Scabiosa*, *Lomelosia*, *Pseudoscabiosa* and *Pterocephalidium*, the columellae are longer at the poles than in the mesocolpia and often appear rhombic in equatorial view. In *Bassecoia*, the columellae become very loose in the polar region. In *Pseudoscabiosa*, *Pterocephalidium diandrum* and *Pterothamnus centennii*, cavae can be found in the polar region (e.g. Erdtman 1952; Diez and Devesa 1981; Verlaque 1986a; Valdez et al. 1987). These have been interpreted as polar sexine detached from the nexine (Erdtman 1952). In *Triplostegia* the margo surrounding the apertures on the inside is lacking; instead, a halo similar to that of pollen of Valerianaceae can be observed (Backlund and Nilsson 1997). This feature in combination with other characters encouraged the latter authors to support the inclusion of *Triplostegia* in Valerianaceae.

Karyology. The dominant chromosome number in Dipsacaceae is $2n = 18$ ($x = 9$), found in the majority of species of *Cephalaria*, *Dipsacus*, *Lomelosia*, *Pterocephalus* and *Pycnocomon*. Also *Triplostegia* has $2n = 18$. Nearly all species are diploids. Exceptions are tetraploid species with $2n = 36$ of the *Cephalaria* sect. *Cephalaria* group, *Dipsacus chinensis* (Kachidze 1929) and *Pterocephalus kurdicus* (Verlaque 1986a). The W

Mediterranean annual *L. stellata* is triploid with $2n = 27$, 28 or rarely 30 chromosomes (Verlaque 1986b). Chromosome numbers are $2n = 20$ ($x = 10$) in *Pseudoscabiosa*, *Succisa* and *Succisella*; in these three genera no polyploids are found. *Knautia* also has the base number $x = 10$ (except for the annual *K. orientalis* with $x = 8$) but, apart from diploids, also tetraploids and hexaploids occur (see Frajman et al. 2015). According to Ehrendorfer (1962), aneuploid individuals can be found in *Knautia* sect. *Sylvaticae*. *Bassecoia siamensis*, *Pterocephalidium*, all species of *Scabiosa* and *Sixalix* are diploids with $2n = 16$, except *S. parviflora* for which peculiar $3x$ (and $4x$) counts have been reported (Verlaque 1986a: 50). The annual *P. brevis* deviates from all other *Pterocephalus* species with $2n = 16$ (Díaz Lifante et al. 1992). The annual *Lomelosia persica* has $2n = 16$, 17 or 18 chromosomes and, for the annual *L. calocephala*, $2n = 16$ ($x = 8$) has been reported. *Lomelosia brachiata* has $2n = 14$ ($x = 7$; Verlaque 1977).

Pollination and Reproductive System. The single flowers are usually hermaphroditic and protandrous, gynodioecy also occurs. Flower number does not differ between female and hermaphroditic capitula (Larsson 2005). According to Müller (1873), the hairs on the inner side of the corolla prevent the intrusion of raindrops. The filaments elongate during the male stage and expose the anthers ca. 4–6 mm outside the flower tube. The versatile anthers act like a 'seesaw' stamping pollen onto pollinators (Larsson 2005). The flowers are insect-pollinated, offer nectar and pollen as reward and are considered to exhibit a generalized pollination system. They are pollinated mainly by bees and bumblebees, but also visited by many other insect groups: long-tongued hoverflies (Syrphidae; Kugler 1955), moths (Willis and Burkill 1895–1908 cited in Proctor et al. 1996), Lepidoptera (e.g. Lycaenidae) and Diptera (Müller 1873). Fertilization is predominantly allogamous. Some *Knautia* and *Scabiosa* species are reported to be self-compatible (Ehrendorfer 1962; Vange 2002).

Fruit and Seed. The fruits of Dipsacaceae are single-seeded dry nutlets. The pericarp consists of an outer and inner epidermis and residues of the parenchymatous tissue (Mayer 1995; Mayer and Ehrendorfer 1999, 2000). However, the most distinctive feature of Dipsacaceae s.str. is the ± lignified epicalyx, which encloses the fruits completely (Fig. 29B) and functions in ovary protection, seed dispersal and germination (Ehrendorfer 1964, 1965; Mayer 1995). The epicalyx tube is apically open in *Bassecoia*, *Pseudoscabiosa* and in some species of *Cephalaria*, *Dipsacus*, *Pterocephalus* and *Succisa*. In all *Lomelosia*, *Pycnocomon*, *Scabiosa* and *Sixalix* species, a thin ventral outgrowth (diaphragm) closes the tube apically (Fig. 29B, di). In the remaining species, ± prominent bulges are formed. A ± large dorsal membranous outgrowth (corona) is found in all *Lomelosia*, *Pycnocomon* and *Scabiosa* species, as well as in some *Bassecoia*, *Cephalaria* and *Pseudoscabiosa* species (Fig. 29B, co). Between diaphragm and corona an "epidiaphragm" is developed which varies in length and position (e.g. short in *Scabiosa*, elongated and upright in *Sixalix* and *Pycnocomon*, elongated and horizontal in *Lomelosia*; see also Mayer and Ehrendorfer 1999, 2000) (Fig. 29B, ed). A detailed developmental study of the diaphragm, corona and pits was provided by Hilger and Hoppe (1984). The tube itself can be strongly modified (Figs. 30, 31): 4-angled with an entire margin or lobed (*Dipsacus*, *Cephalaria*, *Succisa*), 4–8-angled with 4 or more apical setae (*Cephalaria*); ovoid to elliptical with a white elaiosome at the base (*Knautia*); cylindrical with conspicuous ribs and furrows, sometimes hairy (*Pterocephalus*, *Scabiosa*, *Succisella*); cylindrical with 8 deep pits (= foveoles), often densely hairy (*Lomelosia*, two species of *Pseudoscabiosa*); 4-angled with 8 deep pits (*Pycnocomon*). Heterocarpic diaspores are found in *Sixalix semipapposa* (Fig. 31E), *S. rhizantha* and *Pycnocomon rutifolium* (Fig. 30N). In the diaspores of *Pterocephalidium diandrum* one epicalyx tooth is developed into a flattened and curved seta up to 15 mm long (Fig. 30E). The epicalyx may become very large. In annual species of *Lomelosia* it constitutes up to 70% of the whole fruit mass due to an increasing sclerification of the tube. In addition to the primary ring of longitudinal lignified fibres, a secondary inner ring of horizontal fibres is formed. A correlation of this feature with the habitat of the plants could be established (Mayer 1995). In some genera (*Bassecoia*, *Lomelosia*, *Pterocephalus*, *Pterothamnus*, *Pseudoscabiosa*, *Scabiosa* and *Sixalix*), the calyx is persistent and modified into spiny or plumose setae. The dispersal unit consists of fruit s.str. plus

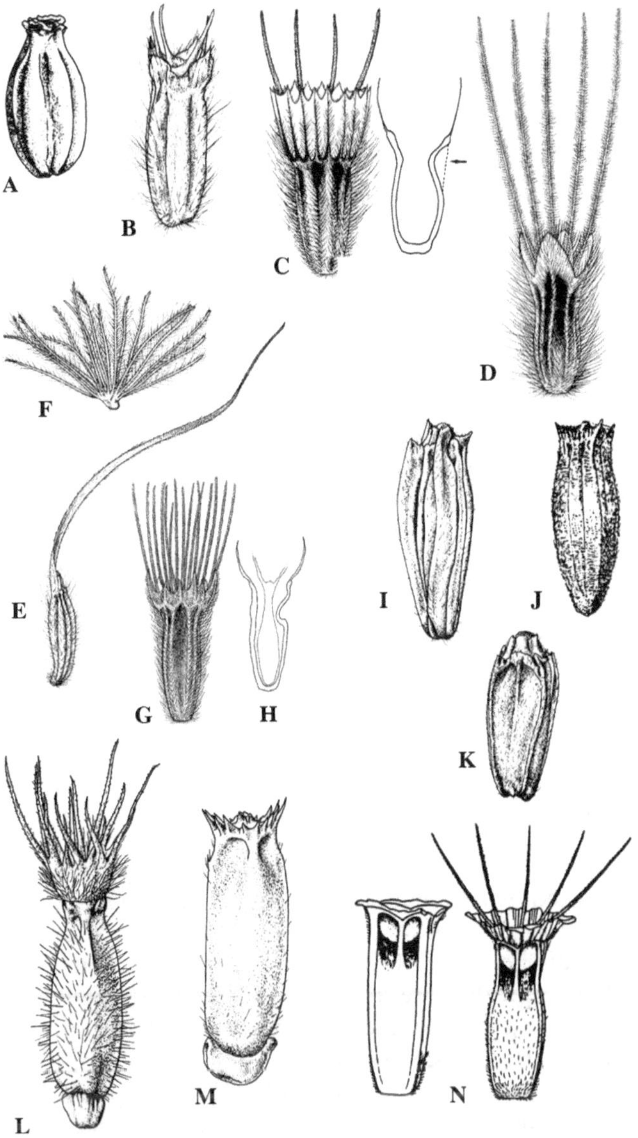

Fig. 30. Dipsacaceae. Diaspores. **A** *Succisella inflexa*. **B** *Succisa pratensis*. **C, D** *Pseudoscabiosa*. **C** *P. saxatilis*, diaspore habit and schematic longitudinal section. **D** *P. grossii*. **E, F** *Pterocephalidium*. **E** *P. diandrum*, diaspore with long seta. **F** *P. diandrum*, shed calyx. **G, H** *Bassecoia*. **G** *B. siamensis*, diaspore habit. **H** *B. siamensis*, schematic longitudinal section. **I, J** *Cephalaria*. **I** *C. uralensis*. **J** *C. leucantha*. **K** *Dipsacus pilosus*. **L, M** *Knautia*. Diaspores with elaiosome at the base. **L** *K. dipsacifolia*. **M** *K. orientalis*. **N** *Pycnocomon rutifolium*, heterocarpic fruits. **O–R** *Lomelosia*. **O** *L. brachiata*, diaspore habit and schematic longitudinal section. **P** *L. caucasica*. **Q** *L. cretica*. **R** *L. calocephala*. (C, D, N–R from Mayer and Ehrendorfer 1999; E, F from Mayer and Ehrendorfer 2013; L, M from Mayer and Svoma 1998)

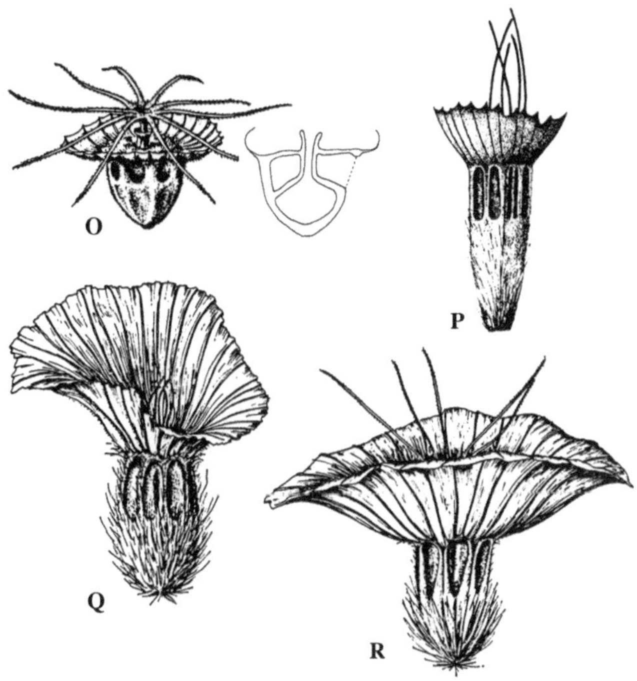

Fig. 30. (continued)

epicalyx and, if persistent, the calyx. The diaspore of *Triplostegia* has a double epicalyx. The outer whorl consists of four conspicuous, persistent, basally connate bract-like epicalyx lobes, the inner, also persistent epicalyx is 8-ribbed and urceolate. The seed coat is vascularised, composed of the outer epidermis and a hypodermis, the rest of the tissue having been crushed.

Dispersal. Different modes of dispersal are found in Dipsacaceae: (1) The most obvious one is anemochory. The large membranous corona of the epicalyx (*Lomelosia*, some *Pterocephalus*, *Scabiosa*) and the plumose calyx setae (*Pseudoscabiosa grosii*, *Pterocephalidium* and all species of *Pterocephalus*) are adaptations to wind dispersal. Apart from a significantly reduced sinking velocity which allows anemochory, the fruits have been observed to bounce along the ground driven by wind in open habitats. (2) Due to the long, spiny calyx setae which adhere to the fur of animals, epizoochorous dispersal occurs in *Bassecoia*, *Lomelosia*, *Pseudoscabiosa*, *Scabiosa* and

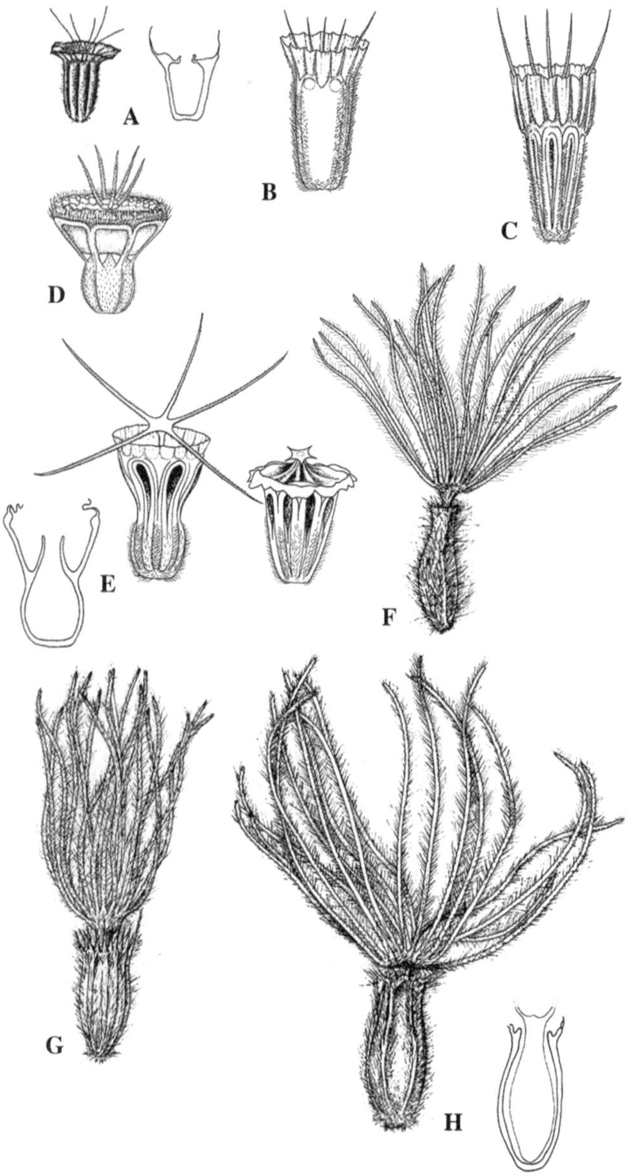

Fig. 31. Dipsacaceae. Diaspores. **A–C** *Scabiosa*. **A** *S. columbaria*, diaspore and schematic longitudinal section. **B** *S. lachnophylla*. **C** *S. africana*. **D, E** *Sixalix*. **D** *S. eremophila*. **E** *S. semipapposa*, heterocarpic diaspores and schematic longitudinal section. **F–J** *Pterocephalus*. **F** *P. depressus*. **G** *P. perennis*. **H** *P. frutescens*, diaspore habit and schematic longitudinal section. **I** *P. wendelboi*, diaspore habit and schematic longitudinal section. **J** *P. plumosus*, diaspore habit and schematic longitudinal section. (A–E from Mayer and Ehrendorfer 1999; F–J from Mayer and Ehrendorfer 2000)

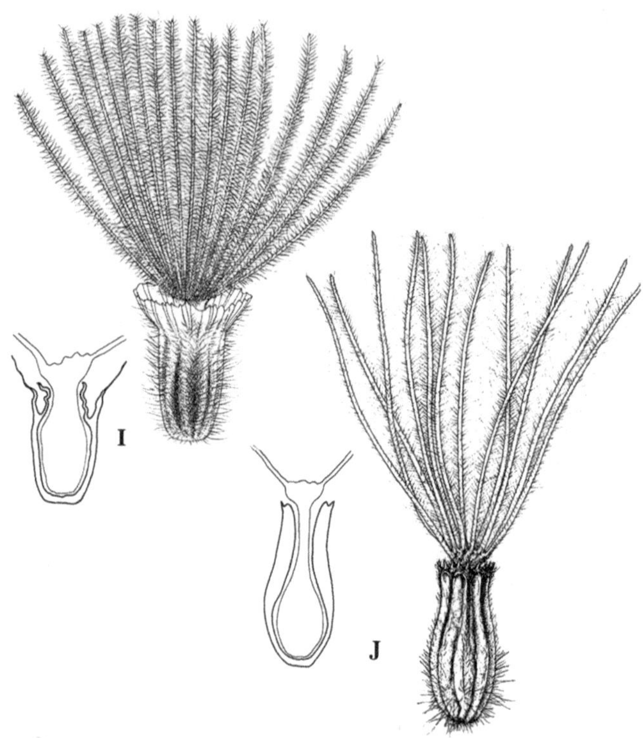

Fig. 31. (continued)

Sixalix. The setae and the corona margins are adorned with small teeth which may increase dispersal efficiency. In some oriental species of *Cephalaria*, most of them annuals, the main ribs of the epicalyx are apically elongated into four awns longer than the epicalyx tube itself (e.g. *Cephalaria dichaetophora*, *C. setosa*, *C. syriacaca*). Epizoochory is very probable in these species but has never been demonstrated experimentally. (3) Myrmecochory is observed in *Knautia*. The presence of an elaiosome in *Knautia* was already pointed out by Sernander (1906) and his experiments demonstrated ant dispersal of these species. The elaiosome is formed by the epicalyx (Mayer and Svoma 1998). (4) The stiff, spiny and downward-curving involucral bracts and receptacular scales in species of *Dipsacus* sect. *Dipsacus* constitute a catapult mechanism. (5) Additionally, hydrochory in

Succisa pratensis (Ridley 1930) and some *Cephalaria* species (Szabó 1940) has been reported. The diaspores of many *Cephalaria*, *Dipsacus*, *Succisa* and *Succisella* species are very simple and unspecialised (Fig. 30A, B, I–K). After detachment they fall to the ground.

Phytochemistry. Dipsacaceae frequently contain iridoids, glycosides and seco-iridoids (Jensen et al. 1975; Jensen 1991; Graikou et al. 2002; Christopoulou et al. 2008; Sarıkahya et al. 2011). Mono- and bis-iridoid compounds such as loganin, sweroside, cantleyoside, sylvestroside, laciniatoside, picein and loganic acid have been isolated (e.g. Jensen et al. 1975; Kocsis et al. 1993; Tomassini et al. 2004; Sarıkahya et al. 2011). Triterpenoid glycosides (based on hederagenin, oleanolic acid, ursolic acid) were found in all species investigated so far (*Bassecoia*, *Cephalaria*, *Dipsacus*, *Pterocephalus*, *Scabiosa*; Akimaliev et al. 1971; Kırmızıgül et al. 1995; Yang et al. 2007; Liu et al. 2011). Within the large group of phenylpropanoids, flavonoid composition is heterogeneous in the species studied (*Cephalaria gigantea*, *Scabiosa*, *Lomelosia*) and luteolin, galangin, apigenin, isovitexin, quercetin, diosmetin, rutin and their derivates could be detected (Perdetzoglou et al. 1994; Movsumov et al. 2006). Coumarins sometimes occur, also phenolic compounds like vanillin, vanilloloside and vanillic acid (Christopoulou et al. 2008), caffeoyl quinic acid derivatives (Hung et al. 2006) and lignan glycoside (Pasi et al. 2002). Monoterpenoid glucoindole alkaloids were detected in *Pterocephalus* (Gülcemal et al. 2010). Dipsacaceae s.str. lack valeric acid and thus valepotriates, which are typical for the sister group Valerianaceae. The chemistry of *Triplostegia* illustrates its problematic placement in either Dipsacaceae or Valerianaceae. *Triplostegia grandiflora* contains the iridoid glucosides loganin, loganic acid, sweroside and a number of triterpenoid saponins which are frequently found in Dipsacaceae (Ma et al. 1991, 1992). On the other hand, Backlund and Moritz (1998) detected a valepotriate in *T. grandiflora* so far known only from Valerianaceae.

Distribution and Habitats. In its natural distribution the Dipsacaceae are an Old World family basically distributed in the temperate and subtropical northern hemisphere. The vast majority of species occurs in the Mediterranean (mainly E Mediterranean) and Irano-Oriental regions. Most genera show a disjunct Mediterranean/S to SE African and/or Mediterranean/C to E Asian distribution area. Disjunct relic areas in the C European and Mediterranean regions are common in *Pseudoscabiosa*, *Succisa* and *Succisella*. All annuals have a strictly Mediterranean distribution. Approximately 26 species extend to the C African mountains and S Africa, most of them belong to *Cephalaria* (16 spp.) and *Scabiosa* (7 spp.), only three species to other genera (*Succisa trichotocephala*, *Dipsacus narcisseanus*, *D. pinnatifida*). The monotypic genus *Pterothamnus* is endemic to SE Africa (Mozambique), one *Pterocephalus* species is found in E Africa (Ethiopia; *P. frutescens*). *Pterocephalus wendelboi*, endemic to a 2300 m high mountain in S Iran at the Strait of Hormuz (Persian Gulf), could be a relic from progenitors marking the migration route of *P. frutescens* to its present distribution in Yemen and E Africa. In Asia *Dipsacus* is the most widespread genus (8 spp.), *Scabiosa* occurs with ca. 11 locally distributed species in Korea, Mongolia, N China and Taiwan. *Bassecoia* has a distinctly E Asian distribution from N Thailand to Yunnan and Szechuan. *Triplostegia* occurs in the alpine E Himalayas and the mountainous regions of SW China, with extensions to Celebes (Sulawesi) and New Guinea.

Most Dipsacaceae species are xerophilic and prefer dry, open and stony habitats like uncultivated fields, stony places, roadsides, and montane habitats; in *Sixalix*, even desert annuals in sandy habitats evolved. Some species of *Dipsacus*, *Knautia*, *Succisa* and *Succisella* prefer fresh, humid habitats in pastures or along forest edges (e.g. *Knautia dipsacifolia*). Dipsacaceae have a large altitudinal range occurring from sea level to up to 4000 m altitude. *Dipsacus fullonum* and *D. laciniatus* have become invasive in North America and meanwhile are regarded as noxious weeds in the local flora of some states (Rector et al. 2006).

Palaeobotany. Fossil records of Dipsacaceae are fairly poor. Fruits of *Scabiosa* cf. *columbaria* were described from the Pliocene of Poland (Szafer 1954). *Scabiosa* type pollen has been recorded from the Hungarian Middle Miocene and the upper Miocene of France, *Dipsacus* type pollen has been recorded from the upper Miocene of Spain (Muller 1981). The overall picture from the fossil record is consistent with an origin of the family in the Tertiary, followed by rapid diversification and dispersal. Biogeographical and divergence time reconstruction for *Scabiosa* indicate that this genus originated during the Miocene (Carlson et al. 2012). The initiation of the Mediterranean climate with increasing aridity may have played an important role in the diversification of *Scabiosa* and other Dipsacaceae lineages (Bell and Donoghue 2005).

Uses and Economic Importance. Dipsacaceae are used in traditional medicine of many countries—e.g. the traditional Chinese medicine "Xuduan" is made of roots of *Dipsacus* species (especially *D. asper*). It is used for treatment of pain and swellings in the lower back and knees, of fractures and osteoporosis, and of external disorders such as sores and uterus bleeding during pregnancy (Bensky and Gamble 1993; Liu et al. 2011). In South Africa, leaves or roots of *Scabiosa columbaria* are used as a remedy for colic and heartburn. Dried roasted roots are made into a wound-healing ointment, the powdered roots are also used as baby powder (Van Wyk et al. 1997). *Pycnocomon rutifolium*, *Pterocephalus perennis* and some *Cephalaria* species have anti-inflammatory and strong antibacterial activity (Herrera et al. 1990; Kırmızıgül et al. 1996; Graikou et al. 2006; Liu et al. 2011; Sarıkahya et al. 2011). Diuretic, cytotoxic and antioxidant activities have been reported for *Cephalaria*, *Dipsacus* and *Lomelosia* (Garcia Gímenez et al. 1988; Hung et al. 2006; Liu et al. 2011; Sarıkahya et al. 2011). *Dipsacus sativus* (teasel) is cultivated for its heads. The stiff bracts of the capitula are used to carde or "tease" fibre ends from costly wool, flannel and velvet (Ryder 1994). Some species are popular as ornamental plants, especially those with large showy marginal flowers like *Lomelosia caucasica*, *Cephalaria gigantea*, *Knautia macedonica* and *Lomelosia graminifolia*.

Subdivisions and Relationships Within the Family. For a long time, the subdivision of Dipsacaceae has been difficult due to the morphological similarity among the genera and the lack of clear distinguishing characters. The most valuable characters to distinguish genera are found at the level of the diaspores, which led Van Tieghem (1909) to divide the family into the three tribes Dipsaceae, Scabioseae and Knautieae. Morphological, anatomical, palynological, karyological, taxonomic and finally recent DNA analytical studies (i.a. Szabó 1940; Baksay 1952, 1955; Ehrendorfer 1962; Devesa 1984; Verlaque 1985a, 1985b, 1986a, 1986b; López González 1987; Burtt 1999; Mayer and Ehrendorfer 1999, 2000, 2013; Caputo et al. 2004; Avino et al. 2009; Carlson et al. 2009) allowed to identify characters often used as differential markers as homoplasious at the generic level (e.g. the membranous corona, shape of diaphragm, existence of foveoles, number and indumentum of calyx setae). The clarification of the phylogeny of Dipsacaceae made it necessary to modify its traditional classification. At present Dipsacaceae s.str. with 14 genera are classified into two subfamilies (Dipsacoideae, Scabiosoideae), eight tribes (Bassecoieae, Dipsaceae, Knautieae, Lomelosioideae, Pseudoscabioseae, Pterocephalidieae, Scabioseae, Succiseae) and *Triplostegia* as debated member. Clear diagnostic characters for the two subfamilies are only few: in Dipsacoideae (except in *Knautia*) floral bracts are well-developed, the corollas usually 4-merous, ± actinomorphic and in spherical flowering heads, whereas Scabiosoideae have only small or no floral bracts, and the flowers are usually 5-merous with zygomorphic corollas. The epicalyx of Dipsacoideae diaspores has a lignified and thick diaphragm—if present, in Scabiosoideae this is not lignified. Subfamily Dipsacoideae comprises six tribes (Bassecoieae, Dipsaceae, Knautieae, Pseudoscabioseae, Pterocephalidieae and Succiseae). With regard to the DNA phylogenetic trees, subfamily Dipsacoideae is paraphyletic when Bassecoieae are included. The latter tribe appears to be sister to the rest of the Dipsacaceae in DNA phylogenies (Avino et al. 2009; Carlson et al. 2009), but morphology and differentiation pattern of the three *Bassecoia* species show striking similarities with the four relic W Mediterranean species of *Pseudoscabiosa* and justify the inclusion of Bassecoieae in subfamily Dipsacoideae (Mayer and Ehrendorfer 2013). *Succisa* and *Succisella* are sister taxa in most analyses, only in the cpDNA analysis *Succisella* is resolved as weakly supported sister to *Pseudoscabiosa* (Carlson et al. 2009). They share the chromosome base number $x = 10$ and all are centred in the W Mediterranean, but differences in habit, pollen characters, epicalyx anatomy and calyx morphology contradict a close relationship (Mayer and Ehrendorfer 2013). *Cephalaria* and *Dipsacus* are closely related in morphological as well as DNA analyses (Verlaque 1985b; Caputo et al. 2004;

Avino et al. 2009; Carlson et al. 2009; Mayer and Ehrendorfer 2013) and are grouped in tribe Dipsaceae. Dipsaceae are sister to the monogeneric Knautieae plus Pterocephalidieae (Caputo et al. 2004; Avino et al. 2009; Carlson et al. 2009; Mayer and Ehrendorfer 2013). Pterocephalidieae consist of *Pterocephalidium diandrum*, a Spanish endemic and *Pterothamnus centennii*, an E African endemic. Both share pollen features and epicalyx characters. *Knautia*, however, is the most aberrant genus in terms of its extensive trend to polyploidisation and the presence of an elaiosome at the base of the diaspore.

Subfamily Scabiosoideae consists of two tribes: (1) Scabioseae including *Pterocephalus*, *Sixalix* and *Scabiosa*; (2) Lomelosieae with *Lomelosia*, *Pycnocomon* and *Tremastelma*. In Scabioseae, *Pterocephalus* is probably basal and the stepwise development of diaphragm, epidiaphragm and corona can be observed. *Sixalix* is sister to *Scabiosa*. Lomelosieae are the most apomorphic tribe in the Scabiosoideae, with a second sclerenchyma ring in the epicalyx tube, triporate and operculate pollen, and a highly evolved epicalyx. *Pycnocomon* appears either nested within *Lomelosia* or sister to *Lomelosia* (Caputo et al. 2004; Avino et al. 2009; Carlson et al. 2009; Mayer and Ehrendorfer 2013). *Lomelosia brachiata* (the former *Tremastelma palaestinum*), an Eastern Mediterranean annual which differs in calyx morphology and the chromosome base number $x = 7$, appears in the DNA phylogeny as sister to *Lomelosia* + *Pycnocomon*.

The position of *Triplostegia* is not easy to be solved. In the past, it was pushed back and forth between Morinaceae and Dipsacaceae or Valerianaceae and Dipsacaceae (see also Backlund and Bremer 1998) or raised to the separate family Triplostegiaceae by Airy Shaw (1964). Inflorescence structure, pollen morphology and phytochemistry, for example, indicate affinities to Valerianaceae (Ma et al. 1991, 1992; Backlund and Nilsson 1997; Backlund and Moritz 1998); other characters like the pseudomonomerous ovary and the presence of an epicalyx indicate affinities to Dipsacaceae. Whereas morphological characters and combined datasets (morphology and DNA) suggest an intermediate position between Valerianaceae and Dipsacaceae (Pyck and Smets 2004), DNA phylogenies indicate with high statistical support that *Triplostegia* is more closely related to Dipsacaceae than to Valerianaceae, and that the genus is sister group to Dipsacaceae (Bell 2004; Pyck and Smets 2004; Hidalgo et al. 2004; Soltis et al. 2011). Therefore, *Triplostegia* is here treated together with Dipsacaceae but not integrated as a separate subfamily into the Dipsacaceae s.str.

Affinities of the Family. In recent years several phylogenetic analyses have supported Dipsacaceae-*Triplostegia* as sister to Valerianaceae (Bell et al. 2001; Zhang et al. 2003; Bell 2004; Winkworth et al. 2008; Soltis et al. 2011). The split between Dipsacaceae-*Triplostegia* and Valerianaceae may date to the Palaeocene followed by a rather recent radiation of Dipsacaceae in semiarid areas around the Mediterranean basin (Bell and Donoghue 2005). The Dipsacaceae–*Triplostegia*–Valerianaceae clade is linked to Morinaceae. Dipsacaceae, *Triplostegia* and Morinaceae share the presence of an epicalyx from fused bracts, which in the past prompted several authors to include *Morina* s.l. in Dipsacaceae (e.g. Coulter 1823; Moore 1976; Cronquist 1981). Morinaceae, Valerianaceae, *Triplostegia* and Dipsacaceae form a clade which is connected with the Linnaeeae of the traditional Caprifoliaceae (Winkworth et al. 2008; Soltis et al. 2011). APG III (2009) unites Caprifoliaceae s.s., Morinaceae, Dipsacaceae and Valerinaceae into Caprifoliaceae s.l., which is sister to Adoxaceae s.l.

Classification of Dipsacaceae

Triplostegia Wall. (1830).

Dipsacaceae s.str.

I. Subfam. Dipsacoideae A. Eaton, Bot. Dict. Ed. 4: 36 (1836).
1. Bassecoieae V. Mayer & Ehrend. (2013).
2. Dipsaceae Rchb., Fl. Germ. Excurs. 1(3): 192 (1831).
3. Knautieae Tiegh., Ann. Sci. Nat. IX, Bot. 10: 168 (1909).
4. Pseudoscabioseae V. Mayer & Ehrend. (2013).
5. Pterocephalidieae V. Mayer & Ehrend. (2013).
6. Succiseae V. Mayer & Ehrend. (2013).

II. Subfam. Scabiosoideae Burnett, Outl. Bot. 918 (1835).
7. Scabioseae s.str. DC., Prodr. 4: 645 (1830).
8. Lomelosieae V. Mayer & Ehrend. (2013).

Key to the Genera of Dipsacaceae incl. *Triplostegia*

1. Inflorescence thyrsoidal, not a capitulum. Bracts of the epicalyx covered with glandular hairs, epicalyx double **Triplostegia**
– Inflorescence a capitulum. Bracts of the epicalyx not covered with glandular hairs, epicalyx simple (Dipsacaceae s.str.) 2
2. Corolla 4-merous; prominent receptacular bracts present 3
– Corolla 5-merous, rarely 4- and 5-merous flowers in one capitulum; no prominent receptacular bracts 11
3. Receptacular bracts coriaceous or rigid and spine-tipped 4
– Receptacular bracts herbaceous, hairy or missing 5
4. Stem often densely prickly, leaves connate or sessile; involucral bracts equal to or longer than receptacular bracts **3. *Dipsacus***
– Stem never prickly, leaves not connate; involucral bracts smaller than receptacular bracts **2. *Cephalaria***
5. Capitulum globose 6
– Capitulum hemispherical 7
6. Calyx setae absent, calyx 4-lobed, epicalyx ± cylindrical **9. *Succisella***
– Calyx setae 4–5, epicalyx 4-angled **8. *Succisa***
7. Calyx cupuliform or with 10–16 bristly teeth, epicalyx flat, ovoid, rarely cylindrical, with a white elaiosome at the base **4. *Knautia***
– Calyx with setae; epicalyx cylindrical, without an elaiosome at the base 8
8. Calyx setae more than 10, plumose or smooth, epicalyx at the apex either 4-lobed or with a membranous corona 9
– Calyx setae 4–6, plumose or smooth, when plumose, epicalyx 4-lobed at the apex, if smooth, with a membranous corona and pits, densely hairy **5. *Pseudoscabiosa***
9. Epicalyx with a long, flattened, sickle-shaped lamina and a solid diaphragm; plants annual **6. *Pterocephalidium***
– Epicalyx without a long lamina; diaphragm membranous if present; plants perennial 10
10. Epicalyx 4-lobed at apex and with 20–24 plumose calyx setae, if smooth, 12–16 setae, membranous corona and pits, no diaphragm **1. *Bassecoia***
– Epicalyx with small membranous corona at the apex, calyx setae 18–20, plumose, with small diaphragm **7. *Pterothamnus***
11. Calyx with 5 stiff or 10 fringed setae; receptacular bracts small, herbaceous or hairy 12
– Calyx always with more than 10 plumose setae; receptacular bracts lacking **10. *Pterocephalus***
12. Epicalyx without prominent pits in the upper part of the tube 13
– Epicalyx with 8 prominent pits in the upper half 14
13. Epicalyx with 8 prominent ridges, horizontal diaphragm and conspicuous membranous corona **11. *Scabiosa***
– Epicalyx with conically expanded infundibuliform tube and diaphragm, corona tiny **12. *Sixalix***
14. Involucral bracts connate, epicalyx 4-angled, corona short **14. *Pycnocomon***
– Involucral bracts not connate, entire and radiant, epicalyx cylindrical, corona ± large **13. *Lomelosia***

Triplostegia Wall. ex DC. Fig. 28

Triplostegia Wall. ex DC., Prodr. 4: 642 (1830).
Hoeckia Engl. & Graebn. (1901).

Perennial herbs with thin rhizome. Inflorescence thyrsoidal. Corolla 5(–4)-merous, actinomorphic; two epicalyx whorls (= double epicalyx); bracts of the outer epicalyx covered with glandular hairs; stigma relatively large, disc-like. Achene 8-ribbed, crowned by the persistent calyx limb with 5 prominent triangular calyx lobes and several smaller teeth; fruit surrounded by the persistent 4-lobed epicalyx, consisting of linear-lanceolate leaves, connected at their bases and with hook-like tips. $2n = 18$.

Two spp., *T. glandulifera* Wall. ex DC. and *T. grandiflora* Gagnepain, in China and Taiwan, Celebes and New Guinea.

Dipsacaceae s.str.

I. Subfam. Dipsacoideae A. Eaton (1836).

Capitula globose, receptacular bracts prominent. Corolla 4-merous, mostly actinomorphic; calyx in most species caducous.

I.1. Tribe Bassecoieae V. Mayer & Ehrend. (2013).

Perennials, slightl y woody at base, with basal leaf rosettes. Peduncles ± leafless with single capitula. Calyx > 10 setae, plumose or bristly; epicalyx tube simple without diaphragm; pollen tricolpate, poles with loose columellae, exine with rather loosely spread monomorphic echinae.

1. ***Bassecoia*** B.L. Burtt Fig. 30G, H

Bassecoia B.L. Burtt, in Biodiversity, Taxonomy and Ecology. Scientific Publishers (India): 137 (1999).

Pterocephalodes Mayer & Ehrendorfer, Bot. J. Linn. Soc. 132: 247–278 (2000); Mayer & Ehrendorfer, Taxon 62: 112–126 (2013), rev.

Leaves entire or irregularly divided by deep incisions. Receptacular bracts herbaceous and prominent, corolla mauve, pale pink to lavender-rose or cream-white to yellowish. Calyx with 20–24 plumose or 12–16 bristly setae, persistent; epicalyx tube with 8 broad ribs, densely hairy, without diaphragm and either terminating in 4 lobes or in a prominent corona, in the latter case also with conspicuous pits (= foveoles). $2n = 16, 18$.

Three spp., *B. breitschneideri* (Batalin) B.L. Burtt from NW Yunnan and Szechwan, *B. hookeri* (C.B. Clarke) Mayer & Ehrend. from Nepal, Sikkim and SW China, and *B. siamensis* (Craib) B.L. Burtt from N Thailand. All in open, rocky habitats, limited to the montane and alpine zone above 2000 m.

I.2. Tribe Dipsaceae Rchb. (1831).

Annual, biennial or perennial herbs with basal leaf rosettes. Capitula small and globose or large and ovoid. Corolla 4-merous, ± actinomorphic; calyx sessile, cupuliform, caducous; epicalyx tube often quadrangular, diaphragm thick and lignified; pollen 3-colporate, with a wide halo surrounding the aperture, aperture membrane with long branched echinae.

2. ***Cephalaria*** Schrad. ex Roem. & Schult. Fig. 30I, J

Cephalaria Schrad. ex Roem. & Schult., Syst. Veg. 3: 1 (1818); Szabó, Magyar. Tud. Akad., Budapest: 1–351 (1940), rev.; Napper, Kew Bull. 21: 463–470 (1968), rev.

Mostly herbs, rarely shrubs; stems usually glabrous or pilose, rarely setose, often ribbed, mostly branched. Leaves entire, pinnate or deeply pinnatisect, sometimes coriaceous. Capitula ovoid or globose, sometimes radiate; involucral bracts coriaceous or scarious, receptacular bracts larger than involucral bracts. Corolla white, cream or yellow, rarely bluish; flowers of *Cephalaria* subg. *Phalacrocarpus* (Boiss.) Szabó zygomorphic; epicalyx tube either quadrangular or 8-ridged, apex usually with 4 or 8 teeth. $2n = (10, 14), 18$.

Approximately 80 species. Perennials mostly with small, disjunct distribution areas, annuals with large ranges. Mediterranean, extending to E and S Africa, Russia and Persia; open, dry, often rocky places; 0–2000 m.

Four subgenera:

(1) Subgenus *Lobatocarpus* Szabó. Perennials. Epicalyx quadrangular with 4 ribs ending in 4 apical teeth. Ca. 16 spp. endemic to E and S Africa.
(2) Subgenus *Fimbriatocarpus* Szabó. Perennials. Epicalyx quadrangular with 4 ribs, ending in a short membranous corona. Three spp., C and W Mediterranean.
(3) Subgenus *Cephalaria*. Perennials or annuals. Epicalyx ovoid with 8 ribs ending in 8 apical teeth (often 4 long, 4 short) or ± prominent setae. Ca. 55 spp.; mostly E Mediterranean and S Europe.
(4) Subgenus *Phalacrocarpus* (Boiss.) Szabó. Annuals. Epicalyx small, edentate, no corona. Four spp., E Mediterranean, Anatolia and Caucasus.

3. ***Dipsacus*** L. Figs. 29A, 30K

Dipsacus L., Sp. Pl. 1: 97 (1753); Hedberg & Hedberg, Bot. Notiser 129: 383–389 (1976), rev.
Virga Hill (1763).

Stems prickly or hirsute. Leaves entire with serrate to laciniate margin, petiolate or sessile, sometimes connate at the base. Receptacular bracts ending in acuminate spines. Corolla whitish to lilac; calyx caducous, either bristly or entire or cupuliform; epicalyx either quadrangular or with eight ribs, grooved at the apex, tube apically either with four lobes or a narrow plicate margin continuing the main ribs. $2n = 18$

Disjunct between Europe, Africa and Asia; all species in open weedy places; in the Himalayas up to 4000 m.

Two sections:

(1) Section *Dipsacus*. Biennials; stems prickly along entire length. Cauline leaves connate forming a cup-shaped conceptacle often filled with water. Capitula large, ovoid, with long lignified recurved acuminate bracts. Seven spp.; mainly Mediterranean and C and N Europe; dry places, below 1000 m.
(2) Section *Sphaerodipsacus* Lange. Biennials or annuals; stems smooth or hirsute, not prickly; prickles only in the leaf angles. Cauline leaves not connate. Capitula small, spherical, with small herbaceous spiny bracts. 13 spp., Asia (8 ssp.), Africa (2 spp.), Eurasia (2 spp.) and Turkey (1 sp.); mostly fresh humid places, up to 4000 m.

I.3. Tribe Knautieae Tiegh. (1909).

Predominantly perennials, few annuals; often with basal leaf rosette; stems often hollow.

Capitula radiate, receptacle with hairs instead of receptacular bracts. Corolla 4-merous, mostly ± actinomorphic; calyx caducous, cupulate, the margin with minute awns or setae; epicalyx tube flat, ovoid, green to light brownish, often with small teeth at the apical margin, at the base with a white globular elaiosome; pollen triporate, trioperculate and breviaxial.

4. ***Knautia*** L. Fig. 30L, M

Knautia L., Sp. Pl. 101 (1753). Szabó, Bot. Jahrb. Syst. 36: 389–442 (1905), rev.; Ehrendorfer, Oesterr. Bot. Z. 109: 276–343 (1963), key.

Leaves entire to pinnate, pubescent to subglabrous. Receptacular bracts in most species lacking, receptacle hairy instead (except for *K. orientalis*). Corolla rarely zygomorphic, marginal flowers often enlarged, bluish-lilac, purple, cream or yellow; calyx with (4–)8–16(–24) minute awns or setae, setae sometimes ciliate. $2n = 16, 20$. $2n = 4x = 40$, $2n = 6x = 60$.

Ca. 60 species. The species are rather variable in their morphology, and taxonomic treatments are therefore different. Distributed in Europe and Eastern Asia Minor; open woods, wood margins, mesophilous to dry grassland; 0–2000 m.

Three sections:

(1) Section *Knautia* (Fig. 30M). Annuals. Capitula with only 5–10 flowers per head. Elaiosome slightly sickle-shaped. These species differ markedly from the others by their few-flowered *Lychnis*-like capitula and strongly zygomorphic flowers. Two species (*K. orientalis* L., *K. degenii* Borbás) in the E Mediterranean; thermophilous (sub-)Mediterranean forests; below 1000 m. $2n = 16$.

(2) Section *Trichera* (Schrad.) DC. (Fig. 30L). Perennials. $2n = 20$, polyploidy common. About 50 spp. Distributed all over Europe, centred on the Balkan Peninsula; up to 2000 m. $2n = 16, 20$. $2n = 4x = 40$, $2n = 6x = 60$.

(3) Section *Tricheroides* DC. Annuals. Two species (*K. integrifolia* (L.) Bert., *K. byzantina* Fritsch), Mediterranean and E Anatolia; roadsides, fields, slopes; 300–1500 m. $2n = 20$.

I.4. Tribe Pseudoscabioseae V. Mayer & Ehrend. (2013).

Perennial half-shrubs, woody at base, with basal leaf rosettes. Leaves entire; branched stellate hairs on stems and leaves. Capitula globose, receptacular bracts prominent. Corolla 4-merous, actinomorphic; calyx 4–6-setate, plumose or bristly, persistent; pollen tricolpate, longiaxial, with cavae at the pollen poles.

5. ***Pseudoscabiosa*** Devesa Fig. 30C, D

Pseudoscabiosa Devesa, Lagascalia 12: 216 (1984); Romo et al. 1997, Feddes Repert. 108: 31–38 (1997), rev.
Scabiosa sect. *Asterothrix* P. Font Quer (1944).

Leaves often glabrous or tomentose on the upper side, densely covered with stellate hairs below. Corolla tube in marginal flowers slightly longer than in central ones; lilac or white; calyx either with 4–6 relatively long and slightly plumose setae or with 4 much shorter bristly setae; epicalyx a simple tube, densely hairy, without diaphragm, either terminating in 4 broadly triangular lobes or in a membranous corona, in the latter case also with prominent pits. $2n = 20$.

Four W Mediterranean spp. with disjunct relic distribution in the Mediterranean and N Africa; limestone rocks, rock crevices.

I.5. Tribe Pterocephalidieae V. Mayer & Ehrend. (2013).

Capitula globose, receptacular bracts prominent, inner and outer flowers ± equal. Corolla 4-merous; calyx with more than 14 setae, plumose, caducous; epicalyx ± round with eight broad flat ridges; pollen triporate, subequiaxial, with cavae at the poles.

6. ***Pterocephalidium*** G. López González Fig. 30E, F

Pterocephalidium G. López González, Anales Jard. Bot. Madrid 43: 251 (1987).

Annual, stem pubescent. Leaves pinnatisect. Corolla blue or lilac; stamens 2; calyx of 14–27 plumose setae, caducous; epicalyx with a long, flattened and sickle-shaped lamina representing an outgrowth of one of the epicalyx teeth, prominent lignified diaphragm. $2n = 16$.

Monotypic with *Pterocephalidium diandrum* G. López González endemic to C Spain and Portugal; dry places.

7. ***Pterothamnus*** (M.J. Cannon) V. Mayer & Ehrend.

Pterothamnus (M.J. Cannon) V. Mayer & Ehrend., Taxon 62: 112–126 (2013).

Perennial half-shrub. Leaves lanceolate, toothed. Receptacular bracts broadly ovate. Corolla whitish; calyx of 18–20 plumose setae; epicalyx with a rather short diaphragm, apically with a slightly lobed margin sometimes ending in a rudimentary membranous corona.

Monotypic with *P. centennii* (M.J. Cannon) V. Mayer & Ehrend. known only from its type locality in Mozambique; ca. 2000 m.

I.6. Tribe Succiseae V. Mayer & Ehrend. (2013).

Perennial herbs with stolons and basal leaf rosettes. Leaves entire to serrate. Capitula hemispherical, receptacular bracts prominent. Corolla 4-merous, marginal and central flowers ± equal; epicalyx tapered or quadrangular, with parenchyma cells only, no fibrous sheaths in the region of the ridges; pollen tricolpate, angulaperturate.

8. ***Succisa*** Haller Fig. 30B

Succisa Haller, Hist. Stirp. Helv. 1: 87 (1768); Baksay, Ann. Hist.-Nat. Mus. Natl. Hung., n.s. 2: 237–260 (1952), rev.

Corolla lilac to dark violet blue; calyx of 4–5 persistent setae; epicalyx quadrangular, apically ending in four lobes, tiny diaphragm. $2n = 20$.

Three spp., two of them with small relic areas (*S. pinnatifida* Lange in Iberian Peninsula, *S. trichotocephala* Baksay endemic to Mt. Cameron, W Africa); both growing on fresh, moist soil; 500–1100 m.

9. ***Succisella*** G. Beck Fig. 30A

Succisella G. Beck, Fl. Nieder-Österreich 2: 1145 (1893); Baksay, Ann. Hist.-Nat. Mus. Natl. Hung., n.s. 6: 167–176 (1955), rev; Amich, Devesa & Bernardos, Bot. J. Linn. Soc. 144: 351–364 (2004), rev.

Corolla pale lilac to blue; calyx setae absent, only a plateau with four lobes; epicalyx tapered, with prominent ribs, diaphragm lignified. $2n = 20$.

Five spp., four of them with rather small disjunct relict areas in the Mediterranean, only *S. inflexa* (Kluk) G. Beck with a larger distribution (S and E Europe from Italy to the Baltic States and the Caucasus). Except for *S. microcephala* (Willk.) G. Beck, all species prefer wet places; up to 1000 m.

II. Subfam. Scabiosoideae Burnett (1835).

Capitula mostly radiant, floral bracts small and herbaceous, rudimentary or absent. Corolla 5-merous, ± zygomorphic; calyx persistent, 5 bristly setae or more than10 plumose setae. diaphragm (if present) thin and hyaline.

II.7. Tribe Scabioseae DC. s.str. (1830).

Perennials, biennials or annuals. Epicalyx various, a simple tube without diaphragm and corona or elaborate with diaphragm, epidiaphragm and corona; pollen tricolpate, subequiaxial to longiaxial, angulaperturate, non-operculate.

10. ***Pterocephalus*** Adans. Fig. 31F–J

Pterocephalus Adans., Fam. Pl. 2: 152, 595 (1763).

Predominantly perennial half-shrubs; usually stems hairy. Leaves entire to 1- to 2-pinnatisect. Receptacular bracts rudimentary or absent. Corolla mostly 5-merous, in some species 5- and 4-merous flowers in one capitulum; lilac, pinkish, cream or rarely yellow; calyx 12–24 plumose setae. $2n = 18$, the annual *P. brevis* deviates with $2n = 16$.

Approximately 30 spp.; the centre of diversity is the Irano-Oriental and E Mediterranean region, with outposts in SE Spain, Morocco, the Canary Islands and E Africa; open, relatively dry habitats; 0–2500 m.

11. ***Scabiosa*** L. Figs. 27, 31A–C

Scabiosa L., Sp. Pl.: 98 (1753); Devesa, Lagascalia 12: 143–212 (1984), rev.
Asterocephalus Zinn (1757).

Perennials, chamaephytes and hemicryptophytes, only two species annual; stem often shortly pubescent. Leaves often in basal rosettes, simple or pinnate. Capitula long-pedunculate, hemispherical to cylindrical; receptacular bracts linear-lanceolate. Corolla pale blue, pink, yellowish or cream; calyx with five stiff setae originating from a basal plateau, only in *S. triandra* setae partly or sometimes completely reduced, epicalyx with 8 prominent ridges, horizontal diaphragm and conspicuous membranous corona (Fig. 31A); in a few species from N China, Korea and Japan, epicalyx quadrilateral with four prominent ridges (Fig. 31B). $2n = 16$, all diploid except *S. parviflora* with $3x$ (and $4x$).

More than 30 spp. ranging from Europe and the Mediterranean to E Asia and to E and S Africa. Dry and rocky habitats, uncultivated fields, alpine meadows; most species below 1000 m, few up to 2600 m.

12. ***Sixalix*** Raf. Fig. 31D, E

Sixalix Raf., Fl. Tellur. 4: 95 (1838).
Scabiosa sect. *Cyrtostemma* Mert. & Koch (1823).

Perennial (4 spp.), biennial (1 sp.) or annual (3 spp.); stems usually branched. Leaves entire, dentate or pinnatifid, often becoming pinnatisect towards apex of stem. Corolla white, pale lilac or purplish; calyx with five rough, sometimes reduced setae originating from a basal plateau; epicalyx with eight ribs, funnel-shaped in upper half, pits elongated, diaphragm conical, with tiny undulate membranous corona at the apical rim. Diaspores of some species heterocarpic. $2n = 16$.

Eight spp., centred in NW Africa (five of the eight spp.), *S. atropurpurea* (L.) Greuter & Burdet circum-Mediterranean; sandy desert soils, rocky slopes and meadows; up to 2000 m.

II.8. Tribe Lomelosieae V. Mayer & Ehrend. (2013).

Perennial or annual herbs with basal leaf rosettes. Capitula globose or flat, radiant, small herbaceous receptacular bracts. Calyx usually with five stiff and rough setae, persistent; epicalyx in all species with a prominent ±large membranous corona at the apical rim and a flat diaphragm closing the epicalyx tube; epidiaphragm elongated and horizontal, 8 prominent pits in upper half of the tube; second sclerenchyma ring present; pollen triporate, breviaxial to subequiaxial, angulaperturate and operculate.

13. ***Lomelosia*** Raf. Fig. 30O–R

Lomelosia Raf., Fl. Tellur. 4: 95 (1838).
Scabiosa sect. *Trochocephalus* Mert. & Koch (1823).

Perennials or annuals, perennials often woody at the base. Leaves entire, upper leaves pinnatisect with linear lobes. Corolla white, cream, pale yellow, blue, pink; calyx usually persistent, only in *L. stellata* caducous, in *L. brachiata* setae 8–10, broad, ciliate; epicalyx tube smooth or densely hairy. $2n = 18$, rarely $2n = 16$, *L. brachiata* $2n = 14$; polyploidy in few species.

One of the largest genera of the family with more than 50 species; mainly distributed in the E Mediterranean and Middle East, few species in the W Mediterranean and C Asia; dry, stony places, rocks, steppes; up to 2500 m.

14. ***Pycnocomon*** Hoffmanns. & Link Fig. 30N

Pycnocomon Hoffmanns. & Link, Fl. Portug. 2: 93, t. 88 (1820).

Perennial or annual. Basal leaves entire to pinnatifid, cauline leaves 1–2-pinnatifid. Involucral bracts linear-lanceolate, connate at the base or up to half the length. Corolla pink to yellowish-white or pale lilac; calyx with 5(–8) plumose or bristly setae; epicalyx 4-angled or rounded, terminating in a short membranous corona, with prominent pits. Diaspores homo- or heterocarpic; $2n = 18$.

Two spp., the annual *P. intermedium* (Lag.) Greuter & Burdet endemic to the Iberian Peninsula, *P. rutifolium* (Vahl) Hoffmanns. & Link with circum-Mediterranean distribution; dry sandy ground.

Selected Bibliography

Airy Shaw, H.K. 1964. Diagnoses of new families, new names, etc., for the seventh edition of Willis's 'Dictionary'. Kew Bull. 18: 249–271.

Akimaliev, A., Alimbaeva, P.K., Mzhel'skaya, L.G., Abubakirov, N.K. 1971. Triterpene glycosides of *Scabiosa soongorica* II. The structure of songorosides C, G and I. Chem. Nat. Comp. 12: 415–418.

Alvarado, S. 1927. Der morphologische Aufbau des Hüllkelches der Dipsacaceen. Bot. Jahrb. Syst. 61, Beiheft 138: 10–21.

APG III 2009. An update of the Angiosperm Phylogeny Group classification for the orders and families of flowering plants: APG III. Bot. J. Linn. Soc. 161: 105–121.

Avino, M., Tortoriello, G., Caputo, P. 2009. A phylogenetic analysis of Dipsacaceae based on four DNA regions. Plant Syst. Evol. 279: 69–86.

Backlund, A., Bremer, B. 1998. To be or not to be. Principles of classification and monotypic plant families. Taxon 47: 391–400.

Backlund, A., Moritz, T. 1998. Phylogenetic implications of an expanded valepotriate distribution in the Valerianaceae. Biochem. Syst. Ecol. 26: 309–335.

Backlund, A., Nilsson, S. 1997. Pollen morphology and the systematic position of *Triplostegia* (Dipsacales). Taxon 46: 21–31.

Baksay, L. 1952. Monographie der Gattung *Succisa*. Ann. Hist.-Nat. Mus. Natl. Hung., n.s. 2: 237–260.

Baksay, L. 1955. Anatomische und systematische Untersuchungen über die Gattung *Succisella*. Ann. Hist.-Nat. Mus. Natl. Hung., n.s. 6: 167–176.

Behnke, H.-D. 1994. Sieve-element plastids: Their significance for the systematics and evolution of the order. In: Behnke, H.-D., Mabry, T.J. (eds.) Caryophyllales: Evolution and systematics. Berlin: Springer, pp. 87–121.

Bell, C.D. 2004. Preliminary phylogeny of Valerianaceae (Dipsacales) inferred from nuclear and chloroplast DNA sequence data. Mol. Phylog. Evol. 31: 340–350.

Bell, C.D., Donoghue, M.J. 2005. Dating the Dipsacales: Comparing models, genes and evolutionary implications. Amer. J. Bot. 92: 284–296.

Bell, C.D., Edwards, E J., Kim, S.-T., Donoghue, M.J. 2001. Dipsacales phylogeny based on chloroplast DNA sequences. Harvard Pap. Bot. 6: 481–499.

Bensky, D., Gamble, A. 1993. Chinese herbal medicine: materia medica. Seattle, Washington: Eastland Press.

Burtt, B.L. 1999. The importance of some far Eastern species of Dipsacaceae in the history of the family. In: Tandon, R.K., Prithipalsingh (eds.) India, Scientific Publications: 131–139.

Caputo, P., Cozzolino, S. 1994. A cladistic analysis of the Dipsacaceae (Dipsacales). Plant Syst. Evol. 189: 41–63.

Caputo, P., Cozzolino, S., Moretti, A. 2004. Molecular phylogenetics of Dipsacaceae reveals parallel trends in seed dispersal syndromes. Plant Syst. Evol. 246: 163–175.

Carlquist, S. 1982. Wood anatomy of Dipsacaceae. Taxon 31: 443–450.

Carlson, S.E., Mayer, V., Donoghue, M.J. 2009. Phylogenetic relationships, taxonomy, and morphological evolution in Dipsacaceae (Dipsacales) inferred by DNA sequence data. Taxon 58: 1075–1091.

Carlson SE, Linder HP, Donoghue MJ. 2012. The historical biogeography of *Scabiosa* (Dipsacaceae): Implications for Old World plant disjunctions. Biogeogr. 39: 1086–1100.

Čelakovsky, L. 1893. Über den Blütenstand von *Morina* und den Hüllkelch (Außenkelch) der Dipsacaceen. Bot. Jahrb. Syst. 17: 395–418.

Christopoulou, C., Graikou, K., Chinou, I. 2008. Chemosystematic value of chemical constituents from *Scabiosa hymettia* (Dipsacaceae). Chem. Biodiv. 5: 318–322.

Clarke, G., Jones, M.R. 1981. The Northwest European pollen flora, 21: Dipsacaceae. Rev. Palaeobot. Palynol., Netherl. 33: 1–25.

Coulter, T. 1823. Mémoire sur les Dipsacées. Mém. Soc. Phys. Genève 2: 13–60.

Cronquist, A. 1981. An integrated system of classification of flowering plants. New York: Columbia University Press.

Devesa, J.A. 1984. *Pseudoscabiosa*, genero nuevo de Dipsacaceae. Lagascalia 12: 213–221.

Díaz Lifante, Z., Luque, T., Santa Bárbara, C. 1992. Chromosome numbers of plants collected during Iter Mediterraneum II in Israel. Bocconea 3: 229–250.

Diez, M.J., Devesa, J.A. 1981. Contribution al estudio palinologica del genero *Scabiosa* L. (Dipsacaceae) en la Peninsula Iberica y Baleares. Bot. Macaronés., IV Ci 8–9: 129–146.

Donoghue, M.J., Bell, C.D., Winkworth, R.C. 2003. The evolution of reproductive characters in Dipsacales. Int. J. Plant Sci. 164: S453–S464.

Ehrendorfer, F. 1962. Beiträge zur Phylogenie der Gattung *Knautia* (Dipsacaceae), I. Cytologische Grundlagen und allgemeine Hinweise. Österr. Bot. Z. 109: 276–343.

Ehrendorfer, F. 1964. Über stammesgeschichtliche Differenzierungsmuster bei den Dipsacaceen. Ber. Deutsch. Bot. Ges. 77 (Sondernummer): 83–94.

Ehrendorfer, F. 1965. Evolution and karyotype differentiation in a family of flowering plants: Dipsacaceae. Genetics Today. Proc. XI Int. Congr. Genetics, The Hague, The Netherlands, 1963, 2: 399–407.

Erdtman, C. 1952. Pollen morphology and plant taxonomy. Stockholm: Almquist & Wiksell.

Frajman B, Rešetnik I, Weiss-Schneeweiss H, Ehrendorfer F, Schönswetter P. 2015. Cytotype diversity and genome size variation in *Knautia* (Caprifoliaceae, Dipsacoideae). BMC Evolutionary Biology 15: 140–157.

Frey, L. 1976. Embryological studies on *Scabiosa lucida* Vill. and *Scabiosa ochroleuca* L. in Poland. Acta Biol. Cracov., Ser. Bot. 19: 173–179.

Garcia Gímenez, M.D., Marhuenda Requena, E., Saenz Rodrigez, M.T. 1988. Contribution to the pharmacodynamic study of *Scabiosa atropurpurea* L.: III. Diuretic activity. Pl. Méd. Phytothérap. 22: 175–179.

Goebel, K. 1924. Die Entfaltungsbewegungen der Pflanzen und deren teleologische Deutung. Jena: Gustav Fischer.

Graikou, K., Aligiannis, N., Ioanna B. Chinou, I.B., Harvala, C. 2002. Cantleyoside-dimethyl-acetal and other iridoid glucosides from *Pterocephalus perennis* - Antimicrobial activities. Z. Naturforsch. 57c: 95–99.

Graikou, K., Aligiannis, N., Chinou, I.B. 2006. Chemical constituents from *Pterocephalus perennis* subsp. *perennis* (Dipsacaceae). Biochem. Syst. Ecol. 34: 438–441.

Greuter; W., Raus, Th. (eds.) 1985. Med-Checklist Notulae, 11. Willdenowia 15: 72–76.

Gülcemal, D., Masullo, M., Alankus-Calıskan, O., Karayıldırım, T., Senol, S.G., Piacente, S., Bedir, E. 2010. Monoterpenoid glucoindole alkaloids and iridoids from *Pterocephalus pinardii*. Magn. Reson. Chem. 48: 239–243.

Günthart, A. 1904. Blütenbiologische Untersuchungen. Beiträge zur Blütenbiologie der Dipsacaceen. Flora 93: 199.

Hedberg, I., Hedberg, O. 1976. The genus *Dipsacus* in tropical Africa. Bot. Not. 129: 383–389.

Herrera, M.D., Marhuenda, E., Saenz, M.T. 1990. Contribution to the pharmacodynamic study of *Pycnocomon rutifolium* (Vahl.) Hoffmansegg & Link: Antiinflammatory activity and antibacterial activity. Ann. Pharm. Franç. 48: 81–86.

Hidalgo, O., Garnatje, T., Susanna, A., Mathez, J.L. 2004. Phylogeny of Valerianaceae based on *matK* and ITS markers, with reference to *matK* individual polymorphism. Ann. Bot. 93: 283–293.

Hilger, H.H., Hoppe, M. 1984. Die Entwicklung des Außenkelchs von *Scabiosa* Sekt. *Sclerostemma* und

Trochocephalus (Dipsacaceae). Beitr. Biol. Pflanzen 59: 55–73.
Höck, F. 1897. Dipsacaceae. In: Engler, A., Prantl, K. (eds.) Die natürlichen Pflanzenfamilien 22 (IV, 4–5). Leipzig: Wilhelm Engelmann, pp. 183–189.
Höck, F. 1902. Verwandtschaftsbeziehungen der Valerianaceae und Dipsacaceae. Bot. Jahrb. Syst. 31: 405–411.
Hung, T.M., Na, M.K., Thuong, P.T., Su, N.D., So, D.E., Song, K.S., Seong, Y.H., Bae, K.H. 2006. Antioxidant activity of caffeoyl quinic acid derivatives from the roots of *Dipsacus asper* Wall. J. Ethnopharmacol. 108: 188–192.
Jacobs, B., Geuten, K., Pyck, N., Huysmans, S., Jansen, S., Smets, E. 2011. Unraveling the phylogeny of *Heptacodium* and *Zabelia* (Caprifoliaceae): An interdisciplinary approach. Syst. Bot. 36: 231–252.
Jensen, S.R. 1991. Plant iridoids, their biosynthesis and distribution in Angiosperms. In: Harborne & Barbaran (eds.) Annual Proceedings of the Phytochemical Society of Europe. Ecological Chemistry and Biochemistry of Plant Terpenoids. Oxford: Oxford University Press, pp. 133–158.
Jensen, S.R., Nielsen, B.J., Dahlgren, R. 1975. Iridoid compounds, their occurrence and systematic importance in the Angiosperms. Bot. Not. 128: 148–180.
Johri, B.M., Ambegaokar, K.B., Srivastava, P.S. 1992. Comparative embryology of Angiosperms, vol. 2. Berlin: Springer, pp. 782–798.
Kachidze, N. 1929. Karyologische Studien über die Familie der Dipsacaceae. Planta 7: 482–502.
Kamelena, O.P. 1980. Sravnitel'naya embriologiya semeistv Dipsacaceae i Morinaceae. Leningrad: Nauka (in Russian).
Kırmızıgül, S., Anil, H., Rose, M.E. 1995. Triterpenoid glycosides from *Cephalaria transsilvanica*. Phytochemistry 39: 1171–1174.
Kırmızıgül, S., Anıl, H., Uçar, F., Akdemir, K. 1996. Antimicrobial and antifungal activities of three new triterpenoid glycosides. Phytother. Res. 10: 274–276.
Kocsis, Á., Szabó, L.F., Podányi, B. 1993. New bis-iridoids from *Dipsacus laciniatus*. J. Nat. Prod. (Lloydia) 56: 1486–1499.
Kugler, H. 1955. Einführung in die Blütenökologie. Stuttgart: Gustav Fischer.
Larsson, M. 2005. Higher pollinator effectiveness by specialist than generalist flower-visitors of unspecialized *Knautia arvensis* (Dipsacaceae). Oecologia 146: 394–403.
Liu, J.-J., Wang, X.-L., Guo, B.-L., Huang, W.-H., Xiao, P.-G., Huang, C.-Q., Zheng, L.-Z., Zhang, G., Qin, L., Tu, G.-Z. 2011. Triterpenoid saponins from *Dipsacus asper* and their activities in vitro. J. Asian Nat. Prod. Res. 13: 851–860.
López González, G. 1987. *Pterocephalidium*, un nuevo género ibérico de la familia Dipsacaceae. Anales Jard. Bot. Madrid 43: 245–252.
Ma, W.G., Wang, D.Z., Zeng, Y.L., Yang C.H. 1991. Three triterpenoid saponins from *Triplostegia grandiflora*. Phytochemistry 30: 3401–3404.
Ma, W.G., Wang, D.Z., Zeng, Y.L., Yang C.H. 1992. Triterpenoid saponins from *Triplostegia grandiflora*. Phytochemistry 31: 1343–1347.
Mayer, V. 1995. The epicalyx in fruits of *Scabiosa* and *Tremastelma* (Dipsacaceae): Anatomy and ecological significance. Bot. Jahrb. Syst. 117: 211–238.
Mayer, V., Ehrendorfer, F. 1999. Fruit differentiation, palynology, and systematics in the *Scabiosa* group of genera and *Pseudoscabiosa* (Dipsacaceae). Plant Syst. Evol. 216: 135–166.
Mayer, V., Ehrendorfer, F. 2000. Fruit differentiation, palynology, and systematics in *Pterocephalus* Adanson and *Pterocephalodes*, gen. nov. (Dipsacaceae). Bot. J. Linn. Soc. 132: 247–278.
Mayer, V., Ehrendorfer, F. 2013. The phylogenetic position of *Pterocephalidium* and the new African genus *Pterothamnus* within an improved classification of Dipsacaceae. Taxon 62: 112–126.
Mayer, V., Svoma, E. 1998. Development and function of the elaiosome in *Knautia* (Dipsacaceae). Bot. Acta 111: 402–410.
Metcalfe, C.R., Chalk, L. 1950. Anatomy of dicotyledons, vol. 2: 780–781. Oxford: Clarendon Press.
Molliard, M. 1895. Recherches sur les Cécidies florales. Ann. Sci. Nat., Bot. 8, 1: 67–245.
Moore, D.M. 1976. Dipsacaceae. In: Tutin, T.G. et al. (eds.) Flora Europea 4: 56–74. Cambridge: Cambridge University Press.
Movsumov, I.S., Garaev, E.A., Isaev, M.I. 2006. Flavonoids from *Cephalaria gigantea* flowers. Chem. Nat. Comp. 42: 677–680.
Muller, J. 1981. Fossil pollen records of extant Angiosperms. Bot. Rev. 47: 1–142
Müller, H. 1873. Die Befruchtung der Blumen durch Insekten und die gegenseitige Anpassung beider. Leipzig: Wilhelm Engelmann.
Neubauer, H.F. 1978. On nodal anatomy and petiolar vascularization of some Valerianaceae and Dipsacaceae. Phytomorphology 28: 431–436.
Pasi, S. Aligiannis, N., Skaltsounis, A.L., Chinou, I.B. 2002. A new lignan glycoside and other constituents from *Cephalaria ambrosioides*. Nat. Prod. Lett. 16: 365–370.
Peng, C.-I., Tobe, H., Takahashi, M. 1995. Reproductive morphology and relationships of *Triplostegia* (Dipsacales). Bot. Jahrb. Syst. 116: 505–516.
Perdetzoglou, D.K., Skaltsa, H., Tzakou, O., Harvala, C. 1994. Comparative phytochemical and morphological study of two species of the *Scabiosa* L. genus. Feddes Repert. 105: 157–165.
Proctor, M., Yeo, P., Lack, A. 1996. The natural history of pollination. London: Harper Collins.
Pyck, N., Smets, E. 2004. On the systematic position of *Triplostegia* (Dipsacales): a combined molecular and morphological approach. Belgian J. Bot. 137: 125–139.
Rector, B.G., Harizanova, V., Sforza, R., Widmer, T. Wiedemann, R.N. 2006. Prospects for biological control of teasels, *Dipsacus* spp., a new target in the United States. Biolog. Control 36: 1–14.
Reichenbach, H.G.L. 1850. Icones florae Germanicae et Helveticae. Vol. XII. Leipzig: F. Hofmeister.
Ridley, H.N. 1930. The dispersal of plants throughout the world. Ashford (Kent): L. Reeve.
Romo, A., Cirujano, S. Peris, J.B., Stuebing, G. 1997. Contributions to the knowledge of the genus *Pseudoscabiosa* (Dipsacaceae). Feddes Repert. 108: 31–38.

Ryder, M.L. 1994. Fascinating Fullonum. Circaea 11: 23–31.

Sarıkahya, N.B., Pekmez, M., Arda, N., Kayce, P., Yavaşoğlu, N.U.K., Kırmızıgül, S. 2011. Isolation and characterization of biologically active glycosides from endemic *Cephalaria* species in Anatolia. Phytochem. Lett. 4: 415–420.

Sernander, R. 1906. Entwurf einer Monographie der europäischen Myrmekochoren. Kungl. Svenska Vetenskapsakad. Handl. 41: 1–409.

Solereder, H. 1899. Systematische Anatomie der Dicotyledonen. Stuttgart: Ferdinand Enke.

Soltis, D.E., Smith, S.A., Cellinese, N., Wurdack, K.J., Tank, D.C., Brockington, S.F., Refulio-Rodriguez, N. F., Walker, J.B., Moore, M.J., Carlsward, B.S. et al. 2011. Angiosperm phylogeny: 17 genes, 640 taxa. Amer. J. Bot. 98: 704–730.

Szabó, Z. 1905. Monographie der Gattung *Knautia*. Bot. Jahrb. Syst. 36: 389–442.

Szabó, Z. 1922. A *Cephalaria* genus rendszere. Math. Term. Értes. 39: 267–275.

Szabó, Z. 1923. The development of the flower of the Dipsacaceae. Ann. Bot. 37: 325–334.

Szabó, Z. 1930. Entwicklungsgeschichtliche Deutung des Blütenstandes der Dipsacaceen. Szent István Akadémia Ertesítöje 1: 55–72.

Szabó, Z. 1940. Monographie der Gattung *Cephalaria*. Math. Természettud. Közlem 38. 352 pp.

Szafer, W. 1954, Plioceriska flora okolic Czorsztyna i jej stosunek do Plejstocenu. Warsaw: Inst. Geol. Publ., v. 11, 238 pp.

Tomassini, L., Foddai, S., Nicoletti, M. 2004. Iridoids from *Dipsacus ferox* (Dipsacaceae). Biochem. Syst. Ecol. 32: 1083–1085.

Valdez, B., Diez, M.J., Fernandez, I. 1987. Atlas polinico de Andalucia occidental. Spain: Instituto de desarrollo regional no. 43, Universidad de Sevilla. EXCMA. Diputacion de Cadiz.

Vange, V. 2002. Breeding system and inbreeding depression in the clonal plant species *Knautia arvensis* (Dipsacaceae): implications for survival in abandoned grassland. Biol. Cons. 108: 59–67.

Van Tieghem, P. 1909. Remarques sur les Dipsacées. Ann. Sci. Nat. Bot., ser. 9, 10: 148–200.

Van Wyk, B.-E., van Oudtshoorn, B., Gericke, N. 1997. Medical plants of South Africa. Pretoria: Briza Publications.

Verlaque, R. 1977. Rapports entre les Valerianaceae, les Morinaceae et les Dipsacaceae. Bull. Soc. Bot. France 124: 475–482.

Verlaque, R. 1981. Utilisation des caractères du pollen pour la réorganisation taxinomique de la famille des Dipsacaceae. Compt. Rend. Hebd. Séances Acad. Sci. 293: 351–354.

Verlaque, R. 1984. Etude biosystématique et phylogénétique des Dipsacaceae. I. Délimination des Dipsacaceae à l'intérieur des Dipsacales, rapports avec les autres familles de l'ordre. Rev. Gen. Bot. 91: 81–121.

Verlaque, R. 1985a. Étude biosystématique et phylogénétique des Dipsacaceae. II. Caractères généraux des *Dipsacaceae*. Rev. Cytol. Biol. Vég. Botaniste 8: 117–168.

Verlaque, R. 1985b. Étude biosystématique et phylogénétique des Dipsacaceae. III. Tribus des Knautieae et Dipsaceae. Rev. Cytol. Biol. Vég. Botaniste 8: 171–243.

Verlaque, R. 1986a. Étude biosystématique et phylogénétique des Dipsacaceae. IV. Tribu des Scabioseae (phylum No 1,2,3). Rev. Cytol. Biol. Vég. Botaniste 9: 5–72.

Verlaque, R. 1986b. Étude biosystématique et phylogénétique des Dipsacaceae. IV. Tribu des Scabioseae (phylum No 4). Rev. Cytol. Biol. Vég. Botaniste 9: 97–176.

Wagenitz, W., Laing, B. 1984. Die Nektarien der Dipsacales und ihre systematische Bedeutung. Bot. Jahrb. 104: 483–507.

Weberling, F. 1961. Die Infloreszenzen der Valerianaceen und ihre systematische Bedeutung. Abh. Akad. Wiss. Lit. Mainz, math.-naturw. Kl. 5: 153–281.

Winkworth, R.G., Bell, C.D., Donoghue, M.J. 2008. Mitochondrial sequence data and Dipsacales phylogeny: Mixed models, partitioned Bayesian analyses, and model selection. Mol. Phylog. Evol. 46: 830–843.

Yang, P., Li, Y., Liu, X., Jiang, S. 2007. Determination of free isomeric oleanolic acid and ursolic acid in *Pterocephalus hookeri* by capillary zone electrophoresis. J. Pharm. Biomed. Anal. 43: 1331–1334.

Zhang, W.-H., Chen, Z.-D., Ki, J.-H., Chen, Y.-C., Anndtang, H.-B. 2003. Phylogeny of the Dipsacales s.l. based on chloroplast *trnL* and *ndhF* sequences. Mol. Phylog. Evol. 26: 176–189.

Ehretiaceae

Ehretiaceae Mart., Nov. Gen. Sp. Pl. 2: 136, 138 (1827), nom. cons.,
including Cordiaceae R.Br. ex Dumort., Anal. Fam. Pl.: 25 (1829), nom. cons.

M. Gottschling, M. Weigend, and H.H. Hilger

Trees or shrubs, rarely subshrubs, lianas or herbs, mostly evergreen; indumentum sericeous or hispid, plants rarely glabrous, trichomes usually simple, unicellular, eglandular, sometimes additionally glandular, rarely branched or dolabriform. Leaves mostly alternate, sometimes fasciculate, generally simple, petiolate or sessile, mostly exstipulate, coriaceous or membranaceous, mostly bifacial. Inflorescences terminal or axillary, thyrsoidal, sometimes congested (appearing pseudoumbellate, subglobose or spicate), usually ebracteose, rarely flowers solitary. Flowers mostly perfect, occasionally unisexual, actinomorphic, tetracyclic, chasmogamous, sometimes dimorphic and dichogamous or cleistogamous, sessile or shortly pedicellate, often fragrant; perianth heterochlamydeous, usually pentamerous; calyx mostly synsepalous, persistent, sometimes accrescent and more or less enclosing the fruit, aestivation valvate or imbricate (quincuncial), lobes usually five, rarely four or more than five, equal, occasionally unequal and of varying shape; corolla sympetalous, usually white, rarely red, orange, yellow or blue, aestivation mostly imbricate (quincuncial), tube usually cylindrical or funnel-shaped, mostly porrect, lobes usually five or rarely more, spreading or reflexed; androecium usually haplostemonous, antesepalous and epipetalous, stamens five, rarely four or more than five, filaments linear, more or less adnate to corolla tube, anthers tetrasporangiate, included or exserted, dorsifixed, opening with longitudinal slits; gynoecium superior, mostly bicarpellate, syncarpous, ovules four, orthotropous or anatropous, unitegmic, style terminal, stigmatic lobes one, two or four; nectary disk often present. Fruits usually four-seeded, mostly indehiscent and drupaceous, rarely schizocarpous, one or more seeds occasionally abortive, endocarp entire or endocarpids (pyrenes) two or four, in total enclosing four, two or one seeds, sometimes with additional sterile chambers including placental tissue. Testa with transfer cells, embryo straight to curved, endosperm copious or absent, cotyledons flat or plicate.

Comprising 10 genera with approximately 500 species in the tropical and subtropical zones of all continents from tropical rain forest to dry savannah, most diverse in at least seasonally dry habitats. Ehretiaceae are here treated to include Cordiaceae (*Cordia*, *Coldenia*, *Hoplestigma*) and to exclude Lennoaceae (see below), which are described separately.

Vegetative Morphology. The vast majority of species are woody plants, mostly trees or shrubs, with the exceptions of the herbaceous *Coldenia procumbens* and *Tiquilia* p.p. Some species are lianescent or subscandent shrubs (*Bourreria* p.p., *Rochefortia* p.p., *Varronia* subclade of *Cordia* p.p.). The plants are mostly evergreen, but deciduous species growing in seasonally dry or temperate regions are found in, for example, *Cordia* and *Ehretia*. The primary root is always well developed and persistent. In species from very arid climates (e.g. *Tiquilia*), an extensive horizontal root system is developed. Underground rhizomes and xylopodia are occasionally found in *Bourreria* and *Ehretia*. The bark is mostly dark, occasionally grey to almost white (e.g. *Ehretia rigida*). Three species of *Cordia* produce domatia in plant-ant mutualisms, most likely with protective value for the plants (Tillberg 2004; Edwards et al. 2007). Ehretiaceae have linear to widely ovate leaves, and shape may vary even within species (Gaviria 1987; Miller 1989). Phyllotaxy is typically alternate (Fig. 32A, D, Fig. 33A, G), sometimes fasciculate (e.g. *Bourreria* p.p., *Cordia* (≡ *Patagonula*) *americana*, *Ehretia microphylla*, *Rochefortia* p.p.). Many

J.W. Kadereit and V. Bittrich (eds.), *Flowering Plants. Eudicots*, The Families and Genera of Vascular Plants 14,
DOI 10.1007/978-3-319-28534-4_12, © Springer International Publishing Switzerland 2016

species of the *Collococcus* subclade of *Cordia* have two opposite leaves in each branch axil, one of which is much smaller and different in shape from the other, larger leaf. The lamina is generally undivided (Fig. 32A, D, F, Fig. 33A–E, G) except in a few *Cordia* species with compound leaves. Usually, the leaves are flat, but plicate and marginally lobed leaves are found in *Coldenia* (Fig. 33C) and *Tiquilia*. Few species have highly divergent leaf shapes such as strongly coriaceous leaves with one lobe on each side (*Cortesia*, *Ehretia microphylla*, Fig. 33B, D). The margin is usually entire (flat or revolute), but crenate, serrate or dentate (Fig. 32F) in some species of *Cordia* and *Ehretia*. The leaf apex is acute, acuminate or rounded, and the base is subcordate, cuneate or decurrent.

Vegetative Anatomy. Wood anatomy in Ehretiaceae is relatively uniform. Bast arises superficially from the cambium (Metcalfe and Chalk 1950) and is generally well developed. Annual growth rings are absent to very distinct. Vessels are mostly diffuse, occasionally arranged in rings, and both arrangements may be found in the same species. Vessels are solitary or in radial multiples, latewood vessels are solitary or in clusters (*Ehretia* p.p.) and are arranged as wavy bands or in a dendritic pattern. Vessel-ray and vessel-parenchyma pits are similar, half-bordered, enlarged or even perforated (in *Cordia* p.p.). Vasicentric tracheids accompany some narrow vessels of *Ehretia* species, then with helical thickenings. The fibres have minutely bordered pits and are

Fig. 32. Cordioideae. **A, D, F** Habit. **A** *Cordia sebestena*, **D** *Cordia saccellia* (≡ *Saccellium lanceolatum*), **F** *Cordia* (≡ *Varronia*) *verbenacea*. **B–C, I–K** Flowers. **B** *Cordia* (≡ *Patagonula*) *americana*, **C** *Cordia saccellia*, **I** and **J** *Cordia sebestena*, showing a distinct dimorphism, **K** *Cordia africana*. **E, L–N** Fruits. **E** *Cordia americana*, **L** *Cordia faulknerae*, **M** *Cordia balanocarpa*, **N** *Cordia somaliensis*. **G** Seedling of *Cordia dentata*. **H** Transverse section of the fruit of *Cordia sebestena*. (Data sources: **A** Sargent et al. 1894, **B–E** Pérez-Moreau and Cabrera 1983, **F** Uhlarz and Weberling 1977, with permission: John Wiley and Sons, **G** Gaviria 1987, with permission: copyright owner, **H–J** Tomlinson and Fawcett 1980, with permission: copyright owner, **K–N** Verdcourt 1991, with permission: courtesy of the Board of the Royal Botanic Gardens Kew.)

mostly non-septate (septate fibres are found in *Cordia* p.p., *Ehretia* p.p.). Parenchyma is abundant and paratracheal to banded (*Cordia* p.p.), predominantly apotracheal and diffuse-in-aggregates in *Ehretia* p.p. The rays are four- to six-seriate, heterocellular, with few rows of square to upright marginal cells, to homocellular in some species of *Cordia* (Baas in Riedl 1997). Vestured pits have been reported for a number of Ehretiaceae (Miller 1977; Rabaey et al. 2010). Calcium oxalate crystals are universally present in the wood: Crystal sand and prismatic crystals are found in *Cordia*, while the other Ehretiaceae have crystal druses as solitary structures or arranged in rows. Pyramidal crystals are restricted to the *Sebestena* subclade of *Cordia* and columnar crystals to the subordinate *Cordia sebestena* species group (Heubl et al. 1990; Al-Shehbaz 1991; Gottschling et al. 2005). Nodal anatomy has only been reported from *Cordia myxa* (Neubauer 1977). The nodes are unilacunar with one main and two lateral traces. Leaves are usually bifacial, equifacial leaves are found occasionally in *Tiquilia*. Palisade tissue is mostly one-layered, but more than one layer is found in some *Cordia* species and in *Tiquilia* so that the upper surface becomes bullate. Stomata are usually anomocytic. Leaf venation pattern appears to be ancestrally brochidodromous, and craspedodromous venation is mostly restricted to species with serrate leaf margins predominantly found in the *Varronia* subclade of *Cordia*. Vascular bundles are embedded in the mesophyll or are vertically transcurrent through sclerenchymatous bundle sheath extensions (in *Cordia* p.p., *Ehretia*). Sclereids are sometimes present in *Cordia*. The indumentum is variously pubescent to hispid, rarely absent and usually consists of unicellular trichomes frequently containing cystoliths (also in adjacent subsidiary cells) causing the typical roughness of the leaf surface. Density and composition of indumentum may vary within species

Fig. 33. Ehretioideae. **A, E, G** Habit. **A** *Ehretia tinifolia*, **E** *Halgania cyanea*, **G** *Bourreria succulenta*. **B–D** Unusual leaf shapes. **B** *Cortesia cuneifolia*, **C** *Coldenia procumbens*, **D** *Ehretia microphylla*. **F, H, I** Flowers. **F** *Halgania erecta* (connate anthers surrounding the style), **H** *Bourreria succulenta*, **I** *Ehretia cymosa*. **J** Fruit of *Ehretia cymosa*. **K–N** Transverse sections of the fruit. **K** *Ehretia cymosa*, **L** *Ehretia acuminata*, **M** *Ehretia microphylla*, **N** *Bourreria teitensis*. (Data sources: **A** Browne 1756, **B–D, L, M** original, **E** M. Fagg, Australian Plant Image Index ANBG Photo

Fig. 33. (continued) No: a.2634, **F** Toelken and Craven 1986, with permission: courtesy of the Board of the Botanic Gardens and State Herbarium Adelaide, **G** Sargent et al. 1933, **H** Tomlinson and Fawcett 1980, with permission: copyright owner, **I–K, N** Verdcourt 1991, with permission: courtesy of the Board of the Royal Botanic Gardens Kew, **L** Pitot 1939, with permission: copyright owner.)

according to climate or habitat (Gaviria 1987; Miller 1989). Unicellular and uniseriate trichomes without cystoliths are also found, as are branched, uniseriate trichomes (in the *Sebestena* and *Varronia* subclades of *Cordia*, *Halgania* p.p., *Rochefortia stellata*), dolabriform trichomes (*Cordia* p.p., *Halgania* p.p.) and uniseriate, gland-tipped trichomes with a single head cell (e.g. *Coldenia*, *Halgania*, on immature anthers in the *Varronia* subclade of *Cordia*, on mature anthers in the *Myxa* subclade of *Cordia*). Frequently, trichomes have a slightly bulbous, multicellular base with cystolith-like structures in trichomes and adjacent subsidiary cells and vary in shape and surface structure (Mez 1890).

Inflorescence Structure. The basically thyrsoidal inflorescence architecture of Ehretiaceae is rather uniform. A terminal, sometimes axillary, acropetalous, ebracteose thyrsoid with mono- or dichasial paracladia (Fig. 32A, D, Fig. 33A, E, G) is most common (e.g. *Bourreria*, *Cordia*, *Ehretia*). The flowers are pedicellate (Fig. 32A) or subsessile (Fig. 33I), rarely sessile (*Rochefortia* p.p.). Thyrsoids with dichasial paracladia and pedicellate flowers appear to represent the ancestral condition, while monochasial paracladia and sessile flowers are probably derived. The more divergent inflorescence types found in some taxa (e.g. *Varronia* subclade of *Cordia*) can be readily derived from this basic structure (Uhlarz and Weberling 1977; Gaviria 1987). In some species of *Cordia*, morphologically terminal inflorescences are displaced into a lateral, internodal position (metatopic displacement). The usually elongated internodes in the inflorescence can be congested, resulting in pseudocapitate synflorescences (*Cordia* p.p., *Ehretia* p.p.). Also, inflorescences can be reduced to solitary, axillary flowers, a development that took place several times independently within the respective taxa such as *Bourreria*, *Cordia*, *Ehretia*, *Halgania* or *Rochefortia*. In the *Varronia* subclade of *Cordia*, complex patterns of growth lead to strongly congested (capitate or glomerulate) architectures, and pseudospicate inflorescences (Fig. 32F) with basipetal anthesis (concaulescent-syndesmic) are derived from these congested inflorescences (Uhlarz and Weberling 1977; Miller and Gottschling 2007). Capitate inflorescences are also found in some *Tiquilia* species (Richardson 1977).

Flower Structure. Detailed studies of floral anatomy are rare (Rao and Rani 1987; Gottschling 2004; Gottschling et al. 2014b). The flowers usually are <15 mm in diameter, but large-flowered species are found in, for example, *Bourreria* p.p. and the subclades *Collococcus* p.p. and *Sebestena* p.p. of *Cordia*. *Cordia mairei* from the *Myxa* subclade of *Cordia* probably has the largest flower of Ehretiaceae (up to 10 cm in diam.). The calyx is synsepalous, divided almost to the base (Fig. 33I) or forming a long tube (e.g. *Bourreria*, *Cordia*, Fig. 32A, I–K, Fig. 33H). The calyx tube is cylindrical, urceolate, campanulate or funnel-shaped and sometimes ribbed in *Cordia* p.p. (Miller 2013). The lobes are lanceolate, triangular or narrowly ovate. Calyx aestivation is sometimes valvate (*Bourreria*, *Cordia*), otherwise quincuncial-imbricate (e.g. *Ehretia*, *Lepidocordia*, *Rochefortia*). Occasionally, the calyx is irregularly ruptured by the developing corolla, leading to indistinct calyx lobes (in some species of the *Myxa* subclade of *Cordia*, *Hoplestigma*). In *Cortesia*, flowers are unusual in having five spoon-shaped sepal lobes and a unique involucre composed of 10 to 15 teeth (derived from bracts) surrounding the entire flower (Fries 1910). The calyx persists in fruit and encloses it at least when young. The corolla is always sympetalous, with a cylindrical or funnel-shaped tube and spreading or reflexed lobes. Corolla aestivation is imbricate (quincuncial), occasionally contort. Usually, the corolla tube exceeds the calyx in length (rarely shorter in *Ehretia* p.p. and the *Varronia* p.p. subclade of *Cordia*). Five corolla lobes are predominantly present, sometimes only four (e.g. *Coldenia*, *Cordia tetrandra*, *Tiquilia* p.p.) or more than five (*Hoplestigma* with 10–15 lobes, some species of *Cordia*). The corolla lobes are mostly shorter than the corolla tube, but exceptions are found in most of the larger taxa. They are ovate to suborbicular, sometimes long and narrow with entire to repand margins, sometimes involute (Fig. 32B–C, I–K, Fig. 33H, I). The androecium is haplostemonous in most taxa, but *Hoplestigma* is highly unusual in having 20–35 stamens arranged in three irregular series. Filaments insert on the corolla tube at different levels (Fig. 32B–C, I, J). A trichome tuft frequently is present at the points of filament attachment in *Bourreria* p.p. and *Cordia* p.p. Veins below the points of attachment are swollen or winged in some species of *Tiquilia*. Anthers are oblong, ovate or cordate, introrse, usually exserted beyond the stigma, rarely shorter and included in the corolla. In *Halgania*, anthers are connate into a cone around the style (Fig. 33F, *Solanum*-type flowers: Fægri 1986). The four microsporangia open with longitudinal slits. A nectary disk is more or less developed (a voluminous disk is found in, for example, *Bourreria*: Gottschling 2004, and *Cordia* p.p.), and a short gynophore is formed in *Tiquilia dichotoma* (Gottschling et al. 2014b). The ovary is bicarpellate, whereas the internal architecture is characterised by the development of several additional septa (basal septa, apical septa, false septa), thus leading mostly to an imperfectly four-locular ovary. Only two locules and ovules are found in some species of *Halgania*, probably after loss of one carpel (Junell 1938). The style is terminal and bifid in most Ehretiaceae, but twice bifid in *Cordia* (with four stigmatic lobes) and undivided in some species of *Bourreria*. The relative length of the two style branches varies from shortly divided apically to divided almost to the base (e.g. *Ehretia microphylla*, *Hoplestigma*, *Rochefortia* p.p.). Some species of *Cordia* have distylous flowers. The placentae are well (e.g. *Bourreria*, *Ehretia* p. p.) to poorly developed (e.g. *Cordia*, *Tiquilia*). Placentation is basal (e.g. *Sebestena* subclade of *Cordia*) to axile (most Ehretiaceae).

Embryology. Embryology of Ehretiaceae is poorly studied (Briechle and Hilger 1988). The anther wall usually has one middle layer, in some species it comprises two to four (sometimes five) layers and fibrous thickenings in the endothecium. The tapetum is multinucleate and glandular. The archaespore functions directly (*Cordia*), or a parietal cell is formed (remainder of Ehretiaceae). Microsporogenesis is simultaneous. The initial microspore is tetrahedral and decussate-isobilateral. Ovules are generally anatropous, but orthotropous ovules are found, for example, in the *Sebestena* subclade of *Cordia* (Svensson 1925; Rao and Rao 1984; Khaleel 1985; Gottschling 2004). In Ehretiaceae, the ovules are mostly crassinucellate, while tenuinucellate ovules are found in *Cordia*. A placental obturator is documented for some species (Khaleel 1982), as well as a suspensor (Gottschling 2004), which is uniseriate, filamentous and consists of four to six cells in *Ehretia acuminata* (Khaleel 1977). Polar nuclei fuse prior to fertilisation. Three ephemeral antipodial cells are formed, which do not proliferate. The synergids are hooked and disintegrate during the entry of the pollen tube. Fertilisation is porogamous. The endosperm develops ab initio cellular and corresponds to the *Myosotis*-type. Both chalazal and micropylar haustoria are found. In some species, one of them is more active than the other, while in others both are equally well developed (Khaleel 1985). Embryogenesis corresponds to the crucifer- or chenopodiad-type, and embryo sac development follows the *Polygonum*- or *Allium*-type (Rao and Rao 1984).

Pollen Morphology. Ehretiaceae are eurypalynous with many heteromorphic pollen types. Sticky pollen kitt is abundant in, for example, the *Myxa* subclade of *Cordia* and may obscure the exine structure. Pollen of *Cordia* is oblate-spheroidal to subprolate in outline, usually 3–(4)-colpor(oid)ate and 20–60 μm in diameter. The *Varronia* subclade of *Cordia* has 3-porate pollen as does, for example, *Cordia lutea* (*Myxa* subclade of *Cordia*). Exine ornamentation is irregularly striate [*Cordia* (≡ *Auxemma*) *glazioviana*)], rugulose [*Cordia* (≡ *Auxemma*) *oncocalyx*), *Cordia americana*], reticulate with scabrae (*Varronia* subclade of *Cordia*), striate-reticulate (*Sebestena* subclade p.p. of *Cordia*) or variously spinulose and clavate (subclades *Sebestena* p.p. and *Myxa* of *Cordia*). Borhidi et al. (1988) distinguished three main types, Heubl et al. (1990) five types and Nowicke and Miller (1990) at least seven types of pollen in *Cordia*, some of which correspond to clades retrieved by molecular data (Gottschling et al. 2005). The (2)3(4)-colpor(oid)ate pollen grains of other Ehretiaceae (except *Lepidocordia*) can usually be distinguished from those of *Cordia* by the presence of pseudocolpi or pseudocolpoid depressions alternating with the colpi (Sahay 1979; Miller and Nowicke 1990; Nowicke and Miller 1990). Pollen shape is prolate-spheroidal to subprolate in outline with a diameter of 15–35 μm. Exine ornamentation is granulate (e.g. *Bourreria*, *Hoplestigma*), scrobiculate (e.g. *Coldenia*), punctate (e.g. *Halgania*), finely reticulate (e.g. *Ehretia* p.p., *Lepidocordia*), distinctively reticulate (*Rochefortia*) or faintly striate (e.g. *Ehretia* p.p.). Both size and shape dimorphisms have been reported from distylous species of *Cordia* (Al-Shehbaz 1991).

Karyology. Chromosome numbers of Ehretiaceae are poorly studied (Al-Shehbaz 1991), and records of only about 15 % of the species are available. Although some reports are conflicting (even within species), the following generalised patterns emerge (Heubl et al. 1990): *Cordia* has haploid chromosome numbers of n = 7 (*Myxa* subclade of *Cordia*), n = 8 (*Sebestena* subclade of *Cordia*) or n = 9 (*Cordia americana*, *Varronia* subclade of *Cordia*). In some species of the *Sebestena* subclade of *Cordia*, n = 15 occurs, which can be interpreted as being derived from n = 7 or 8. Two haploid chromosome numbers are readily recognised in other Ehretiaceae: n = 8 (*Ehretia*, *Tiquilia* p.p.) and n = 9 (*Cortesia*, *Tiquilia* p.p.; Richardson 1977; Heubl et al. 1990). For *Bourreria*, n = 19 has been reported (Al-Shehbaz 1991). Polyploidisation (mostly tetraploidy) is a common feature in Ehretiaceae (e.g. in subclades *Myxa* and *Sebestena* of *Cordia*, *Ehretia*, *Tiquilia*).

Pollination. Flowering can precede leaf development in deciduous species of the *Myxa* subclade of *Cordia* (Gaviria 1987). The breeding systems are diverse (Opler et al. 1975; Gaviria 1987; Askins et al. 1987; Warfa 1988), and protandry is a common feature. Some examples of transitions between homostyly and distyly (Fig. 32I, J) and monoecy and dioecy, respectively, are found in *Cordia* (*Myxa* subclade: Lloyd 1979; Verdcourt 1991; Gottschling et al. 2005). Distyly is further known from the *Varronia* subclade of *Cordia* (partly with a high degree of specialisation: Milet-Pinheiro and Schlindwein 2010; Machado et al. 2010) and *Ehretia saligna* (Randell 1993), and dioecy has been reported from *Lepidocordia* (Miller and Nowicke 1990) and *Rochefortia* (Urban 1908; Irimia et al. 2015). Flower morphology and scent suggest that Ehretiaceae are typically zoophilous, but few pollinator observations have been published (di Fulvio 1965; Percival 1974; Opler et al. 1975; Gaviria 1987; Milet-Pinheiro and Schlindwein 2010). For small, white flowers of many Ehretiaceae, insects such as Hymenoptera, Diptera, Lepidoptera, Coleoptera or Thysanoptera have been recorded. The 'solanoid' blue flowers of *Halgania* are buzz-pollinated (Fægri 1986) and may be visited by UV-sensitive insects. The large, red flowers of subclades *Collococcus* p.p. and *Sebestena* p.p. of *Cordia* suggest bird pollination. However, at least *C. sebestena* suffers from flower destruction and nectar depletion by avian robbers (i.e. no pollination: Askins et al. 1987). Furthermore, the frequent occurrence of exserted anthers and stigmas indicates wind pollination.

Fruit and Seed. Fruits are usually small (≤10 mm in diameter), but sometimes larger (up to 5 cm in diameter) in *Bourreria* p.p. (Miller 2003) and the *Sebestena* p.p. subclade of *Cordia*. Fruits often are coloured. They are orange or red in *Bourreria*, *Ehretia*, *Rochefortia* and the *Varronia* subclade of *Cordia*, white in some species of *Cordia* and in *Ehretia latifolia*, rarely yellow or black in some species of *Bourreria*, *Cordia* and *Ehretia*. Usually, the globose to ovoid fruit is drupaceous (Fig. 32L–N, Fig. 33G, J), sometimes nut-like in *Cordia* p.p. (Miller 2013). Schizocarps occur in *Bourreria* p.p., *Coldenia* and *Tiquilia*. In *Bourreria* p.p., they are morphologically unique by having a carpophore (derived from the ventral vascular bundle), which fixes the apex of the schizocarpids to the persistent gynobase (columella; Gottschling 2004; Gottschling and Miller 2007). The calyx (and sometimes also the corolla: Miller 2013) persists (Fig. 32L–N, Fig. 33J) and encloses the fruit at least when young. It is accrescent in many species of *Cordia*, ultimately more or less enclosing the fruit (fruits completely enclosed in most species of the *Sebestena* subclade of *Cordia* (Fig. 32D). In some species, the calyx lobes are elongated in fruit (e.g. *Cordia americana*: Fig. 32E, *Ehretia microphylla*). The style (at least its base) persists in fruit (Fig. 33J) in most Ehretiaceae and in *Cordia* p.p., the fruits are capped by the persistent base of the style (Miller 2013). Fruit anatomy provides important taxonomic characters (Miller 1989; Al-Shehbaz 1991; Gottschling and Hilger 2001, 2004b). The exocarp is thin and coriaceous, and the mesocarp is mostly fleshy to mucilaginous; only in some cases (e.g. *Coldenia*, *Sebestena* subclade p.p. of *Cordia*, *Tiquilia*: Gottschling et al. 2014b), it is dry and/or spongy (*Bourreria* p.p.: Gottschling 2004). Typically, each of four seeds is enclosed in a sclerenchymatic, multi-layered endocarp. Cells of the endocarp may be relatively thin-walled (*Cordia* p.p., *Halgania*, *Lepidocordia*, *Tiquilia*). The endocarp is either undivided (*Cordia*, species of *Ehretia*, *Halgania* p.p.,

Hoplestigma, *Lepidocordia*, Fig. 32H, Fig. 33M) or divided into two (*Ehretia* p.p., *Halgania* p.p., Fig. 33L) or four parts (remainder of Ehretiaceae, Fig. 33K, N), termed endocarpids (pyrenes). The four-parted endocarp appears ancestral in Ehretiaceae, and fusion of pyrenes (synmericarpy) has evolved multiple times and more frequently than assumed to date (Gottschling et al. 2014a). The abaxial surface of the pyrenes is glabrous, rugose-sulcate, ridged or distinctly lamellate (*Bourreria*, *Ehretia* p.p.: Gottschling and Hilger 2001, 2004a; Gottschling 2004; Gottschling and Miller 2007). Occasionally, some pyrenes and/or seeds abort (*Cordia* p.p., *Ehretia* p.p., *Halgania*, *Tiquilia* p.p.), or additional sterile chambers (e.g. *Ehretia microphylla*, *Hoplestigma*, *Lepidocordia*, Fig. 33M) containing placental tissue are found (Gottschling 2004). The testa epidermis consists of transfer cells with tangential and radial protuberances of the cell walls (Diane et al. 2002). The embryo is curved or straight. Endosperm is absent in mature seeds of *Coldenia*, *Cordia* and *Hoplestigma* but copious and fleshy in the remainder of Ehretiaceae. Seed germination is epigeal and phanerocotylar. In many Ehretiaceae, cotyledons are flat, while the plicate cotyledons of *Cordia* (Fig. 32G) may represent the derived character state (Gottschling et al. 2005). The first pair of foliage leaves is alternate or subopposite (Gaviria 1987; Warfa 1988).

Dispersal. Usually, the dispersal unit is the entire fruit, sometimes with the persistent calyx attached (e.g. *Sebestena* subclade of *Cordia*). Fruit parts are dispersed in *Bourreria* p.p., *Coldenia* and *Tiquilia*. Diaspores are zoochorous, with endozoochory and dysochory as most common mechanisms. Birds have been reported as vectors (Ridley 1930; Scott and Martin 1984; Al-Shehbaz 1991), but mammals probably also play a role (Babweteera 2012). Fruits of some species of the *Sebestena* subclade of *Cordia* are wind-dispersed (Al-Shehbaz 1991; Miller 2013), which may also be the case in those species of *Bourreria* with dry and light schizocarps (mesocarp spongiose rather than juicy as in the other species). Hydrochory is found in, for example, *Cordia subcordata* and is mediated by a suberose mesocarp (Johnston 1951).

Phytochemistry. Species of Ehretiaceae have a vast array of secondary metabolites including triterpenes, sesquiterpenes, flavonoids and saponins in all parts of the plant (Chen et al. 1983; Khattab et al. 2001; Kuroyanagi et al. 2001; Costa et al. 2005; Santos et al. 2005). They frequently deposit calcium carbonate and/or silicon dioxide in trichomes and epidermis cells. Pyrrolizidine alkaloids and aromatic organic compounds such as quinonoids and phenolic derivates (PAs; e.g. ehretinine) are typical for Ehretiaceae (Hegnauer in Riedl 1997; Stegelmeier et al. 1999; Dai et al. 2010; Li et al. 2010). Pyrrolizidine alkaloids (PAs) all belong to the ester-type and are similar to, or identical with, alkaloids also occurring in Asteraceae. PAs are frequently concentrated in young leaves and inflorescences or flowers. They are a part of the defence against herbivores and are used to increase floral constancy of visiting pollinators when present in nectar (Hegnauer in Riedl 1997). Red alkannin is—via various pathways—the precursor of a range of quinonoids and/or phenolic constituents such as the allergenic alliodorin, cordiachromones and glaziovianols (Pessoa et al. 1995; Costa et al. 1999). These latter compounds show antibiotic activities and have been isolated from stems and roots. Cordiaquinones, ehretianones and microphyllones are probably also derivates of hydroquinone (Silva Filho et al. 1993; Selvanayagam et al. 1996). Nitrile glucosides (ehretiosides) are found in the stem bark of *Ehretia philippinensis*, and caffeic and rosmarinic acid appear to be ubiquitous (Simpol et al. 1994; Ticli et al. 2005). The seeds are often very rich in allantoin (presumably for nitrogen storage), and the seed oil is rich in oleic acid (Hegnauer 1989). Major components of essential oil are humulene, caryophyllene, cyclohexane and eudesmol (Fernandes et al. 2007; Hernandez et al. 2007).

Phylogeny. In their most recent circumscription (Weigend et al. 2014), Boraginales comprise the sister groups Boraginales I with Codonaceae, Wellstediaceae and Boraginaceae and Boraginales II with Hydrophyllaceae (monophyly doubtful), Heliotropiaceae, Cordiaceae and Ehretiaceae (including Lennoaceae; Ferguson 1999; Gottschling et al. 2001, 2014a; Judd and Olmstead 2004; Moore and Jansen 2006; Luebert and Wen 2008; Weigend and Hilger 2010). Cordiaceae, Ehretiaceae and Heliotropiaceae together are monophyletic based on both molecular and morphological data such as the multilayered endocarp that is otherwise rare in the asterids (Gottschling et al. 2001; Diane et al. 2002; unified as Primarily Woody Boraginales: Gottschling 2003). Different from Weigend et al. (2014), Ehretiaceae are here treated to include Cordiaceae and to exclude the parasitic Lennoaceae, which are described separately. Ehretiaceae sensu Weigend et al. (2014) comprise *Bourreria* (including *Hilsenbergia*), *Cortesia*, *Ehretia* (including *Carmona* and *Rotula*), *Halgania*, *Lennoa*, *Lepidocordia*, *Rochefortia* and *Tiquilia* (Gottschling et al. 2014a). Cordiaceae sensu Weigend et al. (2014) comprise *Cordia*, *Coldenia* and *Hoplestigma* and may be characterized by the absence of endosperm in mature seeds as a derived character state.

Distribution and Habitats. Ehretiaceae are largely restricted to tropical or subtropical regions. Only a few species occur in more temperate regions, and this is correlated with certain morphological features such as serrate leaves in, for example, *Ehretia* p.p. (Gottschling and Hilger 2001). Both *Cordia* and *Ehretia* have centres of diversity both in the New and in the Old World (particularly Africa and East Asia), whereas *Bourreria* has an unusual disjunction between the Neotropics and eastern Africa (Gottschling et al. 2004, 2005, 2014a). The other taxa of the Ehretiaceae have more restricted distributions, mostly limited to the New World (i.e. *Cortesia*, *Lepidocordia*, *Rochefortia*, *Tiquilia*). *Halgania* is endemic to Australia and *Hoplestigma* to western Africa. The habitats range from seasonally flooded localities (*Ehretia aquatica*), tropical rainforest (*Cordia*, *Ehretia* p.p.) to arid savannahs (*Bourreria*, *Ehretia* p.p., *Varronia* subclade of *Cordia*) and deserts (*Tiquilia*). Some taxa are adapted to extremely poor (*Tiquilia*) or toxic soils (e.g. some species of *Bourreria* and *Rochefortia* on ultramafic soils in Cuba).

Palaeobotany. Fossilised pyrenes of *Ehretia* p.p. are known from the Eocene to Pliocene of Europe (England, France, Germany; Chandler 1964; Mai 1995), closing the present-day holarctic distribution gap in western Eurasia (Gottschling et al. 2002, 2004). A fossil endocarp of *Cordia* has been reported from the Miocene of Tasmania (Ettingshausen 1888), as well as some leaf imprints of *Cordia* from the Palaeocene of North America (Ward 1887) and the Eocene of Russia (Chelebajeva 1984). There exist a few pollen records for *Cordia* from the Miocene of Spain and the Micronesian Marshall Islands (Muller 1981). Quaternary seeds, fragmentary wood (Brea and Zucol 2006), and twigs of *Cordia* have been reported from the Caribbean (Iturralde-Vinent et al. 2000). The fossil record shows that Ehretiaceae were more widely distributed in both hemispheres in the geological past than they are today.

Economic Importance. Several species of *Cordia* and *Ehretia* are valuable timber trees in East Asia and Central and South America. Their wood is used for construction and carpentry, flooring, tool handles and musical instruments (Al-Shehbaz 1991). The fleshy mesocarp of the fruit of many species is edible (Watt and Breyer-Brandwijk 1962; Usher 1971; Facciola 1990). Furthermore, bark and leaves are sometimes used for the preparation of herbal teas or as condiments. Flowers and leaves are occasionally eaten as vegetables. Some species are cultivated as ornamentals (e.g. *Bourreria huanita*, *Cordia lutea*, *Cordia saccellia* (≡ *Saccellium lanceolatum*), *Tiquilia* p. p.). Alkannin present in wood can be used as a dye. Several uses in traditional medicine have been reported for wound-healing and to treat urinary infections, lung diseases and leprosy because of astringent, anti-inflammatory, antihelminthic, antimalarial and diuretic properties of their secondary metabolites (Quisumbing 1951; Watt and Breyer-Brandwijk 1962; Lewis 1991; Sertié et al. 1991; Ficarra et al. 1995; Costa et al. 1999; Nayar et al. 1999; Dong et al. 2000). Some essential oils of Ehretiaceae have larvicidal activity (Costa et al. 2004; Santos et al. 2006), other compounds have psychoactive (Nayar et al. 1999), antiallergic or spasmolytic activity (Hegnauer in Riedl 1997). In Africa, the roots of *Cordia gharaf* and *Cordia quarensis* are used as an abortive (Watt and Breyer-Brandwijk 1962).

Key to the Genera

(Parasitic Lennoaceae are treated separately)

1. Androecium with 20–35 stamens **7. *Hoplestigma***
– Androecium with usually 5, never more than 10 stamens 2

2. Annual herbs with adventitious roots; lamina asymmetrical; flowers tetramerous **2. *Coldenia***
- Woody plants or rarely perennial herbs; lamina usually symmetrical; flowers mostly pentamerous 3
3. Flower with involucre composed of 10–15 teeth formed by bracts, sepals free and spoon-shaped **4. *Cortesia***
- Flower without involucre, calyx synsepalous 4
4. Style twice bifid (if once bifid as in *Cordia saccellia*, then fruit completely enclosed in accrescent calyx); endocarp undivided; cotyledons plicate **3. *Cordia***
- Style entire or once bifid; endocarp rarely undivided, mostly two- or four-parted; cotyledons flat 5
5. Herbs or shrubs up to 1 m tall; inflorescences congested, rarely few-flowered; fruits dry, separating into four ovoid nutlets shorter than 5 mm; leaves pinnately veined, upper surface bullate **10. *Tiquilia***
- Trees or shrubs; inflorescences branched and usually not congested, rarely few-flowered; fruits drupaceous at least when young, rarely schizocarpous (*Bourreria*), then each mericarpid longer 5 mm; leaves sometimes pinnately veined, surface never bullate 6
6. Anthers connate in a cone surrounding the style; flowers mostly blue, rarely white, somewhat tilted; Australia **6. *Halgania***
- Anthers free; flowers usually white 7
7. Flowers unisexual 8
- Flowers bisexual 9
8. Stems without spines **8. *Lepidocordia***
- Stems with spines **9. *Rochefortia***
9. Calyx aestivation valvate; fruits with 4 triangular pyrenes with an additional sterile chamber and abaxially distinctly lamellate **1. *Bourreria***
- Calyx aestivation imbricate; fruit with 1–4 pyrenes, abaxially usually not lamellate **5. *Ehretia***

Genera of Ehretiaceae

1. ***Bourreria*** P. Br. Fig. 33G, H, N

Bourreria P. Br., Civ. Nat. Hist. Jam.: 168–169, tab. 15 (1756), nom. cons.; Schulz in Urb., Symb. Antill. 7: 45–71 (1911), rev.; Gottschling & Hilger, Bot. Jahrb. Syst. 123: 249–268 (2001), phylog.; Miller, Adansonia III, 25: 151–189 (2003), rev. (*Hilsenbergia*); Gottschling, Flora 199: 409–423 (2004), anat.; Gottschling & Miller, Ann. Missouri Bot. Gard. 94: 734–744 (2007), rev. of S Am spp.
Hilsenbergia Tausch ex Meisn. (1840).

Trees or shrubs, occasionally woody climbers. Leaves alternate or fasciculate, sometimes microphyllous, mostly smooth or scabrid, margin usually entire. Thyrsoids terminal, rarely axillary, with dichasial paracladia, sometimes few-flowered. Flowers bisexual, small to large; calyx cylindrical or campanulate, aestivation valvate, lobes sometimes irregular in shape; corolla white, rarely yellow, red or blue, tube often broadened and with trichome tuft in throat, lobes spreading or reflexed; anthers included or exserted, free; style bifid or entire. Drupes small to large, not enclosed in calyx, endocarp four-parted, endocarpids abaxially distinctly lamellate, with an additional sterile chamber containing placental tissue. n = 19.

50 species in Central and northern South America and the Caribbean, Africa, Madagascar, the Comores and the Mascarenes.

Bourreria shows a peculiar disjunction between the Neotropics and eastern Africa (Gottschling et al. 2004), and the Old World species were segregated as *Hilsenbergia* (Miller 2003). However, the monophyly of the New World species is questionable (Gottschling et al. 2014a) and, therefore, *Bourreria* is treated here in a broad sense.

2. ***Coldenia*** L. Fig. 33C

Coldenia L., Sp. Pl.: 125 (1753); Richardson, Rhodora 79: 467–572 (1977), rev.

Procumbent annual herb with slender, branched stems, often with adventitious roots. Leaves clustered, small, numerous, bullate, margin crenate. Flowers bisexual, solitary, small, tetramerous; calyx deeply lobed; corolla white or yellow, tube cylindrical, glabrous, lobes spreading; anthers included; style bifid. Fruits small, dry, ovoid-conical, four-lobed, separating into four one-seeded apiculate nutlets, mericarpids without sterile chambers.

One species, *Coldenia procumbens* L., coastal, Old World tropics and introduced elsewhere.

3. ***Cordia*** L. Fig. 32

Cordia L., Sp. Pl.: 190 (1753); Johnston, Contr. Gray Herb. 92: 5–65 (1930), rev.; Johnston, J. Arnold Arbor. 30: 85–104 (1949), rev.; Borhidi et al., Acta Bot. Hung. 34: 375–423, rev.; Gottschling et al., Ann. Missouri Bot. Gard.

92: 425–437 (2005), phylog.; Gottschling & Miller, Syst. Bot. 31: 361–367 (2006), rev. (*Auxemma*, *Patagonula*, *Saccellium*); Miller & Gottschling, Taxon 56: 163–169 (2007), rev. (*Varronia*); Miller, J. Bot. Res. Inst. Texas 7: 55–83 (2013), rev. (*Cordia* sect. *Gerascanthus*).
Patagonula L. (1753).
Varronia P. Br. (1756).
Saccellium Humb. & Bonpl. (1806).
Auxemma Miers (1875).

Trees, shrubs or woody climbers. Leaves alternate, rarely subopposite or fasciculate, ovate to lanceolate, margin entire, crenate, dentate or serrate, leaves rarely compound. Thyrsoids mostly terminal, occasionally axillary, basically with dichasial or monochasial paracladia, but also capitate, spicate or very shortly cymose umbellate to globose, sometimes in corymbo-thyrsoids. Flowers bisexual or sometimes unisexual and dioecious, often dichogamous or functionally unisexual; calyx cylindrical, lobes occasionally thickened at the margin or appendaged; corolla sometimes persistent in fruit, white, yellow, orange or red, tube often with trichome tuft in the throat, lobes spreading or reflexed; anthers included or exserted, free; style twice bifid. Drupes small to large, largely or almost completely enclosed by the accrescent calyx (this sometimes cupuliform) at maturity, endocarp undivided, without sterile chambers, often one-seeded, the other ovules abortive. n = 7, 8, 9.

Approximately 350 species distributed pantropically with centres of diversity in the Caribbean, northern South America (including Galapagos) and Africa.

Cordia is monophyletic (Gottschling et al. 2005; Moore and Jansen 2006; Miller and Gottschling 2007) based on both molecular data and apomorphic traits such as plicate cotyledons, an undivided endocarp and the presence of four stigmatic lobes. It is treated here in a broad sense and comprises the four subclades *Collococcus*, *Myxa*, *Sebestena* (including *Auxemma*, *Patagonula* and *Saccellium*; together approx. 250 spp.) and *Varronia* (approx. 100 species). Nuclear molecular data show that *Cordia* (with subclades *Collococcus*, *Myxa*, *Sebestena*) and *Varronia* are sister groups (Gottschling et al. 2005), but this topology is not corroborated by chloroplast data (Weigend et al. 2014).

4. ***Cortesia*** Cav. Fig. 33B

Cortesia Cav., Icones 4: 53, tab. 377 (1797); Gottschling & Hilger, Taxon 53: 919–923 (2004), anat.

Small shrubs. Leaves alternate or fasciculate, long-cuneiform with 2–3 lobes, lobe margin entire. Flowers bisexual, mostly solitary, terminal or axillary, small, with an involucre with 10–15 teeth formed by bracts, persistent in fruit; calyx composed of free sepals, sepals spoon-shaped, caducous, aestivation imbricate; corolla equalling involucre; anthers exserted; style bifid. Drupes small, endocarp two-parted, without sterile chambers. n = 9.

One species, *Cortesia cuneifolia* Cav., Argentina.

5. ***Ehretia*** P. Br. Fig. 33A, D, I–M

Ehretia P. Br., Civ. Nat. Hist. Jam.: 168, tab. 16 (1756); Miller, Ann. Missouri Bot. Gard. 76: 1050–1076 (1989), rev. (New World species); Gottschling & Hilger, Bot. Jahrb. Syst. 123: 249–268 (2001), phylog.; Miller, Adansonia 24: 137–157 (2002), rev. (Madagascan species); Gottschling & Hilger, Blumea 49: 145–153 (2004), anat.
Rotula Lour. (1790).
Carmona Cav. (1799).

Trees or shrubs. Leaves alternate or fasciculate, ovate to lanceolate, smooth or scabrid, margin entire, crenate or dentate. Thyrsoids terminal or axillary, with dichasial or monochasial paracladia, rarely few-flowered. Flowers bisexual, sometimes dimorphic; calyx cylindrical, shallowly to deeply lobed, rarely accrescent, aestivation imbricate; corolla white, pale yellow, pale pink or blue, tube very short or exceeding lobes in length, glabrous, lobes spreading or reflexed; anthers usually exserted, sometimes included, free; style scarcely divided into two branches or divided to base. Drupes small, endocarp four- or two-parted or undivided, rarely abaxially lamellate, rarely with additional sterile chambers including placental tissue. n = 9.

About 40 species in Old and New World tropics, with centres of diversity in Africa and East Asia.

6. ***Halgania*** Gaudich. Fig. 33E–F

Halgania Gaudich. in Freyc., Voy. Aut. Monde (Bot.): 448, tab. 59 (1829).

Shrubs. Leaves alternate, lanceolate, margin entire or toothed to serrate, glandular hairs abundant. Thyrsoids with dichasial paracladia terminal, rarely few-flowered. Flowers bisexual, small, somewhat tilted; calyx campanulate, aestivation imbricate; corolla white or blue to purple, tube short, glabrous; anthers exserted, connate and forming a yellow cone around the style, tapering into five terminal appendages which form a straight beak; gynoecium with four or two locules (by abortion of one carpel), style shortly bifid. Drupes small, one- or two-seeded by abortion, endocarp undivided or rarely two-parted, without sterile chambers.

About 20 species in Australia.

7. ***Hoplestigma*** Pierre

Hoplestigma Pierre, Bull. Mens. Soc. Linn. Paris II: 116 (1899).

Trees. Leaves alternate, ovate, margin entire. Thyrsoids terminal, with dichasial paracladia. Flowers bisexual; calyx irregularly lobed; corolla yellowish, tube short, glabrous, lobes 10–15, imbricate; stamens 20–35, in three irregular series, anthers included; style bifid. Drupes large, endocarp undivided, with 2 sterile chambers.

Two species in western Africa.

8. ***Lepidocordia*** Ducke

Lepidocordia Ducke, Archiv. Jard. Bot. Rio de Janeiro 4: 170, tab. 22 (1925); Miller & Nowicke, Amer. J. Bot. 77: 543–551 (1990), rev.

Trees. Leaves alternate, ovate, margin entire. Thyrsoids terminal with dichasial paracladia. Flowers unisexual, dioecious, small; calyx campanulate, aestivation imbricate, lobes nearly free to base, slightly unequal; corolla white, tube glabrous, lobes spreading or reflexed; staminate flowers with functional anthers and viable pollen, style absent, stigma reduced; pistillate flowers with degenerated anthers and an elongated bifid style. Drupes small, endocarp undivided, with sterile chambers.

Two species in Central America and northern South America.

9. ***Rochefortia*** Sw.

Rochefortia Sw., Prod. Veg. Ind. Occ.: 53 (1788); Irimia et al., Plant Syst. Evol. 301: 1509–1516 (2015), phylog.

Shrubs, with axillary spines. Leaves alternate or fasciculate, mostly ovate, margin entire. Thyrsoids often axillary or terminal, then with dichasial paracladia, occasionally flowers solitary. Flowers unisexual; calyx campanulate, aestivation imbricate; corolla white, tube short, glabrous; anthers exserted; style bifid. Drupes small, endocarp four-parted, without sterile chambers.

Nine species in Central America, northern South America and the Caribbean.

10. ***Tiquilia*** Pers.

Tiquilia Pers., Syn. Pl. 1: 157 (1805); Richardson, Rhodora 79: 467–572 (1977), rev.; Moore & Jansen, Mol. Phylogenet. Evol. 39: 668–687 (2006), phylog.; Gottschling et al., Biol. J. Linn. Soc. 112: 520–534 (2014), anat.

Shrubs or herbs, prostrate, spreading or erect, densely branched. Leaves microphyllous, often clustered near branch apices, lamina bullate, margins entire or crenate. Inflorescences congested. Flowers bisexual, small, sessile; calyx campanulate; corolla white to blue, throat often yellow, glabrous; anthers included or exserted; style bifid. Fruits drupaceous when young, later separating into four (or fewer by abortion) dry nutlets, mericarpids without sterile chambers. $n = 8, 9$.

About 28 species in arid western North and South America (including Galapagos Islands).

Selected Bibliography

Al-Shehbaz, I.A. 1991. The genera of Boraginaceae in the southeastern United States. J. Arnold Arb. Suppl. Ser. 1: 1–64.

Askins, R.A., Ercolino, K.M., Walter, J.D. 1987. Flower destruction and nectar depletion by avian nectar robbers on a tropical tree, *Cordia sebestena*. J. Field Ornithol. 58: 345–349.

Babweteera, F. 2012. Seed dispersal and tree spatial recruitment patterns in secondary tropical rain forests. In: Sudarshana, P., Nageswara-Rao, M., Soneji, J.R. (eds.) Tropical forests. Rijeka: InTech, pp. 157–176.

Borhidi, A.E., Gondár, E., Orosz-Kovács, Z.S. 1988. The re-consideration of the genus *Cordia* L. Acta Bot. Hung. 34: 375–423.

Brea, M., Zucol, A.F. 2006. Leños fósiles de Boraginaceae de la Formación Peñas Coloradas (Paleoceno superior), Puerto Visser, Chubut, Argentina. Ameghiniana 43: 139–146.

Briechle, M. Hilger, H.H. 1988. Die Embryogenese von *Microparacaryum intermedium* und die embryologische Klassifizierung der Boraginaceae nach Souèges. Flora 181: 45–59.

Browne, P. 1756. The civil and natural history of Jamaica. London: by author.

Chandler, M.E.J. 1964. The Lower Tertiary floras of Southern England 4. London: British Museum.

Chelebajeva, A.I. 1984. The genus *Cordia* (Boraginaceae) from the Palaeogene of Kamchatka (Russian SFSR, USSR) and adjacent territories. Bot. Zhurn. (St. Petersburg) 69: 605–615.

Chen, T.K., Ales, D.C., Baenziger, N.C., Wiemer, D.F. 1983. Ant-repellent triterpenoids from *Cordia alliodora*. J. Org. Chem. 48: 3525–3531.

Costa, J.G.M. da, Lemos, T.L.G. de, Pessoa, O.D.L., Monte, F.J.Q., Braz-Filho, R. 1999. Glaziovianol, a new terpenoid hydroquinone from *Auxemma glazioviana*. J. Nat. Prod. 62: 1044–1045.

Costa, J.G.M. da, Pessoa, O.D.L., Menezes, E.A., Santiago, G.M.P., Lemos, T.L.G. de 2004. Composition and larvicidal activity of essential oils from heartwood of *Auxemma glazioviana* Taub. (Boraginaceae). Flavour Frag. J. 19: 529–531.

Costa, J.G.M. da, Pessoa, O.D.L., Monte, F.J.Q., Menezes, E.A., Lemos, T.L.G. de 2005. Benzoquinones, hydroquinones and sesquiterpenes of *Auxemma glazioviana*. Quim. Nova 28: 591–595.

Dai, J., Sorribas, A., Yoshida, W.Y., Williams, P.G. 2010. Sebestenoids A–D, BACE1 inhibitors from *Cordia sebestena*. Phytochemistry 71: 2168–2173.

di Fulvio, T.E. 1965. Sobre la embriología de *Cortesia cuneifolia* (Boraginaceae). I. - Anatomía floral, esporogénesis y gametogénesis. Kurtziana 2: 7–25.

Diane, N., Hilger, H.H., Gottschling, M. 2002. Transfer cells in the seeds of Boraginales. Bot. J. Linn. Soc. 140: 155–164.

Dong, M., Oda, Y., Hirota, M. 2000. (10*E*,12*Z*,15*Z*)-9-Hydroxy-10,12,15-octadecatrienoic acid methyl ester as an anti-inflammatory compound from *Ehretia dicksonii*. Biosci. Biotechnol. Biochem. 64: 882–886.

Edwards, D.P., Arauco, R., Hassall, M., Sutherland, W.J., Chamberlain, K., Wadhams, L.J., Yu, D.W. 2007. Protection in an ant-plant mutualism: An adaptation or a sensory trap? Anim. Behav. 74: 377–385.

Ettingshausen, C. von 1888. Contributions to the Tertiary flora of Australia. Mem. Geol. Surv. New South Wales 2: 1–189.

Facciola, S. 1990. Cornucopia. A source book of edible plants. Vista: Kampong.

Fægri, K. 1986. The solanoid flower. Trans. Bot. Soc. Edinburgh 45 Suppl. 1: 51–59.

Ferguson, D.M. 1999. Phylogenetic analysis and relationships in Hydrophyllaceae based on *ndh*F sequence data. Syst. Bot. 23: 253–268.

Fernandes, E.S., Passos, G.F., Medeiros, R., da Cunha, F. M., Ferreira, J., Campos, M.M., Pianowski, L.F., Calixto, J.B. 2007. Anti-inflammatory effects of compounds alpha-humulene and (−)-trans-caryophyllene isolated from the essential oil of *Cordia verbenacea*. Eur. J. Pharmacol. 569: 228–236.

Ficarra, R., Ficarra, P., Tommasini, S., Calabrò, M.L., Ragusa, S., Barbera, R., Rapisarda, A. 1995. Leaf extracts of some *Cordia* species: Analgesic and anti-inflammatory activities as well as their chromatographic analysis. Il Farmaco 50: 245–256.

Fries, R.E. 1910. Über den Bau der *Cortesia*-Blüte, ein Beitrag zur Morphologie und Systematik der Borragineen. Ark. Bot. 9: 1–13.

Gaviria, J.C. 1987. Die Gattung *Cordia* in Venezuela. Mitt. Bot. Staatssamml. München 23: 1–279.

Gottschling, M. 2003. Phylogenetic analysis of selected Boraginales. PhD Thesis, Berlin: FU Berlin (http://www.diss.fu-berlin.de/2003/30/).

Gottschling, M. 2004. Floral ontogeny in *Bourreria* (Ehretiaceae, Boraginales). Flora 199: 409–423.

Gottschling, M., Hilger, H.H. 2001. Phylogenetic analysis and character evolution of *Ehretia* and *Bourreria* (Ehretiaceae, Boraginales) and their allies based on ITS1 sequences. Bot. Jahrb. Syst. 123: 249–268.

Gottschling, M., Hilger, H.H. 2004a. Characterisation of a novel fruit type found in *Ehretia* (Ehretiaceae, Boraginales). Blumea 49: 145–153.

Gottschling, M., Hilger, H.H. 2004b. The systematic position of *Ehretia cortesia* nom. nov. (≡ *Cortesia cuneifolia*: Ehretiaceae, Boraginales) inferred from molecular and morphological data. Taxon 53: 919–923.

Gottschling, M., Miller, J.S. 2007. A revision of *Bourreria* (Ehretiaceae, Boraginales) in South America. Ann. Missouri Bot. Gard. 94: 734–744.

Gottschling, M., Hilger, H.H., Wolf, M., Diane, N. 2001. Secondary structure of the ITS1 transcript and its application in a reconstruction of the phylogeny of Boraginales. Plant Biol. 3: 629–636.

Gottschling, M., Mai, D.H., Hilger, H.H. 2002. The systematic position of *Ehretia* fossils (Ehretiaceae, Boraginales) and implications for character evolution. Rev. Palaeobot. Palynol. 121: 149–156.

Gottschling, M., Hilger, H.H., Diane, N., Weigend, M. 2004. Testing hypotheses on disjunctions present in the Primarily Woody Boraginales: Ehretiaceae, Cordiaceae, and Heliotropiaceae, inferred from ITS1 sequence data. Int. J. Plant Sci. 165 (4 Suppl.): S123–S135.

Gottschling, M., Miller, J.S., Weigend, M., Hilger, H.H. 2005. Congruence of a phylogeny of Cordiaceae (Boraginales) inferred from ITS1 sequence data with morphology, ecology, and biogeography. Ann. Missouri Bot. Gard. 92: 425–437.

Gottschling, M., Luebert, F., Hilger, H.H., Miller, J.S. 2014a. Molecular delimitations in the Ehretiaceae (Boraginales). Mol. Phylogenet. Evol. 72: 1–6.

Gottschling, M., Nagelmüller, S., Hilger, H.H. 2014b. Generative ontogeny in *Tiquilia* (Ehretiaceae,

Boraginales) and phylogenetic implications. Biol. J. Linn. Soc. 112: 520–534.

Gürke, M. 1893. Borraginaceae. In: Engler, A., Prantl, K. (eds), Natürl. Pflanzenfam. IV, 3a. Leipzig: Engelmann, pp. 71–131.

Hegnauer, R. 1989. Chemotaxonomie der Pflanzen. Nachträge zu Band 3 und 4 (Acanthaceae bis Lythraceae). Basel: Birkhäuser.

Hernandez, T., Canales, M., Teran, B., Avila, O., Duran, A., Garcia, A. M., Hernandez, H., Angeles-Lopez, O., Fernandez-Araiza, M., Avila, G. 2007. Antimicrobial activity of the essential oil and extracts of *Cordia curassavica* (Boraginaceae). J. Ethnopharm. 111: 137–141.

Heubl, G.R., Gaviria, J.C., Wanner, G. 1990. A contribution to the taxonomy and evolution of *Cordia* (Boraginaceae) and allied genera. Chromosome numbers, pollen morphology and crystal pattern in wood. Bot. Jahrb. Syst. 112: 129–165.

Irimia, R.-E., Pérez-Escobar, O.A., Gottschling, M. 2015. Strong biogeographic signal in the phylogenetic relationships of *Rochefortia* Sw. (Ehretiaceae, Boraginales). Plant Syst. Evol. 301: 1509–1516.

Iturralde-Vinent, M.A., MacPhee, R.D.E., Díaz-Franco, S., Rojas-Consuegra, R., Suárez, W., Lomba, A. 2000. Las breas de San Felipe, a Quaternary fossiliferous asphalt seep near Martí (Matanzas Province, Cuba). Caribbean J. Sci. 36: 300–313.

Johnston, I.M. 1951. Studies in the Boraginaceae, XX. Representatives of three subfamilies in Eastern Asia. J. Arnold Arb. 32: 1–26, 99–109.

Judd, W.S., Olmstead, R.G. 2004. A survey of tricolpate (eudicot) phylogenetic relationships. Amer. J. Bot. 91: 1627–1644.

Junell, S. 1938. Über den Fruchtknotenbau der Borraginazeen mit pseudomonomeren Gynäzeen. Svensk Bot. Tidskr. 32: 261–73.

Khaleel, T.F. 1977. Embryology of *Ehretia acuminata* R. Br. Proc. Mont. Acad. Sci. 37: 35–53.

Khaleel, T.F. 1982. Embryology of *Cordia sebestena* (Boraginaceae). Plant Syst. Evol. 139: 303–311.

Khaleel, T.F. 1985. A review of endosperm and the taxonomic position of Boraginaceae. J. Plant Sci. Res. 1: 117–133.

Khattab, A.M., Grace, M.H., el-Khrisy, E.A. 2001. A new flavone derivative from *Ehretia ovalifolia* leaves. Pharmazie 56: 661–662.

Kuroyanagi, M., Seki, T., Hayashi, T., Nagashima, Y., Kawahara, N., Sekita, S., Satake, M. 2001. Anti-androgenic triterpenoids from the Brazilian medicinal plant, *Cordia multispicata*. Chem. Pharm. Bull. 49: 954–957.

Lewis, W.H. 1991. Leaves of *Ehretia cymosa* (Boraginaceae) used to heal fractures in Ghana increase bone remodeling. Econ. Bot. 45: 281–282.

Li, L., Li, M.-H., Xu, L.-J., Guo, N., Wu-Lan, T.-N., Shi, R.-B., Peng, Y., Xiao, P.-G. 2010. Distribution of seven polyphenols in several medicinal plants of Boraginaceae in China. J. Med. Plants Res. 4: 1216–1221.

Lloyd, D.G. 1979. Evolution towards dioecy in heterostylous populations. Plant Syst. Evol. 131: 71–80.

Luebert, F., Wen, J. 2008. Phylogenetic analysis and evolutionary diversification of *Heliotropium* sect. *Cochranea* (Heliotropiaceae) in the Atacama Desert. Syst. Bot. 33: 390–402.

Machado, I.C., Lopes, A.V., Sazima, M. 2010. Contrasting bee pollination in two co-occurring distylic species of *Cordia* (Cordiaceae, Boraginales) in the Brazilian semi-arid Caatinga: Generalist in *C. globosa vs.* specialist in *C. leucocephala*. An. Acad. Bras. Cienc. 82: 881–891.

Mai, D.H. 1995. Tertiäre Vegetationsgeschichte Europas. Jena: Fischer.

Metcalfe, C.R., Chalk, L. 1950. Anatomy of the dicotyledons. Oxford: Clarendon.

Mez, C. 1890. Morphologische und anatomische Studien über die Gruppe der Cordieae. Bot. Jahrb. Syst. 12: 526–586.

Milet-Pinheiro, P., Schlindwein, C. 2010. Mutual reproductive dependence of distylic *Cordia leucocephala* (Cordiaceae) and oligolectic *Ceblurgus longipalpis* (Halictidae, Rophitinae) in the Caatinga. Ann. Bot. 106: 17–27.

Miller, J.S. 1989. A revision of the New World species of *Ehretia* (Boraginaceae). Ann. Missouri Bot. Gard. 76: 1050–1076.

Miller, J.S. 2003. Classification of Boraginaceae subfam. Ehretioideae: Resurrection of the genus *Hilsenbergia* Tausch ex Meisn. Adansonia, ser. 3 25:151–189.

Miller, J.S. 2013. A revision of *Cordia* section *Gerascanthus* (Boraginales: Cordiaceae). J. Bot. Res. Inst. Texas 7: 55–83.

Miller, J.S., Gottschling, M. 2007. Generic classification in the Cordiaceae (Boraginales): Resurrection of the genus *Varronia* P.Br. Taxon 56: 163–169.

Miller, J.S., Nowicke, J.W. 1990. Dioecy and a reevaluation of *Lepidocordia* and *Antrophora* (Boraginaceae: Ehretioideae). Amer. J. Bot. 77: 543–551.

Miller, R.B. 1977. Vestured pits in Boraginaceae. IAWA Bull. n.s. 3: 43–48.

Moore, M.J., Jansen, R.K. 2006. Molecular evidence for the age, origin, and evolutionary history of the American desert plant genus *Tiquilia* (Boraginaceae). Mol. Phylogenet. Evol. 39: 668–687.

Muller, J. 1981. Fossil pollen records of extant angiosperms. Bot. Rev. 47: 1–142.

Nayar, T.S., Anil Kumar, E.S., Pushpangadan, P. 1999. *Rotula aquatica*, Boraginaceae - First record on its psychoactive property. Econ. Bot. 53: 115–117.

Neubauer, H.F. 1977. Über den Knotenbau und Blattgrund von *Cordia myxa* L., *Anchusa officinalis* L. und *Borago officinalis* L. (Boraginaceae). Bot. Jahrb. Syst. 98: 362–371.

Nowicke, J.W., Miller, J.S. 1990. Pollen morphology of the Cordioideae (Boraginaceae): *Auxemma*, *Cordia*, and *Patagonula*. Plant Syst. Evol. (Suppl.) 5: 103–121.

Opler, P.A., Baker, H.G., Frankie, G.H. 1975. Reproductive biology of some Costa Rican *Cordia* species (Boraginaceae). Biotropica 7: 234–247.

Percival, M. 1974. Floral ecology of coastal scrub in Southeastern Jamaica. Biotropica 6: 104–129.

Pérez-Moreau, R.L., Cabrera, Á.L. 1983. Boraginaceae. In: Cabrera, Á.L. (ed.) Flora de la provincia de Jujuy 8. Buenos Aires: Colección Científica del INTA, pp. 247–261.

Pessoa, O.D.L., Lemos, T.L.G. de, Carvalho, M.G. de, Braz-Filho, R. 1995. Cordiachromes from *Auxemma oncocalyx*. Phytochemistry 40: 1777–1786.

Pitot, A. 1939. Le fruit d'"*Ehretia serrata*" Roxb. Bull. Soc. Bot. Fr. 86: 310–314.

Quisumbing, E. 1951. Medical plants of the Philippines. Manila: Bureau of Printing.

Rabaey, D., Lens, F., Smets, E., Jansen, S. 2010. The phylogenetic significance of vestured pits in Boraginaceae. Taxon 59: 510–516.

Randell, B.R. 1993. New taxa and combinations in the Boraginaceae. J. Adelaide Bot. Gard. 15: 93–99.

Rao, B.H., Rani, S.S. 1987. Floral anatomy of *Ehretia*. Acta Bot. Indica 15: 50–54.

Rao, B.H., Rao, P.S.P. 1984. Embryology of three species of *Ehretia*. Proc. Indian Acad. Sci., Plant Sci. 93: 57–65.

Richardson, A.T. 1977. Monograph of the genus *Tiquilia* (*Coldenia*, sensu lato), Boraginaceae: Ehretioideae. Rhodora 79: 467–572.

Ridley, H.N. 1930. The dispersal of plants throughout the world. Ashford: Reeve.

Riedl, H. 1997. Boraginaceae. Flora Malesiana 13. Den Haag: Nijhoff, pp. 43–144.

Sahay, S.K. 1979. On the pollen morphology of Ehretiaceae with reference to taxonomy. IV Int. Palynol. Conf., Lucknow (1976–77) 1: 471–479.

Santos, R.P. dos, Silveira, E.R., Lemos, T.L.G. de, Viana, F. A., Braz-Filho, R., Pessoa, O.D.L. 2005. Characterization of two minor saponins from *Cordia piauhiensis* by ^{1}H and ^{13}C NMR spectroscopy. Magn. Reson. Chem. 43: 494–496.

Santos, R.P. dos, Nunes, E.P. Nascimento, R.F., Santiago, G.M.P., Menezes, G.H.A., Silveira, E.R., Pessoa, O.D. L. 2006. Chemical composition and larvicidal activity of the essential oils of *Cordia leucomalloides* and *Cordia curassavica* from the northeast of Brazil. J. Braz. Chem. Soc. 17: 1027–1030.

Sargent, C.S., Faxon, C.E., Picart, Ph., Picart, E. 1894. The silva of North America. Cambridge: Riverside Press (Houghton Mifflin).

Sargent, C.S., Faxon, C.E., Gill, M.W. 1933. Manual of the trees of North America. Boston: Riverside Press (Houghton Mifflin).

Scott, P.E., Martin, R.F. 1984. Avian consumers of *Bursera*, *Ficus*, and *Ehretia* fruit in Yucatán. Biotropica 16: 319–323.

Selvanayagam, Z.E., Gnanavendhan, S.G., Balakrishna, K., Rao, R.B., Sivaraman, J., Subramanian, K., Puri, R., Puri, R.K. 1996. Ehretianone, a novel quinonoid xanthene from *Ehretia buxifolia* with antisnake venom activity. J. Nat. Prod. 59: 664–667.

Sertié, J.A., Basile A.C., Panizza, S. Oshiro, T.T., Azzolini, C.P., Penna, S.C. 1991. Pharmacological assay of *Cordia verbenacea*. III: Oral and topical antiinflammatory activity and gastrotoxicity of a crude leaf extract. J. Ethnopharmacol. 31: 239–247.

Silva Filho, A.A. da, Lima, R.M.O.C., Nascimento, S.C. do, Silve, E.C., Andrade, S.A.S., Lins, S.C.N., Bieber, L.W. 1993. Biological activity of cordiaquinones A and B, isolated from *Cordia corymbosa*. Fitoterapia 64: 78–80.

Simpol, L.R., Otsuka, H., Ohtani, K., Kasai, R., Yamasaki, K. 1994. Nitrile glucosides and rosmarinic acid, the histamine inhibitor from *Ehretia philippinensis*. Phytochemistry 36: 91–95.

Stegelmeier, B.L., Edgar, J.A., Colegate, S.M., Gardner, D. R., Schock, T.K., Coulombe, R.A., Molyneux, R.J. 1999. Pyrrolizidine alkaloid plants, metabolism and toxicity. J. Nat. Toxins 8: 95–116.

Svensson, H.G. 1925. Zur Embryologie der Hydrophyllaceen, Borraginaceen und Heliotropiaceen mit besonderer Rücksicht auf die Endospermbildung. Uppsala Univ. Årsskr. 2: 3–175.

Ticli, F.K., Hage, L I.S., Cambraia, R.S., Pereira, P.S., Magro, Â.J., Fontes, M.R.M., Stábeli, R.G., Giglio, J. R., França, S.C., Soares, A.M., Sampaio, S.V. 2005. Rosmarinic acid, a new snake venom phospholipase A_2 inhibitor from *Cordia verbenacea* (Boraginaceae): Antiserum action potentiation and molecular interaction. Toxicon 46: 318–327.

Tillberg, C.V. 2004. *Cordia gerascanthus* (Boraginaceae) produces stem domatia. J. Trop. Ecol. 20: 355–357.

Toelken, H.R., Craven, L.A. 1986. Family Boraginaceae. In: Jessop, J.P., Toelken, H.R. (eds.) Flora of South Australia 3. Adelaide: Woolman, pp. 1146–1163.

Tomlinson, P.B., Fawcett, P. 1980. The biology of trees native to Tropical Florida. Cambridge, MA: Harvard University Press.

Uhlarz, H., Weberling, F. 1977. Ontogenetische Untersuchungen an *Cordia verbenacea* DC. (Boraginaceae), ein Beitrag zur Kenntnis der Syndesmien. Ber. Deutsch. Bot. Ges. 90: 127–134.

Urban, I. 1908. Borraginaceae. In: Symbolae antillanae seu fundamenta florae indiae occidentalis 5. Leipzig: Bornträger, pp. 474–483.

Usher, G. 1971. A dictionary of plants used by man. London: Constable.

Verdcourt, B. 1991. Boraginaceae. In: Polhill, R.M. (ed.) Flora of Tropical East Africa. London: Crown Agents for the Colonies, pp. 1–77.

Ward, L.F. 1887. Types of the Laramie flora. Bull. U.S. Geol. Surv. 37: 1–354.

Warfa, A.M. 1988. *Cordia* (Boraginaceae) in NE Tropical Africa and Tropical Arabia. Acta Univ. Ups. 174: 1–80.

Watt, J.M. Breyer-Brandwijk, M.G. 1962. The medical and poisonous plants of Southern and Eastern Africa. Edinburgh: Livingstone.

Weigend, M., Hilger, H.H. 2010. Codonaceae—A newly required family name in Boraginales. Phytotaxa 10: 26–30.

Weigend, M., Luebert, F., Gottschling, M., Couvreur, T.L. P., Hilger, H.H., Miller, J.S. 2014. From capsules to nutlets—phylogenetic relationships in the Boraginales. Cladistics 30: 508–518.

Yatskievych, G., Zavada, M. 1984. Pollen morphology of Lennoaceae. Pollen et Spores 26: 19–30.

Eremosynaceae

Eremosynaceae Dandy in Hutch., Fam. Fl. Pl., ed. 2, 1: 460 (1959).

J.G. Conran and T.D. Macfarlane

Annual herbs covered with glandular and eglandular trichomes; root a secondarily thickened taproot with fibrous secondary roots; stems branched, non-woody, fibrous. Leaves basal, rosulate, alternate, exstipulate, obovate-spathulate, entire; scape bracts (stem leaves) leaf-like, alternate, deeply palmately 5–7-lobed, lobes minutely toothed. Inflorescences axillary, scapose, branched, short-lived, erect. Flowers numerous in terminal dichasia, hermaphrodite, minute; perianth actinomorphic, perigynous; sepals 5, shortly connate basally, valvate, adnate to the basal third of the ovary, persistent, densely hairy; petals 5, free, rotate, white, imbricate in bud; stamens 5, alternipetalous, perigynous; filaments free; anthers dorsifixed, 2-thecate, 4-sporangiate, introrse, opening by lateral slits; gynoecium of two united carpels; ovary half-inferior, hairy, laterally compressed 1,2-locular with axile placentation; ovules single in each locule, sub-basally attached, erect, campylotropous, unitegmic, tenuinucellate with a well-developed endothelium; stylodia two, shortly fused basally, terminal, filiform-tapering, recurved, stigma minutely capitate, papillate. Fruit a hairy, laterally compressed, loculicidal, cartilaginous capsule. Seeds small, ovoid-turbinate, brown, finely reticulate-sculptured, exarillate at maturity; endosperm hard, copious, non-starchy, oily; embryo small, linear.

One genus, *Eremosyne* with a single species, *E. pectinata* Endl., native to far SW Western Australia.

Vegetative Morphology. The root consists of a primary taproot with well-developed and extensive lateral secondary branching. The stem base at the position of the rosette leaves is short and thick, but there is no rhizome. Flowering stems are numerous, erect and sympodially branched towards the apex. Leaves are arranged in a basal rosette (Figs. 34A, 35A), large, numerous, spirally inserted, more or less fleshy to herbaceous, obovate-spathulate, entire, dorsiventral, non-sheathing and exstipulate. The lower leaves in the rosette are distinctly petiolate by narrowing of the lamina, but the upper rosette leaves tend to become sessile. Venation in the rosette leaves is acrodromous with a central primary, two basal and two medially divergent secondaries which all merge at the leaf apex. Higher order venation is reticulate-random with irregular to incomplete areoles and looped marginal veins. The leaf-like inflorescence bracts (also termed stem leaves) are small, palmatifid-dissected, with 5–7 minutely toothed lobes and actinodromous venation (Gornall and Al-Shammary 1998; Conran and Macfarlane, in prep.).

Germination in the family is epigeal phanerocotylar. The seedlings have a well-developed taproot and the hypocotyl is smooth. The cotyledons are linear, glabrous, eglandular and leaf-like. The first leaves are alternate, and resemble the later ones and are glandular hairy.

Vegetative Anatomy. Calcium oxalate crystals are absent. Vessels occur in the roots and stems and have annular, spiral or scalariform thickening, sometimes with alternating pits, and simple more or less horizontal perforation plates (Gornall and Al-Shammary 1998).

The primary roots are tetrarch and have large, well-developed metaxylem vessels filling the centre of the stele. The phloem and

J.W. Kadereit and V. Bittrich (eds.), *Flowering Plants. Eudicots*, The Families and Genera of Vascular Plants 14,
DOI 10.1007/978-3-319-28534-4_13, © Springer International Publishing Switzerland 2016

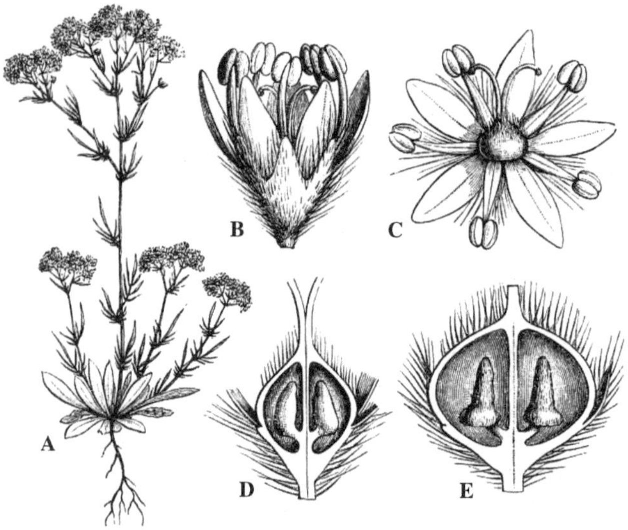

Fig. 34. Eremosynaceae. **A–E** *Eremosyne pectinata.* **A** Habit. **B** Flower, lateral view. **C** Flower, view from above. **D** Longitudinal section of ovary. **E** Longitudinal section of developing fruit. (Engler 1890, image courtesy of Biodiversity Heritage Library)

endodermis are more or less parenchymatous and there is no obvious pericycle. The cortex parenchyma is lignified angular and brick-like. The epidermis appears to be more or less suberised with thickened outer walls and the root hairs are short-lived. The roots soon undergo secondary thickening with a considerable expansion of the secondary xylem and phloem, both of which are lignified. Secondary medullary rays are not clearly distinguishable. There is a clearly defined 4–6 layered suberised phelloderm (Conran and Macfarlane, in prep.).

The primary vascular eustele in the aerial scapose stems forms a ring of closely spaced irregular collateral bundles surrounded externally by a 5–6-layered sclerenchymatous pericycle. The 1–3 layers of cortical cells are tanniferous and often crushed in older stems. The central pith of the primary stem consists of large, isodiametric-circular parenchymatous cells with large, irregular intercellular spaces. Secondary growth is absent. Epidermal hairs on the scapes are simple and eglandular and the stomata resemble those of the leaves (Gornall and Al-Shammary 1998). Nodal anatomy of the scapose aerial shoots is unilacunar with a single trace.

The leaf has a thin cuticle on both surfaces with furrows along the anticlinal cell walls but the

Fig. 35. Eremosynaceae. **A–E** *Eremosyne pectinata.* **A** Habit and basal leaf rosette. **B** Scape with divided bracts (stem leaves). **C** Part of inflorescence. **D** Flower at anthesis. **E** SEM photo of seed. (J.G. Conran)

cuticle is smooth to striate over the periclinal walls. The epidermal cells have sinuous anticlinal walls and are larger on the adaxial surface. The anomocytic stomata are located on both surfaces and are not sunken, and possess polar t-pieces and elliptical guard cells. Epidermal trichomes consist of unicellular eglandular hairs with a peg-like base and warty cuticle, and multiseriate glandular hairs with radially arranged multicellular head cells (Al-Shammary and Gornall 1994). The mesophyll is uniform and spongy, more densely packed on the adaxial side, with irregular-sized air-spaces abaxially. The leaf vascular bundles are small, poorly lignified and collateral, and terminate at the leaf or leaf lobe apices in a glandular hydathode (Gornall and Al-Shammary 1998; Conran and Macfarlane, unpubl. obs.).

Inflorescence Structure. The numerous axillary aerial stems in *Eremosyne* represent the scapes of the indeterminate thyrses with dichasial branching of the partial inflorescences.

Flower Morphology (Figs. 34B, C, 35C, D). Nectaries are not reported. The perigynous flowers are pentamerous in two whorls and the sepals are shortly fused at the base and adnate to the basal third of the ovary, deeply and bluntly lobed, regular, valvate in bud and persistent in fruit. The petals are free, white, regular and deciduous, sessile or shortly clawed.

Androecium of 5 free, isomerous fertile stamens in a single alternipetalous whorl. The filaments are thin, subulate and the anthers are small, dorsifixed and versatile, dehiscing by introrse to latrorse longitudinal slits.

Gynoecium bicarpellary, syncarpous, bilocular and partly inferior. The stylodia are apically inserted, shortly fused basally, subulate-attenuate and divergent-recurved. The stigmas are dorsal-apical on the stylodia. There is a single, erect, sub-basal, axile, funiculate, campylotropous ovule per loculus. The ovule appears to be arillate, but there is no aril development in the seed.

Embryology. Microsporogenesis is simultaneous resulting in tetrahedral pollen tetrads. The glandular tapetum is binucleate and degenerates in situ, the epidermis is crushed at maturity and the endothecium is spirally thickened.

The unitegmic ovule is campylotropous and tenuinucellate. There is a well-developed endothelium, but no clearly differentiated or lignified hypostase. The archaesporal cell seems to function directly as the megaspore mother cell and divides to produce a linear tetrad of which only the chalazal megaspore survives. Embryology is incompletely known, but the embryo sac development appears to be of the Polygonum type and the embryo sac is 8-nucleate at maturity. The synergids are elongated and form micropylar haustoria, the polar nuclei fuse prior to fertilisation and the antipodals seem to degenerate at fertilisation. Endosperm development appears to be nuclear (Conran, unpubl. obs.).

The systematic embryology of Eremosynaceae is the subject of ongoing research by Tobe et al. (pers. comm.).

Pollen Morphology. Pollen grains are trichotomosulcate, binucleate when shed, 20–24 μm in diameter, tricolporate with a coarsely reticulate, continuous, partial tectum and complex apertures (Hideaux and Ferguson 1976; Conran, unpubl. obs.).

Karyology. The chromosome number for the family is apparently $2n = 18$ with very small chromosomes (Conran and Macfarlane, in prep.).

Pollination. The flowers are visited by small bees (Halictidae and Anthophoridae), flies (Syrphidae, Bombyliidae, Muscidae) and beetles (Chrysomelidae: Hispinae and Curculionidae). The bees were seen collecting pollen actively, whereas the other visitors were in situ pollen feeders (Conran, unpubl. obs.).

Fruit and Seed Morphology. The fruit is a hirsute, laterally compressed, apically dehiscent loculicidal capsule. The seeds are small, brown and with finely reticulate sculpturing. There is a small, linear embryo and the endosperm lacks starch and contains copious oil. No special adaptations for dispersal are evident.

Phytochemistry. Basically unknown, but no alkaloids were detected by Aplin and Cannon (1971).

Affinities. The Eremosynaceae have traditionally been associated with the Saxifragaceae (e.g. Cronquist 1981), although they have been long recognised as distinctive and isolated within the family, with Engler (1928) placing it within its own subfamily. Takhtajan (1997) recognised them as a distinct family, but within the Saxifragales.

Hideaux and Ferguson (1976) considered that the pollen differed from core Saxifragaceae, but not enough to warrant removal to another family. Trichome anatomy studies of Saxifragaceae concluded that *Eremosyne* is probably allied to subfamily Escallonioideae (= Escalloniaceae) or at least parts thereof, but *Eremosyne* and Escallonioideae are otherwise not closely related to the remainder of the Saxifragaceae (Al-Shammary and Gornall 1994).

Subsequent analyses of 18S rDNA (Soltis and Soltis 1997) and *rbc*L sequence data (Backlund and Bremer 1997; Hibsch-Jetter et al. 1997) suggested that *Eremosyne* is only distantly related to core members of Saxifragaceae s.l. and that it is allied instead with members of Asteridae, in particular *Escallonia*. The Angiosperm Phylogeny Group (APG I 1998) placed it in the Asteridae: Euasterid II clade.

Bremer et al. (2001) found that a combined molecular and morphological analysis placed *Eremosyne* with the Escalloniaceae in a clade with the Tribeliaceae and this clade as sister to the Bruniaceae but still part of the Euasterids II. Although that study placed *Eremosyne* within Escalloniaceae, the resultant paraphyly of Escalloniaceae without *Eremosyne* lacks bootstrap support, while a sister group relationship of *Eremosyne* with *Escallonia* + *Forgesia* has only moderate bootstrap support. Similarly, Albach et al. (2001) using a combined 4-gene analysis found that *Eremosyne* + *Escallonia* and *Berzelia* (Bruniaceae) were sister to the Apiales + Dipsacales. The 6-gene study of Bremer et al. (2002) placed Eremosynaceae with Escalloniaceae (*Escallonia* only), Polyosmaceae, and Tribelaceae, but only with moderate bootstrap support. Very long branch length differences between *Eremosyne* and *Escallonia* and their position within the campanulid Euasterids were considered to be still unresolved and requiring further study. Similarly, multigene studies of the campanulids by Winkworth et al. (2008) and Tank and Donoghue (2010) placed it in a phylogenetically isolated Escalloniaceae, but with uncertain affinities within the family.

Eremosynaceae were retained by APG II (2003), but returned to Escalloniaceae in APG III (2009). However, the expanded Escalloniaceae appear to be a monophyletic, but phylogenetically isolated and heterogeneous group of genera of uncertain position and with relatively poor support (Winkworth et al. 2008; APG III 2009; Tank and Donoghue 2010).

The morphological characters examined in Bremer et al. (2001) distanced *Eremosyne* from Escalloniaceae, albeit without any discussion of differentiating features. Comparison of the two families shows that they share leaves that are simple, usually alternate, often gland-toothed or tipped and exstipulate flowers that are hermaphrodite, actinomorphic, with differentiated calyx and corolla, equal numbers of sepals, petals and stamens, sepals valvate, persistent, joined at least basally and usually also with the ovary, petals free though sometimes coherent, and an ovary usually partly or fully inferior. The diversity of this group is expressed in individual exceptions for many characters and variability in others such as numbers of floral parts and carpels, ovary locules and ovules per locule. However, Eremosynaceae differ from Escalloniaceae in numerous features: the annual herbaceous habit and lack of any features associated with a woody habit, glandular hairs, acrodromous leaf venation, small flowers, dorsifixed anthers, lack of a nectary, and loculicidal capsule dehiscence. The pollen structure also differs between the two families, but there are still similarities, indicating a possible transitional series from a perforate (*Eremosyne*) to complete tectum (*Escallonia*; Hideaux and Ferguson 1976).

Given the limited molecular sampling within Escalloniaceae sensu APG III and the extremely long sequence-derived branch lengths for *Eremosyne* in recent analyses, the position and affinities of the genus still require further study.

Distribution and Habitats. The family is endemic to the far southwest of Western Australia. *Eremosyne* occurs in isolated pockets in swampland and is most abundant in swamps which were burnt 1–2 years previously. The plants are locally very abundant on seasonally damp peaty white-siliceous soils in full sun to light shade, growing with *Anarthria scabra* (Anarthriaceae) and *Lomandra* spp. (Laxmanniaceae/Asparagaceae).

Economic Importance. Eremosynaceae have no economic importance.

One genus:

Eremosyne Endl. Figs. 34, 35

Eremosyne Endl., Enum. Pl. 53 (1837).

Characters as for the family.

One species, *E. pectinata* Endl. from far SW Western Australia.

Selected Bibliography

Albach, D.C., Soltis, P.S., Soltis, D.E., Olmstead, R.G. 2001. Phylogenetic analysis of Asterids based on sequences of four genes. Ann. Missouri Bot. Gard. 88: 163–212.

Al Shammary, K.I.A., Gornall, R.J. 1994. Trichome anatomy of the Saxifragaceae *s.l.* from the southern hemisphere. Bot. J. Linn. Soc. 114: 99–131.

APG (Angiosperm Phylogeny Group) 1998. An ordinal classification for the families of flowering plants. Ann. Missouri Bot. Gard. 85: 531–553.

APG (Angiosperm Phylogeny Group) II 2003. An update of the Angiosperm Phylogeny Group classification for the orders and families of flowering plants: APG II. Bot. J. Linn. Soc. 141: 399–436.

APG (Angiosperm Phylogeny Group) III 2009. An update of the Angiosperm Phylogeny Group classification for the orders and families of flowering plants: APG III. Bot. J. Linn. Soc. 161: 105–121.

Aplin, T.E.H., Cannon, J.R., 1971. Distribution of alkaloids in some Western Australian plants. Econ. Bot. 25: 366–380.

Backlund, A., Bremer, B. 1997. Phylogeny of the Asteridae *s. str.* based on *rbc*L sequences, with particular reference to the Dipsacales. Plant Syst. Evol. 207: 225–254.

Bremer, K., Backlund, A., Sennblad, B., Swenson, U., Andreason, K., Hjertson, M., Lundberg, J., Backlund, M., Bremer, B. 2001. A phylogenetic analysis of 100+ genera and 50+ families of euasterids based on morphological and molecular data with notes on possible higher level morphological synapomorphies. Plant Syst. Evol. 229: 137–169.

Bremer, B., Bremer, K., Heidari, N., Erixon, P., Olmstead, R.G., Anderberg, A.A., Källersjö, M., Barkhordarian, E. 2002. Phylogenetics of asterids based on 3 coding and 3 non-coding chloroplast DNA markers and the utility of non-coding DNA at higher taxonomic levels. Mol. Phylogenet. Evol. 24: 274-301.

Cronquist, A. 1981. An integrated system of classification of flowering plants. New York: Columbia Univ. Press.

Engler, A. 1890. Saxifragaceae. In: Natürl. Pflanzenfam. III, 2a. Leipzig: W. Engelmann, pp. 46, 65–66.

Engler, A. 1928. Saxifragaceae. In: Natürl. Pflanzenfam., ed. 2, 18a. Leipzig: W. Engelmann, pp. 74–226.

Gornall, R.J., Al-Shammary, K.I.A. 1998. Eremosynaceae. In: Cutler, D.F., Gregory, M. (eds.) Anatomy of the Dicotyledons, 2nd edn. Vol. 4. Saxifragales. Oxford: Clarendon Press, pp. 236–237.

Hibsch-Jetter, C., Soltis, D.E., Macfarlane, T.D. 1997. Phylogenetic analysis of *Eremosyne pectinata* (Saxifragaceae *s. l.*) based on *rbc*L sequence data. Plant Syst. Evol. 204: 225–232.

Hideaux, M., Ferguson, I.K. 1976. The stereostructure of the exine and its evolutionary significance in Saxifragaceae *sensu lato*. In: Ferguson, I.K., Muller, I. (eds.) The Evolutionary Significance of the Exine. Linnean Society Symposium Series 1. London: Academic Press, pp. 327–377.

Soltis, D.E., Soltis, P.S. 1997. Phylogenetic relationships in Saxifragaceae *sensu lato* – a comparison of topologies based on 18s rDNA and rbcL sequences. Amer. J. Bot. 84: 504–522.

Takhtajan, A. 1997. Diversity and Classification of Flowering Plants. New York: Columbia Univ. Press.

Tank, D.C., Donoghue, M.J. 2010. Phylogeny and phylogenetic nomenclature of the Campanulidae based on an expanded sample of genes and taxa. Syst. Bot. 35: 425–441.

Winkworth, R.C., Lundberg, J., Donoghue, M.J. 2008. Toward a resolution of campanulid phylogeny, with a special reference to the placement of Dipsacales. Taxon 57: 53–65.

Escalloniaceae

Escalloniaceae R. Br. ex Dumort., Anal. Fam. Pl.: 35, 37 (1829), nom. cons.

J. Lundberg

Subshrubs (*Valdivia*), shrubs or small trees. Leaves alternate, exstipulate, simple, entire, crenulate, or serrate to biserrate. Inflorescence axillary or terminal racemes, sometimes panicles, many-flowered (up to 800 flowers in some *Escallonia*), few-flowered (*Anopterus*, *Forgesia*, *Valdivia* and some *Escallonia*), or flowers solitary (some *Escallonia*). Flowers bisexual, rarely unisexual, actinomorphic, hypogynous to epigynous; sepals fused, 5(−9), persistent; petals free, 5(−9), imbricate, rarely valvate; nectary disk present, epigynous; stamens 5(−9), alternipetalous; filaments free; anthers basifixed or versatile, bisporangiate, introrse to extrorse; ovary of 2 (−5) united carpels, inferior to superior, 1–3 (−5)-locular, placentation parietal; ovules numerous to few; styles simple or two more or less separate stylodia (*Anopterus*, *Forgesia*); stigma capitate, 2–5-lobed. Fruits capsular, septicidal, opening from the top, rarely from the base (*Escallonia*), or indehiscent (*Valdivia*), with many to few seeds.

Four genera with ca. 45 species in South America, especially the Andes (*Escallonia*, *Valdivia*), southeastern Australia and Tasmania (*Anopterus*), and Isle de Réunion (*Forgesia*).

Vegetative Morphology. With the exception of *Valdivia gayana*, a low subshrub, Escalloniaceae are shrubs or small trees. The leaves are alternate, although in *Anopterus* they can appear as almost whorled. Most Escalloniaceae have trichomes (Al-Shammary and Gornall 1994) on their vegetative parts (only *Forgesia* is entirely glabrous except for the flowers), although sometimes trichomes are restricted to the inflorescences.

Vegetative Anatomy. Venation is semicraspedodromous in species with serrate leaves, except for *Anopterus glandulosus* and *Forgesia* with a venation that is intermediate between eucamptodromous and semicraspedodromous (Al-Shammary 1991; Gornall et al. 1998). Most species with serrate leaves have dilated vein-endings in the teeth (*Anopterus*, *Escallonia* and *Forgesia*), while in *Valdivia* the veins taper to a point (Ramamonjiarisoa 1980; Hils 1985; Al-Shammary 1991; Gornall et al. 1998). The stomata are mainly anomocytic (Gornall et al. 1998).

The nodes are unilacunar in *Escallonia* (Stern 1974), but trilacunar in the other genera (Gornall et al. 1998). Among the trilacunar genera, the vascular bundles enter the leaf lamina fused (*Forgesia*, *Anopterus*). Apart from *Escallonia angustifolia*, the leaves are dorsiventral. Holle (1893) reported a two-layered hypodermis for *Forgesia* and a one- to several-layered hypodermis for some *Escallonia* species. The palisade parenchyma is uniseriate in most *Escallonia* species (Stern 1974) and *Anopterus* (Hils 1985), but biseriate or triseriate in *Forgesia* (Ramamonjiarisoa 1980). Crystals are present as druses, or absent (Holle 1893; Sterns 1974; Ramamonjiarisoa 1980; Hils 1985; Al-Shammary 1991; Gornall et al. 1998).

The vessel element perforation plates are scalariform (simple and scalariform perforation plates are present in *Escallonia millegrana* and *E. micrantha*), often with many bars (*Escallonia* has between 3 and 22 bars per plate). The vessel end wall angle is very steep in *Forgesia* (60° to 85°) and *E. millegrana* (up to 64°). Crystals are not present in the wood, although druses are

J.W. Kadereit and V. Bittrich (eds.), *Flowering Plants. Eudicots*, The Families and Genera of Vascular Plants 14,
DOI 10.1007/978-3-319-28534-4_14,

abundant in the cortex of *Forgesia* (Holle 1893; Ramamonjiarisoa 1980; Gornall et al. 1998). The axial parenchyma is scanty paratracheal in *Forgesia*, whereas in the other genera it is diffuse apotracheal (or sometimes also diffuse-in-aggregates: *E. millegrana* and *E. micrantha*). Tracheids are the only imperforate elements in *Anopterus* and *Escallonia*, while also fibre-tracheids are present in *Forgesia*. Dark deposits (possibly tannins) are present in the parenchyma cells and the imperforate elements of *Forgesia*. Tanninoid substances are also present in the ray cells in all investigated *Escallonia* species; there is also a red-brown-staining tanninoid substance reported from the mesophyll of the leaves from some *Escallonia* species (Stern 1974).

In most Escalloniaceae the trichomes are unicellular or, in *Anopterus* and *Valdivia*, uniseriate and multicellular. *Escallonia* itself has both multiseriate glandular trichomes with multicellular heads and unicellular trichomes.

Inflorescence. There is some variation in the structure of the generally racemose inflorescences. *Forgesia* has rather lax, terminal panicles with few (10–25) flowers. The inflorescence of *Valdivia* is more compact, and the few flowers are born in axillary racemes or few-branched panicles. The few-flowered racemes of *Anopterus* are terminal with non-persistent, leaf-like bracts. *Escallonia* shows somewhat greater variation in the morphology of the inflorescences. The flowers can be arranged in sometimes rather dense terminal racemes or panicles or, more rarely, solitary in the axils of the upper leaves.

Flower Structure. The flowers of Escalloniaceae are regular, often pentamerous and usually bisexual. There are five fused sepals, except for *Anopterus* (6–9). The calyx is hairy (with the same kind of trichomes as the vegetative parts of the plant; Al-Shammary and Gornall 1994) in most species (only *Forgesia* has an entirely glabrous calyx, while the calyx of *Valdivia* only has rather few multicellular uniseriate hairs). The petals are also usually five, but variable in number in *Anopterus* (6–9), and *Valdivia* (5–7). In bud the petals are either imbricate (*Anopterus*, *Escallonia*) or valvate (*Forgesia*, *Valdivia*). The petals of *Escallonia*, *Valdivia* and *Anopterus glandulosus* are glabrous, while unicellular trichomes are present on the adaxial surface of the petals of *Forgesia* and *Anopterus macleayanus* (Al-Shammary and Gornall 1994). The petals of *A. macleayanus* also have uniseriate multicellular trichomes. A nectar disk is present in all genera, and at least *Escallonia* and *Valdivia* produce large amounts of nectar. The stamens are isomerous with the petals, and inserted at the base of the petals below the margin of the nectar disk. The filaments are often well developed. Trichomes are present on the filaments of *Forgesia* (only unicellular trichomes) and *Anopterus macleayanus* (both unicellular and uniseriate multicellular; Al-Shammary and Gornall 1994). The gynoecium is composed of 2(3) united carpels. The ovary is partly bi(−multi)-locular with incomplete septa in its lower part. The often numerous ovules are borne on strongly protruding and more or less bilobed parietal placentas.

In *Anopterus* and *Forgesia* the two stylodia are partly to entirely distinct, while they are connate in the other genera. Stylar trichomes similar to the filament trichomes are present only in *Anopterus macleayanus* and *Forgesia* (Al-Shammary and Gornall 1994). The stigmata are capitate, and more or less distinctly lobed with as many lobes as there are carpels.

Floral Anatomy. The floral anatomy of *Escallonia* has been investigated by Bensel and Palser (1975). From a central vascular cylinder, 10 compound peripheral bundles diverge. The strands remaining in the centre after this initial divergence provide the ventral carpel supply. Bundles from the two ventral strands supply both locules but ovules of different placentae (and never combine in supplying one ovule). The ovular traces are dorsal. Bundles of the ventral and the dorsal systems combine at the top of the ovary and both systems contribute to the vasculature of the stylodia or style. The nectar disk is supplied by anastomosing branches from the carpel wall bundles. This has been described as forming "a strongly developed reticulate collar below the style" by Philipson (1967), although Bensel and Palser (1975) considered this misleading, as they found only few branches and a low number of anastomoses. Gustafsson (1995) described the petal venation for two of the genera (*Anopterus* and *Escallonia*): marginal veins are absent, and the single vein entering the petal has few branches with no or few anastomoses.

EMBRYOLOGY. Only *Escallonia* is known embryologically (Jönsson 1881; Kamelina 1984, 1988). The formation of the anther wall is dicotyledonous, and it consists of epidermis, endothecium, and 2–3 middle layers. The irregularly thickened tapetum is glandular, and microsporogenesis is simultaneous. Pollen grains are 2-celled when shed. The ovules are anatropous, unitegmic and tenuinucellate (Kamelina 1984, 1988). Anatropous, unitegmic ovules are also found in *Anopterus*, *Forgesia* and *Valdivia* (Al-Shammary 1991). Embryo sac development is of the Polygonum type. The endosperm is cellular, and micropylar haustoria are present.

POLLEN MORPHOLOGY. Pollen of Escalloniaceae has been described by Erdtman (1952), Agababyan (1961, 1964), Sleumer (1968), Wakabayashi (1970), Pastre and Pons (1973), Hideux and Ferguson (1976) and Al-Shammary (1991). The pollen grains are normally 3-colporate, with the exception of *Anopterus* with 3(4)-colporate pollen. The columella layer is usually narrow to almost absent. The pollen grains of *Anopterus* have erect and irregular columella layers. The pollen of *Forgesia* (Straka and Friedrich 1988; Al-Shammary 1991) has a thin endexine and a rather narrow columella layer. In both *Escallonia* and *Valdivia* the columella layer is erect and more or less sponge-like in the upper part (Al-Shammary 1991). The endexine is very thin in *Escallonia*. The tectum is perforate in all investigated species. The endoaperture is simple-complex in *Anopterus* and *Escallonia*, but simple in *Forgesia* and *Valdivia*.

KARYOLOGY. *Escallonia* has a haploid chromosome number of n = 12 and relatively small chromosomes (Zielinski 1955).

POLLINATION. *Escallonia* is well known for its rich production of nectar, and planted individuals often attract large numbers of honeybees, bumblebees, butterflies and hummingbirds. Only one species, *E. myrtoidea*, has been studied in nature (Valdivia and Niemeyer 2006) and, for most species, the pollinators are unknown. However, many species form natural hybrids where they meet, indicating cross-pollination (Sleumer 1968). *Valdivia* also produces much nectar, and is visited by various insects as well as the hummingbird Chilean Fire Crown (*Eustephanus galeritus*; Gunckel 1931).

FRUITS AND SEEDS. The fruits of Escalloniaceae are generally dry capsules with septicidal dehiscence, apart from *Valdivia* with dry indehiscent fruits. The fruits of *Escallonia* dehisce from the base, but from the apex in *Forgesia* and *Anopterus*. The dehiscence of *Forgesia* is minimal; the fruits only open below and between the two stylodia. The capsules of *Anopterus* split almost to the base into two halves. Both *Escallonia* and *Valdivia*, with inferior ovaries, have fruits crowned by the persistent calyx. With the exception of *Anopterus macleayanus* the seeds have a very simple exotestal seed-coat consisting of 2–3 cell layers (Al-Shammary 1991; Nemirovich-Danchenko and Lobova 1998). The seeds of *Escallonia*, *Valdivia* and *Forgesia* are small (less than 1 mm long), dark and linear-oblong to linear-ovoid with almost unbranched longitudinal ridges (Nemirovich-Danchenko 2000). *Anopterus* has seeds with large (up to ca. 5 mm) wings and, in *A. macleayanus*, a seed coat with up to seven cell layers (Krach 1976, 1977; Al-Shammary 1991). The seeds of *Anopterus* are somewhat similar to the seeds of *Polyosma* (Polyosmaceae) in having an outer epidermis with discontinuous thickenings on the inner periclinal walls (Al-Shammary 1991).

PHYTOCHEMISTRY. Iridoids have been reported from all genera of Escalloniaceae, and are of two types. Undetermined seco-iridoids are found in *Anopterus* (Al-Shammary 1991), while 10-hydroxylated compounds are present in *Escallonia* (as daphylloside, deacetyl-daphylloside, geniposide and 6'-O-beta-D-glucosylasperuloside; Plouvier 1956; Al-Shammary 1991; Tomasini et al. 1993) as well as in *Valdivia* and *Forgesia* (as deacetyl-daphylloside; Al-Shammary 1991). The pentacyclic triterpene ursolic acid has been detected in *Escallonia tortuosa* (Goodson 1938). Saponins are present in *Valdivia gayana* var. *robusta*, but absent from most *Escallonia* species. Diterpene alkaloids have been detected in *Anopterus* (Hegnauer 1990), where they are present mainly in the form of anopterin. Ellagic acid is absent from the iridoid-containing *Escallonia* as well as from *Anopterus*. Other phenolic compounds detected in Escalloniaceae include the widely distributed flavonols kaempferol and quercetin (but not myricetin, which seems to be absent), the simple phenylpropane caffeic acid and the anthocyanidins cyanidin (detected in *Anopterus*, but not from *Escallonia*) and delphinidin (small amounts present in *Escallonia*, absent from *Anopterus*).

Affinities and Intrafamilial Relationships. The precise phylogenetic affinities of Escalloniaceae are not known with certainty. Traditionally, the family has been allied with Saxifragaceae and contained many more genera (e.g. Engler 1928), but recent phylogenetic analyses have indicated that Escalloniaceae are a much more restricted lineage of the Campanulids (sensu APG III 2009). Recent phylogenetic analyses have shown that Escalloniaceae are non-monophyletic even in their present, restricted sense (Lundberg 2001; Winkworth et al. 2008; Tank and Donoghue 2010). The monophyletic core Escalloniaceae are restricted to *Escallonia*, *Forgesia* and *Valdivia*, while *Anopterus* is member of a clade consisting of the core Escalloniaceae, *Tribeles australis* (Tribelaceae), *Eremosyne pectinata* (Eremosynaceae) and *Polyosma* (Polyosmaceae). In view of these relationships, it has been argued that Escalloniaceae, in addition to *Escallonia*, *Forgesia* and *Valdivia*, also should include *Anopterus*, Eremosynaceae, Tribelaceae and Polyosmaceae (Lundberg 2001; APG III 2009; Tank and Donoghue 2010; Sede et al. 2013). Despite this and for practical reasons, the present account treats Escalloniaceae in a slightly more traditional way, but with *Anopterus* treated separately in the generic descriptions below.

Distribution and Habitats. Escalloniaceae have a disjunct but predominantly south hemispheric distribution. The largest genus, *Escallonia*, has its main distribution area in the montane rainforests of South America, ranging from the Venezuelan Cordilleras through most of the Andes south to Tierra del Fuego (Sleumer 1968; Zapata 2010). In the southern part it reaches sea level in the littoral rainforest, although it reaches an altitude of up to 4000 m in Bolivia and further north. Other species are found in coastal mountains of south-eastern Brazil and Uruguay, where they grow in the mist forest (up to 2800 m) or even reach the subalpine region. The westernmost species (*E. callcottiae*) is endemic to the Juan Fernández Islands, where it grows in scrubs or montane forests up to ca. 800 m. The most widely distributed species, *E. myrtilloides*, reaches as far north as the mountainous areas of Costa Rica and Panama, and also grows in Colombia, Ecuador, Peru, Bolivia and Argentina. *Valdivia* is also South American, with only three known populations in moist and shady calcareous maritime habitats close to the city of Valdivia, Chile (Ramírez and Sempe 1981). The other two genera have a mainly palaeotropical distribution. *Forgesia* is locally very common in hygrophilic forests at 600–2300 m on the Isle de Réunion in the western Indian Ocean (Scott 1997), while *Anopterus* has a disjunct distribution in Australia: *A. glandulosus* grows in eucalypt forests and temperate rain forests of Tasmania, and *A. macleayanus* in moist and cool areas in rainforests of north-eastern New South Wales and southern Queensland.

Palaeobotany. Macrofossils (flowers, fruits, and leaves) associated with *Escallonia* have been described from the middle to late Eocene and early Miocene in Chile (Troncoso and San Martin 1999).

Uses. Several species and hybrids of *Escallonia* (e.g. *E. rubra* and *E. virgata* and their hybrids; Sleumer 1968) are cultivated as ornamentals and hedge plants in warm-temperate regions. A few high-altitude species of *Escallonia* are used locally because of their hard wood (Sleumer 1968).

Key to the Genera

1. Petals valvate — 2
– Petals imbricate — 3
2. Shrub or small tree; flowers white or pink; stylodia 2, distinct — **2. *Forgesia***
– Small subshrub; flowers red; style 1, simple — **3. *Valdivia***
3. Petals and stamens 5; ovary inferior — **1. *Escallonia***
– Petals and stamens 6–9; ovary superior — **4. *Anopterus***

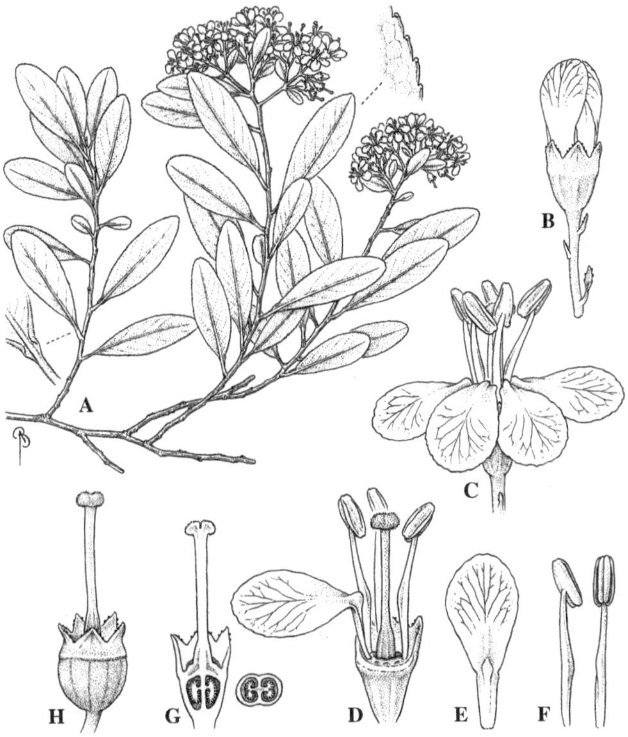

Fig. 36. Escalloniaceae. *Escallonia bifida*. **A** Stem with leaves and inflorescences. **B** Flower bud near anthesis. **C** Lateral view of anthetic flower. **D** Same, but part of calyx, four petals and two stamens removed. **E** Adaxial view of petal. **F** Stamens in lateral and adaxial view. **G** Medial section of gynoecium (left) and transverse section of ovary (right). **H** Immature fruit. (Drawn by Bobbi Angell, with gracious permission from the artist.)

1. ***Escallonia*** Mutis ex L.f. Fig. 36

Escallonia Mutis ex L.f., Suppl. pl.: 21 (1781); Sleumer, Verh. Kon. Ned. Akad. Wetensch., Afd. Natuurk. 58(2): 1-146 (1968), rev.
Stereoxylon Ruiz & Pav. (1794).
Vigieria Vell. (1829) ('1825').

Evergreen glabrous or glandular-pubescent medium-sized trees to small shrubs, often 'resinous'. Leaves with more or less serrate margins. Flowers in terminal racemes or panicles, rarely single and axillary. Calyx tube adnate to the ovary, 5-lobed; petals 5, imbricate; stamens 5, inserted below the margin of the epigynous nectar disk; ovary inferior, only partly 2-3-locular, style simple; stigma capitate, 2-3-lobed; ovules numerous, on subglobose parietal placentas. Fruit dry, 2-3-locular, septicidal, 2-3-valved and opening from the base. Seeds minute. n = 12.

About 40 spp. mostly in montane Central and South America distributed from Costa Rica and Panama along the Andes to Chile and Argentina (16 spp.), but also in south-eastern Brazil and Uruguay.

2. ***Forgesia*** Comm. ex Juss.

Forgesia Comm. ex Juss., Gen. Pl.: 164 (1789).
Defforgia Lam. (1793).

Glabrous shrub or small tree, up to 8 m high. Leaves with glandular-serrate margins. Calyx tube turbinate, 5-lobed; petals 5, valvate, caducous, inserted at the base of the perigynous nectar disk, white or rose, apically with thick, white indumentum; stamens 5; filaments papillose; ovary semi-inferior, only partly 2-locular, stylodia 2, persistent; ovules numerous, on thick, 2-lobed parietal placentas. Fruit dry, partly 2-locular, and septicidally opening between the stylodia, with numerous seeds.

One species, *F. racemosa* J.F. Gmel., Isle de Réunion.

3. ***Valdivia*** J. Rémy

Valdivia J. Rémy in Gay, Fl. Chil. 3: 43, t. 29 (1848).

Subglandular-pilose subshrub. Leaves with biserrate margins. Flowers in short, few-flowered axillary racemes or little-branched panicles, red. Calyx tube adnate to the ovary, 5-lobed, persistent; petals 5-7, valvate, perigynous, barbate on the inner surface towards the base; stamens 5-7, inserted below the margin of the epigynous nectar disk; filaments subulate; anthers linear; ovary inferior, only partly 2-3-locular, style filiform, persistent; stigma capitate, 2-3-lobed; ovules numerous, on globose, 2-lobed, parietal placentas. Fruit membranous, indehiscent, with numerous seeds. Seeds minute; embryo small.

One rare and endangered species, *V. gayana* J. Rémy, around Valdivia in south-central Chile.

Genus not included in monophyletic Escalloniaceae:

4. ***Anopterus*** Labill.

Anopterus Labill., Nov. Holl. pl. 1: 85, t. 112. (1805).

Glabrous shrubs or small trees. Leaves oblanceolate, serrate. Flowers in terminal racemes, white or pink, bracteate; bracts leafy, deciduous. Calyx tube inconspicuous, adnate to the ovary base; calyx lobes 6–9, short, persistent; petals 6–9, imbricate; stamens 6–9; filaments dilated at the base; ovary superior, 1-locular, style short, partly 2-parted; ovules on parietal placentas. Capsule coriaceous, 1-locular, septicidal, 2-valved. Seeds flattened, with a large membranous wing; embryo very small.

Two species, one in eastern Australia (*A. macleyanus* F. Muell.), one in Tasmania (*A. glandulosus* Labill.).

Selected Bibliography

Agababyan, V.S. 1961. Materialy k palinosistematicheskoma izucheniya semeistva Saxifragaceae s.l. Izv. Akad. Nauk Armyansk. S.S.R., Biol. Nauki 14: 45–61.

Agababyan, V.S. 1964. Evolyutsia pyltsy v poryadkak Cunoniales i Saxifragales v svyazi s nekotoymi voprosami ikh sistematiki i filogenii. Izv. Akad. Nauk Armyansk. S.S.R., Biol. Nauki 17: 59–75.

Al-Shammary, K.I.A. 1991. Systematic studies of the Saxifragaceae s.l., chiefly from the southern hemisphere. Ph.D. Thesis, University of Leicester, Leicester, UK.

Al-Shammary, K.I.A., Gornall, R.J. 1994. Trichome anatomy of the Saxifragaceae s.l. from the southern hemisphere. Bot. J. Linn. Soc. 114: 99–131.

Angiosperm Phylogeny Group III (AGP III) 2009. An update of the Angiosperm Phylogeny Group classification for the orders and families of flowering plants: APG III. Bot. J. Linn. Soc. 161: 105–121.

Bensel, C.R., Palser, B.F. 1975. Floral anatomy in the Saxifragaceae sensu lato. III. Kirengeshomoideae, Hydrangeoideae and Escallonioideae. Amer. J. Bot. 62: 676–687.

Engler, A. 1928. Saxifragaceae. In: Natürl. Pflanzenfam., 2nd edn, 18a: 74–226. Leipzig: W. Engelmann.

Erdtman, G. 1952. Pollen morphology and plant taxonomy. Angiosperms. Stockholm: Almqvist & Wiksell.

Goodson, J.A. 1938. The occurrence of ursolic acid in *Escallonia tortuosa*. Conversion of ursolic acid into alpha-amyrin. J. Chem. Soc. 1938: 999–1001.

Gornall, R.J., Al-Shammary, K.I.A., Gregory, M. 1998. Escalloniaceae. In: Cutler, D.F., Gregory, M. (eds.) Anatomy of the Dicotyledons. Ed. 2. Oxford: Clarendon Press, pp. 41–86.

Gunckel, H. 1931. Contribución al conocimiento de la flora valdiviana (quinta comunicación) *Valdivia gayana* Rémy. Revista Univ. (Santiago) 16: 510–517.

Gustafsson, M.H.G. 1995. Petal venation in the Asterales and related orders. Bot. J. Linn. Soc. 118: 1–18.

Hegnauer, R. 1990. Chemotaxonomie der Pflanzen. Vol. 9. Basel: Birkhäuser.

Hideux, M.J., Ferguson, I.K. 1976. The stereostructure of the exine and its evolutionary significance in Saxifragaceae sensu lato. In: Ferguson, I.K., Muller, J. (eds.) The evolutionary significance of the exine. Linnean Society Symposium Series, no. 1. London: Academic Press, pp. 327–377.

Hils, M.H. 1985. Comparative anatomy and systematics of twelve woody Australasian genera of the Saxifragaceae. Ph.D. Thesis, University of Florida, Gainesville, U.S.A.

Holle, G. 1893. Beiträge zur Anatomie der Saxifragaceen und deren systematische Verwerthung. Bot. Centralbl. 53: 1–9, 33–41, 65–70, 97–102, 129–136, 161–169, 209–222.

Jönsson, B. 1881. Om embryosäckens utveckling hos Angiospermerna. Acta Univ. Lund 16: 1–88.

Kamelina, O.P. 1984. K embriologii roda *Escallonia* (Escalloniaceae). Bot. Zhurn. 69: 1304–1316.

Kamelina, O.P. 1988. Sporo-, gametogenesis and fertilization of *Escallonia* and *Brexia* with comments on their taxonomy. In: Cresti, M., Gori, P., Pacini, E. (eds.) Sexual reproduction in higher plants. Berlin: Springer, pp. 431–435.

Krach, E. 1976. Samenanatomie der Rosifloren I. Die Samen der Saxifragaceae. Bot. Jahrb. Syst. 97: 1–60.

Krach, E. 1977. Seed characters in and affinities among the Saxifragineae. Plant Syst. Evol., Suppl. 1: 141–153.

Lundberg, J. 2001. Phylogenetic studies in the Euasterids II with particular reference to Asterales and Escalloniaceae. Ph.D. Thesis, Uppsala University, Uppsala, Sweden.

Nemirovich-Danchenko, E.N. (ed.) 2000. Sravnitel'naja anatomija semjan. Vol. 6. Dvudol'nye. Rosidae II. St. Petersburg: Academia Scientiarum Rossica, Institutum Botanicum Nomine V. L. Komarovii NAUKA.

Nemirovich-Danchenko, E.N., Lobova, T.A. 1998. Stroenie semennoy kozhury nekotorykh predstaviteley poryadka Hydrangeales. Bot. Zhurn. 83: 1–9.

Pastre, A., Pons, A. 1973. Quelques aspects de la systématique des Saxifragacées à la lumière des données de la palynologie. Pollen Spores 15: 117–133.

Philipson, W.R. 1967. *Griselinia* Forst. fil. - anomaly or link. New Zealand J. Bot. 5: 134–165.

Plouvier, V. 1956. Sur la présence d'aspéruloside chez les *Escallonia* et de dulcitol chez le *Brexia madagascariensis* Thou. (Saxifragacées). Compt. Rend. Hebd. Séan. Acad. Sci. (Paris) 242: 1643–1645.

Ramamonjiarisoa, B.A. 1980. Comparative anatomy and systematics of African and Malagasy woody Saxifragaceae sensu lato. Ph.D. Thesis, University of Massachusetts, Amherst, U.S.A.

Ramírez, C., Sempe, J. 1981. *Valdivia gayana* als Beispiel einer im subantarktischen Bereich von Südamerika endemischen Pflanzenart. Oberhess. Naturwiss. Z. 46: 75–80.

Sede, S.M., Dürnhöfer, S.I., Morello, S., Zapata, F. 2013. Phylogeny of *Escallonia* (Escalloniaceae) based on plastid DNA sequence data. Bot. J. Linn. Soc. 173: 442–451.

Scott, A. J. 1997. 84. Escalloniacées. In: Bosser, J., Cadet, T., Guého, J., Marais, W. (eds.) Flore des Mascareignes La Réunion, Maurice, Rodrigues. The Sugar Industry Research Institute, Mauritius.

Sleumer, H. 1968. Die Gattung *Escallonia* (Saxifragaceae). Verh. Kon. Ned. Akad. Wetensch., Afd. Natuurk. 58 (2): 1–146.

Stern, W.L. 1974. Comparative anatomy and systematics of woody Saxifragaceae. *Escallonia*. Bot. J. Linn. Soc. 68: 1–20.

Straka, H., Friedrich, B. 1988. Familien 65 bis 97. In: Straka, H. (ed.) Palynologica Madagassica et Mascarenica. Stuttgart: Steiner-Verl. Wiesbaden, pp. 1–117.

Tank, D.C., Donoghue, M.J. 2010. Phylogeny and phylogenetic nomenclature of the Campanulidae based on an expanded sample of genes and taxa. Syst. Bot. 35: 425–441.

Tomassini, L., Foddai, S., Nicoletti, M., Giuffra, S.E., Garcia, M.R., Bravo, F.H. 1993. Iridoid glycosides from *Escallonia* species. Biochem. Syst. Ecol. 21: 621–623.

Troncoso, A.A., San Martin, A.J. 1999. Presencia del genero *Escallonia* (Magnoliopsida, Escalloniaceae) en el terciario de Chile central. Bol. Mus. Nac. Hist. Nat. 48: 29–36.

Valdivia, C.E., Niemeyer, H.M. 2006. Do floral syndromes predict specialisation in plant pollination systems? Assessment of diurnal and nocturnal pollination of *Escallonia myrtoidea*. New Zealand J. Bot. 44: 135–141.

Wakabayashi, M. 1970. On the affinity in Saxifragaceae s. lato with special reference to the pollen morphology. Acta Phytotax. Geobot. 24: 128–145.

Winkworth, R.C., Lundberg, J., Donoghue, M.J. 2008. Toward a resolution of Campanulid phylogeny, with special reference to the placement of Dipsacales. Taxon 47: 53–65.

Zapata, F. 2010. Phylogenetics and diversification of *Escallonia* (Escalloniaceae). Ph.D. Thesis, University of Missouri-St. Louis, U.S.A.

Zielinski, Q.B. 1955. *Escallonia*: the genus and its chromosomes. Bot. Gaz. 117: 166–172.

Eucommiaceae

Eucommiaceae Engl., Syllabus (ed. 5): 139 (1907), nom. cons.

L.-B. Zhang

Deciduous trees. Leaves alternate, simple, pinnately veined, serrate, petiolate, without stipules. Flowers unisexual, dioecious, borne near base of current year's branchlets and appearing from the scale-like buds before or together with leaves, without perianth, wind-pollinated. Flowers in the axil of bracts. Male flowers fascicled, on short pedicel; stamens 5–10, linear, filaments very short, anthers tetrasporangiate and longitudinally dehiscent; female flowers solitary in lower part of branchlets, on short pedicel; ovaries unilocular, flat, stipitate, of two fused carpels, bifid at apex, 2-ovulate; ovules anatropous, pendulous from the top of the ovary. By abortion of one ovule the ovary develops into one-seeded samara; samara oblong-elliptic, non-dehiscent, flat, bifid at apex, pericarp thinly coriaceous. Seeds compressed-oblong, pendulous, seed coat membranous; endosperm copious; embryo erect, of same length as endosperm; cotyledons fleshy, flat.

One species, *Eucommia ulmoides* Oliv., distributed in C, W, SW and NW China and cultivated extensively in China.

Vegetative Morphology. *Eucommia ulmoides* is a deciduous tree with a height of up to 20 m and a girth of up to 1.2 m. The trunk is gray, smooth when young and rough when old. There are white thread-like strands between the two portions when sapwood or bark is pulled apart. Young branches are covered with brown hairs and become glabrous with age, old branches with conspicuous whitish lenticels. An annual flowering twig is 1.5–10 cm long with 1–6 leaves above the flowering portion (Fig. 37A). Winter buds are brown and conic-ovoid, covered by 6–8 imbricate ciliate scales. The leaves are exstipulate and petiolate. The petioles are 1–2 cm long, 1–1.5 mm thick, sparsely hairy. Leaf blades are elliptic, ovate or oblong, thinly coriaceous, 6–15 cm long, 3.5–6.5 cm wide, rounded or broadly cuneate at the base, acuminate at the apex, the acumen 5–15 mm long, margin dentate with 20–40 teeth on each side of the blades. The upper surface of the blade is dark green, with brown hairs when young, glabrous later, rugose when old. The lower surface is pale green, hairy when young, later only sparsely hairy on the nerves. There are 6–9 pairs of lateral nerves.

Vegetative Anatomy (Including Ultrastructure). According to Hu (1979), the cork tissue has many layers of compressed cells with the inner walls thickened and suberized, the cork cambium has 2–3 layers of small compressed cells with conspicuous nuclei, the cortex consists of 10 or more layers of parenchyma cells and stone cells mixed with some fibres, forming a ring near the phloem, the phloem consists of thin-walled cells and rays 2–3 cell-layers in thickness and there are irregular lumps of gutta-percha in the phloem and near the stone cell bands. All the parenchyma cells are devoid of calcium oxalate crystals, but some of them contain starch grains. Old bark contains partially degenerated cortex, thicker phloem with 5–6 bands of thick-walled stone cells, and much gutta-percha material. In longitudinal section, cork cells have thick lignified walls and small pores. In the cork there exist many stone cells of irregular shape with prominent lumen which are of various size and up to 120 μm long and 30 μm wide. The white thread-like gutta-percha material is about 14 μm thick, twisted or straight. Fibres are short and rare. Articulated laticifers are present in

J.W. Kadereit and V. Bittrich (eds.), *Flowering Plants. Eudicots*, The Families and Genera of Vascular Plants 14,
DOI 10.1007/978-3-319-28534-4_15,

Fig. 37. Eucommiaceae. *Eucommia ulmoides*. **A** Flowering branch. **B** Fruiting branch. **C** Male flower. **D** Female flower. **E** Stamen in lateral view. **F** Ovary in longitudinal section. (from Zhang et al. 2003, fig. 25, artists: Feng Zhongyuan and Cai Shuqin, with permission from Missouri Botanical Garden Press, St. Louis, and Science Press, Beijing)

the leaves and in the pith and bark of the stem. The wood is diffuse-porous. Its vessel members have simple perforation plates, with end walls being oblique, lateral walls with bordered pits and spiral thickenings. Leaves have one leaf trace, sieve-tube plastids are of the S-type, and diffuse sclerenchyma is present in the secondary phloem (Zhang et al. 1990).

Inflorescence Structure. Male flowers occur in clusters and female flowers are solitary in the axils of bracts.

Flower Structure and Anatomy. The flowers are small, without perianth, and unisexual with dioecious distribution. The male flowers have a glabrous pedicel of ca. 3 mm length, their are spoon-shaped, 6–8 mm long, rounded at the apex, ciliate on the margin, and caducous. Stamens are linear, 5–10 per male flower, ca. 1 cm long, glabrous, with ca. 1 mm long filaments. Anthers are ca. 8 mm long, acuminate and mucronate at the apex. The female flowers are solitary in the axils of ovate bracts with ca. 8 mm long pedicels. Ovaries are glabrous, unilocular, flat and narrow, bifid at the apex and with very short stalk. There are two stigmas which are decurrent and reflexed spreading, and two anatropous pendant ovules. One of the ovules is aborted so that each ovary develops into a 1-seeded samara (Fig. 37B).

Embryology. The glandular tapetal cells of the anthers are usually multinuclear. Cell formation after meiosis of pollen mother cells is simultaneous. Microspores are arranged tetrahedrally. Ovules are anatropous, weakly crassinucellate, and unitegmic. Parietal cells are of 2–3 layers. The embryo sac is of the monosporic Polygonum type and the proembryo is of the Solanad type (Zhang et al. 1990). The embryo consists of a large, stout, white radicle, two cotyledons ca. 8 mm long, and is embedded in an endosperm of uniform thickness beneath a delicate structure which is hardly separable from the endosperm. Endosperm is cellular. Both the cotyledons and the endosperm are rich in oil (Hu 1979).

Pollen Morphology. The pollen grains are tricolporate, prolate, the polar axis 32.7 (30.5–54.8) μm and the equatorial axis 29.3 (27.8–31.1) μm long. The colpi are narrow, unequal in length, often two long and one short or two short and one long. The exine is psilate under LM, granulate under SEM, and shortly baculate under TEM. The tectum is thin and with dense and small granules, the columellae layer consists of short bacula, and the foot layer is very thick (Zhang et al. 1988).

Karyology. The chromosome number of *Eucommia* is $2n = 34$ (Steshina in Fedorov 1974; Tanaka and Oginuma 1983). Chromosomes vary in length from about 0.8 μm to 1.6 μm. The chromosome complement was categorized as symmetric in terms of arm ratio. Satellites were found in the distal regions of the short arms of one chromosome pair (Tanaka and Oginuma 1983).

Phenology, Pollination and Reproductive Systems. *Eucommia* flowers in early spring, with or slightly before the young leaves appear. The genus is anemophilous, and individuals first flower when seven years old.

Fruit and Seed. The fruits of *Eucommia* are samaras which are compressed, oblong-elliptic, 3.5–4 cm long, 1–2 cm across the middle, cuneate at the base, notched at the apex, and contain a central elongated seed. The wing of the samara is papery. The pericarp is very rich in gutta-percha. The seed is compressed, linear, 1.4–1.5 mm long, ca. 3 mm across the broad side, rounded at both ends, and with a conspicuous raphe along the entire length of one side. It contains ca. 27 % oil. Fruits are ripe in late fall.

Dispersal. The fruits may be wind-dispersed.

Phytochemistry. *Eucommia* contains alkaloids, cyanidin, flavonols (e.g. kaempferol, quercetin), iridoids (e.g. aucubin, eucommioside, eucommiol), loliolide, proanthocyanidins and saponins/sapogenins, but lacks arbutin, ellagic acid and verbascosides. *Eucommia* further contains 2.5 % gutta-percha in the leaves, and 27 % of gutta-percha in the fruit. The threads in the bark of *Eucommia* consist of caoutchouc. *Aucuba*, *Garrya* and *Eucommia* share the chemical character of not being able to synthesise catalpol (Grayer et al. 1999). The unusual iridoids eucommioside and eucommiol occur only in *Aucuba* and *Eucommia* (Hegnauer 1989).

Affinities. On the basis of the apiculate anthers and free carpels, Oliver (1895) placed *Eucommia* in Trochodendraceae together with *Cercidiphyllum*, *Euptelea*, *Tetracentron* and *Trochodendron*, and suggested a closer relationship to *Euptelea* and *Cercidiphyllum* than to *Tetracentron* and *Trochodendron*. He also placed Trochodendraceae between Saxifragaceae and Hamamelidaceae. Solereder (1899) separated *Eucommia* and *Cercidiphyllum* from the other genera of Trochodendraceae and placed them in Hamamelidaceae based on the 2-carpellate connate ovary. Later, van Tieghem (1900) suggested that *Eucommia* should be placed as only genus in its own family, which was described by Engler (1907). Tippo (1940) suggested that Eucommiaceae form a connecting link between Urticales and Hamamelidales, on the basis of the anatomical structure of the simple perforation plates of the vessels, the alternate intervascular pitting, the laticiferous cells, and flower morphology. Varossieau (1942) considered that Eucommiaceae should be placed in Urticales, near Ulmaceae. Hu (1979) and others placed it in the Rosales between Hamamelidaceae and Rosaceae. Dahlgren (1983) considered it close to Cornales. Zhang et al. (1988) considered that Eucommiaceae are related to Hamamelidales because Eucommiaceae have tricolporate pollen grains which are similar to some members of Hamamelidales. Later, according to anatomical and embryological characters, Zhang et al. (1990) further suggested relatively close relationships with Hamamelidales. APG I (Angiosperm Phylogeny Group 1998) placed Aucubaceae, Eucommiaceae, Garryaceae and Oncothecaceae together in Garryales based on molecular data. Also based on molecular data and unisexual flowers and apical placentation, APG II (2003) and III (2009) re-circumscribed Garryales to consist of Eucommiaceae and Garryaceae (incl. Aucubaceae), and suggested that family limits require further phylogenetic work.

Distribution and Habitats. *Eucommia ulmoides* is endemic to China and naturally distributed in the Anhui, Gansu, Guangdong, Guangxi, Guizhou, Henan, Hubei, Hunan, Sichuan, Shaanxi, Yunnan and Zhejiang provinces, at altitudes of 300–500 m.

Palaeobotany. In the Tertiary, species of *Eucommia* were widely distributed in the northern Hemisphere. The earliest fossil records of *Eucommia* were found in Atlantic North America and dated to the Eocene and Oligocene. Five species of *Eucommia* were recognized in North America by Call and Dilcher (1997): *E. constans* from Neogene rocks in central Mexico, *E. eocenica* from middle Eocene strata of the Mississippi Embayment in Missouri, Tennessee and Mississippi, *E. jeffersonensis* from the latest Eocene or earliest Oligocene of the John Day Formation of Oregon, *E. montana* from early Eocene to early Oligocene localities in British Columbia, Washington, Oregon, Utah, Colorado and Montana, and *E. rolandii* from Eocene localities in British Columbia, Mississippi and Oregon. Fruit evolution in *Eucommia* may have involved increases in samara size and symmetry and reduction in seed number from two to one, perhaps in adaptation to wind dispersal (Call and Dilcher 1997).

Economic Importance. Apart from its importance in landscape engineering, *Eucommia* is used as an antihypertensive drug with pinoresinol di-β-D-glucoside as pharmacologically active principle (Hu 1979). Further, *Eucommia* contains 3 % gutta-percha which is used for insulation of wire-ropes, submarine cables, and for dental supplies. Additionally, *Eucommia* fruit contains 27 % oil, and this oil has various industrial uses. The wood is used for making wooden clogs and shoes (Hu 1979). Usually, the bark of larger trunks is collected for medicinal uses, and smaller trunks, branches and leaves are used for the production of gutta-percha.

Conservation. Because of its economic value *E. ulmoides* is widely cultivated in China. There is no threat of extinction but its natural habitats are under threat.

One monospecific genus:

Eucommia Oliv. Fig. 37

Eucommia Oliv. in Hook. f., Hooker's Icon. Pl. 20, t. 1950 (1890).

Characters as for family.

Selected Bibliography

APG (Angiosperm Phylogeny Group) 1998. An ordinal classification for the families of flowering plants. Ann. Missouri Bot. Gard. 85: 531–553.
APG (Angiosperm Phylogeny Group) II 2003. An update of the Angiosperm Phylogeny Group classification for the orders and families of flowering plants: APG II. Bot. J. Linn. Soc. 141: 399–436.
APG (Angiosperm Phylogeny Group) III 2009. An update of the Angiosperm Phylogeny Group classification for the orders and families of flowering plants: APG III. Bot. J. Linn. Soc. 161: 105–121.
Call, V.B., Dilcher, D.L. 1997. The fossil record of *Eucommia* (Eucommiaceae) in North America. Amer. J. Bot. 84: 798–814.
Dahlgren, R. 1983. General aspects of angiosperm evolution and macrosystematics. Nord. J. Bot. 3: 119–149.
Fedorov, A. 1974. Chromosome numbers of flowering plants (reprinted in English). Koenigstein: Koeltz.
Grayer, R.J., Chase, M.W., Simmonds, M.S.J. 1999. A comparison between chemical and molecular characters for the determination of phylogenetic relationships among plant families: An appreciation of Hegnauer's "Chemotaxonomie der Pflanzen". Biochem. Syst. Ecol. 27: 369–393.
Hegnauer, R. 1989. Chemotaxonomie der Pflanzen, vol. 8. Basel: Birkhäuser.
Hu, S.Y. 1979. A contribution to our knowledge of Tu-chung--*Eucommia ulmoides*. Amer. J. Chin. Med. 7: 5–37.
Solereder, H. 1899. Zur Morphologie and Systematik der Gattung *Cercidiphyllum* Sieb. et Zucc., mit Berücksichtigung der Gattung *Eucommia* Oliv. Ber. Deutsch. Bot. Gesell. 17: 387–406.
Tanaka, R., Oginuma, K. 1983. Karyomorphological study on *Eucommia ulmoides*. Chromosome Inform. Serv. 35: 21–22.
Tieghem, M.P. van 1900. Sur les Dicotylédones du groupe des Homoxylées. J. Bot. (Morot) 14: 259–276, 277–297, 330–361.
Tippo, O. 1940. The comparative anatomy of the secondary xylem and the phylogeny of the Eucommiaceae. Amer. J. Bot. 27(11): 832–838.
Varossieau, W.W. 1942. On the taxonomical position of *Eucommia ulmoides* Oliv. (Eucommiaceae). Blumea 5: 81–92, 1 folded table.
Zhang, Y.L., Wang, F.H., Chien, N.F. 1988. A study on pollen morphology of *Eucommia ulmoides* Oliver. Acta Phytotax. Sin. 26: 367–370.
Zhang, Z.Y., Lu, A.M., Pan, K.Y., Wen, J. 1990. The anatomy, embryology and systematic relationships of Eucommiaceae. Acta Phytotax. Sin. 28: 430–441.
Zhang, Z.Y., Zhang, H., Chang, H., Turland, N.J. 2003 (text), 2004 (illustrations). Eucommiaceae. In: C.Y. Wu, P.H. Raven, D.Y. Hong (eds.) Fl. China, vol. 9: 43, fig. 25. Beijing: Science Press & St. Louis: Missouri Botanical Garden Press.

Garryaceae

Garryaceae Lindl., Edwards's Bot. Reg. 20: ad t. 1686 (1834), nom. cons.

A. Liston

Dioecious evergreen trees and shrubs with decussate branching. Leaves simple, petiolate, opposite-decussate, exstipulate; blades coriaceous, with entire, variably undulate margins; upper surfaces sparingly pubescent to glabrous, often lustrous; lower surfaces glabrate to densely hairy; epidermal papillae common and well-developed. Inflorescences terminal, catkin-like, solitary, fasciculate or branched at the base; flowers solitary or ternate, borne in the axils of decussate bracts connate only at the base or fused into cup-like structures. Staminate flowers pedicellate, actinomorphic; a vestigial nectary disk sometimes present; sepals absent, petals 4, distinct, apices united by intertwined hairs forming a semi-enclosed chamber with four openings; stamens 4, distinct, alternating with petals; anthers basifixed, tetrasporangiate, dehiscing with longitudinal slits; filaments short; pistillate flowers short-pedicellate; perianth segments absent or bract-like, 2, rarely 4, adnate to the ovary, the free tips opposite or alternate with the stylodia; ovary inferior, syncarpous, bicarpellate or rarely tricarpellate, unilocular; placentation apical, ovules 2, rarely 3, pendant, anatropous, unitegmic and crassinucellar; stylodia 2, rarely 3, distinct, persistent and spreading, stigmatic through much of their length. Fruit a berry, dry and thin-walled at maturity, long persistent, indehiscent. Seeds 2, rarely 3, subglobose to ovoid-elliptical, with a translucent arilloid sarcotesta and copious oily endosperm; embryo minute, straight.

A monogeneric family with 15 species, distributed in western North America, Central America, and the Greater Antilles.

Vegetative Morphology and Anatomy. Plants vary from small shrubs to large trees, often with well-developed crown sprouts indistinguishable from the main stem axis. Wood anatomy has been studied by Moseley and Beeks (1955) and Noshiro and Baas (1998). Characteristic features include distinct growth rings (rarely indistinct in subtropical species of subg. *Fadyenia*), vessels with scalariform perforation plates (3–8 bars) and helical thickenings, fibres with distinctly bordered pits and helical thickenings, and heterogeneous rays from 1–11 cells wide. Crystals are absent. The sieve-tube plastids are of the S-type (Behnke 1991).

Trichomes typically occur on the stems, leaves, flowers and fruits. Trichomes are unicellular, with a counter-clockwise orientation of ridges and furrows; protuberances are frequently present and restricted to ridges (Dahling 1978). Foliar trichomes vary in size, shape and density, and are considered diagnostic of species and subspecies.

Leaves of *Garrya* appear opposite, but the trilacunar vascular traces for one leaf become distinct from the stele slightly lower than the traces to the opposite leaf (Moseley and Beeks 1955). Leaves are thick, coriaceous and persistent. They have two layers of lower epidermis, the cells of the outer layer are markedly papillose (Paliwal and Kakkar 1970). There is a well-developed hypodermis below the upper epidermis. Sclereids are present throughout the leaves (Paliwal and Kakkar 1970). The stomata are paracytic and confined to the lower leaf surface (Dahling 1978).

Inflorescence and Flower Structure. The inflorescence is composed of an axis bearing decussate bracts which subtend the unisexual flowers (Fig. 38A). The inflorescence typically becomes pendulous and can be considered a catkin. In some species "nascent inflorescences" (cf.

J.W. Kadereit and V. Bittrich (eds.), *Flowering Plants. Eudicots*, The Families and Genera of Vascular Plants 14,
DOI 10.1007/978-3-319-28534-4_16, © Springer International Publishing Switzerland 2016

Arctostaphylos) are initiated in the summer and then remain dormant until flowering in the winter. In subg. *Garrya* the inflorescences are unbranched (but often fasciculate). Paired bracts are sclerified and fused into cup-like structures subtending ternate flowers. In subg. *Fadyenia* the inflorescences are branched at the base. The paired bracts are foliaceous, but generally smaller than leaves, connate only at the base, and subtend solitary flowers (Fig. 38B). In both subgenera, the staminate flowers are pedicellate and have four regular perianth parts which are abaxially pubescent and adaxially glabrous (Fig. 38C). Based on homology with the flowers of *Aucuba*, these can be considered petals (see also Baillon 1877). The distinct petals are joined at their tips by intertwined hairs. The four stamens alternate with the petals. The anthers are tetrasporangiate, introrse, and dehiscent with longitudinal slits. They are basifixed to short filaments. The mature anthers protrude slightly from the chambers created by the apically connate petals. A small mass of undifferentiated tissue is variably present internal to the filaments. This has sometimes been considered a rudimentary pistil (Wangerin 1906; Horne 1914; Kapil and Mohana Rao 1966; Takhtajan 1997) but more likely represents a vestigial nectary disk (Fig. 2A, B of Liston 2003). Pistillate flowers (Fig. 38D) have been variously interpreted as epigynous or hypogynous (reviewed in Hallock 1930; Eyde 1964). Small appendages are found near the apex of the ovary and alternate with the stylodia in subg. *Garrya* (Fig. 38E). Pistillate flowers in subg. *Fadyenia* generally lack appendages, but the subtending bracts are sometimes adnate to the ovary in the terminal flowers of an inflorescence. These bracts are opposite the styles. Eyde (1964) used patterns of floral vasculature to further document the distinctiveness of these two situations. He interpreted the paired ovarian appendages of subg. *Garrya* as adnate perianth remnants, and considered the flowers to be epigynous. Stylodia in subg. *Garrya* are elongate, slender and terete; in subg. *Fadyenia* they are shorter and thicker.

Fig. 38. Garryaceae. **A, C–E** *Garrya elliptica*. **B** *G. wrightii*. **A** Branch and staminate inflorescence. **B** Pistillate inflorescence. **C** Staminate flowers. **D** Pistillate flowers. **E** Pistillate flower, median section. (Wangerin 1910, modified)

Embryology. Embryology has been studied by Hallock (1930), Eyde (1964), and Kapil and Mohana Rao (1966). Microsporogenesis is simultaneous and the mature pollen grains are 2-celled. The tapetum has been described as periplasmodial with binucleate cells in *G. elliptica* (Hallock 1930) and glandular with multinucleate cells in *G. flavescens* and *G. veatchii* (Kapil and Mohana Rao 1966). Placentation is apical, the two (rarely three) ovules pendant, anatropous, unitegmic and crassinucellate. Embryo sac development is of the Polygonum type and embryogenesis conforms to the Solanad type. The embryo is characterized by a well-developed placental obturator, a hypostase, persistent antipodal cells, and nuclear endosperm development. Suspensor polyembryony has been reported in *G. veatchii* (Kapil and Mohana Rao 1966). The minute embryo is embedded in a copious, oily endosperm. Hemicellulose is also present as a storage compound (Dahlgren 1991).

Pollen Morphology, Pollination. Pollen grains are tricolporate and subspheroidal ranging from oblate-spheroidal to prolate-spheroidal. The exine is thick and roughened, the sexine is reticulate with simpli- or duplibaculate muri (Dahling 1978). The grains are partially tectate (Eyde 1988). The pollen is small to medium in size and produced in large quantities. Pollen abundance and flower phenology and morphology are indicative of wind pollination (Hallock 1930; Dahling 1978).

Karyology. A chromosome number of $n = 11$ has been reported from three species representing both subgenera.

Reproductive Systems. All species are dioecious and obligately outcrossing. There exists morphological evidence for interspecific hybridization between four species pairs (Dahling 1978). In addition, six "problematic taxa" were reduced to subspecies by Dahling on the basis of their intergradation with other species.

Fruit, Seed and Dispersal. The fruit is a dry, thin-walled berry with two (rarely three) seeds. The flesh of the immature fruit is mostly derived from the outer layer of the single, massive ovule integument, comprised of large, thin-walled, palisade cells. It dries at maturity and forms a thin, wrinkled, translucent layer surrounding the seed. This "arilloid structure", which can rehydrate and increase in size, likely plays a role in seedling establishment. The fruits are anatomically indehiscent and long-persistent on the parent plant. The thin fruit wall does gradually break down due to abiotic weathering or insect burrowing. *Neotoma* spp. (woodrats and packrats) collect the fruits and apparently disperse the seeds (Hallock 1930), and fruits have been found in fossil packrat middens.

Phytochemistry. Aucubin, a C4-decarboxylated (route II) iridoid (Jensen 1992), is characteristic of *Garrya*. Toxic diterpenoid alkaloids commonly occur in leaves and bark (Pelletier and Mody 1980). Gallic and ellagic acids are absent (Bate-Smith et al. 1975) but caffeic acid is present (Bate-Smith 1962). The endosperm is rich in petroselinic acid (Kleiman and Spencer 1982). Gutta-percha (*trans*-1,4-polyisoprene) has been reported from two species of the genus (Roth et al. 1985).

Affinities. Dahling (1978) reviewed the tortuous taxonomic history of *Garrya*. Most taxonomists have either placed it with the "Amentiferae" (e.g. Wangerin 1906), considered it a distinct family allied to Cornaceae (e.g. Hallock 1930), or included it in a broadly circumscribed Cornaceae (e.g. Eyde 1964). Despite conspicuous differences in gross morphology, an association between *Garrya* and *Aucuba* (Aucubaceae) has long been suspected, as demonstrated in early reports of successful grafting of the two genera (Baillon 1877; Horne 1914). Shared features include dioecy, decussate evergreen leaves, and the presence of aucubin, petroselinic acid and caffeic acid (but tannins absent). Interpreting the intrastaminal tissue observed in *Garrya* staminate flowers as a vestigial nectary disk (Liston 2003) provides an additional link between *Garrya* and *Aucuba*. Chloroplast and nuclear ribosomal DNA sequences strongly support a close relationship between *Garrya* and *Aucuba*, and remove the two from Cornaceae (Soltis et al. 2000). DNA sequence data also ally *Eucommia* (Eucommiaceae) with Garryaceae and Aucubaceae (Soltis et al. 2000). These three monotypic families comprise an apparently monophyletic Garryales characterized by dioecy, unitegmic crassinucellate ovules, and the presence of aucubin. The presence of gutta-percha (found in *Garrya* and *Eucommia*, unknown in *Aucuba*) may represent an additional synapomorphy.

Distribution and Habitats. *Garrya* is distributed in western North America (Pacific Northwest to Panama); one species occurs in the Greater Antilles. *Garrya* typically occurs in evergreen, sclerophyllous shrublands or as an understory in coniferous (usually *Pinus*) forests. Two species of subg. *Garrya* occur on serpentine soils (*G. buxifolia* and *G. congdonii*) while several species of subg. *Fadyenia* are restricted to limestone. Most species grow in habitats with pronounced dry seasons and, consistent with this, many of their morphological and anatomical characteristics are decidedly xeromorphic. Examples include coriaceous leaves with stomata restricted to the lower surface (Paliwal and Kakkar 1970) and wood with helical thickenings in vessels and fibres and relatively short vessel elements (Carlquist 2001). In contrast, the presence of scalariform perforation plates is more typical of plants occurring in mesic habitats (Carlquist 2001). Other features associated with seasonally dry habitats include resprouting after grazing and fire in *G. ovata* (Lloret et al. 1999), and enhanced seed germination after simulated fire in *G. flavescens* (Keeley 1987).

Economic Importance and Conservation. *Garrya elliptica* has been cultivated in Europe since its introduction by David Douglas in 1828. This and several other species are grown as ornamentals. California Native Americans used the leaves of *G. elliptica* in a tea to induce menstruation (Goodrich et al. 1996). In Mexico, *G. laurifolia* has been used in treating chronic diarrhea (Martinez 1959). No species of *Garrya* are of immediate conservation concern; however, many grow in habitats which are subject to anthropogenic degradation.

Only one genus:

Garrya Dougl. ex Lindl. Fig. 38

Garrya Dougl. ex Lindl., Edwards's Bot. Reg. 20: ad t. 1686 (1834).
Fadyenia Endlicher, Gen. Suppl. 2: 30. Mar-Jun 1842 (non W.J. Hooker 1840).

Two subgenera: subg. *Garrya*, seven spp., western United States, northern Baja California, Mexico, and disjunct in central Guatemala; subgen. *Garrya* is characterized by unbranched inflorescences, floral bracts fused into cup-like structures, ternate flowers and long, slender stylodia. Subgenus *Fadyenia* Dahling, eight spp., southwestern United States, northern Mexico, disjunct to Costa Rica and Panama, Greater Antilles; subg. *Fadyenia* is characterized by inflorescences which are branched at the base, foliaceous floral bracts, solitary flowers and short, thick stylodia. A phylogenetic study using nuclear ribosomal DNA sequences supports the distinctiveness of the two subgenera (Burge 2011).

Selected Bibliography

Baillon, H.E. 1877. Organogénie florale des *Garrya*. Compt. Rend. Assoc. Franç. Avancem. Sci. 1877: 561–566. Reprinted as Adansonia 12: 262–269. 1879.

Bate-Smith, E.C. 1962. The phenolic constituents of plants and their taxonomic significance. J. Linn. Soc., Bot. 58: 95–173.

Bate-Smith, E.C., Ferguson, I.K., Hutson, K., Jensen, S.R., Nielsen, B.J., Swain, T. 1975. Phytochemical interrelationships in the Cornaceae. Biochem. Syst. Ecol. 3: 79–89.

Behnke, H.-D. 1991. Distribution and evolution of forms and types of sieve-element plastids in the dicotyledons. Aliso 13: 167–182.

Burge, D.O. 2011. Molecular phylogenetics of *Garrya* (Garryaceae). Madroño 58: 249–255.

Carlquist, S. 2001. Comparative wood anatomy: systematic, ecological, and evolutionary aspects of dicotyledon wood. Berlin: Springer.

Dahlgren, G. 1991. Steps toward a natural system of the dicotyledons: embryological characters. Aliso 13: 107–165.

Dahling, G.V. 1978. Systematics and evolution of *Garrya*. Contrib. Gray Herb. Harvard Univ. 209: 1–104.

Eyde, R.H. 1964. Inferior ovary and generic affinities of *Garrya*. Amer. J. Bot. 51: 1083–1092.

Eyde, R.H. 1988. Comprehending *Cornus*: Puzzles and progress in the systematics of the dogwoods. Bot. Rev. 54: 233–351.

Goodrich, J., Lawson, C., Lawson, V.P. 1996. Kashaya Pomo Plants. Berkeley: Heyday Books.

Hallock, F.A. 1930. The relationship of *Garrya*. The development of the flowers and seeds of *Garrya* and its bearing on the phylogenetic position of the genus. Ann. Bot. 176: 771–812.

Horne, A.S. 1914. A contribution to the study of evolution of the flower with special reference to the Hamamelidaceae, Caprifoliaceae, and Cornaceae. Trans. Linn. Soc. London, II, 8: 239–309.

Jensen, S.R. 1992. Systematic implications of the distribution of iridoids and other chemical compounds in the Loganiaceae and other families of the Asteridae. Ann. Missouri Bot. Gard. 79: 284–302.

Kapil, R.N., Mohana Rao, P.R. 1966. Studies of the Garryaceae II. Embryology and systematic position of *Garrya* Douglas ex Lindley. Phytomorphology 16: 564–578.

Keeley, J.E. 1987. Role of fire in seed germination of woody taxa in California chaparral. Ecology 68: 434–443.

Kleiman, R., Spencer, G.F. 1982. Search for new industrial oils. XVI. Umbelliflorae – seed oils rich in petroselinic acid. J. Amer. Oil Chem. Soc. 59: 29–38.

Liston, A. 2003. A new interpretation of floral morphology in *Garrya* (Garryaceae). Taxon 52: 271–276.

Lloret, F., Verdú, M., Flores-Hernández, N., Valiente-Banuet, A. 1999. Fire and resprouting in Mediterranean ecosystems: insights from an external biogeographical region, the mexical shrubland. Amer. J. Bot. 86: 1655–1661.

Martinez, M. 1959. Las plantas medicinales de Mexico. Mexico, D.F.: Ediciones Botas.

Moseley, M.F. Jr., Beeks, R.M. 1955. Studies of the Garryaceae I. The comparative morphology and phylogeny. Phytomorphology 5: 314–346.

Noshiro, S., Baas, P. 1998. Systematic wood anatomy in Cornaceae and allies. J. Int. Assoc. Wood. Anat. 19: 43–97.

Paliwal, G.S., Kakkar, L. 1970. Leaf anatomy of some *Garrya* species. J. Linn. Soc., Bot. 63: 81–90.

Pelletier, S.W., Mody, N.V. 1980.Developments in the chemistry of diterpenoid alkaloids. J. Natural Prod. 43: 41–71.

Roth, W.B., Carr, M.E., Davis, E.A., Bagby, M.O. 1985. New sources of gutta-percha in *Garrya flavescens* and *Garrya wrightii*. Phytochemistry 24: 183–184.

Soltis, D.E., Soltis, P.S., Chase, M.W., Mort, M.E., Albach, D.C., Zanis, M., Savolainen, V., Hahn, W. H., Hoot, S.B., Fay, M.F., Axtell, M., Swensen, S.M., Price, L.M., Kress, W.J., Nixon, K.C., Farris, J.S. 2000. Angiosperm phylogeny inferred from 18S rDNA, *rbc*L, and *atp*B sequences. J. Linn. Soc., Bot. 133: 381–461.

Takhtajan, A. 1997. Diversity and Classification of Flowering Plants. New York: Columbia Univ. Press.

Wangerin, W. 1906. Die Umgrenzung und Gliederung der Familie der Cornaceae. Bot. Jahrb. 18: 1–88.

Wangerin, W. 1910. Garryaceae. In: Engler, A., Das Pflanzenreich. IV(56a): 1–18. Leipzig: W. Engelmann.

Heliotropiaceae

Heliotropiaceae Schrad., Commentat. Soc. Regiae Sci. Gott. Recent. 4: 192 (1819), nom. cons.

N. Diane, H.H. Hilger, H. Förther, M. Weigend, and F. Luebert

Annual or perennial herbs, subshrubs, shrubs, lianas or small trees; hermaphrodite or rarely unisexual and dioecious; indumentum of simple, unicellular, eglandular trichomes, sometimes with glandular cells. Leaves alternate, petiolate or sessile, exstipulate; lamina linear to suborbicular, margin usually entire, very rarely crenate or dentate, often revolute; apex mostly acute or acuminate; base cuneate or decurrent. Inflorescences thyrsoids, terminal or axillary; partial inflorescences scorpioid cymes. Flowers perfect, rarely unisexual, actinomorphic, chasmogamous, sessile or shortly pedicellate; calyx tube usually short, mostly campanulate, calyx lobes often unequal, linear to ovate, glabrous to densely hairy, mostly persistent after anthesis; corolla sympetalous, with subcircular to linear lobes; androecium haplostemonous, stamens antesepalous and borne on corolla tube, usually included, filaments short, linear; anthers dorsifixed, usually linear, tetrasporangiate; gynoecium superior, syncarpous, bicarpellate, usually 4-loculate with one ovule in each locule; ovule anatropous to hemitropous, unitegmic, tenuinucellate; style terminal with a conical stigmatic head with a basal ring-shaped stigma and a sterile, sometimes two-lobed apex; nectary disk at the base of the ovary. Fruit dry or fleshy, usually 4-seeded, rarely 1–2-seeded, falling apart into 1–4 mericarpids with 1–2 seeds each. Embryo minute, straight to curved, embedded in thin endosperm, cotyledons linear to ovoid.

Four genera with ca. 450 species in the tropical, subtropical and temperate zones of all continents, most diverse in seasonally dry habitats.

Vegetative Morphology. The family comprises predominantly woody species, i.e. shrubs or subshrubs, but annual or perennial herbs are also common. Tall shrubs and small trees are common in Neotropical *Heliotropium* species. Tall, woody lianas are typical of *Myriopus* and rare in *Heliotropium*. Some species in both *Euploca* and *Heliotropium* are very small (<10 cm), ephemeral herbs in semiarid areas. The leaves are usually evergreen in perennial taxa. Phyllotaxy is alternate, rarely pseudo-opposite or pseudo-alternate in some *Heliotropium* species (Figs. 39A, G, N, 40A, I). The primary root is always well developed and persistent, sometimes secondarily thickened into a root tuber (Fig. 39G). Rhizomes and underground stolons are occasionally found in *Heliotropium*. Adventitious roots are profusely developed in some *Euploca*.

Vegetative Anatomy. The leaves are usually bifacial. Isobilateral leaves occur in *Ixorhea*, *Euploca* and *Heliotropium*. Kranz chlorenchyma organisation is found in *Euploca* (Diane et al. 2003). Calcium oxalate crystals in the leaves are widespread (Diane et al. 2003). The indumentum consists of mostly unicellular, eglandular trichomes, rarely interspersed with uniseriate gland-tipped trichomes with a few-celled stalk which can secrete resinous exudates covering the surface of stems and leaves (DiFulvio 1982). Trichome surface varies from smooth to sculptured with papillae. The primary cortex is composed of chlorenchymatic, collenchymatic and parenchymatic layers, frequently with isolated sclerenchymatic strands. The xylem is mostly dif-

J.W. Kadereit and V. Bittrich (eds.), *Flowering Plants. Eudicots*, The Families and Genera of Vascular Plants 14,
DOI 10.1007/978-3-319-28534-4_17, © Springer International Publishing Switzerland 2016

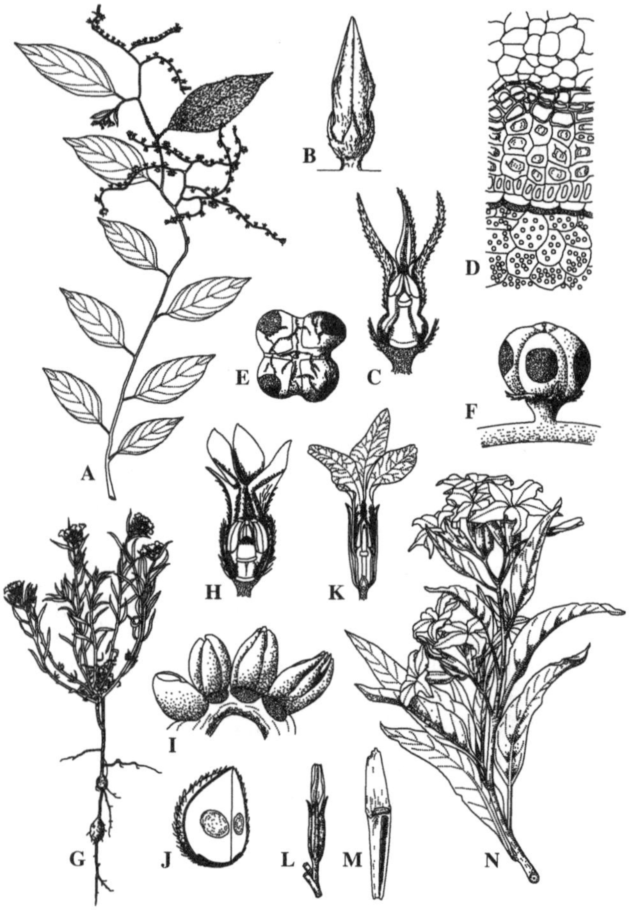

Fig. 39. Heliotropiaceae. **A–F** *Myriopus volubilis*. **A** Habit (after Nowicke 1969). **B** Bud. **C** Dissected flower. **D** Transverse section of endocarp. **E** Fruit in top view. **F** Fruit in lateral view. **G** *Euploca mendocina*. Habit, with root tuber. **H–J** *Euploca procumbens*. **H** Dissected flower. **I** Buds; the three right buds with dissected calyx. **J** Mericarpids with pits. **K–N** *Ixorhea tschudiana* (L–N after Di Fulvio 1978). **K** Dissected flower. **L** Bud. **M** Winged mericarpid. **N** Habit. (orig., drawn by Horst Lünser)

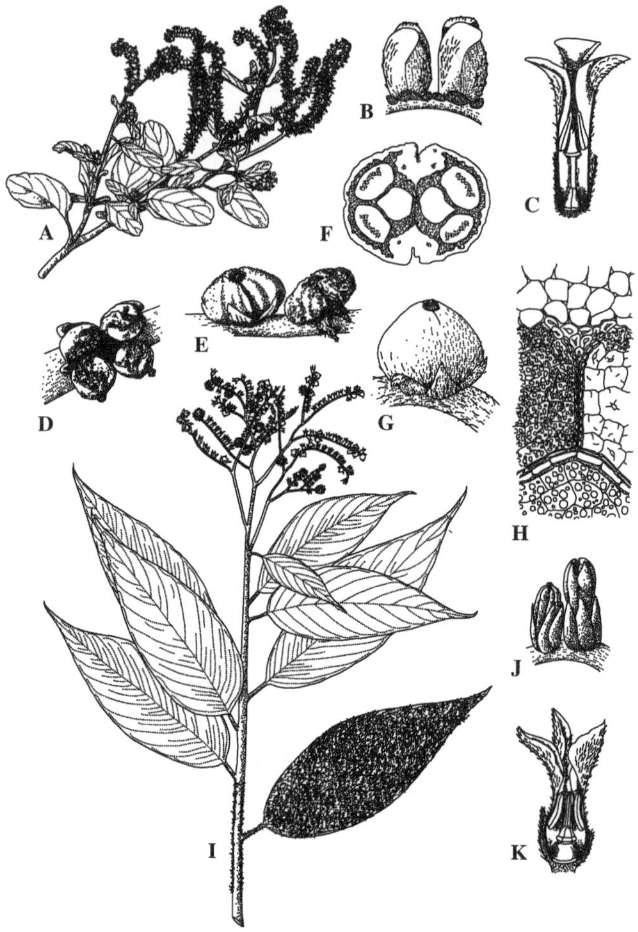

Fig. 40. Heliotropiaceae. **A** *Heliotropium europaeum*. Habit. **B, C** *Heliotropium nelsonii*. **B** Buds with dissected calyx. **C** Dissected flower. **D** *Heliotropium curassavicum*. Fruit. **E** *Heliotropium angiospermum*. Fruits. **F** *Heliotropium nicotianifolium*. Transverse section of fruit. **G** *Heliotropium verdcourtii*. Fruit in lateral view. **H** *Heliotropium angiospermum*. Transverse section of endocarp. **I–K** *Heliotropium verdcourtii*. **I** Habit (after Nowicke 1969). **J** Buds. **K** Dissected flower. (orig., drawn by Horst Lünser)

fuse-porous to semi-ring porous. Vessels are clustered in groups of two or three, and have simple perforation plates. The ground tissue of the wood is composed of tracheids, fibre-tracheids, and libriform fibres. Axial parenchyma is apotracheal diffuse, diffuse-in-aggregates, and paratracheally banded. The rays are one to three cells wide and usually less than twenty cells high, heterogeneous, and composed of square to upright marginal cells. Suberization occurs in the hypodermal layer of *Heliotropium* and *Myriopus*.

Inflorescence Structure. Inflorescences are mostly terminal, rarely axillary thyrsoids with 1–∞ monochasial branches. The monochasial branches are scorpioid cymes, so-called boraginoids (Buys and Hilger 2003; Figs. 39A, G, N, 40A, I). Bracts are usually absent, the flowers are well-spaced (especially in *Myriopus*, Fig. 39A), or more or less condensed (most *Heliotropium*, Fig. 40A), but then internodes sometimes elongate postanthetically. In *Euploca* the terminal thyrsoids are usually reduced either to monochasial or dichasial cymoids or to an axillary monochasium with only one flower developing.

Flower Structure. The flowers are mostly sessile to short-pedicellate, rarely long-pedicellate (*Ixorhea*). The calyx is synsepalous, mostly divided nearly to base and shows imbricate quincuncial aestivation. Corolla aestivation varies among genera: *Myriopus* is characterized by subvalvate aestivation with involute corolla lobes (Fig. 39B); *Euploca* by usually apert aestivation (rarely indistinctly cochlear) and involute corolla lobes (Fig. 39I). *Ixorhea* and some *Heliotropium* are characterized by apert-duplicative aestivation sometimes with spirally twisted corolla lobes (Figs. 39L, 40J). *Heliotropium* shows various aestivation types, cochlear aestivation being the most frequent type (Fig. 40B). The anthetic corolla is always sympetalous with usually spreading corolla lobes (Figs. 39C, H, K, 40C, K) which are mostly shorter than the tube, but can be elongated in some taxa (e.g. *H. zeylanicum*). *Myriopus* differs from all other taxa in having long and narrow corolla lobes with involute margins (Fig. 39C). A peculiar character is the presence of compact scales inside the corolla tube (e.g. *H. hirsutissimum*, *H. arbainense*). The androecium is haplostemonous, the stamens are antesepalous, the short and linear filaments insert at different levels on the inside of the corolla tube. The stamens are usually included in the corolla tube (Figs. 39C, H, K, 40C, K). The anthers have 4 pollen sacs and are basifixed, usually linear, rarely ovoidal to cordate, and open with introrse longitudinal slits. Anthers with conspicuously protracted connectives are found in all genera, with the apex smooth in *Ixorhea* (Fig. 39K), papillose in *Myriopus* (Fig. 39C), and pubescent in *Euploca* (Fig. 39H) and some *Heliotropium*. The connectives are apically coherent and close the corolla tube above the style-stigma complex, except in *Ixorhea*, where connectives do not surpass the apex of the stigmatic head. In *Heliotropium* (except section *Heliothamnus*) the anthers are apically free (Fig. 40C, K). The gynoecium is bicarpellate, syncarpous, superior, and usually 4-loculate with one anatropous to hemitropous ovule in each locule. Altogether, four mostly basal septa form chambers which completely surround the ovules and separate them from each other (Hilger 1992). Placentation is axile. Only *Myriopus* has a well-developed placenta consisting of transfer cells (Diane et al. 2002). The style is terminal and terminates in a conical stigmatic head with a basal ring-shaped stigma and a sterile, sometimes two-lobed apex (Khaleel 1978; Figs. 39C, H, L, 40C, K); the sterile apex is usually glabrous, rarely pubescent. Style and stigma morphology are taxonomically useful for differentiation and characterisation of species. The conical stigmatic head with its sterile apex is the key synapomorphy of Heliotropiaceae within the Boraginales.

Embryology. Detailed embryological studies are available only for some species of *Heliotropium* (e.g. Khaleel 1978, 1985) and *Ixorhea* (DiFulvio 1978): the ovules are unitegmic and have been described as tenuinucellar with ab initio cellular endosperm and distinctive micropylar and chalazal haustoria. The functional archesporial cell or the nucellar epidermis undergo periclinal divisions forming a parietal layer capping the archaesporial cell which functions as megaspore mother cell. Thus, the ovule appears to be crassinucellate at the beginning of megasporogenesis. Embryo-sac development is of the Polygonum type. The innermost layer of the integument forms an integumentary tapetum. Three types of embryology have been reported so far: *Ixorhea* conforms to the Chenopodiad type, and *Heliotropium* to the Chenopodiad, Solanad or Onagrad type. The anther wall is 4-layered with bi- or multinucleate tapetal cells, and a small amount of microsporogenous tissue. Microsporogenesis is simultaneous, producing tetrahedrally arranged pollen tetrads. The pollen grains are binucleate at anthesis.

Pollen Morphology. Pollen grains are spheroidal to prolate in outline, mostly subprolate with a circular or 6-lobed, very rarely a subtriangular polar shape. The most common pollen type is 6-heterocolporate. The sporoderm is always tectate. The exine is mostly psilate to rugulose, sometimes reticulate along the pseudocolpi. This typical pollen morphology is generally found in *Myriopus*, *Ixorhea* and *Heliotropium*. Various modifications of this pollen type can be found (Nowicke and Skvarla 1974; Perveen et al. 1995; Scheel et al. 1996). The pollen of *Euploca* is always spheroidal to prolate with a circular, rarely a quadrangular polar shape and a psilate to rugulose exine, but showing a vast range of aperture types and numbers (Perveen et al. 1995).

Karyology. Only few chromosome counts are available. Base numbers apparently range from $x = 7$ to $x = 13$, both polyploidy and aneuploidy seem to be common. *Euploca* has base chromosome numbers of $x = 7$ and $x = 8$, and the ploidy levels vary from $2x$ to $8x$. Old World *Heliotropium* has a base number of $x = 8$ and the species are usually di- or tetraploid, rarely hexaploid; exceptions are $2n = 26$ in *H.* (= *Argusia*) *sibiricum* and $2n = 28$ in *H. zeylanicum* and *H. ophioglossum*. In *Heliotropium* sect. *Heliothamnus*, $n = 9$ was reported for *H. arborescens* (Britton 1951). In other New World *Heliotropium*, base numbers are higher and tetraploids exist at least in *Heliotropium veronicifolium* and *H. curassavicum* (DiFulvio 1969). One chromosome count each is available for *Myriopus* ($n = 12$) and *Ixorhea* ($2n = 10$), and no count is available for *Heliotropium* sect. *Cochranea*.

Pollination. The frequent occurrence of yellow nectar guides in the flowers, UV-patterns (Frohlich 1976), the secretion of nectar, and an often intense floral scent in *Euploca* and *Heliotropium* suggest insect pollination (Knuth 1899, 1905). Our own observations document that the flowers of *Heliotropium* sect. *Heliothamnus* and sect. *Cochranea* are visited by insects, especially Lepidoptera. The anthers are positioned above the stigma and self-fertilization is thus very likely when the plants are self-compatible.

Fruit and Seed. The fruits are usually 4-seeded, rarely 1–2-seeded, falling apart into 1- or 2-seeded mericarpids (Figs. 39E, F, J, M, 40D, E, G). Dry mericarpids with a dry, brown, glabrous exocarp and a thin and dry mesocarp are the most common type. Fleshy fruits seem to have evolved several times independently within Heliotropiaceae. Typical drupes have a white to bright orange exocarp and a juicy mesocarp (all species of *Myriopus* and some *Heliotropium*, Figs. 39E, F, 40G). Some taxa have a brown to black exocarp and a more or less dry to spongiose mesocarp (some species of *Euploca* and some *Heliotropium*). The endocarp is homogenous, and sclerified with ligneous, strongly thickened cell walls filling the cell lumina (Fig. 40H). Sometimes one or more empty cells are found next to the locules in some species-groups of *Heliotropium* (Johnston 1928; Fig. 40F). Calcium oxalate crystals (mostly druses and crystal sand) are usually embedded in the cell walls of the endocarp. All Heliotropiaceae except *Myriopus* have a clear, longitudinal dehiscence line in the endocarp (Fig. 40H). The testa consists of an epidermis of transfer cells with protuberances of the inner tangential and radial cell walls (Diane et al. 2002) and a few inner layers of thin-walled cells, persisting or not. *Myriopus* has a 4-layered endocarp (Fig. 39D). All species investigated have a long suspensor. The embryo is straight in *Ixorhea* and *Heliotropium*, and curved in *Euploca* and *Myriopus*.

Dispersal. The dry or drupaceous fruits seem to reflect alternative mechanisms of dispersal, though only anecdotal evidence is available. Endozoochory can be assumed for drupaceous fruits present in *Myriopus* and *Heliotropium*. Air-filled empty chambers in mericarpids (several species of *Heliotropium*), a spongiose mesocarp (e.g. *H. foertheri*, *H. gnaphalodes*) or inflated calyces (*H. supinum*, *H. drepanophyllum*) suggest adaptations to hydrochory. In some species of *Euploca*, elaiosomes are found in two pits on the adaxial side of each mericarpid (Fig. 39J) and myrmecochory has been inferred (Craven 1996). *Euploca serpylloides* shows postfloral elongation of the pedicel, and *E. hypogaea* is geocarpic. The long-winged mericarpids of *Ixorhea* (Fig. 39M) and some *Heliotropium* species such as *H. pterocarpum* are probably anemochorous.

Phytochemistry. Pyrrolizidine alkaloids (PAs) are present in Heliotropiaceae (Culvenor 1978; Boppré 2011). The PAs are composed of necines (aminoalcohols) and necic acid and are also occasionally found in the form of nitrogen oxides. All hepatotoxic PAs consist of unsaturated necines (with a 1,2 double bond) and show esterification of the CH_2OH groups in the side chains. The most frequently found compounds are mono- and diester PAs. Flavonoids and aromatic geranyl derivatives have been reported to be present in the resinous exudates of some species of *Heliotropium* sect. *Cochranea* (Urzúa et al. 2000).

Subdivision and Affinities. Generic limits have been drawn very differently in the past (De Candolle 1845; Gürke 1893; Johnston 1935; Förther 1998; Hilger and Diane 2003). Förther (1998) recognized ca. 450 species in eight genera, i.e. *Argusia*, *Ceballosia*, *Heliotropium*, *Hilgeria*, *Ixorhea*, *Nogalia*, *Schleidenia* and *Tournefortia*. Most species were placed in the two large genera *Heliotropium* and *Tournefortia*, while the other genera mostly had only one to three species. Molecular studies (Hilger and Diane 2003; Luebert and Wen 2008; Luebert et al. 2011) indicate that Heliotropiaceae consist of two large clades with the position of *Ixorhea* still uncertain. The first clade is composed of *Myriopus* (=*Tournefortia* sect. *Cyphocyema*) and *Euploca* (incl. *Hilgeria*, *Schleidenia*, *Heliotropium* sect. *Orthostachys*), and is morphologically characterized by curved embryos and involute corolla aestivation. *Ixorhea* seems to be sister to these two genera (Weigend et al. 2014), but its position still needs to be confirmed. The remaining groups of the family constitute the second large clade, treated here as the genus *Heliotropium*, which in turn consists of four well-supported subclades: (i) *Heliotropium* sect. *Heliothamnus*, (ii) Old World *Heliotropium* (incl. *Argusia*, *Ceballosia*, *Nogalia*), (iii) *Heliotropium* sect. *Cochranea* and (iv) all remaining New World sections of *Heliotropium* plus *Tournefortia* sect. *Tournefortia*. The inclusion of *Tournefortia* sect. *Tournefortia*, as well as the genera *Argusia*, *Ceballosia* and *Nogalia*, in *Heliotropium* (Craven 2005) is in accordance with molecular phylogenetic evidence (Hilger and Diane 2003; Luebert and Wen 2008; Luebert et al. 2011) and is supported by morphology (Hilger and Diane 2003; Diane et al. 2003). Heliotropiaceae have mostly been treated as a subfamily of Boraginaceae s.l. (together with the subfamilies Cordioideae, Ehretioideae and Boraginoideae). Molecular phylogenetic analyses show that this Boraginaceae s.l. is paraphyletic in relation to Hydrophyllaceae and Lennoaceae, and contains clear subclades (Ferguson 1999; Gottschling et al. 2001). Gottschling et al. (2001), based on ITS sequences, and our own unpublished results based on plastid sequence data suggest that Heliotropiaceae are sister to the family Ehretiaceae (incl. Cordioideae and Lennoaceae), but other studies based only on the plastid gene *ndhF* (e.g. Ferguson 1999; Luebert and Wen 2008) fail to resolve clear sister relationships.

Distribution and Habitats. Heliotropiaceae are distributed worldwide, mainly in tropical and subtropical zones. *Heliotropium* is most diverse in temperate and subtropical zones such as the Mediterranean, the Irano-Turanian region and southern South America, frequently dominating the vegetation of dry, disturbed areas, but the genus is also diverse in montane and alpine environments of the Neotropics. *Euploca* has centres of diversity in Africa, Australia, and the tropical and subtropical regions of the Americas where it is most abundant and diverse in semiarid habitats. *Myriopus* is essentially restricted to dry forest in the Caribbean and Central America south to central Peru and northern Argentina. *Ixorhea* is a xerophyte endemic to north-western Argentina.

Palaeobotany. Heliotropiaceae fossils are poorly documented, and only pollen grains of *Heliotropium* have been reported (as *Tournefortia*). Pollen ascribed to the *Tournefortia bicolor* type (Muller 1981 = pollen type III of Nowicke and Skvarla 1974) has been recorded by Graham and Jarzen (1969) from the Oligocene of Puerto Rico and by Graham (1976) and Graham and Dilcher (1998) from the middle Pliocene of Mexico.

Economic Importance. Reviews of the economic use of Heliotropiaceae have been provided by Al-Shehbaz (1991) and Förther (1998). Extracts from some species show significant antitumor activity. The resin of some species of *Heliotropium* sect. *Cochranea* have compounds with antioxidant, antibacterial and antiviral properties. Various species of Heliotropiaceae are cultivated as ornamentals (e.g. *Heliotropium arborescens*, *H. amplexicaule*). *Heliotropium arborescens* is also used in the perfume industry. *Heliotropium* species which contain unsaturated pyrrolizidine alkaloids are noxious weeds poisonous to livestock if they contaminate hay and are also dangerous to humans if food grain is contaminated with their fruits (Boppré 2011).

Key to the Genera

1. Shrubs with glandular trichomes only, covered with strongly scented resin, especially after drying; flowers long-pedicellate; corolla limb broader than 2 cm in diameter; fruit with 4 apically winged mericarpids; endemic to north-western Argentina **1. *Ixorhea***

- Annuals to trees or lianas with numerous eglandular and with or without glandular trichomes, usually without strong scent after drying; flowers sessile or sub-sessile; corolla limb not broader than 1.5 cm in diameter; fruit with 1–4 wingless or laterally winged mericarpids 2
2. Anthers apically glabrous, connectives not protracted **4. *Heliotropium***
- Anthers with pubescent or papillose apex, connectives protracted 3
3. Corolla lobes subulate; anthers apically papillose; fruit with fleshy mesocarp; lianas or subscandant shrubs **2. *Myriopus***
- Corolla lobes rounded; anthers with pubescent apex; fruit dry, without fleshy mesocarp; annuals, perennial herbs or erect shrubs 4
4. Stigmatic apex pubescent; anthers apically coherent; calyx lobed to 1/3–1/2; embryo curved **3. *Euploca***
- Stigmatic apex glabrous; anthers rarely coherent apically; calyx lobed nearly to base; embryo straight **4. *Heliotropium***

Genera of Heliotropiaceae

1. ***Ixorhea*** Fenzl Fig. 39K–N

Ixorhea Fenzl, Verh. K. K. Zoolog.-Bot. Ges. Wien 36: 287 (1886).
Oxyosmyles Speg. (1901).

Shrub; entire plant, except corolla limb, densely glandular, covered with an aromatic, varnish-like, glutinous resin. Leaves alternate to pseudo-opposite, lanceolate, margin entire. Inflorescences branched, ebracteose, few-flowered. Calyx tubular, lobed to 1/3 to 1/4; corolla whitish to violet-rose, lobes triangular; aestivation apert-duplicative; anthers long protracted, connectives not surpassing the apex of the stigmatic head; style long, stigmatic head long protracted. Fruit dry, separating into one-seeded, apically winged mericarpids. Embryo straight. $2n = 10$.

One species, *Ixorhea tschudiana* Fenzl, endemic to north-western Argentina (provinces Salta and Tucumán).

2. ***Myriopus*** Small Fig. 39A–F

Myriopus Small, Manual Southeast. Fl. 1131 (1933); Johnston, Contr. Gray Herb. 92: 66–89 (1930), reg. rev. as *Tournefortia* sect. *Cyphocyema*; Johnston, J. Arnold Arbor. 16: 46–56 (1935), reg. rev. as *Tournefortia* sect. *Cyphocyema*; Macbride, Fl. Peru 5(2): 540–553 (1960), reg. rev.; Miller, Ann Missouri Bot. Gard. 75 456–521 (1988), reg. rev. as *Tournefortia* sect. *Cyphocyema*; Feuillet, J. Bot. Res. Inst. Texas 2(1): 263–265 (2008), reg. rev. *Tournefortia* L. sect. *Cyphocyema* I.M. Johnst. (1930).

Lianas, rarely subscandant shrubs. Leaves alternate to pseudo-opposite, lanceolate to ovate, margin entire, glabrous to sericeous. Inflorescence 1-∞-branched, ebracteose. Calyx lobed to 1/2 or nearly to base; corolla orange, yellow, greenish, lobes subulate, margins involute, base of corolla tube inflated; anthers long protracted with papillose apex, apically coherent, closing the corolla tube; style long, stigmatic head truncate-cylindrical. Fruit fleshy, deeply 4-lobed, white or bright orange, with four one-seeded mericarpids. Embryo curved. $2n = 24$.

About 25 spp. in the Caribbean, Central and South America.

3. ***Euploca*** Nutt. Fig. 39G–J

Euploca Nutt., Trans. Amer. Phil. Soc. II, 5: 189 (1836); Johnston, Contr. Gray Herb. 81: 3–83 (1928), reg. rev. as *Heliotropium* sect. *Orthostachys*; Verdcourt, Fl. Trop. E. Africa, Boraginaceae: 51–77 (1991), reg. rev. as *Heliotropium* sect. *Orthostachys*; Craven, Aust. Syst. Bot. 9: 521–657 (1996), reg. rev. as *Heliotropium* sect. *Orthostachys*; Förther, Sendtnera 5: 35–241 (1998), rev. as *Heliotropium* sect. *Orthostachys*, *Hilgeria* and *Schleidenia*.
Heliotropium L. sect. *Orthostachys* (R. Br.) G. Don (1831).
Preslaea Mart. (1827) nom. illeg. non *Preslia* Opiz (1824).
Schleidenia Endl. (1839).
Hilgeria Förther (1998).

Annual or perennial herbs or small shrubs. Leaves alternate, rarely pseudo-opposite, linear to (ob)ovate. Indumentum variable, rarely glabrous. Inflorescences unbranched or 1–4-branched, ebracteose, bracteose, frondose, or reduced to solitary, axillary flowers. Calyx lobed 1/3 to 1/2 of length; corolla white to yellow, lobes subcircular to ovate-acuminate, lobes involute in bud; anthers long protracted with pubescent apex, apically coherent, closing the corolla tube. Fruit dry, separating into one-seeded mericarpids each with two pits on the abaxial side, rarely dry drupes. Embryo curved. $2n = 14, 24, 28, 32, 42, 48$.

About 100 species, cosmopolitan with centres of diversity in Africa, Australia, and tropical America.

The name *Preslaea* Mart. (1827) is older than *Euploca* Nutt. (1836). However, there is established practice in considering *Preslaea* Mart. and the earlier name *Preslia* Opiz (1824, Lamiaceae) as homonyms, which makes *Preslaea* Mart. illegitimate. This practice dates back to Endlicher (1839: 646), who noticed the homonymy and created the replacement name *Schleidenia*. This has been followed in treatments by *Heliotropium* specialists (Johnston 1928: 46; Förther 1998: 133), in general works (Meissner 1839: 187; De Candolle 1845: 557; Post and Kuntze 1904: 459) as well as in the treatment of the Brazilian Heliotropiaceae (Fresenius [in Martius] 1857: 33), from where the type species (*Preslaea paradoxa* Mart. ≡ *Euploca paradoxa* (Mart.) J.I.M. Melo & Semir ≡ *Schleidenia paradoxa* (Mart.) DC.) comes. *Preslaea* Mart. has not been accepted or even used in any other recent work.

4. ***Heliotropium*** L. Fig. 40A–K

Heliotropium L., Sp. Pl. 1: 130 (1753); Johnston, Contr. Gray Herb. 81: 3–83 (1928), reg. rev.; Johnston, Contr. Gray Herb. 92: 66–89 (1930), reg. rev. as *Tournefortia* sect. *Eutournefortia*; Johnston, J. Arnold Arbor. 16: 46–56 (1935), reg. rev. as *Tournefortia* sect. *Eutournefortia*; Macbride, Fl. Peru 5(2): 540–568 (1960), reg. rev.; Johnston in Lundell et al., Fl. Texas 1: 138–160 (1964), reg. rev.; Gibson, Fieldiana, Bot. 24: 138–149, 157–167 (1970), reg. rev.; Brummit, Fl. Eur. 3: 84–86 (1972), reg. rev.; Popov, Fl. URSS 19: 86–113 (1974), reg. rev.; Miller, Ann Missouri Bot. Gard. 75: 456–521 (1988), reg. rev.; Nowicke and Miller, Revis. Handb. Fl. Ceylon 7: 18–20, 29–33 (1991), reg. rev.; Verdcourt, Fl. Trop. E. Africa, Boraginaceae: 48–77 (1991), reg. rev; Akhani and Förther, Sendtnera 2: 187–276 (1994), reg. rev.; Ge-ling et al., Fl. China 16: 329–427 (1995), reg. rev.; Craven, Aust. Syst. Bot. 9:521–657 (1996), reg. rev.; Förther, Sendtnera 5: 35–241 (1998), rev.

Argusia Böhm. (1760).
Ceballosia Kunkel ex Förther (1994).
Messerschmidia L. ex Hebenstr. (1763).
Nogalia Verdc. (1987).
Pittonia Mill. (1754).
Tiaridium Lehm. (1818).
Tournefortia L. (1753), nom. cons.

Annual or perennial herbs, subshrubs, shrubs or small trees, rarely scandent or lianescent. Leaves alternate to pseudo-opposite, rarely pseudoternate, linear to broadly ovate or obovate, margin entire to repand, rarely crenate, dentate, or pinnatisect, sometimes revolute, membranous to chartaceous, rarely succulent; indumentum variable, rarely glabrous. Inflorescence 1–∞-branched, ebracteose, many-flowered. Calyx lobed to 1/2 or nearly to base; corolla mostly white, green, rarely yellow, orange or purple; lobes subcircular, oblong to lanceolate; anthers sometimes protracted, apex usually glabrous, rarely pubescent; style short to long, sometimes not visible; stigmatic head obtuse to long protracted. Fruit dry or fleshy; dry fruits separating into one- or two-seeded mericarpids, rarely dry drupes, fleshy fruits unlobed, white, composed of usually two 2-seeded mericarpids, often with empty chambers. Embryo straight. $2n = 14$, 16, 18, 22, 24, 26, 28, 32, 44, 48, 50, 52, 64.

About 300 species. Cosmopolitan, with centres of diversity in the Irano-Turanian region and in the Neotropics.

The genus, as defined here, is monophyletic. *Heliotropium* sect. *Heliothamnus* I.M. Johnst. forms a well-supported monophylum sister to all other sections. The Old World sections of *Heliotropium* (*Heliotropium* sects. *Heliotropium*, *Chamaetropium* Griseb., *Monimantha* M.A. Franch., *Odontotropium* Griseb., *Pleurolasia* Bunge, *Pseudocoeloma* Förther, *Pterotropium* (DC.) Bunge, *Rutidotheca* (A.DC.) Verdc. and *Zeylanica* Förther) together with the synonymized genera *Argusia* and *Ceballosia* form another monophyletic group (Hilger and Diane 2003); relationships among these sections and their delimitation have been discussed in detail by Förther (1998) but still need to be clarified through more detailed phylogenetic studies. *Heliotropium* sect. *Cochranea* (Miers) Kuntze, endemic to the Atacama Desert, is also monophyletic (Luebert and Wen 2008), and is sister to a large clade comprising *Tournefortia* sect. *Tournefortia* and the remaining New World sections of *Heliotropium* (*Heliotropium* sects. *Coeloma* (DC.) I.M. Johnst., *Heliotrophytum* G. Don, *Hypsogenia* I.M. Johnst., *Plagiomeris* I.M. Johnst., *Platygyne* Benth., *Schobera* (Scop.) I.M. Johnst. and *Tiaridium* (Lehm.) Griseb., Johnston 1928; Förther 1998; Luebert et al. 2011). Several nomenclatural changes are still needed to transfer all *Tournefortia* to *Heliotropium* but should be preceded by a critical taxonomic revision of the former. The taxonomy of *Tournefortia* is chaotic due to plasticity, widespread variable species and weak character differentiation.

Selected Bibliography

Akhani, H., Förther, H. 1994. The genus *Heliotropium* L. (Boraginaceae) in Flora Iranica Area. Sendtnera 2: 187–276.

Al-Shehbaz, I.A. 1991. The genera of Boraginaceae in the Southeastern United States. J. Arnold Arbor. Suppl. Ser. 1: 1–169.

Boppré, M. 2011. The ecological context of pyrrolizidine alkaloids in food, feed and forage: an overview. Food Additives Contaminants 28: 60–281.

Britton, D.M. 1951. Cytogenetic studies on the Boraginaceae. Brittonia 7: 233–266.

Buys, M.H., Hilger, H.H. 2003. Boraginaceae cymes are exclusively scorpioid and not helicoid. Taxon 52: 719–724.

Craven, L.A. 1996. A taxonomic revision of *Heliotropium* (Boraginaceae) in Australia. Austral. Syst. Bot. 9: 521–657.

Craven, L.A. 2005. Malesian and Australian *Tournefortia* transferred to *Heliotropium* and notes on delimitation of Boraginaceae. Blumea 50: 375–381.

Culvenor, C.C.J. 1978. Pyrrolizidine alkaloids - occurrence and systematic importance in angiosperms. Bot. Notiser 131: 473–486.

De Candolle, A.P. 1845. Prodromus systematis naturalis regni vegetabilis, vol. 9. Paris: Treuttel & Würtz, pp. 513–558, 565–566.

Diane, N., Hilger, H.H., Gottschling, M. 2002. Transfer cells in the seeds of Boraginales. Bot. J. Linn. Soc. 140: 155–164.

Diane, N., Jakob, C., Hilger, H.H. 2003. Leaf anatomy and foliar trichome types in Heliotropiaceae and their systematic relevance. Flora 198: 468–485.

DiFulvio, T.E. 1969. Recuentos cromosómicos en *Heliotropium* (Boraginaceae). Kurtziana 5: 89–95.

DiFulvio, T.E. 1978. Sobre la vasculatura floral, embryología y chromosomas de *Ixorhea tschudiana* (Heliotropiaceae). Kurtziana 11: 75–105.

DiFulvio, T.E. 1982. Sobre la anatomía foliar y estomatogenesis de *Ixorhea tschudiana*. (Heliotropiaceae). Kurtziana 15: 9–18.

Endlicher, S. 1839. Genera plantarum secundum ordines naturales disposita, part 9. Vienna: Apud F. Beck, pp. 641–728.

Ferguson, D.M. 1999. Phylogenetic analysis and relationships in Hydrophyllaceae based on ndhF sequence data. Syst. Bot. 23: 253–268.

Förther, H. 1998. Die infragenerische Gliederung der Gattung *Heliotropium* L. und ihre Stellung innerhalb der subfam. Heliotropioideae (Schrad.) Arn. (Boraginaceae). Sendtnera 5: 35–241.

Fresenius, G. 1857. Cordiaceae, Heliotropieae et Borragineae. In: Martius, C.F.P. (ed.) Flora Brasiliensis 8(1). Leipzig: F. Fleischer, pp. 1-64.

Frohlich, M.W. 1976. Appearance of vegetation in ultraviolet light: absorbing flowers, reflecting backgrounds. Science 194: 839–841.

Gottschling, M., Hilger, H.H., Wolf, M., Diane, N. 2001. Secondary structure of the ITS1 transcript and its application in a reconstruction of the phylogeny of Boraginales. Plant Biol. 3: 629–636.

Graham, A. 1976. Studies in neotropical paleobotany. II. The Miocene communities of Veracruz, Mexico. Ann. Missouri Bot. Gard. 63: 787–842.

Graham, A., Dilcher, D.L. 1998. Studies in neotropical paleobotany. XII. A palynoflora from the Pliocene Rio Banano Formation of Costa Rica and the Neogene vegetation of Mesoamerica. Amer. J. Bot. 85: 1426–1438.

Graham, A., Jarzen D.M. 1969. Studies in neotropical paleobotany. I. The Oligocene communities of Puerto Rico. Ann. Missouri Bot. Gard. 56: 308–357.

Gürke, M. 1893. Boraginaceae. In: Natürl. Pflanzenfam. IV, 3a: 49–96. Leipzig: W. Engelmann.

Hilger, H.H. 1992. Morphology of *Heliotropium* (Boraginaceae) dispersal units. Bot. Acta 105: 387–393.

Hilger, H.H., Diane, N. 2003. A systematic analysis of Heliotropiaceae (Boraginales) based on trnL and ITS1 sequence data. Bot. Jahrb. Syst. 125: 19–51.

Johnston, I.M. 1928. Studies in the Boraginaceae. VII. 1. The South American species of *Heliotropium*. Contr. Gray Herb. 81: 3–83.

Johnston, I.M. 1930. Studies in the Boraginaceae. VIII. I. Observations on the species of *Cordia* and *Tournefortia* known from Brazil, Paraguay, Uruguay and Argentina. 3. Treatment of *Tournefortia*. Contr. Gray Herb. 92: 66–89.

Johnston, I.M. 1935. Studies in the Boraginaceae. XI. 1. The species of *Tournefortia* and *Messerschmidia* in the Old World. J. Arnold Arbor. 16: 145–168.

Khaleel, T.F. 1978. Embryology of *Heliotropium scabrum* and *H. strigosum* (Boraginaceae). Plant Syst. Evol. 129: 45–62.

Khaleel, T.F. 1985. A review of endosperm and the taxonomic position of Boraginaceae. J. Plant Sci. Res. 1: 117–133.

Knuth, P. 1899. Handbuch der Blütenbiologie II/2. Familie Boraginaceae. Leipzig: W. Engelmann, pp. 97–125.

Knuth, P. 1905. Handbuch der Blütenbiologie III/2. *Heliotropium*. Leipzig: W. Engelmann, p. 64.

Luebert, F., Wen, J. 2008. Phylogenetic analysis and evolutionary diversification of *Heliotropium* sect. *Cochranea* (Heliotropiaceae) in the Atacama Desert. Syst. Bot. 33: 390–402.

Luebert, F., Brokamp, G., Wen, J., Weigend, M., Hilger, H. H. 2011. Phylogenetic relationships and morphological diversity in Neotropical *Heliotropium* (Heliotropiaceae). Taxon 60: 663–680.

Luque, T. 1996. Karyological studies on Spanish Boraginaceae. VII. *Heliotropium* L. Willdenowia 25: 659–668.

Meissner, C.F. 1839. Plantarum vascularium genera, pars 1, fasc. 6. Leipzig: Libraria Weidmannia, pp. 177–220.

Muller, J. 1981. Fossil pollen records of extant angiosperms. Bot. Rev. 47: 1–142.

Nowicke, J.W. 1969. Boraginaceae. In: Woodson R.E., Schery, R.W. et al. (eds.) Flora of Panama. Ann. Missouri Bot. Gard. 56: 33–69.
Nowicke, J.W., Skvarla, J.J. 1974. A palynological investigation of the genus *Tournefortia* (Boraginaceae). Amer. J. Bot. 61: 1021–1036.
Perveen, A., Qureshi, U.S., Qaiser, M. 1995. Pollen flora of Pakistan. 4. Boraginaceae. Pak. J. Bot. 27: 327–360.
Popov, M.G. 1953. Boraginaceae. In: Shishkin, B.K. (ed.) Flora SSSR 19. Leningrad: Nauka, pp. 97–691.
Post, T., Kuntze, O. 1904. Lexicon generum phanerogamarum. Stuttgart: Deutsche Verlags-Anstalt, 715 pp.
Scheel, R., Ybert, J.-P., Barth, O.M. 1996. Pollen morphology of the Boraginaceae from Santa Catarina State (southern Brazil), with comments on the taxonomy of the family. Grana 35: 138–153.
Urzúa, A., Modak, B., Villarroel, L., Torres, R., Andrade, L., Mendoza, L., Wilkens, M. 2000. External flavonoids from *Heliotropium megalanthum* and *H. huascoense* (Boraginaceae). Chemotaxonomic considerations. Bol. Soc. Chil. Quím. 45: 23–29.
Verdcourt, B. 1991. Boraginaceae. In: Polhill, R.M. (ed.) Flora of Tropical East Africa. Kew: Royal Botanic Gardens, pp. 1–124.
Weigend, M., Luebert, F., Gottschling, M., Couvreur, T.L. P., Hilger, H.H., Miller, J.S. 2014. From capsules to nutlets—phylogenetic relationships in the Boraginales. Cladistics 30(5): 508–518.

Helwingiaceae

Helwingiaceae Decne, Ann. Sci. Nat. II, 6: 69 (1836).

Q.-Y. Xiang

Dioecious evergreen or deciduous shrubs, rarely small trees up to 8 m. Leaves simple, pinnately veined, alternate, petiolate and stipulate; blade chartaceous, sub-coriaceous or coriaceous; margin glandular-serrate; stipules 2, deciduous, thread-like, divided or not. Flowers small, 3–4-merous, rarely 5-merous, green or purple green; calyx small, limbs rudimentary; petals triangular to ovate, valvate; floral disk fleshy and flat; male flowers 3–20 in an umbel; stamen filaments short; anthers short, with two nearly separate locules; female flowers 1–4 in an umbel born on leaf blades, rarely on petioles; style short, with 3–4 (5) distal stigmatic branchlets; ovary inferior, 3–4 (5)-locular; ovules 1 per locule, pendant, apotropous, with dorsal raphe. Fruit drupaceous, with 3–4(5)-lobed stone covered by a fleshy layer, ovate or oblong, green when young, turning red and eventually black when mature, 1–4(5) obtusely angular when dry. Embryo straight.

A family with the single genus *Helwingia* comprising four species distributed from the eastern Himalayas eastward to Japan. Often growing at elevations of 800–3300 m in subtropical evergreen broad-leaved forests or in subalpine coniferous forests, in shady and moist habitats.

Vegetative Anatomy. Leaf cuticular layer varies in thickness among species, thin in *H. japonica*, *H. himalaica*, and well developed in *H. omeiensis* and *H. chinensis* (3.5–5 μm thick in *H. chinensis* and 5–7 μm in *H. omeiensis*). In *H. chinensis* and *H. omeiensis* epidermal cells are sclerified, and lateral veins are not prominently elevated, in contrast to *H. japonica* and *H. himalaica* in which epidermal cells are not sclerified and lateral veins are conspicuously elevated. Lower surface of leaves varies within species from smooth, slightly wrinkled, to rugose and papillose. Stomata are elliptic in *H. japonica* (also larger in polyploid races of the species), but roundish in *H. himalaica* (Hara and Kurosawa 1975). Three species of *Helwingia*, *H. japonica*, *H. chinensis*, and *H. himalaica* have been examined for wood anatomy (Adams 1949; Li and Chao 1954; Noshiro and Baas 1998). According to these studies, wood of *Helwingia* is diffuse-porous with distinct growth rings marked by vessel size and 2–3 layers of flattened fibers. Vessels are mostly solitary or in multiples of 2 or 3. Perforations of vessels are scalariform with 17–41 bars. Intervessel pits are opposite to scalariform, and without vestures. Helical thickenings are absent in vessel elements, or occasionally tertiary spiral thickenings are present. The length of vessel elements varies from 760 μm to 940 μm, with means from 616 μm in *H. himalaica* to 729 μm in *H. chinensis*. Vessel diameter varies from 20 μm to 43 μm. Tyloses are occasionally present in vessels. Wood fibers in all species are septate and axial parenchyma is scanty paratracheal, with 4–8 cells per strand. Wood rays are heterocellular, composed of procumbent, square, and longitudinally elongated body cells. Silica grains are present in the ray cells and axial parenchyma of all species studied, and in septate fibers of *H. himalaica*.

Inflorescence and Floral Morphology. The epiphyllous inflorescence in *Helwingia* is a result of changes in the position of primordium initiation and of intercalary growth (Dickinson and Sattler 1975). Development of the epiphyllous inflorescence is initiated adjacent to the leaf axil on the adaxial side of the base of a leaf primordium. The

J.W. Kadereit and V. Bittrich (eds.), *Flowering Plants. Eudicots*, The Families and Genera of Vascular Plants 14,
DOI 10.1007/978-3-319-28534-4_18,

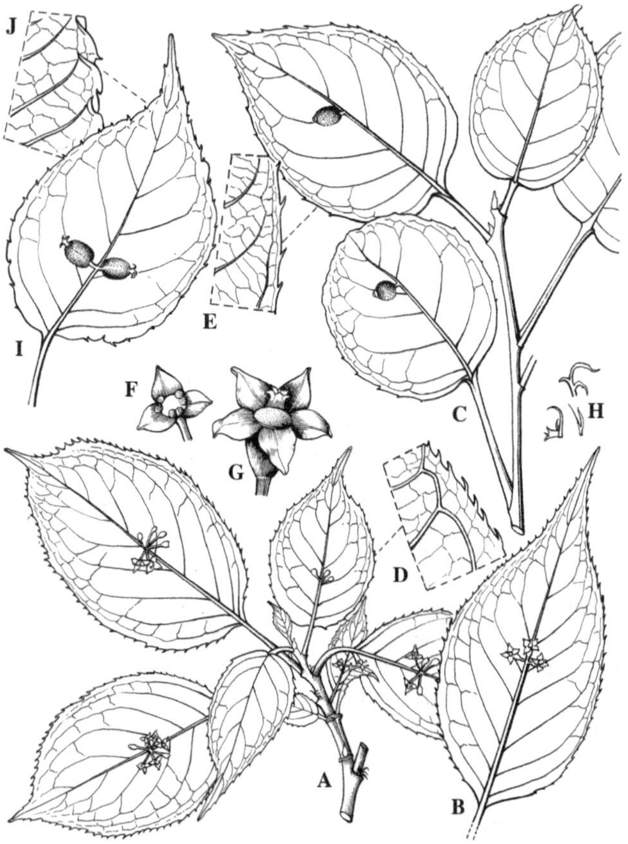

Fig. 41. Helwingiaceae. **A–H** *Helwingia japonica* var. *japonica*. **A** Male flowering branch. **B** Female inflorescence and leaf. **C** Fruiting branchlet. **D, E** Portions of leaf abaxial surface. **F** Male flower. **G** Female flower. **H** Stipules. **I, J** *H. japonica* var. *hypoleuca*. **I** Inflorescence and leaf. **J** Portion of leaf abaxial surface. (From Xiang and Boufford 2006, with permission from Missouri Botanical Garden Press, St. Louis, and Science Press, Beijing)

resulting primordium develops into an inflorescence situated on the midrib of the mature leaf through the action of a basal, intercalary meristem (Fig. 41). A separate, adaxial vascular bundle departs from the leaf trace at the base of the petiole and leads to the inflorescence on the mature leaf. In sterile leaves, this adaxial vascular bundle is absent. Recent study by Ao and Tobe (2015) suggested that the flowers of *Helwingia* lack petals (floral leaves are sepals) and are characterized by obhaplotamous stamens, large recurved stigma, and poorly developed disc nectar.

Embryology. The ovules of *Helwingia* have a single integument enclosing a tenuinucellate nucellus and a micropyle directed inwardly toward the point of attachment of the funiculus. The nucellus is thin and ephemeral. The funicle is short, and no obturator is present (Sato 1976; Ao and Tobe 2015). The development of the embryo sac (based on observations in *H. japonica*) is of the monosporic eight-nucleate type (Sato 1976). The megaspore mother cell undergoes two successive divisions to form a linear arranged tetrad of megaspores. The chalazal megaspore becomes functional and divides to form a two-nucleate embryo sac. The two nuclei are separated by a vacuole. After two successive divisions, an 8-nucleate embryo sac is formed, comprising four micropylar and four chalazal nuclei. One polar nucleus of the chalazal quartet moves upward to fuse with the micropylar nucleus below the egg apparatus. At the beginning of this movement, the embryo sac divides conspicuously into three parts comprising an egg apparatus, a central cell, and three antipodals. At maturity, the central vacuole degrades, and the embryo sac becomes conspicuously rich in cytoplasmic contents (Sato 1976).

Pollen Morphology. Pollen grains of *Helwingia* are subprolate or prolate (21–32 μm x 16–26 μm), tri-colporate, and finely spinulate on surface. Endoapertures are diffuse and the tectum is perforate (Ferguson 1977; Ferguson and Hideux 1978).

Karyology. The base chromosome number of Helwingia is $x = 19$, diploid in *H. himalaica* ($2n = 38$), but highly polyploid in *H. japonica* ($2n = 38$, 76, 74, 114, 120, 144; see Hara 1972; Hara and Kurosawa 1975; Peng et al. 1986).

Phytochemistry. No iridoids, tannins or procyanidins were found in the genus (Bate-Smith et al. 1975). Flavonoid glycosides were detected in leaves of *H. japonica* (Iwashina et al. 1997).

Affinities. *Helwingia* has often been treated as an unusual genus of Cornaceae or as a monotypic family allied with Araliaceae (see review in Eyde 1988 and Xiang et al. 1993). The genus is distinct from other members of Cornaceae in morphological, anatomical, and palynological characters. A possible close relationship between *Helwingia* and *Phyllonoma* (Grossulariaceae or

Phyllonomaceae) was discussed by Rodriguez (1971) based on the epiphyllous inflorescences shared by the two genera. This hypothesis was supported by DNA sequence data from several genes (*rbc*L, 18S rDNA, *ndh*F, *atp*B; Morgan and Soltis 1993; Xiang et al. 1993; Soltis et al. 1997; Xiang and Soltis 1998; Albach et al. 2001). These data suggest that *Helwingia* is distantly related to *Cornus*, but closely related to *Phyllonoma* and *Ilex* (Aquifoliaceae) in the Aquifoliales clade of Euasterids II. Recent phylogenetic studies resolved *Helwingia* as the sister of *Phyllonoma* (Soltis et al. 2011, their Fig. 2), which is also supported by epiphyllous inflorescences shared by the two taxa.

Economic Uses. Pith, leaves, and fruits of *Helwingia* are often used medicinally in China (Hu and Soong 1990). All parts of *H. japonica* are used together as an antipyretic and disinfecting agent, to invigorate blood circulation, and to treat swellings.

One genus:

Helwingia Willd. Fig. 41

Helwingia Willd., Sp. Pl. 4: 716 (1806) ('1805'), nom. cons.; Xiang & Boufford, Helwingiaceae, Flora of China 14: 227–229 (2005).

See family description.

Selected Bibliography

Adams, J.E. 1949. Studies in the comparative anatomy of the Cornaceae. J. Elisha Mitchell Sci. Soc. 65: 218–244.

Albach, C.D., Soltis, P.S., Soltis, D.E., Olmstead, R.G. 2001. Phylogenetic analysis of asterids based on sequences of four genes. Ann. Missouri. Bot. Gard. 88: 163–212.

Ao, C., Tobe, H. 2015. Floral morphology and embryology of *Helwingia* (Helwingiaceae, Aquifoliales): Systematic and evolutionary implications. J. Plant Res. 128: 161–175.

Bate-Smith, E.C., Ferguson, I.K., Hutson, K., Jensen, S.R., Nielsen, B.J., Swain, T. 1975. Phytochemical interrelationships in the Cornaceae. Biochem. Syst. Ecol. 3: 79–89.

Dickinson, T.A., Sattler, R. 1975. Development of the epiphyllous inflorescence of *Helwingia japonica* (Helwingiaceae). Amer. J. Bot. 62: 962–973.

Eyde, R.H. 1988. Comprehending *Cornus*: puzzles and progress in the systematics of the dogwoods. Bot. Rev. 54: 233–351.

Ferguson, I.K. 1977. World pollen and spore flora 6: Cornaceae Dum. Stockholm: Almquist and Wiksell.

Ferguson, I.K., Hideux, M.J. 1978. Some aspects of the pollen morphology and its taxonomic significance in Cornaceae sens. lat. Proc. IV Int. Palynol. Conf., Lucknow 1: 240–249. Headed 1978, printed 1980.

Hara, H. 1972. Corresponding taxa in North America, Japan and the Himalayas. In: Valentine, D.H. (ed.) Taxonomy, phytogeography and evolution. London, New York: Academic Press, pp. 61–72.

Hara, H., Kurosawa, S. 1975. A revision of the genus *Helwingia*. In: Ohashi, H. (ed.) The flora of eastern Himalayas, vol. 3. Tokyo: University of Tokyo Press, pp. 393–413.

Harms, H. 1898. Cornaceae. Nat. Pflanzenfam., Teil III, Abteilung 8. Leipzig: Engelmann, pp. 250–270.

Hu, W.K., Soong, T.P. 1990. Cornaceae. In: Fang, W.P., Hu, W.K. (eds.) Flora reipublicae popularis sinicae 56: 3–108. Beijing: Science Press.

Iwashina, T., Kamenosono, K., Hatta, H. 1997. Flavonoid glycosides from leaves of *Aucuba japonica* and *Helwingia japonica* (Cornaceae): Phytochemical relationship with the genus *Cornus*. J. Japan. Bot. 72: 337–346.

Li, H.-L., Chao, C.-Y. 1954. Comparative anatomy of the woods of the Cornaceae and allies. Quat. J. Taiwan Mus. 7: 119–136.

Morgan, D.R., Soltis, D.E. 1993. Phylogenetic relationships among members of Saxifragaceae sensu lato based on rbcL sequence data. Ann. Missouri Bot. Gard. 80: 631–660.

Noshiro, S., Baas, P. 1998. Systematic wood anatomy of Cornaceae and allies. IAWA J. 19: 43–97.

Peng, C.-I., Yen, S.-F., Guo, J.-Y. 1986. Notes on the chromosome cytology of some rare, threatened, or endangered plants of Taiwan (I). Bot. Bull. Acad. Sin. (Taipei). 27: 219–235.

Rodriguez, R.L. 1971. The relationships of the Umbellales. In: Heywood, V.H. (ed.) The biology and chemistry of the Umbelliferae. London, New York: Academic Press, pp. 63–91.

Sato, Y. 1976. Embryological studies of some cornaceous plants. Sci. Rep. Tohoku Univ. Ser. IV (Biol.) 37: 117–130.

Soltis, D.E., Soltis, P.S., Nickrent, D.L., Johnson, L.A., Hahn, W.J., Hoot, S.B., Sweere, J.A., Kuzoff, R.K., Kron, K.A., Chase, M.W., Swensen, S.M., Zimmer, E.A., Chaw, S.-M., Gillespie, L.J., Kress, W.J., Sytsma, K.J. 1997. Angiosperm phylogeny inferred from 18S ribosomal DNA sequences. Ann. Missouri Bot. Gard. 84: 1–49.

Soltis, D.E., Smith, S.A., Cellinese, N., Wurdack, K.J., Tank, D.C., Brockington, S.F., Refulio-Rodriguez, N. F., Walker, J.B., Moore, M.J., Carlsward, B.S. et al. 2011. Angiosperm phylogeny: 17 genes, 640 taxa. Amer. J. Bot. 98: 704–730.

Xiang, Q.-Y., Boufford, D.E. 2005 (Text), 2006 (Illustrations). Helwingiaceae. In: Wu, Z.Y., Raven, P.H., Hong, D.Y. (eds.) Flora of China, vol. 14. St. Louis: Missouri Bot. Garden Press, Beijing: Science Press, pp. 227–229, fig. 415.

Xiang, Q.-Y., Soltis, D.E. 1998. RbcL sequence data define a cornaceous clade and clarify relationships of Cornaceae sensu lato. In: Boufford, D.E., Ohba, H. (eds.) Sino-Japanese flora - its characteristics and diversification. Univ. Mus. Bull. 37: 123–137.

Xiang, Q.-Y., Soltis, D.E., Morgan, D.R., Soltis, P.S. 1993. Phylogenetic relationships of *Cornus* L. sensu lato and putative relatives inferred from rbcL sequence data. Ann. Missouri Bot. Gard. 80: 723–734.

Hydroleaceae

Hydroleaceae Edwards, Bot. Rev. (Lancaster) 7: Tab. 566 (1821).

V. Bittrich and M.C.E. Amaral

Perennial, rarely annual, herbs or small shrubs, often semi-aquatic, erect, prostrate or decumbent, stems succulent or woody, often with aerenchyma in the primary cortex, glabrous to hispid, glandular trichomes present or not, 1–2 axillary thorns per node, or thorns absent. Leaves alternate, sessile or petiolate, exstipulate, blade linear to ovate, apex acuminate to rounded, base attenuate to cordate, margin entire or serrulate, glabrous to glandular-pubescent. Inflorescence terminal or axillary, cymose or racemose, rarely flowers solitary. Flowers bisexual, actinomorphic or slightly zygomorphic; sepals 5, nearly free to base, usually lanceolate, persistent, glabrous, puberulent or glandular-pubescent; petals 5, imbricate, united at base, blue or white, corolla rotate or campanulate; stamens 5, filaments white or blue, adnate to corolla tube, usually conspicuously dilated at base, anthers sagittate, basifixed; nectary disk absent or present; ovary superior, globose to ellipsoid, oblique to median flower plane, usually 2-locular, ovules very numerous, placentation parietal at the apex, otherwise axillary with enlarged, sometimes bilobed placentas; stylodia 2(−5), free, spreading, persistent, stigma slightly funnelform or capitate, papillose. Capsule with loculicidal, septicidal, or irregular dehiscence, globular to cylindrical. Seeds numerous, ovoid to cylindrical, symmetrical or occasionally asymmetrical, usually with longitudinal ridges. $n = 9, 10, 12$; $2n = 18, 24, 20, 40$.

A single genus with about 11 spp. in tropical or warm-temperate regions, southern U.S.A. to Argentina, Africa, southern Asia to northern Australia.

Vegetative Morphology and Vegetative Anatomy. Typically *Hydrolea* plants are ascendant or erect, but procumbent or dwarf forms are also found, depending on ecological conditions. Plants can be glabrous (rarely, e.g. *H. palustris*) or pubescent or glandular-pubescent; various different trichome types were described for *H. zeylandica*; the glandular trichomes can have a uniseriate or multicellular head (Inamdar and Patel 1973). While in hairs of various genera of Hydrophyllaceae cystolith-like incrustations were found (cf. Solereder 1899), these were not reported for *Hydrolea*. Thorns at the nodes, if present, occur single or in pairs. According to Kainradl (1927), they may have small leaves apically, showing that they are not spines but modified branches. Some species (e.g. *H. quadrivalvis*) may have thorns or not, while other constantly have or lack them. Of the 11 species recognized by Davenport (1988), five never have thorns, two always have them, and four may or may not have them. Leaves are petiolate or sessile, their margin entire or serrulate. Anatomy is poorly studied. The primary cortex has schizogenous air-canals, their neighbouring cells show clustered crystals. Carlquist and Eckhart (1984) included *Hydrolea paraguayensis* (= *H. spinosa* var. *par.*) in a comparative study of the wood anatomy of the Hydrophyllaceae. Growth-rings are basically absent; vessel element perforations are mostly simple and, of the species studied, only *H. spinosa* var. *par.* occasionally also had scalariform perforation plates. Vessel element diameter is 44 µm and length is 458 µm, i.e. notably long. Fibre-tracheids are present and sometimes septate, their walls are gelatinous; diffuse axial parenchyma is absent. Vascular rays with erect cells predominate. Rabaey et al. (2010) reported the presence of vestured pits in the tracheary elements of *Hydrolea*, another difference to Hydrophyllaceae, which lack these. Stomata are present on both leaf surfaces in

J.W. Kadereit and V. Bittrich (eds.), *Flowering Plants. Eudicots*, The Families and Genera of Vascular Plants 14,
DOI 10.1007/978-3-319-28534-4_19, © Springer International Publishing Switzerland 2016

H. spinosa, and druses of calcium oxalate are present in the leaf veins. Patel and Inamdar (1971) described the stomata of *Hydrolea zeylanica* as anomocytic and found a few with a single subsidiary cell situated at right angle to the guard cell. The ontogeny is perigenous.

Inflorescence, Flower Structure and Pollen Morphology (Fig. 42). The flowers are arranged in cymose or racemose, simple or branched inflorescences, sometimes in clusters, corymbose or subcapitate forms (Davenport 1988).

A detailed study of flower ontogeny and anatomy of *Hydrolea palustris* was undertaken by Erbar et al. (2005). Starting adaxially in the diagonal plane of the flower, sepal primordia appear in a 2/5-sequence, followed by the alternating petal primordia, which appear simultaneously. The corolla development shows late sympetaly resulting in a short basal tube, the aestivation of the lobes is quincuncially imbricate. The stamens are alternipetalous, the filaments broaden abruptly at the base in most species (according to Brand 1913, only slightly so in *H. multiflora* (= *H. elatior*)) and are inserted on the corolla tube. Below the insertion points a low corolla-stamen tube is formed. Anthers are versatile, basifixed and sagittate. The ovary is glabrous or the upper half pubescent, usually 2 carpels are developed, rarely more; carpel number varies within species (in *H. spinosa* 2–4-carpellate ovaries are found). In a bicarpellate ovary in *H. palustris*, the two carpels are not positioned exactly in the median plane but slightly oblique. The gynoecium is synascidiate in its larger part, but symplicate apically, where the septa fuse only postgenitally (*H. palustris*). Placentation either is axile throughout, or the septa in the upper region only protrude deeply into the ovary cavity (*H. spinosa*, pers. obs.). According to Brand (1913), the placentas are entire (sect. *Attaleria*) or bilobed (sect. *Hydrolea*) in cross-section, mostly thick and spongiose (Brand mentions *H. ovata*), rarely membranaceous (no species mentioned by Brand). The U-shaped placenta with numerous ovules almost fills the locules. The stylodia are distinct, the stigma papillose and slightly funnelform or capitate.

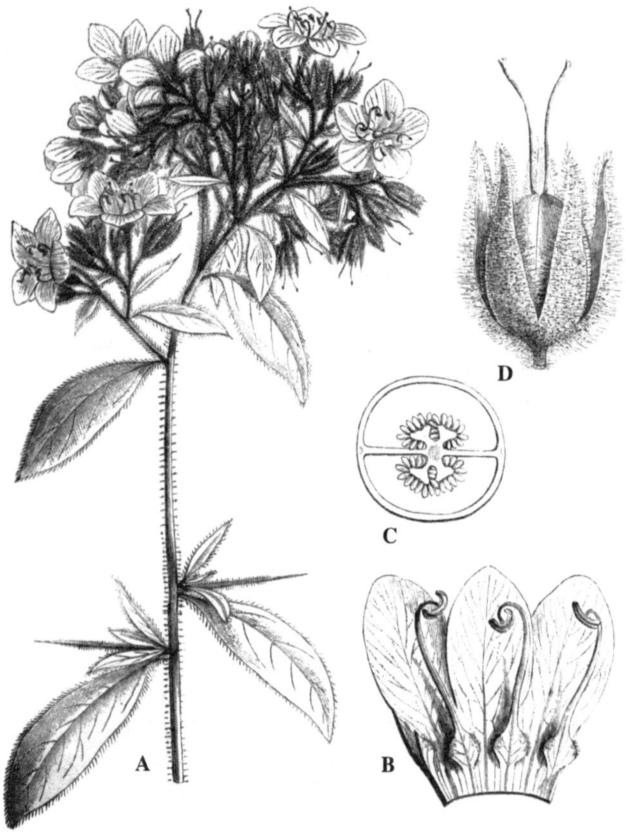

Fig. 42. Hydroleaceae. *Hydrolea spinosa.* **A** Flowering branch. **B** Part of corolla with stamens. **C** Ovary in transversal section. **D** Fruit (closed) with persistent calyx. (From Peter 1897, his Fig. 31)

Although Willis (1894) had described a disk in *Hydrolea spinosa* (sect. *Hydrolea*) producing nectar, Davenport (1988) described flowers of *Hydrolea* as lacking a nectary. Erbar et al. (2005) again reported such a disk with nectary slits and with 5 hump-like swellings alternating with the stamens in *Hydrolea palustris* (sect. *Attaleria*). As the filaments are usually strongly dilated at the base, access to the nectary is only possible for flower visitors at the gaps between the filament bases (Willis 1894; Erbar et al. 2005). Willis (1894) also described the selfing of flowers of *H. spinosa*, cultivated in England.

Pollen is tricolporate with a reticulate tectum without supratectal processes, lumina of various sizes and forms, the larger showing numerous small bacula (Constance and Chuang 1982).

Embryology. Ovules are anatropous, tenuinucellate, unitegmic, the embryo-sac is of the Polygonum-type. No vascular bundles were observed in the ovule, these end at the base of the funicle, and the nucellus disappears early (Kainradl

1927).The polar nuclei degenerate as free nuclei. The mature embryo-sacs in various Hydrophyllaceae studied by Svensson (1925) are rich in starch, in contrast to that of *Hydrolea spinosa.* Svensson (1925) also observed that, in the absence of pollination, the ovary of this species can develop to the size of an immature fruit due to a hypertrophic embryo-sac, but without production of an embryo. Di Fulvio de Basso (1990) also studied *H. spinosa* and reported that endosperm formation is cellular and heteropolar micropylar, while in Hydrophyllaceae it is isopolar. Different from reports by Mitra (1947, for *Hydrolea zeylanica*) and Svensson (1925, for *H. spinosa*), she observed 2 uniseriate pluricellular branched haustoria in *Hydrolea spinosa.* This basically confirms the observations of Kainradl (1927) for the same species, who also reported two strongly developed polar branched haustoria growing mycelia-like into the integument. Kainradl (1927), however, regarded these haustoria as unicellular with one nucleus each. Di Fulvio de Basso (1990) also reported that in Hydrophyllaceae s.str. haustoria are either absent or inconspicuous and unbranched (Phacelieae) or unicellular (Hydrophylleae). According to Mitra (1947), embryo development of *H. zeylanica* is of the solanaceous type.

Fruit, Seeds and Dispersal. Fruits are 2(3–5)-locular dry capsules, dehiscing septicidally, loculicidally or irregularly; regular and irregular deshiscence can occur in the same species. The capsules are glabrous or puberulent or glandular-pubescent in the upper half, and contain numerous seeds. The seeds are small and typically have longitudinal, more rarely transversal ridges, and a surface with prominent reticulations (Davenport 1988). The ridges are formed by enlarged epidermis cells; the other thick epidermis cells contain oil (Netolitzky 1926). The integument is 6–7 cell layers thick. According to Kainradl (1927), several narrow conical structures grow from the inner periclinal wall into the lumen of the epidermal cells. Below the epidermis a 4–5 cell-layered pigmented middle-layer is present; this is mostly absorbed during embryo development. The innermost layer forms an endothelium. The embryo is small and straight. The endosperm contains oil and aleuron and is completely absorbed in the ripe seed (Kainradl 1927 for *H. spinosa*). Hydrochory was suggested as seeds can float for several says and may germinate when still floating, as observed in *H. spinosa* (Kainradl 1927).

Karyology. Chromosome counts for section *Hydrolea* ($2n = 20, 40$) indicate a base number of 10, with *H. spinosa* apparently being diploid or tetraploid (Davenport 1988). The counts reported for *H. zeylanica* (sect. *Attaleria*) are ambiguous with $n = 9$, $2n = 18$ and 24. The data suggest, however, that the base number is different from that in sect. *Hydrolea.*

Affinities. Robert Brown (1818, p. 32) was the first to suggest to move *Hydrolea* into an own family together with the genus *Nama.* As he probably considered *Nama* in the sense of Linnaeus (1759) and thus as a genus of Hydrophyllaceae s.s., he did not completely separate *Hydrolea* from that family. Following Brown (1818), Hydroleaceae subsequently normally included *Nama* and perhaps other related genera. Choisy (1846) and Brand (1913), who included *Hydrolea* in an own tribe (Hydroleae) of Hydrophyllaceae, and Constance (1963) emphasized the distinctiveness of *Hydrolea* with its completely bilocular capsule, axile placentation, and differences in the texture of the placentas: fleshy or cartilaginous in the Hydrophyllaceae s.str. and spongiose (rarely membranaceous) in *Hydrolea.* Also the aerenchyma in the primary cortex (Solereder 1899), the presence of vestured pits in the tracheary elements (Rabaey et al. 2010), and the heteropolar endosperm formation with pluricellular haustoria (Di Fulvio de Basso 1990) are characteristic for *Hydrolea* but unknown from Hydrophyllaceae. Di Fulvio de Basso (1990) therefore proposed to exclude *Hydrolea* from Hydrophyllaceae and put it into a monogeneric family of its own. Probably most taxonomists preferred to keep *Hydrolea* included in the Hydrophyllaceae (or Boraginaceae s.l.) because of its similarity with *Nama*, until phylogenetic analyses based on DNA sequences (Cosner et al. 1994; Ferguson 1998) definitely showed that the genus had to be excluded from the Boraginales. The monogeneric Hydroleaceae belong to the Solanales as sistergroup to Sphenocleaceae (Cosner et al. 1994; Soltis et al. 2011). Different from the late sympetaly of most Solanales incl. *Hydrolea*, however,

Sphenoclea shows early sympetaly (Erbar et al. 2005).

Distribution and Habitats. The species of section *Hydrolea* occur in the Americas while those of sect. *Attaleria* occur in Africa, Australia and Asia. The most common habitats are swamps, lakes and margins of waterways, but various species are also encountered in other moist but non-inundated habitats.

Economic Importance. *Hydrolea spinosa* is invasive. A few species are of minor horticultural importance.

Only one genus:

Hydrolea L. Fig. 42

Hydrolea L., Sp. Pl. ed. 2: 328 (1762), nom. cons.; Davenport, Rhodora 90: 169–208 (1988), rev.
Nama L. (1753) nom. rej. non *Nama* L. (1759), nom. cons.

Characters as for family.

Brand (1913) recognized 2 sections, sect. *Hydrolea* (as sect. *Sagonea* (Aubl.) Brand) and sect. *Attaleria* (Poir.) Brand, both also accepted by Davenport (1988). The species of sect. *Hydrolea* occur in the Americas and their placentas are divided into 2 lobes, thorns are present or absent. Those of sect. *Attaleria* occur in Africa, Asia and Australia, and have entire placentas; thorns are always absent.

Selected Bibliography

Brand, A. 1913. Hydrophyllaceae. In: Engler A. (ed.) Das Pflanzenreich IV, 251: 1–210. Leipzig: Engelmann.
Brown, R. 1818. Observations, systematical and geographical, on Professor Christian Smith's collection of plants from the vicinity of the river Congo. In: Tuckey J.K. (ed.) Narrative of an expedition to explore the river Zaire. London: John Murray, pp. 420–485.
Carlquist, S., Eckhart, V.M. 1984. Wood anatomy of Hydrophyllaceae. II. Genera other than *Eriodictyon*, with comments on parenchyma bands containing vessels with large pits. Aliso 10: 527–546.
Choisy, J.D. 1846. Hydroleaceae. In: A. De Candolle (ed.) Prodromus systematis naturalis regni vegetabilis . Paris: Treuttel and Würtz, pp. 179–185.
Constance, L. 1963. Chromosome numbers and classification in Hydrophyllaceae. Brittonia 15: 273–285.
Constance, L., Chuang, T.I. 1982. SEM survey of pollen morphology and classification in Hydrophyllaceae (waterleaf family). Amer. J. Bot. 69: 40–53.
Cosner, M.E., Jansen, R.K., Lammers, T.G. 1994. Phylogenetic relationships in the Campanulales based on *rbcL* sequences. Plant Syst. Evol. 190: 79–95.
Davenport, L.J. 1988. A monograph of *Hydrolea* (Hydrophyllaceae). Rhodora 90: 169–208.
Di Fulvio de Basso, T.E. 1990. Endospermogenesis y taxonomía de la familia Hydrophyllaceae y su relación con las demas Gamopetalas. Acad. Nac. Cienc. Exact. Física. Natur. Buenos Aires 5: 73–82.
Erbar, C., Porembski, S., Leins, P. 2005. Contributions to the systematic position of *Hydrolea* (Hydroleaceae) based on floral development. Plant Syst. Evol. 252: 71–83.
Ferguson, D.M. 1998. Phylogenetic analysis and relationships in Hydrophyllaceae based on ndhF sequence data. Syst. Bot. 23: 253–268.
Inamdar, J.A., Patel, R.C. 1973. Structure, ontogeny and classification of trichomes in some Polemoniales. Feddes Repert. 83: 473–488.
Kainradl, E. 1927. Beiträge zur Biologie von *Hydrolea spinosa* L. mit besonderer Berücksichtigung von Fruchtwand und Samenentwicklung. Akad. Wiss. Wien, Sitzungsver. Math.-Naturwiss. Kl., Abt. 1, 136: 167–193.
Linnaeus, C. 1759. *Nama*. Systema Naturae, ed. 10, 2. Holmiae: Impensis Direct. Laurentii Salvii, p. 950.
Mitra, J.N. 1947. A contribution to the life-history of *Hydrolea zeylanica* Vahl. J. Indian Bot. 26: 51–61.
Netolitzky, F. 1926. Anatomie der Angiospermen-Samen. In: Linsbauer, K. (ed.) Handbuch der Pflanzenanatomie, II. Abt., III. Teil, Pteridophyten und Anthophyten, Band 10. Berlin: Gebrüder Borntraeger.
Patel, R.C., Inamdar, J.A. 1971. Structure and ontogeny of stomata in some Polemoniales. Ann. Bot. 35(2): 389–409.
Peter, A. 1897. Hydrophyllaceae. In: Engler, A., Prantl, K. (eds.) Die natürlichen Pflanzenfamilien IV, 3a. Leipzig: Engelmann, pp. 54–71.
Rabaey, D., Lens, F., Smets, E., Jansen, S. 2010. The phylogenetic significance of vestured pits in Boraginaceae. Taxon 59: 510–516.
Solereder, H. 1899. Systematische Anatomie der Dicotyledonen. Stuttgart: F. Enke.
Soltis, D.E., Smith, S.A., Cellinese, N., Wurdack, K.J., Tank, D.C., Brockington, S.F., Refulio-Rodriguez, N. F., Walker, J.B., Moore, M.J., Carlsward, B.S., Bell, C. D., Latvis, M., Crawley, S., Black, C., Diouf, D., Xi, Z., Rushworth, C.A., Gitzendanner, M.A., Sytsma, K.J., Qiu, Y.L., Hilu, K.W., Davis, C.C., Sanderson, M.J., Beaman, R.S., Olmstead, R.G., Judd, W.S., Donoghue, M.J., Soltis, P.S. 2011. Angiosperm phylogeny: 17 genes, 640 taxa. Amer. J. Bot. 98: 704–730.
Svensson, H.G. 1925. Zur Embryologie der Hydrophyllaceen, Borraginaceen und Heliotropiaceen mit besonderer Rücksicht auf die Endospermbildung. Uppsala Univ. Årsskrift Matem. Naturvetensk. 2: 1–176, pl. 1–3.
Willis, J.C. 1894. Contributions to the natural history of the flower. Part II. Fertilization methods of various flowers; Cleistogamy in *Salvia Verbenaca*. Bot. J. Linn. Soc. 30: 284–298.

Hydrophyllaceae

Hydrophyllaceae R. Br., Bot. Reg. 3: ad t. 242 (1817), as Hydrophylleae, nom. cons.

M. Hofmann, G.K. Walden, H.H. Hilger, and M. Weigend

Annual, biennial or perennial herbs, subshrubs, shrubs, rarely small trees (*Wigandia*), often aromatic; primary root usually persistent and developed as strong taproot, sometimes roots tuberous; stems erect, rarely prostrate to ascending, sometimes forming simple or few- to many-branched rhizomes; indumentum usually present and strongly developed on whole plant, usually scabrid to hispid, often densely glandular, sometimes distinct stinging hairs present. Leaves alternate or opposite, basal and/or cauline, exstipulate, lamina linear, narrowly ovate to subcircular, simple to compound, margins entire or variously lobed, sessile to petiolate. Inflorescences terminal or axillary, frondose, bracteose or ebracteose, paraclades monochasial or dichasial, lax or very dense, usually scorpioid and contracted into boragoids, these paraclades present as simple terminal inflorescences or combined into complex thyrsoids, sometimes with extensive accessory paraclades; sometimes inflorescences congested into terminal "heads", or strongly reduced to axillary or terminal single flowers. Flowers pentamerous, usually hypogynous, bisexual, commonly protandrous; perianth biseriate, sepals united at base or nearly to apex, usually radially symmetrical, sometimes slightly or distinctly unequal with some lobes much larger than others, with or without appendages between the lobes, persistent and usually accrescent in fruit, spreading or closing; corolla sympetalous, mostly campanulate, more rarely rotate or infundibuliform, tube internally often with scales, glands or hair lines near base of each filament; lobes spreading or porrect, rarely reflexed, narrowly triangular to (usually) oblong or subcircular; aestivation usually quincuncial, rarely contorted; stamens epipetalous and antesepalous, inserted at the same or at different heights in corolla tube, anthers included or exserted, free from each other, dorsifixed, dithecous, tetrasporangiate, opening by longitudinal slits; gynoecium 2-carpellate, syncarpous, superior to half-inferior, pubescent, often glandular, usually with basal nectar disk; stylodia 2 or connate into a usually bifid or deeply bifurcate, rarely entire, slender style; stigma punctate or capitate, dry; ovules (1–)4–∞, anatropous, unitegmic, tenuinucellate; placentation parietal or intrusive-parietal. Fruits 1- to many-seeded capsules, 2-4-valved, with loculicidal or loculicidal and septicidal dehiscence, rarely indehiscent. Seeds ovoid, globose or angular, with dark, often reticulate testa, endosperm copious, oily, rarely with elaiosome; embryo small, straight.

A family comprising 15 genera and 280–300 species, distributed mainly in western North America, several species in Central America and western South America (*Phacelia*, *Nama*, *Wigandia*), the West Indies and Hawaiian Islands (only *Nama*).

Vegetative Morphology (Figs. 43, 44). Hydrophyllaceae are a mostly herbaceous family, including winter- or spring-flowering annuals, some species with life spans of only a few weeks. Most genera consist solely (*Ellisia*, *Pholistoma*, *Emmenanthe*, *Eucrypta*, *Nemophila*) or predominantly (*Phacelia*, *Nama*) of annuals. *Howellanthus* and *Hesperochiron* are acaulescent perennials with more or less distinctly thickened primary roots, although *Howellanthus* has long flowering stems from the caudex axils. *Romanzoffia* has a short vegetative stem, and sometimes elongate flowering stems, and also has thick primary roots and tuberous or bulbous bases in the

J.W. Kadereit and V. Bittrich (eds.), *Flowering Plants. Eudicots*, The Families and Genera of Vascular Plants 14,
DOI 10.1007/978-3-319-28534-4_20, © Springer International Publishing Switzerland 2016

Fig. 43. Hydrophyllaceae. *Nama sericea.* **A** Habit. **B** Flower. **C** Open corolla. **D** Gynoecium. **E** Fruit longisection. (From Brand 1913)

perennial species, with the annual species forming only a slender taproot (*R. thompsonii*). *Romanzoffia tracyi* and *R. californica* are spring-flowering tuberous geophytes, a growth form not otherwise known in the family. The tubers of *Romanzoffia* have not been critically studied, but likely originate from a thickened primary root. Species of *Hydrophyllum*, *Tricardia* and *Draperia* together with some *Nama* and *Phacelia* are short-lived perennial herbs. A biennial life history seems to have evolved several times independently in the family, occurring in *Hydrophyllum* (*H. appendiculatum*) and *Phacelia* (e.g. *P. bipinnatifida*, *P. formosula*, *P. franklinii*, *P. heterophylla* var. *virgata*). Underground structures in these genera have not been described in detail, but are likely pleiocorms or root-pleiocorms in *Nama*, *Phacelia*, *Tricardia* and *Draperia*, and are

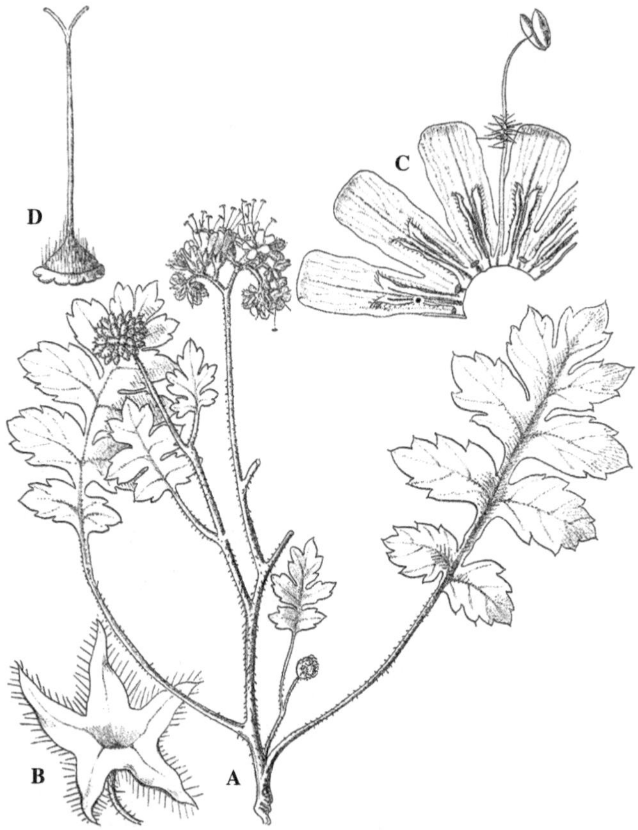

Fig. 44. Hydrophyllaceae. *Hydrophyllum macrophyllum.* **A** Habit. **B** Calyx. **C** Open corolla. **D** Gynoecium. (From Brand 1913)

clearly rhizomes in *Hydrophyllum*. Some species of *Nama* are subshrubs, *Eriodictyon* includes perennial herbs, subshrubs and shrubs, and *Wigandia* is a tall shrub or small tree. All of these taxa have soft-wooded and brittle stems. An extensive, horizontal root system may be present in perennial desert taxa (e.g. *Eriodictyon*, *Nama rothrockii*) and shoots may arise directly from this diffuse root system (Carlquist et al. 1983; Carlquist and Eckhart 1984). The majority of *Eriodictyon* individuals in any given locality may ultimately result from roots rather than seed germination (Carlquist et al. 1983). *Wigandia* can also produce aerial shoots from roots, but clonal growth in *Wigandia* is usually not as conspicuous as in *Eriodictyon*. Phyllotaxy on vegetative parts is mostly alternate, but opposite leaves may occur in the inflorescence. Some taxa, such as *Draperia* and *Eucrypta* and some species of *Phacelia*, have opposite leaves throughout or opposite leaves at least at the base in *Pholistoma* and some species of *Nemophila* and *Phacelia*. Leaves are mostly distinctly petiolate, rarely subsessile or sessile. The lamina is often widely ovate, but sometimes obovate, triangular-ovate or narrowly ovate, usually with a cordate base, but sometimes with the lamina long-decurrent on the petiole. *Draperia*, *Hesperochiron*, *Howellanthus*, *Tricardia* and many species of *Nama* and *Phacelia* have simple leaves with an entire margin, *Eriodictyon*, *Wigandia*, and some species of *Nama* and *Phacelia* have entire leaves with a serrate to lobulate-serrate margin. The remaining species of *Phacelia*, but also *Ellisia*, *Eucrypta*, *Pholistoma*, *Emmenanthe* and *Nemophila* have variously pinnatisect to bipinnate or tripinnate leaves. *Romanzoffia* is aberrant in producing long-petiolate leaves with a subcircular, subpalmately lobed lamina otherwise only found in a few species of *Phacelia*. A peculiarity of *Eriodictyon californicum* is the production of two different sets of leaves each year: drought-deciduous leaves in late winter and spring, and drought-resistant leaves in spring and summer (Johnson et al. 1985). This leaf dimorphism is also found in other species of *Eriodictyon*, with vegetative stem leaves narrow and green, and flowering stem leaves white-tomentose in *E. lobbii*.

Vegetative Anatomy. In many taxa the entire plants, and especially the leaves, are densely glandular or resinous, especially in *Eriodictyon*, but also in *Emmenanthe*, *Eucrypta* and many species of *Phacelia* (Di Fulvio and Dottori 1995). The trichomes can be variable within individual plants and species, and include simple unicellular and multicellular trichomes, and unicellular or multicellular gland-tipped trichomes (Di Fulvio and Dottori 1995). Very little is known about leaf anatomy: the lamina is bifacial or aequifacial, stomata are anomocytic and confined to one surface or present on both surfaces. The mesophyll in *Eriodictyon* has transverse, septate palisade cells. Minor leaf veins have phloem transfer cells (five genera) or not (*Wigandia*). Druses or solitary crystals of Ca-oxalate have been found in idioblasts of stems and leaves in several genera. Crystal form and distribution in vegetative and generative organs might be useful characters for systematic purposes (Hofmann 1999). Nodes are unilacunar. Wood anatomy of Hydrophyllaceae has been studied by Carlquist et al. (1983), Carlquist and Eckhart (1984), and Carlquist (1992). Perforation plates of the vessel elements are simple. Fibriform vessel elements are universally present and particularly abundant in *Wigandia*, fibre tracheids are common, with fully bordered pits to libriform fibres with simple pits, vessel end-walls are simple; the axial parenchyma is diffuse (apotracheal) rather than vasicentric (Carlquist 1992). The wood of *Eriodictyon* is ring-porous with earlywood tracheids and vessels distinctly wider than latewood tracheids and vessels, with growth rings well-demarcated in inland and montane taxa. Growth rings are very indistinct in *Wigandia*. Vessels are variously grouped in *Eriodictyon*, but virtually solitary in *Wigandia*. Vessels are narrow in the stem wood, showing distinctly xeromorphic adaptations. Multiseriate vascular rays are present in most genera, but only form in later stages of wood accumulation in *Phacelia* (e.g. *P. ramosissima*). Secondary thickening takes place from a conventional cambial ring, and an initially superficial cork cambium is present. The xylem has tracheids rather than fibre tracheids or libriform fibres. Vessel end-walls are simple. Root wood in *Eriodictyon* has much wider vessels than stem wood, and also more extensive multiseriate rays and more copious axial parenchyma indicating root succulence.

Inflorescence Structure. Detailed investigations of the inflorescences have not been published. In general, Hydrophyllaceae share the basic architecture of other Boraginales, with exclusively "closed" (monotelic) inflorescences and all paraclades cymose (prophylls are the only leaf type present on their branches). Overall, inflorescences are usually frondose, frondose-bracteose, bracteose or ebracteose. Bract morphology often changes abruptly within the inflorescence, and the pherophylls of (mostly frondose) paraclades may be morphologically strongly different from those subtending dichasial ramifications (mostly bracteose) and these in turn may differ strongly from the bracts in monochasial portions (distinctly bracteose, or absent). As in Boraginaceae, the basic unit of the inflorescence is usually a scorpioid (= coiled) cincinnus, the so-called boragoid (Fig. 44A): the mostly short-pedicellate and often ebracteose flowers are separated by initially very short internodes and are arranged in two alternating rows. Flowers usually all face in the same direction (= secund).

The boragoids are often paired and arise from a first dichasial division (from the axils of a pair of opposite bracts preceding an often well-developed terminal flower). These "paired boragoids" or "double boragoids" are a common basic unit of Boraginales inflorescences. In Hydrophyllaceae, the mono- or dichasia themselves are usually ebracteose. An individual boragoid or double boragoid may make up the entire terminal inflorescence, in which case the inflorescence can be classified as a mono- or dichasial cymoid. Commonly, the boragoids, and more often the double boragoids are found as paraclades of more or less complex thyrsoids, rarely di- or pleiothyrsoids. Perennial taxa generally have quite straightforward thyrsoids with several double boragoids as paraclades. The overall inflorescences are often regular (mesotonic) and cylindrical in outline, with all paraclades more or less evenly developed (e.g. many *Eriodictyon*, most perennial *Phacelia*, *Wigandia*). In several taxa the internodes in the inflorescence are extremely shortened, which is usually combined with an elongation of the naked peduncle. This condensation of the inflorescence can be relatively moderate, leading to a superficially corymbose arrangement of the flowers (*Draperia*, some *Hydrophyllum* and *Phacelia*), or very strong with all flowers situated in a more or less compact head (e.g. *Nama rothrockii*, *Eriodictyon tomentosum*, *Hydrophyllum capitatum*, *Phacelia capitata*). In *Tricardia* and *Romanzoffia* inflorescences are weakly branched, typically monochasial, and additional paraclades are rarely present. Inflorescence structure is obscure in rosulate *Hesperochiron*, *Howellanthus*, and the rosulate species of *Phacelia*, where they are probably reduced to cymoids with obsolete internodes and frondose bracts. Annual taxa generally have a very complex inflorescence structure, which has not been analyzed in detail. Complex synflorescence types, including both homoeocladic and heterocladic thyrsoids with often a mixture of frondose/bracteose portions and ebracteose terminal elements, are found in *Emmenanthe*, *Pholistoma*, *Nemophila*, *Eucrypta*, annual *Nama* and *Phacelia*. Metatopic displacement of pherophylls is common in these taxa. In some taxa the main axis of the inflorescence is extremely shortened with the primary flower situated at or near ground level and sometimes abortive. It is overtopped by paraclades and renewal shoots with elongated proximal internodes and short distal internodes flowering simultaneously with the terminal inflorescence. The overall inflorescence is then often flat on the ground and densely frondose, while the rosette leaves degenerate early, leading to an essentially anthocladal architecture. The derivations in inflorescence architecture in annual Hydrophyllaceae thus closely parallel those in annual Boraginaceae (especially American *Cryptantha*, *Plagiobothrys*; Schwarzer 2007).

Floral Structure and Anatomy. Flowers are usually erect in the inflorescences, sometimes horizontal (*Hydrophyllum*) or nodding (*Emmenanthe*). Sepals, petals and stamens originate centripetally as five separate primordia each in quincuncial order and in alternate and aequidistant position, the first sepal arising is median adaxial. Carpel primordia are in a median position. The pentamerous, bisexual flowers have sepals united only at the base, sometimes divided 3/4 of their length in *Romanzoffia*. Tetramerous flowers are rare (e.g. *Phacelia tetramera*), or inflorescences on individual plants can contain flowers ranging from 4- to 6-merous (*Hesperochiron*, *Howellanthus*, *Phacelia* spp.). The calyx is persistent and usually accrescent in fruit, sometimes strongly accrescent. Calyx lobes are usually all equal, but sometimes slightly different in size resulting in unequal calyx lobes, which have been called heteromorphic in the literature at times (some *Phacelia*), rarely differentiated into two linear and three large, cordate sepals (*Tricardia*). The corolla is commonly coloured in various shades of blue or purple, but sometimes yellow or white, and may be concolourous or bicolourous. The corolla lobes are always well-differentiated and mostly ovate to subcircular in outline. The corolla is deciduous in most taxa, but may be tardily deciduous or persistent in fruit—e.g. in some species of *Phacelia*, and in *Emmenanthe*. The corolla is often campanulate with a porrect, half-spreading or spreading limb. Some groups (*Hesperochiron*, *Nemophila*, some *Phacelia*) have subrotate (bowl-shaped) corollas, and groups in *Phacelia* have infundibuliform or tubular or tubular-infundibuliform corollas. The corolla mouth is always wide and open, and rarely corolla lobes are erect or slightly incurved at anthesis (some species of *Phacelia*), while the tube may be constricted in some species of *Phacelia*. At the point of filament attachment there is usually a differentiation of the inside of the corolla tube, which can be scale-shaped ("corolla scales"), or a hairy line or fold is present (V-shaped folds on the corolla tube in *Eucrypta*, *Romanzoffia*). "Corolla scales" may restrict pollinator access to the nectar reward, or may be a restriction point forcing pollinators to contact the stigmatic surface or anthers. Presence, modification and absence of corolla scales are variable in the family, and have been used as a diagnostic character (Gray 1875). Even within a genus (e.g. *Eucrypta*, *Phacelia*), there are species with and without corolla scales. Homology of these corolla scales to the faucal or basal scales of Boraginaceae is likely, but detailed investigations of homology are needed. Filaments are mostly filiform, sometimes abruptly widened above or below in *Hesperochiron*, *Eriodictyon*, *Nama*. They are attached near the base or near the mouth of the tube and are often equal, but may be quite different in size in some genera (e.g. *Hesperochiron*, *Tricardia*, *Phacelia* spp.). The anthers are often included, but sometimes exserted from the corolla (*Hydrophyllum*, some *Wigandia* and *Phacelia*). The ovary is mostly superior, rarely semi-inferior (e.g. *Nama stenocarpa*). The connate portion of the calyx becomes postflorally adnate to the ovary and is persistent in fruit, sometimes becoming indurate in fruit. In these cases the initially superior ovary appears semi-inferior in fruit. The ovary is surrounded by a ring-shaped, circular or five-lobed nectary with secretary stomata; sometimes the nectary is fragmented into five nectar glands ("hypogynous disk" of *Phacelia* sect. *Miltitzia*; Halse 1981). Secretory stomata are dispersed on the nectary or arranged in clusters. The ovary is usually ovoid and pubescent and sometimes glandular. It is variously subdivided by false septa from the intrusive parietal placentae and usually 2- or 4-locular, rarely unilocular or 5-celled. It may be divided at the bottom by a cross zone and a false septum, and at the top by an apical septum. The apical style is often pubescent and included or exserted from the corolla. It may be deeply forked into two long stigmatic lobes (e.g. *Eriodictyon*) to nearly undivided with two marginally separate stigmas (e.g. *Emmenanthe*). Heterostyly has been reported from *Eriodictyon*, *Nama* and *Phacelia* (Brand 1913), but these reports have not been substantiated by recent critical studies. However, gynodioecy has been reported in a series of studies in *Phacelia dubia* (del Castillo 1993, 1994, 1998; del Castillo and Trujillo 2009), *Nemophila menziesii* (Ganders 1978), *Phacelia distans* (Horner 1977), *Phacelia linearis* (Eckhart 1991, 1992) and *Phacelia ranunculacea* (Sewell and Vincent 2009), and is likely more prevalent than reported in the family. The reports of heterostyly may thus go back to a misinterpretation of gynodioecious taxa. In general, comprehensive studies of the breeding system across the family have not been carried out in Hydrophyllaceae.

Embryology. Ovules are unitegmic and usually tenuinucellate [crassinucellate in *N. jamaicense* (Di Fulvio 1989) or weakly pseudocrassinucellate in *N. californicum* (Di Fulvio 1993)]. The cells of the inner integument epidermis function as an endothelium. After fertilization, the endothelium shows intense secretory activity; its cells acquire thick inner tangential walls with wedge-like projections in the radial walls. Embryo sac development is of the Polygonum type. Polar nuclei fuse prior to fertilization. Antipodal cells are formed (three, not proliferating, ephemeral). Endosperm formation is ab initio cellular [belonging to the isopolar type in tribe Hydrophylleae (Di Fulvio 1987)], or nuclear, or cellular to nuclear. Svensson (1925) described four different endosperm types and the presence of micropylar and chalazal haustoria. These are unicellular with or without lateral branches in *Ellisia*, *Hydrophyllum*, *Nemophila* and *Pholistoma*, bicellular micropylar in *Eucrypta* and multicellular in other cases. The mature endosperm is ruminated or not. Rumination is produced by "giant cells" of the testa epidermis with tubular extensions into the endosperm (Berg 1985, 2009) or by endothelial cells that push inwards against the storage tissue (Di Fulvio 1989). Embryogeny conforms to the Solanad type.

Pollen Morphology. A major survey of pollen morphology of the family was provided by Constance and Chuang (1982): pollen grains are prolate to nearly spherical, more rarely oblate with a triangular outline. Grains are tricolpate to tricolporate, rarely 5–6-colpate or zonosulcate. The pollen grains are isopolar or heteropolar, the true colpi extending to both poles, to one pole but not the other, or not reaching either pole. The tectum is often reticulate or microreticulate, but may also be rugulose or verrucose. The tectum is mostly uniform, but differentiated colpus margins are also found, with widely reticulate mesocolpial regions alternating with microreticulate or punctate colpus margins. Similarly, the apocolpial region sometimes is less deeply sculptured. Hexacolpate pollen is common in *Phacelia*, and zonosulcate pollen is common in *Nama* and *Pholistoma*. Ferguson (1998) found that in tribe Phacelieae there were two clades, first identified by Constance and Chuang (1982), one clade having both three pseudocolpi and three true colpi, and the other three true colpi only.

Karyology. Haploid chromosome numbers vary from $n = 5$ (*Phacelia dubia*, *P. maculata*) to $n = 38$ (*Wigandia kunthii*), but no base chromosome number is discernible, except for *Phacelia* with $x = 11$ (Mayrose et al. 2011; GKW, unpubl. data). In *Phacelia*, the genus best investigated, both ascending and descending dysploidy occur. The perennial *P. magellanica* group ($n = 11, 22, 33$) has been traditionally considered to have arisen by allopolyploidy from a diploid ancestor ($n = 11$; Kruckeberg 1956; Heckard 1956, 1960; Constance 1963; Shelly 1989), but annual polyploids also exist in the genus. Polyploidy is also evident in *Nama* ($n = 7$, 14, 17), *Wigandia* ($n = 19$, 38) and *Eucrypta* ($n = 10$, 20; $n = 6$, 12). *Eucrypta micrantha* ($n = 10$, 20) is more similar in chromosome number to *Ellisia* ($n = 10$) than to *Eucrypta chrysanthemifolia* ($n = 6$, 12; Ferguson 1998; Walden 2010). Polyploidy appears to be absent in *Draperia* ($n = 9$), *Emmenanthe* ($n = 18$), *Ellisia* ($n = 10$), *Hesperochiron* ($n = 8$), *Howellanthus* ($n = 8$), *Hydrophyllum* ($n = 9$), *Nemophila* ($n = 9$), *Pholistoma* ($n = 9$) and *Romanzoffia* ($n = 11$), but *Emmenanthe* is a candidate for a possibly relictually polyploid genus, with a hypothesized diploid ancestor of $n = 9$ (Cave and Constance 1950; Ferguson 1998; Walden 2010).

Reproductive Biology incl. Pollination. Flowers are generally protandrous, often self-compatible. Some degree of selfing has been reported in *Hydrophyllum* (Wolfe and Shore 1992) and *Phacelia*, notably in the annuals (sect. *Whitlavia*, sect. *Gymnobythus* by Gillett 1955; Levy 1988; *Phacelia* cf. *viscosa*, *Nama dichotoma*, *Emmenanthe* by MW, pers. obs.). The degree of interspecific isolation varies widely: some studies identified interfertility between all members of supraspecific taxa in *Phacelia* (e.g. *Phacelia* subsect. *Whitlaviae*); interfertility between some, but not all, members of supraspecific taxa (e.g. *Phacelia* subsect. *Campanulariae*); or intersterility between all members of supraspecific taxa (e.g. *Phacelia* sect. *Gymnobythus*; Gillett 1955; Hansen et al. 2009; Walden 2010). Some species in *Eriodictyon* and *Nama* produce cleistogamous flowers (Brand 1913). Conversely, the perennial taxa are probably predominantly outcrossing and sometimes self-incompatible [e.g. clonal *Eriodictyon capitatum* is self-incompatible (Elam 1994)]. Halse (1979) studied UV absorbance and reflectance in flowers of *Phacelia* sect. *Miltitzia*, and found no specialized patterns, speculating that the species were likely pollinated by generalists. Generalist hymenopteran pollinators have indeed been widely reported, with species of *Bombus* likely the single most important group: 18 species of *Bombus* have been reported as visitors of representatives of *Eriodictyon*, *Hesperochiron*, *Hydrophyllum*, *Nama* and *Phacelia* in California by Thorp et al. (1983). Similarly, the most frequent pollinators observed on *Hydrophyllum* by Beckmann (1979)) were *Apis mellifera*, *Bombus pennsylvanicus*, *B. vagans*, *B. grisecollis*, *B. nevadensis*, *Osmia* spp., *Hoplitis* spp. and four *Anthophora* spp. *Phacelia argentea* is pollinated by honey bees, bumble bees, and a locally native leafcutter bee (*Anthidium palliventre*; Rittenhouse 1993). *Nemophila menziesii* is also frequently visited by generalist bee pollinators such as *Habropoda miserabilis*, *Bombus vosnesenskii* and *Andrena* spp. (McCall 2006). *Eriodictyon* is known to be pollinated by wasps, butterflies and various bees of the genera *Anthophora*, *Bombus*, *Chelostoma*, *Hylaeus*, *Osmia* and *Nomadopsis* (Moldenke 1976). More or less specialized pollination is found as asymmetrical specialization—the bees are oligolectic on Hydrophyllaceae, but the plants have a much wider spectrum of flower visitors (Cruden 1972a, 1972b). Thus, individual species of oligolectic bees of the genera *Osmia* and *Pseudomasaris* favour *Eriodictyon*, *Hydrophyllum*, *Nemophila* and *Phacelia* (Ganders 1978; Anon. 2007). *Pseudomasaris edwardsii* is apparently locally specialized on *Phacelia leucophylla* and five spp. of Masarinae are more or less specialized on species of *Phacelia* and *Eriodictyon* (O'Neill 2001). Bird pollination is not usually reported for the family, but Calliope hummingbirds (*Stellula calliope*) have been reported to feed on *Wigandia urens* in California (Des Granges 1979), whereas carpenter bees of the genus *Xylocopa* are the primary visitors of that plant in South America and bats are the likely pollinators of *Wigandia ecuadoriensis* (MW and HH, pers. obs.).

FRUIT AND SEED. The fruits are loculicidally dehiscent capsules, or loculicidally and septicidally dehiscent in *Eriodictyon*. The dehiscence tissue develops from the median carpel vascular bundles and the small false septa. The sclerified endocarp acts as opening tissue; it consists of both the inner epidermis and adjacent subepidermal layers (Hilger 1987, for *Wigandia caracasana*). Dispersal is usually by gravity or wind, or the seeds move secondarily downslope or upslope on soils (Armstrong 1992). Sometimes the accrescent calyx encloses the capsule and is dispersed as a unit (Horner-Till 1982). Pedicels often elongate slightly postflorally (e.g. *Phacelia* sect. *Whitlavia*, *Phacelia* subg. *Cosmanthus*), but generally the capsules retain the same orientation as the flowers. *Nemophila* represents an exception—here pedicels elongate strongly postflorally and position the capsule directly on the ground, sometimes half burying it (MW, own observation, e.g. *Nemophila pedunculata*). This corresponds to the assumed ant-dispersal of the seed: seeds of *Nemophila* (tribe Hydrophylleae) are unique in the family in having a fleshy appendage (elaiosome) derived from the outer epidermis of the integument and attached to the chalazal end of the mature seed (Constance 1941; Pemberton and Irving 1990; Chuang and Constance 1992; Berg 2009). Crystals are present mainly in the inner subepidermal layer of the carpel wall and, to a lesser extent, in the placental ground tissue and the outer parts of the parenchymatous mesocarp or the nectary.

Seeds vary greatly in shape and ornamentation and seed sculpturing is important in differentiating genera, and in distinguishing infrageneric taxa (Brand 1913; Gray 1875). Trends towards reduction in ovule number as a derived character have been proposed for *Phacelia*, but remain to be tested comparatively across the family (Constance 1963; Atwood 1975; Ferguson 1998; Walden 2010). Seed morphology and sculpturing are rather uniform in tribes Nameae (Bacon et al. 1986a) and Hydrophylleae (Chuang and Constance 1992, with the exception of *Eucrypta*). Some species of *Eucrypta* (tribe Hydrophylleae) differ from all other Hydrophyllaceae in the production of ovules on both sides of each placenta ("bifacial placentation") and in usually having a pair of seeds retained behind one of the false valves or septa; in the rest of the genus, ovules are produced on one side of the placenta and all seeds are shed at maturity (Constance 1938). All seeds are homomorphic (corrugated) in *Eucrypta micrantha*, whereas in *E. chrysanthemifolia* the seeds are dimorphic (the proximal elliptic or round and smooth, the distal oblong-ovoid and wrinkled). Seed chemistry is largely unknown; *Hydrophyllum* seems to store mainly starch (82 %, Morgan 1971).

PHYTOCHEMISTRY. Phenolics (prenylated hydroquinones, e.g. geranyl hydroquinone) are exuded from the heads of glandular trichomes and can cause contact dermatitis (Munz 1932; Berry et al. 1962; Reynolds and Rodriguez 1979; Reynolds et al. 1980, 1986). This reaction is known from *Wigandia caracasana*, *Eriodictyon parryi* and several species of *Phacelia* (Reynolds and Rodriguez 1981a, 1981b, 1986; Reynolds et al. 1980, 1985, 1989). The glandular trichomes containing phenols are thought to be an anti-herbivory defence and an adaptation to arid environments (Rodriguez 1983, 1985; Klocke et al. 1986). Flavonols are present and varied, and have been most extensively studied in *Eriodictyon* for chemotaxonomic characters and anti-herbivory properties (Hadley and Gisvold 1944; Johnson 1983; Bacon 1986; Bacon et al. 1986a, 1986b; Bohm and Constant 1990). Bacon et al. (1986b) suggested that pubescence is correlated with presence and expression of classes of flavonoids, and proposed that there may be groups of flavonoids common across the genus, and some found only in individual species, which may suggest that the family is more "biosynthetically diverse" than currently described. Thus, alkanes were highly variable across sampled *Phacelia* (Horner and Scora 1983). Anthocyanins in petals have been studied, sometimes in concert with UV absorbance and reflectance in pollinator studies, with identification of phacelianin as responsible for the blue flower colour in *Phacelia campanularia* (Mori et al. 2006). Alkaloids are absent (six genera), iridoids have not been detected.

Subdivision and Affinities. The last family-wide monograph was published by Brand (1913). He recognized 18 genera in three tribes (Hydrophylleae, Phacelieae, Hydroleeae). Hydrophyllaceae has been called a "collection of morphological odds-and-ends and one may ask if this really is a natural family" (Constance 1963). Since then, the situation has improved with the removal of two particularly aberrant genera: *Hydrolea* (Hydroleeae) has been placed in the monogeneric Hydroleaceae in an expanded Solanales (Davenport 1988; Ferguson 1998; Erbar et al. 2005; APG III 2009). *Codon*, as the only Old World genus, is isolated both morphologically and geographically from core-Hydrophyllaceae. DNA-based systematics justifies its separation as a monogeneric Codonaceae related more to Boraginaceae s.str. and Wellstediaceae than to Hydrophyllaceae (Weigend and Hilger 2010; Weigend et al. 2014). Hydrophyllaceae are doubtfully monophyletic, but clearly nested in Boraginales (Ferguson 1998; APG III 2009) and retrieved in "Boraginales II" (i.e. Heliotropiaceae, Cordiaceae; Weigend et al. 2014) rather than in Boraginaceae s.str. Hydrophyllaceae are an entirely New World group and fall into two major clades (Ferguson 1998; Weigend et al. 2014): a clade of annual and perennial herbs including tribes Hydrophylleae and Phacelieae (*Draperia*, *Ellisia*, *Emmenanthe*, *Eucrypta*, *Hesperochiron*, *Howellanthus*, *Hydrophyllum*, *Nemophila*, *Phacelia*, *Pholistoma*, *Romanzoffia*, *Tricardia*), and a clade containing *Nama* and the shrubby genera *Eriodictyon* (incl. *Turricula*) and *Wigandia*. This second group corresponds to tribe Nameae sensu Gray, which Brand (1913) included in tribe Phacelieae in his monograph. *Eucrypta*, as currently circumscribed, is doubtfully monophyletic (Ferguson 1998). Additional studies of cryptic diversity and character evolution require a much broader sampling across species, especially in the major genera (Constance 1938; Walden 2010).[1]

Distribution and Habitats. Hydrophyllaceae are restricted to the Americas, with the majority of taxa concentrated in south-western North America. A few genera reach eastern North America (*Hydrophyllum* and *Phacelia*) and north-western North America (*Romanzoffia* and *Phacelia*), representatives of *Nama* are widespread, but not very species-rich in the Caribbean and Central America, with a single, annual species in Hawaii (Hitchcock 1933a, 1933b). Hydrophyllaceae are most diverse in semiarid areas. *Wigandia* and *Eriodictyon* are found in subtropical and tropical, rarely Mediterranean-type areas with a distinct arid season. *Wigandia* is the only truly tropical genus from Central and South America and is found in semiarid situations, e.g. in the Andes up to elevations of ca. 3000 m (Gibson 1967). South America otherwise harbours few Hydrophyllaceae. Apart from *Wigandia*, there are nine annual and perennial representatives of *Phacelia*, most of them restricted to Argentina and Chile and only three to four species reaching Bolivia and Peru (Deginani 1982). *Nama* is present in Chile, Argentina and Bolivia with three species (Hitchcock 1933a, 1933b).

The bulk of annual species are found in semi-desert and desert habitats. Perennial taxa are found on montane or alpine scree slopes (*Phacelia* and *Nama*), in more or less arid pine forests (*Hesperochiron*, some *Phacelia*), mesic forests (*Hydrophyllum*) or even humid forests and on wet cliffs (*Romanzoffia*). Overall, the family is represented from sea level to altitudes exceeding 4000 m (in the Andes), a common element in a wide variety of habitats. Edaphic endemism is found throughout Hydrophyllaceae with granite, limestone, gypsum or serpentine endemics documented (Taylor and Levy 2002). Four species of *Phacelia* plus monospecific *Howellanthus* are endemic to ultramafic (serpentine) soils, three species of *Phacelia* and *Emmenanthe penduliflora* var. *rosea* are broadly endemic on serpentine, with additional six species of *Phacelia*, *Emmenanthe penduliflora* var. *penduliflora* and *Eriodictyon californicum* tolerant of ultramafic soils (Shelly 1989; Safford et al. 2005; Walden and Patterson 2010). Gypsovags and gypsophiles are also found in *Phacelia* as well as in *Nama* (Hitchcock 1933a, 1933b, 1939; Bacon 1974; Atwood 1975).

[1] Hydrophyllaceae including tribe Nameae (*Eriodictyon* Benth., *Nama*, L., *Wigandia* Kunth) are doubtfully monophyletic. Segregation of tribe Nameae as a separate family could resolve this issue.

Economic Importance. Some Hydrophyllaceae with showy flowers are important ornamentals (e.g. species of *Nemophila*, *Phacelia*, *Wigandia*). *Phacelia tanacetifolia* is grown worldwide as an important pollinator attractant in agriculture, as a green manure crop, and in apiculture. Other species are only locally used, especially in folk medicine (with several medical applications). *Eriodictyon californicum* [yerba santa] has a long history of traditional use in folk medicine such as bitters (Ley et al. 2005).

Natural History. Hydrophyllaceae in general are not intensively browsed, due to the presence of stinging hairs (*Wigandia*) or glandular, often malodorous pubescence and the presence of secondary metabolites (Klocke et al. 1986). However, some specialized butterflies preferentially feed on their leaves and especially some ethmiid moths are specialized on *Eriodictyon* (Powell 1973; Murphy 1982). The seeds of Hydrophyllaceae sometimes germinate rapidly and without any special treatment (*Wigandia*, some *Phacelia*, own observation), but many species have erratic germination, requiring either stratification (e.g. *Phacelia franklinii*) or a variety of other seed treatments (Quick 1947; Heckard 1960; Gillett 1968; Serrato-Valenti et al. 2000). Often only some seeds of a given crop germinate at the next opportunity after dispersal (spring in perennials, autumn in winter annuals), and germination is spread over several years in some *Hydrophyllum* (Morgan 1971) and *Phacelia* (Baskin and Baskin 1973, 1983). A particular sequence of high and low temperature appears to be required for germination to take place in several species (Gillett 1960, 1968). More importantly, many species of *Phacelia*, *Nemophila*, *Eriodictyon*, *Turricula* and especially *Emmenanthe* are distinct pyrophytes and are abundant only after fire in California chaparral vegetation (Biswell and Gilman 1961; Wicklow 1977; Quinn and Keeley 2006; Walden 2010). Germination here occurs almost only after exposure to smoke. Perennial *Eriodictyon* species also resprout from perennial underground organs after fires (Sampson and Jespersen 1963), and the seedlings are very weak competitors (Schultz and Biswell 1952).

Key to the Tribes

1. Ovary unilocular, placentae enlarged, lining or filling ovary; capsule appearing 2- or 5-celled; aestivation of petals contorted **1. Tribe Hydrophylleae**
– Ovary unilocular or bilocular, placentae narrow, partially or completely dividing ovary; capsule appearing 2-celled; aestivation of petals imbricate 2
2. Stylodia 2, distinct to base or connate 3/4 of length; ovary bilocular, placentae membranaceous or cartilaginous, completely dividing ovary **2. Tribe Nameae**
– Style single; ovary unilocular or bilocular, placentae cartilaginous, partially or completely dividing ovary **3. Tribe Romanzoffieae**

Key to the Genera in Tribe Hydrophylleae

1. Plants perennial or biennial; leaves usually basal, alternate; stamens exserted **5. *Hydrophyllum***
– Plants annual; leaves usually cauline, lower usually opposite; stamens included 2
2. Herbage viscid-glandular and scented; ovules borne on both surfaces of the placentae **4. *Eucrypta***
– Herbage pubescent to hispid, but neither viscid-glandular nor strongly scented; ovules borne on adaxial surface of the placentae 3
3. Plants climbing or sprawling; stems and leaves uncinate-setose, soft and succulent; capsules hispid-prickly **6. *Pholistoma***
– Plants prostrate to erect; stems and leaves glabrate, minutely prickly to retrorse-hispid, not succulent; capsules not prickly 4
4. Stem angles glabrate to minutely prickly; calyx sinus appendaged with sepaloid auricle, sometimes reduced or rarely absent; seeds with elaiosome, rarely without **7. *Nemophila***
– Stem angles retrorsely hispid; calyx sinus appendaged without sepaloid auricle, rarely sepaloid tooth present; seeds without elaiosome **8. *Ellisia***

Key to the Genera in Tribe Nameae

1. Perennials, shrubs or small trees, 1–6 m, stamens included or exserted, equal, equally inserted on corolla tube **1. *Wigandia***
– Annuals or perennials, mat-forming, subshrubs or shrubs, to 3 m; stamens included, subequal or unequal in length, equally or unequally inserted on corolla tube 2
2. Flowers many in terminal thyrsoids, these sometimes capitate or open thyrsoidal; capsules cartilaginous, dehiscing loculicidally and septicidally with 4 valves **3. *Eriodictyon***

- Flowers apparently solitary in axils (frondose inflorescences) or in few-flowered cymoids or thyrsoids, or rarely flowers many in terminal capitate thyrsoid; capsules membranaceous or semi-cartilaginous, dehiscing loculicidally with 2 valves **2. *Nama***

Key to the Genera in Tribe Romanzoffieae

1. Style entire or slightly 2-lobed at apex; plants usually bulbous or tuberous at base, rarely bulbous base absent; leaves simple, reniform to subcircular to ovate **12. *Romanzoffia***
- Style clearly bifid or divided to base; plants neither bulbous nor tuberous at base; leaves simple to compound, linear to (ob-)ovate, rarely subcircular 2
2. Plants acaulescent, sometimes flowering stems with few reduced leaves; inflorescence one- or few-flowered reduced cymoids 3
- Plants mostly caulescent; stems more or less leafy; inflorescence few- to many-flowered cymoids 4
3. Flowers apparently solitary from basal rosette, long-pedicellate; style bifid; seeds many **14. *Hesperochiron***
- Inflorescence one- to few-flowered, peduncle present, sometimes short, with usually two bracts; style divided to base; seeds 2–4 **15. *Howellanthus***
4. Leaves opposite throughout; stamens unequally inserted on corolla tube **11. *Draperia***
- Leaves alternate, rarely lowermost leaves opposite; stamens equally inserted on corolla tube 5
5. Calyx lobes strongly unequal, the outer 3 conspicuously enlarged, cordate and conspicuously veined in fruit, the inner 2 linear **13. *Tricardia***
- Calyx lobes equal or unequal, if unequal then outer lobes not cordate 6
6. Flowers nodding, long-pedicellate; corolla white, yellow, or pink, persistent, becoming papery and enclosing the capsule; ovary with emergences; seeds flattened, oblong-elliptic, strongly alveolate-reticulate **10. *Emmenanthe***
- Flowers usually erect, subsessile to shortly pedicellate; corolla white, yellow or blue or purple, usually deciduous, rarely persistent, not papery in fruit; ovary variously glandular- and/or eglandular-pubescent; seeds usually oblong to spherical, reticulate and/or transverse rugulate, or excavated on each side of a salient ridge **9. *Phacelia***

Tribes and Genera of Hydrophyllaceae

1.**Tribe Nameae** Choisy (1846) in A.P. de Candolle and A.L.P.P. de Candolle, Prodr. 10: 182 (1846).

Leaves cauline, usually alternate, simple. Aestivation of petals imbricate; corolla scales absent; styles 2, distinct to base or connate 3/4 of length; ovary bilocular, placentae narrow, membranaceous or cartilaginous, completely dividing ovary. Capsule appearing 2-celled, dehiscing loculicidally or loculicidally and septicidally, by 2 or 4 valves.

1. ***Wigandia*** Kunth

Wigandia Kunth in Humboldt et al., Nova Gen. Sp. 3: ed. fol. 98 (1819); ed. qu. 126 (1819), nom. cons.

Perennials, shrubs or small trees, with extensive root system, usually resprouting; variously pubescent and hispid (including coarse stinging hairs) and densely glandular throughout. Leaves cauline, alternate, long-petiolate, lamina simple, ovate to oblong to elliptic, entire or margins crenately lobed, net-veined. Flowers few to many in terminal thyrsoids with boragoids as branches, rarely reduced to cymoids. Corolla campanulate, greenish-white, yellow or lavender to purple, deciduous; stamens included or exserted, equal and filaments equally inserted on corolla tube; gynoecium superior or half-inferior. Capsule cartilaginous, dehiscing loculicidally by 2 valves. Seeds many. $n = 19, 38$.

Six spp., variable and widely distributed. Tropical Mexico, Central America, South America, in disturbed habitats, often weedy, cultivated as an ornamental elsewhere and usually encountered as a persistent relic from former cultivation in western North America.

2. ***Nama*** L. Fig. 43

Nama L., Syst. Nat. ed. 10, 2: 950 (1759), nom. cons.; Hitchcock, Am. J. Bot. 20: 415–430 (1933), part. rev.; Hitchcock, Am. J. Bot. 20: 518–534 (1933), part. rev.; Hitchcock, Am. J. Bot. 26: 341–347 (1936), rev.; Bacon, Brittonia 26: 101–105 (1974), rev.; Bacon, Aliso 12: 327–333 (1989), rev.; Taylor, in Jeps. Man. ed. 2: 480 (2012), reg. rev.

Annuals, perennials, herbs or subshrubs, with slender or stout taproot, sometimes rhizomatous or woody at base; stems prostrate or mat-forming or erect, pubescent and glandular. Leaves cauline, alternate, rarely opposite, lamina narrowly (ob-) ovate or oblong. Flowers apparently solitary (frondose cymes) or in few-flowered (sometimes dichotomous) lateral or terminal cymoids or thyrsoids, or rarely flowers many in terminal capitate thyrsoid (*N. rothrockii*). Corolla tubular to infundibuliform, white to purple or lavender, deciduous; stamens included, subequal or unequal in length, subequally or unequally inserted on corolla tube, staminal appendages present or filament bases dilated; stylodia 2, distinct to base or style connate 3/4 of length (sect. *Conanthus*); gynoecium superior (half-inferior in *N. stenocarpa*). Capsule membranaceous or semicartilaginous, dehiscing loculicidally by 2 valves. Seeds 2–∞. $n = 7$, 14 or 17.

About 50 spp., arid regions of the south-western USA and northern Mexico, Hawaiian Islands, West Indies, south-western South America. Some species gypsophiles or gypsovags.

3. ***Eriodictyon*** Benth.

Eriodictyon Benth., Bot. Sulph. 35 (1844); Abrams and Smiley, Bot. Gaz. 60: 115–133 (1915), rev.; Bacon et al., Sida 11: 271–381 (1986), part. rev.; Hannan, Am. J. Bot. 74: 579–588 (1988), rev.; Hannan, in Jeps. Man. ed. 2: 470–472 (2012), reg. rev.
Turricula J.F. Macbr. (1917).

Erect perennial herbs, subshrubs or shrubs, with stout taproot and usually rhizomatous, woody at base, with shredding bark; glabrate to variously pubescent and densely viscid-glandular, scented. Leaves cauline, alternate, lower short-petiolate or sessile, upper sessile; simple, entire or dentate, abaxial surface with netted veins, some leaves wooly-tomentose throughout (*E. crassifolium*, *E. tomentosum*), or leaves glabrate and adaxially glutinous-resinous (*E. californicum*), lamina (narrowly) ovate. Flowers many in terminal thyrsoids with numerous, boragoids as branches (capitate in *E. capitatum*, open thyrsoid in *E. parryi*). Corolla infundibuliform to campanulate, white or lavender or purple, deciduous; stamens included, equal or unequal in length, equal or unequally inserted on corolla tube, staminal appendages reduced or obsolete; stylodia 2, distinct to base; gynoecium superior. Capsule cartilaginous, dehiscing loculicidally and septicidally with 4 valves (rarely indehiscent). Seeds 1–20. $n = 13$, 14.

Eleven spp., south-western USA, northern Mexico and Baja California, disturbed areas, chaparral, dry or rocky soils, sometimes on serpentine soils, sometimes abundant following fires.

2.**Tribe Hydrophylleae** Rchb., Fl. Germ. Excurs. 1 (3): 347 (1831); Constance, Madroño 5: 38–33 (1939), rev.

Petal aestivation contorted; corolla scales reduced or absent, stamens of equal length, filaments equally inserted on the corolla tube; style 1; ovary unilocular, the two large, fleshy parietal placentae filling the locule at anthesis, ovary thus appearing 2- or 5-celled; nectary at ovary base reduced to glands. Capsule membranaceous, dehiscing loculicidally with 2 valves. Mesophytes preferring shaded sites, confined to North America.

4. ***Eucrypta*** Nutt.

Eucrypta Nutt., Proc. Acad. Nat. Sci. Philadelphia 4: 12 (1848); Constance, Lloydia 1: 143–152 (1938), rev.
Ellisia L. subg. *Eucrypta* (Nutt.) A. Gray (1875).
Phacelia Juss. sect. *Helminthospermum* Torr. ex A. Gray (1875).

Annual herbs, with slender taproots; stems ascending to erect, diffusely to many-branched; variously pubescent and viscid-glandular, scented; stems terete or angled. Leaves cauline, lower opposite and petiolate, petioles narrowly winged, upper alternate and sessile, base clasping; pinnatifid, lamina oblong to ovate. Flowers rarely solitary, usually many in loose terminal or axillary thyrsoids or cymoids; pedicels thread-like, elongating in fruit. Calyx divided 2/3, sinus appendaged with sepaloid auricle; corolla campanulate, white or blue-purple, tube yellow, deciduous; corolla scales reduced and minute (*E. chrysanthemifolia*), or corolla scales absent with v-shaped folds present in upper corolla tube (*E. micrantha*), stamens included. Capsule 5-celled by large parietal placentae lining capsule, ovules borne on both surfaces of the placentae. Seeds 5–15, dimorphic in *E. chrysanthemifolia*. $n = 6$, 12, 10, 20.

Two spp., south-western USA, northern Mexico and Baja California, usually in understory or in shade of rocks.

5. ***Hydrophyllum*** L. Fig. 44

Hydrophyllum L., Sp. Pl. 1: 146 (1753); Constance, Am. Midl. Nat. 27: 710–731 (1942), rev.

Perennial or biennial herbs, roots fleshy-fibrous or tuberous, with horizontal rhizomes (except *H. appendiculatum* with taproot); erect, simple or branched; glabrate or pubescent, sometimes retrorse-hispid, unscented; stems terete or angled. Leaves basal, alternate, cauline leaves few and reduced, long-petiolate, petioles widened at base or clasping, pinnatifid or pinnate, margins usually incised or serrate, secondary veins pinnate or palmate, lamina oblong to ovate to suborbicular. Flowers several to many in terminal, lax to compact, 1–few-branched cymoids to thyrsoids. Calyx divided to base, sinus appendaged with sepaloid auricle and usually minute, conspicuous in *H. appendiculatum*; corolla campanulate, green-white or white or purple to violet, or white with purple markings, deciduous; corolla scales linear, pair bordering but free from filaments; stamens exserted, filaments hairy at the middle; style simple, distally bifid. Capsule dehiscing loculicidally with 2 valves, 2-celled by large parietal placentae lining capsule, ovules borne on adaxial surface of placentae. Seeds 1–3. $n = 9$.

Eight spp., western and eastern USA, Canada, northern Mexico; mesophytic, shaded habitats, or moist soils in the shade of rocks in semiarid habitats.

6. ***Pholistoma*** Lilja

Pholistoma Lilja in Lindblom, Bot. Not.1: 40 (1839); Constance, B. Torrey Bot. Club 66: 341–352 (1939), rev.

Annual herbs, with taproot, climbing or sprawling, many-branched, variously pubescent, sometimes glaucous in *P. membranaceum*, unscented. Stems terete and angled, angles glabrate or uncinate-prickly, succulent. Leaves cauline, lower opposite, upper alternate, petiolate, petioles clasping; lamina simple to pinnatifid, margins lobed, lamina ovate to deltoid. Flowers few to several in terminal or axillary thyrsoids or cymoids, rarely flowers apparently solitary in axils. Calyx divided to base, sinus appendaged with sepaloid auricle, rarely absent in *P. membranaceum*; corolla rotate or open campanulate, throat constricted in *P. auritum*, white with a purple spot on each lobe, white, blue or purple, deciduous; corolla scales present or reduced to pairs of glands bordering each filament, nectaries yellow and conspicuous in *P. auritum*; stamens included; style branched for 1/2 of length. Capsule dehiscing loculicidally with 2 valves, 2-celled by large parietal placentae lining capsule, ovules borne on adaxial face of the placentae. Seeds 1–8. $n = 9$.

Three spp., south-western USA, northern Mexico and Baja California, on hillsides, ravines or ocean bluffs.

7. ***Nemophila*** Nutt.

Nemophila Nutt., J. Acad. Nat. Sci. Philadelphia 2: 179 (1822), nom. cons.; Constance, Univ. Calif. Publ. Bot. 19: 341–399 (1941), rev.

Annual herbs, with slender taproot, prostrate to erect, simple to diffusely branched; glabrate or pubescent, unscented; stems terete and angled, angles glabrate or minutely prickly (e.g. *N. breviflora*). Leaves cauline, all opposite or all alternate, or the lower opposite and the upper alternate, petiolate; lamina simple, entire or lobed or incised, lamina ovate or oblong or spathulate. Flowers in lax 1–few-branched cymoids to thyrsoids with frondose bracts, or sometimes flowers few in a short, terminal cymoid with straight, elongated internodes (e.g. *N. phacelioides*); pedicels elongating in fruit. Calyx divided to base, sinus appendaged with sepaloid auricle, sometimes reduced (*N. microcalyx*) or absent (*N. pulchella*); corolla campanulate to rotate, white or blue or purple, sometimes spotted or marked, deciduous; corolla scales present as pairs at base of each filament, sometimes reduced; stamens included or exserted; style bifid to branched nearly to base. Capsule dehiscing loculicidally with 2 valves, 2-celled by large parietal placentae lining capsule, ovules borne on adaxial surface of the placentae. Seeds 1–20, elaiosome present or absent (e.g. *N. kirtleyi*), deciduous or sometimes persistent (e.g. *N. breviflora*). $n = 7, 9$.

Eleven spp., generally of western North America, to British Columbia and Baja California, with two species occurring in south-eastern USA, roadsides, grassland or woodland.

8. *Ellisia* L.

Ellisia L., Sp. Pl., ed. 2: 1662 (1763), nom. cons.; Constance, Rhodora 42: 33–39 (1940), rev.

Annual herbs, with slender taproot; ascending to erect, simple or diffusely branched, pubescent, unscented; stems terete and angled, retrorsely hispid. Leaves cauline, all opposite or lower opposite and upper alternate, petiolate, petioles narrowly winged or clasping; lamina simple, entire or margins lobed, lamina oblong to ovate. Flowers solitary in axils or opposite the alternate leaves, or flowers few in terminal cymoid with straight, elongated internodes; pedicels elongating considerably in fruit. Calyx divided to base, strongly accrescent in fruit, sinus not appendaged , rarely sepaloid tooth present; corolla narrowly campanulate, white or lavender, deciduous; corolla scales present as reduced pairs at base of each filament; stamens included; style branched for 1/2 of length. Capsule dehiscing loculicidally with 2 valves, 2-celled by large parietal placentae lining capsule, ovules borne on adaxial face of the placentae. Seeds 4–8. $n = 10$.

One species, *E. nyctelea* (L.) L., confined to the eastern and central USA, in moist, shaded sites, roadsides, edges of prairies or plains, alluvial soils, usually on cultivated or disturbed soil.

3.**Tribe Romanzoffieae** Dumort., Anal. Fam. Pl. 26 (1829).

Phacelieae Benth. ex A. Gray (1875).

Corolla aestivation imbricate; corolla scales present, reduced or absent; stamens equal or unequal in length, equally or unequally inserted on corolla tube; style 1; ovary unilocular or bilocular, placentae narrow, cartilaginous, partially or completely dividing ovary, capsule appearing 2-celled. Capsule dehiscing loculicidally with 2 valves. Nectar disk present.

9. *Phacelia* Juss.

Phacelia Juss., Gen. Pl. (Jussieu) 129 (1789); Voss, B. Torrey Bot. Club 64: 81–96 (1937), part. rev.; Voss, B. Torrey Bot. Club 64: 133–144 (1937), part. rev.; Howell, Am. Midl. Nat. 29: 1–26 (1943), part. rev.; Howell, Am. Midl. Nat. 30: 6–18 (1943), part. rev.; Howell, Am. Midl. Nat. 33: 460–494 (1945), part. rev.; Howell, Am. Midl. Nat. 36: 381–411 (1946), part. rev.; Constance, Contr. Gray Herb. 168: 1–48 (1949), part. rev.; Gillett, Univ. Calif. Publ. Bot. 28: 19–78 (1955), part. rev.; Gillett, Rhodora 62: 205–222 (1960), part. rev.; Heckard, Univ. Calif. Publ. Bot. 32: 1–126 (1960), part. rev.; Constance, Brittonia 15: 273–285 (1963), part. rev. cytol.; Gillett, Brittonia 20: 368–374 (1968), part. rev.; Atwood, Great Basin Nat. 35: 127–190 (1975), part. rev.; Halse, Madroño 28: 121–132 (1981), part. rev.; Lee, Am. J. Bot. 73: 230–235 (1986), part. rev.; Ferguson, Syst. Bot. 23: 253–268 (1998), part. rev., mol. syst.; Gilbert et al., Syst. Bot. 30: 627–634 (2005), mol. syst.; Hansen et al., Syst. Bot. 34: 737–746 (2009), mol. syst.
Eutoca R. Br. 1823.

Annual, biennial, or perennial herbs, with slender or thickened taproot or with thick woody root pleiocorm, prostrate to spreading to erect; glabrate or variously pubescent and usually glandular, unscented or scented. Leaves in basal rosette or cauline, mostly alternate (the lower sometimes opposite), petiolate or sessile; lamina reduced upwards, simple or compound, entire to pinnatifid, lamina ovate to suborbicular. Flowers few to many in dense, simple or variously branched terminal cymoids or thyrsoids, sometimes contracted and capitate (in *P. hydrophylloides* and *P. capitata*), virgate (in *P. heterophylla* var. *virgata* and *P. grandiflora*), or open thyrsoidal (e.g. species group *Franklinii*). Corolla tubular, campanulate or rotate; white or yellow or blue to purple, deciduous, sometimes persistent (e.g. *Phacelia* sect. *Miltitzia*), perianth tetramerous in *P. tetramera*; corolla scales present, reduced or absent; stamens included or exserted, equal or unequal, equally inserted on corolla tube; style bifid or divided to base; gynoecium superior or half-inferior (*P. tanacetifolia*). Seeds 1 to many. $n = 5, 6, 7, 8, 9, 10, 11, 12, 13, 14, 22, 23, 24, 33$. Western N America and northern Mexico, eastern USA (*Phacelia* subg. *Cosmanthus*), South America; wide range of habitats, from alkaline and gypsum soils to alpine scree, some cultivated as ornamentals.

About 200 spp., currently divided into three subgenera, five sections and species groups (Constance 1963; Ferguson 1998).

10. *Emmenanthe* Benth.

Emmenanthe Benth., Trans. Linn. Soc. London 17: 281 (1835).

Annual herbs, with slender taproots; erect, simple to many-branched; pubescent and viscid-glandular, scented; stems terete. Leaves basal and cauline, alternate, lower short-petiolate, upper sessile, petiole bases clasping; lamina simple to pinnatifid, ovate to oblong. Flowers many in branched, lax, scorpioid terminal thyrsoids or cymoids; long-pedicellate, pedicels filiform, nodding. Calyx divided to base; corolla campanulate, white or yellow or pink, persistent; corolla scales absent; stamens included, subequal, equally inserted on corolla tube; style bifid. Capsule bilocular, dehiscing loculicidally by 2 valves. Seeds 6–15, flattened, elaiosomes absent. $n = 18$.

One species, *E. penduliflora* Benth., south-western USA to Baja California, usually understory, common after fires, on decomposed granite, serpentine soils.

11. *Draperia* Torr.

Draperia Torr., Proc. Am. Acad. Arts 7: 401 (1868).

Perennial herbs, with stout, woody taproot or horizontal rhizomes; decumbent to erect, diffusely branched; hirsute, unscented; stems terete. Leaves cauline, opposite; lower petiolate, upper sessile; lamina simple and entire, blades ovate. Flowers many, crowded in ebracteate, pedunculate, terminal, mostly geminate scorpioid cymoid, subsessile. Calyx divided to base; corolla tubular-infundibuliform, white to pink or pale violet, deciduous; corolla scales absent; stamens included, unequal, unequally inserted on corolla tube; style bifid. Capsule bilocular, dehiscing loculicidally by 2 valves. Seeds 1–4, elaiosomes absent. $n = 9$.

One species, *D. systyla* Torr., endemic to the Californian Sierra Nevada and North Coast Ranges in woodlands, or shaded hillsides, on dry, rocky soils.

12. *Romanzoffia* Cham.

Romanzoffia Cham. in Nees, Horae Phys. Berol. 71: 1–8 (1820); Greene, Pittonia 5: 34–42 (1902), rev.; Marttala, Madroño 43: 404–414 (1996), part. rev., key.

Perennial herbs with bulbose base or annuals without root tubers, roots fibrous; erect, simple or branched; glabrate or variously pubescent, glandular, unscented. Leaves mainly from a basal rosette, cauline leaves few, reduced and alternate (lower leaves opposite in *R. thompsonii*), long-petiolate; lamina simple, entire or shallowly lobed, lamina reniform, subcircular or ovate. Flowers one to many in raceme-like, compact or lax, terminal secund cymoids, flowers sometimes replaced by bulbils in *R. californica* and *R. tracyi*. Calyx divided to base, lobes equal, slightly accrescent in fruit; corolla campanulate to infundibuliform, white, tube yellow, deciduous; corolla scales absent, sometimes v-shaped folds present on corolla tube; stamens included, unequal, equally inserted on corolla tube; style undivided or shortly bilobed at apex. Capsule bilocular, dehiscing loculicidally by 2 valves. Seeds 2–many. $n = 11$.

Five spp., temperate western N America, extending northward to the coasts of Alaska, British Columbia and Unalaska Island; on ocean bluffs, road banks, wet cliffs, or moist rocky areas.

13. *Tricardia* Torr.

Tricardia Torr., Botany (Fortieth Parallel): 258, pl. 24 (1871).

Perennial herbs, with stout, woody taproot; erect, simple or branched; hirsute below, glabrate above, unscented. Leaves mainly in basal rosette, cauline leaves few and reduced, lower shortly petiolate, upper sessile; lamina simple and entire, blade (narrowly) ovate. Flowers few in loose, terminal cymoids or thyrsoids, pedicellate. Calyx lobes divided to base, very unequal, outer 3 cordate, inner 2 linear, strongly accrescent in fruit; corolla campanulate to rotate, white to cream, generally marked lavender or purple, deciduous; corolla scales linear, paired, free from filaments; stamens included, unequal, equally inserted on corolla tube; style bifid. Capsule unilocular, dehiscing loculicidally by 2 valves. Seeds 4–8. $n = 8$.

One species, *T. watsonii* Torr., south-western USA; on sandy or gravelly slopes, desert mountains, generally in shelter of shrubs.

14. *Hesperochiron* S. Watson

Hesperochiron S. Watson, Botany (Fortieth Parallel): 281, pl. 30 (1871), nom. cons.

Acaulescent perennial herbs, with short, vertical pleiocorm; erect, hirsute, unscented. Leaves in basal rosette, petiolate; simple and entire, blade (narrowly) ovate. Flowers solitary from axils, long-pedicellate. Calyx divided to base, lobes unequal, accrescent in fruit; corolla campanulate or infundibuliform or rotate, white or pale blue, generally tinged or marked with lavender or purple, deciduous; corolla scales absent; stamens included, subequal or unequal, equally inserted on corolla tube, filaments widened at base; style bifid. Capsule unilocular, dehiscing loculicidally by 2 valves. Seeds many. $n = 8$.

Two spp., south-western USA, Baja California, northern Mexico; springs, wet meadows, and moist stream banks.

15. ***Howellanthus*** (Constance) Walden & R. Patt.

Howellanthus (Constance) Walden & R. Patt., Madroño 57: 270 (2010).
Phacelia subg. *Howellanthus* Constance (1953).

Acaulescent perennial herbs, with short pleiocorm; erect, hirsute and viscid-glandular, unscented. Leaves in basal rosette, flowering stem leaves few and reduced, petiolate; simple and entire, blade oblong to elliptic. Flowers pedicellate, solitary or few flowered sessile cymoids in leaf axils. Calyx divided to base, lobes unequal, accrescent in fruit; corolla campanulate to rotate, white, purple-marked, deciduous; corolla scales semicircular, paired, free from filament base; stamens included or slightly exserted, equal, equally inserted on corolla tube, filaments narrow at base; stylodia 2, distinct. Capsule unilocular, dehiscing loculicidally by 2 valves. Seeds 2–4, elaiosomes absent. $n = 8$.

One species, *H. dalesianus* (J.T. Howell) Walden & R. Patt., endemic to northern California; ultramafic soils, forest meadows or flats.

Selected Bibliography

Anonymous. 2007. Status of pollinators in North America. National Research Council (U.S.). Committee on the Status of Pollinators in North America. National Academy of Science, Washington DC, 307 pp.

APG III 2009. An update of the Angiosperm Phylogeny Group classification for the orders and families of flowering plants. Bot. J. Linn. Soc. 161: 105–121.

Armstrong, V.A. 1992. Site characteristics and habitat requirements of the endangered clay phacelia (*Phacelia argillacea* Atwood, Hydrophyllaceae) and aspects of the biology of *Phacelia argillacea* Atwood (Hydrophyllaceae), an annual. M.S. Thesis, Biology, Brigham Young University, Provo UT.

Atwood, N.D. 1975. A revision of the *Phacelia crenulata* group (Hydrophyllaceae) for North America. Great Basin Nat. 35: 127–191.

Bacon, J.D. 1974. Chromosome numbers and taxonomic notes on the genus *Nama* (Hydrophyllaceae). Brittonia 26: 101–105.

Bacon, J.D. 1986. Chemosystematics of the Hydrophyllaceae: flavonoids of three species of *Eriodictyon*. Biochem. Syst. Ecol. 14: 591–595.

Bacon, J.D., Bragg, L.H., Hannan, G.L. 1986a. Systematics of *Nama* (Hydrophyllaceae): Comparison of seed morphology of sects. *Arachnoidea* and *Cinerascentia* with five species of *Eriodictyon* and *Turricula parryi*. Sida 11: 271–281.

Bacon, J.D., Fang, N., Mabry, T.J. 1986b. Systematics of *Nama* (Hydrophyllaceae): flavonoids and phyletic position of sect. *Arachnoidea* and sect. *Cinerascentia*. Plant Syst. Evol. 151: 223–228.

Baskin, J.M., Baskin, C.C. 1973. Delayed germination in seeds of *Phacelia dubia* var. *dubia*. Can. J. Bot. 51: 2481–2486.

Baskin, J.M., Baskin, C.C. 1983. Germination ecophysiology of eastern deciduous forest herbs: *Hydrophyllum macrophyllum*. Am. Midl. Nat. 109: 63–71.

Beckmann, R.L. 1979. Biosystematics of the genus *Hydrophyllum* L. (Hydrophyllaceae). Am. J. Bot. 66: 1053–1061.

Berg, R.Y. 1985. Gynoecium and development of embryo sac, endosperm, and seed in *Pholistoma* (Hydrophyllaceae) relative to taxonomy. Am. J. Bot. 72: 1775–1787.

Berg, R.Y. 2009. Embryo sac, endosperm, and seed of *Nemophila* (Boraginaceae) relative to taxonomy, with a remark on embryogeny in *Pholistoma*. Am. J. Bot. 96: 565–579.

Berry, C.Z., Dahlen, R.F., Shapiro, S.I. 1962. Dermatitis venenata from *Phacelia crenulata*. Arch. Dermatol. 85: 737.

Biswell, H.H., Gilman, J.H. 1961. Brush management in relation to fire and other environmental factors on the Tehama deer winter range. Calif. Fish Game 47: 357–389.

Bohm, B.A., Constant, H. 1990. Leaf surface flavonoids of *Eriodictyon trichocalyx*. Biochem. Syst. Ecol.18: 491–492.

Brand, A. 1913. Hydrophyllaceae. In: Pflanzenreich 59, IV. 251. Leipzig: W. Engelmann. 210 pp.

Carlquist, S. 1992. Wood anatomy of sympetalous dicotyledon families: A summary, with comments on systematic relationships and evolution of the woody habit. Ann. Missouri Bot. Gard.79: 303–332.

Carlquist, S., Eckhart, V.M. 1984. Wood anatomy of Hydrophyllaceae. II. Genera other than *Eriodictyon*, with comments on parenchyma bands containing vessels with large pits. Aliso 10: 527–546.

Carlquist, S., Eckhart, V.M., Michener, D.C. 1983. Wood anatomy of Hydrophyllaceae. I. *Eriodictyon*. Aliso 10: 397–412.

Cave, M.S., Constance, L. 1950. Chromosome numbers in the Hydrophyllaceae: IV. Univ. Calif. Publ. Bot. 23: 363–382.

Chuang, T.I., Constance, L. 1992. Seeds and systematics in Hydrophyllaceae: Tribe Hydrophylleae. Am. J. Bot. 79: 257–264.

Constance, L. 1938. The genus *Eucrypta* Nuttall. Lloydia 1: 143–152.

Constance, L. 1941. The genus *Nemophila* Nuttall Univ. Calif. Publ. Bot. 19: 341–398.

Constance, L. 1963. Chromosome number and classification in Hydrophyllaceae. Brittonia 15: 273–285.

Constance, L., Chuang, T.I. 1982. SEM survey of pollen morphology and classification in Hydrophyllaceae (waterleaf family). Am. J. Bot. 69: 40–53.

Cruden, R.W. 1972a. Information on chemistry and pollination biology relevant to the systematics of *Nemophila menziesii* (Hydrophyllaceae). Madroño 21: 505–516.

Cruden, R.W. 1972b. Pollination biology of *Nemophila menziesii* with comments on the evolution of oligolectic bees. Evolution 26: 373–389.

Davenport, L.J. 1988. A monograph of *Hydrolea* (Hydrophyllaceae). Rhodora 90: 169–208.

Deginani, N.B. 1982. Revisión de las especies argentinas del género *Phacelia* (Hydrophyllaceae). Darwiniana 24: 405–496.

del Castillo, R.F. 1993. Consequences of male sterility in *Phacelia dubia*. Evol. Trend Plants 7: 15–22.

del Castillo, R.F. 1994. Factors influencing the genetic structure of *Phacelia dubia*, a species with a seed bank and large fluctuations in population size. Heredity 72: 446–458.

del Castillo, R.F. 1998. Fitness consequences of maternal and nonmaternal components of inbreeding in the gynodioecious *Phacelia dubia*. Evolution 52: 44–60.

del Castillo, R.F., Trujillo, S. 2009. Evidence of restoration cost in the annual gynodioecious *Phacelia dubia*. Evol. Biol. 22: 306–313.

Des Granges, J. 1979. Organization of a tropical nectar feeding bird guild in a variable tropical environment. Living Bird 17: 199–236.

Di Fulvio, T.E. 1987. La endospermogénesis en Hydrophylleae (Hydrophyllaceae) con relación a la taxonomía. Kurtziana 19: 13–34.

Di Fulvio, T.E. 1989. Embriología de *Nama jamaicense* (Phacelieae, Hydrophyllaceae). Kurtziana 20: 9–31.

Di Fulvio, T.E. 1993. Embriología de *Lemmonia californica* (Phacelieae). Kurtziana 22: 19–30.

Di Fulvio, T.E., Dottori, N. 1995. Contribución al conocimiento de tricomas y emergencias en Hydrophyllaceae. Clasificación y consideraciones taxonómicas. Kurtziana 24: 19–24.

Eckhart, V.M. 1991. The effects of floral display on pollinator visitation vary among populations of *Phacelia linearis* (Hydrophyllaceae). Evol. Ecol. 5: 370–384.

Eckhart, V.M. 1992. The genetics of gender and the effects of gender on floral characters in gynodioecious *Phacelia linearis* (Hydrophyllaceae). Am. J. Bot. 79: 792–800.

Elam, D.R. 1994. Genetic variation and reproductive output in plant populations: effects of population size and incompatibility (*Lilium parryi*, *Lilium humboldtii*, *Raphanus sativus*, *Eriodictyon capitatum*). Ph.D. Dissertation, University of California, Riverside CA, 221 pp.

Erbar, C., Porembski, S., Leins, P. 2005. Contributions to the systematic position of *Hydrolea* (Hydroleaceae) based on floral development. Plant Syst. Evol. 252: 71–83.

Ferguson, D.M. 1998. Phylogenetic analysis and relationships in Hydrophyllaceae based on *ndhF* sequence data. Syst. Bot. 23: 253–268.

Ganders, F.R. 1978. The genetics and evolution of gynodioecy in *Nemophila menziesii* (Hydrophyllaceae). Can. J. Bot. 56: 1400–1408.

Gibson, D.N. 1967. Hydrophyllaceae. In: Macbride, J.F. (ed.) Flora of Peru. Publ. Field Mus. Nat. Hist., Bot. Ser. 13, part V–A, no. 2: 101–131.

Gilbert, C., Dempcy, J.M., Ganong, C.K., Patterson, R., Spicer, G.S. 2005. Phylogenetic relationships within *Phacelia* subgenus *Phacelia* (Hydrophyllaceae) inferred from nuclear rDNA ITS sequence data. Syst. Bot. 30: 627–634.

Gillett, G.W. 1955. Variation and genetic relationships in the *Whitlavia* and *Gymnobythus* phacelias. Univ. Calif. Publ. Bot. 28: 19–78.

Gillett, G.W. 1960. A systematic treatment of the *Phacelia franklinii* group. Rhodora 62: 205–222.

Gillett, G.W. 1968. Systematic relationships in the *Cosmanthus* phacelias (Hydrophyllaceae). Brittonia 20: 368–374.

Gray, A. 1875. A conspectus of the North American Hydrophyllaceae. Proc. Am. Acad. Arts Sci. 10: 312–332.

Greene, E.L. 1902. Revision of *Romanzoffia*. Pittonia 5: 34–42.

Hadley, W.J., Gisvold, O. 1944. A phytochemical study of *Eriodictyon angustifolium*, Nutt. J. Am. Pharm. Assoc. Sci. Ed. 33: 275–277.

Halse, R.R. 1979. Taxonomy of *Phacelia* sect. *Miltitzia* (Hydrophyllaceae). Ph.D. Dissertation, Biology, Oregon State University, Corvallis OR.

Halse, R.R. 1981. Taxonomy of *Phacelia* sect. *Miltitzia* (Hydrophyllaceae). Madroño 28: 121–132.

Hansen, D.R., Spicer, G.S., Patterson, R. 2009. Phylogenetic relationships between and within *Phacelia* sections *Whitlavia* and *Gymnobythus* (Boraginaceae). Syst. Bot. 34: 737–746.

Heckard, L.R. 1956. The subspecies of *Phacelia nemoralis* Greene. Leafl. W. Bot. 8: 29–32.

Heckard, L.R. 1960. Taxonomic studies in the *Phacelia magellanica* polyploid complex. Univ. Calif. Publ. Bot. 32: 1–126.

Hilger, H.H. 1987. Flower and fruit development in *Wigandia caracasana* (Hydrophyllaceae). Am. J. Bot. 74: 250–259.

Hitchcock, C.L. 1933a. A taxonomic study of the genus *Nama* I. Am. J. Bot. 20: 415–430.

Hitchcock, C.L. 1933b. A taxonomic study of the genus *Nama* II. Am. J. Bot. 20: 518–534.

Hitchcock, C.L. 1939. The perennial Mexican namas. Am. J. Bot. 26: 341–347.

Hofmann, M. 1999. Flower and fruit development in the genus *Phacelia* (Phacelieae, Hydrophyllaceae): Characters of systematic value. Syst. Geogr. Plants 68: 203–212.

Horner, P.F. 1977. Biosystematics of the *Phacelia distans* – *P. tanacetifolia* complex. Ph.D. Dissertation, Botany, University of California Riverside, Riverside, CA.

Horner, P., Scora, R.W. 1983. Leaf hydrocarbons of *Phacelia* species (Hydrophyllaceae). Phytochemistry 22: 2489–2491.

Horner-Till, S. 1982. Experimental insights on factors effecting the distribution of *Phacelia cookei* Constance & Heckard, a narrow endemic. M.S. Thesis, Biology, Utah State University.

Johnson, N.D. 1983. Flavonoid aglycones from *Eriodictyon californicum* resin and their implications for herbivory and UV screening. Biochem. Syst. Ecol. 11: 211–215.

Johnson, N.D., Brain, S.A., Ehrlich, P.R. 1985. The role of leaf resin in the interaction between *Eriodictyon californicum* (Hydrophyllaceae) and its herbivore, *Trirhabda diducta* (Chrysomelidae). Oecologia 66: 106–110.

Klocke, J.A., Vanwagenen, B., Balandrin, M.F. 1986. Biologically active constituents of North American plants. 2. The ellagitannin geraniin and its hydrolysis products as insect growth inhibitors from semiarid land plants. Phytochemistry 25: 85–91.

Kruckeberg, A.R. 1956. Notes on the *Phacelia magellanica* complex in the Pacific Northwest. Madroño 13: 209–221.

Levy, F. 1988. Effects of pollen source and time of pollination on seed production and seed weight in *Phacelia dubia* and *P. maculata* (Hydrophyllaceae). Am. Midl. Nat. 119: 193–198.

Ley, J.P., Krammer, G., Reinders, G., Gatfield, I.L., Bertram, H.J. 2005. Evaluation of bitter masking flavanones from Herba Santa (*Eriodictyon californicum* (H.&A.) Torr., Hydrophyllaceae). J. Agric. Food Chem. 53: 6061–6066.

Mayrose, I., Zhan, S.H., Rothfels, C.J., Magnuson-Ford, K., Barker, M.S., Rieseberg, L.H., Otto, S.P. 2011. Recently formed polyploid plants diversify at lower rates. Science 333(6047): 1257.

McCall, A.C. 2006. Natural and artificial floral damage induces resistance in *Nemophila menziesii* (Hydrophyllaceae) flowers. Oikos 112: 660–666.

Moldenke, A.R. 1976. California pollination ecology and vegetation types. Phytologia 34: 304–361.

Morgan, M.D. 1971: Life history and energy relationships of *Hydrophyllum appendiculatum*. Ecol. Monogr. 41: 329–349.

Mori, M., Kondo, T., Toki, K., Yoshida, K. 2006. Structure of anthocyanin from the blue petals of *Phacelia campanularia* and its blue flower color development. Phytochemistry 67: 622–629.

Munz, P.A. 1932. Dermatitis produced by *Phacelia* (Hydrophyllaceae). Science 76(1965): 194.

Murphy, D.D. 1982. On the status of *Euphydryas editha baroni* with a range extension of *E. editha luesterae*. J. Res. Lepidoptera 21: 194–198.

O'Neill, K.M. 2001. Solitary Wasps: Behaviour and Natural History. Ithaca, London: Comstock Publishing Associates. 406 pp.

Pemberton, R.W., Irving, D.W. 1990. Elaiosomes on weed seeds and the potential for myrmecochory in naturalized plants. Weed Sci. 38: 615–619.

Powell, J.A. 1973. A Systematic Monograph of New World Ethmiid Moths (Lepidoptera: Gelechioidea). Washington DC: Smithsonian Institution Press. 302 pp.

Quick, C.R. 1947. Germination of *Phacelia* seeds. Madroño 9: 17–20.

Quinn, R.D., Keeley, S.C. 2006. Introduction to California Chaparral. Berkeley, CA: University of California Press, 304 pp.

Reynolds, G., Rodriguez, E. 1979. Geranylhydroquinone: A contact allergen from trichomes of *Phacelia crenulata*. Phytochemistry 18: 1567–1568.

Reynolds, G., Rodriguez, E. 1981a. Prenylated hydroquinones: contact allergens from trichomes of *Phacelia minor* and *P. parryi*. Phytochemistry 20: 1365–1366.

Reynolds, G., Rodriguez, E. 1981b. Prenylated phenols that cause contact dermatitis from trichomes of *Phacelia ixodes*. Planta Med. 43: 187–193.

Reynolds, G., Rodriguez, E. 1986. Dermatotoxic phenolics from glandular trichomes of *Phacelia campanularia* and *P. pedicellata*. Phytochemistry 25: 1617–1619.

Reynolds, G., Epstein, W., Terry, D., Rodriguez, E. 1980. A potent contact allergen of *Phacelia* (Hydrophyllaceae). Contact Dermatitis 6: 272–274.

Reynolds, G.W., Proksch, P., Rodriguez, E. 1985. Prenylated phenolics that cause contact dermatitis from glandular trichomes of *Turricula parryi*. Planta Med. 51: 494–498.

Reynolds, G.W., Epstein, W., Rodriguez, E. 1986. Unusual contact allergens from plants in the family Hydrophyllaceae. Contact Dermatitis 14: 39–44.

Reynolds, G.W., Gafner, F., Rodriguez, E. 1989. Contact allergens of an urban shrub *Wigandia caracasana*. Contact Dermatitis 21: 65–68.

Rittenhouse, B. 1993. Observation on the pollination of silvery phacelia. Bull. Native Plant Soc. Oregon 26: 10.

Rodriguez, E. 1983. Cytotoxic and insecticidal chemicals of desert plants. ACS Symp. Ser. 208: 291–302.

Rodriguez, E. 1985. Insect feeding deterrents from semiarid and arid land plants. ACS Symp. Ser. 276: 447–453.

Safford, H.D, Viers, J.H., Harrison, S.P. 2005. Serpentine endemism in the California flora: a database of serpentine affinity. Madroño 52: 222–257.

Sampson, A.W. Jespersen, B.S. 1963. California range brushlands and browse plants. Berkeley, CA: University of California, Division of Agricultural Sciences, California Agricultural Experiment Station, Extension Service, 162 pp.

Schultz, A.M., Biswell, H.H. 1952. Competition between grasses reseeded on burned bushlands in California. J. Range Manage. 5: 338–345.

Schwarzer, C. 2007. Systematische Untersuchungen an den peruanischen Vertretern der Gattungen *Pectocarya* DC. ex Meisn., *Amsinckia* Lehm., *Plagiobothrys* Fisch. & C.A. Mey. und *Cryptantha* Lehm. ex G. Don (Boraginaceae). Diploma Thesis, Systematic Botany, FU Berlin.

Serrato-Valenti, G., Cornara, L., Modenesi, P., Piana, M., Mariotti, M.G. 2000. Structure and histochemistry of embryo envelope tissues in the mature dry seed and early germination of *Phacelia tanacetifolia*. Ann. Bot. 85: 625–634.

Sewell, M., Vincent, M.A. 2009. Biosystematics of the *Phacelia ranunculacea* complex (Hydrophyllaceae). Castanea 71: 192–209.

Shelly, J.S. 1989. Biosystematic studies of *Phacelia capitata* (Hydrophyllaceae), a species endemic to

serpentine soils in southwestern Oregon. Madroño 36: 232–247.
Svensson, H.G. 1925. Zur Embryologie der Hydrophyllaceen, Borraginaceen und Heliotropiaceen mit besonderer Rücksicht auf die Endospermbildung. Uppsala Univ. Årsskr. 2: 1–176.
Taylor, S.I., Levy, F. 2002. Responses to soils and a test for preadaptation to serpentine in *Phacelia dubia* (Hydrophyllaceae). New Phytol. 155: 437–447.
Thorp, R.W., Horning, D.S., Dunning, L.L. 1983. Bumble bees and cuckoo bumble bees of California (Hymenoptera, Apidae). Berkeley, CA: University of California Press, 79 pp.
Walden, G.K. 2010. Phylogeny of infrageneric relationships within *Phacelia* (Boraginaceae) inferred from chloroplast sequence data. M.S. Thesis, Biology, San Francisco State University, San Francisco.
Walden, G.K., Patterson, R. 2010. *Howellanthus dalesianus*, recognition of a new genus and species in tribe Phacelieae (Boraginaceae). Madroño 57: 268–273.
Weigend, M., Hilger, H.H. 2010. Codonaceae - a newly required family name in Boraginales. Phytotaxa 10: 26–30.
Weigend, M., Luebert, F., Gottschling, M., Couvreur, T.L. P., Hilger, H.H., Miller, J. 2014. From capsules to nutlets—phylogenetic relationships in the Boraginales. Cladistics 30: 508–518.
Wicklow, D.T. 1977. Germination response in *Emmenanthe penduliflora* (Hydrophyllaceae). Ecology 58: 201–205.
Wolfe, L.M., Shore, J.S. 1992. The mating system of *Hydrophyllum appendiculatum,* a protandrous species. Sex. Plant Reprod. 5: 239–245.

Icacinaceae

Icacinaceae (Benth.) Miers, Ann. Mag. Nat. Hist. II, 8:174 (1851), nom. cons.

M.J. Potgieter and R. Duno

Trees, shrubs or lianas, indumentum generally present. Large tubers rarely present. Leaves alternate, rarely opposite, simple, rarely lobed, exstipulate; petioles canaliculate; domatia present or absent, when present pits or pockets; blades coriaceous to sub-coriaceous, rarely membranaceous; margins entire or dentate, slightly revolute; penninerved, rarely palmatinerved. Inflorescences axillary, terminal or cauliflorous, generally cymose or racemose, with one to many bracts, rarely ebracteate; pedicels generally articulated with the flowers. Flowers hypogynous, commonly actinomorphic, rarely zygomorphic; bisexual or unisexual, at least functionally; flowers mostly pentamerous, but in *Polycephalium* and some *Pyrenacantha* 3-merous, in *Polyporandra* up to 7-merous; sepals (3–)5(6), mostly connate below to a varying degree and 4–6-lobed or -toothed, rarely free, imbricate or valvate, fleshy, generally glabrous; petals 4–6, free or united below to a varying degree, pubescent or glabrous, valvate, rarely subimbricate, generally reflexed; apex inflexed, rarely extended into an appendage; stamens 4–6, alternating with the petals; filaments filiform, fleshy, or sometimes broad and flattened; anthers 2(−4)-locular, dorsi- or basifixed, longitudinally dehiscent, introrse, less often latrorse or more or less extrorse; connective sometimes laterally expanded or extended into an appendage; staminodes sometimes present in pistillate flowers; disk present or absent, either annular or cup-like, free or adnate to the ovary, sometimes reduced to a unilateral fleshy scale; pistil pubescent or glabrous, 1(2–3)-locular, rarely pubescent inside, rudimentary in staminate flowers; ovules 2 per locule, pendant from near the apex, collateral or superposed, anatropous, apotropous, unitegmic; style simple, very rarely 2 or 3 stylodia (*Casimirella*); stigma capitate, punctiform, subcapitate or peltate, entire, 2–5-lobed or crenate or papillose. Fruit a drupe, ovoid, oblong, globose or rarely flattened and with fleshy appendage on one side; endocarp thin and crustaceous to thick and woody, sometimes spongious or fibrous, often veined or ribbed longitudinally. Seed 1, exarillate; embryo commonly minute; endosperm copious, usually non-ruminate.

Thirty five genera with about 245 species, distributed mainly in the tropics of both the Old and New World, rapidly decreasing in number towards the subtropics.

Vegetative Morphology. Most members of the family are trees, or sometimes shrubs with scandent branches or lianas. Branchlets and petioles are puberulous or glabrous. Leaves are simple, alternate, rarely opposite to subopposite (*Cassinopsis*, *Iodes*, *Mappianthus*), petiolate and exstipulate; the petiole sometimes is prehensile in *Pyrenacantha*. Leaves of *Pyrenacantha* vary markedly in shape, ranging from relatively large and circular to palmately lobed or elliptic (Potgieter 1994). The lamina is penninerved or rarely palmatinerved (*Natsiatopsis*, *Natsiatum*, *Polycephalium*, some *Pyrenacantha*, *Phytocrene*), and the leaf margin is entire to dentate, rarely the juvenile leaves are spinose-tipped or -margined. Secondary veins fuse in a series of marginal loops. The leaves have pocket domatia in some genera, either with or without hairs (*Alsodeiopsis*; Kårehed 2001). The leaves of *Mappia* commonly have pores but never with internal hairs (Kårehed 2001). Bifid tendrils at or close to the nodes are present in *Iodes* (Villiers 1973) and *Mappianthus*, and extra-axillary or intrapetiolar tendrils in *Polyporandra*.

J.W. Kadereit and V. Bittrich (eds.), *Flowering Plants. Eudicots*, The Families and Genera of Vascular Plants 14,
DOI 10.1007/978-3-319-28534-4_21,

Vegetative Anatomy. The vegetative anatomy of the family is extremely diverse and apparently no character of leaf, branch, wood or bark anatomy is consistent or even typical of the family at least in its traditional circumscription (Sleumer 1971). Indeed, Potgieter (1994) found that supposedly species-specific characters of the epidermis and lamina in the southern African species of *Apodytes* and *Pyrenacantha* are rather variable.

A mucilaginous epidermis is present in *Apodytes* (Metcalfe and Chalk 1950; Napp-Zinn 1973). Potgieter and van Wyk (1992) reported the presence of intercellular pectic protuberances in species of *Apodytes* and *Cassinopsis*. Minor leaf veins lack phloem transfer cells in *Cassinopsis*, and small, irregularly-shaped, yellowish cell inclusions of unknown chemical composition and function are present beneath the abaxial epidermis in some species of this genus (Potgieter and van Wyk 1999).

Stomatal types include anisocytic, anomocytic, cyclocytic, helicocytic, paracytic and several intermediate types. For example, *Cassinopsis*, *Iodes* (van Staveren and Baas 1973) and *Pyrenacantha* have cyclocytic stomata as opposed to the anomocytic type of *Alsodeiopsis* and *Apodytes*. A peristomal rim may be present in *Cassinopsis*. Stomata occur on the abaxial leaf surface, but in some genera are found near veins on the adaxial surface. The stomatal index can be extremely variable within species (van Staveren and Baas 1973; Baas 1974). Hydathodes are present in some species of *Chlamydocarya*, *Polycephalium* and *Pyrenacantha*. Ten types of hairs are found in the family and include, amongst others, normal hairs which occur in nine genera, glandular hairs (*Platea* only), and peltate-stellate hairs, also present in *Platea* but not found in any other genus. There is no variation of indumentum within genera but, among genera, indumentum type may have diagnostic value. The indumentum is found on all parts of the plant and may persist or plants are glabrescent. The normal hairs are unicellular and usually solitary or clustered in some genera (Heintzelman and Howard 1948). In *Sleumeria* the indumentum is distinctly bright yellow (Utteridge et al. 2005).

Crystals are abundant in all genera. Rhombic crystals are less common but found in *Alsodeiopsis*, *Desmostachys*, *Icacina*, *Lavigeria* and *Rhaphiostylis*. Crystal sand in *Lavigeria* (Edelhoff 1887) is usually associated with the midrib (Heintzelman and Howard 1948). Wreathing of the sclerenchyma in the bark phloem by cells containing calcium oxalate crystals occurs in *Apodytes* where also cells with prismatic crystals were found. *Cassinopsis* has druses and chambered crystalliferous strands (Potgieter and van Wyk 1994d).

Nodes are trilacunar or unilacunar. Most of the trilacunar representatives of the family have an arborescent habit, and several of the unilacunar genera have a tendency towards a scrambling, twining or climbing habit. The trilacunar genera have vessels either with scalariform (*Apodytes*, *Cassinopsis*) or with porous perforations (*Alsodeiopsis*, *Chlamydocarya*, *Sleumeria*). *Apodytes* and *Cassinopsis* have large, conspicuously bordered pits. Unilacunar genera have porous perforations in the vessels of the secondary xylem. The wood parenchyma is a mixture of apotracheal and paratracheal types (*Alsodeiopsis*, *Mappia*, *Merrilliodendron*, *Nothapodytes*; Bailey and Howard 1941c).

Secondary thickening starts from a conventional cambial ring or is anomalous: *Pyrenacantha* has alternating wedges of xylem and phloem, and *Chlamydocarya* has interxylary phloem. Successive cambia are present in certain species of *Icacina*. The xylem is with or without tracheids and fibre tracheids. Tracheid walls are very thick in *Apodytes* (Carlquist 1992). Aggregate rays are present in some species of *Alsodeiopsis*, multiseriate rays in *Apodytes*, and uniseriate rays in *Desmostachys*. Bark fibres associated with sclereids or sclereids only are present in *Apodytes*, and fibres are abundant or sparse in *Cassinopsis* (Potgieter and van Wyk 1994d).

Inflorescence Structure. The inflorescence is generally axillary, sometimes terminal. It is very variable and complex, ranging from one or two flowers (*Ottoschulzia*, *Hosiea*, *Sleumeria*) to racemes, thyrses, panicles, spikes or cymes. Inflorescences are cauliflorous in *Chlamydocarya*, *Stachyanthus* and some members of *Merrilliodendron*. The dichasially branched axillary inflorescences of *Cassinopsis* alternate between nodes.

Flower Structure. Flowers are small, mostly odourless, regular and tetracyclic. The pedicel is generally articulated at the base of the flower (not in *Pleurisanthes*), and sometimes the flowers are sessile in *Desmostachys* and *Lavigeria*. They are generally bisexual, but sometimes functionally unisexual or completely unisexual with only stamens or pistils (*Calatola*, *Iodes*, *Platea*, *Ryticaryum*).

The calyx is distinct (absent in *Pyrenacantha*), (3)4–5(6)-merous, connate and tubular or with small imbricate lobes. It rarely is accrescent as in *Oecopetalum* and *Sarcostigma*. Petals are free, rarely sympetalous to a varying degree, usually valvate; the petal apex is usually inflexed. Petals are pubescent on the abaxial side in *Desmostachys*, adaxially in *Icacina* [and others], and persistent in fruit in *Pyrenacantha*.

A hypogynous disk is usually absent but sometimes present (*Icacina*, *Mappia* and others) and either extra- or intrastaminal.

The androecium consists of (3)4–5(6) fertile stamens. Stamens are free or adnate to the corolla tube, all equal, and epipetalous or rarely episepalous (*Rhaphiostylis*). The filaments are either filiform or often fleshy, flattened, subulate or thick in *Apodytes*, and sometimes pubescent. Anthers dehisce by longitudinal slits. They usually are introrse and unappendaged; they are minute in *Pyrenacantha*. The anther connectives touch and form a ring around the ovary in *Sleumeria*. In *Polyporandra* the thick anthers are many-celled with numerous small pollen-bearing alveoles, each of which is covered with a caducous membrane.

The gynoecium consists of (2–)3(–5) carpels, is syncarpous and superior, glabrous or pubescent in *Alsodeiopsis*, *Cassinopsis*, *Chlamydocarya*, *Desmostachys* p.p., *Lavigeria* and *Pyrenacantha*. The ovary is 1-locular, rarely 2- or 3-locular, with two narrow swellings at the top in *Rhaphiostylis*. Usually there are 2, rarely one ovule per locule; ovules are pendant, anatropous and suspended from the apex of the ovary. One of the two ovules of *Apodytes*, *Cassinopsis* (Potgieter 1994) and *Oecopetalum* is situated higher in the locule. There is one apical style attenuate from the ovary. It is usually short (long in *Desmostachys*), ± eccentric in *Apodytes*, persistent in *Cassinopsis* and *Apodytes*, and sometimes pubescent in *Icacina*. Stigma capitate, punctiform, subcapitate or peltate, entire, 2–5-lobed or crenate or papillose.

Embryology. The information provided here is mainly from Mauritzon (1936) and Fagerlind (1945). The endothecium has fibrous thickenings, microsporogenesis is simultaneous, microspore tetrads are tetrahedral, the anther wall initially has one or more middle layers, and the tapetum is glandular. Pollen grains are binucleate.

One of the two ovules occasionally is aborted. Ovules are unitegmic and tenuinucellate (Mauritzon 1936). *Apodytes* has unitegmic-crassinucellate, and *Cassinopsis* unitegmic-semicrassinucellate ovules. The integument is more than 10 cells thick in *Cassinopsis*, and 7–10 cells in *Pyrenacantha*. The micropyle is long. Placentation usually is apical but axile in plurilocular ovaries. A small obturator develops from the placenta (Johri et al. 1992).

Embryo sac development is of the Polygonum type. Polar nuclei fuse prior to fertilization, 3 antipodal cells are formed which do not proliferate and are ephemeral or persistent. Synergids are hooked (with filiform apparatus).

Pollen Morphology. Pollen grains are ellipsoidal, spheroidal or oblately spheroidal and oblate, rarely prolate, sometimes triangular from polar view (especially in the oblate grains) (Erdtman 1966). Their size can be up to approximately 100 μm in *Desmostachys* and *Nothapodytes*. Pollen grains mostly are 3(–7)-aperturate (colpate, colporoidate, colporate, pororate, porate, forate) or nonaperturate (*Stachyanthus*). The exine varies from foveolate (*Cassinopsis*), foveate (*Lavigeria*), foveate-reticulate (*Apodytes, Icacina*) to echinate (*Iodes*, *Polycephalium*; Dahl 1952). An endexine can be absent or present (Lobreau-Callen 1975). Lobreau-Callen (1969) stated that the endexine, when present, is cracked (*Mappianthus*, *Miquelia*), structured (most genera) or irregular (*Alsodeiopsis*, *Miquelia*). Lobreau-Callen (1975) also noted that icacinaceous genera with dioecious or polygamous-dioecious flowers (e.g. *Pyrenacantha*) have pollen grains with very thin and long and often unequal columellae, or with bacula. Genera with bisexual flowers have pollen with more or less short to long columellae of one type. The tectum of the latter type of pollen grain is very often perforated (Lobreau-Callen 1973).

Karyology. Of the few chromosome counts available, most come from Old World genera:

Apodytes $n = 12$, *Calatola* $n = 14$ (Vera Caletti 1999), *Chlamydocarya* $n = 11$, *Iodes* $n = 10$, *Icacina* $n = 24$, *Pyrenacantha* $n = 20$, *Rhaphiostylis* $n = 20$, *Stachyanthus* $n = 20$.

Pollination and Reproductive Systems. Little is known about the pollination process and reproductive system. Plants are usually hermaphrodite, more rarely andromonoecious, gynomonoecious, dioecious (especially the lianas), androdioecious, gynodioecious or polygamomonoecious. In *Nothapodytes amamianus*, three species of Diptera and five species of Lepidoptera visit flowers to collect nectar (Nagamasu and Kato 2004), and in *Nothapodytes nimmoniana* flies and bees were frequently observed taking nectar, but only bees carried pollen (Sharma et al. 2010). The flowers of *Icacina oiliviformis* stay open for several days. They are pollinated by a few species of bees, most collecting only pollen, *Meliponula bocandei* also collecting nectar (Fay 1993).

Fruit and Seed. The fruit is a subglobose or globose, obliquely ellipsoid (*Apodytes*) or orbicular to ovoid drupe which is fleshy or non-fleshy and pubescent (e.g. *Sleumeria*) or glabrous (e.g. *Alsodeiopsis*, *Chlamydocarya*, *Pyrenacantha*). It is usually stalked but sessile in *Polycephalium*. Fruits of *Apodytes* have a large lateral fleshy appendage (Potgieter and van Wyk 1994a). The drupes contain one seed. The exocarp is usually thin and often coloured; the mesocarp is variably fleshy and sometimes rich in oils. The endocarp is hard and smooth but in some genera several longitudinal ribs may be present. In *Pyrenacantha* the endocarp is verrucose-papillose inside, the papillae projecting into the interstices of the deeply ruminate and pitted endosperm (Potgieter and van Wyk 1994b). In *Merrilliodendron* the inner layer becomes fibrous-corky and is responsible for the buoyancy of the fruit. The style is persistent in *Apodytes*, *Cassinopsis* (Potgieter and van Wyk 1994c) and *Rhaphiostylis*. Endosperm is nuclear, oily and abundant. There are two usually flat or folded (*Lavigeria* and other genera) cotyledons which are fleshy in *Sleumeria*. The embryo is chlorophyllous and straight to curved. Seeds mostly albuminous (Corner 1976) and exarillate. Germination is phanerocotylar or cryptocotylar (Sleumer 1942).

Dispersal. Seeds are dispersed mainly by birds, with records of at least 18 bird species removing fruits (Palmer and Pitman 1972; Frost 1980). Fruits are also dispersed by bushpigs (Breytenbach and Skinner 1982) and vervet monkeys (Nichols 1991). Fay (1993) observed in the northern Central African Republic that ca. 90 % of the fruits of *Icacina oliviformis* were collected by baboons (*Papio cynocephalus anubis*), but often when the seeds were not yet fully mature. Some fruits appear to be able to float, especially in *Merrilliodendron*.

Phytochemistry. Secondary compounds found in the family include monoterpenoids, diterpenoids and sesquiterpenoids, with a strong diversification of mevalonate-derived metabolites (Gershenzon and Mabry 1983; Jensen 1992). *Icacina* produces diterpenoid derivatives, the alkaloids icaceine, de-N-methylicaceine and icacine as well as the additional C_{20}-compound icacinol and C_{19}-compound icacinone (Kaplan et al. 1991). Aguwa and Okunji (1986) identified triterpenoid saponins in the leaves of some species of *Pyrenacantha*. Rasoanaivo et al. (1990) showed that leaves of Malagasy *Cassinopsis* contain the glycoside cassinopin, a kaempferol trirhamnoside. Verbascosides have also been detected in *Cassinopsis* (Rasoanaivo et al. 1990). Oleanolic acid occurs in the bark of some species of *Apodytes* (Sleumer 1971). Carbocyclic iridoids have been reported in *Apodytes*, with *Poraqueiba* containing emmotins, a class of sesquiterpenes based on a rearranged eudesmane skeleton. In species of *Merriliodendron* and *Nothapodytes* camptothecine (quinoline alkaloids) were found (Kaplan et al. 1991).

Subdivisions and Relationships Within the Family. Engler (1896) recognised three subfamilies, Icacinoideae, Lophopyxidoideae and Cardiopterygoideae, of which only Icacinoideae are still treated as Icacinaceae. Subfamily Icacinoideae was divided into four tribes: Icacineae, Iodeae, Sarcostigmateae and Phytocreneae, mainly on the basis of wood anatomical features. Sleumer (1969) noted that this classification was based on insufficient material and therefore inaccurate. The four tribes proposed by Engler (1896) had come under increasing criticism due to the incongruence of macromorphological and anatomical characters.

Bailey and Howard (1941a, b, d) did not propose a new intrafamilial classification but recognised three groups or specialization based on wood and nodal anatomy. Their first group has trilacunar nodes and scalariform vessel perforations, the second trilacunar nodes and a mixture of simple and scalariform vessel perforations, and the third unilacunar nodes and simple vessel perforation. They also suggested that the Icacineae can be divided into two sections with either trilacunar or unilacunar genera. The unilacunar Icacineae resemble Iodeae, Sarcostigmateae and Phytocreneae in having unilacunar nodes and a tendency towards a scrambling, twining or climbing habit. These authors suggested that the unilacunar Icacineae may be transitional in form and structure between the non-scandent, trilacunar Icacineae and the unilacunar Iodeae, Sarcostigmateae and Phytocreneae. Lobreau-Callen (1973), in agreement with the suggestion by Bailey and Howard (1941a–d), also distinguished three levels of specialization based on pollen morphology.

van Staveren and Baas (1973) noted a strong resemblance between the Iodeae and Phytocreneae in leaf epidermal characters. They also supported Melchior and Werdermann's (1964) view that Sarcostigmateae should be regarded as a rather isolated tribe. According to van Staveren and Baas (1973), Icacineae are heterogeneous in leaf epidermal characters, secondary xylem, node and pollen features.

Based on the analysis of DNA sequences from four genes and morphological data, Kårehed (2001) came to the conclusion that Icacinaceae (s.s.) should be divided into three groups: the *Icacina* group, the *Emmotum* group and the *Apodytes* group. Kårehed (2001) furthermore suggested that *Cassinopsis* may constitute a subgroup within the *Icacina* group. Other genera, once included in Icacinaceae s.l. and discussed in the above papers, will not be discussed here (see treatments in this volume: Stemonuraceae (12 genera), Cardiopteridaceae (5 genera), both Aquifoliales, and Metteniusaceae (*Metteniusa* and *Oncotheca*, with *Dendrobangia* tentatively included) probably close to Icacinaceae s.s.; *Pennantia* (Pennantiaceae) belongs to Apiales treated in another volume).

Affinities. In the past, Icacinaceae had been associated with various orders such as Celastrales (Takhtajan 1980; Cronquist 1981), Santalales, Theales (Dickison and Baas 1977; Keller 1992; Savolainen et al. 1994) and Rhamnales. The classification systems of Cronquist (1981), Dahlgren (1980), Takhtajan (1980) and Thorne (1992) place the family in either Celastrales or Cornales.

The large-scale molecular studies by Savolainen et al. (2000a, 2000b) and Soltis et al. (2000) first indicated that Icacinaceae should be subdivided. *Icacina* and related genera formed a monophyletic group with a rather uncertain position in the euasterids, whereas other genera grouped with Cardiopteridaceae and, together with that family, constituted the sister group of Aquifoliales.

On the basis of an analysis of DNA sequences and morphological data, Kårehed (2001) showed that the family is not monophyletic and should be split into at least four families in three different orders (Garryales, Aquifoliales and Apiales), i.e. Cardiopteridaceae, Icacinaceae s.s., Pennantiaceae and Stemonuraceae.

Until recently, the Icacinaceae have been regarded as a rosid taxon related to Aquifoliaceae. In the Angiosperm Phylogeny Group (APG) system (APG 1998), Icacinaceae were unassigned to order but listed under euasterids II. Characters supporting this placement are the unitegmic ovules, the presence of iridoids, and features of wood anatomy. In the updated APG III system (APG 2009), the Icacinaceae s.s. are placed in euasterids I without having been assigned to an order. Weak affinities to the Garryales have been suggested by Kårehed (2001) and Soltis et al. (2011). New data suggest that several genera included here in the Icacinaceae (e.g. *Apodytes*, *Emmotum*, *Poraqueiba*) rather belong into Metteniusaceae s.l. (Greg Stull, pers. comm.). More research is needed to resolve the circumscription and relationships of Icacinaceae s.s.

Distribution and Habitats. The family is largely confined to the tropics of both the Old and the New World, with centres of diversity in Africa and SE Asia (Hua and Howard 2008). Genera are mostly endemic to major phytogeographical realms. An exception is *Apodytes* which is widespread and found in tropical and subtropical Africa, Madagascar, Southeast Asia and tropical Australia. In general, the number of genera decreases rapidly towards the subtropics. The largest genera occur in the Paleotropics, especially in S and SE Asia and Africa. These are *Iodes* (30 species) and *Pyrenacantha* (23 species).

Icacinaceae occupy a variety of habitats excluding deserts and very cold high mountain areas. Genera are most abundant in rain forests of the tropical lowlands, but some reach elevations of 3000 m in cloud forests. No species has been reported to be dominant in any vegetation type, although occasionally species may be locally abundant.

Palaeobotany. Fossil wood similar to *Icacina* has been recovered from the Lower Cretaceous in North America (Utah and Colorado; Thayn et al. 1985). Fossil pollen of the family is known from the Palaeocene of North America (Oregon) and the Eocene of Europe (Manchester and Tiffney 1993). Fossil fruits are known from the Upper Cretaceous in Egypt (Gregor and Hagn 1982) and, with less certainty, from the Upper Cretaceous of New York (Scott and Barghoorn 1957). Fossils are also known from the Eocene of London, England (Lobreau-Callen and Srivastava 1974). Soudry and Gregor (1997) found a late Cretaceous fruit of *Iodes israelii* in southern Israel. Fossil records of *Pyrenacantha* have been reported from the Eocene of Oregon and California (USA), from the Oligocene of Egypt, and from the Palaeogene of Japan together with *Merilliodendron*, *Phytocrene* and two extinct genera (*Goweria*, *Huziokaea*; Tanai 1990). These occurrences indicate that *Pyrenacantha*, today disjunct between Asia and Africa, may be a remnant of the Eocene boreotropical forest with biogeographical ties to Africa (Manchester and Tiffney 1993).

Extinct genera include *Echiperiporites* from the Lower Eocene to Miocene of Cameroon (Muller 1981), *Icacinicarya* from the Upper Cretaceous and Tertiary of Nigeria (Chesters 1957), *Icacinicaryites* from the late Palaeocene of western North America (Pigg et al. 2008), *Croomiocarpon* from the Eocene of SE North America (Stull et al. 2011), and *Palaeophytocrene* from the Eocene of, for example, Vancouver Island (Rankin et al. 2008). *Stizocaryopsis barthelii* (Mid-Danian) from Egypt and England is most probably related to *Chlamydocarya* and *Pyrenacantha* (Gregor and Hagn 1982).

Economic Importance. Members of Icacinaceae are not widely utilized commercially in Africa at present, although in the early parts of the last century the wood of *Apodytes* was used quite extensively for felloes in the ox wagon trade (Palmer and Pitman 1972). Presently, a number of genera are used locally for various purposes.

In Africa, *Pyrenacantha* plays an important role in the ethnobotany of many tribes, having a variety of medicinal and traditional uses (Watt and Breyer-Brandwijk 1962; Mabogo 1990). A water extract of the leaves of *Pyrenacantha* is used in Nigeria for the treatment of various gastro-intestinal ailments and high blood pressure. It appears to act directly as a muscle relaxant (Akube et al. 1983). In South Africa, members of *Pyrenacantha* are used by the Vhavenda for luck, to avoid trouble and in the treatment of malaria (Mabogo 1990), while the fruits are used by the Zulu as an ingredient in stews, and the leaves as a vegetable (Cunningham 1985). The leaves of *Apodytes dimidiata* are used by the Zulu in the treatment of ear inflammation (Coates Palgrave 1983). Leaves of *Apodytes* are highly toxic to snails of the genus *Schistostoma* and are used as a molluscicide (Pretorius et al. 1991). Species of *Cassinopsis* are used in Madagascar as a traditional remedy against malaria (Rasoanaivo et al. 1990). Infants are fed the fruits of *Alsodeiopsis*, while the roots of this genus are used as an aphrodisiac by adults. In Cameroon the leaves and roots of *Desmostachys* are applied to wounds. The fruits of *Lavigeria* are consumed and used as an aphrodisiac (Villiers 1973). *Icacina oliviformis* is a suffrutescent savannah pyrophyte used for its edible seeds, tuberous roots and fruits in many areas of west and north central Africa (Fay 1987).

In Asia some genera are used for construction purposes and as firewood (*Platea*) or for medicinal purposes (*Sleumeria auriculata*). Young leaves of *Ryticaryum* and *Polyporandra* are cooked and eaten as a vegetable by local people. Several genera have edible fruits. Campothecin, an indole alkaloid promising as anticancer drug, is extracted from wood-chips of *Nothapodytes nimmoniana*, a practice threatening the survival of this species (Sharma et al. 2010). In the New World, only two species have economic importance: *Poraqueiba sericea* has fleshy fruits rich in oils which are eaten fresh; the seed also provides flour and the fruit is widely traded in the Amazon basin; *Casimirella ampla* has a tuber (5–20 kg) rich in starch which is edible after removal of the bitter compounds.

KEY TO THE OLD WORLD GENERA

1. Climbing shrubs or lianas, with or without tendrils 2
– Trees or non-climbing shrubs 17
2. Leaves opposite, sometimes with tendrils 3
– Leaves alternate, tendrils absent 4
3. Anthers broadly club-shaped to subglobular, pollen sacs subdivided, with numerous pollen-bearing alveolus **21. *Polyporandra***
– Anthers with only 2 fertile pollen sacs **7. *Iodes***
4. Flowers hermaphrodite 5
– Flowers unisexual, plants dioecious 8
5. Roots not tuberous 6
– Roots tuberous; fruit ellipsoid to oblongoid 7
6. Flowers in supraaxillary fascicles **35. *Rhaphiostylis***
– Flowers in simple or compound racemes **25. *Sleumeria***
7. Inflorescences cauline; petals glabrous inside; anthers appendiculate; disk absent **8. *Lavigeria***
– Inflorescences axillary; petals pubescent inside; anthers not appendiculate; disk present **6. *Icacina***
8. Calyx absent in male flowers 9
– Calyx present in male flowers 11
9. Fruit with accrescent corolla or an apical appendage; flowers 3–4-merous 10
– Fruit without accrescent organ; flowers 5-merous **11. *Mappianthus***
10. Leaves penninerved; flowers 4-merous **3. *Chlamydocarya***
– Leaves palmatinerved; flowers 3-merous **20. *Polycephalium***
11. Petals free or fused only at base 12
– Petals connate into tube with free lobes 15
12. Roots tuberous; leaves dentate and palmatinerved with 3–5 pairs lateral veins; stamens with a flat, bottle-like apical appendage **15. *Natsiatum***
– Roots not tuberous; leaves entire and penninerved or palmatinerved; stamens without appendage 13
13. Flowers 3–4-merous; fruits in globose heads **17. *Phytocrene***
– Flowers 4–5(–6)-merous; fruits not in globose heads 14
14. Inflorescences cauline; stamens included; filaments slightly fleshy at base **26. *Stachyanthus***
– Inflorescences axillary; stamens exserted; filaments filiform **24. *Sarcostigma***
15. Anthers longer than filaments **13. *Miquelia***
– Anthers shorter than filaments 16
16. Flowers 5-merous **5. *Hosiea***
– Flowers 4-merous **14. *Natsiatopsis***
17. Leaves opposite **27. *Cassinopsis***
– Leaves alternate 18
18. Flowers unisexual; dioecious trees or shrubs 19
– Flowers hermaphrodite; trees or shrubs 24
19. Filaments very short (less than 1 mm) or shorter than petals 20
– Filaments long (more than 1 mm) or as long as petals 22
20. Calyx absent **22. *Pyrenacantha***
– Calyx present 21
21. Ovary thick-cylindric to subconical, glabrous **32. *Platea***
– Ovary conical to ovoid, densely covered with erect hairs **23. *Ryticaryum***
22. Flowers sessile **4. *Desmostachys***
– Flowers not sessile 23
23. Ovary oblique with a unilateral swelling; fruit ± unilaterally developed; lateral appendage large and succulent **34. *Apodytes***
– Ovary with 2 narrow swellings on top; fruit without appendage **35. *Rhaphiostylis***
24. Disk present 25
– Disk absent, if present then very small 27
25. Roots tuberous; stamens with filiform filaments, curved, calyx not persistent on fruit **6. *Icacina***
– Roots not tuberous; stamens with slightly fleshy and somewhat flattened filament; calyx persistent on fruit 26
26. Leaf margins sinuate, petals spathulate, glabrous inside; connective protruding; peduncles with small bracts; branches not angular **18. *Pittosporopsis***
– Leaves entire; petals linear-oblong, hairy inside (at base), connective not protruding; peduncles ebracteate; branches usually angular **16. *Nothapodytes***
27. Stigma capitates **1. *Alsodeiopsis***
– Stigma punctiform **12. *Merrilliodendron***

KEY TO THE NEW WORLD GENERA

1. Lianas or climbing shrubs; petiole with a distinct abscission line at the base 2
– Trees or shrubs, sometimes with climbing branches; petiole generally without abscission line, if present then leaves with domatia on the abaxial surface 4
2. Generally all vegetative parts covered by malpighiaceous hairs (two-armed, one reduced, the other well-developed) **9. *Leretia***
– Generally all vegetative parts covered with an indument, but hairs never malpighiaceous 3
3. Lianas; generally all vegetative parts with hispid hairs; inflorescence an axillary or terminal spike; flowers not articulate at their base **19. *Pleurisanthes***

- Lianas or climbing shrubs; generally hairy but never with hispid hairs, hairs fasciculate, rarely simple and then leaves almost glabrous; inflorescence a terminal, rarely axillary cyme; flowers articulated at their base **2. *Casimirella***
4. Leaves without domatia on the abaxial surface, drying black; plants dioecious, male plants with flowers in long pseudo-spikes and female plants with 1–3(–10) flowers in racemes; flowers 4-merous; fruit up to 60 mm long **28. *Calatola***
- Leaves often with domatia on the abaxial surface, never drying black; plants bisexual or functionally dioecious, but flowers not morphologically different; flowers 5-merous; fruit less than 40 mm long 5
5. Leaves membranaceous or subcoriaceous, domatia present; flowers bisexual **10. *Mappia***
- Leaves subcoriaceous to coriaceous; domatia absent; plants dioecious 6
6. Hairs malpighiaceous 7
- Hairs never malpighiaceous 8
7. Leaves more than 10 cm long, with (4–)7–10(–12) pairs of secondary nerves; inflorescences dichasial; petals free; calyx accrescent **30. *Oecopetalum***
- Leaves less than 10 cm long, with 4–7(–8) pairs of secondary nerves; inflorescences paniculate, reduced to 1–3 flowers; petals fused at base; calyx not accrescent **31. *Ottoschulzia***
8. Leaves generally densely hairy on the abaxial surface; petals reflexed and barbed **29. *Emmotum***
- Leaves generally glabrous; petals not reflexed, glabrous or hairy but never barbed **33. *Poraqueiba***

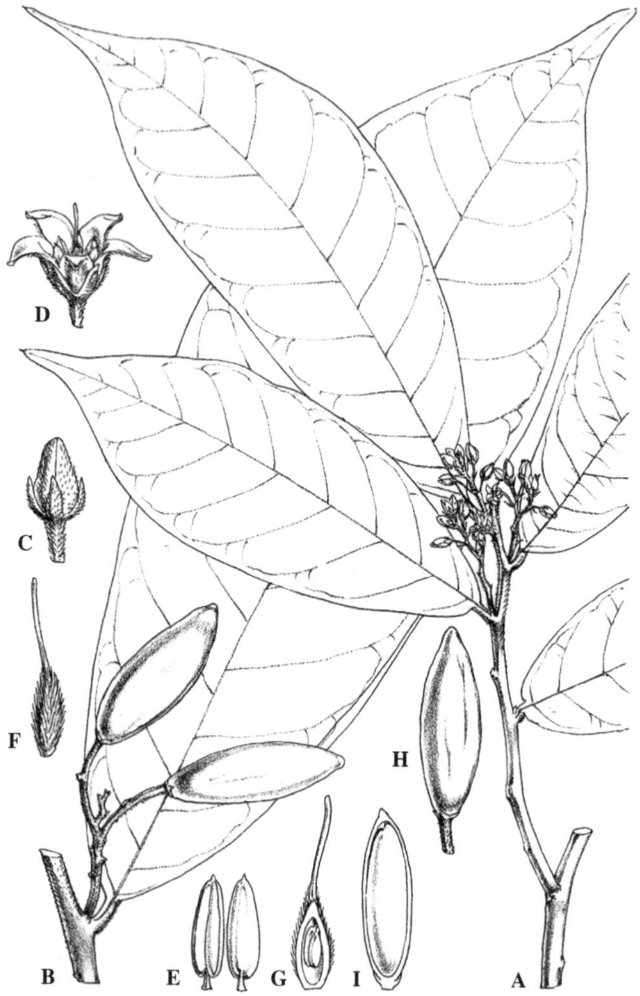

Fig. 45. Icacinaceae. **A–I** *Alsodeiopsis mannii*. **A** Flowering branch. **B** Fruiting branch. **C** Bud. **D** Open flower. **E** Stamens. **F** Gynoecium. **G** Gynoecium in longitudinal section. **H** Fruit. **I** Fruit, opened. (From Hooker 1867, modified, courtesy Missouri Botanical Garden)

Genera of Icacinaceae

Icacina group

Leaves alternate or opposite, penninerved or palmatinerved, entire or serrate; sometimes with domatia. Flowers hermaphrodite or unisexual; petals free or connate, with or without a keel; stamens free or epipetalous; anthers introrse; disc absent or cup-shaped, columnar or lobed; ovary without a fleshy appendage; ovules two.

1. ***Alsodeiopsis*** Oliver ex Benth. & Hook.f. Fig. 45

Alsodeiopsis Oliver ex Benth. & Hook.f., Gen. Pl. 1(3): 996 (1867).
Alsodeiidium Engl. (1895).

Small trees or shrubs; branchlets tomentose. Leaves alternate, membranous to coriaceous, margin entire; domatia absent. Inflorescences axillary, umbelliform or corymbiform cymes or racemes. Flowers bisexual, 5-merous; sepals free or partially united at base; petals free or united up to half their length; stamens very short, free, alternating with petals; filaments flattened, relatively short; disk absent; ovary ellipsoid, hirsute; style filiform; stigma small, capitate. Drupe oblong-ellipsoid, pilose or glabrous.

Eleven species, tropical Africa.

2. ***Casimirella*** Hassl.

Casimirella Hassl., Feddes Repert. Spec. Nov. Regni Veg. 12: 249, fig. 251 (1913); Howard, Brittonia 44: 167 (1992).

Humirianthera Huber (1914).

Lianas or shrubs with scandent branches; large tubers present; generally hairy, hairs fasciculate, rarely simple and then leaves almost glabrous. Leaves alternate. Inflorescences umbelliform panicles, axillary, supraaxillary or terminal. Flowers bisexual, 5(−6)-merous, articulated at their base; sepals almost free; petals free, villose inside; stamens free, filaments filiform; disk absent; ovary conical; style short, thickly filiform; stigma small, capitate. Drupe ovoid, endocarp woody, pubescent inside.

Seven species, tropical S America to Bolivia and Paraguay.

3. ***Chlamydocarya*** Baill.

Chlamydocarya Baill., Adansonia 10: 276 (1872).

Lianas; plants dioecious. Leaves sometimes with hydathodes; margins entire; domatia absent. Inflorescences panicles, axillary or cauliflorous. Flowers 4-merous; sepals absent; petals fused at their bases; staminate flowers: stamens free, filaments short and flattened; pistillate flowers: staminodes alternipetalous, disk present, ovary pubescent. Drupe with a hirsute or echinulate appendage at the top, endocarp verrucose inside. $2n = 22$.

Six species, tropical Africa.

4. ***Desmostachys*** Planch. ex Miers

Desmostachys Planch. ex Miers, Ann. Mag. Nat. Hist., ser. II, 9: 398–399 (1852).

Small trees or shrubs, sometimes scandent. Leaves alternate; margins entire; domatia absent. Inflorescences racemose, spike-like or glomerulose. Flowers bisexual, sessile, (4–)5-merous; sepals almost free; petals free or fused (*D. vogelii*), pubescent on abaxial surface; stamens free or attached to the corolla; ovary asymmetric, glabrous or pubescent; style long, glabrous or pubescent, with a slightly lobed stigma; disk absent or very small. Drupe ellipsoid.

Seven species, tropical Africa and Madagascar.

5. ***Hosiea*** Hemsl. & Wilson

Hosiea Hemsl. & Wilson, Bull. Misc. Inform. Kew 1906: 154 (1906).

Lianas; plants dioecious. Leaves long-petiolate, with palmate venation and long-dentate margins; domatia absent. Inflorescences cymes or compound panicles, often reduced to a few flowers, axillary. Flowers functionally unisexual, 5-merous; petals much longer than calyx, coherent at base, outside pubescent, inside puberulent, lobes recurved with caudate apex; stamens free; filaments shorter than petals; disk present as glands between the stamens; ovary ovoid. Drupe slightly flattened, ellipsoid.

Two species, E Asia (China, Japan).

6. ***Icacina*** A. Juss.

Icacina A. Juss., Mém. Soc. Hist. Nat. Paris 1:174 (1823).
Thollonia Baill. (1886).

Trees or shrubs with ascending, somewhat twining branches, or lianas; subterranean tubers present. Leaves alternate, margins entire. Inflorescences axillary, cymose. Flowers bisexual, 5-merous; sepals united with short or long lobes; petals free, pubescent on inner surface; stamens free; anthers not appendiculate; disk present; ovary ovoid, often asymmetrical; style long, often eccentric and bifid, sometimes pubescent. Drupe ellipsoid, tomentose or tomentellous. $2n = 48$.

Six species, tropical Africa.

7. ***Iodes*** Blume

Iodes Blume, Bijdr. Fl. Ned. Ind. 29 (1825) ('Iödes'); Sleumer, Flora Males. I, 7(1): 67–73 (1971), rev. reg.

Lianas, plants dioecious; branchlets hairy, generally with short bifid tendrils at or close to the nodes (but not axillary). Leaves opposite or subopposite, entire, penninerved, with hairs on the veins; domatia absent. Inflorescences axillary panicles, rarely terminal umbels. Flowers small, fragrant, (4–)5-merous; sepals and petals partially united at base; staminate flowers: stamens alternating with petals; filaments very short, usually flattened; anthers introrse; pistillate flowers: staminodes sometimes present; ovary subsessile; disk absent; style short; stigma thick, peltate, apex concave, sometimes slightly oblique. Drupe ellipsoid, laterally compressed and longitudinally grooved. $2n = 20$.

About 28 species, tropical Africa, Madagascar and Australasia.

8. ***Lavigeria*** Pierre

Lavigeria Pierre, Fl. Forest. Cochinch. 4 [or Fasc. 17], t. 267 (1892).

Lianas or climbing shrubs; branchlets pubescent; generally with subterranean tubers. Leaves pubescent with clustered hairs on nerves on abaxial surface; margins entire; domatia absent. Inflorescences panicles clustered on the main stem, tomentellous. Flowers bisexual, 5-merous, sessile; sepals almost free; petals free, glabrous inside, nearly glabrous outside; stamens free, anthers not appendiculate; ovary ovoid, pubescent. Drupe large, ellipsoid or oblongoid, corrugated.

One species, *L. macrocarpa* Pierre, tropical Africa.

9. ***Leretia*** Vell.

Leretia Vell., Fl. Flumin. 99 (1829) ['1825'].

Lianas or shrubs with scandent branches; generally all vegetative parts covered with malpighiaceous hairs. Leaves alternate, margins entire; petiole with a distinct abscission line at the base. Inflorescences much-branched cymes, axillary. Flowers bisexual, 5-merous, pedicles articulate below the calyx; sepals almost free; petals free, pilose inside; stamens free; filaments filiform; ovary cylindric; disk absent; style glabrous; stigma capitate, apex obliquely umbonate. Drupe ovoid-ellipsoidal, slightly flattened.

Two species, Panama, northern South America to Ecuador, Peru and Brazil.

10. ***Mappia*** Jacq.

Mappia Jacq., Pl. Rar. Horti Schoenbr. 1: 22, t. 47 (1797), nom. cons.

Trees or shrubs. Leaves membranaceous or subcoriaceous, generally with domatia; margins entire, slightly revolute. Inflorescences dichasial, axillary. Flowers bisexual, (4–)5(–6)-merous; sepals fused; petals free, pilose inside; stamens free, filaments filiform; disk fleshy, hirsute on the edge and inside, lobed; ovary ovoid; style short, apex umbonate; stigma capitate, occasionally bilobed. Drupe subovoid or ellipsoidal.

Three or four species, Central America and West Indies.

11. ***Mappianthus*** Hand.-Mazz.

Mappianthus Hand.-Mazz., Anz. Akad. Wiss. Wien, Math. Naturw. Kl. 58: 150 (1921); Sleumer in Engl. & Prantl, Nat. Pflanzenfam., 2nd edn, 20b: 379 (1942), rev.

Lianas; plants dioecious. Leaves opposite or subopposite; margins entire; domatia absent. Inflorescences short, few-flowered axillary cymes. Flowers rather small, 5-merous; staminate flowers: calyx cupular, small, shallowly 5-lobed; corolla campanulate to funnel-shaped, large, lobes valvate, hairy; anthers ovoid-triangular; disk absent; pistillate flowers: as staminate flowers but slightly smaller; ovary subglobose, style very short or absent, stigma discoid, rounded, 5-lobed; disk absent. Drupe narrowly ellipsoid, compressed.

Two species, China and northern Vietnam.

12. ***Merrilliodendron*** Kaneh.

Merrilliodendron Kaneh., Bot. Mag. (Tokyo) 48: 920, f. 7. Ic. 956 (1934); Sleumer, Fl. Malesiana 7(1): 49–51 (1971), rev.
Peekeliodendron Sleum. (1937).

Trees or shrubs. Leaves alternate; margins entire; domatia absent. Inflorescences lax cymes to elongate panicles, in foliate or defoliate axils, or on older branches. Flowers bisexual, (3)4- or 5-merous; sepals fused; petals free; stamens free, filaments dilated; disk absent; ovary ovoid, tapering to a thick-filiform style; stigma punctiform. Drupe large, oblongoid to ellipsoid, becoming corky.

One species, *M. megacarpum* (Hemsl.) Sleum., Micronesia, Melanesia and east Malesia.

13. ***Miquelia*** Meisn.

Miquelia Meisn., Pl. Vasc. Gen. 1: 152 (Sept. 1838), nom. cons., non Blume (June 1838), nom. rej.
Jenkinsia Gruff. (1843), nom. illegit., non Hook. (1841).

Climbing or twining shrubs; dioecious. Leaves alternate, entire or coarsely lobed, penninerved or palmatinerved; domatia absent. Inflorescences heads or umbels, axillary. Flower (4–)5-merous;

sepals almost free; staminate flowers: petals united to various degrees, stamens free, filaments linear, anthers longer than filaments, ovary rudimentary; pistillate flowers: sepals and petals either separated by the elongated flower axis or close together, staminodes flattish, very short, ovary thick cylindric, hirsute; stigma sessile, disc-shaped, its centre slightly depressed; disk absent. Drupe oblongoid to subovoid to ellipsoid, laterally compressed, sometimes narrowed into a stalk-like base.

About eight species, SE Asia and Malesia (Sumatra, Borneo, Philippines and Sulawesi).

14. ***Natsiatopsis*** Kurz

Natsiatopsis Kurz, J. Asiat. Soc. Bengal, Pt. 2, Nat. Hist. 44 (2): 201 (1876) ('1875'); Ma He Teng, Fl. China 11: 514 (2008), rev.

Lianas; plants dioecious. Leaves either entire or toothed, palmatinerved; domatia absent. Inflorescences lax spikes, axillary. Flowers 4-merous; staminate flowers: sepals fused, petals fused into a tube; stamens 4, free, filaments linear; ovary rudimentary, densely hairy; pistillate flowers: calyx and corolla as in male flowers; anthers small; ovary ovoid, hirsute; disk absent. Fruit an ovoid drupe, compressed, polygonously reticulate.

One species, *N. thunbergiifolia* Kurz, China and Myanmar.

15. ***Natsiatum*** Buch.-Ham. ex Arn.

Natsiatum Buch.-Ham. ex Arn., Edinburgh New Philos. J. 16: 314 (1834); Sleumer, Blumea 17: 231 (1969), rev.

Climbing semi-herbaceous shrubs with tuberous roots; plants dioecious. Leaf margins mostly serrulate, palmatinerved, 7–9-veined; domatia absent. Inflorescences pendulous spike-like racemes, slightly supraaxillary. Flowers 4–6-merous; sepals almost free; petals free or connate at base; staminate flowers: stamens free, with a flat, bottle-like apical appendage; filaments short; pistillate flowers: staminodes present; ovary sessile; style short, apex 2- or 3-lobed; stigma capitate; disk absent. Drupe obliquely ovoid, laterally compressed.

One species, *N. sinense* Oliv., tropical Asia (east India, Nepal, Myanmar, north Thailand, north Vietnam).

16. ***Nothapodytes*** Blume

Nothapodytes Blume, Mus. Bot. 1: 248 (1851) ('1850'); Howard, J. Arnold Arbor. 23: 66 (1942); Sleumer, Fl. Malesiana 7(1): 53–56 (1971), reg. rev.

Mappia Jacq. sect. *Trichocrater* Miers (1852).

Mappia Jacq. sect. *Nothapodytes* (Blume) O. Ktze, Post & Kuntze (1904).

Neoleretia Baehni (1936).

Trees or treelets; branches usually angular, with large leaf-scars. Leaves partly subopposite, entire, appearing very finely pellucid-punctate against strong light; domatia absent. Inflorescences corymbose cymes, terminal, rarely axillary, ebracteose. Flowers bisexual, 5-merous; sepals almost fused; petals linear-oblong, slightly cohering at base, villose inside; stamens free; filaments fleshy, flattened, anthers with a pulviniform appendage; disk cup-shaped, fleshy; ovary ovoid to subconical, symmetrical, hirsute; upper half of style glabrous; stigma obliquely truncate and slightly concave. Drupe ellipsoid, slightly flattened; calyx persistent.

Eleven species, S Asia (India, Sri Lanka, Myanmar, Thailand, Malaysia, Cambodia, China, Vietnam).

17. ***Phytocrene*** Wall.

Phytocrene Wall., Pl. Asiat. Rar. 3:11, t. 16 (1831), nom. cons.; Sleumer, Fl. Males. I, 7(1): 79–87 (1971), reg. rev.

Gynocephalum Blume (1825), nom. rej.

Climbing shrubs or lianas, dioecious. Leaves entire or palmately 3–7-lobed, subpalmately nerved; domatia absent. Staminate flowers in globose clusters, spikes or panicles; pistillate flowers in large solitary either cauline or mostly axillary pedunculate globose heads. Flower 3–5(6)-merous; sepals free, 3–6; petals united below, 4–6; staminate flowers: stamens free, 4–6, filaments filiform, connate at base into a column with the hirsute rudiment of the ovary; pistillate flowers: staminodes absent; ovary subovoid; style short, columnar; stigma bilobed, forming a funnel. Drupes numerous, arranged in generally large, globose heads, each one tipped by the style, more or less reversely setulose with brittle irritant hairs.

About 11 species, SE Asia and in Malesia from Sumatra to New Guinea.

18. ***Pittosporopsis*** Craib

Pittosporopsis Craib, Bull. Misc. Inform. Kew 1911: 28 (1911); Hua & Howard, Flora of China 11: 508–509 (2008), rev.

Shrubs or trees. Leaves alternate, margins sinuate, cartilaginous, glabrescent. Inflorescences cymose, axillary. Flowers bisexual, 5-merous. Peduncles with small bracts. Sepals almost fused. Petals free. Stamens free, filaments fleshy, flattened, abruptly contracted upward. Connective protruding. Disk connate with ovary. Ovary ovoid-ellipsoid. Style straight at first, becoming geniculate, persistent. Stigma minute. Drupe ovoid, slightly flattened, with a prominent rib on one side, smooth, crowned by the style. Calyx persistent on fruit.

One species, *P. kerrii* Craib, SE Asia (Myanmar, Thailand, Laos, Vietnam, China).

19. ***Pleurisanthes*** Baill.

Pleurisanthes Baill., Adansonia 11: 201 (1874).
Martia Valet. (1886).
Valetonia T. Durand (1888).

Lianas or shrubs with scandent branches. All parts with hispid hairs. Leaves sometimes dentate. Domatia absent. Inflorescences spikes, axillary, supraaxillary or terminal. Flowers bisexual, 3–5(−6)-merous, non-articulated. Sepals almost free. Petals free. Stamens free, filaments filiform. Ovary hirsute. Disk absent. Style glabrous, as long as ovary. Stigma minute. Drupe oblongoid.

Six or seven species, northern South America to Ecuador, Peru and Brazil.

20. ***Polycephalium*** Engl.

Polycephalium Engl., Nat. Pflanzenfam., ed. 1, Nachtr. 227 (1897).

Lianas, plants dioecious. Leaves palmatinerved or penninerved, lobed or entire. Domatia absent, hydathodes present. Male inflorescences in heads arranged in spikes or panicles, axillary. Staminate flowers: 3-merous, sepals absent, petals fused into a tube; stamens 3, free; ovary absent; pistillate flowers: sepals absent, petals fused into a tube, ovary ellipsoid, style long, pubescent, stigma plumose or laciniate. Drupe sessile, with accrescent corolla, endocarp verrucose inside.

Two species, tropical Africa.

21. ***Polyporandra*** Becc.

Polyporandra Becc., Malesia 1:125 (1877); Sleumer, Fl. Males. I, 7(1): 65–67 (1971), rev.

Lianas; plants dioecious. Leaves opposite, sometimes with extra-axillary or intrapetiolar tendrils. margins entire; domatia absent. Inflorescences panicles, axillary. Flowers 5 or 6(−7)-merous; staminate flowers: sepals fused into a lobed tube, petals valvate, free, stamens free, sessile, anthers thick, clavate-subglobose, many-celled, with numerous small pollen-bearing alveoles, each of which is covered with a caducous membrane; ovary rudimentary; pistillate flowers: sepals almost free, petals very short, connate for ca. 1/3 of length, stamens absent, ovary subglobose to cylindric, stigma large, disk-shaped, its centre depressed. Drupe compressed-ellipsoid.

One species, *P. scandens* Becc., Solomon Islands, New Guinea, Moluccas.

22. ***Pyrenacantha*** Wight

Pyrenacantha Wight in Hook., Bot. Misc. 2: 107 (1830) suppl. t. 9 & 10, nom. cons.
Cavanilla Thunb. (1792) nom. illegit., non Cavanilla J.F. Gmel. 1791 [1792].
Moldenhauera Spreng. (1824), nom. illegit. non *Moldenhawera* Schrad. (1802).
Adelanthus Endl. (1840).
Endacanthus Baill. (1892).
Freeria Merr. (1912).
Monocephalium S. Moore (1920).

Lianas or scandent shrubs, dioecious or monoecious. Leaves alternate, hispid, entire to deeply lobed, glandular-repand-dentate, penninerved or palmatinerved; domatia absent. Inflorescences spikes or racemes. Flowers unisexual by abortion, (3)4(5)-merous; sepals absent; petals free or fused at base, persistent in fruit; staminate flowers: filaments very short; pistillate flowers: vestigial staminodes alternating with petals or absent, ovary pubescent, ovoid; stigma capitate, sessile; disk absent. Drupe spherical to broadly ovoid, usually laterally compressed and emarginate, pubescent; punctate outside, endocarp verrucose-papillose inside, the papillae projecting into the interstices of the deeply ruminate and pitted endosperm. $2n = 40$.

About 25 species, four species not described, Africa, Madagascar, SE Asia, Philippines.

23. ***Ryticaryum*** Becc.

Ryticaryum Becc., Malesia 1:120, t. 5, f. 18–25 (1877), Malesia 1: 256 (1878), corr. '*Rhyticaryum*'; K. Schum. & Laut., Fl. Schutzgeb. Südsee 415 (1900) ('*Rhytidocaryum*'); Sleumer, Fl. Males. I, 7(1): 35–41 (1971), reg. rev.
Pocillaria Ridl. (1916).

Shrubs or trees, dioecious. Leaves alternate, margins entire. Inflorescences spikes, rarely panicles or fascicles, axillary. Flowers (4)5-merous; sepals fused into a lobed or subentire tube; petals connate to ca. middle or beyond; staminate flowers: stamens adnate to the petals, filaments short, ovary rudimentary, hairy; pistillate flowers: staminodes present or absent, ovary conical to ovoid, densely covered by erect hairs; stigma sessile, flattened; disk absent. Drupe generally ovoid-ellipsoid, compressed laterally and almond-shaped, with a sharp crest and sometimes a second, less distinct crest on the flattened side.

About 12 species, tropical NE Australia (Queensland: Cape York Peninsula), Melanesia (Solomon, Bismarcks, New Britain, New Ireland, Admiralty Is.), East Malesia (New Guinea, Moluccas).

24. ***Sarcostigma*** Wight & Arn.

Sarcostigma Wight & Arn., Edinburgh New Philos. J. 14: 299 (1833); Sleumer, Fl. Malesiana, ser. I, 7(1): 73–75 (1971), reg. rev.

Lianas; plants dioecious. Leaves alternate, margins entire. Inflorescences of clustered or solitary flowers along the axis of long, pendulous, interrupted or spike-like racemes. Flowers (4)5(6)-merous; sepals fused; petals shortly connate at base; staminate flowers: stamens free, filaments filiform, ovary rudimentary; pistillate flowers: staminodes present or absent, ovary ovoid-oblongoid; stigma almost sessile, thick, umbonate; disk absent. Drupe unequally subovoid to ellipsoid or oblongoid, laterally compressed and almond-shaped. Seeds without endosperm, embryo with fleshy cotyledons.

Two species, SE Asia (India, Myanmar, Vietnam, Malaysia) and Malesia (Sumatra, Java, Borneo, Philippines).

25. ***Sleumeria*** Utteridge, Nagam. & Teo

Sleumeria Utteridge, Nagam. & Teo, Syst. Bot. 30: 638 (2005).

Scrambler or woody climber. Leaves alternate, coriaceous, reticulation prominent abaxially, hispid abaxially, base rounded, often unequal; margin entire; domatia absent. Inflorescences simple or compound racemes, axillary. Flowers bisexual, 4–(5)-merous; calyx cup-shaped, sepals fused; petals fleshy, shortly connate at base; connectives swollen, auriculate, anthers introrse; ovary spherical, densely strigose, style filiform, stigma unlobed; disk absent. Drupe laterally compressed.

One species, *S. auriculata* Utteridge, Nagam. & Teo, northern Borneo.

26. ***Stachyanthus*** Engl.

Stachyanthus Engl., Nat. Pflanzenfam., Nachtr. 1: 227 (1897), nom. cons.
Neostachyanthus Exell & Mendonça (1951).

Lianas; plants dioecious. Leaves alternate; margins entire; domatia absent. Inflorescences fascicles, cauliflorous. Flowers 5(6)-merous; staminate flowers: stamens free, filaments rather long, slightly fleshy at base; calyx distinct in pistillate flowers; ovary asymmetrical, stigma discoid. Drupe ellipsoid, endocarp ± smooth on the internal surface. $2n = 40$.

Six species, western tropical Africa (Ghana to Angola).

27. ***Cassinopsis*** Sond.

Cassinopsis Sond. in Harv. & Sond., Fl. Cap. 1: 473 (1860).
Hartogia Hochst. (1844), nom. illegit.
Tridianisia Baill. (1879).

Shrubs or small trees. Leaves opposite; margins entire or serrate; domatia absent. Inflorescences dichasially branched, axillary, alternating between nodes. Flowers bisexual, 5-merous; sepals united at base; petals slightly united at base, imbricate; filaments subulate; ovary pubescent; disk absent. Drupe subglobose or ovoid, sometimes compressed, fleshy with style persistent.

Six species, S Africa and Madagascar.

Emmotum group

Leaves alternate, penninerved, entire; domatia absent. Flowers hermaphrodite or unisexual; petals free, with a keel; stamens free; anthers latrorse; disc absent; ovary without a fleshy appendage; ovules two.

28. ***Calatola*** Standl.

Calatola Standl., Contrib. U.S. Natl. Herb. 23: 688 (1923); Howard, Contr. Gray Herb. 142: 13–20 (1942).

Trees; plants dioecious. Leaves papery to coriaceous, margins entire, revolute or sinuate-arcuate, slender, drying blackish; domatia absent. Inflorescences pseudo-spikes, axillary or extra-axillary; female inflorescences few-flowered (2–10) or flowers solitary. Flowers 4-merous; staminate flowers: sepals free, petals fused at base, pilose inside on midrib; stamens free, erect; disk absent; ovary absent; pistillate flowers: sepals fused at base; petals free; stamens absent; ovary cylindrical. Fruit an obovoid drupe, up to 60 mm long, with several longitudinal ribs. $2n = 28$.

Six or seven species, southern Mexico, Central and South America (Colombia, Venezuela, Ecuador, Peru, Bolivia, Brazil).

29. ***Emmotum*** Desv. ex Ham.

Emmotum Desv. ex Ham., Prodr. Pl. Indiae Occid. 29 (1825); De Stefano & Fernández-Concha, Ann. Missouri Bot. Gard. 98: 1–27 (2011), rev.
Pogopetalum Benth. (1840).

Trees or shrubs. Leaves generally densely hairy on abaxial surface, if glabrous then secondary veins diverging at right angles; margins entire; domatia absent. Inflorescences panicles, axillary. Flowers bisexual, 5-merous; sepals fused into a lobed tube; petals free, reflexed, inside lanate on raised, midrib fleshy, barbed; stamens free, filaments fleshy, flattened, connective fleshy, anthers longitudinally dehiscent, more or less extrorse; ovary ovoid with conspicuous differentiated area at base, 2- or more often 3-locular; style terminal or eccentric. Drupe ovoid with a heavy stony endocarp.

Twelve species, northern South America to Peru and Bolivia.

30. ***Oecopetalum*** Greenm. & C.H. Thomps.

Oecopetalum Greenm. & C.H. Thomps., Ann. Missouri Bot. Gard. 1: 408 (1915).

Trees. Leaves with (4–)7–10(–12) pairs of secondary veins diverging at right angles, drying brownish; margins entire; domatia absent. Inflorescences dichasial cymes, axillary or supraaxillary. Flowers bisexual, 5-merous; calyx accrescent, sepals almost fused, persistent; petals free; stamens inserted at base of petals, filaments thick, fleshy, connective very thick, anthers dehiscing laterally; disk absent; ovary conical, ovules superposed. Drupe ovoid, slightly flattened along longitudinal axis, with 2 ribs.

Two species, S Mexico to Guatemala and Costa Rica.

31. ***Ottoschulzia*** Urb.

Ottoschulzia Urb., Symb. Antill. 7: 272 (1912).

Trees or shrubs. Leaves with 4–7(8) pairs of secondary veins generally diverging at right angles, drying blackish, sometimes bicoloured, penninerved; margins entire; domatia absent. Inflorescences few-flowered (1–3). Flowers bisexual, 5-merous, sometimes functionally unisexual; sepals almost completely fused; petals fused at the base; stamens epipetalous, filaments short and fleshy, connective protruding beyond the laterally dehiscing pollen sacs; ovary conical. Drupe oblongoid.

Three or four species, southern Mexico, Guatemala, West Indies.

32. ***Platea*** Blume

Platea Blume, Bijdr. Fl. Ned. Ind. 646 (1826); Sleumer, Fl. Males. I, 7(1): 9–14 (1971), reg. rev.

Trees, rarely shrubs; dioecious. Leaves ± conduplicate in bud; young leaves with rust-coloured stellate scales or simple hairs; margins entire; domatia absent. Staminate flowers in axillary subspicate racemes or short-branched panicles; pistillate flowers in few-flowered cymes, axillary. Flowers (4)5-merous; sepals free or united at base; petals shortly connate at base, absent in female flowers; male flowers: stamens adnate to the petal tube, filaments short, ovary rudimentary or absent; female flowers: ovary thick-cylindric to subconical, blunt; stigma sessile, discoid. Drupe elongate-ovoid, crowned by a large discoid stigma, with 1 or 2 shallow longitudinal grooves.

About five species, SE Asia (China, India, Sumatra to New Guinea, New Britain).

33. ***Poraqueiba*** Aubl. Fig. 46

Poraqueiba Aubl., Hist. Pl. Guiane 1: 123, t. 47 (1775).
Barreria Scop., Introd. 182 (1777) non Linnaeus (1753).
Meisteria J.F. Gmel. (1791), non Scop. (1777) nec Sieb. & Zucc. (1846).

Trees. Leaf veins patent and fused near the margin; margin entire; domatia absent. Inflorescences panicles, axillary. Flowers bisexual, 5-merous; sepals fused; petals free, fleshy, with three pubescent ridges inside; stamens free, filaments fleshy, flattened, connective broad and protruding beyond the laterally dehiscing pollen sacs; ovary ovoid; style terminal, short; stigma minute, capitate; disk absent. Drupe ovoid to oblongoid, more or less oblique.

Three species, northern South America.

Fig. 46. Icacinaceae. **A–G** *Poraqueiba sericea*. **A** Flowering branch. **B** Floral diagram. **C** Petal adaxial. **D** Calyx. **E** Flower in longitudinal section. **F** Gynoecium. **G** Fruit. **H** *Poraqueiba guianensis*, fruiting branch. (From Engler 1872, modified, courtesy Missouri Botanical Garden)

Apodytes group

Leaves alternate, penninerved, entire; domatia absent. Flowers hermaphrodite; petals free, keeled or not; stamens free; anthers introrse; disc absent; ovary with a fleshy appendage or not; ovules two.

34. ***Apodytes*** E. Mey. ex Arn.

Apodytes E. Mey. ex Arn., J. Bot. (Hooker) 3: 155 (1840); Sleumer, Fl. Males. I, 7(1): 47–49 (1971), reg. rev.
Hemilobium Welwitsch (1862).
Anisomallon Baill. (1874).
Jobaboa Chiov. (1935).

Trees or shrubs. Leaves often shiny, drying blackish; margins entire. Inflorescences terminal or axillary racemes or panicles, sometimes corymbose. Flowers bisexual, 5-merous; sepals almost completely united, persistent; petals free or united at base, with short inflexed apex; filaments short, subulate, or thick, anthers sagittate; ovary oblique, with a unilateral swelling, style eccentric; stigma subcapitate, capitate or truncate; disk absent. Fruit a unilaterally developed drupe, obliquely ellipsoid or orbicular, ± compressed with a large, lateral fleshy appendage, style often persistent, lateral to almost basal at maturity. $2n = 24$.

Five species, tropical Australia (Queensland), SE Asia, subtropical Africa, Madagascar, Mauritius.

35. ***Rhaphiostylis*** Planch. ex Benth.

Rhaphiostylis Planch. ex Benth. in Hook., Niger Fl. 259, t. 28 (1849).
Chelonocarya Pierre (1896).

Shrubs or lianas. Leaves alternate, drying blackish; margins entire; domatia absent. Inflorescences supraaxillary fascicles. Flowers bisexual, 5-merous; petals free or united at base; filaments alternipetalous; ovary asymmetrical, with 2 narrow swellings at the top; style long, eccentric, filiform, persistent. Drupe reticulate-wrinkled, oblique on the stipe, broad-ovoid. $2n = 40$.

Ten species, tropical Africa.

Selected Bibliography

Aguwa, C.N., Okunji, C.O.1986. Gastrointestinal studies of *Pyrenacantha staudtii* leaf extracts. J. Ethnopharmacol. 15: 45–56.
Akube, P.I., Mittal, G.C., Aguwe, C.N. 1983. Preliminary pharmacological study of some Nigerian medicinal plants. J. Ethnopharmacol. 8: 53–64.
Angiosperm Phylogeny Group (APG) 1998. An ordinal classification for the families of flowering plants. Ann. Missouri Bot. Gard. 85: 531–553.
Angiosperm Phylogeny Group III 2009. An update of the Angiosperm Phylogeny Group classification for the orders and families of flowering plants: APG III. Bot. J. Linn. Soc. 161:105–121.
Baas, P. 1974. Stomatal types in Icacinaceae. Additional observations on genera outside Malesia. Acta Bot. Neerl. 23: 193–200.
Bailey, I.W., Howard, R.A. 1941a. The comparative morphology of the Icacinaceae. I. Anatomy of the node and internode. J. Arnold Arbor. 22: 125–131.
Bailey, I.W., Howard, R.A. 1941b. The comparative morphology of the Icacinaceae. II. Vessels. J. Arnold Arbor. 22: 171–187.
Bailey, I.W., Howard, R.A. 1941c. The comparative morphology of the Icacinaceae. III. Imperforate tracheary elements and xylem parenchyma. J. Arnold Arbor. 22: 432–442.
Bailey, I.W., Howard, R.A. 1941d. The comparative morphology of the Icacinaceae. IV. Rays of the secondary xylem. J. Arnold Arbor. 22: 556–568.
Breytenbach, G.J., Skinner, J.D. 1982. Diet, feeding and habitat utilization by bushpigs *Potamochoerus porcus* L. S.A. J. Wildlife Res. 12:1–7.
Carlquist, S. 1992. Wood anatomy of sympetalous dicotyledon families: A summary, with comments on systematic relationships and evolution of the woody habit. Ann. Missouri Bot. Gard. 79: 303–332.
Chesters, K.I.M. 1957. Some plant remains from the Upper Cretaceous and Tertiary of West Africa. Ann. Mag. Nat. Hist. 8: 498–504
Coates Palgrave, K. 1983. Trees of Southern Africa. 2nd impression. Cape Town: Struik.
Corner, E.J.H. 1976. The seeds of dicotyledons. 2 vols. Cambridge: Cambridge University Press.
Cronquist, A. 1981. An integrated system of classification of flowering plants. New York: Colombia University Press.
Cunningham, A.B. 1985. The resource value of indigenous plants to rural people in a low agricultural area. Vol. 1. Ph.D. Thesis, University of Cape Town, Cape Town.
Dahl, A.O. 1952. The comparative morphology of the Icacinaceae. 4. The pollen. J. Arnold Arbor. 33: 252–291.
Dahlgren, R. 1980. A revised system of classification of the angiosperms. J. Linn. Soc. Bot. 80: 91–124.
Dickison, W.C., Baas, P. 1977. The morphology and relationships of *Paracryphia* (Paracryphiaceae). Blumea 23: 417–438.
Edelhoff, E. 1887. Vergleichende Anatomie des Blattes der Familie der Olacineen. Bot. Jahrb. 8: 100–153.
Engler, A. 1872. Icacineae. In: Martius, C.F.P., Eichler, A. W., Urban, I. (eds.) Flora brasiliensis, vol. 12 (2). Leipzig: Frid. Fleischer, pp. 41–62, tab. 9–12.
Engler, A. 1896. Icacinaceae. In: Nat. Pflanzenfam. ed. 1. Bd. 3/5: 233–257.
Erdtman, G. (ed.) 1966. Pollen morphology and plant taxonomy. Angiosperms. New York: Hafner.
Fagerlind, F. 1945. Bau des Gynoceums, der Samenlage, und des Embryosackes bei einigen Repräsentanten der Familie Icacinaceae. Svensk Bot. Tidskr. 39: 346–364.
Fay, J.M. 1987. *Icacina oliviformis* (Icacinaceae): A close look at an under exploited crop. 1. Overview and ethnobotany. Econ. Bot. 41: 512–522.
Fay, J.M. 1993. *Icacina oliviformis* (Icacinaceae): A close look at an underexploited food plant. 3. Ecology and production. Econ. Bot. 47(2): 163–170.
Frost, P.G.H. 1980. Fruit-frugivore interactions in a South African coastal dune forest. In: Nohring, R. (ed.) Acta XVII Congr. lnt. Omithologici 2: 1179–1184. Berlin.
Gershenzon, J., Mabry, T.J. 1983. Secondary metabolites and the higher classification of angiosperms. Nord. J. Bot. 3: 5–34.
Gregor, H.-J., Hagn, H. 1982. Fossil fructification from the Cretaceous-Palaeocene-boundary of SW-Egypt (Danian, BirAbu Munqar). Tertiary Res. 4: 121–147.
Heintzelman, C.E., Howard, R.A. 1948. The comparative morphology of the Icacinaceae. V. The pubescence and the crystals. Am. J. Bot. 35: 42–52.
Hooker, J.D. 1867. *Alsodeiopsis mannii*. In: Hooker's Icones Plantarum, vol. 11: t. 1008. London: Williams and Norgate.
Hua, P., Howard, R.A. 2008. Icacinaceae. In: Wu, Z.Y., Raven, P.H., Hong, D.Y. (eds.) Flora of China 11: 505–514. Beijing: Science Press and St. Louis: Missouri Botanical Garden Press.
Jensen, S.R. 1992. Systematic implications of the distribution of iridoids and other chemical compounds in

the Loganiaceae and other families of the Asteridae. Ann. Missouri Bot. Gard. 79(2): 284–302.

Johri, B.M., Ambegaokar, P.S., Srivastava, P.S. 1992. Comparative embryology of angiosperms. 2 vols. Berlin: Springer.

Kaplan, M.A.C., Ribeiro, J., Gottlieb, O.R. 1991. Chemogeographical evolution of terpenoids in Icacinaceae. Phytochemistry 30: 2671–2676.

Kårehed, J. 2001. Multiple origin of the tropical forest tree family Icacinaceae. Am. J. Bot. 88: 2259–2274.

Keller, R. 1992. Les familles de Dicotylédones ligneuses tropicales et leur identification à l'aide d'une clef fondée sur les seuls caractères végétatifs. Thèse Doct. Sc., Univ. du Languedoc, Montpellier.

Lobreau-Callen, D. 1969. Les limites de l'ordre des Célastrales d'après le pollen. Pollen et Spores 11: 499–555.

Lobreau-Callen, D. 1973. Le pollen de Icacinaceae: 2. Observations en microscopie électronique, Corrélations, Conclusions. Pollen et Spores 15: 47–89.

Lobreau-Callen, D. 1975. Le pollen des Célastrales et groupes apparentés. Thèse Doct. Sc., Univ. du Languedoc, Montpellier.

Lobreau-Callen, D., Srivastava, S.K. 1974. Paleocene occurrence of *Platea* (Icacinaceae) pollen in U.S.A. Pollen et Spores 16: 411–423.

Mabogo, E. 1990. Ethnobotany of the VhaVenda. M.Sc. Thesis, University of Pretoria, Pretoria.

Manchester, S.R., Tiffney, B.H. 1993. Fossil fruits of *Pyrenacantha* and related Phytocreneae (Icacinaceae) in the Paleogene of North America, Europe and Africa. Am. J. Bot. (Abstracts) 80: 91.

Mauritzon, J. 1936. Embryologische Angaben über Stackhousiaceae, Hippocrateaceae und Icacinaceae. Svensk Bot. Tidskr. 30: 541–550.

Melchior, H., Werdermann, E. (eds.) 1964. Engler's Syllabus der Pflanzenfamilien. Ed. 12. Angiospermen. 2 vols. Berlin: Gebr. Bornträger.

Metcalfe, C.R., Chalk, L. 1950. Anatomy of the dicotyledons. 2 vols. Oxford: Clarendon Press.

Muller, J. 1981. Fossil pollen records of extant angiosperms. Bot. Rev. 47: 1–142.

Nagamasu, H., Kato., M. 2004. *Nothapodytes amamianus* (Icacinaceae), a new species from the Ryukyu Islands. Acta Phytotax. Geobot. 55(2): 75–78.

Napp-Zinn, K. 1973. Handbuch der Pflanzenanatomie. Berlin: Gebr. Bornträger.

Nichols, G. 1991. Pancakes and fritters. Plant Life - Newsletter for amateur botanists and indigenous plant growers 4: 4.

Palmer, E., Pitman, N. 1972. Trees of Southern Africa. 2 vols. Cape Town: A.A. Balkema.

Pigg, K.B., Manchester, S.R., DeVore, M.L. 2008. Fruits of Icacinaceae (Tribe Iodeae) from the Late Paleocene of western North America. Am. J. Bot. 95: 824–833.

Potgieter, M.J. 1994. Contributions to the systematics of Southern African Icacinaceae. M.Sc. Thesis, University of Pretoria, Pretoria.

Potgieter, M.J., van Wyk, A.E. 1992. Intercellular pectic protuberances in plants: Their structure and taxonomic significance. Bot. Bull. Acad. Sin. 33: 295–316.

Potgieter, M.J., van Wyk, A.E. 1994a. Fruit structure of the Southern African species of *Apodytes* E. Mey. ex Arn. (Icacinaceae). Bot. J. Linn. Soc. 115: 221–233.

Potgieter, M.J., van Wyk, A.E. 1994b. Fruit structure of the genus *Pyrenacantha* Hook. (Icacinaceae) in Southern Africa. Bot. Bull. Acad. Sin. 35: 105–113.

Potgieter, M.J., van Wyk, A.E. 1994c. Fruit structure of the genus *Cassinopsis* Sond. (Icacinaceae) in Africa. S. Afr. J. Bot. 60: 122–126.

Potgieter, M.J., van Wyk, A.E. 1994d. Bark structure of the Southern African Icacinaceae. IAWA J. 15: 161–170.

Potgieter, M.J., van Wyk, A.E. 1999. Leaf anatomy of the Southern African Icacinaceae and its taxonomic significance. S. Afr. J. Bot. 65: 153–162.

Pretorius, S.J., Joubert, P.H., Scott, A.H., Fourie, T.G. 1991. The mollucicidal properties of *Apodytes dimidiata* subsp. *dimidiata* (Icacinaceae). J. Trop. Med. Hyg. 94: 159–165.

Rankin, B.D., Stockey, R.A., Beard, G. 2008. Fruits of Icacinaceae from the Eocene Appian Way locality of Vancouver Island, British Columbia. Int. J. Plant Sci. 169: 305–314.

Rasoanaivo, P., Ratsimamango-Urverg, S., Messana, I., De Vincente, Y., Galeffi, C. 1990. Cassinopin, a kaempferol trirhamnoside from *Cassinopsis madagascariensis*. Phytochemistry 29: 2040–2043.

Savolainen, V., Manen, J.-F., Douzery, E., Spichiger, R. 1994. Molecular phylogeny of the Celastrales related families based on *rbc*L 5'flanking sequences. Mol. Phylogenet. Evol. 3: 27–37.

Savolainen, V., Chase, M.W., Hoot, S.B., Morton, C.M., Soltis, D.E., Bayer, C., Fay, M.F., de Bruijn, A.Y., Sullivan, S., Qiu, Y.-L. 2000a. Phylogenetics of flowering plants based upon a combined analysis of plastid *atpB* and *rbcL* gene sequences. Syst. Biol. 49: 306–362.

Savolainen, V. et al. 2000b. Phylogeny of the eudicots: a nearly complete familial analysis based on *rbcL* gene sequences. Kew Bull. 55: 257–309.

Scott, R.A., Barghoorn, E.S. 1957. *Phytocrene microcarpa* - a new species of Icacinaceae based on cretaceous fruits from Kreischerville, New York. The Palaeobotanist 6: 25–28.

Sharma, M.V., Uma Shaanker, R.U., Leather, S.R., Vasudeva, R., Shivanna, K.R. 2010. Floral resources, pollinators and fruiting in a threatened tropical deciduous tree. J. Plant Ecol. 4: 259–264.

Sleumer, H. 1942. Icacinaceae. In: Engler, A. (ed.) Nat. Pflanzenfam., 2nd ed., 20b: 322–396.

Sleumer, H. 1969. Materials towards the knowledge of the Icacinaceae of Asia, Malesia, and adjacent areas. Blumea 17: 181–264.

Sleumer, H. 1971. Icacinaceae. In: Van Steenis, C.G.G.J. (ed.) Flora Males. I, 7(1): 1–87. Leyden: Noordhoff.

Soltis, D.E., Soltis, P.A., Chase, M.W., Mort, M.E., Albach, D.C., Zanis, M., Savolainen, V., Hahn, W. H., Hoot, S.B., Fay, M.F., Axtell, M., Swensen, S.M., Prince, L.M., Kress, W.J., Nixon, K.C., Farris, J.S. 2000. Angiosperm phylogeny inferred from 18S rDNA, *rbcL*, and *atpB* sequences. Bot. J. Linn. Soc. 133: 381–461.

Soltis, D.E., Smith, S.A., Cellinese, N., Wurdack, K.J., Tank, D.C., Brockington, S.F., Refulio-Rodriguez, N. F., Walker, J.B., Moore, M.J., Carlsward, B.S., Bell, C. D., Latvis, M., Crawley, S., Black, C., Diouf, D., Xi, Z., Rushworth, C.A., Gitzendanner, M.A., Sytsma, K.J., Qiu, Y.L., Hilu, K.W., Davis, C.C., Sanderson, M.J., Beaman, R.S., Olmstead, R.G., Judd, W.S., Donoghue, M.J., Soltis, P.S. 2011. Angiosperm phylogeny: 17 genes, 640 taxa. Am. J. Bot. 98: 704–730.

Soudry, D., Gregor, H.-J. 1997. *Jodes israelii* sp. nov.: a huge phosphate-mineralized icacinacean fructification from the late Cretaceous of the Negev, southern Israel. Cretaceous Res. 18: 161–178.

Stull, G.W., Moore, R.W., Manchester, S.R. 2011. Fruits of Icacinaceae from the Eocene of southeastern North America and their biogeographic implications. Int. J. Plant Sci. 172: 935–947.

Takhtajan, A.L. 1980. Outline of the classification of flowering plants (Magnoliophyta). Bot. Rev. 46: 225–359.

Tanai, T. 1990. Euphorbiaceae and Icacinaceae from the Paleogene of Hokkaido, Japan. Bull. Nat. Sci. Mus., Ser. C, 16: 91–118.

Thayn, G.F., Tidwell, W.D., Stokes, W.L. 1985. Flora of the Lower Cretaceous Cedar Mountain formation of Utah and Colorado. Part III: *Icacinoxylon pittiense* n. sp. Am. J. Bot. 77: 175–180.

Thorne, R.F. 1992. Classification and geography of the flowering plants. Bot. Rev. 58: 225–348.

Utteridge, T., Nagamasu, H., Teo, S.P., White, L.C., Gasson, P. 2005. *Sleumeria* (Icacinaceae): A new genus from northern Borneo. Syst. Bot. 30: 635–643.

van Staveren, M.G.C., Baas, P. 1973. Epidermal leaf characters of the Malesian Icacinaceae. Acta Bot. Neerl. 22: 329–359.

Vera Caletti, P. 1999. Revisión del Género *Calatola* (Icacinaceae) en México. Tésis de Maestro, Instituto de Enseñanza e Investigación en Ciencias Agricolas, Texcoco, México. 107 pp.

Villiers, J.-F. 1973. Icacinaceae. In: Aubréville, A., Leroy, J.-F. (eds.) Flora du Cameroun. Mus. Natl Hist. Nat. 15: 3–100.

Watt, J.M., Breyer-Brandwijk, M.G. 1962. The medicinal and poisonous plants of southern and eastern Africa. 2nd edn. Edinburgh: Livingstone.

Lennoaceae

Lennoaceae Solms, Abh. Naturf. Ges. Halle 11: 174 (1870), nom. cons.

V. Bittrich

Annual or perennial succulent herbaceous holorhizoparasites with completely or nearly subterranean stems. Glandular hairs on stems, leaves and calyx. Leaves much reduced, scale-like, alternate, spirally arranged, exstipulate, simple, with entire margin. Inflorescences cymose, capitate or spicate, terminal. Flowers bracteate, pedicellate or sessile, bisexual, actinomorphic to slightly zygomorphic, perianth and the androecium usually (4)5–9(10)-merous, calyx lobes and stamens sometimes fewer than corolla lobes. Sepals united at least at the base, puberulent to tomentose, persistent; petals united for most of their length, lobes induplicate valvate, persistent; stamens adnate to the corolla tube, included, in 1 or 2 series alternating with corolla lobes, anthers dehiscing with longitudinal slits; nectary disk absent; ovary superior, 6–17-carpellate, syncarpous, each of the 6–17 primary locules divided by a false septum into 2 uniovulate chambers; placentation axile; style simple. Fruit irregularly circumscissily dehiscent. Seeds embedded in hardened ovary tissue forming 1-seeded pyrenes, tiny, exarillate, with minute globose embryo and copious starchy endosperm. $x = 9$.

A New World family of 2 genera with 4 species.

Vegetative Morphology. The Lennoaceae are annual (*Lennoa*) or perennial (*Pholisma*) obligatory parasitic herbs. They are generally considered achlorophyllous but Cummings and Welschmeyer (1998) detected the presence of chlorophyll a (no traces of chlorophyll b) in *Pholisma arenarium*. Annual or perennial habit is only discernable by field studies because perennial plants die back after flowering and survive as undifferentiated tissue at the infection site of the host plant (Yatskievych 1985). The succulent white to brownish stems are largely subterranean, up to 1.5 m long and 5 cm in diameter; their terminal part is sometimes enlarged, forming mushroom-like bodies on which the flowers arise. "Pilot" roots spread horizontally and give rise to short lateral haustorial roots (Kuijt 1966). The pilot roots also serve vegetative propagation; this explains the sometimes linear arrangement of inflorescences observed in *Pholisma*. The haustorial root grows straight towards a host root and its apical meristem is directly transformed into the haustorial organ (Kuijt 1966). Infections of Lennoaceae are localized, with minimal development of the haustorium beyond the infection site (Yatskievych 1985).

The spirally arranged entire leaves are reduced to linear to broadly deltoid scales up to 25 mm long. Suessenguth (1927) suggested a Fibonacci series of at least 5/13 for *Lennoa*. The leaves around the terminal inflorescences can form an involucrum. Stem, leaves and calyx are covered with glandular uniseriate multicellular trichomes; rarely part of these lack the glandular end cell (*Pholisma culiacanum*).

Vegetative Anatomy. Root anatomy of *Lennoa* was studied by Suessenguth (1927). According to him, the pilot roots lack an endodermis and a calyptra, the vascular bundles are collateral, and the haustorial roots are of exogenous origin. Kuijt (1966), on the other hand, described the origin of the haustorial roots as endogenous. Suessenguth (1927) rejected an interpretation of the pilot roots as modified stems because no leaves of any kind exist and the haustorial roots are arranged in four rows as is typical for tetrarchic roots.

Stems and leaves are covered with anomocytic stomata. The parenchyma of the stem and pith is rich in starch grains. Vascular bundles are

J.W. Kadereit and V. Bittrich (eds.), *Flowering Plants. Eudicots*, The Families and Genera of Vascular Plants 14,
DOI 10.1007/978-3-319-28534-4_22, © Springer International Publishing Switzerland 2016

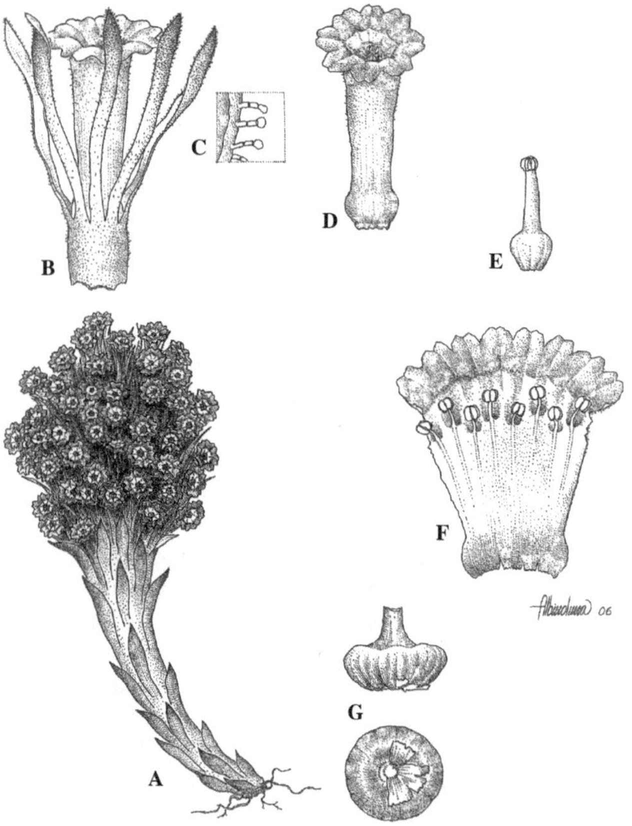

Fig. 47. Lennoaceae. *Lennoa madreporoides.* **A** Habit. **B** Flower. **C** Detail of calyx indument. **D** Corolla. **E** Gynoecium. **F** Corolla spread out showing insertion of the stamens. **G** Fruit (after Albino Luna in Alvarado-Cárdenas (2007), Fig. 1, with permission of the editor)

arranged in an irregular eustele with 7–12 major bundles in *Lennoa* and 15–25 in *Pholisma* in the more basal parts of the stems (Yatskievych and Mason 1986). Cortical bundles are present and secondary thickening is absent; vessel end-walls were described as simple by Solereder (1899), but Suessenguth could not find vessels in *Lennoa*, nor typical sieve plates in the phloem. The leaves of *Lennoa* have several reduced vascular bundles without a bundle sheath (Suessenguth 1927).

Inflorescence and Flower Structure. According to Copeland (1935) and Yatskievych and Mason (1986), the inflorescences are basically cymose (generally monochasial) giving rise to spicate and capitulum-like types. In *Pholisma sonorae* the flowers open in successive circles on the upper side of a shallowly concave receptacle. The bisexual, actinomorphic or slightly zygomorphic flowers are sessile or pedicellate and subtended by a bract. Calyx and corolla are persistent and slightly accrescent during fruit development. The calyx is either tubular with 6–8 short triangular lobes or more often divided for most of its length into 4–10 linear lobes, with open aestivation. It is densely tomentose or plumose externally with stipitate glandular trichomes. In *Lennoa* the first sepal develops obliquely to the bract (Suessenguth 1927). The petals are united nearly to the apex forming a tubular or hypocrateriform corolla with a 4–10-lobed limb, the lobes show induplicate valvate aestivation and are frequently emarginate. The petal primordia are separate in early stages of flower ontogeny (Suessenguth 1927; Copeland 1935). The corolla is finely tomentose externally or more rarely completely glabrous, lavender to bluish, sometimes with a white limb margin or with a yellow throat. The stamens are included, adnate to the corolla tube and arranged in 1 (*Pholisma*) or 2 (*Lennoa*) series, generally isomerous with the corolla lobes and alternating with them. Filaments are short, the anthers are introrse, dithecal, tetrasporangiate, and open by longitudinal slits. The anthers are incurved and the thecae basally spreading in *Lennoa*, in *Pholisma* anthers are erect and the thecae parallel. The anther wall is 3- or 4-layered and, in contrast to some reports in the literature, the endothecium has typical fibrous thickenings. The gynoecium is syncarpous, superior, globose to subturbinate, and composed of 5–17 carpels (carpel number can be highly variable within species), with a single terminal solid or hollow style and a capitate, entire or slightly lobed moist stigma. There are 2 horizontal epitropous ovules per locule originating from a thick axile placenta; each locule is later divided into 2 chambers by a false dorsal septum. A nectary disk is wanting.

Embryology. Anthers are tetrasporangiate with a persistent epidermis. The tapetum is glandular with 1- or 2-nuclear cells, the pollen grains are two-celled when shed. Ovules are tenuinucellate, unitegmic, anatropous or later campylotropous. The embryo sac is of the Polygonum type, endosperm formation is exclusively cellular, and embryogeny is of the caryophyllad type but with a much reduced suspensor. A small antipodial haustorium was observed in *Lennoa* by Suessenguth (1927).

Pollen Morphology. Pollen is 3–4(5)-colporate (Yatskievych and Zavada 1984), or colpate and colporate, often with alternating aperturate and inaperturate colpi (pseudocolpi). The exine sculpturing is psilate or finely reticulate in the polar region and adjacent to the colpi with the tectum being perforate or imperforate, while the pseudocolpi are scabrate to rugulate and perforate. The infrastructural layer is granular with few irregular short and stout columellae in *Lennoa* and columellate in *Pholisma*, the columellae are underlain by a thin foot layer.

Karyology. Meiotic chromosome counts suggest a base number of $x = 9$ for the family, with *Lennoa* being diploid with $n = 9$ and all three *Pholisma* species tetraploid with $n = 18$ (Yatskievych and Mason 1986).

Fruit and Seed. The fruit is slightly compressed and dry at maturity. It is included in the persistent calyx and the slightly accrescent corolla. The fruit (and the corolla) dehisces capsule-like by an irregular circumscissile ring below the middle of the fruit. Due to the hardening of the ovary tissue around the developing seeds, 1-seeded pyrenes are formed that are arranged peripherally in the lower part of the open fruit (Suessenguth 1927). The wall of the pyrenes is formed by stone cells. The seeds are tiny (0.5–1 mm long), laterally flattened, and protruding in the region of the raphe. The testa is thin, tan or dark brown, and has a reticulate surface. Like in Ehretiaceae and Heliotropiaceae, transfer cells are present in the seed coat, funicle and placenta (Diane et al. 2002). The embryo is poorly differentiated when the fruit opens, globose, embedded in copious starchy endosperm that also contains oil and protein crystals, the outer endosperm layer contains aleuron (Netolitzki 1926; Suessenguth 1927).

Dispersal. In *Pholisma sonorae*, infructescences were observed to break off and be blown away by the wind, breaking apart when tumbling across the ground. Harvester ants are recorded to carry away infructescence parts to their nests (Yatskievych 1985).

Reproductive Systems. The flowers contain small amounts of nectar, suggesting pollination by small insects. In *Pholisma culiacanum*, visits by flies, beetles and small butterflies were observed. Extremely low seed set found in older inflorescences suggests self-incompatibility (Yatskievych 1985).

Phytochemistry. Unknown.

Affinities. The family was first thought to be related to the Ericales but Hallier (1923), later largely supported by the detailed studies of Suessenguth (1927), suggested a close similarity with Boraginaceae. Hallier (1923) emphasized especially the presence of glandular hairs, vessels with simple pores, cymose inflorescences, the division of ovary locules by a false septum and the similarity of the fruits. He concluded that the Lennoaceae should be included in Boraginaceae. Other authors suggested relationships with Verbenaceae or Hydrophyllaceae. The transfer cells in the seed coat, funicle and placenta suggest close relationships to Ehretiaceae and Heliotropiaceae. Hallier's idea was basically followed by most later authors and recently supported by cladistic analyses based on DNA sequences of the plastid gene *rps2*, the mitochondrial gene *cox1* and *atpA*, and the nuclear ITS region (Smith et al. 2000) and using the chloroplast genes *rbcL* and *ndhF* (Olmstead and Ferguson 2001). The latter study found Lennoaceae (represented by the genus *Pholisma*) to be sister to *Tiquilia* (Ehretiaceae, subf. Ehretioideae). The most recent phylogenetic study based on four chloroplast markers confirms a close association of Lennoaceae with Ehretiaceae with strong support for the former to be nested in the Ehretiaceae, likely sister to *Tiquilia* (Weigend et al. 2014). Thus, Lennoaceae should either be included in Boraginaceae s.l. (accepted by APG III 2009) or part of the latter family has to be united with the Lennoaceae under the name Ehretiaceae within an order Boraginales to form monophyletic taxa. Boraginales are sister to the Lamiales (cf. Soltis et al. 2011).

Distribution and Habitats. Detailed information can be found in Yatskievych (1985). Species of *Pholisma* occur in the south-western United States and Mexico, They are found on dunes (inland and coastal) and other sandy areas, sometimes in deserts (Sonora and Mojawe), more rarely in thorn or short-tree forests on mountain slopes. *Lennoa madreporoides* occurs from Mexico to northern Colombia and Venezuela. Typical habitats for *Lennoa madreporoides* include dry or seasonally dry habitats, coastal sand dunes to oak woodland forests at up to 2200 m alt. *Lennoa madreporoides* also grows as a weed in corn or tobacco fields or in avocado groves. No evidence was found that this species is a parasite of these crops. Populations are often small and well-isolated from each other, leading to considerable infraspecific variation (Yatskievych and Mason 1986). The distribution of the species is probably highly correlated with the occurrence of suitable host plants, as all four species show host specificity (Yatskievych and Mason 1986). Host plants include members of Asteraceae, Boraginaceae, Nyctaginaceae, Polygonaceae and Zygophyllaceae. For *Lennoa madreporoides* 11, for *Pholisma culiacanum* 4, for *P. sonorae* 5, and for *P. arenarium* 8–9 host species have been documented. It is noteworthy that hosts include two species of the genus *Tiquilia* (Boraginaceae, Ehretioideae) found to be sister group of Lennoaceae in the cladistic study of Olmstead and Ferguson (2001). Plants of individual populations, however, generally parasitize only one or two host species even when other potential hosts are present (Yatskievych 1985). The fact that the parasitized host plants look perfectly healthy has spurred speculations that the parasites may be able to absorb moisture not through their roots that lack root hairs, but through their numerous stomata, and that part of this moisture is transferred via the haustoria to their hosts (Yatskievych 1985). No experiments have been reported, however, to support this hypothesis that would implicate a partly symbiotic relationship between parasite and host.

Economic Importance. The fleshy, semisweet underground stems of *Pholisma* and *Lennoa* species were collected by native tribes and eaten raw, cooked, baked or dried for later use. Especially important was *Pholisma sonorae*, the "Sonoran sandfoot", called "camote de los médanos" (sweet-potato of the dunes) by local Mexicans. It has also been suggested, however, that the poor dental health of the Papago indians might be due to a diet with high proportions of "Sonoran sandfoot", possibly due to the encrusted sand grains (Yatskievych 1985). A flour of a mixture of ground dried pieces of *Pholisma sonorae* and mesquite beans (*Prosopis juliflora* and *P. pubescens*) was used as a trail-food ("pinole"). Attempts to cultivate this species as a crop failed.

Conservation. According to the IUCN red list of threatened list, two species of Lennoaceae must be considered as threatened. This is due to clearing of native dune plants, agriculture, and use of off-road vehicles.

Key to the Genera

1. Flowers 8-merous; stamens in two series, thecae basally spreading **1. *Lennoa***
– Flowers 4–10-merous; stamens in 1 series, thecae basally parallel **2. *Pholisma***

Genera of Lennoaceae

1. ***Lennoa*** Lex. (Fig. 47).

Lennoa Lex. in Llave & Lex., Nov. Veg. Descr. 1: 7 (1824); Yatskievych & Mason, Syst. Bot. 11: 539–543 (1986), rev.
Corallophyllum Kunth (1825).

Annual herbs, stems up to 15(30) cm long and 2 cm in diameter. Leaves 3–20 mm long, lanceolate to narrowly deltoid. Inflorescences thyrsoid, globose to elongate. Flowers basically 8-merous; calyx lobed nearly to the base, with glandular trichomes; corolla lavender or bluish with yellow throat, sometimes externally puberulent; stamens in two series, anthers slightly incurved, thecae basally spreading; ovary (7)8(9)-carpellate. $n = 9$.

One species, *L. madreporoides* Lex., with two forms based on differences in corolla length, from Mexico to coastal Colombia and Venezuela.

2. ***Pholisma*** Nutt. ex Hook.

Pholisma Nutt. ex Hook., Icon. pl. 7: t. 626 (1844); Yatskievych & Mason, Syst. Bot. 11: 543–547 (1986), rev.
Ammobroma Torrey ex A. Gray (1854).

Perennial herbs, stems up to 1.5 m long and 5 cm in diameter. Leaves 5–25 mm long, linear to broadly deltoid. Inflorescences variable, cymes, spikes, sometimes capitulum-like. Flowers (4)5–9(10)-merous; calyx tubular or lobed nearly to the base, with glandular trichomes; corolla lilac to bluish-purple or pinkish and mauve, margin of the limb sometimes white, externally puberulous or glabrous; stamens in one series, anthers erect, thecae basally parallel; ovary 5–17-carpellate. $n = 18$.

Three species, south-western United States (California, Arizona) to Mexico.

Selected Bibliography

Alvarado-Cárdenas L.O. 2007. Lennoaceae. In Lemos, R. M. et al. (eds.) Flora del Valle de Tehuacán-Cuicatlán, fasc. 50. Mexico D.F.: Universidad Nacional Autonoma de Mexico, Instituto de Biologia.

Angiosperm Phylogeny Group (APG) 2009. An update of the Angiosperm Phylogeny Group classification for the orders and families of flowering plants: APG III. Bot. J. Linn. Soc. 161: 105–121.

Copeland, H.F. 1935. The structure of the flower of *Pholisma arenarium*. Amer. J. Bot. 22: 366–383.

Cummings, M.P., Welschmeyer, N.A. 1998. Pigment composition of putatively achlorophyllous angiosperms. Plant Syst. Evol. 210: 105–111.

Diane, N., Hilger, H.H., Gottschling, M. 2002. Transfer cells in the seeds of Boraginales. J. Bot. Linn. Soc. 140: 155–164.

Drude, O. 1891. Lennoaceae. In: A. Engler & K. Prantl (eds.) Die natürlichen Pflanzenfamilien, ed. 1, Teil 4, Abt. 1, pp. 12–15.

Hallier, H. 1923. Über die Lennoen, eine zu Linne's Bicornes verirrte Sippe der Borraginaceae. Beih. Bot. Centralbl. 40, 2. Abt.: 1–19.

Kuijt, J. 1966. Parasitism in *Pholisma* (Lennoaceae). I. External morphology of subterranean organs. Amer. J. Bot. 53: 82–86.

Netolitzki, F. 1926. Anatomie der Angiospermen-Samen. In: Linsbauer, K. (ed.) Handbuch der Pflanzenanatomie, II. Abt., III. Teil, Pteridophyten und Anthophyten, Band 10. Berlin: Gebrüder Borntraeger.

Olmstead, R., Ferguson, D. 2001. A molecular phylogeny of the Boraginaceae/Hydrophyllaceae. BSA Congress Botany 2001, Albuquerque (New Mexico), systematic section, abstract 521.

Smith, R.A., Ferguson, D.M., Barkman, T.J., Depamphilis, C.W. 2000. Molecular evidence for the origin of Lennoaceae: A case of adelphoparasitism in the angiosperms? Amer. J. Bot. 85 (suppl. 6): 157, BSA Congress Botany 2000, Portland (Oregon), systematic section, abstract 148.

Solereder, H. 1899. Systematische Anatomie der Dicotyledonen. Stuttgart: F. Enke.

Soltis, D.E., Smith, S.A., Cellinese, N., Wurdack, K.J., Tank, D.C., Brockington, S.F., Refulio-Rodriguez, N. F., Walker, J.B., Moore, M.J., Carlsward, B.S., Bell, C. D., Latvis, M., Crawley, S., Black, C., Diouf, D., Xi, Z., Rushworth, C.A., Gitzendanner, M.A., Sytsma, K.J., Qiu, Y.L., Hilu, K.W., Davis, C.C., Sanderson, M.J., Beaman, R.S., Olmstead, R.G., Judd, W.S., Donoghue, M.J., Soltis, P.S. 2011. Angiosperm phylogeny: 17 genes, 640 taxa. Amer. J. Bot. 98: 704–730.

Steyermark, J.A. 1968. El genero *Lennoa*. Acta Bot. Venez. 3: 213–237.

Suessenguth, K. 1927. Über die Gattung *Lennoa*. Ein Beitrag zur Kenntnis exotischer Parasiten. Flora 122: 264–305.

Weigend, M., Luebert, F., Gottschling, M., Couvreur, T.L. P., Hilger, H.H., Miller, J. 2014. From capsules to nutlets—phylogenetic relationships in the Boraginales. Cladistics 30: 508–518.

Yatskievych, G. 1985. Notes on the biology of the Lennoaceae. Cact. Succ. J. (U.S.A.) 57(2): 73–79.

Yatskievych, G., Mason, C.T. Jr. 1986. A revision of the Lennoaceae. Syst. Bot. 11(4): 531–548.

Yatskievych, G., Zavada, M. 1984. Pollen morphology of Lennoaceae. Pollen Spores 26(1): 19–30.

Metteniusaceae (incl. *Dendrobangia* incert. sed.)

Metteniusaceae Schnizl., Iconogr. Fam. Regni Veg. 2: ad t. 142 (1860–70).
Oncothecaceae Kobuski ex Airy Shaw, Kew Bull. 18: 264 (1965).

W.C. Dickison and V. Bittrich

Shrub or more commonly small to medium-sized trees with glabrous (*Oncotheca*) or hairy (*Metteniusa*), exstipulate, alternate leaves; leaves simple, margin entire or minutely glandular-denticulate. Flowers pentamerous, borne in axillary thyrsoid inflorescences, bisexual, actinomorphic; sepals 5, imbricate, persistent; petals 5, united into sympetalous corolla, lobes imbricate or valvate, glabrous or inside with secretory trichomes; stamens 5, alternate with petals, inserted on corolla tube, filaments short, anthers are dithecate, basically tetrasporangiate becoming polysporangiate by transversal septa (*Metteniusa*) or dithecate and bisporangiate (*Oncotheca*) by reduction of the dorsal pollen-sacs, extrorse, opening by longitudinal slits, connectives prolonged or not, when prolonged forming cover on superior ovary; ovary syncarpous, 5-carpellate, 5-locular with apical-axile placentation or unilocular with parietal placentation; ovules 2 per locule, pendant, anatropous, unitegmic; stylodia 5, distinct or connate for most of their length or style simple, filiform. Fruit drupaceous, with hard endocarp, 5-locular or 1-locular, with 1(2) seeds per locule. Seed albuminous, embryo straight or curved.

Two small genera, *Oncotheca* endemic to New Caledonia, *Metteniusa* restricted to NW South America and southern Central America.

Vegetative Morphology and Anatomy. Leaves glabrous or with an indument of unicellular T-shaped (or similar forms) or globular hairs below. Petiole anatomy is similar in the two genera with five or (*Metteniusa* spp.) up to nine collateral traces in a more or less U-shaped configuration at the petiole base. In *Oncotheca*, however, these fuse distally into one single arc which is invaginated at both sides, while in *Metteniusa* they remain separate, and the central one shows invaginated sides. The leaf mesophyll is bifacial, a hypodermis was found in *Oncotheca balansae* and *Metteniusa cogolloi*, but not in the other species of *Metteniusa*. Stomata of *Oncotheca* are of an unusual type, possibly to be interpreted as a modified paracytic or anisocytic type with subdivided subsidiary cells (Baas 1975; Carpenter and Dickison 1976); the stomata of *O. humboldtiana* have encircling striae. Stomata of *Metteniusa* were reported as cyclocytic (Lozano-Contreras and Lozano 1994). Mesophyll is densely tanniniferous and with clustered crystals of calcium oxalate. Highly branched foliar sclereids are present in leaves of *O. humboldtiana*. Baas (1975) and Carpenter and Dickison (1976) studied the stem anatomy of *O. balansae* in detail. Nodal anatomy is pentalacunar and 5-traced; petioles are vascularized proximally by five vascular bundles forming an arch in transversal section. Cork arises subepidermally, but also deeper in the secondary phloem; wood is hard, and its ground tissue is composed of very thick-walled fibre-tracheids, vessel elements have scalariform perforation plates; wood rays are heterocellular, uniseriate and multiseriate; wood axial parenchyma is sparse, diffuse and scanty. Nodal anatomy of *Metteniusa* was also found to be 5-lacunar 5-traced (González, pers. comm.). Wood anatomy of *Metteniusa* was investigated by Lens et al. (2008) in a comparative study of the polyphyletic Icacinaceae s.l. They considered the anatomy similar to genera of Cardiopteridaceae. Growth rings are visible. Vessels are mainly solitary, and perforation plates scalariform with numerous (up to 50) bars; intervessel pits are opposite to slightly alternate. Fibres have distinctly bordered pits. The axial parenchyma usually is diffuse-in-

Dickison (deceased).

J.W. Kadereit and V. Bittrich (eds.), *Flowering Plants. Eudicots*, The Families and Genera of Vascular Plants 14,
DOI 10.1007/978-3-319-28534-4_23, © Springer International Publishing Switzerland 2016

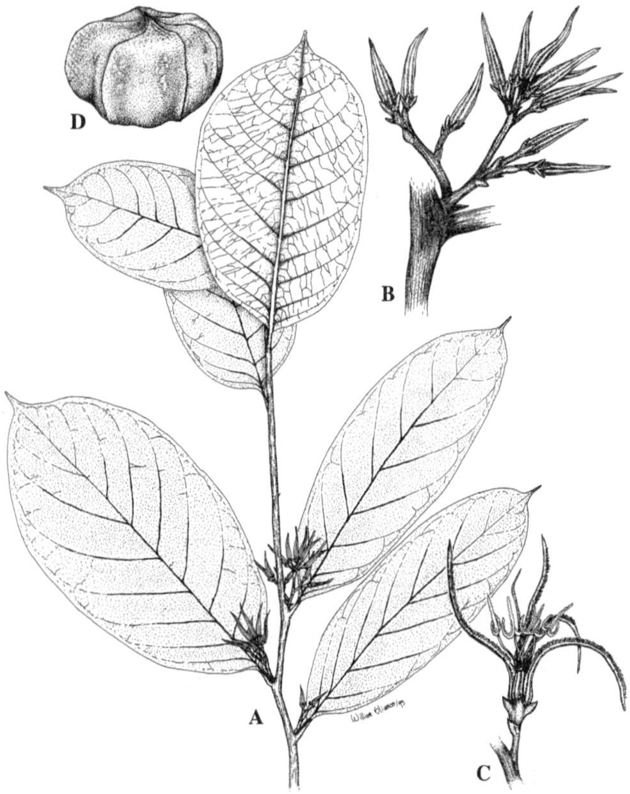

Fig. 48. Metteniusaceae. A–D *Metteniusa cogolloi*. **A** Flowering branch. **B** Detail of the inflorescence. **C** Open flower. **D** Fruit in dorsal view. (from Lozano-Contreras and Lozano 1994, with permission)

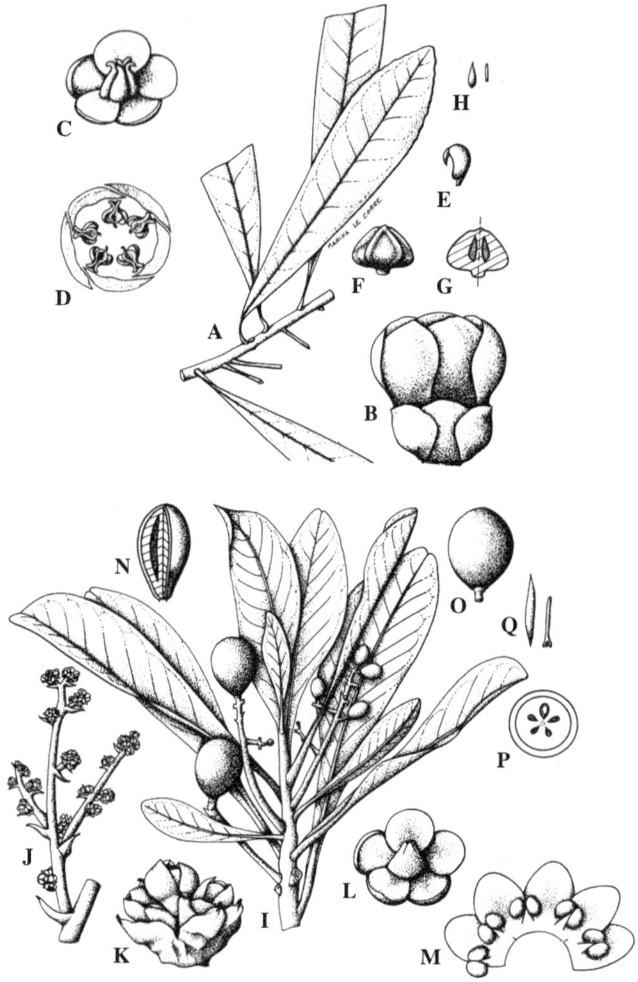

Fig. 49. Metteniusaceae. **A–H** *Oncotheca balansae*. **A** Vegetative. **B** Flower bud. **C** Flower showing calyx and gynoecium. **D** Flower, transverse section showing stamen attachment. **E** Stamen. **F** Fruit. **G** Fruit, longitudinal view. **H** Seed (left) and embryo (right). **I–Q** *Oncotheca humboldtiana*. **I** Fruiting branch. **J** Inflorescence. **K** Detail of part of inflorescence. **L** Flower showing calyx and gynoecium. **M** Corolla with attached androecium. **N** Fruit, partial longitudinal section. **O** Mature fruit. **P** Fruit, transverse section. **Q** Seed (left) and embryo (right). (from Morat and Veillon 1988, with permission)

aggregates plus scanty paratracheal, having a tendency to form short narrow bands. Uniseriate rays are present, multiseriate rays are 5–11-seriate, combining procumbent and square body cells, prismatic crystals are present in non-chambered cells.

Floral Morphology and Anatomy (Figs. 48, 49). Flowers are borne in axillary cymose inflorescences. The two species of *Oncotheca* can be separated on the basis of the presence (*O. balansae*) or absence (*O. macrocarpa*) of a prolonged stamen connective and either distinct ventrally stigmatic stylodia (*O. balansae*) or simple styles with distally free lobes (*O. macrocarpa*). The sepals are only slightly fused at their bases, the petals are fused in their lower half, forming a tube around the ovary, and the lobes are quincuncially imbricate. *Oncotheca* deviates from the majority of angiosperms in having dithecate bisporangiate stamens instead of the normal tetrasporangiate, dithecal pattern (Endress and Stumpf 1990). The mature anther wall contains a subepidermal endothecium composed of cells with prominent fibrous thickenings. The thecae open extrorsely by longitudinal slits. Each locule of the syncarpous ovary contains 2 pendant anatropous, epitropous crassinucellate unitegmic ovules with long funicles on apical-axile placentas. *Oncotheca balansae* has a single central ovary bundle while *O. macrocarpa* has free ventral carpel bundles. A detailed anatomical description of the flowers of *Oncotheca* was presented by Carpenter and

Dickison (1976) and Dickison (1986). The flowers of *Metteniusa* were studied in detail by González and Rudall (2010). The sepals are imbricate and connate at the base, the corolla shows late sympetaly, and its lobes are valvate. In the unusual polysporangiate anthers, no endothecial differentiation was observed. Both thecae are primarily bisporangiate and secondarily divided by transverse septa into secondary microsporangia. This results in the moniliform appearance of the anthers, which dehisce basically by 4 longitudinal slits, but each secondary microsporangium opens individually. The gynoecium is 5-carpellate, monosymmetric, pseudomonomerous, unilocular with parietal placentation; two ovule primordia develop in the two smaller carpels, and the three larger carpels (including the median one) are sterile; the one locule is bi-ovulate, and the one viable ovule is unitegmic, tenuinucellate and on a long funicle, whereas the abortive ovule has a short funicle (for details, see González and Rudall 2010).

Embryology. The mature anther wall may contain remnants of an internal tapetum. The ovules in *Oncotheca* are unitegmic and crassinucellate, the integument four to seven cells thick (Dickison 1986). Small druses and crystal fragments occur in peripheral integumentary cells. The single integument is massive and vascularized in *Metteniusa*, the nucellus strongly reduced, probably tenuinucellate (González and Rudall 2010). Nothing is known about embryo sac development.

Pollen. Pollen of *Oncotheca* has been described at the SEM and TEM levels by Lobreau-Callen (1975) and Dickison (1986). Pollen grains are small (P = 10–13 μm, E = 9–11 μm), 3-colporate and spheroidal or subprolate. The exine shows clear stratification into tectum, columellae, foot layer and endexine. The tectum is pitted or with microperforations. According to Lozano-Contreras and Lozano (1988), the pollen of *Metteniusa* is 3-colporate, scabrate, 31–42 μm long.

Reproductive Biology. *Oncotheca balansae* and *O. humboldtiana* are described as autogamous, and the stigmas can be covered by pollen masses without opening of the corolla (Morat and Veillon 1988). The fruits are eaten by birds. No data seem to exist for *Metteniusa*.

Fruit and Seeds. The fruit is a drupe with thin flesh and a very thick-walled, 5-locular stone. In *Oncotheca* each of the five locules has one, rarely two, fertile seeds, in *Metteniusa* the drupe is 1-seeded. The embryo is embedded in abundant endosperm and has a long hypocotyl and two short cotyledons. The seeds of *Metteniusa edulis* were used as food by indigenous people in Colombia (Karsten 1860).

Infrafamilial Relationships and Affinities. *Metteniusa* and *Oncotheca* were recently placed in their own families and even orders. Before, *Metteniusa* had been included in Icacinaceae, Cardiopteridaecae (cf. morphological cladistic analysis of Kårehed 2001, but with *Oncotheca* not included) or Opiliaceae among others. For *Oncotheca* an inclusion in Aquifoliaceae and Ebenaceae had been suggested. More recent anatomical and pollen morphological studies by Baas (1975), Carpenter and Dickison (1976), Dickison (1982, 1986) and Shiklina (1977) seemed to support the placement of Oncothecaceae near Theaceae; highly branched foliar sclereids as present in leaves of *O. humboldtiana* are also a characteristic feature of many theaceaous plants (Dickison 1982). The pentalacunar nodes and unitegmic ovules, however, did not fit with this placement. Evidence from DNA sequence data supported an isolated position of Metteniusaceae and Oncothecaceae in the Lamiidae close to a clade of (Boraginaceae + Gentianales + Lamiales + Solanales + Vahliaceae) according to González et al. (2007), while unpublished data by R. Olmstead (pers. comm.) suggest that the clade of Oncothecaceae + Metteniusaceae is sister to Icacinaceae s.s. Characters supporting the close relationship of Metteniusaceae and Oncothecaceae are pentalacunar nodes and a 5-carpellate gynoecium, both very rare in Lamiidae. Further characters suggesting a close relationship between the two families include inflorescence architecture, thick triangular bracts, 2 ovules per carpel, long funicles, a drupaceous fruit with persistent calyx, vessel elements with scalariform perforation plates, druses and tanniferous cells in most tissues (González and Rudall 2010). Recent studies suggest that several genera currently included in Icacinaceae (e.g. *Apodytes*, *Emmotum*, *Poraqueiba*) will need to be moved to Metteniusaceae (Greg Stull, pers. comm.), which would considerably alter the circumscription of this family. The new data also suggest that *Oncotheca* is more closely related to the reduced Icacinaceae s.s. than to the expanded Metteniusaceae.

Distribution and Habitat. *Oncotheca* is endemic to New Caledonia. *Oncotheca balansae* grows on stony, alluvial banks along rivers, on organically rich forest soils, and on open shrub plains. The species is often subject to flooding and prolonged submersion. It occurs to up to 800 m elevation and is known to accumulate nickel (Morat and Veillon 1988). *Oncotheca humboldtiana* grows in humid forests from sea level to up to 800 m elevation. *Metteniusa* occurs in montane cloud forests of the northern Andes and adjacent Central America. Long-distance dispersal and vicariance events were invoked to explain the disjunct distribution pattern of the supposed sister taxa *Oncotheca* and *Metteniusa* (e.g. Heads 2010).

Key to the Genera

1. Plants with indument; petals hairy adaxially, anthers moniliform; ovary 1-locular, style simple; fruit 1-seeded (Neotropics) **1. *Metteniusa***
– Plants glabrous; petals glabrous, anthers not moniliform; ovary 5-locular, stylodia or style branches 5; fruit with 1(2) seeds per locule (New Caledonia) **2. *Oncotheca***

1. ***Metteniusa*** H. Karst. Fig. 48

Metteniusa H. Karst., Fl. Columb. (H. Karst.) 1: 79, t. 39 (1860); Lozano-Contreras and de Lozano, Fl. Colombia 11: 1–53 (1988), rev.

Trees 6 to 18 m high. with simple and T-shaped hairs. Leaves simple, alternate, margin entire. Inflorescences axillary, few-flowered, cymose, bracts thick, triangular; pedicel articulate close to the flower. Flowers bisexual, sepals 5, imbricate, persistent; petals 5, fused at base forming a tube up to 2 cm long, lobes valvate, hairy inside; stamens alternate to the petals, anthers locellate by transverse septa (moniliform), sagittate, with free parts recurved at maturity; ovary unilocular, bi-ovulate (1 ovule fertile), style one, filiform, stigma punctate. Fruit a drupe, seed one with copious endosperm.

Seven species from Costa Rica to Peru, mainly in montane cloud forests.

2. ***Oncotheca*** Baillon Fig. 49

Oncotheca Baillon, Bull. Mens. Soc. Linn. Paris 2: 931 (1891).

Trees up to 30 m high, glabrous in all parts. Leaves alternate, margin minutely glandular-denticulate. Flowers small, sepals and petals fused at base, petals white, lobes imbricate; anthers dithecate, bisporangiate, connectives prolonged and covering the ovary or not; ovary 5-locular, 5-lobed, each carpel with a short stylodium and one biovulate locule. Drupe with 5 lateral grooves, 1(2) seeds per locule.

Two species, endemic to New Caledonia.

Genus incertae sedis:

Dendrobangia Rusby

Dendrobangia Rusby, Mem. Torrey Bot. Club 6: 19 (1896); Duno de Stefano, R., Candollea 62: 91–103 (2007).
Asterolepidion Ducke (1922).

Trees up to 40 m, with buttresses, most plant parts covered with lepidote-stellate hairs. Leaves alternate, simple, margin entire, venation brochidodromous, exstipulate. Inflorescence axillary panicles, flowers sessile, glomerulate. Flowers actinomorphic, hermaphrodite, pedicel articulated apically, bracteoles 2–6, sepals 5, imbricate, petals 5, united basally into a tube, lobes with strap-like, apically broadened acumen; stamens 5, adnate to the corolla tube and alternate to the lobes, anthers basifixed; ovary pubescent, unilocular, uniovulate. Fruit a drupe, yellow to orange. Seed one, embryo minute, endosperm copious. Wood anatomy (Lens et al. 2008): Vessels solitary, perforation plates scalariform with numerous bars + reticulate portions; intervessel pits opposite. Fibres with distinctly bordered pits. Axial parenchyma usually diffuse-in-aggregates, scanty paratracheal having a tendency to form short narrow bands, chambered axial parenchyma cells with prismatic crystals. Uniseriate and 8-seriate rays present combining procumbent and square body cells, with few small prismatic crystals (often in chambered cells) and small styloids.

Two species, Costa Rica to Bolivia and Brasil.

The position of this genus is presently uncertain. It seems to belong neither into Icacinaceae s. str. nor Cardiopteridaceae. It is here annexed to the Metteniusaceae s.lat. treatment but considered as *incertae sedis*. New data suggest that *Dendrobangia* is closely related to genera such as *Apodytes* and *Rhaphiostylis* and should be included in the expanded Metteniusaceae (Greg Stull, pers. comm.).

Selected Bibliography

Airy Shaw, H.K. 1965. Diagnoses of new families, new names, etc., for the seventh edition of Willis's "Dictionary". Kew Bull. 18: 249–273.

Baas, P. 1975. Vegetative anatomy and the affinities of Aquifoliaceae, *Sphenostemon*, *Phelline*, and *Oncotheca*. Blumea 22: 311–407.

Cameron, K. 2002. On the phylogenetic position of the New Caledonian endemic families Paracryphiaceae, Oncothecaceae, and Strasburgeriaceae: a comparison of molecules and morphology. Bot. Rev. 68(4): 428–443.

Carpenter, C.S., Dickison, W.C. 1976. The morphology and relationships of *Oncotheca balansae*. Bot. Gaz. 137: 141–153.

Dickison, W.C. 1982. Vegetative anatomy of *Oncotheca macrocarpa*, a newly described species of Oncothecaceae. Bull. Mus. Natl. Hist. Nat. Paris, IV. B, Adansonia 3-4: 177–181.

Dickison, W.C. 1986. Further observations on the floral anatomy and pollen morphology of *Oncotheca* (Oncothecaceae). Brittonia 38: 247–259.

Duno de Stefano, R. 2007. Tratamiento taxonómico del género *Dendrobangia* Rusby (Cardiopteridaceae o Icacinaceae). Candollea 62: 91–103.

Endress, P.K., Stumpf, S. 1990. Non-tetrasporangiate stamens in the angiosperms: structure, systematic distribution and evolutionary aspects. Bot. Jahrb. Syst. 112: 193–240.

González, F., Rudall, P.J. 2010. Flower and fruit characters in the early-divergent lamiid family Metteniusaceae, with particular reference to the evolution of pseudomonomery. Amer. J. Bot. 97: 191–206.

González, F., Betancur, J., Maurin, O., Freudenstein, J.V., Chase, M.W. 2007. Metteniusaceae, an early-diverging family in the lamiid clade. Taxon 56: 795–800.

Heads, M. 2010. The endemic plant families and the palms of New Caledonia: a biogeographical analysis. J. Biogr. 37: 1239–1250.

Kårehed, J. 2001. Multiple origin of the tropical forest tree family Icacinaceae. Amer. J. Bot. 88: 2259–2274.

Karsten, H. 1860. *Metteniusa*. In: Florae Colombiae, vol. 1, pt. 2. Berlin: Ferdinand Duemmler, pp. 79–80.

Lens, F., Kårehed, J., Baas, P., Jansen, S., Rabaey, D., Huysmans, S., Hamann, T., Smets, E. 2008. The wood anatomy of polyphyletic Icacinaceae s.l., and their relationships within asterids. Taxon 57: 525–552.

Lobreau-Callen, D. 1975. Les pollens des Célastrales et groups apparentés. Thesis, Univ. du Languedoc, Montpellier.

Lozano-Contreras, G., de Lozano, N.B. 1988. Metteniusaceae. In: Pinto, P., Lozano-Contreras, G. (eds.) Flora de Colombia, vol. 11. Bogotá: Instituto de Ciencias Naturales, Universidad Nacional de Colombia, pp. 1–53.

Lozano-Contreras, G., de Lozano, N.B. 1994. Una nueva especie de *Metteniusa* Karsten Metteniusaceae) de Colombia. Novon 4: 266–270.

Morat, P., Veillon, J-M. 1988. Oncothecaceae. In: Morat, P., Mackee, H.S. (eds.) Flore de la Nouvelle-Calédonie et Dépendances, 15, pp. 90–98. Mus. Natl. d'Histoire Naturelle, Paris.

Shiklina, I.A. 1977. The comparative anatomy of the wood of the genus *Oncotheca* (order Theales). Bot. Zhurn. 62: 1273–1275 (in Russian).

Montiniaceae

Montiniaceae Nakai, Chosakuronbun Mokuroku: 243 (1943), nom. cons.

Louis P. Ronse De Craene

Small trees, lianas, or shrubs, dioecious; stems erect with libriform fibres; indumentum of eglandular unicellular trichomes, present only in leaf axils. Leaves simple, entire, opposite or alternate, exstipulate, margins entire. Staminate inflorescences axillary, rarely terminal, paniculate; each flower subtended by a bract or with several bracts of reduced flowers; bracteoles absent. Flowers tri-, tetra-, or rarely pentamerous; calyx reduced to a toothed persistent rim or shortly three- to five-lobed; petals white or greenish, free, imbricate to valvate, deciduous; stamens alternipetalous, inserted at the base of a well-developed nectary, filaments short, dilated at the base, anthers extrorse, rarely introrse, opening by longitudinal slits; pistillode absent or present as central conical rudiment. Female inflorescence terminal, consisting of a single terminal flower, occasionally with a few flowers below, or a few-flowered axillary cyme. Perianth similar to staminate flowers, or petals absent; alternipetalous staminodes with abortive anthers present or absent; ovary bicarpellate, inferior to half-inferior; placentation intruding-parietal to axile, style bifid or two decurrent stylodia; stigma bilobed or decurrent on the stylodia, persistent; ovules tenuinucellate, anatropous, one or up to 12 per locule. Fruits capsular with winged seeds, dry indehiscent, or a compressed orange two-seeded drupe. Endosperm fleshy, scanty or absent, embryo straight, with large, compressed, orbiculate cotyledons and well developed radicle.

A family comprising three genera and five species confined to equatorial and southern Africa and Madagascar.

Vegetative Morphology and Anatomy. Montiniaceae are shrublets or large shrubs, rarely lianas. The vegetative anatomy of *Montinia*, *Kaliphora* and *Grevea* has been described by Milne-Redhead (1955), Ramamonjiarisoa (1980), Carlquist (1989) and Gregory (1998). Common features include vessels with simple oblique perforation plates, solitary or grouped into radial multiples, vestured pits (vessel walls vestured in *Grevea*), libriform non-septate or one- to two-septate fibres, axial parenchyma scanty paratracheal (absent or scarce vasicentric in *Kaliphora*), and multiseriate to uniseriate rays. *Montinia* and *Kaliphora* wood is very similar, but *Grevea* wood characters generally fit in Montiniaceae. Main differences of *Grevea* include vestured vessel walls vs. vestured pits, mostly absence of uniseriate rays, wood storied vs. non-storied, pith bundles present vs. absent, and rays homogeneous vs. heterogeneous. Trichomes are uniseriate and eglandular (Al-Shammary and Gornall 1994), mainly confined to the base of the petiole and leaf axil. Leaves are alternate (Fig. 50A) or (sub)opposite in *Grevea* (Fig. 51A), coriaceous or membranous (*Grevea*), persistent (deciduous in *Grevea*), exstipulate, with unilacunar nodes (Ramamonjiarisoa 1980). The number of traces differs from one (*Kaliphora*) to three in *Montinia* and 4–11 in *Grevea* (Gregory 1998). Stomata are restricted to the abaxial surface (amphistomatic in *Montinia*), anomocytic and occasionally anisocytic (*Grevea*).

Floral Morphology and Anatomy. Flowers are small, unisexual and dioecious. Flower structure follows a tetramerous plan, although there is variation from trimery to pentamery in *Montinia* and *Grevea*; *Kaliphora* is always tetramerous (Milne-Redhead 1955; Keraudren 1958; Ronse De Craene et al. 2000). Staminate flowers are grouped in axillary or terminal (*Montinia*) panicles

J.W. Kadereit and V. Bittrich (eds.), *Flowering Plants. Eudicots*, The Families and Genera of Vascular Plants 14,
DOI 10.1007/978-3-319-28534-4_24, © Springer International Publishing Switzerland 2016

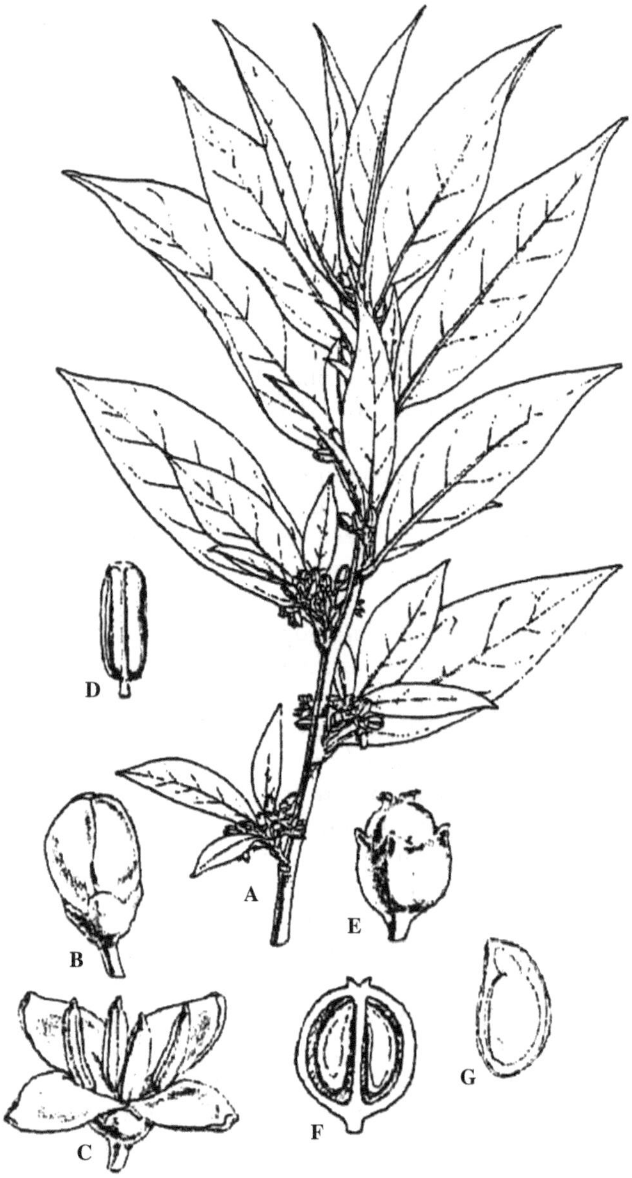

Fig. 50. *Kaliphora madagascariensis.* **A** Male branch. **B** Male flower bud. **C** Mature male flower. **D** stamen, adaxial side. **E** Pistillate flower at anthesis. **F** Longisection of young fruit. **G** Longisection of seed. (**A–C**, **E**, **G** after Hooker 1867a; **D**, **F** after Keraudren 1958)

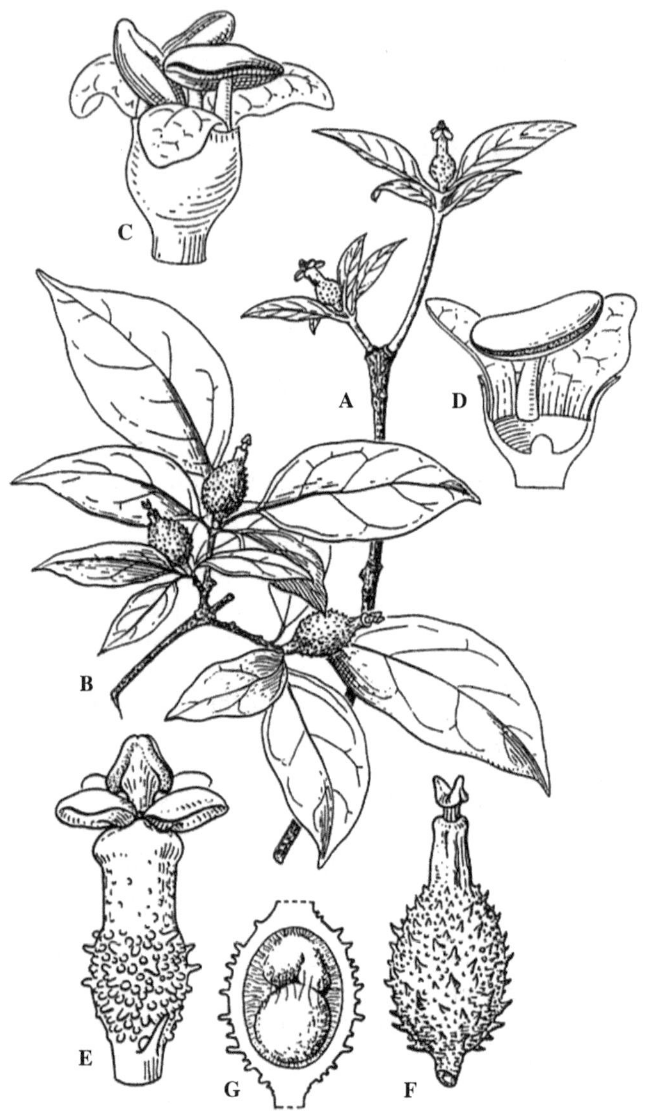

Fig. 51. *Grevea eggelingii.* **A** Female branch at anthesis. **B** Female fertile branch at fruiting. **C** Staminate flower. **D** Longisection of male flower with pistillode. **E** Pistillate flower at anthesis. **F** Pistillate flower at fruiting. **G** Longisection of fruit with two seeds visible. (After Milne-Redhead 1955)

(Fig. 50A), rarely racemes (*Grevea bosseri*). Sepals small, basally confluent into a rim (Figs. 50B, 51C). Petals broad, fleshy, free and imbricate (valvate in *Kaliphora* at least at the base; Capuron 1969a). Stamens alternipetalous, arising below bulging nectary. Filaments short, thickened. Anthers basifixed (*Kaliphora*) to dorsifixed, becoming extrorse at maturity in *Montinia* and *Grevea* (Ronse De Craene et al. 2000; Figs. 50D, 51C, D); thecae with longitudinal dehiscence; connective occasionally inflated, apically prolonged in *Kaliphora*. Rudimentary gynoecium absent or present as a conical rudiment confluent with nectary (Fig. 51D). Nectary thick, cushion-like. Pistillate flowers solitary and terminal (Fig. 51A, B), or in axillary clusters (Fig. 50A), sometimes reduced to a single flower (*Kaliphora*). Sepals and petals similar to staminate flowers or petals absent (Figs. 50E, 51E). Calyx lobes adnate to ovary with four to five teeth. Petals 4–5, deciduous. Staminodes absent (*Kaliphora*), or with aborted anthers (Fig. 51E). Gynoecium two-carpellate. Style persistent, deeply cleft with bilobed recurved stigma, each lobe being bilobed again (Fig. 51E, F), or two short recurved stylodes with decurrent stigmas (*Kaliphora*: Fig. 50E, F). Ovary inferior (Fig. 51F) to semi-inferior (*Kaliphora*: Fig. 50E), covered by broad nectary, bilocular or unilocular by placentas not fusing in the centre (Capuron 1969a, 1969b; Verdcourt 1973b; Ronse De Craene et al. 2000). Ovules ascending and anatropous, linear with inferior micropyle and short funicle, unitegmic. A single ovule per locule inserted about mid-level in the cavity (*Kaliphora*: Fig. 50F), or four to twelve ovules in four rows (Fig. 51G). Floral anatomy has only been studied for *Montinia* (Ronse De Craene et al. 2000). Transverse sections of pistillate flowers show two narrow septa linked to a broad, central, rectangular placenta on which the ovules are attached in four parallel rows. The placenta is divided into two equal parts in the upper part of the ovary. The stylar lobes are supplied by two pairs of marginal traces while the dorsals end below the nectary.

Pollen Morphology. The three genera are very similar in pollen morphology and share spheroidal tricolporate, isopolar, coarsely reticulate pollen grains (Milne-Redhead 1955; Pastre and Pons 1973; Hideux and Ferguson 1976). The three colpi have transverse furrows with “nexinous thickenings” (Hideux and Ferguson 1976). In *Grevea* and *Kaliphora* simple bacula occur in the mazes between the columellae.

Karyology. A chromosome number of $n = 16$ has been recorded in *Kaliphora* (Rabakonandrianina and Carr 1987) and of $n = 34$ in *Montinia* (Goldblatt 1976).

Fruit and Seed. Fruits are berry-like drupes with flattened pyrenes (*Kaliphora*), dry indehiscent (*Grevea*) or capsular (*Montinia*), with persistent calyx, style and nectary and the flattened seeds pressed against placentas. Fruits are crowned by the persistent styles (Figs. 50F, 51B, F). Fruits of *Grevea* are either strongly ribbed or the ribs are obscure (Verdcourt 1973a, 1973b, 1996). Seeds of *Grevea* and *Montinia* share several anatomical similarities (Krach 1976). In *Montinia* seeds are apically prolonged into a flattened wing serving dispersal, while seeds are globular in *Grevea* (Fig. 51G); exotesta lignified and thickened in *Montinia*, thin in *Grevea* and *Kaliphora*; endosperm abundant and fleshy (*Grevea*), thin (*Kaliphora*), or absent (*Montinia*); the embryo is large and straight, with a short axis (curved in *Kaliphora*) and broad *Tilia*-like cotyledons.

Phytochemistry. The presence of iridoid compounds (Montinioside) has been reported in *Montinia* (Dahlgren et al. 1977). Capuron (1969a) reported an acrid smell similar to pepper in crushed leaves of the three genera. *Montinia* scores negative for aluminium accumulation (Ronse De Craene et al. 2000).

Distribution and Habitats. *Montinia* is endemic to southern Africa, from the Cape Peninsula north to southern Angola, Botswana and Namibia, and east to Port Elisabeth, where it occurs in fynbos or dry savannah in either dry or more humid conditions. *Kaliphora* is restricted to the central part of Madagascar, growing in submontane, evergreen forest, while *Grevea* has three species found in western Madagascar (growing in dry forest), East Africa (growing in riverine thickets), and the Congo basin (growing in forest).

Uses. Leaves of Montiniaceae have an acrid, peppery smell, but uses are limited. Pungent leaves of *Grevea madagascariensis* are used to sooth colds and headaches (Capuron 1969b).

Family Status and Affinities. In earlier classifications the three genera have been classified widely apart, *Kaliphora* in the Cornaceae (Hooker 1867a, 1867b; Keraudren 1958), and *Grevea* and *Montinia* either in the Myrtales (Onagraceae or Oliniaceae: Hooker 1867b; Milne-Redhead 1955), in the Saxifragaceae (Baillon 1884; Engler 1928), Grossulariaceae (Cronquist 1981), or in the Escalloniaceae (Hutchinson 1967; Capuron 1969a, 1969b). Milne-Redhead (1955) accommodated the single species of *Montinia* and two species of *Grevea* in Montiniaceae. *Kaliphora* has sometimes been associated with *Melanophylla*, sharing comparable pollen morphology (Takhtajan 1973), or was placed in a family of its own (Takhtajan 1997). Thorne (1992) even included

Melanophylla in Montiniaceae but there are several differences, such as amplexicaulous leaves, bisexual pentamerous flowers, glandular trichomes, the presence of bracteoles, contorted petals, obscure or absent nectary, a trimerous gynoecium with two sterile locules, and a single pendent ovule (Hutchinson 1967; Capuron 1969a; Takhtajan 1997). *Melanophylla* is now placed in Torricelliaceae (Apiales). Despite some differences, several characters including anatomy also supported placement of *Kaliphora* with Montiniaceae (Carlquist 1989; Gregory 1998). Distinct characters supporting Montiniaceae are: dioecious plants with simple leaves without stipules and absence of trichomes, except for tufts of hairs in the leaf axils; tetramerous, haplostemonous flowers with free petals and broad nectary; pistillate flowers with (semi-) inferior ovary and bifid style or two persistent stylodes; seeds with large embryos having a short radicle and large cotyledons; similar pollen morphology. In addition, the three genera are reported to emit a spicy, burning odour when crushed. Leaves turn black when dried in *Kaliphora* and *Grevea* (Schatz 2001), but not in *Montinia*. Al-Shammary and Gornall (1994) discussed a possible affinity of Montiniaceae to *Vahlia*, especially based on vegetative characters. *rbc*L sequence data (Morgan and Soltis 1993) position *Montinia* in the Solanales (Asterids), in close relationship to *Sphenoclea* (Sphenocleaceae) and *Hydrolea* (Hydroleaceae; Cosner et al. 1994). Bremer et al. (2002) found strong support for inclusion of *Kaliphora* and *Grevea* in Montiniaceae and Soltis et al. (2011) found strong support for a sister-group relationship with Sphenocleaceae + Hydroleaceae based on a study of multiple genes. A possible synapomorphy for this group of families is the presence of vestured pits (Stevens 2001). Ronse De Craene et al. (2000) found little morphological and floral ontogenetic evidence supporting inclusion in Solanales (e.g. inferior ovaries, absence of stamen-petal tubes), although a stamen-petal tube may be secondarily lost in Montiniaceae.

Key to the Genera

1. Leaves alternate, petals valvate or absent (female flowers), anthers introrse at anthesis, ovary semi-inferior with two stylodia, one ovule per locule **1. *Kaliphora***
– Leaves opposite or alternate, petals imbricate, anthers extrorse at anthesis, ovary inferior with bifid style, several ovules per locule 2
2. Leaves opposite, membranous, flowers tri- to tetramerous, ovules 4 to 12, seeds globular **2. *Grevea***
– Leaves alternate, coriaceous, flowers mostly tetramerous, ovules 8 to numerous, seeds winged **3. *Montinia***

Genera of Montiniaceae

1. ***Kaliphora*** Hook. f. Fig. 50

Kaliphora Hook. f., Hook. Icon. Pl. 11, pl. 1023 (1867).

Evergreen tree or shrub, strongly branched; dioecious. Leaves entire, alternate, anisophyllous with reduced bract-like leaves subopposite full-sized leaves, elliptic-lanceolate, leathery, asymmetric at the base of the blade with abaxially protruding main vein. Flowers grouped in axillary panicles subtended by bract-like leaves; male flowers in pendent few- to multi-flowered panicles (up to 30 flowers); pedicel without bracteoles. Calyx 4-parted, teeth-like; petals 4, greenish, linear-oblong, valvate; 4 alternipetalous stamens inserted below margin of nectary, with short, dilated filament and elongated, basifixed anthers; conical stylar rudiment occasionally present; nectary strongly developed, cushion-like, four-angled. Pistillate flowers similarly arranged, but without corolla or staminodes; ovary semi-inferior with two decurrent stylodia; stigmatic surface papillate on adaxial side of stylodia; one ovule per locule inserted in upper half of locule. Fruit a berry-like drupe with 2 compressed pyrenes. Embryo large, with flat, thickened cotyledons and short curved radicle, endosperm abundant.

One species, *K. madagascariensis* Hook. f., Madagascar.

2. ***Grevea*** Baill. Fig. 51

Grevea Baill., Bull. Mens. Soc. Linn. Paris 1: 420 (1884).

Shrub or liana with (sub)opposite, membranous leaves with abaxially thickened main vein; glabrous except for tuft of hairs in axil of leaves; plants dioecious. Male flowers in axillary, long-pedunculate cymes with small bracts or in racemes. Sepals 3–4, short; petals 3–4, greenish, imbricate; stamens 3–4, extrorse, with short dilated filaments, connective broadly dilated in *G. bosseri*; gynoecium none or rudimentary, inserted on broad nectary; female flowers terminal and solitary; sepals 4; petals 4, imbricate; staminodes 4, well developed; ovary inferior, with tuberculate surface in *G. eggelingii* , bicarpellate with 2–12 ovules each, (in)completely septate; large bilobed stigma with each lobe bilobuled. Fruit dry and indehiscent. Seed globular, with copious endosperm.

Three species, *G. eggelingii* Milne-Redh. in E. Africa (Verdcourt 1996), *G. madagascariensis* Baill. in Madagascar (Verdcourt 1973a, 1973b), and *G. bosseri* R. Letouzey, a liana from Congo-Brazzaville (Letouzey 1977).

3. ***Montinia*** Thunb.

Montinia Thunb., Phys. Saellsk. Handl. 1: 108 (1776).

Erect shrublets with coriaceous, entire, alternate leaves; trichomes restricted to leaf axil; plants dioecious. Staminate inflorescence terminal, basically paniculate, with lateral flowers subtended by several bracts of reduced flowers. Flowers mostly tetramerous; calyx reduced to 3–5 teeth; petals 3–5, imbricate; stamens 3–5, alternipetalous, anthers extrorse; pistillode present or absent; pistillate inflorescence terminal, single-flowered; ovary bicarpellate, inferior, with massive bilobed, decurrent style and bilobed stigma; ovary synascidiate below, symplicate in upper half, with narrow septal connection; placentation basically intruding-parietal, ovules 8–24 superposed, with funicle absorbed in the ovular tissue. Seeds adaxially winged, without endosperm.

One species, *M. caryophyllacea* Thunb., in the Cape and Southwest Africa.

Selected Bibliography

Al-Shammary, K.I.A., Gornall, R.J. 1994. Trichome anatomy of the Saxifragaceae s.l. from the southern hemisphere. Bot. J. Linn. Soc. 114: 99–131.

Baillon, H. 1884. *Grevea madagascariensis*. Bull. Soc. Linn. Paris 1: 420, 477.

Bremer, B., Bremer, K., Heidari, N., Erixon, P., Olmstead, R.G., Anderberg, A.A., Källersjö, M., Barkhordarian, E. 2002. Phylogenetics of asterids based on 3 coding and 3 non-coding chloroplast DNA markers and the utility of non-coding DNA at higher taxonomic levels. Mol. Phylogenet. Evol. 24: 274–301.

Capuron, R. 1969a. Contribution à l'étude de la flore forestière de Madagascar. Sur la présence du genre *Kaliphora* Hook.f. Adansonia II, 9: 395–397.

Capuron, R. 1969b. Observation sur le *Grevea madagascariensis* Baillon. Adansonia II, 9: 511–514.

Carlquist, S. 1989. Wood anatomy and relationships of *Montinia*. Aliso 12: 369–378.

Cosner, M.E., Jansen, R.K., Lammers, T.G. 1994. Phylogenetic relationships in the Campanulales based on *rbc*L sequences. Plant Syst. Evol. 190: 79–95.

Cronquist, A. 1981. An integrated system of classification of flowering plants. New York: Columbia University Press.

Dahlgren, R., Jensen, S.R., Nielsen, B.J. 1977. Seedling morphology and iridoid occurrence in *Montinia caryophyllacea* (Montiniaceae). Bot. Not. 130: 329–332.

Engler, A. 1928. Saxifragaceae. In: Natürlich. Pflanzenfam., 2nd edn, 18a: 74–236. Leipzig: W. Engelmann.

Goldblatt, P. 1976. New or noteworthy chromosome records in the angiosperms. Ann. Missouri Bot. Gard. 63: 889–895.

Gregory, M. 1998. Montiniaceae. In: Cutler, D.F., Gregory, M. (eds.) Anatomy of the Dicotyledons, ed. 2, 4. Saxifragales (sensu Armen Takhtajan 1983). Oxford: Clarendon Press, pp. 110–116.

Hideux, M.J., Ferguson, I.K. 1976. The stereostructure of the exine and its evolutionary significance in Saxifragaceae sensu lato. In: Ferguson, I.K., Muller, J. (eds.) The evolutionary significance of the exine. London: Academic Press, pp. 327–378.

Hooker, J.D. 1867a. *Kaliphora madagascariensis* Hook.fil. Icones Plantarum 11, ser. 16–17: pl. 1023.

Hooker, J.D. 1867b. *Montinia, Kaliphora*. In: Bentham, G., Hooker, J.D. (eds.) Genera Plantarum. London: L. Reeve and Co., pp. 794, 951–952.

Hutchinson, J. 1967. The genera of flowering plants. Vol. I, Dicotyledons, 3rd edn. Oxford: Clarendon Press.

Keraudren, M. 1958. Cornacées. In: Humbert, H. (ed.) Flore de Madagascar et des Comores 158. Paris: Muséum national d'Histoire naturelle, pp. 1–17.

Krach, J.E. 1976. Samenanatomie der Rosifloren I. Die Samen der Saxifragaceae. Bot. Jahrb. Syst. 97: 1–60.

Letouzey, R. 1977. Présence du genre *Grevea* Baill. en Afrique Centrale Occidentale. Adansonia II, 2, 17: 119–123.

Milne-Redhead, E. 1955. Montiniaceae. Hooker's Icones Plantarum 6, V: 1–16, tab. 3541–3544.

Morgan, D.R., Soltis D.E. (1993). Phylogenetic relationships among members of Saxifragaceae sensu lato based on *rbc*L sequence data. Ann. Missouri Bot. Gard. 80: 631–660.

Pastre, A., Pons, A. 1973. Quelques aspects de la systématique des Saxifragacées à la lumière des données de la palynologie. Pollen Spores 15: 117–133.

Rabakonandrianina, E., Carr, G.D. 1987. Chromosome numbers of Madagascar plants. Ann. Missouri Bot. Gard. 74: 123–125.

Ramamonjiarisoa, B. 1980. Comparative anatomy and systematics of African and Malagasy woody Saxifragaceae sensu lato. PhD Thesis, University of Massachusetts, Amherst, 241 pp.

Ronse De Craene, L.P., Linder, H.P., Smets, E.F. 2000. The questionable relationship of *Montinia* (Montiniaceae): evidence from a floral ontogenetic and anatomical study. Amer. J. Bot. 87: 1408–1424.

Schatz, G.E. 2001. Generic tree flora of Madagascar. Richmond: Royal Botanic Gardens, Kew, 477 pp.

Soltis, D.E., Smith, S.A., Cellinese, N., Wurdack, K.J., Tank, D.C., Brockington, S.F., Refulio-Rodriguez, N. F., Walker, J.B., Moore, M.J., Carlsward, B.S., Bell, C. D., Latvis, M., Crawley, S., Black, C., Diouf, D., Xi, Z., Rushworth, C.A., Gitzendanner, M.A., Sytsma, K.J., Qiu, Y.L., Hilu, K.W., Davis, C.C., Sanderson, M.J., Beaman, R.S., Olmstead, R.G., Judd, W.S., Donoghue, M.J., Soltis, P.S. 2011. Angiosperm phylogeny: 17 genes, 640 taxa. Amer. J. Bot. 98: 704–730.

Stevens, P.F. 2001 onwards. Montiniaceae. Angiosperm Phylogeny Website. Version 9, June 2008 [and more or less continuously updated since]. http://www.mobot.org/MOBOT/research/APweb/.

Takhtajan, A. 1973. Evolution und Ausbreitung der Blütenpflanzen. Stuttgart: G. Fischer.

Takhtajan, A. 1997. Diversity and classification of flowering plants. New York: Columbia University Press.

Tahkhtajan, A., Trifonova, V. 1999. Fruit and seed anatomy of the genus *Kaliphora* (Kaliphoraceae) in relation to its taxonomical position. Bot. Zhurn. 84(4): 1–7.

Thorne, R.F. 1992. An updated phylogenetic classification of the flowering plants. Aliso 13: 365–389.

Verdcourt, B. 1973a. Notes on the genus *Grevea* Baill. (Montiniaceae). Kew Bull. 28: 147–150.

Verdcourt, B. 1973b. Montiniaceae. In: Polhill, R.M. (ed.) Flora of Tropical East Africa. London: Crown Agents, pp. 1–6.

Verdcourt, B. 1996. A reappraisal of the *Grevea* (Montiniaceae) growing on the Kenya coast. Kew Bull. 51: 771–773.

Morinaceae

Morinaceae Raf., Ann. Gén. Sci. Phys. Bruxelles 6: 88 (1820).

U. Hofmann and V. Bittrich

Shrubs or perennial herbs with basal leaf rosettes and initial taproot, basal part of axis ("rootstock") mostly covered with the remains of old leaf bases. Leaves of vegetative and flowering shoots opposite or whorled, exstipulate, simple, lanceolate, entire or mostly deeply lobed, margin entire, dentate or with spiny bristles; lamina basally tapering into a rather indistinct petiole, petioles in most species basally connate to form cylindrical sheaths. Flowers hermaphrodite, white, yellow or pink to purple, in terminal compact thyrses with 1–3-flowered cymes, or cymes in the axils of showy opposite or verticillate bracts, forming "verticillasters", these forming a head-like or cylindrical thyrse. Involucel (= epicalyx) 12-veined, of 4 fused bracteoles, cylindrical-campanulate, rim with few to numerous spiny teeth or seta 2. Calyx 4–5-merous, cupular or tubular, zygomorphic, with oblique, weakly 5-lobed opening (*Acanthocalyx*) or 2-lipped with lips 2- or 3-lobed or emarginate (*Morina*). Corolla of (4)5 petals, with a narrow tube, often curved, often villous outside, limb scarcely to markedly 2-lipped, lobes (4)5, spreading; in some species of *Morina* the corolla is shorter than the calyx and sometimes not or scarcely opening at anthesis (flowers possibly cleistogamous). Stamens 4 in two pairs, borne in the corolla tube, all fertile, or the 2 posterior fertile and the anterior sterile. Anthers introrse, 2-celled, locules subequal or markedly unequal, opening by longitudinal slits. Nectary at abaxial inside base of the corolla tube, covered with hairs, either cushion-like and uni- or trilobed, rarely bilobed, or nectary indistinctive. Ovary inferior, 3-locular with two locules containing sterile ovules only or 1-locular, pseudomonomerous, formed by the fusion of one lateral, fertile carpel and two solid, sterile carpels, ovule solitary, pendant. Style simple, slender, stigma disk-shaped. Fruit dry, 1-seeded, indehiscent (achene), enclosed in the involucel, when present, and surmounted by the ± persistent calyx, the fertile carpel elongated into a prominent tip. Endosperm copious, embryo large or small.

The family comprises two genera and 12 species from the Balkans to the Himalayas and SW China, with the centre of distribution in East Asia.

Vegetative Morphology. The seedling of *Morina longifolia* develops a taproot and a basal leaf rosette. The upper leaves in this rosette are joined in whorls of three or four, forming a short basal sheath which breaks irregularly between leaves. Therefore, the leaves in the rosette seem to be arranged irregularly. The rosette leaves are small, lanceolate, tapering into the indistinct petiole, the decurrent blade margin has coarse double-toothed triangular teeth, terminating in a spiny seta. In the second vegetation period, the main axis elongates into the flowering shoot with few whorls of leaves. The shoot leaves are short-petiolate with a short sheath. Innovation shoots, also with a basal rosette, are situated in the axils of basal rosette leaves of the flowering shoot. The first leaves of the lateral rosettes are opposite and decussate, later there are whorls composed of three or four leaves. Sometimes, there are bifurcated leaves in the transition whorl (Hofmann and Göttmann 1990). In fruit, the basal rosette leaves of the flowering shoot wither and finally the flowering axis dies off as does the core of the taproot. The peripheral parts split off longitudinally. Such a rootstock is crowned by several innovation rosettes. The other *Morina* species follow the same pattern. In *Acanthocalyx*, the

Hofmann Ursula (deceased).

J.W. Kadereit and V. Bittrich (eds.), *Flowering Plants. Eudicots*, The Families and Genera of Vascular Plants 14,
DOI 10.1007/978-3-319-28534-4_25, © Springer International Publishing Switzerland 2016

flowering shoots arise from below the leaf rosettes; all stem leaves are opposite, entire, small, lanceolate and have petioles and long sheaths.

VEGETATIVE ANATOMY. The mesomorphic, bifacial leaves of *Morina longifolia* have a two-layered palisade parenchyma and a spongy mesophyll. The collateral median vascular bundle is embedded in mostly chlorophyll-free parenchyma, sclerenchyma is lacking. The epidermis of both sides has anomocytic stomata and short multicellular, sunken, glandular hairs. Mostly on the adaxial surface, there are rather big isodiametric, sunken epidermal cells with a globular inclusion of unknown nature. Possibly the "mucilaginous epidermis" mentioned by Watson and Dallwitz (1992) refers to these cells. The leaves are not aromatic and are not mucilaginous when wet. Nodal anatomy of *Morina longifolia* is 5-lacunar 5-traced, rarely 3:3 (Neubauer 1978). Like in the species of Valerianaceae and Dipsacaceae studied by Neubauer (1978), the outer branches of the bifurcating lateral traces of opposite leaves join in the nodal cortex, forming a bridge. Different from the situation in these two families, however, in *M. longifolia* bundles starting at this bridge and leading to the petioles are lacking.

The flowering axis of *Morina* and *Acanthocalyx* is composed of a narrow cortex, a closed ring of conducting tissue and a dead, lignified parenchymatic pith about 2/3 the size of the axis. The epidermis has few stomata and the same isodiametric cells with globular inclusions as the leaf epidermis. Periderm is lacking. Adjacent to the phloem there are small groups of thin-walled, weakly lignified fibres. Sclerenchyma is lacking. Only small groups of primary xylem indicate separate primary vascular bundles. The adjacent inner parts of the secondary xylem consist of vessels with simple perforation, and the larger outer main part is composed of lignified prosenchymatic or fibrous cells. Parenchyma of cortex and pith contains oxalate druses.

INFLORESCENCE. The inflorescences are thyrses. The bracts, decussate in *Acanthocalyx*, mostly in whorls of 3–6 or more in *Morina*, each support a cymose partial inflorescence and together form a so-called verticillaster (Fig. 52E). In *Acanthocalyx* few verticillasters are condensed into a head-like inflorescence at the end of the flowering shoots, in *Morina* there are more and clearly distant verticillasters, or these form a cylindrical inflorescence. The bracts of the cymes are broadly ovate at the base, free or connate, with a long-acuminate spiny tip and spiny seta on the margin, mainly at the base. In *Morina longifolia*, the cymes have two basal dichasial ramifications, and additional flowers are arranged monochasially. Bracteoles as pherophylls of flowers of the next order are lacking. There are up to 14 flowers per partial inflorescence; early flowers are shortly pedicellate, later flowers sessile. Other species contain only up to 20 flowers per verticillaster (Cannon and Cannon 1984).

FLOWER ONTOGENY AND STRUCTURE. The zygomorphic flowers have a persistent, cylindrical-campanulate, truncate, slightly laterally compressed involucel or epicalyx in *Acanthocalyx* and *Morina*. In *Morina longifolia*, the involucel is formed by two median and two transversal bract primordia, the joint bases of which soon form a disk-like structure developing into the tube of the involucel (Hofmann and Göttmann 1990). The transverse rim has spiny bristles; the longest bristles in the median plane develop late from the tips of the median primordia. After formation of the primordium of the involucel, two additional pentagonal edges arise simultaneously on the flower primordium, of which the lower develops into the calyx. The median abaxial corner of this calyx primordium soon discontinues growth; the lateral ones grow up in pairs together with the lateral, earlike lips of the calyx. In *Acanthocalyx*, the calyx tube is cylindrical; the opening is oblique, with the longest part on the adaxial side and a fissure on the abaxial side, the margin forming 2 lateral and 3 posterior teeth, often with spiny bristles. Its ontogeny is unknown. In *Morina longifolia*, the corolla develops from the uppermost edge of the flower primordium. Its corners alternate with those of the calyx. Soon the common corolla base exceeds the flower apex. On the inner side of the corolla tube primordium, the adaxial pair of stamen primordia develops first, followed by the abaxial pair. The corolla lobes grow up and soon close the bud, the middle lobe of the lower lip covers the lobes of the upper lip and the latter the lateral lobes of the lower lip.

Fig. 52. Morinaceae. **A–C** *Morina chinensis*. **A** Leaf. **B** Involucel and calyx. **C** Calyx. **D–G** *M. kokonorica*. **D** Lower portion of plant. **E** Upper portion of plant. **F** Involucel and calyx. **G** Corolla spread out. **H–I** *M. chlorantha*. **H** Leaf. **I** Involucral bract. (from Hong and Barrie 2012, Fig. 505, with permission from Missouri Botanical Garden Press, St. Louis, and Science Press, Beijing)

In *Acanthocalyx*, both pairs of stamens are fertile, scarcely emergent from the corolla tube, and inserted with a short free filament near the throat of the corolla. In *Morina* only the adaxial stamens are fertile and similarly arranged as in *Acanthocalyx*, the cordate staminodes are inserted at the base of the corolla tube or shortly above. The staminodes are either conspicuously smaller than the fertile stamens (*Morina* s.s.) or of equal size (former genus *Cryptothladia*). The conspicuous abaxial nectary at the base of the corolla tube (normally abaxially gibbous here) is 3-lobed in *Morina longifolia*, contains xylem and phloem, and is densely covered with the nectar-secreting trichomes characteristic for the Dipsacales (Wagenitz and Laing 1984). The nectary of some other *Morina* species and that of *Acanthocalyx* are 1-lobed. In some species, there is one abaxial nectary, in others two separate ones at the base of the staminodes.

The symmetry axis of the inferior unilocular ovary is perpendicular to that of the calyx and corolla. The ovary can be interpreted as pseudomonomerous, composed of two sterile, solid carpels and a ± lateral fertile one, similar to, for example, *Centranthus* of Valerianaceae. The fertile carpel contains one pendant ovule and elongates into a prominent tip. The slender style contains three commissural vascular bundles, but the disk-like stigma does not provide evidence for the number of carpels.

Embryology. In *Morina longifolia*, many but not all cells of the endothecium in the outer anther wall divide periclinally to form a partially double (or triple) fibrous layer. Equally, some of the tapetum cells divide periclinally. The tapetum remains intact until the middle layer degenerates. Finally, the inner cell walls break down and the cell contents spread between the pollen grains as was observed in other families of Dipsacales. Thus, the tapetum is periplasmodial or amoeboid. The description of the tapetum as secretory (Vijayaraghavan and Sarveshwari 1968) is incorrect.

The anatropous, unitegmic and tenuinucellar ovule is very similar to that of, for example, *Centranthus* and has the same kind of vascular bundle in the integument. In the mature seed, the conspicuous embryo is embedded in copious endosperm, which is not ruminate (contra Vijayaraghavan and Sarveshwari 1968).

Pollen Morphology. Pollen of all species of *Acanthocalyx* and *Morina* were studied in detail by Blackmore and Cannon (1983), Verlaque (1983), and Jacobs et al. (2011). Morinaceae pollen grains are rather large. The pollen grains of *Acanthocalyx* are inconspicuously tricolporate, triangular in polar view (equatorial axis 100–125 μm), with the colpi in the rounded angles, subrhomboidal in equatorial view (polar axis 120–150 μm) and the endoapertures are zonorate. Pollen of *Morina* (incl. *Cryptothladia*, treated separately by Blackmore and Cannon 1983) is tripororate and has three strikingly long and prominent apertural protrusions, already seen by Mohl (1835), Fritzsche (1837), and Kerner (1898). Species of the former genus *Cryptothladia* have an equatorial axis of 55–70 μm, incl. the prominent domed protrusion and a polar axis of 100–120 μm, the endoapertures are zonorate. *Morina* s.s. pollen grains have an equatorial axis of 120–160 μm, incl. the funnel-shaped protrusion and a polar axis of 150–270 μm, the endoapertures are porate. Observing fresh pollen of *Morina longifolia*, Blackmore and Cannon (1983) found an enormous elasticity of the pollen wall and the production of pollen tube-like structures when the grains were hydrated or after physical pressure. Pollen grains of *Morina* and *Acanthocalyx* show an endocingulum at the inner nexine wall layer, a structure which in Dipsacales otherwise is found only in *Zabelia*. These three genera also share a psilate ornamentation of the pollen surface (Verlaque 1983; Jacobs et al. 2011).

Karyology. The chromosomes are much smaller than those of the Valerianaceae and Dipsacaceae (Verlaque 1983). Chromosome counts of three *Morina* species are available, all with $2n = 34$ (Cannon and Cannon 1984; IPCN).

Pollination. The showy white or pink to red, rarely yellow flowers of *Acanthocalyx and Morina* with a long, narrow corolla tube seem to be adapted to pollinators with a long tongue. The flowers of *Morina longifolia* open in the evening and are first white and may be visited by moths (Kerner 1898). They become red later. Self-pollination occurs by recurving of the style bringing the stigma into contact with the anthers. Bagged flowers produce well-developed fruits. The short corollas of some *Morina* spp. (*Cryptothladia*) are hidden between the lateral calyx lobes. As the corolla lobes do not open, these flowers are possibly cleistogamous. Direct observations of pollinators in Morinaceae are lacking.

Fruit and Seed, Dispersal. The rootstocks of *Morina* species may have several leaf-rosettes, but proper vegetative propagation seems to be absent. The achenes simply fall out of the involucel, at least in *Morina*. The large, bilobed calyx of *Morina* may act as a sort of sail in wind dispersal, but it soon falls off. The calyx in *Acanthocalyx* enlarges in fruit and becomes ± leathery (Hong and Barrie 2011).

Phytochemistry. Records of chemical compounds in Morinaceae concern mostly *Morina* species. Alkaloids, saponins and polyphenols were found. The characteristic iridoids of Dipsacales are lacking in *Morina* (Hegnauer 1966, 1986, 1989).

Economic Importance. Two *Morina* species are ornamentals. Seeds of *Morina persica* are eaten like rice by peasants of Iran (Cannon and Cannon 1984).

Affinities. The characteristic nectary suggests that Morinaceae is a member of a monophyletic order Dipsacales s.str (= excl. Adoxaceae s.l.) according to Wagenitz and Laing (1984), and this has been confirmed by molecular data (e.g. Backlund and Pyck 1998). There are morphological similarities with Dipsacaceae (involucel) and also with Valerianaceae (details of ovule and ovary, cymose parts of inflorescence), but morphological differences as well. Form and number of chromosomes and the special pollen type suggested that Morinaceae s.str. (*Morina* and *Acanthocalyx*) is a monophyletic family, confirmed by Bell and Donoghue (2003) using DNA sequence data. In their phylogeny the family appears as sister group to a clade containing Valerianaceae and Dipsacaceae plus *Triplostegia*. APG III (2009) unites Caprifoliaceae s.s., Morinaceae, Dipsacaceae and Valerinaceae into Caprifoliaceae s.l., which is sister to Adoxaceae s.l. The data of Bell and Donoghue (2003) also depict *Cryptothladia* and *Acanthocalyx* as monophyletic, but *Morina* as paraphyletic after the exclusion of *Cryptothladia*. A close relationship of *Morina* and *Cryptothladia* was suggested also by Blackmore and Cannon (1983) in a cladogram based on palynological data. Hong (2010) discussed the morphological characters used to separate the two genera and also proposed to re-include *Cryptothladia* in *Morina*. The inclusion of *Zabelia* (formerly thought to belong into Caprifoliaceae (tribe Linnaeeae) or even in *Abelia*) was suggested by recent phylogenetic studies based on DNA sequences (Pyck 2001; Jacobs et al. 2010, 2011; Soltis et al. 2011). Palynological data provided support to this idea with pollen being tricolporate, psilate and especially having an endocingulum (Verlaque 1983; Jacobs et al. 2011). Most characters, however, like unarmed leaves, lacking involucel, actinomorphic calyx and achene morphology, did not fit with Morinaceae. Landrein et al. (2012) re-examined the relationships of *Zabelia* with improved sampling and additional DNA markers. Their data, although still not conclusive, did not support an inclusion of *Zabelia* in Morinaceae, but rather a sister-group relationship to the (Morinaceae (Dipsacaceae + Valerianaceae)) clade. The genus is included in this volume in the Caprifoliaceae and treated as *incertae sedis*.

Key to the Genera

1. Stem leaves always opposite; inflorescence (sub)capitate; calyx tube oblique at opening; fertile stamens 4 **1. *Acanthocalyx***
– Stem leaves opposite or mostly in whorls; inflorescence cylindric or with well-separated verticillasters; calyx 2-lipped; fertile stamens 2 plus 2 staminodes **2. *Morina***

Genera of Morinaceae

1. ***Acanthocalyx*** (DC.) Tiegh.

Acanthocalyx (DC.) Tiegh., Ann. Sci. Nat. Bot. IX, 10: 199 (1909); Hong & Barrie, Fl. China 19: 649–653 (2011), rev.
Morina sect. *Acanthocalyx* DC. (1830).

Low perennial herbs; stem leaves always decussate. Lamina linear to lanceolate, margin often with yellow, spreading spiny bristles, petioles connate into long sheaths. Flowering shoots arising from below the basal rosette, up to 35 (50) cm, with 2–4 leaf pairs; inflorescence head-like; involucel cylindrical-campanulate, persistent, rim with spiny teeth. Calyx tube cylindrical, opening oblique, fissured ventrally, usually with 2 inconspicuous lateral and 3 posterior teeth and sometimes with spiny bristles; corolla white, cream or yellowish to pink or purple, exceeding the calyx, with spreading lobes; stamen 4, all fertile, scarcely emergent from the corolla tube; nectary one, 1-lobed; style with a tuft of hairs just below the stigma. Achene rugose, sometimes puberulent.

Two species in a wide range of habitats from woodland to alpine meadows, Sino-Himalaya and Hengduan Mts. (Nepal, Sikkim, Bhutan, China (Xizang, Gansu, Sichuan, Yunnan), 2500–4900 m.

2. ***Morina*** L. Fig. 52

Morina L., Sp. Pl. 28 (1753); Hong & Barrie, Fl. China 19: 649–653 (2011), reg. rev.
Asaphes Spreng. (1827) non DC. (1825).
Cryptothladia (Bunge) M.J. Cannon (1984).

Robust thistle-like herbs; rosette leaves and lower stem leaves opposite or in whorls of up to 4–6, glabrous, simple or 2–5-lobed, spinose, petioles fused to form short sheaths; flowering stems up to 125 cm long, with ca. 4 whorls of leaves; inflorescence cylindrical or composed of 7–9 well-separated verticillasters with up to 50 flowers; bracts of cymes showy, spinose; involucel ± cylindrical, limb spiny-toothed. Calyx with a short tube and 2 entire or shallowly to deeply bilobed or trilobed lips, sometimes spinose; corolla either showy, the long tube exceeding the calyx lobes or ± hidden in the calyx, yellow or white, becoming pink or purple, limb 2-lipped, upper one 2-lobed, lower one 3-lobed, spreading; fertile stamens 2, inserted just below the corolla throat or halfway down the tube, staminodes 2, minute and inserted near the stamens or well-developed and inserted near the base of the tube; nectary 1(2), irregular or 3-lobed, at the abaxial, sometimes gibbous base of the corolla tube. Achene rugose, embryo large. $2n = 34$.

Ten species in a wide range of habitats, mainly alpine, Balkan to E Himalaya, in China eight spp., four endemic, 500–4700 m. *Morina longifolia* Wall. and *M. persica* L. are sometimes cultivated as ornamentals (whorl flower).

Selected Bibliography

APG III 2009. An update of the Angiosperm Phylogeny Group classification for the orders and families of flowering plants: APG III. Bot. J. Linn. Soc. 161: 105–121.

Backlund, A., Pyck, N. 1998. Diervillaceae and Linnaeaceae, two new families of caprifolioids. Taxon 47: 657–661.

Bell, C.D., Donoghue, M.J. 2003. Phylogeny and biogeography of Morinaceae (Dipsacales) based on nuclear and chloroplast DNA sequences. Organisms Divers. Evol. 3(3): 227–237.

Blackmore, S., Cannon, M.J. 1983. Palynology and systematics of Morinaceae. Rev. Palaeobot. Palynol. 40: 207–226.

Bunge, A. 1852. Beitrag zur Kenntnis der Flora Russlands. 321–323. St. Petersburg: Kaiserliche Akademie der Wissenschaften.

Cannon, M.J., Cannon, J.F.M. 1984. Revision of the Morinaceae. Bull. Br. Mus. nat. Hist. (Bot.) 12(1): 1–35.

Fritzsche, J., 1837. Über den Pollen. Mém. Sav. étrang. Acad. St. Petersburg 3: 649–769.

Hegnauer, R. 1966, 1986, 1989. Chemotaxonomie der Pflanzen, vols. IV, VII, VIII. Basel: Birkhäuser.

Hofmann, U., Göttmann, J. 1990. *Morina* L. und *Triplostegia* Wall. ex DC. im Vergleich mit Valerianaceae und Dipsacaceae. Bot. Jahrb. Syst. 111: 499–553.

Hong, D.Y. 2010. Nomenclatural notes on the Morinaceae and Valerianaceae in China. Novon 20(4): 418–419.

Hong, D.Y., Barrie, F.R. 2011 (text), 2012 (illustrations). Morinaceae. In: Wu, Z.Y., Raven, P.H., Hong, D.Y. (eds.) Flora of China 19. Beijing: Science Press; St. Louis: Missouri Bot. Garden Press, pp. 649–653.

Index to Plant Chromosome Numbers (IPCN). http://www.tropicos.org/Project/IPCN.

Jacobs, B., Pyck, N., Smets, E. 2010. Phylogeny of the Linnaea clade: Are *Abelia* and *Zabelia* closely related? Mol. Phylogenet. Evol. 57: 741–752.

Jacobs, B., Geuten, K., Pyck, N., Huysmans, S., Jansen, S., Smets, E. 2011. Unraveling the phylogeny of *Heptacodium* and *Zabelia* (Caprifoliaceae): An interdisciplinary approach. Syst. Bot. 26: 231–252.

Kerner von Marilaun, A. 1898. Pflanzenleben, Bd. 2, Leipzig, Wien: Bibliographisches Institut; pp. 88, 316.

Landrein, S., Prenner, G., Chase, M.W., Clarkson, J.J. 2012. *Abelia* and relatives: Phylogenetics of Linnaeeae (Dipsacales–Caprifoliaceae s.l.) and a new interpretation of their inflorescence morphology. Bot. J. Linn. Soc. 169: 692–713.

Mohl, H. 1835. Sur la structure et les formes des grains de pollen. Ann. Sci Nat. ser. 2, 3: 148–180, 220–236, 304–346.

Neubauer, H.F. 1978. On nodal anatomy and petiole vascularization of some Valerianaceae and Dipsacaceae. Phytomorphology 28: 431–436.

Pyck, N. 2001. Phylogenetic relations within Dipsacales: a combined molecular and morphological approach. Ph.D. Dissertation. Leuven: Katholieke Universiteit Leuven.

Qiner, Y., Landrein, S. 2011. Linnaeaceae. In: Wu, Z.Y., Raven, P.H., Hong, D.Y. (eds.) Flora of China 19. Beijing: Science Press; St. Louis: Missouri Bot. Garden Press, pp. 642–648.

Rehder, A. 1911. Caprifoliaceae. In: Sargent, C.S. (ed.) Plantae Wilsonianae, vol. 1. Cambridge: Cambridge Univ. Press, pp. 106–144.

Soltis, D.E., Smith, S.A., Cellinese, N., Wurdack, K.J., Tank, D.C., Brockington, S.F., Refulio-Rodriguez, N. F., Walker, J.B., Moore, M.J., Carlsward, B.S., Bell, C. D., Latvis, M., Crawley, S., Black, C., Diouf, D., Xi, Z., Rushworth, C.A., Gitzendanner, M.A., Sytsma, K.J., Qiu, Y.L., Hilu, K.W., Davis, C.C., Sanderson, M.J., Beaman, R.S., Olmstead, R.G., Judd, W.S., Donoghue, M.J., Soltis, P.S. 2011. Angiosperm phylogeny: 17 genes, 640 taxa. Amer. J. Bot. 98: 704–730.

Verlaque, R. 1983. Contribution à l'étude du genre *Morina* L. Pollen et Spores 25: 143–162.

Vijayaraghavan, M.R., Sarveshwari, G.S. 1968. Embryology and systematic position of *Morina longifolia*. Bot. Not. 121: 383–402.

Wagenitz, G., Laing, B. 1984. Die Nektarien der Dipsacales und ihre systematische Bedeutung. Bot. Jahrb. Syst. 104: 483–507.

Watson, L., Dallwitz M.J. 1992 onwards. The families of flowering plants: Morinaceae. http://delta-intkey.com/angio/www/morinace.htm.

Xu, L., Lu, L., Li, D.-Z., Wang, H. 2011. Evolution of pollen in Dipsacales. Plant Divers. Resources 33: 249–259.

Paracryphiaceae

Paracryphyiaceae Airy Shaw, Kew Bull. 18: 265 (1965).

W.C. Dickison and J. Lundberg

Shrubs to medium-sized trees, some vines. Leaves alternate to almost verticillate (*Paracryphia*), simple, margins finely serrate or sometimes entire; stipules absent; dense pubescence on young leaves, absent on mature foliage. Flowers in axillary or terminal racemes or compound spikes, bisexual or unisexual (plants andromonoecious); perianth differentiated into 4–5 sepals and 4–5 white, free, deciduous petals (*Quintinia*), or with undifferentiated perianth of 4 caducous, decussate, concave, free, imbricate segments (*Paracryphia*); stamens 4–5 (*Quintinia*) or ca. 8 (*Paracryphia*) in a single whorl; anthers basifixed, tetrasporangiate, with longitudinal dehiscence; ovary superior (*Paracryphia*) or inferior (*Quintinia*); 8–15- (*Paracryphia*) or 3–5-locular (*Quintinia*), ovules 4 per locule (*Paracryphia*) or numerous; style elongated with 3–5-lobed stigma (*Quintinia*), or absent (*Paracryphia*). Fruit capsular, septicidal; seeds small, winged in *Paracryphia* and most *Quintina*, copiously endospermic.

Two genera with about 25 spp. distributed in New Guinea, the Philippines, Vanuatu, New Caledonia, tropical Australia and New Zealand.

Vegetative Morphology. Paracryphiaceae mostly consist of shrubs or small to medium high trees (with some *Quintinia* species up to 40 m), more rarely vines. The leaves are alternate to subverticillate (*Paracryphia*). The venation is semicraspedodromous in species with serrate leaves or, in the *Quintinia* spp. with entire leaf margins, brochidodromous. *Paracryphia* has somewhat dilated vein-endings in the teeth, while *Quintinia* has veins with a constant diameter (Dickison and Baas 1977; Al-Shammary 1991).

Vegetative Anatomy. Much information can be found in Patel (1973; *Quintinia*) and Dickison and Baas (1977; *Paracryphia*). The nodes are trilacunar and 3-traced (with the exception of one species of *Quintinia* with pentalacunar nodes; Al-Shammary 1991). Styloid crystals are scattered in the mesophyll of the leaves and in young stems of *Paracryphia*, but seem to be absent from *Quintinia*. The leaf palisade parenchyma is biseriate or triseriate (Holle 1893; Dickison and Baas 1977). Vessel elements are very long, up to 1500 µm or more in *Paracryphia*. Vessel element perforation plates are scalariform with many bars (*Quintinia* has an average of 62 bars per plate, while *Paracryphia* has more than 110 bars per plate) that are at least sometimes reticulate and, at least in *Paracryphia*, with very steep (nearly vertical) vessel end walls and nearly intact pit membranes (Carlquist 1988, 1992). Tracheids are the only imperforate elements. Rays are heterogeneous, and both uniseriate and multiseriate rays are present. Axial parenchyma is diffuse and sometimes also diffuse-in-aggregates and scanty.

The trichomes of *Quintinia* are specialized (Al-Shammary and Gornall 1994; Gornall et al. 1998): the peltate, multicellular glandular hairs, which are found on all parts of the plant except the petals, the androecium, and the gynoecium, have a short, multiseriate stalk that bears a flattened or concave, multicellular, peltate, glandular 'shield' with radially arranged apical cells. *Paracryphia* lacks these trichomes, but has unbranched unicellular trichomes forming a dense pubescence on young leaves (absent from mature foliage), young stems and inflorescences.

Dickison (deceased).

J.W. Kadereit and V. Bittrich (eds.), *Flowering Plants. Eudicots*, The Families and Genera of Vascular Plants 14,
DOI 10.1007/978-3-319-28534-4_26, © Springer International Publishing Switzerland 2016

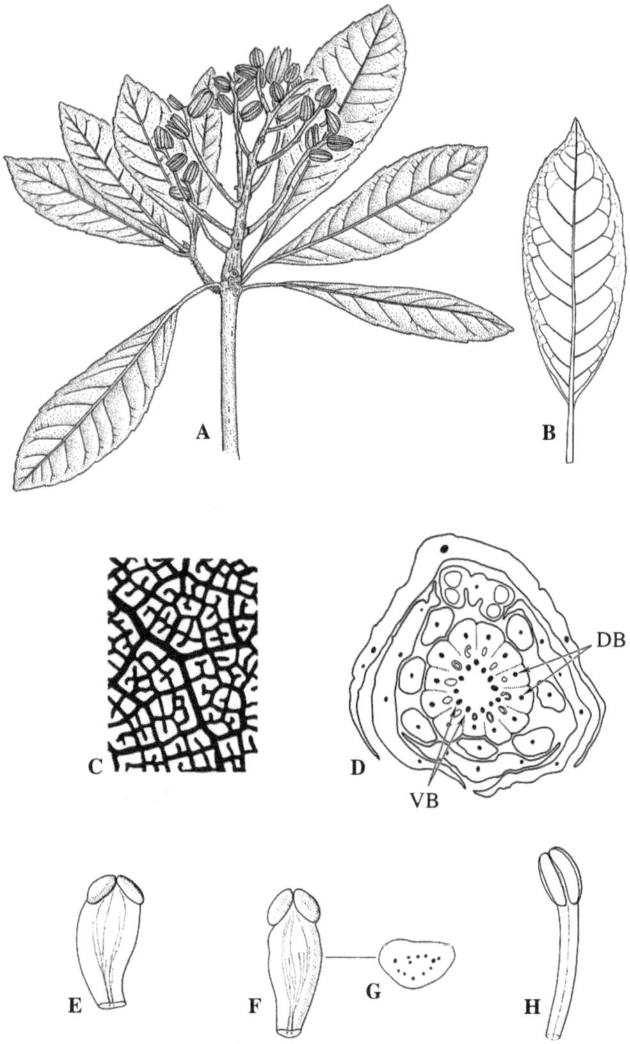

Fig. 53. Paracryphiaceae. **A–H** *Paracryphia alticola*. **A** Fruiting branch. **B** Leaf. **C** Leaf venation. **D** Flower cross-section showing dorsal carpellary bundles (DB) and ventral carpellary bundles (VB). **E, F** Stamens from male flower showing expanded filament and branched venation system. **G** Stamen filament in cross-section. **H** Stamen from bisexual flower. (A original; B–H Dickison and Baas 1977, with permission)

Inflorescence, Floral Morphology and Anatomy (Fig. 53). The inflorescences of *Paracryphia* are basically paniculate, with the sessile flowers forming a terminal compound spike. *Quintinia* bears its pedicillate flowers in terminal racemes or panicles (rarely solitary and axile). Both *Quintinia* and *Paracryphia* have bisexual as well as unisexual flowers. In *Paracryphia*, both bisexual and male flowers occur in the same inflorescence, and the staminate flowers completely lack a gynoecium. *Quintinia*, on the other hand, has rudimentary ovaries in the male flowers, and sterile anthers in the female flowers. In both *Paracryphia* and *Quintinia*, bisexual flowers are more common than unisexual ones. Floral morphology differs greatly between the two genera. *Quintinia* has a differentiated, biseriate perianth, with 4–5 fused sepals and 4–5 free petals that are imbricate in bud. The petals are either glabrous or have unicellular trichomes on the adaxial surface (Al-Shammary and Gornall 1994). Petal marginal veins are absent; the single vein entering the petal has few branches with few anastomoses (Gustafsson 1995). *Paracryphia* has sometimes been described as having no perianth but a series of bracts and bracteoles subtending the androecium and gynoecium (Airy Shaw 1965). Other authors refer to the presence of imbricate, caducous perianth segments (Baker 1921; Bausch 1938; Dickison and Baas 1977). *Quintinia* has 4–5 stamens, while *Paracryphia* has about twice that number (often 8 stamens). The filaments are well developed, but short in *Quintinia*, and conspicuously swollen in the unisexual flowers of *Paracryphia*. The anthers are basifixed, tetrasporangiate and dehiscence is longitudinal. The gynoecium is inferior and 3–5-locular in *Quintinia*, and superior, somewhat oblique and 8–15-locular in *Paracryphia*. *Paracryphia* and some species of *Quintinia* (e.g. *Q. ledermannii* and *Q. fawkneri*) are completely multilocular, while other species have incomplete septa in the lower part of the ovary. *Paracryphia* has four ovules per locule on axile placentas, while *Quintinia* has numerous ovules per locule. The parietal placentas of *Quintinia* are in some species closely adpressed and thus appearing axial, but are always delimited by an epidermis (Bensel and Palser 1975). The floral vascular anatomy was described by Dickison and Baas (1977) for *Paracryphia*, and by Dravitzki (1967; summarized in Philipson 1967) for *Quintinia*. The stamens receive one single trace, which is undivided in *Quintinia* and in the stamens of bisexual flowers of *Paracryphia*, but undergoes a basal dichotomy and subsequent three-dimensional branching in the enlarged filament of stamens in male flowers of *Paracryphia*. The carpel vascularization is simple, in *Quintinia* consisting of one dorsal and two ventral strands, in *Paracryphia* only of a single dorsal and a single ventral strand (interpreted as resulting from the fusion of two ventrals from adjacent carpels) except for distal parts, where the fused ventrals separate resulting in one dorsal and two ventral strands per carpel. The ovules are vascularized by traces from the ventral strands. In *Quintinia*, each ventral strand supplies the ovules in one locule. The nectar disk of *Quintinia* is supplied by branches from the carpel wall bundles.

Pollen Morphology. Pollen of *Paracryphia* has been described by Agababyan and Zavaryan (1971) and Dickison and Baas (1977), and of *Quintinia* by Agababyan (1961, 1964) and Hideux and Ferguson (1976). *Quintinia* has small 4–6-colporate pollen grains with, at least in some species, a very thin endexine. The columella layer is usually short to almost absent, and the tectum varies from smooth or perforated to variously striate-rugulose or radiate-rugulose in most species. The endoaperture is simple-complex. The pollen of *Paracryphia* is small, tricolporate, and spherical to oblate-spheroidal in shape (ca. 24 × 22 µm). The exine is stratified into endexine, foot layer, columellae and tectum. Columellae are short. Sculpturing is reticulate or reticulate-rugulate.

Embryology. Neither *Paracryphia* nor *Quintinia* are well known embryologically. Nothing is known about the anthers of *Quintinia*, but in *Paracryphia* the anther wall contains a well-developed endothecium with lignified thickenings in form of bands (Dickison and Baas 1977). The endothecium is interrupted in the region of the connective and locular partition. Philipson (1974) reported that the ovules of *Quintinia* are tenuinucellate and bitegmic, although the latter condition could not be confirmed by Nemirovich-Danchenko (2000). The ovules of *Paracryphia* have been reported to be unitegmic with a 3–4-layered integument where the outer cells have thicker walls and are more densely staining, and with a nucellus that is 2–3 layers thick (Dickison and Baas 1977).

Karyology. The haploid chromosome number of *Quintinia* has been reported as $n = 22$ (Dawson 1995).

Fruit and Seed. The fruits are at least initially septicidal capsules. In *Paracryphia*, the capsules dehisce longitudinally along the ventral suture of each carpel. In order for this to occur, each carpel separates from the central column along its entire ventral margin at maturity, although a distal connection is maintained. The mature fruit of *Paracryphia* has massive sclerification of cells lining the locules. In the fruit of *Quintinia* dehiscence only reaches the level of the persistent calyx after having separated the fused styles into as many parts as there are carpels. Seeds are small (2.0–2.5 mm in length in *Paracryphia*, ca. 1.5 mm in *Quintinia*), exarillate, albuminous and, in all but one species of *Quintinia*, the testa forms small wings. The endosperm is cellular, and the embryo straight.

Phytochemistry. Undetermined seco-iridoids are found in *Quintinia quatrefagesii* and *Q. verdonii* (Al-Shammary 1991). Saponins are present in at least two *Quintinia* species. Ellagic acid has been reported from *Quintinia serrata* (Gardner 1976, who failed to detect iridoids). Other phenolic compounds detected in *Quintinia* include the widely distributed flavonols kaempferol and quercetin, and the simple phenylpropane caffeic acid. Cyanidin has been detected in *Quintinia*. Some species of *Quintinia* furthermore show aluminium hyperaccumulation (Jansen et al. 2002). The phytochemistry of *Paracryphia* is unknown.

Affinities. The peculiar wood anatomy of *Paracryphia*, showing many features often considered "primitive" (e.g. very long vessel elements, and scalariform and very oblique end plates with numerous bars), has prompted some authors to regard *Paracryphia* as a "primitive" and "ancient" although isolated member of the Angiosperms (e.g. Takhtajan 1997). In contrast to the wood, the reproductive organs are advanced or specialized, leading to a variety of suggested affinities for *Paracryphia* (summarized in Dickison and Baas 1977), including Trochodendraceae and Eucryphiaceae. No support was found for these affinities, and Dickison and Baas (1977) instead suggested an affinity to Actinidiaceae, Theaceae, and/or Sphenostemonaceae. Molecular-based phylogenetic analyses (Savolainen et al. 2000; Lundberg 2001; Bremer et al. 2002) support a close relationship between *Paracryphia*, *Quintinia* and *Sphenostemon* (Sphenostemonaceae). Based on this finding, it has been suggested that *Paracryphia* together with *Quintinia* and *Sphenostemon* should be included in one family, Paracryphiaceae (Lundberg 2001; APG III 2009; *Sphenostemon* is treated as a separate family in this series mainly on practical grounds). Later analyses (Tank and Donoghue 2010) have confirmed this relationship, and also a placement of the expanded family Paracryphiaceae as sister to the order Dipsacales. There are so far no unequivocal morphological synapomorphies known for this clade.

Distribution and Habitats. *Paracryphia* is endemic to New Caledonia, where it occurs in humid forests between 570–1500 m altitude. *Quintinia* also occurs in New Caledonia (4 spp.), but has a wider distribution from the Philippines (1 sp.) through New Guinea (12 spp.), Vanuatu (1 sp.) and Australia (4 spp.) to New Zealand (1 sp.). Most species inhabit montane rainforests, although several New Guinean species reach the subalpine scrubs or even the alpine grasslands up to about 3900 m (*Q. montiswilhelmii*; van Royen 1983).

Palaeobotany. A fossil cuticle from the Miocene of New Zealand has been identified as *Paracryphia* (Pole 2010). Pollen of the *Quintinia* type has been reported from the Upper Eocene of New Zealand (Mildenhall 1980) and the Oligocene-Pliocene of Australia (Martin 1973), and *Quintinia* macrofossils thought to be related to *Q. serrata* (New Zealand) from the early Pleistocene of Tasmania (Jordan 1997). *Silvianthemum suecicum*, charcoalified floral structures from the late Cretaceous of southern Sweden, has been hypothesized to be a close relative of *Quintinia*, thus extending the palaeodistribution of Paracryphiaceae into the northern hemisphere (Friis 1990).

Key to the Genera

1. Perianth undifferentiated, of 4 segments; stamens 8–15; ovary superior with sessile stigma — 1. ***Paracryphia***
– Perianth of sepals and petals; stamens 4–5; ovary inferior, style present — 2. ***Quintinia***

1. ***Paracryphia*** Baker f. Fig. 53

Paracryphia Baker f., J. Linn. Soc. Bot. 45: 306 (1921).

Shrub to medium-sized tree. Leaves alternate to almost verticillate, margins finely serrate; dense pubescence on young leaves, absent on mature foliage. Flowers in a terminal, compound spike, bisexual or occasionally exclusively staminate. Perianth undifferentiated, of 4 caducous, decussate, concave, free, imbricate segments, cochleate with one large helmet-shaped segment that encloses and covers the others; stamens usually 8, rarely up to 15; filaments in male flowers conspicuously swollen; ovary superior, 8-15-locular; stigma sessile, ovules small, anatropous, born in a single row on axile placentas, 4 in each locule. Fruit capsular, septicidal but later schizocarpic. Seeds with abundant endosperm, winged.

One species, *Paracryphia alticola* (Schltr.) Steenis, endemic to New Caledonia.

2. ***Quintinia*** A.DC.

Quintinia A.DC., Monogr. Campan.: 92 (1830).
Dedea Baill. (1879).
Curraniodendron Merrill (1910).

Shrubs, small to tall trees (up to 40 m high), or vines. Leaves entire or somewhat serrate. Flowers in axillary or terminal racemes or many-flowered panicles, small, white, bisexual or unisexual. Calyx tube adnate to the ovary, 4–5-lobed; petals 4–5, deciduous; stamens 4–5; filaments subulate; ovary inferior, 3–5-locular; style elongated, persistent; stigma capitate, 3–5-lobed; ovules numerous, 2-seriate, on parietal (or sometimes seemingly axile) placenta. Fruit dry, small, becoming 1-locular, opening from the top. Seeds small, slightly winged.

About 25 spp., New Guinea (12 spp.), New Caledonia (6 spp.), Australia (4 spp.), Vanuatu, Philippines, and New Zealand (1 sp. each).

Selected Biography

Agababyan, V.S. 1961. Materialy k palinosistematichesko-maizucheniyasemeistva Saxifragaceae s.l. Izv. Akad. Nauk Armyansk. S.S.R., Biol. Nauki 14: 45–61.
Agababyan, V.S. 1964. Evolyutsia pyltsy v poryadkak Cunoniales i Saxifragales v svyazi s nekotoymi voprosami ikh sistematiki i filogenii. Izv. Akad. Nauk Armyansk. S.S.R., Biol. Nauki 17: 59–75.
Agababyan, V.S., Zavaryan, E.L. 1971. Palynotaxonomy of the genus *Paracryphia* Bak.f. Biol. Zh. Arm. 24: 35–40.
Airy Shaw, H.K. 1965. Diagnoses of new families, new names, etc., for the seventh edition of Willis's "Dictionary". Kew Bull. 18: 249–273.

Al-Shammary, K.I.A. 1991. Systematic studies of the Saxifragaceae s.l., chiefly from the southern hemisphere. Ph.D. Thesis, University of Leicester, Leicester, UK.

Al-Shammary, K.I.A., Gornall, R.J. 1994. Trichome anatomy of the Saxifragaceae s.l. from the southern hemisphere. Bot. J. Linn. Soc. 114: 99–131.

Angiosperm Phylogeny Group III (APG III) 2009. An update of the Angiosperm Phylogeny Group classification for the orders and families of flowering plants: APG III. Bot. J. Linn. Soc. 161: 105–121.

Baker, E.G. 1921. Systematic account of the plants collected in New Caledonia and the Isle of Pines by Prof. R.H. Compton, M.A., in 1914. Part 1. Dicotyledons. Polypetalae. J. Linn. Soc. Bot. 45: 264–325.

Bausch, J. 1938. A revision of the Eucryphiaceae. Kew Bull. 1938: 317–349.

Bensel, C.R., Palser, B.F. 1975. Floral anatomy in the Saxifragaceae sensu lato. III. Kirengeshomoideae, Hydrangeoideae and Escallonioideae. Amer. J. Bot. 62: 676–687.

Bremer, B., Bremer, K., Heidari, N., Erixon, P., Olmstead, R.G., Anderberg, A.A., Källersjö, M., Barkhordarian, E. 2002. Phylogenetics of asterids based on 3 coding and 3 non-coding chloroplast DNA markers and the utility of non-coding DNA at higher taxonomic levels. Mol. Phylogenet. Evol. 24: 274–301.

Carlquist, S. 1988. Comparative wood anatomy. Berlin: Springer.

Carlquist, S. 1992. Pit membrane remnants in perforation plates of primitive dicotyledons and their significance. Amer. J. Bot. 79: 660–672.

Dawson, M.I. 1995. Contributions to a chromosome atlas of the New Zealand flora – 33. Miscellaneous species. New Zeal. J. Bot. 33: 477–487.

Dickison, W.C., Baas, P. 1977. The morphology and relationships of *Paracryphia* (Paracryphiaceae). Blumea 23: 417–438.

Dravitzki, P.V. 1967. A comparative study of the wood anatomy and floral vascular systems of the New Zealand genera of the Escalloniaceae, *Carpodetus*, *Ixerba* and *Quintinia*. M.Sc. Thesis, University of Canterbury, Christchurch, New Zealand.

Friis, E.M. 1990. *Silvianthemum suecicum* gen. et sp. nov., a new saxifragalean flower from the Late Cretaceous of Sweden. Biol. Skrift. 36: 1–35.

Gardner, R.O. 1976. Studies in Alseuosmiaceae. Ph.D. Thesis, University of Auckland, Auckland, New Zealand.

Gornall, R.J., Al-Shammary, K.I.A., Gregory, M. 1998. Escalloniaceae. In: Cutler, D.F., Gregory, M. (eds.) Anatomy of the Dicotyledons. Ed. 2. Oxford: Clarendon Press, pp. 41–86.

Gustafsson, M.H.G. 1995. Petal venation in the Asterales and related orders. Bot. J. Linn. Soc. 118: 1–18.

Hideux, M.J., Ferguson, I.K. 1976. The stereostructure of the exine and its evolutionary significance in Saxifragaceae sensu lato. In: Ferguson, I.K., Muller, J. (eds.) The evolutionary significance of the exine. Linnean Society Symposium Series, no. 1. London: Academic Press, pp. 327–377.

Holle, G. 1893. Beiträge zur Anatomie der Saxifragaceen und deren systematische Verwerthung. Bot. Centralbl. 53: 1–9, 33–41, 65–70, 97–102, 129–136, 161–169, 209–222.

Jansen, S., Broadley, M.R., Robbrecht, E., Smets, E. 2002. Aluminium hyperaccumulation in Angiosperms: a review of its phylogenetic significance. Bot. Rev. 68: 235–269.

Jordan, G.J. 1997. Evidence of Pleistocene plant extinction and diversity from Regatta Point, western Tasmania, Australia. Bot. J. Linn. Soc. 123: 45–71.

Lundberg, J. 2001. Phylogenetic studies in the Euasterids II with particular reference to Asterales and Escalloniaceae. Ph.D. Thesis, Uppsala University, Uppsala, Sweden.

Martin, H.A. 1973. The palynology of some Tertiary Pleistocene deposits, Lachlan River Valley, New South Wales. Austral. J. Bot., Suppl. Ser. 6: 1–57.

Mildenhall, D.C. 1980. New Zealand Late Cretaceous and Cenozoic plant biogeography: a contribution. Palaeogeogr. Palaeoclimatol. Palaeoecol. 31: 197–233.

Nemirovich-Danchenko, E.N. (ed.) 2000. Sravnitel'najaanatomijasemjan. Vol. 6. Dvudol'nye. Rosidae II. St. Petersburg: Academia Scientiarum Rossica, Institutum Botanicum Nomine V. L. Komarovii NAUKA.

Patel, R.N. 1973. Wood anatomy of the dicotyledons indigenous to New Zealand. New Zeal. J. Bot. 11: 421–434.

Philipson, W.R. 1967. *Griselinia* Forst. fil.–anomaly or link. New Zeal. J. Bot. 5: 134–165.

Philipson, W.R. 1974. Ovular morphology and the major classification of the dicotyledons. Bot. J. Linn. Soc. 68: 89–108.

Pole, M. 2010. Was New Zealand a primary source for the New Caledonian flora? Alcheringa 34: 61–74.

Savolainen, V., Fay, M.F., Albach, D.C., Backlund, A., van der Bank, M., Cameron, K.M., Johnson, S.A., Lledó, M.D., Pintaud, J.-C., Powell, M., Sheahan, M.C., Soltis, D.E., Soltis, P.S., Weston, P., Whitten, W.M., Wurdack, K.J., Chase, M.W. 2000. Phylogeny of the eudicots: a nearly complete familial analysis based on *rbcL* gene sequences. Kew Bull. 55: 257–309.

Takhtajan, A. 1997. Diversity and classification of flowering plants. New York: Columbia University Press.

Tank, D.C., Donoghue, M.J. 2010. Phylogeny and phylogenetic nomenclature of the Campanulidae based on an expanded sample of genes and taxa. Syst. Bot. 35: 425–441.

van Royen, P. 1983. The alpine flora of New Guinea. Vol. 4. Vaduz, Lichtenstein: J. Cramer.

Phyllonomaceae

Phyllonomaceae Small in Britt., N. Amer. Fl. 22(1): 2 (1905).
Dulongiaceae J.G. Agardh (1858), nom. illeg.

V. Bittrich

Terrestrial or rarely epiphytic shrubs or small trees, completely glabrous except for the stipules. Leaves alternate, petiolate, simple, acuminate, with entire or serrate or dentate margin; stipules small, with glandular hairs on the adaxial surface, caducous. Flowers small, bisexual, green, in epiphyllous inflorescences, actinomorphic, (4)5-merous. Sepals free or very slightly connate at base, persistent; petals free, valvate, late caducous to persistent; stamens free, alternating with the petals, persistent; anthers bilobed, dehiscing by lateral slits, pollen 3-colporate; flat nectary present on top of the ovary; ovary inferior, 2(3)-carpellate, syncarpous, unilocular with protruding parietal placentae; ovules 4–14, described as bitegmic but probably unitegmic, tenuinucellate; style bifid (trifid) or free to the base, recurved, stigmas small, terete. Fruit a small 3–10-seeded berry. Seeds small, with rugose or tuberculate testa, exarillate, endosperm fleshy, containing aleuron and oil; embryo straight, very small.

A Neotropical family with a single genus of four species.

Vegetative Morphology. Most information can be found in Mori and Kallunki (1977). The Phyllonomaceae are small trees (rarely up to 15 m tall) or shrubs, often with drooping young branches. Leaves are spirally arranged, the petiole is canaliculate, the blade elliptic to narrowly ovate (or rarely obovate) with an acuminate to long acuminate tip, sometimes with a constriction between blade and acumen. The margin is variously serrate or dentate, and in some species it varies from entire to serrate in the same population or even on the same branch. The 2 stipules are small, caducous, with numerous glandular hairs on the adaxial surface (Dickinson and Sattler 1974) and enclose the unexpanded sylleptic axillary shoot. The leaves remain folded until maturity, protecting the young inflorescence inserted adaxially on the leaf blade. The lower leaves of a shoot can be sterile, all others are fertile. The inflorescence is inserted between the acumen and the middle of the leaf blade.

Vegetative Anatomy. Most data are found in the study of Dickinson and Sattler (1974) on *Phyllonoma integerrima* and in Metcalfe and Chalk (1950). The vascular bundles form an ectophloic siphonostele, and the vessels have scalariform perforations. The pericycle has a ring of fibres. Cork develops superficially. Nodal anatomy is unilacunar, giving off one trace to the leaf and two to the axillary bud, but no traces to the stipules. The leaf midrib up to the insertion of the inflorescence has a ring-like vascular bundle that may be discontinuous laterally. The bundle is enclosed by a fibre sheath with extensions to the epidermis.

Inflorescence Structure and Flower structure. The inflorescences are fascicled, sometimes pseudoumbellate, the bracts fimbriate, sometimes stipulate (Fig. 54). Stork (1956) described the inflorescences as cymose. Dickinson and Sattler (1974) and Mori and Kallunki (1977) observed considerable variation in the branching pattern. In *P. ruscifolia*, branching is highly irregular, with bracts and flowers not always closely associated, and its inflorescences were classified as irregular monochasial cymes by Mori and Kallunki (1977). In the three other species the inflorescence appears to be a cluster of one to four unbranched racemes, with rather broad bracts which enclose the bases of the pedicels. Detailed anatomical investigations of one of these species, *P. tenuidens*, however, revealed cymose branching of the

J.W. Kadereit and V. Bittrich (eds.), *Flowering Plants. Eudicots*, The Families and Genera of Vascular Plants 14,
DOI 10.1007/978-3-319-28534-4_27,

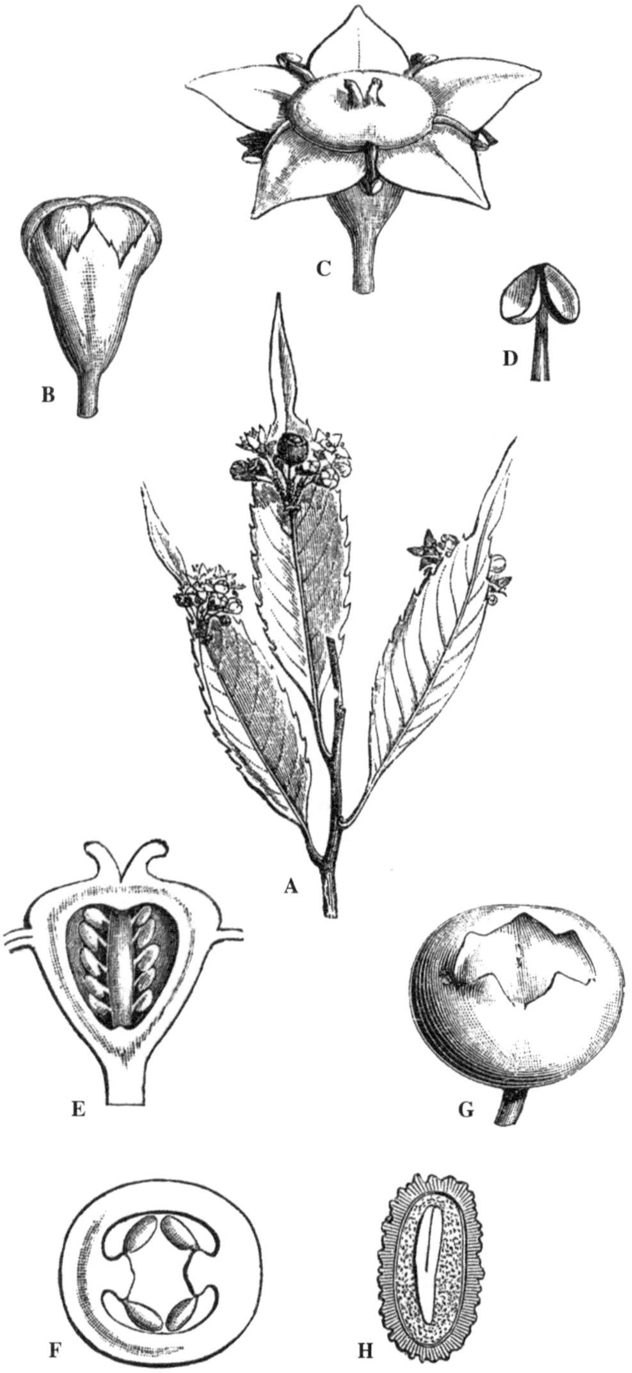

Fig. 54. Phyllonomaceae. **A–H** *Phyllonoma laticuspis*. **A** Flowering branch. **B** Flower bud. **C** Open flower. **D** Anther. **E** Ovary in longitudinal section. **F** Ovary in cross section. **G** Fruit. **H** Seed in longitudinal section. (Engler 1891, modified)

inflorescence resulting in a cincinnus, suggesting that the whole genus is characterized by monochasia (Tobe 2014) as assumed by Stork (1956). The unusual epiphyllous initiation of the inflorescences was studied in detail by Dickinson and Sattler (1974). These authors concluded that inflorescence development is truly epiphyllous as the leaves can neither be interpreted as modified branches nor do they show any evidence for "congenital" fusion with the inflorescence. Weberling (1981) maintained that the data would not exclude the possibility of recaulescence, and Weber (2004, p. 28) supported this view stating: "[. . .] the inflorescence primordium arises as a low, flat meristem extending from the leaf axil upwards. Only the distal part develops into a large meristematic head which later differentiates into bracts and flower primordia". Weber (2004) also emphasized that the epiphyllous inflorescences of *Phyllonoma* and *Helwingia* cannot be regarded as completely different, and that those of *Phyllonoma* only represent a more advanced form.

The bisexual, actinomorphic, greenish flowers show little interspecific differences (Mori and Kallunki 1977). They are small, measuring about 2–4 mm in diameter. The pedicels lack bracteoles. The sepals are free, triangular, with small multicellular glands on the margins (Tobe 2013), persistent, valvate or slightly quincuncially imbricate in bud, and the odd sepal is in an abaxial position. The petals are free, valvate in bud and spreading or reflexed at anthesis. The alternipetalous stamens are always shorter than the petals. They are first erect and become reflexed later. The globose bilobed anthers open by lateral slits. The gynoecium is syncarpous, inferior, and composed of 2(3) carpels in transversal position, with a short bifid (trifid) terminal style (or styles free to the base) and small terete stigmas. There are up to 14 horizontal campylotropous ovules originating from the protruding parietal placentae. An epigynous disk nectary with nectarostomata covers most of the ovary roof (Tobe 2013).

Embryology. Only few observations were reported by Mauritzon (1933). According to him, the nucellus is of a transitional type, tenuinucellate but with the formation of a parietal cell, a situation today normally called weakly crassinucellar. Mauritzon also reported only one integument, while Krach (1976) considered 2 integuments as more probable (see below).

Pollen Morphology. Pollen is prolate or oblate-spheroidal, 3-colporate. The tectum is complete, the exine surface is microspinulose, granulose or smooth (Hideux and Ferguson 1976).

Karyology. Unknown.

Fruit and Seed. The fruit is a small globose or subglobose, 3–10-seeded berry. The seeds are small (up to 2.8 mm long), straight or slightly curved, their surface rugose when dry; an aril is lacking. Seed anatomy was studied in detail by Krach (1976). He described the testa as bitegmic, but with reservations as he did not observe a cuticle between the postulated integuments. Families closely related to Phyllonomaceae (cf. Soltis et al. 2011, see below) are characterized by having unitegmic ovules. According to Krach (1976), the "outer integument" is composed of several layers, with cell size diminishing continuously from the outer to the innermost layer; the "inner integument" is 2-layered. The rather large cells of the outer epidermis of the "outer integument" have thick cellulose walls and contain mucilage. The cells of the inner epidermis of the "inner integument" are filled with mucilage and tannins. The endosperm has unpitted cell walls of hemicelluloses, the cells contain aleuron and fatty oils. The embryo is straight and very small.

Dispersal. The small berries are probably bird-dispersed.

Reproductive Systems. The small nectariferous flowers suggest pollination by small insects, but observations are lacking. Nothing is known about breeding systems in the family.

Phytochemistry. Aluminium accumulation was demonstrated in four species (Hegnauer 1973).

Affinities. The family was formerly often included in the Saxifragaceae s.l. Takhtajan (1997) included it in the Hydrangeales and suggested a closer relationship with Escalloniaceae, despite pollen characters which differ from those of that family (Hideux and Ferguson 1976). Krach (1976) suggested that seed anatomical characters would point to a relationship with *Ribes* (Grossulariaceae), and Cronquist (1981) included *Phyllonoma* in the Grossulariaceae. In the APG classification (1998), the family is included in the Aquifoliales. A study of the "eudicots" (Savolainen et al. 2000) based on *rbcL* sequence data showed Phyllonomaceae and the monogeneric Helwingiaceae as sister taxa, both together forming the sister group to Aquifoliaceae. *Helwingia* Willd., formerly often included in Cornaceae or Araliaceae, occurs in Asia and is noteworthy for having, at least superficially, quite similar epiphyllous inflorescences as *Phyllonoma*, as already noted by de Candolle (1890) and studied in more detail by Dickinson and Sattler (1975) and Weber (2004). *Helwingia* species are dioecious shrubs with stipulate, alternate and serrate or dentate leaves, perianth uniseriate with 3–5, valvate, green petals, a 3–5-loculed ovary with an epigynous disk, and globose drupaceous fruits. In the angiosperm phylogeny of Soltis et al. (2011), based on sequence data of 17 genes, Helwingiaceae and Phyllonomaceae are sister taxa with 99 % BS support and Aquifoliaceae sister to these two families with BS 100 %. A clade formed of Cardiopteridaceae and Stemonuraceae is sister to the clade of the three families and all five families comprise the order Aquifoliales.

Distribution and Habitats. Species of *Phyllonoma* occur from Mexico to Peru and Bolivia in the understorey of mountain cloud forests.

Only one genus:

Phyllonoma Schultes Fig. 54

Phyllonoma Schultes in Roem. & Schultes, Syst. Veg. 6: xx (1820); Mori & Kallunki, Brittonia 29: 69–84 (1977), rev.
Dulongia Kunth (1824), nom. illeg.

Characters as for family. About four species, montane rain forests from Mexico to Bolivia.

Selected Bibliography

Angiosperm Phylogeny Group (APG) 1998. An ordinal classification for the families of flowering plants. Ann. Missouri Bot. Gard. 85: 531–553.

Candolle, C. de. 1890. Recherches sur les inflorescences épiphylles. Mém. Soc. Phys. Hist. Nat. Genève, suppl. 6: 1–37.

Cronquist, A. 1981. An Integrated System of Classification of Flowering Plants. New York: Columbia University Press.

Dickinson, T.A., Sattler, R. 1974. Development of the epiphyllous inflorescence of *Phyllonoma integerrima* (Turcz.) Loes.: Implications for comparative morphology. J. Linn. Soc. Bot. 69: 1–13.

Dickinson, T.A., Sattler, R. 1975. Development of the epiphyllous inflorescence of *Helwingia japonica* (Helwingiaceae). Amer. J. Bot. 69: 962–973.

Engler, H.G.A. 1891. Saxifragaceae. In: Engler, A., Prantl, K. (eds.) Die natürlichen Pflanzenfamilien III, 2a. Leipzig: W. Engelmann, pp. 41–93.

Hegnauer, R. 1973. Chemotaxonomie der Pflanzen, vol. 6. Basel: Birkhäuser.

Hideux, M.J., Ferguson, I.K. 1976. The stereostructure of the exine and its evolutionary significance in Saxifragaceae sensu lato. In: Ferguson, I.K., Muller, J. (eds.) The evolutionary significance of the exine, pp. 327–377. Linn. Soc. Symposium no. 1. London: Academic Press.

Krach, J.E. 1976. Die Samen der Saxifragaceae. Bot. Jahrb. Syst. 97: 1–60.

Mauritzon, J. 1933. Studien über die Embryologie der Familien Crassulaceae und Saxifragaceae. Lund: Hakans Ohlsson.

Metcalfe, R.C., Chalk, L. 1950. Anatomy of the Dicotyledons. Oxford: Clarendon Press.

Mori, S.A., Kallunki, J.A. 1977. A revision of the genus *Phyllonoma* (Grossulariaceae). Brittonia 29(1): 69–84.

Savolainen, V., Fay, M.F., Albach, D.C., Backlund, A., van der Bank, M., Cameron, K.M., Johnson, S.A., Lledó, M.D., Pintaud, J.-C., Powell, M., Sheahan, M.C., Soltis, D.E., Soltis, P.S., Weston, P., Whitten, W.M., Wurdack, K.J., Chase, M.W. 2000. Phylogeny of the eudicots: a nearly complete familial analysis based on *rbcL* gene sequences. Kew Bull. 55: 257–309.

Soltis, D.E., Smith, S.A., Cellinese, N., Wurdack, K.J., Tank, D.C., Brockington, S.F., Refulio-Rodriguez, N. F., Walker, J.B., Moore, M.J., Carlsward, B.S., Bell, C. D., Latvis, M., Crawley, S., Black, C., Diouf, D., Xi, Z., Rushworth, C.A., Gitzendanner, M.A., Sytsma, K.J., Qiu, Y.L., Hilu, K.W., Davis, C.C., Sanderson, M.J., Beaman, R.S., Olmstead, R.G., Judd, W.S., Donoghue, M.J., Soltis, P.S. 2011. Angiosperm phylogeny: 17 genes, 640 taxa. Amer. J. Bot. 98: 704–730.

Stork, H.E. 1956. Epiphyllous flowers. Bull. Torrey Bot. Club 83: 338–341.

Takhtajan, A. 1997. Diversity and Classification of Flowering Plants. New York: Columbia University Press.

Tobe, H. 2013. Floral morphology and structure of *Phyllonoma* (Phyllonomaceae): systematic and evolutionary implications. J. Plant Res. 126: 709–718.

Tobe, H. 2014. Inflorescences of *Phyllonoma* (Phyllonomaceae, Aquifoliales): Anatomical observations. Acta Phytotax. Geobot. 65: 117–126.

Weber, A. 2004. What is morphology and why is it time for its renaissance in plant systematics? In: Stuessy, T.F., Mayer, V., Hörandl, E. (eds.) Deep Morphology: Toward a Renaissance of Morphology in Plant Systematics. Ruggell, Liechtenstein: A.R.G. Gantner, pp. 3–31.

Weberling, F. 1981. Morphologie der Blüten und Blütenstände. Stuttgart: E. Ulmer.

Polyosmaceae

Polyosmaceae Blume, Mus. Bot. 1: 258 (1851).

J. Lundberg

Evergreen shrubs or trees. Leaves opposite to subopposite, exstipulate, petiolate, simple, entire to spinulose-dentate. Inflorescence a terminal, lax raceme, flowers rarely solitary, with 1–3 entire bracteoles or one 3-fid bracteole. Flowers bisexual, actinomorphic, perigynous to epigynous, fragrant; sepals 4, small, fused, persistent; petals 4, valvate in bud, free but sometimes cohering, deciduous, hairy; stamens 4, alternipetalous, free, filaments flattened, hairy; anthers introrse, linear, basifixed; intrastaminal nectar disk present; ovary of 2 united carpels, partly to completely inferior, 1-locular (appearing 2-locular due to intruding placentation), placentation parietal, ovules numerous; style simple, filiform, stigma weakly bilobed. Fruit a drupe. Seed single, large, with thick, smooth testa, endosperm abundant, embryo small.

One genus with about 80 species distributed from north-eastern India to New Caledonia and Australia.

Vegetative Morphology. Polyosmaceae are a family of erect shrubs (0.5–3 m high) and trees up to about 25 m high. Young branches, leaf petioles and leaves are covered with unicellular, pointed hairs (Holle 1893). The leaves are opposite or almost so, petiolate and simple. The leaf margin is variable, from entire through serrate to spinulose-dentate. Leaf venation is brochidodromous (species with entire leaf margin) or semicraspedodromous. Domatia are absent, as are stipules.

Vegetative Anatomy. In certain species of *Polyosma*, a 1–3-layered hypodermis is present on the upper side of the leaf (Holle 1893). This hypodermis can be either complete or partially absent (e. g. *P. integrifolia*). The nodes are unilacunar (Swamy 1954). Large druses are present in the leaves (Holle 1893; Al-Shammary 1991; Gornall et al. 1998). Cork arises in subepidermal layers (Holle 1893). The sclerenchyma in the pericycle is a complete ring of brachysclereids (Solereder 1899). In mature wood the pore distribution is variable, as solitary, in radial multiples, or clustered (Stern et al. 1969; Gornall et al. 1998). Perforation plates are scalariform with 40–70(–125) bars. Intervascular pitting is alternate in all species investigated by Stern et al. (1969), with two exceptions (*P. integrifolia* and *P. serratula*), where the pitting is exclusively opposite; in all species with alternate pitting also opposite pits are present. In most investigated species of *Polyosma* mainly fibre-tracheids but also tracheids are present; in *P. cunninghamii* only tracheids are present (Stern et al. 1969). The apotracheal axial parenchyma is diffuse or, more often, diffuse-in-aggregates (Stern et al. 1969; Boer and Sosef 1998), while the paratracheal axial parenchyma is scanty (Boer and Sosef 1998). Multiseriate rays (3–6 cells wide) are heterocellular and often have uniseriate margins, while the uniseriate rays are homocellular with only upright cells (Stern et al. 1969). Some of the rays are storied. Trichomes are sickle-shaped with their bases perforated and embedded in the epidermis (Al-Shammary and Gornall 1994).

Flower Structure (Fig. 55). The flowers are epigynous and tetramerous, and borne in terminal racemes. Each flower is subtended by 1–3 small entire bracteoles or sometimes one 3-fid bracteole, often caducous. The four sepals are fused into a conspicuous calyx tube with four lobes, and the abaxial surface is usually covered with the unicellular, pointed hairs described above. The calyx is persistent on the fruit. The

J.W. Kadereit and V. Bittrich (eds.), *Flowering Plants. Eudicots*, The Families and Genera of Vascular Plants 14,
DOI 10.1007/978-3-319-28534-4_28, © Springer International Publishing Switzerland 2016

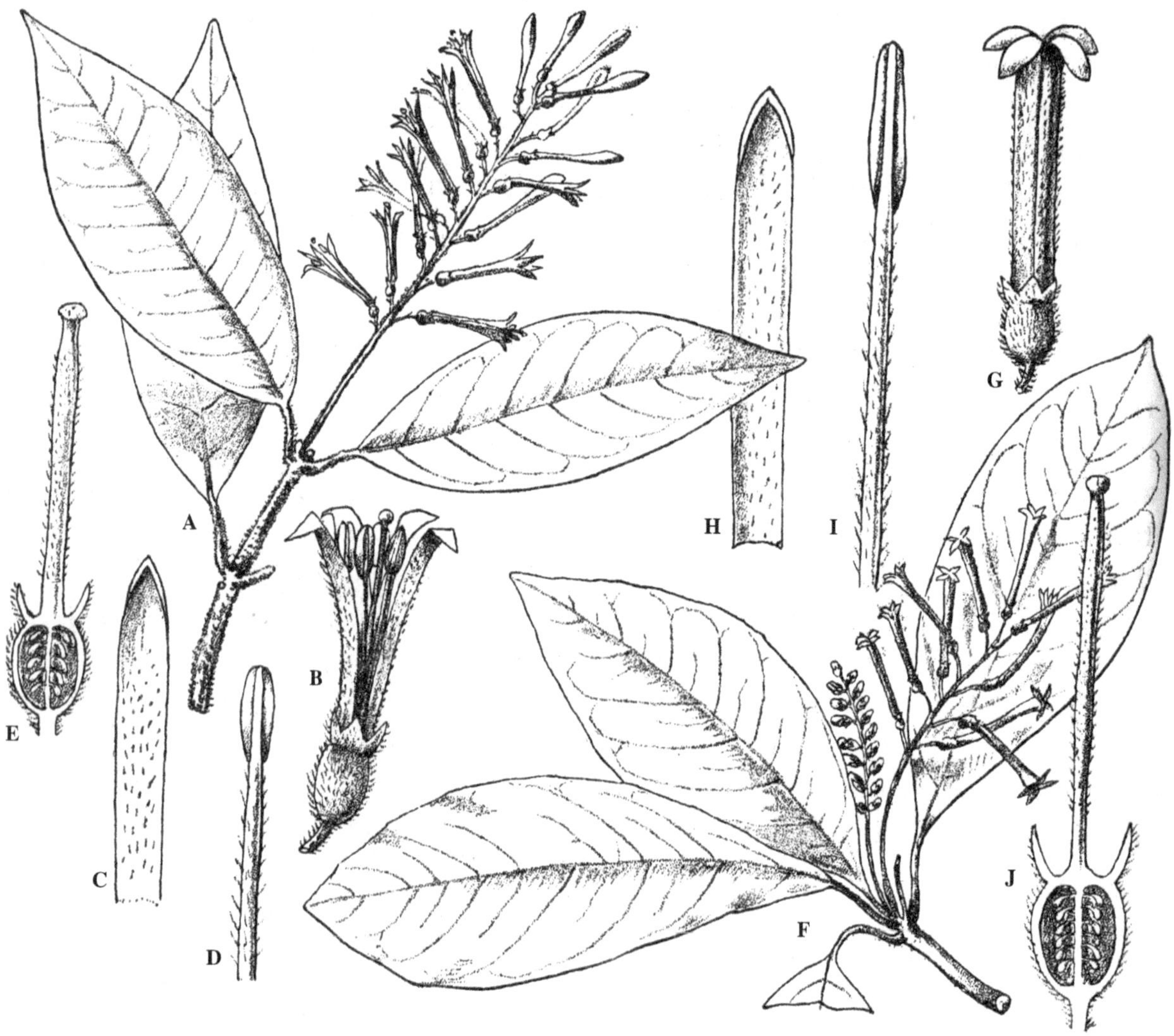

Fig. 55. Polyosmaceae. **A–E** *Polyosma finisterrae*. **A** Flowering branch. **B** Flower. **C** Petal. **D** Stamen. **E** Gynoecium showing style and the ovary with numerous ovules. **F–J** *Polyosma dentata*. **F** Branch with buds and flowers. **G** Flower. **H** Petal. **I** Stamen. **J** Gynoecium showing style and the ovary with numerous ovules. (A–J Schlechter 1914)

petals are free although sometimes coherent and appearing as fused into a tube with four spreading lobes, and have the unicellular, pointed hairs described above on their abaxial surface, in some species also on the adaxial surface (Al-Shammary and Gornall 1994). The stamens are free from each other and also from the perianth. The linear, more or less flattened filaments are hairy. A nectariferous disk is present and positioned on top of the ovary.

The partly to completely inferior, mostly unilocular ovary is formed by two completely connate carpels, with a single, compound and sometimes hairy style with a weakly bilobed stigma. In some species (e.g. *P. occulta*) the ovary is partially bilocular with a septum only in the lower part of the ovary. Placentation is parietal, although intruding deep into the locule, and can thus be mistaken for being partly axial. Several anatropous and unitegmic ovules are borne at the level where the placenta intrudes. Pollinators not known.

Floral Anatomy. The floral vascularisation of an undetermined Australian *Polyosma* species was described by Bensel and Palser (1975). High up in the receptacle, just below the locular base, a flat reticulum of small, anastomosing lateral traces (originating from the vascular cylinder) supplies the ovules through a single ventral carpel bundle that splits into two compound bundles each supplying a bilobed placenta. The dorsal carpel bundles originate from sepal-plane traces, and are the only ones that extend into the style, while the ventral traces disappear near the top of the locule.

Embryology. Ovules are orthotropous and somewhat intermediate between tenui- and crassinucellate (Nemirovich-Danchenko 2000). The single integument is up to 10 cell layers thick.

Pollen Morphology. Pollen of *Polyosma* has been investigated and described by Erdtman (1952), Agababyan (1961, 1964), Wakabayashi (1970), Pastre and Pons (1973), Hideux and Ferguson (1976), and Al-Shammary (1991). The pollen is triporate (occasionally tetraporate), with an unsculptured and sparsely perforated tectum and a slightly reticulate sexine (Hideux and Ferguson 1976). The endexine is thin to almost absent while the columella layer is robust and relatively short (Al-Shammary 1991). Around each aperture is a thickened and convolute ektexine (Agababyan 1964; Hideux and Ferguson 1976; Al-Shammary 1991).

Fruit and Seed. The fruits are drupes (Krach 1976) although often described as berries. The single seed is large, with a smooth and more or less thick testa (up to twelve cell layers; Al-Shammary 1991). The outer epidermis of the seed has discontinuous thickenings on the inner periclinal walls, somewhat similar to those found in *Anopterus* (Escalloniaceae; Al-Shammary 1991). The embryo is undifferentiated, small and found in the upper part of the copious endosperm. Starch is present in the endosperm (Krach 1976; Nemirovich-Danchenko 2000). The fruits seem to be eaten by flying-foxes (*Pteropus* spp.) in New South Wales, Australia (Roberts 2006).

Phytochemistry. The chemistry of Polyosmaceae is not well known. Saponins, triterpenes, alkaloids and aluminium accumulation have been reported (Hegnauer 1973). Iridoids are mainly present as the 10-hydroxylated daphylloside (asperuloside; Kooiman 1971; Al-Shammary 1991).

Affinities. Phylogenies based on molecular data (Lundberg 2001; Winkworth et al. 2008; Tank and Donoghue 2010) have placed *Polyosma* in a well-supported clade with *Anopterus*, *Escallonia*, *Forgesia*, *Tribeles* and *Valdivia*, and it has been suggested that they should all be placed in one family, Escalloniaceae (e.g. APG III 2009). There is, however, some uncertainty regarding the exact placement of *Polyosma* in this clade. Lundberg (2001) and Tank and Donoghue (2010) identified *Polyosma* as sister to the rest of this expanded, monophyletic Escalloniaceae, but only with low support.

Distribution and Habitats. Polyosmaceae are found in tropical rainforests ranging from the Khasi Hills in north-eastern India, through Indochina and China (14 spp.), the Malayan Peninsula (13 spp.), Java (5 spp.), Sumatera (5 spp.), Borneo (13 spp.), the Philippines (14 spp.), New Caledonia (7 spp.), and New Guinea (23 spp.) to tropical Australia (7 spp.). *Polyosma integrifolia* is distributed from Indo-China, China, the Malay Peninsula to the Lesser Sunda Islands and the Moluccas, while *P. mutabilis* is distributed from Thailand through Sumatra and Borneo to the Philippines (H.-J. Esser, pers. comm.), and *P. brachyantha* further south, in New Guinea, Bougainville and the Amboina Island. The other species have more restricted distributions. *Polyosma* inhabit low-altitude forests to subalpine scrubs at altitudes of up to about 3750 m in Papua New Guinea (*P. subalpina*; van Royen 1983), with the highest species diversity in montane forests at 1000–1500 m.

Parasites. Schlechter (1914) reported that disproportionally many flowers of *Polyosma* from New Guinea had galls.

Uses. *Polyosma* is only utilized in local economies, where its wood has been used for house building. The inner bark has been used for treating scurf, and its leaf sap against eye diseases (Boer and Sosef 1998).

Only one genus:

Polyosma Blume Fig. 55

Polyosma Blume, Bijdr. Fl. Ned. Ind.: 658 (1825).

Characters as for family.

The genus is in great need of revision, and the number of species is uncertain.

Selected Bibliography

Agababyan, V.S. 1961. Materialy k palinosistematicheskoma izucheniya semeistva Saxifragaceae s.l. Izv. Akad. Nauk Armyansk. S.S.R., Biol. Nauki 14: 45–61.

Agababyan, V.S. 1964. Evolyutsia pyltsy v poryadkak Cunoniales i Saxifragales v svyazi s nekotorymi voprosami ikh sistematiki i filogenii. Izv. Akad. Nauk Armyansk. S.S.R., Biol. Nauki 17: 59–75.

Al-Shammary, K.I.A. 1991. Systematic studies of the Saxifragaceae s.l., chiefly from the southern hemisphere. Ph.D. Thesis, University of Leicester, Leicester, UK.

Al-Shammary, K.I.A., Gornall, R. J. 1994. Trichome anatomy of the Saxifragaceae s.l. from the southern hemisphere. Bot. J. Linn. Soc. 114: 99–131.

Angiosperm Phylogeny Group III 2009. An update of the Angiosperm Phylogeny Group classification for the orders and families of flowering plants: APG III. Bot. J. Linn. Soc. 161: 105–121.

Bensel, C.R., Palser, B.F. 1975. Floral anatomy in the Saxifragaceae sensu lato. III. Kirengeshomoideae, Hydrangeoideae and Escallonioideae. Amer. J. Bot. 62: 676–687.

Boer, E., Sosef, M.S.M. 1998. *Polyosma* Blume. In: Sosef, M. S.M., Hong, L.T., Prawirohatmodjo, S. (eds.) Plant Resources of South-East Asia 5(3). Timber trees: Lesser-known timbers. Leiden: Backhuys, pp. 464–465.

Erdtman, G. 1952. Pollen morphology and plant taxonomy. Angiosperms. Stockholm: Almqvist and Wiksell.

Gornall, R.J., Al-Shammary, K.I.A., Gregory, M. 1998. Escalloniaceae. In: Cutler, D.F., Gregory, M. (eds.) Anatomy of the Dicotyledons. Ed. 2. Oxford: Clarendon Press, pp. 41–86.

Hegnauer, R. 1973. Chemotaxonomie der Pflanzen, vol. 6. Basel: Birkhäuser.

Hideux, M.J., Ferguson, I.K. 1976. The stereostructure of the exine and its evolutionary significance in Saxifragaceae sensu lato. In: Ferguson, I.K., Muller, J. (eds.) The evolutionary significance of the exine. Linnean Society Symposium Series number 1. London: Academic Press, pp. 327–377.

Holle, G. 1893. Beiträge zur Anatomie der Saxifragaceen und deren systematische Verwerthung. Bot. Centralbl. 53: 1–9, 33–41, 65–70, 129–136, 161–169, 209–222.

Kooiman, P. 1971. Ein phytochemischer Beitrag zur Lösung der Verwandtschaftsproblems der Theligonaceae. Österr. Bot. Zeitschr. 119: 395–398.

Krach, J.E. 1976. Samenanatomie der Rosifloren I. Die Samen der Saxifragaceae. Bot. Jahrb. Syst. 97: 1–60.

Lundberg, J. 2001. Phylogenetic studies in the Euasterids II with particular reference to Asterales and Escalloniaceae. Ph.D. Thesis, Uppsala University, Uppsala, Sweden.

Nemirovich-Danchenko, E.N. (ed.) 2000. Sravnitel'naja anatomija semjan. Vol. 6. Dvudol'nye. Rosidae II. St. Petersburg: Academia Scientiarum Rossica, Institutum Botanicum Nomine V. L. Komarovii NAUKA, Russia.

Pastre, A., Pons, A. 1973. Quelques aspects de la systématique des Saxifragacées à la lumière des données de la palynologie. Pollen Spores 15: 117–133.

Roberts, B.J. 2006. Management of urban Flying-fox camps: Issues relevant to camps in Lower Clarence, NSW. Maclean: Valley Watch.

Schlechter, R. 1914. Die Saxifragaceae Papuasiens. Engl. Bot. Jahrb. 52: 118–138.

Solereder, H. 1899. Systematische Anatomie der Dicotyledonen. Stuttgart: Ferdinand Enke.

Stern, W.L., Brizicky, G.K., Eyde, R.H. 1969. Comparative anatomy and relationships of Columelliaceae. J. Arnold Arb. 50: 36–75.

Swamy, B.G.L. 1954. Morpho-taxonomical notes on the Escallonioideae, part A. Nodal and petiolar vasculature. J. Madras Univ. 24: 299–306.

Tank, D.C., Donoghue, M.J. 2010. Phylogeny and phylogenetic nomenclature of the Campanulidae based on an expanded sample of genes and taxa. Syst. Bot. 35: 425–441.

van Royen, P. 1983. The alpine flora of New Guinea. Vol. 4. Vaduz: J. Cramer.

Wakabayashi, M. 1970. On the affinity in Saxifragaceae s. lato with special reference to the pollen morphology. Acta Phytotax. Geobot. 24: 128–145.

Winkworth, R.C., Lundberg, J., Donoghue, M.J. 2008. Toward a resolution of Campanulid phylogeny, with special reference to the placement of Dipsacales. Taxon 47: 53–65.

Solanaceae

Solanaceae Juss., Gen. pl.: 124 (1789), nom. cons.

G.E. Barboza, A.T. Hunziker, G. Bernardello, A.A. Cocucci, A.E. Moscone, C. Carrizo García, V. Fuentes, M.O. Dillon, V. Bittrich, M.T. Cosa, R. Subils, A. Romanutti, S. Arroyo, and A. Anton

Shrubs or perennial to annual herbs, rarely trees, rosette-forming or ephemerals, sometimes with tuberous or gemmiferous roots, or with tubers or stolons; stems occasionally with heteroblastic growth or with cauline spines; plants glabrous, frequently viscose or slightly or densely wooly-tomentose, with a diverse array of non-glandular and glandular trichomes. Leaves alternate, often in pairs, sometimes becoming opposite in the inflorescence, usually simple, entire, infrequently pinnatifid to deeply dissected or compound, exstipulate. Flowers perfect, rarely functionally unisexual in dioecious or andromonoecious plants, sessile to mostly pedicellate, fragrant or not, solitary or more commonly in axillary, extra-axillary, or terminal multi-flowered lax panicles, cymes or fascicles. Perianth (4)5(6)-merous; calyx actinomorphic, rarely zygomorphic, undivided or slightly to much divided, usually persistent and variously accrescent; corolla actinomorphic or zygomorphic, rotate, stellate, tubular, infundibuliform or hypocrateriform, exceptionally papilionaceous, tube glabrous or hairy inside, lobes longer or shorter than the tube, similar or dissimilar in size, aestivation valvate, valvate-induplicate, valvate-plicate, valvate-conduplicate, valvate-supervolute, cochlear, cochlear-conduplicate, cochlear-plicate, contorted-induplicate, contorted-conduplicate, contorted-plicate, quincuncial, or reciprocative; androecium included or exserted, 5-merous, 6-merous (*Goetzea*), 4-merous (then stamens equal in *Nothocestrum*, otherwise didynamous: 4 fertile or 2 fertile stamens and 2 staminodes; sometimes also with a fifth staminode) or with 2 fertile mobile stamens in lateral or dorsal position and 3 staminodes with vestigial or sometimes without anthers, filaments straight or declinate, inserted at different heights on the corolla tube, distinct or connate in a basal ring fused to the corolla, filament base appendages adnate to petal tube ("stapet") absent, inconspicuous or conspicuous (with or generally without basal auricles), anthers dorsifixed, basifixed, dorsi-basifixed, or ventrifixed, extrorse, introrse or latrorse, frequently connivent (in *Solanum* sect. *Lycopersicon* with sterile apices and joined in a column), thecae generally non confluent apically, equal or unequal, dehiscence longitudinal or by terminal pores or slits, connective inconspicuous, wide, forming a dorsal layer of uniform and slight thickness, or thick with an emerging hump; gynoecium with oblique orientation (except *Nicandra*), usually bicarpellate, 3–5-carpellate in *Jaborosa*, *Nicandra*, *Trianaea*, and up to 30-carpellate in *Nolana*, ovary generally bilocular, sometimes 4-locular due to false septa, superior or sometimes partly inferior, glabrous or with trichomes or prickles; style heteromorphic or homomorphic, straight or declinate, terminal or subterminal, stigma variously shaped, usually wet and papillate (papillae rarely lacking); nectary absent, cryptic, or evident, then annular, invaginated or with 2 prominent lobes. Fruit generally a many-seeded berry or a septifragal, septicidal or septicidal-loculicidal capsule, rarely a pyxidium, drupe, diclesium or schizocarp. Seeds 1 to ca. 5000, compressed, then discoid, lenticular, reniform, irregularly ovoid or not compressed, then generally angular, cuboidal, ovoid, prismatic, polyhedric, subspherical, reniform; embryo straight to coiled, U-shaped only in

Hunziger (deceased).
Fuentes (deceased).

J.W. Kadereit and V. Bittrich (eds.), *Flowering Plants. Eudicots*, The Families and Genera of Vascular Plants 14,
DOI 10.1007/978-3-319-28534-4_29, © Springer International Publishing Switzerland 2016

Duckeodendron, cotyledons incumbent or oblique, less frequently accumbent, usually equal, slightly longer or shorter (up to 12 times shorter in the Australian endemic genera) and as wide or rarely broader than rest of embryo; endosperm rarely absent, generally copious, storing mainly oil and starch, with cellular type of endosperm formation (nuclear only in *Schizanthus*).

Cosmopolitan family of 96 genera and ca. 2400 species, much more diversified in South America than elsewhere.

Vegetative Morphology. Annual, biennial or more commonly perennial plants, succulents in *Sclerophylax* and *Nolana*; shrubs (0.5–6 m tall but occasionally stunted, rarely exceeding 20 cm high), sometimes lianas (up to 30 m high) or small trees (5–10 m tall, rarely up to 30 m), rarely greatly reduced (annuals, hemicryptophytes or chamaephytes); occasionally with fleshy, thick taproots, or with gemmiferous roots, or root tuberosities, or tubers, stolons or rhizomes. Branching sympodial, sometimes little branched, zigzag, or strongly dichasial; stems usually erect, or reclinate to prostrate, sometimes climbing, occasionally with heteroblastic growth, or sometimes with spines or prickles, or leafless, internodes long or short, hollow or usually solid. Leaves entire, infrequently toothed or cleft (pinnatifid, pinnatipartite or pinnatisect), generally simple, or occasionally compound (usually imparipinnate or trifoliolate), always exstipulate, occasionally thick, coriaceous, occasionally highly succulent (especially *Nolana* and *Sclerophylax*), alternate or frequently anisogeminate (the members of a pair both on the same side of the stem), often markedly unequal, sometimes in whorls of three to six, fasciculate or not, sessile or petiolate; basal leaves sometimes in a rosette. In some taxa (especially in *Solanum*), hair tuft type domatia on the leaves (Brouwer and Clifford 1990).

Various systems of vegetative reproduction are known. For example, in *Nierembergia aristata*, aerial stolons arise at the basal tuber-like nodes of the erect plants; the species of *Solanum* sect. *Petota* (*S. tuberosum*, *S. chacoense*, etc.) have subterraneous stolons arising from the underground stem and form a terminal tuber each. In other cases, root buds (Fig. 56C) are responsible for the origin of subterranean organs; in this way the horizontal rhizomes of *Nierembergia rigida* (Cosa 1989) or the vertical rhizomes of some *Jaborosa* spp. are formed (Barboza and Hunziker 1987). In *Bouchetia anomala*, *Leptoglossis linifolia* (Fig. 56B) and *Solanum elaeagnifolium*, there are lateral root buds which show two growth phases: a horizontal followed by a vertical phase, the buds arising at the point of curvature of the roots (Cosa 1989). Finally, the vertical root ends in a prominent tuberosity in *S. juvenale* and *S. hieronymi* (Cosa et al. 1998, 2000).

Vegetative Anatomy. All members of the family are distinctive in having internal phloem and 1-traced, unilacunar nodes. The foliar epidermis (trichomes and stomata) and the internal structure of leaves and stems have relevant features for the delimitation of the different taxa. The primary root is usually diarch (Fig. 56A), such as in *Nicotiana tabacum*, *Solanum juvenale*, *S. elaeagnifolium*, *S. tuberosum*, *S. sisymbriifolium* and *Bouchetia anomala* (Dottori et al. 2000; Hadid et al. 2007), but sometimes triarch types are present such as in *Solanum palinacanthum* (Pericola et al. 2004), or even tetrarch, pentarch and hexarch in other *Solanum* species (Nurit-Silva et al. 2007, 2011). The first phellogen originates from pericyclic layers (*Combera*, *Benthamiella*, *Nicotiana*), or immediately below the epidermis (*Solanum sisymbriifolium*). In some species (e.g. *Solanum hieronymi*) the roots develop into small fleshy storage organs.

In the primary stem, the epidermis consists of a single layer of cells, while the cortex shows distinct features according to the position and kinds of tissues (Metcalfe and Chalk 1950; Cosa 1991, 1993, 1994; Liscovsky et al. 2002; Liscovsky and Cosa 2005; Liscovsky 2007, Figueroa et al. 2008); in some genera, e.g. *Petunia*, *Benthamiella* and *Combera*, the cortex consists exclusively of chlorenchyma; the chlorenchymatic subepidermal layers are arranged palisade-like in *Leptoglossis*, *Salpiglossis* and *Reyesia* (Fig. 56P). In other genera, the cortex consists of chlorenchyma and collenchyma: sometimes, as in *Lycium*, *Jaborosa*, *Sclerophylax* and *Brunfelsia* (Fig. 56D), the collenchyma is superficial, while in Browallia (Fig. 56G), Nicotiana (Fig. 56J), Schizanthus (Fig. 56M), *Cestrum*, *Sessea*, some *Solanum* and the majority of *Fabiana* spp. it is located on the innermost layers of the cortex. Finally, in genera such as *Datura* and *Brugmansia* the cortex consists of chlorenchyma, collenchyma and

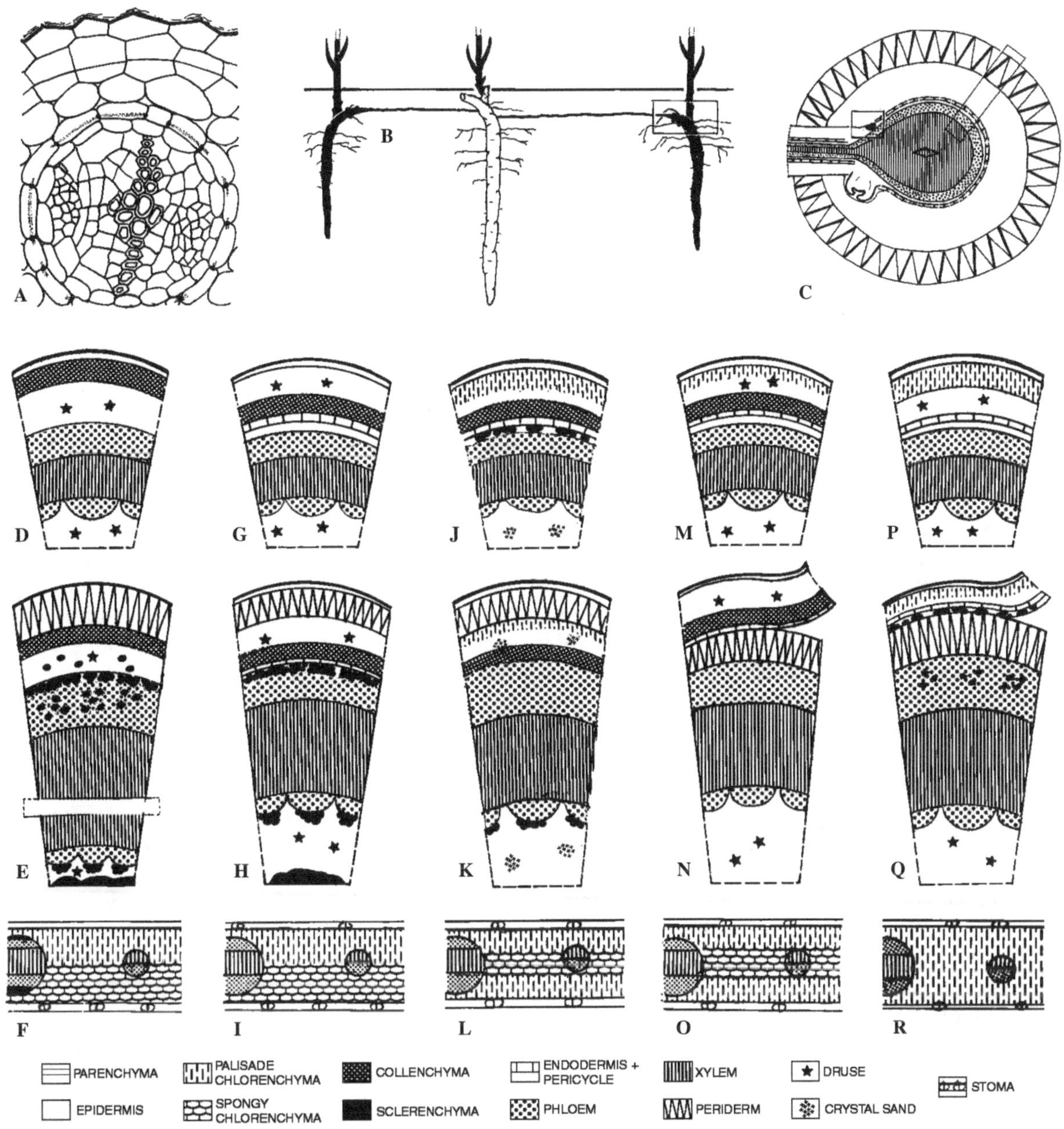

Fig. 56. Solanaceae. Vegetative anatomy. **A–C** *Leptoglossis linifolia*. **A** Diarch primary root. **B** Vegetative propagation. **C** Secondary root with root buds. **D–F** *Brunfelsia* spp.; **G–I** *Browallia* spp.; **J–L** *Nicotiana* spp.; **M–O** *Schizanthus* spp.; **P–R** *Reyesia* spp. **D, G, J, M, P** Primary stem diagrams (observe the different position and kind of tissues). **E, H, K, N, Q** Secondary stem diagrams (note the different position of the periderm). **F, I, L, O, R** Foliar structures. **F, I** Dorsiventral, **L, O** Isolateral, **R** Homogeneous. (Orig.)

parenchyma, with the collenchyma located beneath the superficial chlorenchyma. In some species of *Melananthus*, *Protoschwenckia* and *Schizanthus*, the innermost cortical layer usually contains many large starch grains (starch sheath); in other cases (e.g. *Nierembergia*, *Combera*, *Schwenckia*, *Streptosolen*), these cells have Casparian strips. Two patterns of arrangement of the vascular tissues are characteristic: the most common is an amphiphloic siphonostele, but in some

Solanum spp. and in *Sclerophylax* a eustele with bicollateral bundles is present. In the secondary stem, the origin of the periderm and the presence and distribution of fibres and crystals (druses, solitary crystals and crystal sands) are taxonomically important. The periderm may have a superficial or a deep origin; in the former case, it arises from the epidermal layer (e.g. *Brunfelsia*, Fig. 56E, *Schwenckia*, *Nierembergia* and *Leptoglossis*), or from subepidermal layers (*Browallia*, Fig. 56H, *Cestrum*, *Latua*); in the latter case, the periderm originates at the outer limit of the external phloem, such as in *Salpiglossis* and *Reyesia* (Fig. 56Q) and *Schizanthus* (Fig. 56N). The pericycle consists commonly of isolated groups of fibres (Fig. 56E, H, Q) or, in some cases, a continuous ring of fibres is present, generally with cellulose or with slightly or well-lignified cell walls. Sclerenchyma also occurs in the cortex or in the external and/or internal phloem and/or in the pith, such as in *Brunfelsia* (Fig. 56E), *Cestrum*, *Sessea*, *Browallia* (Fig. 56H), *Streptosolen*, *Nicotiana* (Fig. 56K); sclereids also appear in the epidermis (*Cestrum* and *Sessea*). In *Schizanthus* (Fig. 56N) and *Combera*, there are no fibres at all. *Datura* shows atypical growth in the vascular system by forming an extra-parenchyma as a water storage tissue (Liscovsky et al. 2001).

Wood structure is diverse in Solanaceae: ring porous (e.g. *Fabiana* and *Lycium*) to diffuse porous (*Capsicum rhomboideum*, *Nothocestrum latifolium*), and the vessels have simple perforation plates, or rarely vestiges of bars (Carlquist 1988, 1992); wood of *Duckeodendron* is clearly mesomorphic, that of *Espadaea* is intermediate and that of *Henoonia* is relatively xeromorphic, based on vessel element dimensions, vessel density, and presence of vasicentric tracheids (Carlquist 1988). Carlquist (1987) reviewed the wood anatomy in *Nolana* and suggested it was indicative of paedomorphosis. Both in primary and secondary stems druses (Fig. 56H, N), solitary crystals or crystal sand (Fig. 56K) are present in the cortex and pith and sometimes also in the mesophyll. Stem and leaves in *Salpiglossis* and *Reyesia* contain druses and solitary crystals but *Browallia*, *Streptosolen*, *Brunfelsia* and *Schizanthus* have only druses; the stems of *Benthamiella*, *Combera*, *Pantacantha*, *Nierembergia* and *Bouchetia* completely lack any kind of crystals.

The anatomy of the petiole is uniform. The vascular system consists of a central arc-shaped bicollateral bundle accompanied by 1–4 smaller bundles on each side (e.g. *Browallia*, *Salpiglossis*, *Reyesia*, *Solanum*). Dorsiventral (Fig. 56F, I), isolateral (Fig. 56L, O) and homogenous (Fig. 56R) leaves are present in the family, but the dorsiventral type is the most common. Some genera show only one type of leaf structure, e.g. *Jaborosa*, *Schultesianthus*, *Brunfelsia* and *Schwenckia*, while others (such as *Lycium*, *Calibrachoa*, *Reyesia*, *Solanum* and *Nierembergia*) have two and even all three structural types (Cosa 1991, 1993; Reis et al. 2002; Araújo et al. 2010). The vascular system consists of a main bicollateral bundle from which smaller veins branch. In some species, the phloem is surrounded by collenchymatous cells (*Nicotiana*, *Browallia*, *Streptosolen*) or fibre groups (*Pantacantha*, *Benthamiella*, *Brunfelsia*, some *Solanum*); in *Duckeodendron*, the main vein is surrounded by a sclerenchymatous sheath. Stomata occur either on both sides of the leaf—e.g. some species of *Calibrachoa*, *Datura*, *Schwenckia*, *Protoschwenckia* and *Solanum*—or only on the lower side, as in some *Solanum* spp. (Benítez de Rojas 2007; Nurit-Silva and Agra 2011), *Cestrum*, *Atropa*, *Brunfelsia*, *Latua*, *Streptosolen* and *Duckeodendron*.

The presence of different types of stomata on one leaf is frequent in Solanaceae. The most common types are anomocytic and anisocytic stomata; however, *Brunfelsia* has mostly paracytic stomata and, in *Dyssochroma*, *Juanulloa*, *Withania*, *Physalis*, *Schwenckia* and *Melananthus*, the diacytic type is predominant (Patel and Inamdar 1971; Bessis and Guyot 1979). Apparently, *Fabiana* is the only genus where the stomata are placed on projections of the epidermis. Raphides are present in the mesophyll of some *Sclerophylax* (Di Fulvio 1961).

Trichomes are present on nearly all parts of the plants and are of taxonomic importance (Seithe 1962). The array, density, and variety of trichomes (i.e. morphology, size, and ornamentation of the cell walls) produce a great diversity of indumentum types (Roe 1971). Non-glandular, simple to variously dendritic, uniseriate, multicellular trichomes as well as glandular trichomes (head and stalk uni- or multicellular) are the most widespread in the family. Simple short trichomes result in a puberulose or strigose appearance, as in some *Jaltomata*, *Larnax*, etc., while longer trichomes which are the most common result in a villose (*Hyoscyamus*, some *Aureliana*, *Nolana* and *Solanum*, etc.), tomentose (some Australian genera, *Solanum*, etc.) or viscid indumentum (e. g. *Salpiglossis*, *Reyesia*, *Nicotiana*, *Nolana*, *Petunia*, *Fabiana*, *Calibrachoa*, *Exodeconus*, some *Athenaea*, and *Solanum* spp.). Dendritic trichomes vary from few-branched as in *Leptoglossis*, *Withania*, *Brunfelsia*, etc. to many-branched as in *Solanum* (Mentz and de Oliveira 2004), *Sessea* (Benítez de Rojas and Jáuregui 2005), *Cestrum*, *Nolana*, *Juanulloa*, *Anthotroche* and *Symonanthus*. In general, the surface of the trichomes is smooth or slightly to strongly verrucate, especially within the non-glandular types; rarely striate non-glandular trichomes (*Vestia*, *Leptoglossis*, some *Solanum* spp.) or trichomes with verrucate stalk (*Schwenckia*, *Protoschwenckia*, *Heteranthia*, and some *Jaborosa*, *Capsicum*, and *Lycianthes* spp.) appear within the glandular type. The stellate, echinoid and lepidote trichomes of *Solanum* subgen. *Brevantherum* and subgen. *Leptostemonum* are very variable in structure (Mentz et al. 2000). The stellate trichomes are sessile or stalked (stalk uniseriate or multiseriate, multicellular, emergent or intrusive), the lateral rays (more than 4) are in one plane or multiangulate (arms radiating outwards in all directions) while the short or long central ray is glandular or not. The most conspicuous trichomes are: a) unicellular non-glandular trichomes, b) glandular trichomes with long multiseriate stalk and multicellular head (shaggy hairs) present only in *Schizanthus*, c) glandular trichomes with a multiseriate, multicelled head (more than 20 cells) and a long, multicellular, uniseriate, smooth stalk characteristic of *Salpiglossis* and *Reyesia*, d) whorled branched trichomes mixed with other dendritic trichomes in the majority of the Australian genera, e) glandular scale-like peltate trichomes, sunken in the mesophyll of *Schultesianthus*, f) vesicular, shiny, strongly verrucate trichomes only in *Quincula*, g) viscid glandular trichomes (head uni- or multicelled) in some *Exodeconus*, *Nicotiana*, *Nolana*, *Petunia*, *Calibrachoa*, and oil-secreting trichomes on the inner surface of the corolla in *Nierembergia* and *Atropa*, h) bayonet trichomes (two-celled trichomes, their basal cell very large and thick-walled, the apical cell small and thin-walled) of *Solanum* sect. *Basarthrum*, i) non-glandular and glandular trichomes, present in *Oryctes*, *Lycium* and *Lycianthes*, j) branched-glandular trichomes in *Sclerophylax*, *Physochlaina orientalis*, *Eriolarynx lorentzii* and some *Physalis*, or stellate-glandular ones with many short glandular branches (e.g. *Solanum velleum* and *S. robustum*), and k) dendritic-echinoid (*S. axillifolium*) or paleaceous (*S. castaneum*) trichomes. The presence of prickles is peculiar to *Solanum* subgen. *Leptostemonum*; they may be straight, curved or uncinate, acicular, mammiform or conic, compressed or not, patent to retrorse, glabrous or with non-glandular or glandular trichomes, rarely with stellate trichomes at their base and sometimes with a tuft of unicellular rays at their apex. In *Saracha*, the young stems are covered with brownish, dendriform, multicellular emergences.

Inflorescence Structure. Danert (1958, 1967) and Troll (1969) intensively studied the inflorescence structure in the family; before them, an account had been provided by Eichler (1875, 1878). The flowering shoots are basically of the monotelic (closed) thyrse type. They are often strongly modified by serial (basis to top) differentiation in the degree of branching, by recaulescence (shift of the subtending leaf along their axillary shoots), and concaulescence (fusion of axillary shoot to the primary shoot). These basic types of branching are consistent in the family. However, the diverse patterns found are still difficult to connect with each other. Each flowering shoot ending in a flower commonly continues growth by sympodial branching resulting in cymose systems. These represent the repetition of modules (sympodial units), each one consisting of a flower and its underlying node and internodes (under-structure or Unterbau). The number of nodes and internodes of the under-structure varies considerably. A tendency to the reduction of the under-structure to its minimum is evident, that is, three internodes delimited by the nodes of the first (α) and second (β) prophylls. The buds of the β-prophyll most commonly continue branching. End flowers may recede to their secondary flowers, whereby they are shifted to a lateral position. Thus, sympodial systems, superficially resembling racemes, may arise in which flowers, though lateral in position, lack a subtending leaf and have instead one opposite to it. These are the so-called spur shoots (Child 1979). This pattern is often further modified by recaulescence in such a way that each flower is accompanied by two leaves (the so-called geminate leaves), of which one, usually the smaller, corresponds to the β-prophyll of the same flower and the other to the β-prophyll of the preceding flower. In many Solanaceae, and more often in Solanoideae, these sympodial systems are particularly rich in number of sympodial units in the upper parts of the main shoot. This is the so-called acrotonic development promotion (Troll 1969). If these branching systems continue forming normal leaves, they represent the main photosynthetic part of the whole plant. Such foliose sympodial chains are the so-called anthocladia (Troll 1969). Anthocladia may further become independent from the main shoot's under-structure if they produce adventitious roots. Reduction of the main shoot's under-structure leads to a creeping habit of the anthocladia. Complete loss of the under-structure and independence of the anthocladia is clearly evident in *Lycianthes asarifolia* and *Nierembergia repens* (Troll 1969). Anthocladia may further become zonalized by abortion or caducity of the basal sympodial members (*Przewalskia tangutica*). Genetic control of branching system development is just beginning to be understood (Lippman et al. 2008).

Flower Structure. Perianth usually 5-lobed, rarely 5-parted or unevenly 4-toothed or 4-lobed (*Nothocestrum*, some *Lycium*, etc.), or 6-lobed (*Goetzea*). Calyx synsepalous, actinomorphic, sometimes zygomorphic (e.g. *Brugmansia*, *Nicotiana undulata*), exceptionally asymmetric (*Sclerophylax*, *Nolana*), or circumscissile after anthesis (*Datura*), entire (*Tubocapsicum*, *Capsicum*, etc.) or with 5–10 linear teeth or appendages (*Lycianthes*, *Capsicum*). In some genera tube almost absent (*Athenaea*, *Nectouxia*, *Markea*, *Schizanthus*, etc.). Calyx usually persistent, very frequently accrescent around the berry or capsule, sometimes inflated and the upper sector almost closed (*Physalis*, *Quincula*), or tightly appressed to the pericarp or loosely enclosing it and widely open in the upper sector (*Chamaesaracha*, *Leucophysalis*), or reflexed (*Jaltomata* spp., *Schraderanthus*, *Capsicophysalis*) and exposing the mature berry. *Nicandra* is unique for having auriculate calyx segments.

Corolla gamopetalous, very small (2.5 mm long.) to very large (100–370 mm long in *Solandra*), actinomorphic, slightly (e.g. *Sclerophylax*) or distinctly zygomorphic (e.g. *Browallia*, *Streptosolen*, *Schultesianthus*), in general rotate, stellate, tubular, infundibuliform or hypocrateriform (papilionate in *Schizanthus*); aestivation valvate (*Aureliana*, *Dyssochroma*), valvate-induplicate (*Cestrum*, *Jaltomata*, *Protoschwenckia*), valvate-plicate (*Capsicum*), valvate-conduplicate (*Jaborosa*, *Oryctes*, *Schwenckia*), valvate-supervolute (each lobe has one of its edges rolled inwards and overlapped by its opposite edge, two lobes being dextroinvolute and the other three sinistroinvolute, tribe Anthocercideae), cochlear (*Petunia*, *Nierembergia*, *Lycium*, *Markea*, *Schultesianthus*), cochlear-conduplicate (*Latua*), cochlear-plicate (*Nicandra*, some *Lycium*), contorted-induplicate (*Exodeconus*), contorted-conduplicate (*Sclerophylax*, *Metternichia*, *Datura*), quincuncial (*Duckeodendron*), reciprocative (the anterior induplicative lobe covers the four others which are conduplicate, *Calibrachoa*); exceptionally tube very short (*Athenaea*), lobes usually entire, or deeply laciniate in *Schizanthus*, and generally equal (except *Browallia* with unequal lobes). In *Lycium* and *Cestrum*, the corollas leave, after falling off, an annular persistent basal portion around the ovary base (called cyathium by some authors).

Several filament and anther features are of interest in recognizing genera and suprageneric taxa. These can be summarized as follows. The basic number of stamens in the family is five. This condition is maintained in Solanoideae. There is an exceptional increase or decrease in stamen number due to alteration of the symmetry of the whole flower (Robyns 1931; Knapp 2002a). In Cestroideae and Nicotianoideae, reduction in the androecium by abortion of 1 or 3 stamens is frequent. When 4 stamens are fertile, reduction or loss of the anterior stamen has occurred (Fig. 57A, N, O, Q). Four fertile stamens are found in some genera of tribes Anthocercideae, Salpiglossideae, and Schwenckieae. The lateral pair is normally longer than the dorsal pair (Fig. 57A). If two fertile stamens are present, this condition may be achieved by further reduction of either the lateral, as in *Leptoglossis* (Fig. 57P) and *Hunzikeria*, or the ventral stamens as in *Schwenckia* and also in *Schizanthus*

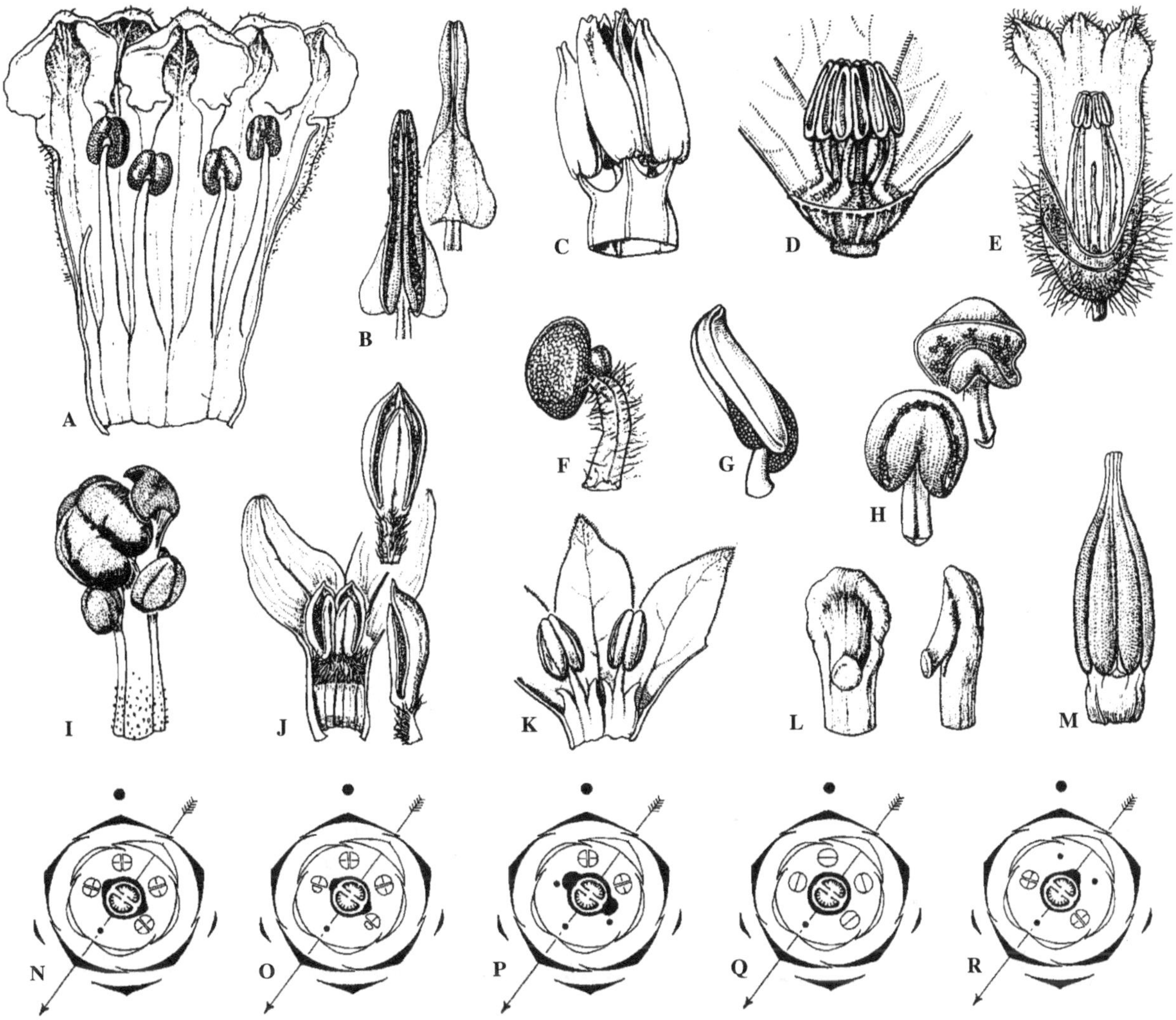

Fig. 57. Solanaceae. Androecium. **A–M** Examples of major stamen modifications. **A** *Bouchetia anomala*, ventrifixed anthers. **B** *Nectouxia formosa*, apical geniculation of the filament. **C** *Solanum herculeum*, filament fusion, basifixed anthers and anther dehiscence by apical slits. **D** *Nicandra physalodes*, basal geniculation of the filaments. **E** *Symonanthus aromaticus*, anther adnation. **F** *Browallia americana*, anther asymmetry. **G** *Solanum sciadostylis*, dorsal and ventral connective outgrowths. **H** *Anthotroche pannosa*. **I** *Nierembergia browallioides*, filament adnation. **J** *Juanulloa wardiana*, confluent thecae. **K** *Capsicum chacoense*, wing-like expansions of the filament. **L** *Jaborosa integrifolia*. **M** *Solanum lycopersicum*. **N–R** Diagrams showing reduction in the androecium. **N** *Bouchetia*, *Leptoglossis* (in part), *Melananthus*, *Plowmania*, *Protoschwenckia*, *Reyesia* (in part), *Salpiglossis*. **O** *Browallia* and *Streptosolen*. **P** *Hunzikeria*, *Leptoglossis* (in part), *Reyesia* (in part). **Q** *Brunfelsia*, *Crenidium*, *Cyphanthera*, *Duboisia*. **R** *Schizanthus* and *Schwenckia*. (A–M from Hunziker 2001, N–R orig.)

(Fig. 57R). Six stamens are exceptional, such as in *Goetzea*.

Fusion at the filament level takes place when an organ complex is formed in which the boundary of the individual stamens cannot be recognized anatomically. This condition is found in *Solanum*, though it may be masked by additional fusion of the filaments to the corolla tube. In some species, however, it is clearly evident because the filaments are fused also above their fusion with the corolla tube. Adnation, i.e. the bonding of stamens without losing anatomical boundaries, may take place at the anther level—e.g. in *Solanum* sect. *Lycopersicon* (Fig. 57M) and *Symonanthus* (Fig. 57E)—and at the filament level, as in *Nierembergia*.

Filaments may have lateral auriculate appendages at their base. In addition, these expanded segments may be fused to the corolla tube to form the so-called stapet (Ritterbusch 1976), which provides important features in recognizing *Capsicum* (Fig. 57K) and allied genera. The filament expansions are functionally related to nectar conduction from the disc to the place of presentation. Geniculation or curvature of the filament may occur near the anther (e.g. *Atropa*, *Mandragora*, *Solandra*, *Sclerophylax* and *Jaborosa*, Fig. 57L). In *Nicandra* (Fig. 57D) and *Cestrum*, filaments are geniculate near the base.

Filament attachment is also variable. The most common conditions are dorsifixed and basifixed. Ventrifixed anthers are present in many genera such as *Cuatresia*, *Jaltomata*, *Withania*, *Trianaea*, *Schwenckia*, *Melananthus*, *Heteranthia*, *Nierembergia*, *Leptoglossis*, *Bouchetia* (Fig. 57A), and in the Australian genera.

Connective outgrowths are of systematic and ecological importance in *Solanum* sect. *Pachyphylla* (ex *Cyphomandra*, Fig. 57G); lateral connective expansions are also present in *Nectouxia* (Fig. 57B). Thecal fusion occurs when the two thecae of an anther share the sporogenous tissue. Thus, anthers have two loculi instead of four and one semicircular dehiscence line instead of two straight lines (Fig. 57H, J, Q). Such synthecous anthers are found in *Brunfelsia*, five genera of Anthocercideae, in *Markea* spp. (ex *Hawkesiophyton*), *Juanulloa* spp. (ex *Rahowardiana*) and some *Fabiana* species. The dehiscence line is longitudinal in most genera but may be restricted to a longitudinal apical slit (Fig. 57C) or pore (*Solanum* in part, *Lycianthes*). In some *Solanum*, in addition to the larger end pore, a longitudinal row of short slits is found which may enlarge and be fused after opening (Carrizo García et al. 2008). Anther asymmetry by the unequal development of the thecae in one anther may result in size differences (*Nierembergia*, Fig. 57I) or in abortion of one theca (*Browallia* and *Streptosolen*, Fig. 57F, O).

The ovary in the Solanaceae is typically superior, bicarpellate, and bilocular (Fig. 58R). *Solandra* stands out with a partly inferior ovary (Fig. 58V), as well as *Nothocestrum* and *Datura*. In *Nicandra* (Fig. 58Y) and *Trianaea*, there are 5- or 4-carpellate ovaries while in some *Jaborosa* species ovaries are 3- to 5-carpellate. In *Nolana*, the development of the carpels is unique in the subdivision of the loculi into locelli (Bondeson 1986; Di Fulvio 1971; Huber 1980).

Regarding the number of ovary locules, some variation is found: *Trianaea* has 8–10 locules, *Solandra* has 4 locules while *Coeloneurum* and *Henoonia* have only one. In some genera, the number of locules varies along the ovary due to the development of false septa; thus, some *Lycium* ovaries are 4-locular in the upper half and bilocular in the lower half, whereas in *Datura* (Fig. 58M, N) and some *Vassobia* species they are 4-locular only in the lower half. The ovules are mostly borne on more or less fleshy axile placentas, or on parietal (*Tsoala*) or basal placentas (*Melananthus*); in *Sclerophylax*, ovules are pendant. Four different types of ovules have been reported so far (Barboza 1991): anatropous, anacampylotropous, hemitropous and hemicampylotropous. There are usually numerous ovules per locule (Fig. 58M, P, R, V, W), but sometimes they can be reduced to a few or one (Fig. 58K).

Styles are mostly cylindrical, solid, and terminal (Fig. 58K, O, U); occasionally, they can be hollow with a stylar canal (Fig. 58Z), and less frequently subterminal. *Browallia* (Fig. 58Q) and *Streptosolen* stand out by having a corrugated style. Stylar heteromorphism was recorded in some *Solanum* spp., *Athenaea* and *Aureliana*. *Leptoglossis* (Fig. 58X) and allied genera are typical for the presence of a distally broadened style with two thin lateral expansions. Stigmata are chiefly capitate and bilobed with an evident slit (Fig. 58O), except for *Schizanthus* with a very small inconspicuous stigma. The receptive surface has small to medium-sized 1- to 4-cellular papillae which are absent in some taxa, e.g. *Nierembergia linariaefolia* and *Solanum nigrum*. The majority of the analyzed stigmata are wet with the exception of *Solandra* where they are dry. Although the stigma has a comparatively large surface in *Browallia* (Fig. 58Q), only two small lateral areas are receptive. Most genera have a floral nectary while a few of them completely lack it. When present, the nectary (Fig. 58K, O, P, T, U, W) is located at the ovary base. Nectar is secreted through modified stomata (Bernardello 2007). There are some variations in structure, shape and colour which are of taxonomic interest: 1. macroscopically inconspicuous and of the same colour as the ovary, but microscopically detectable (e.g. *Browallia*, many *Lycium* spp., Fig. 58T, U, *Vassobia*, etc.); 2. macroscopically conspicuous and orange to red carotinoid-coloured (e.g. some *Lycium*, *Salpichroa*,

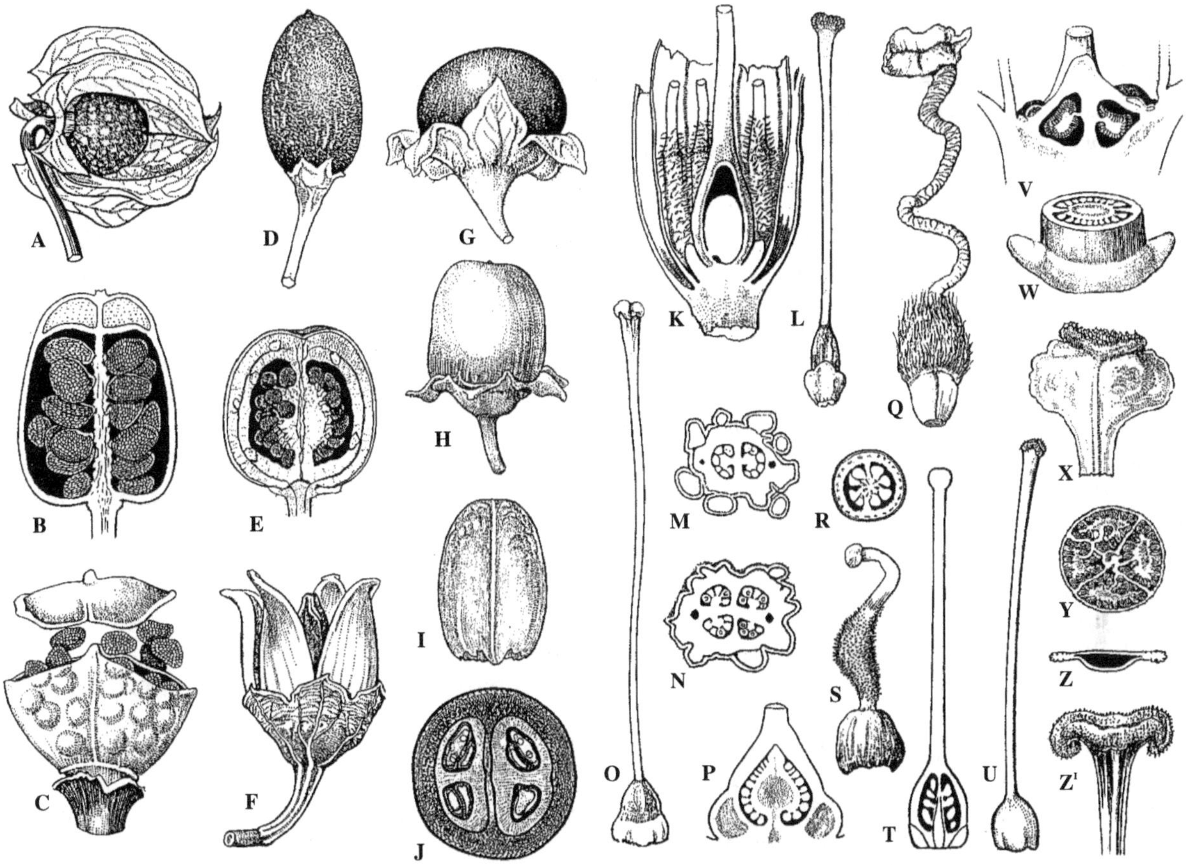

Fig. 58. Solanaceae. Fruit and gynoecium. **A–J** Fruits. **A** Berry with accrescent and inflated calyx (partially removed) of *Nicandra physalodes*. **B** Berry with two macroscopic stone cells of *Lycium bridgesii*. **C** Pyxidium of *Scopolia japonica* with accrescent calyx removed. **D** Berry of *Lycium gilliesianum*. **E** Berry with scattered sclerosomes of *Lycianthes rantonnei*. **F** Capsule of *Vestia foetida*. **G** Berry of *Atropa belladonna*. **H–J** Fruit of *Lycium boerhaviifolium*. **H** Lateral view, **I** Lateral view of pyrene, **J** Cross section showing the two pyrenes. **K–Z′** Gynoecia. **K, L** *Melananthus multiflorus*. **K** Longitudinal section of basal part of flower showing ovary and nectary. **L** Pistil. **M, N** Cross sections of ovary in the upper and basal part of *Datura ferox*. **O, P** *Vassobia iochromoides*. **O** Pistil. **P** Longitudinal section of ovary. **Q** Pistil of *Browallia americana*. **R, S** *Solanum nitidibaccatum*. **R** Cross section of ovary. **S** Pistil. **T, U** *Lycium cestroides*. **T** Longitudinal section of pistil showing cryptic nectary. **U** Pistil. **V** Longitudinal section of ovary *Solandra grandiflora*. **W** Ovary base with nectary of *Mandragora officinarum*. **X** Stigma and upper part of style of *Leptoglossis lomana*. **Y** Cross section of ovary of *Nicandra physalodes*. **Z, Z′** Cross section of upper part of style, and style and stigma of *Bouchetia anomala* respectively. (From Hunziker 2001)

Jaltomata, Fig. 58O); 3. invaginated of two types: invaginated-pelviform (*Schwenckia*) or invaginated-bilobed (e.g. *Fabiana*, *Petunia*). Showy special cases of the second type are *Mandragora* (Fig. 58W) with two opposite commissural lobes, and *Melananthus* with a cup-shaped nectary (Fig. 58K, L).

Extrafloral nectaries occur on the outer surface of the corollas (Anderson and Symon 1985) or on the abaxial surface of the leaves. Extrafloral nectaries on the calyces of *Physalis viscosa* and *Lycium* (sub nom. *Phrodus* and *Grabowskia*) have been reported (Bernardello 1987; Rodríguez 2000).

Embryology. Anther wall formation follows the basic as well as the dicotyledonous type; posterior divisions are frequent in the middle layers. In the mature anther wall, the epidermis may become papillate, the endothecial cells develop lignified thickenings following different patterns (reticulate, helical, annular), and sometimes also middle layer cells remain (Carrizo García 2002a; Barboza and Carrizo García 2005); in species with poricidal dehiscence, the lignified layers usually are restricted to the zone surrounding the pores (Siddiqui and Khan 1988; Carrizo García 2002b). The tapetum is glandular and belongs to the pachymerous type (inner tapetum cells larger and radially more elongated than the outer tapetum cells); a very common feature is the presence of an early placentoid arisen from the connective tissue (Hartl 1963). In many species, the so-called resorption tissue differentiates in the septum, usually in hypodermal position; this tissue is placed in subhypodermal position in poricidal anthers, except in the pore zone (e.g. *Solanum betaceum*, *S. glaucophyllum*, Carrizo García et al. 2008). The resorption tissue is formed by a variable number of cells of which the cytoplasm is gradually replaced by calcium oxalate crystals in the first stage of development (D'Arcy et al. 1996). Regarding tissue thickness, some are well developed, up to 12 cells wide in cross section (e.g. *Nicotiana glauca*), and others are very narrow, only two to four cells wide (e.g. *Solanum*).

Normally, two rows of archaesporial cells occur; cytokinesis of pollen mother cells is simultaneous; the microspore tetrads are mainly tetrahedral, but isobilateral (e.g. *Hyoscyamus* and *Atropa*, Sharma et al. 1987) and decussate tetrads (e.g. *Nicandra physalodes*, Prasad and Singh 1978) can also be found. Pollen is usually 2-celled at shedding but 3-celled pollen has been found in species of *Capsicum* (Davis 1966), *Nicotiana* (Jagannadham 1988), *Jaborosa*, and *Salpichroa* (Barboza 1989).

The ovules are always tenuinucellate and unitegmic, with a well-developed endothelium. The nucellus forms a hypodermal archaesporial cell, which enlarges and functions directly as megaspore mother cell; rarely two archaesporial cells are found (e.g. *Solanum* spp.). The megaspore mother cell usually gives rise to a linear tetrad, T tetrads have been reported in *Nicotiana tabacum* (Goodspeed 1947) and some *Solanum* spp. (Mohan and Singh 1969; Siddiqui and Khan 1986); in both tetrad types, the proximal megaspore is functional. The female gametophyte is usually monosporic and develops according to the Polygonum type; bisporic embryo sacs of either the Allium or the Endymion type (Modilewski 1935; Pavari 1957; Dharamadhaj and Prakash 1978), or of the tetrasporic Adoxa type (Nanetti 1912) have also been observed; in all cases, the antipodals are ephemeral. Embryogenesis conforms to the Solanad type; a few species follow the Onagrad type (e.g. *Przewalskia tangutica*, Lu et al. 1999). Polyembryony has been reported in a few members of the family (e.g. *Solanum torvum*, Mohan and Singh 1969). The development of the endosperm is cellular, exceptionally nuclear (e.g. *Schizanthus pinnatus*, Samuelsson 1913) or helobial (*Hyoscyamus niger*, Svensson 1926).

Pollen Morphology. Pollen grains show a broad range of variation in size, shape, apertures, aggregation and exine ornamentation. Palynological characters have been important in contributing to the delimitation at different taxonomic levels (Barboza 1986; Moncada and Fuentes 1993; Persson et al. 1994; Bernardello and Luján 1997; Knapp et al. 1997) and in providing support for hypotheses of relationships between genera (Stafford and Knapp 2006).

Almost 50% of the genera have medium-sized pollen grains (range: 25–50 μm), the smallest ones have been reported in some *Lycianthes* taxa, i.e. *L. laevis* (mean 10.7 μm, sub nom. *Solanum bigeminatum*) and *L. lysimachioides* (mean 11.6 μm, sub nom. *Solanum lysimachioides*; Sharma 1974), and in *Schwenckia*, *Melananthus* and some *Solanum* where pollen grains measure 12–20 μm. Some *Nierembergia* spp., *Hyoscyamus*, *Salpichroa*, *Tsoala tubiflora* and *Plowmania nyctaginoides* have the largest grains (60–96 μm). The shape is commonly subspheroidal (ranging from suboblate to subprolate), only few genera have also oblate (*Browallia*, *Leucophysalis*, *Tubocapsicum* and *Espadaea*) or prolate (*Exodeconus*, *Atropa*, *Chamaesaracha*, etc.) or exclusively oblate grains (*Markea*, sub nom. *Hawkesiophyton* and *Schultesianthus*; Persson et al. 1994); *Schizanthus* is the only genus with prolate spheroidal to perprolate grains (Fig. 59G). The typical condition is the presence of trizonocolporate apertures. However, tri-, tetracolporate (*Latua*, *Fabiana*, *Witheringia*, *Nothocestrum*, *Schwenckia*, *Plowmania*, *Protoschwenckia* and *Solanum*) or tetra-, pentacolporate (*Physochlaina*) grains may co-occur, and in *Browallia* and *Streptosolen* grains are up to 9-colpate or colporoidate (Fig. 59C). Exclusively tetra- or pentaporate (some *Jaborosa*) or exclusively tricolpate (*Metternichia*, *Coeloneurum*, *Goetzea*, *Henoonia*) or tri- or tetracolpate or

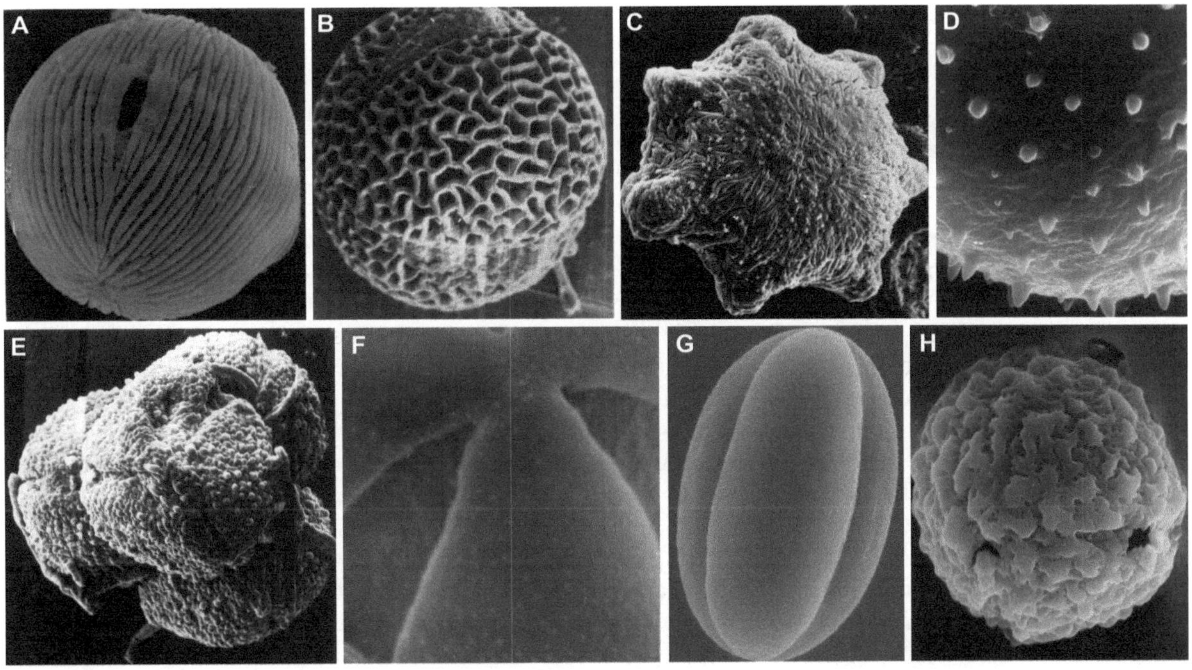

Fig. 59. Solanaceae. Pollen SEM micrographs. **A** *Brugmansia suaveolens*, ectocolpus, endocingulum, striate exine, x 2000. **B** *Leptoglossis ferreyrae*, reticulate exine, x 4000. **C** *Streptosolen jamesonii*, hepta-aperturate pollen, rugulate exine, x 2000. **D** *Metternichia principis*, echinate exine, x 5800. **E** *Salpiglossis sinuata*, tetrad, gemmate exine, x 1200. **F, G** *Schizanthus pinnatus*. **F** Detail of apocolpium, exine microgranulate, x 5400. **G** Equatorial view, x 2000. **H** *Brunfelsia* sp., alveolate exine, x 2000. (Orig.)

colporoidate grains (*Benthamiella*, the Australian genera, some *Brunfelsia*, *Tsoala* and *Espadaea*) are more unusual in Solanaceae. *Anisodus* and *Mandragora* have pantoporate pollen grains. In *Solanum*, inaperturate grains appear in functionally female flowers in all cases of dioecy (*S. appendiculatum*, *S. confertiseriatum*, and many spp. of subgen. *Leptostemonum*). There exist some remarkable peculiarities concerning the morphology of the apertures. One is the case of the cryptaperturate pollen of *Mandragora* where there are no ectoapertures but a symmetrical pattern of endoapertural thinnings occurs on a thick and homogeneous endexine layer (Diez and Ferguson 1984); in *Datura* and *Brugmansia* the endoaperture is a continuous endocingulum with indistinct margins (Fig. 59A; Persson et al. 1999); finally, in some genera the exine is greatly thickened around the ectoapertures forming an annulus as in *Jaborosa* sect. *Jaborosa* (Barboza 1986) or a distinct margo as in *Merinthopodium* (Persson et al. 1994), tribe Benthamielleae, *Petunia*, *Calibrachoa*, *Reyesia chilensis*, and *Salpiglossis spinescens* (Stafford and Knapp 2006) or some *Solanum* spp.

Pollen grains are normally shed as monads; tetrads also occur in *Nierembergia*, *Bouchetia*, *Salpiglossis* (Fig. 59E) and *Reyesia*, and only rarely pollen in massulae is found in some *Nierembergia* spp. (e.g. *N. rigida*, *N. hatschbachii*, etc.). Intra- and intergeneric differences are observed in the ornamentation of the exine (Gentry 1979, 1986; Barboza 1986; Persson et al. 1994, 1999; Bernardello and Lujan 1997; Knapp et al. 2000; Stafford and Knapp 2006). In general, the exine is reticulate (Fig. 59B), scabrate, striate (Fig. 59A), rugulate (Fig. 59H), punctate-foveolate, striate-reticulate or striate-rugulate. The presence of a psilate (some spp. of *Cestrum*, *Nicotiana*, *Physalis* and *Salpichroa*), echinate (*Metternichia*, Fig. 59D, *Exodeconus*, *Markea lopezii*, *Tsoala pubiflora*, *Goetzea*, *Coeloneurum*, *Espadaea* and *Henoonia*), microechinate (*Latua*, *Lycianthes*, some *Physalis* spp.), or faintly granular exine (*Nothocestrum*, *Solanum* spp.) is less frequent, and exceptionally the exine is rugulate-reticulate or rugulate-striate

(*Browallia*), microgranulate (*Schizanthus*, Fig. 59F, *Heteranthia*), scabrate-gemmate (*Anisodus*), fossulate (some *Salpichroa* spp.), or smooth (some *Jaborosa*). In *Mandragora*, many blunt spines (1.5–2 μm high) are distributed on the dense and fine gemmate-clavate exine (Diez and Ferguson 1984).

The structure of the exine has been little studied in Solanaceae. In general, it is completely tectate (*Brunfelsia*, *Cestrum*, *Solanum*, etc.), or tectate in the equatorial area and with bacula not joined to a tectum in the polar region (some spp. in Juanulloeae), or semitectate (e.g. *Lycium*) or intectate (*Mandragora*). The sexine may be up to 6 times thicker or more rarely of the same thickness or thinner than the nexine; however, the nexine is thicker than the sexine in *Vestia* and *Juanulloa* (sub nom. *Rahowardiana*). The imperforate or perforate tectum is usually supported by short and thick (*Brunfelsia*, *Reyesia*, *Salpiglossis*, *Merinthopodium*, etc.) or by short and granular (*Hunzikeria*, *Streptosolen*) or by not well-defined columellae (*Cestrum*, *Solanum*); in *Sclerophylax*, the perforate tectum is on prominent columellae, some of which are bifurcate (Gentry 1986). The foot layer is extremely thin (e.g. *Brunfelsia*), or barely perceptible (*Streptosolen*). *Melananthus*, however, is characterized by an undulating continuous foot layer. Some Australian genera have crowded columellae more or less uniform in length and an irregular and moderately thick foot layer (Knapp et al. 2000). *Schizanthus* stands apart from the other genera and is readily distinguished by the radially and transversely channelled tectum and foot layer. This character is shared only with *Schwenckia* (Gentry 1979).

Karyology. Solanaceae exhibit a dysploid series from $x = 7$ to $x = 14$ (15?), although other basic numbers such as $x = 17$, $x = 19$ and $x = 23$ have been recorded (Moscone 1992; Hunziker 2001). The chromosome number distribution in the subfamilies (fide Olmstead et al. 2008) is as follows (the figures within taxa are arranged according to their decreasing frequency): Goetzeoideae: $x = 12$, 13 (*Espadaea*: $x = 12$; *Metternichia*: $x = 13$); Cestroideae: $x = 12$, 8, 11, 10 (Cestreae: $x = 8$; Browallieae: $x = 11$, 10, 12; Salpiglossideae: $x = 11$); Nicotianoideae: $x = 9$, 10, 12, 8 (*Nicotiana*: $x = 9$, 10, 12, 8; Anthocercideae: $x = 9$, 10); Solanoideae: $x = 12$, 17, 23, 14, 13, 11, 10, 15, 9 (exceptions to $x = 12$ are *Nicandra physalodes*: $x = 10$; Hyoscyameae p.p.: $x = 17$, 14, 11; Physalideae: *Physalis lanceolata*: $x = 10$, *Quincula lobata*: $x = 11$; Capsiceae: *Capsicum* p. p.: $x = 13$; Solaneae p.p.: i.e. *Solanum* subgen. *Archeosolanum*: $x = 23$, one cultivar of *Solanum lycopersicum*: $x = 13$, *Solanum mammosum*: $x = 11$, *Solanum bullatum*: $x = 15$). In tribes and genera not assigned to a subfamily, the chromosome number distribution is: *Schizanthus*: $x = 10$; Benthamielleae: $x = 11$; Petunieae: $x = 8$, 9, 7, 10, 11 (*Fabiana*: $x = 9$; *Calibrachoa*: $x = 9$; *Petunia*: $x = 7$; *Brunfelsia*: $x = 11$; *Leptoglossis*: $x = 10$; *Nierembergia*: $x = 8$, 9; *Bouchetia*: $x = 8$; *Hunzikeria*: $x = 8$); Schwenckieae: $x = 12$. By far the most common base number is $x = 12$ which was found in more than 50% of the species studied, including the large subfam. Solanoideae where it is almost universal.

Twenty-five of the 96 genera of Solanaceae are karyologically unknown, some of them comprising several species [e.g. *Cuatresia* (15), *Juanulloa* (11), *Larnax* (32) and *Markea* (9)], and another 13 being monotypic (*Darcyanthus*, *Duckeodendron*, *Heteranthia*, *Oryctes*, *Plowmania*, *Protoschwenckia*, *Tsoala*, *Calliphysalis*) or unispecific (*Coeloneurum*, *Discopodium*, *Henoonia*, *Nectouxia*). The main gaps are in tribe Schwenckieae where only one count was done and in tribe Juanulloeae where just two reports for one of 4 genera are available. Furthermore, 8 genera belonging to tribe Physalideae remain unexplored regarding their chromosome numbers. Finally, in *Solanum*, less than half of its ca. 1400 spp. have been examined.

A comprehensive hypothesis on possible paths of chromosome number changes during the evolution of the family is still lacking, and even its original base number is a matter to be clarified. Raven (1975) postulated $x = 12$ as being plesiomorphic in Solanaceae; based on more recent chloroplast DNA phylogenetic studies, however, Olmstead et al. (2008) have suggested that this number is apomorphic. Evidence for dysploid changes from $x = 12$ to $x = 13$ via Robertsonian translocations, i.e. centric fissions, has been reported in *Capsicum* (Moscone et al. 1993, 2007) and *Solanum* (sub nom. *Lycopersicon*, Banks 1984).

Polyploidy has occurred at least in 22 genera belonging to subfamilies Cestroideae (tribe

Browallieae), Nicotianoideae (*Nicotiana* and tribe Anthocercideae), Solanoideae (*Mandragora*, *Lycium*, tribes Solaneae and Hyoscyameae), and in tribe Physalideae. The highest chromosome number was found in *Scopolia japonica* ($2n = 14x = 168$). Polyploids are remarkably frequent in *Nicotiana* where ca. 40% of the species are tetraploids often arisen by amphidiploidy, e.g. *N. tabacum* ($2n = 2x + 2x = 48$), as demonstrated by genomic in situ hybridization (Kenton et al. 1993). Furthermore, polyploid series are conspicuous in some genera with $x = 12$, i.e. *Mandragora* (2x, 7x, 8x), *Lycium* (2x, 3x, 4x, 8x, 10x), *Physalis* (2x, 4x, 6x), *Chamaesaracha* (2x, 3x, 4x, 6x), *Withania* (2x, 3x, 4x, 8x), *Solanum* (sect. *Solanum*: 2x, 4x, 6x, 8x; sect. *Petota*: 2x, 3x, 4x, 5x; sect. *Lathyrocarpum*: 2x, 4x, 6x) and *Scopolia* (4x, 7x, 14x).

The appearance of B chromosomes has only been recorded in few species of the following genera (the number of species and Bs is indicated in brackets): *Cestrum* (1: 1–6 Bs), *Combera* (1: 1–5 Bs), *Datura* (1: 1–6 Bs), *Nicandra* (1: 1 B), *Nicotiana* (1: 1 B), *Pantacantha* (1: 1–3, 6 Bs), *Petunia* (2: 1–2 Bs) and *Solanum* (11: 1–3 Bs; Jones and Rees 1982; Moscone 1989).

Karyotype analyses have revealed that the family shows high variation in chromosome size with a range of mean lengths from 1.0–1.5 μm in *Metternichia*, *Solanum* sect. *Basarthrum* and *Atropa*, to 6.5–11.51 μm in *Cestrum* (Las Peñas et al. 2006), and *Solanum* (sect. *Pachyphylla* and sect. *Cyphomandropsis*). Nevertheless, most species have comparatively small or medium-sized chromosomes with average lengths of 1.5 to 5.0 μm (Moscone 1989; Bernardello and Anderson 1990; Berg and Greilhuber 1993; Pringle and Murray 1993; Badr et al. 1997; Acosta et al. 2005; Las Peñas et al. 2006; Chiarini and Barboza 2008; Chiarini et al. 2010). Measurements of the nuclear DNA content, available only for few genera, show that genome size (1C-value) of species with diploid complements varies from 0.6 pg (588 Mbp) in *Solanum chacoense* and *S. tripartitum* to 24.8 pg (24304 Mbp) in *Solanum huilense* (sub nom. *Cyphomandra hartwegii* var. *ramose*, Bennett et al. 2000; Stiefkens and Bernardello 2000).

In general, karyotypes are symmetrical with a majority of metacentric chromosomes of rather similar size; exceptions to this are found in *Nicotiana* and *Solanum* sect. *Acanthophora* where complements mostly composed of subtelocentric and submetacentric chromosomes are frequent. In the latter cases, the chromosomes often have different lengths although bimodal karyotypes are rare. Many genera show a high constancy in chromosome number, shape and size between species. However, in some genera (*Nicotiana*, *Capsicum*, *Solanum*) evidence accumulates that the direction of evolutionary changes due to chromosome rearrangements implies karyotype re-patterning with asymmetry increase in the advanced taxa.

Chromosome banding approaches and gene mapping by fluorescent in situ hybridization have revealed useful markers for chromosome identification to prompt phylogenetic conclusions in some genera such as *Cestrum* (Berg and Greilhuber 1993, etc.), *Capsicum* (Moscone et al. 1993, 1995, 2007, 2011), *Solanum* (Pringle and Murray 1993) and *Nicotiana* (Parokonny and Kenton 1995; Lim et al. 2000). These studies have demonstrated that the heterochromatin is highly variable between species with respect to amount and distribution. Furthermore, in *Cestrum*, it has special properties as it appears in clustered, rather than blocked, bands which in part behave as cold-sensitive regions.

Pollination. Solanaceae often have flowers adapted to specialized types of pollination (Cocucci 1999; Knapp 2010a). Almost every known pollination syndrome is present in the family: bee-, bird-, moth-, butterfly-, fly- and bat-pollination. However, unspecialized pollination systems are also found in, for example, *Lycium*. Reward systems include almost all known kinds of primary attractants, except food-bodies and resins: perfume, oil, nectar and pollen. Pollination by deceit, i.e. without any reward, is present in some blowfly-pollinated species of *Jaborosa*.

Mechanical flower devices for pollen presentation and pollen placement adapted to particular pollinators are noteworthy among butterfly- and bee-pollinated species. These include mobile anthers (rotatory in *Leptoglossis* or balancing in *Browallia* and *Brunfelsia*), explosive pollen release, buzz pollination, and bellow mechanisms.

At the tribal level two or several syndromes commonly occur. At the generic level contrasting patterns are found: *Solanum* is presumably bee-pollinated throughout, whereas *Nicotiana* shows infrageneric syndrome radiation with bee-, moth-, bird- and bat-pollination. In the Physalideae (*Acnistus, Cuatresia, Physalis, Vassobia, Aureliana, Athenaea*), where bee-pollination prevails, flowers attract pollen-collecting females, fragrance-collecting Euglossine males or nectar-feeding bees of both sexes. In Euglossine-pollinated species of *Solanum* (ex *Cyphomandra*), fragrance is produced by the prominent anther connective. Bat-pollination is widespread in the epiphytic genera (*Dyssochroma, Juanulloa, Markea, Trianaea*), while butterfly-pollination is prevalent in *Salpiglossis* and *Reyesia. Nierembergia* is unique for its association with oil-collecting bees. One Coleoptera genus, *Epicante*, feeds exclusively on *Nolana* corollas and may complete pollination, and on at least one occasion *Phodopis vesper* (a northern Chilean hummingbird) has been observed pollinating *Nolana villosa*; *Nolanomelissa toroi* has been found collecting pollen and nectar from *Nolana rostrata* in Chile (Rozen 2003).

Reproductive Systems. In general, Solanaceae have hermaphrodite flowers, wet stigmata and binucleate pollen when shed. The stigma surface consists of a thick lipidic upper layer and a thin aqueous under-layer. Pollen captured on the lipidic layer sinks through the aqueous layer and hydrates (Konar and Linskens 1966a, 1966b). In agreement with the binucleate pollen and the stigma features, the gametophytic type of self-incompatibility system prevails, i.e. genetic recognition and rejection of incompatible pollen takes place during the passage of the pollen tube through the style (Heslop-Harrison and Shivanna 1977).

Breeding systems derived from this general pattern have been described. A number of species has unisexual flowers distributed monoeciously or dioeciously. Andromonoecious species are common in *Solanum* subgen. *Leptostemonum* (Whalen and Costich 1986; Miller and Diggle 2003). The few reported dioecious species are found in *Solanum* (Levine and Anderson 1986; Anderson and Symon 1989; Knapp et al. 1998; Knapp 2010b), *Deprea* (Sawyer and Anderson 2000), *Dunalia* (Hunziker 2001), *Lycium* (Minne et al. 1994; Miller and Venable 2002) and *Withania* (Hepper 1991; Anderson et al. 2006). Gender dimorphism has evolved independently in six lineages, primarily in the Physalideae (sub Solaneae, Sawyer and Anderson 2000). In some Australian and Central American *Solanum* spp., flowers are morphologically hermaphrodite but functionally unisexual: male flowers form abortive ovules and female flowers sterile pollen (Anderson 1979; Anderson and Symon 1989; Knapp et al. 1998; Martine et al. 2006). The two spp. of *Symonanthus* are truly dioecious, and flowers have either rudimentary pistils or stamens (Haegi 1981). *Dunalia solanacea* has long-styled and short-styled flowers with reciprocity in stamen lengths, as expected from distyly; however, the long-styled flowers are male sterile, but it is not known if the short-styled flowers are female sterile (Rodríguez 1998). Finally, four South African *Lycium* spp., *Withania coagulans* from Asia, *W. aristata* from the Canary Islands (Anderson et al. 2006) and *Deprea paneroi* from South America are functionally dioecious as well (Sawyer and Anderson 2000). In *Nolana* stigmas were found to be receptive during a wide range of floral developmental stages, and pollen remained viable throughout the open flower period and into senescence (Douglas and Freyre 2010).

Solanaceae are an experimental model to study gametophytic self-incompatibility systems based on S-RNase (Newbigin et al. 1999; Wang et al. 2001). Self-incompatibility factors of the style, controlled by a single multialeletic locus (S), are glycoproteins with ribonuclease activity. S-RNase from the pistil extracellular matrix is taken up by both incompatible and compatible pollen tubes, but it causes degradation of RNA only in incompatible pollen tubes (McClure and Franklin-Tong 2006). In addition, other factors are also required for pollen rejection (e.g. HT-B). A model of S-RNase compartmentalization in the vacuoles has been suggested recently, in which S-RNase is selectively released later in incompatible pollen tubes (Goldraij et al. 2006); the HT-B factor is involved in the process. The cytotoxic activity of S-RNases causes growth arrest of incompatible pollen tubes by inhibiting translation (McClure and Franklin-Tong 2006).

Fruit and Seed. The family is very diverse in fruit types: berry, capsule, drupe, schizocarp and diclesium; this diversity has been analyzed in a phylogenetic framework (Knapp 2002b). The two most common types are berries or capsules. Regardless of fruit type, the calyx is generally persistent (Fig. 58A, D, G), and rarely circumscissile after anthesis (*Datura*); sometimes, there are different degrees of calyx accrescence which just surrounds the fruit (Fig. 58G) or envelopes it completely; in some cases, it is clearly inflated, as in *Nicandra* (Fig. 58A), *Physalis*, *Przewalskia*, *Withania*, etc. (Hunziker 2001).

Capsules, plesiomorphic in the family (Knapp 2002b), are variable in number of seeds: from one in *Melananthus*, four or five in *Metternichia*, 30–50 in *Fabiana* to ca. 5000 in *Nicotiana*. Capsule dehiscence is almost always longitudinal (Fig. 58F), septicidal, septifragal or septicidal-loculicidal; some genera (*Hyoscyamus*, *Atropanthe*, *Physochlaina*, *Przewalskia* and *Scopolia*) have capsules with circumscissile dehiscence (pyxidium).

Berries are clearly derived in the family and seem to have three separate origins (Knapp 2002b). Some berries show different degrees of sclerification with scattered stone cells variable in size: microscopic, small but macroscopic (Fig. 58E), or conspicuous (Fig. 58B). The highest degree of sclerification is reached in the drupe (i. e. two to eight seeds enclosed in a wooden endocarp surrounded by a fleshy mesocarp) observed in some *Lycium* spp. (Fig. 58H–J), *Lycianthes lycioides* and *Saracha*, which are considered derived (Bernardello 1983; Knapp 2002b). *Duckeodendron* has a very peculiar drupe in which a pericarp with a multi-layered epicarp (epidermis and the loose parenchyma with groups of brachysclereids), a yellow and fibrous mesocarp (long adpressed rough fibres) and a bony dark brown endocarp enclose one seed (Machado et al. 2011). The origin of this fruit type most probably is entirely independent of the drupes found in *Lycium* and *Lycianthes* (Knapp 2002b). The unusual sclerified schizocarp of *Nolana* (Fig. 67J, K) is probably also derived from a berry (Tago-Nakazawa and Dillon 1999). Some *Nolana* species have 3–6 many-seeded mericarps broadly affixed to the receptacle, others have (2–) 5–30 free mericarps.

Some fruits dry after maturity, resulting in the presence of a thin leathery pericarp, as in genera of tribe Hyoscyameae (except *Atropa*), *Oryctes*, and some spp. of *Solanum* subgen. *Leptostemonum*. *Sclerophylax* is very peculiar in having a dry indehiscent fruit, generally with one seed per locule, with a membranous pericarp included in the accrescent and sclerified calyx (= diclesium). The origin of this fruit type inferred from the phylogeny of the family has not yet been explored.

Seeds (Fig. 60) are not compressed or more or less compressed, thick, with various forms: subspherical, reniform, discoid, cuboidal, prismatic, polyhedric, ovoid, or more rarely elongated and thin, lenticular or bacilliform. Some *Datura* spp. have elaiosomes. Seed size varies from 1.2 to 2.5 mm long, but sometimes seeds are smaller (0.3–1 mm long) or larger (4–35 mm long). The testa is variable. It can form a thin (Fig. 60D′) or broad wing, or be smooth, finely reticulate (Fig. 60F‴), tuberculate, honeycombed, etc. The testa cells may be thick-walled, with an undulate or wavy outline (Fig. 60C, H), foveolate or reticulate-foveolate, etc. (Axelius 1992; Freire de Carvalho et al. 1999; Lester et al. 1999).

The embryo has important features for the delimitation of the taxa: form, ratio of its total size and the size of the cotyledons, and how the cotyledons are arranged in relation to the rest of the embryo (Hunziker 2001) are relevant. Regarding their form, embryos are thick (Fig. 60C′, F) or slender (Fig. 60K′), long or short, straight (Fig. 60A″, J″), slightly curved (Fig. 60G″, E), vermiform, annular, coiled (Fig. 60B″, K′) or subcoiled (Fig. 60H′, I), and exceptionally U-shaped. Furthermore, *Cestrum*, *Sessea*, *Vestia*, *Merinthopodium*, *Markea* and *Juanulloa* (Fig. 60F) have embryos with broad cotyledons, unlike all the other taxa where their width is about the same as that of the rest of the embryo. As regards the ratio between total embryo size and cotyledon size, two groups are found: most commonly the cotyledons are equal, slightly longer or up to 2.5–3 times shorter than the rest of the curved or straight embryo; in the endemic Australian genera the cotyledons are between 6–8 times shorter than the rest of the straight embryo (Fig. 60G″). In the majority of the genera (Fig. 60A, B, H, I, K), the embryo has incumbent or oblique cotyledons; some epiphytic genera (*Juanulloa*, *Merinthopodium*, *Schultesianthus*) have accumbent cotyledons (Fig. 60C, D, E, F, J).

There are two main possibilities of seed dispersal: wind dispersal in case of very small seeds

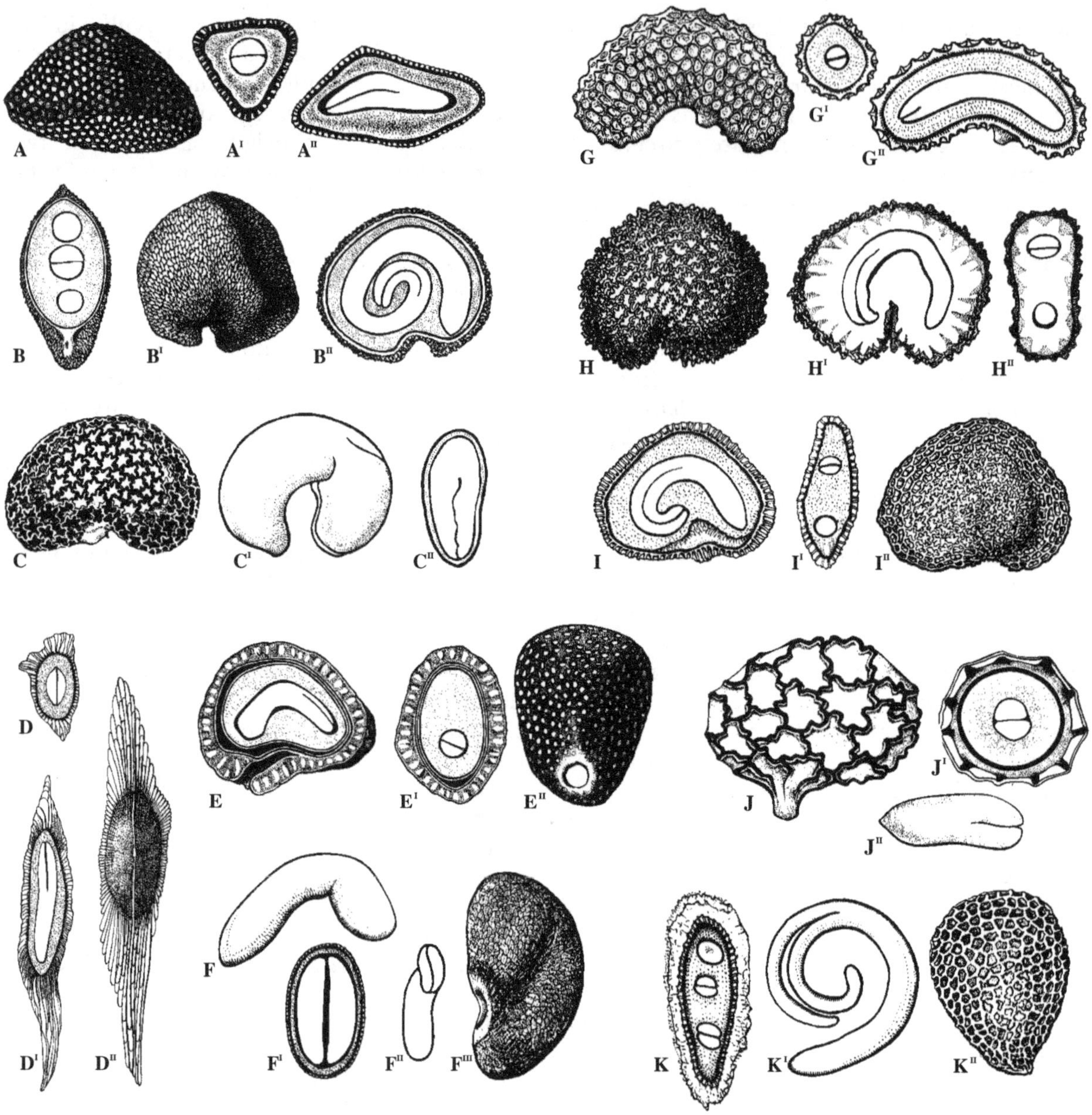

Fig. 60. Solanaceae. Seed. A *Nierembergia linariaefolia*. A lateral view, A′ cross section, A″ longitudinal section. B *Reyesia parviflora*. B cross section, B′ lateral view, B″ longitudinal section. C *Schultesianthus leucanthus*. C lateral view, C′ embryo, C″ cross section. D *Sessea vestioides*. D cross section, D′ lateral view, D″ longitudinal section. E *Latua pubiflora*. E longitudinal section, E′ cross section, E″ hilum view. F *Juanulloa verrucosa*. F lateral view of embryo, F′ seed, cross section, F″ frontal view of embryo, F‴ seed, lateral view. G *Duboisia hopwoodii*. G lateral view, G′ cross section, G″ longitudinal section. H *Schizanthus grahamii*. H lateral view, H′ longitudinal section, H″ cross section. I *Lycium gilliesianum*. I longitudinal section, I′ cross section, I″ lateral view. J *Nicotiana longiflora*. J lateral view, J′ cross section, J″ embryo. K *Solanum nitidibaccatum*. K cross section, K′ embryo, K″ lateral view. (From Hunziker 2001)

(e.g. *Nicotiana*, *Schwenckia*) and winged seeds (e.g. *Sessea*), and animal dispersal mainly by birds eating the berries (as in many *Solanum*) and probably by ants in *Datura* spp. with elaiosomes.

In some genera (*Solanum*, *Tubocapsicum*, *Nicotiana*, *Capsicum*, *Nolana*), germination is phanerocotylar (Dottori et al. 2000; D'Arcy et al. 2001; Perícola et al. 2004; Machado et al. 2008; Cabrera et al. 2010).

PHYTOCHEMISTRY. Hunziker (2001) and Eich (2008) provided a thorough review of the vast phytochemical literature on the secondary compounds found in Solanaceae. Two recent reviews deal with the withanolides and related steroids in subfam. Solanoideae (Chen et al. 2011; Misico et al. 2011). Eich (2008) and Misico et al. (2011) discussed and summarized the chemical information in a phylogenetic framework. Nearly 75% of the genera have been investigated phytochemically at least in part. Tropane alkaloids, steroidal alkaloids/glycoalkaloids, and withanolides (see below) are important distinctive groups of metabolites in Solanaceae.

According to Eich (2008), eight types of alkaloids have been detected in the family: a) simple pyrrolidines (hygrine and its derivates) found mainly in subfam. Solanoideae, and in members of *Schizanthus*, *Brunfelsia*, *Anthocercis*, *Duboisia* and *Nierembergia liniariaefolia*; b) pyrrolidine amides (conioidines), very uncommon and detected only in *Chamaesaracha coniodes*; c) pyrrolydine-type nicotinoids (nicotine and/or its congeners) in large amounts in genera of subfam. Nicotianoideae (*Nicotiana*, *Crenidium*, *Cyphanthera*, *Duboisia*) and in extremely low concentrations in some species of *Petunia* and members of subfam. Cestroideae (*Salpiglossis sinuata*, *Streptosolen jamesonii*, *Cestrum* spp.), and Solanoideae (*Solanum lycopersicum*, *S. melongena*, *Atropa belladonna*, *Capsicum annuum*, *Withania somnifera*, *Brugmansia candida*, *Datura* spp.); d) tropanes (some very poisonous), the most common alkaloids produced by genera of subfam. Nicotianoideae (except *Nicotiana*), Solanoideae (Datureae, Hyoscyameae, *Mandragora*, *Solandra*, *Nicandra*, *Physalis*, *Withania*, *Latua*), and *Schizanthus* and *Nierembergia*; e) calystegines (polyhydroxylated *nor*tropanes), widespread in subfam. Solanoideae (some Hyoscyameae, *Datura* spp., *Solanum* spp., *Capsicum* spp., *Physalis* spp., *Withania* spp., *Lycium chinense*, *Nicandra physalodes*) and in *Duboisia leichhardtii* and *Brunfelsia nitida*; f) pyrrolizidines, rare in the family and present in *Datura stramonium* and *Nicotiana tabacum*; g) β-carbolines, in *Vestia foetida*, *Datura stramonium* and two *Solanum* spp.; h) steroidal alkaloids, common and predominant metabolites of *Solanum* but also of some other genera (*Capsicum*, *Lycianthes*, *Saracha*, *Cestrum*, *Nicotiana*); these alkaloids are pharmacologically and toxicologically very well investigated (see Eich 2008 and bibliography therein). Some alkaloids are unique to one or few species, such as: fabianine (tetrahydroquinoline alkaloid) in *Fabiana imbricata*; 2-methoxy-3-isobutylpyrazine, mainly responsible for the aroma of the fruit of *Capsicum annuum*; nicotianamine in *Nicotiana* spp., *Solanum* spp., *Lycium chinense*, and *Datura metel*; solamines from the roots of two *Solanum* spp. and *Duboisia leichhardtii*; pyrrole alkaloids like the poisonous brunfelsamidine in *Brunfelsia grandiflora* and *Nierembergia lineariaefolia*, and solsodomines in *Solanum* sp.; benzodiazepines and catecholamines in *Solanum tuberosum*; betaines in *Solanum wendlandii* and *Nicotiana tabacum*; *homo*-cholestane glycoside (aethiosides A–C) in *Solanum aethiopicum*; and cardenolides in *Nierembergia rigida*.

Pharmacologically more interesting steroidal lactones are the withanolides (Chen et al. 2011; Misico et al. 2011), i.e. polyoxygenated steroids with a C_{28} ergostane skeleton. Almost 650 withanolides have been isolated so far from 23 genera (ca. 70 spp.) belonging to subfam. Solanoideae, and from only one species of subfam. Cestroideae (*Browallia viscosa*). Some withanolides have been demonstrated to be good chemotaxonomical markers at different levels (Misico et al. 2011); thus, physalins and related withanolides appear only in subtribe Physalidinae (Physalideae); acnistins are exclusive to subtribes Iochrominae and Withaniinae (Physalideae); the withametelin skeleton is exclusive to *Datura*; trechonolides, sativolides, and spiranoid withanolides at C-22 are present only in *Jaborosa* sect. *Lonchestigma* while spiranoid-γ-lactones have been isolated mainly from species of *Jaborosa* sect. *Jaborosa.* The wide biological activity of many withanolides is usually associated with their structural variety. These compounds show insecticidal, phytotoxic, antiparasitic, antimicrobial, anti-inflammatory, antistress, immunosuppressive and immunomodulatory, cancer-related, and CNS-related activities (Chen et al. 2011; Misico et al. 2011).

Flavonoids, especially flavones and flavonols, are common in many *Solanum* species (Sarmento da Silva et al. 2003), and flavonoid aglycones in *Nicotiana*, *Petunia*, *Solanum*, *Datura*, *Physalis*, *Chamaesaracha*, *Iochroma*, *Atropa*, *Hyoscyamus* and *Salpiglossis* (Wollenweber et al. 2005). Anthocyanins are the flower, tuber and fruit pigments best studied in the family, such as in *Petunia/Calibrachoa* (Ando et al. 1999, 2004), *Brunfelsia* (Vaknin et al. 2005), and *Solanum* (Brown 2005; Lachman and Hamouz 2005; Ichiyanagi et al. 2005) species. Monoterpenoids and sesquiterpenoids were studied as volatile compounds in the fragrance of flowers and other plant parts of *Cestrum*, *Nicotiana*, *Brugmansia*, *Datura*, *Solanum*, *Capsicum*, *Fabiana* and *Lycium*, while different diterpenoids were reported in *Nicotiana* spp., *Petunia* x *hybrida*, *Cestrum parqui*, *Physalis coztomatl*, and *Nolana* and *Capsicum* spp. Finally, triterpenoids and phytosterols were detected in many genera of the family (for details, see Eich 2008).

Other compounds are very rare in the family and appear in individual species or genera, such as β-phenylethylamine, tyramine and its N-methyl derivatives in *Nierembergia linariaefolia*; phenylpropanoid acids, phenylethanoids, and phenylmethanoids and others with good phytotoxic activity in *Cestrum parqui*; two unusual glycosides (cesternosides A and B) in *Cestrum nocturnum*; capsaicinoids (ca. 25 compounds, capsaicin the dominating alkaloid) in *Capsicum* spp. fruits; hydroxycoumarins and/or long chain alkyl esters of hydroxycinnamic acids in *Atropa belladonna*, *Fabiana imbricata*, *Nicotiana tabacum*, *Schultesianthus leucanthus*, *Datura stramonium*, and *Brunfelsia* and *Solanum* spp.; some lignans in *Nierembergia rigida* and *Solanum* spp., and lignanamides in *Capsicum annuum* and *Hyoscyamus niger* (Eich 2008).

Distribution and Habitats. Solanaceae are widely distributed throughout the world, with the greatest concentration in the New World. The family ranges from sea level (*Aureliana*, *Athenaea*, *Nolana*, some *Solanum* and *Calibrachoa*) to 4900–5000 m (e.g. *Fabiana bryoides* in South America, and *Przewalskia tangutica* in China). Although the family has many temperate members, the majority of genera and species is tropical; some genera are found in semiarid and arid settings, even deserts, while others are well adapted to montane rainforests. The highest diversity occurs from Central America southwards into South America along the Pacific coast and in the Andean mountains (D'Arcy 1991). The main centre of diversification is South America, where 37 endemic genera occur together with some endemic species of the cosmopolitan genera *Solanum*, *Lycium*, *Lycianthes* and *Nicotiana*. In the uplands of western South America, endemic species of *Saracha*, *Dunalia*, *Salpichroa*, *Iochroma*, some *Capsicum*, *Solanum* sect. *Petota*, *Nicotiana* sect. *Tomentosae* and *Undulatae* and some spp. of sect. *Paniculatae*, etc. are found; several genera have representatives, including most of the lianas and epiphytes, in tropical forests at different elevations, like *Acnistus*, *Streptosolen*, *Browallia*, *Deprea*, *Larnax*, *Duckeodendron*, *Juanulloa*, *Markea*, *Trianaea*, *Schultesianthus*, *Merinthopodium*, most *Solanum*, etc.; the western coast of South America (5–30°S) is dominated by desert conditions except for localities where recurring fogs along the shore allow for the development of plant communities termed lomas formations; in the lomas formations occur 16 genera and ca. 130 species (no fewer than 95 endemic species) belonging mainly to *Exodeconus*, *Jaltomata*, *Leptoglossis*, *Lycium*, *Nicotiana*, *Nolana*, *Reyesia*, *Schizanthus*, *Salpiglossis* and *Solanum* (Dillon 2005). *Nolana*, with 75 endemic lomas species, stands out as the largest and most wide-ranging genus to be encountered in all lomas formations. Towards eastern and southern South America, the diversity is less pronounced; to the north in the subtropics, *Vassobia*, *Eriolarynx*, *Schwenckia*, *Metternichia*, *Protoschwenckia*, etc. are found; from temperate areas in Chile to the south, with some genera reaching Argentina, *Latua*, *Vestia*, *Schizanthus*, *Reyesia*, *Salpiglossis*, *Nierembergia*, *Fabiana*, *Jaborosa*, *Combera*, *Benthamiella*, *Pantacantha*, etc. grow; towards north-eastern South America, mostly in tropical and subtropical lowlands, *Aureliana*, *Athenaea*, *Petunia*, *Calibrachoa*, some *Capsicum*, etc. are centred. *Sclerophylax* is endemic to Argentina and adjacent parts of Paraguay and Uruguay.

Other centres of Solanaceae diversity are 1) Mexico and North America, including *Physalis* and its close relatives (*Chamaesaracha*, *Tzeltalia*, and *Quincula*), *Datura*, and the two endemic genera *Nectouxia* (Mexico) and *Oryctes* (western North America), plus some endemic species of *Cestrum*, *Solanum*, *Lycianthes*, *Lycium* and *Nicotiana*; *Plowmania* and some *Bouchetia* and *Hunzikeria* species are also centred in this area but also reach South America; finally, some species have probably been introduced by man (*Petunia*,

Nierembergia) or arrived by local migrations from the south (e.g. *Jaltomata*); 2) Central America, with centres of upland and lowland diversity, including species of *Cestrum*, *Brunfelsia*, *Witheringia*, *Lycianthes* and *Solanum* sect. *Geminata*; *Goetzea*, *Espadaea*, *Henoonia* and *Coeloneurum* are mostly endemic to the semiarid lowlands of the Greater Antilles; 3) Australia and neighbouring islands, with some endemic taxa (7 genera), *Nicotiana* sect. *Suaveolantes* (exception: *N. africana* of Namibia), and *Solanum* subgen. *Archaeosolanum* and some groups of subgen. *Leptostemonum*; 4) a zone in Eurasia from Europe to Japan, particularly in the Himalayan region where Hyoscyameae are the best represented tribe together with *Mandragora* and *Withania* (Tu et al. 2010).

Solanaceae are little diversified in the rest of the world. For example, in Africa only a few genera are present, namely the two widely ranging genera *Solanum* and *Lycium*, three in common with Europe and Asia (*Mandragora*, *Hyoscyamus* and *Withania*), the extraordinary *Nicotiana africana*, and the endemic *Discopodium*. In Asia the strictly Asian *Tubocapsicum*, species of *Solanum* and *Lycianthes*, and species of the European genera *Atropa*, *Mandragora*, *Hyoscyamus* and *Withania* are found. Finally, *Nothocestrum* is a geographically isolated genus on Hawaii.

Economic Importance. The economic importance of Solanaceae is well documented (Hawkes 1999 and references therein), as well as the role they have played in the development of traditional cultures and civilizations. The family contains not only major food plants such as potato, chili and tomato, but also others locally important and little known outside their ranges.

Among the food plants, the potato (*Solanum tuberosum*) has a remarkable position, being the world's fourth most important crop after wheat, rice and maize (Hawkes 1990); its wild close relatives have been used in programs aimed at breeding for disease resistance, environmental tolerances, and other agronomic traits of interest (Hijmans et al. 2002). *Solanum lycopersicum* (= *Lycopersicon esculentum*), the only domesticated tomato, is a very popular vegetable crop in the modern human diet worldwide, with significant phenotypic variation in fruit size and shape; wild tomato species all produce fruits which are almost invariably round and small; mutations at six genetic loci seem to have been essential in transforming the small, inconspicuous berries of wild tomatoes to the extremely large fruits now found in modern cultivars (Tanksley 2004). Some other *Solanum* spp. are cultivated at a commercial scale for their fruits, especially *S. betaceum* (= *Cyphomandra betacea*), the "tree tomato" or "tamarillo". Brazil is the native region of this and New Zealand the main country of fruit production; the interest in the "tamarillo" as a potential new crop is increasing in many frost-free climatic areas of the world (Prohens and Nuez 2000). *Solanum melongena* ("eggplant", "berengena", "aubergine", "brinjal"), indigenous to Indochina and domesticated in the Indo-Burma region, has medicinal, culinary and even ornamental uses. In India, eggplants are still used for medicinal purposes, but they are cultivated worldwide mostly for their edible berries (Daunay et al. 2007); *S. macrocarpon*, the "gboma eggplant", is cultivated throughout much of Africa, especially in Ghana, for its edible fruits and leaves as is the "scarlet eggplant", *S. aethiopicum*, widely planted in many parts of tropical Africa, less in Brazil and occasionally in S Italy and the Black Sea Russian republics (Daunay et al. 2001). In South America, *S. quitoense* ("naranjilla", "lulo"), *S. muricatum* ("pepino", "pepino dulce") and *S. sessiliflorum* are reputed for their sweet or fragrant fruits; some cultivars have been developed with a higher sugar content in their fruits (Prohens and Nuez 2001) or plants resistant to root knot nematodes, such as the *S. quitoense* 'Lulo la Selva' cultivar (Heiser 2001).

Considering their uses as vegetables and condiments, peppers are superb culinary plants. Cultivars commercially grown throughout the world belong to *Capsicum annuum* and allied species (*C. frutescens*, *C. chinense*, *C. pubescens*, *C. baccatum* var. *pendulum*); some—known as "sweet pepper", "bell pepper", "pimiento"—produce non-pungent fruits while the plants with pungent fruits are called "chili", "yellow chilli", "ají", "red pepper", "paprika", etc.; pungency is due to the presence of a group of metabolites unique to *Capsicum* spp. known as capsaicinoids, a mixture of phenolic amides in which capsaicin is prevalent. These substances have also shown a wide range of pharmacological properties, such as antigenotoxic, antimutagenic and anticarcinogenic effects (Manirakiza et al. 2003).

Of the great number of *Physalis* species, very few are used for their fruit. The Old World species *P. alkekengi* is grown as an ornamental and is popular for its large, bright orange to red husks. Among New World *Physalis*, some have been domesticated mainly in Mexico from where they were taken to Europe and other parts of the world (Montes Hernández and Aguirre Rivera 1992); the edible fruits of the "tomatillo" or "husk-tomato" (*P. philadelphica*), the "cape gooseberry" (*P. peruviana*), and the "ground cherries" (*P. pubescens*, *P. subglabrata*, etc.) are consumed raw or cooked in salsas, gazpacho and guacamole (Morton 1987; Montes Hernández and Aguirre Rivera 1992) or are preserved and canned to be exported. In many countries (South Africa, Australia, Greece, Turkey), these species are commercially cultivated at different scale.

Nicotiana is very well known as "tobacco"; economically important spp. are *N. tabacum* and *N. rustica*. *Nicotiana tabacum* is a natural amphidiploid hybrid; its maternal S-genome donor is an ancestor of *N. sylvestris* and the paternal T-genome donor is derived from a particular lineage of *N. tomentosiformis* (Murad et al. 2002; Chase et al. 2003). Only a few species are ornamental, although some are cultivated as garden plants for their flowers or their foliage, such as *N. alata*, while many varieties of diverse flower colour have been produced experimentally (Japan Tobacco Inc. 1994).

The roots of the Mediterranean *Mandragora* spp., which contain tropane alkaloids, were highly praised as amulets or for their medicinal properties (narcotic, anti-inflammatory, disinfectant, aphrodisiac), and used for all kinds of magic purposes due to their human shape in the Middle Ages. Furthermore, they furnished the most ancient curative drug for humanity. Highly esteemed by the Egyptians, it was mentioned in the Bible, and was also well known to Dioscorides, Theophrastus, Plinius and many other classical authorities. On account of the extraordinary role which "mandrake" played in the popular superstitions and folklore of Europe, it is undoubtedly the most relevant magic plant (Evans 1979).

Atropa belladonna has been known since ancient times for its potent toxic properties, due to the presence of tropane alkaloids in roots, stems, leaves and fruits. It was involved in witchcraft practices in the Middle Ages. Its generic name refers to the goddess Atropos, one of the three Fates in Greek and Roman mythology; this goddess, together with Lachesis and Cloto, was in charge of spinning the thread of human destiny, which Atropos could cut off with her shears whenever she pleased. Nowadays, *A. belladonna* ("beautiful lady") and *A. baetica* are cultivated as a source of atropine, a drug used for therapeutical purposes on account of its depressant effect on the parasympathetic nervous system (Evans 1979).

Hyoscyamus niger ("henbane", "beleño negro", "judquiame noire", "schwarzes Bilsenkraut") is well known in Europe for its extreme toxicity and has been valued medicinally since very early times. Dioscorides, as well as Plinius, wrote about it and it is thought to have been used in witches' brews and other toxic preparations during the Middle Ages, together with other species, above all *H. muticus*. The "white henbane" (*H. albus*) was used medicinally for its narcotic, anti-inflammatory and disinfectant effects (Daunay et al. 2007). *Hyoscyamus niger* has been included in the pharmacopoeia of most European countries, being used in general as an external analgesic.

The alkaloid content (especially scopolamine) of the "daturas" (*Datura* spp.) explains their use as hallucinogens in healing, initiation or divining rites in traditional cultures of Africa, India, the south-western USA, Mexico and South America. The species of greatest importance for these ceremonies and uses are *D. inoxia*, *D. discolor*, *D. wrightii*, *D. quercifolia*, *D. leichhardtii* and *D. ferox*. The "chamico" (*D. ferox*) in South America and, worldwide, the "jimson weed" or "thorn apple" (*D. stramonium*) are toxic weedy plants; their toxic seeds contaminate commercial seeds and grains and also invade all sorts of crops (40 crops in almost 100 countries according to Holm et al. 1997).

Solanum glaucophyllum grows in southern Brazil, Uruguay, Paraguay and northern Argentina, and is one of the most important poisonous plants of Argentina ("duraznillo blanco"), being responsible for a serious disease characterized by wasting and calcification of soft tissues in cattle; the toxicity is caused by high levels of 1,25-dihydroxyvitamin D3 (Costa et al. 1998). Another

South American toxic species is *Nierembergia linariaefolia*, known as "chuscho" or "chucho", causing severe disease in cattle, sheep, goats, horses and rabbits (Gallo 1987; Botha et al. 1999); pyrrole-3-carbamidine is responsible for its lethality (Buschi and Pomilio 1987).

The rare *Latua pubiflora* ("latúe", "palo mato", "palo de bruja"), endemic of S Chile, is a species with narcotic and toxic properties due to its content of atropine and hyosciamine and, to a lesser extent, of scopolamine in stems, seeds and leaves (Schultes 1979). Also some *Cestrum* spp. (*C. sendtnerianum*, *C. laevigatum*, *C. dumetorum*, *C. nocturnum*, *C. diurnum*, *C. parqui*, *C. mathewsii*) have poisonous properties or are used as medicinal plants in many countries of the American continent. *Cestrum parqui* is considered a major problem because of its toxicity to livestock, causing death after ingestion of small quantities of leaves, twigs or fruit, and even dry leaves from plants which have been cut down or sprayed (Gallo 1987).

As regards ornamentals, "petunia" (breeding lines and commercial hybrid varieties of *Petunia* x *hybrida*) is a popular ornamental species, a model for flower colour (anthocyanin biosynthesis), and first among bedding plants (Sink 1984). Of some Argentinean *Calibrachoa* spp. (ex *Petunia* spp.), genotypes suitable for breeding pot plants were selected for their ornamental value (Facciuto et al. 2006). Species of *Brugmansia* ("trumpet flower", "floripondio"), *Brunfelsia* ("jazmín del Paraguay", "galán de moda"), *Capsicum* ("peppers"), *Cestrum* ("dama de noche"), *Physalis* ("cape gooseberry", "winter cherry", "chinese lantern-plant"), *Schizanthus* ("butterfly flower", "fringe flower", "flor de pajarito"), *Solandra* ("chalice vine", "tetona", "copa de oro"), some *Nierembergia* ("cup flower") and *Nicotiana*, and a few *Datura* spp., such as *D. wrightii*, are ornamentals. *Nolana paradoxa* ("bluebird" and "snowbird") and *N. humifusa* (referred to as "*N. prostrata*") have found their way into the horticultural trade and wider usage of artificial hybrids amongst *Nolana* species has been investigated (Freyre et al. 2005).

Finally, the timber of *Duckeodendron cestroides* is of good quality and similar in texture and properties to 'marupá' (*Simaruba amara*; Record and Hess 1942). In Brazil, where this species is endemic, it is used to make fine wooden furniture as well as smaller products (e.g. matches, Freitas da Silva et al. 1977).

Conservation. Habitat destruction due to deforestation and clearing for agriculture and urban development is the main threat to various species of Solanaceae, especially given the small population size and the restrictive distribution of some. This problem is largest in Central and South American forests, both regions with endemic genera at risk, such as *Duckeodendron* (Cramer et al. 2007), *Latua*, *Plowmania*, *Darcyanthus*, as well as many threatened species of other genera (Barboza et al. 2010, 2011; Stehmann and Greppi 2011). Some species are listed in the IUCN Red List (2007) in different categories: *Solanum diploconos* and *Aureliana fasciculata*, both from Brazil (Lower Risk); some *Nothocestrum* spp. from the Hawaiian Islands (Critically Endangered), and *Nicotiana africana* (Least Concern). Australia has provided a list of ca. 35 rare and threatened Solanaceae species (Electronic Flora of South Australia, http://www.flora.sa.gov.au/lucid_keys/Solanaceae/Solanaceae_rare_and_threatened.shtml). Within *Solanum*, many species are also listed in different risk categories (Queensland Nature Conservation Act 1992; Threatened Species Protection Act 1995; Bañares et al. 2004), while others have been considered in the literature as Critically Endangered (Knapp 2010b), Endangered (Knapp 2007, 2010c; Granados-Tochoy et al. 2007; Knapp and Nee 2009; Giacomin and Stehmann 2011), or Vulnerable species (Farruggia and Bohs 2010; Stern and Bohs 2012). Finally, *Solanum fernandezianum* is an endemic species in danger of extinction on the Robinson Crusoe Island, Chile (Stuessy et al. 1998; Ricci 2006) where some conservation measures have been taken (Solano et al. 2011).

Classification and Phylogeny. D'Arcy (1979, 1991) compiled in detail the development of thought on the classification of the Solanaceae from classical times to the Third International Solanaceae Congress held in Colombia in July 1988. Since then, Solanaceae have been divided into two subfamilies: Solanoideae and Cestroideae (D'Arcy 1979, 1991; Hunziker 1979; Olmstead and Palmer 1992). The last formal classification was presented by Hunziker (2001). This author recognised 6 subfamilies

(Cestroideae, Juanulloideae, Solanoideae, Salpiglossideae, Schizanthoideae and Anthocercidoideae), 27 tribes and 92 genera, excluding 15 genera considered members of other families. The main achievement of this classification was a comprehensive morphological analysis of all genera, revealing unknown characters for many taxa. He also provided palynological, karyological and phytochemical evidence to support his system. Seed shape, cotyledon features, and corolla aestivation were the main characters used to circumscribe the subfamilies. From a phylogenetic point of view, Hunziker's system recognizes many non-monophyletic groups (i.e. subfam. Cestroideae, tribe Salpiglossidae, tribe Solaneae, etc.; cf. Stafford and Knapp 2006; Olmstead et al. 2008). In 1999, Olmstead et al. proposed a phylogenetic and provisional classification based on chloroplast DNA sequences; their analysis included 52 genera (79 spp.), some of them previously excluded from the Solanaceae, such as *Nolana* and the Antillean genera of the Goetzeaceae. Additional recent studies strongly support *Duckeodendron* (Fay et al. 1998; Santiago-Valentín and Olmstead 2003), the four Antillean genera (Santiago-Valentín and Olmstead 2003), and *Sclerophylax* (Olmstead and Bohs 2007) as nested within Solanaceae. Olmstead et al. (2008) published a new classification of the family based on the chloroplast regions *ndhF* and *trnLF* of 89 genera (190 spp.); this classification recognised four subfamilies (Goetzeoideae, Cestroideae, Nicotianoideae and Solanoideae), two unranked informal clades within Solanoideae (Atropina and Salpichroina), and many genera and some tribes unassigned to subfamily or within a more inclusive clade. Olmstead et al.'s (2008) work is the most comprehensive and complete for the family so far, with nearly 94% of the genera examined. The data produced an estimate of the phylogeny of Solanaceae with a number of well-supported clades (i.e. Goetzeoideae, Petunieae, Physalideae, Hyoscyameae) which accommodated many genera, and others with little resolution (i.e. basal clades in Solanoideae).

Olmstead et al.'s (2008) main modifications of the Hunziker (2001) classification are: 1) *Nolana* and *Sclerophylax* (in subfam. Solanoideae), *Goetzea*, *Espadaea*, *Henoonia*, *Coeloneurum* and *Tsoala* (in subfam. Goetzeoideae), and *Duckeodendron* (unassigned to a subfamily) are nested in Solanaceae; 2) reorganization of subfam. Solanoideae with 7 tribes also including re-accommodated genera (and other genera not assigned to any tribe or subtribe); 3) a much-reduced subfam. Cestroideae (3 tribes); 4) acceptance of subfam. Nicotianoideae including the Australian endemic genera (tribe Anthocercideae) + *Nicotiana*; 5) recognition of a reorganized tribe Physalideae with three subtribes (Withaniinae, Iochrominae and Physalidinae); 6) incorporation of *Metternichia* in subfam. Goetzeoideae; 7) acceptance of *Petunia*, *Calibrachoa*, *Tzeltalia* and *Tsoala* as valid genera; 8) inclusion of *Cyphomandra*, *Lycopersicon*, *Normania* and *Triguera* in *Solanum*. Olmstead et al. (2008) conclude: "Although significant advancements have been made in the understanding of relationships within Solanaceae, a complete phylogenetic classification has not been yet provided for the whole family. The classification we summarize below is the current framework for Solanaceae based on molecular evidence".

Solanales, Solanaceae and Related Families. Within Lamiidae, Solanales are one of the four major clades together with Boraginales, Lamiales and Gentianales, among which there is no well-supported resolution (Soltis et al. 2011). Solanaceae are assigned to Solanales along with Convolvulaceae, Hydroleaceae, Montiniaceae and Sphenocleaceae (APG III 2009). These families comprise altogether ca. 165 genera and 4080 species. Solanaceae and Convolvulaceae are the larger families with an almost worldwide distribution. Authors differ in the dating of the stem group of Solanales. According to Wikström et al. (2001, *Sphenoclea* not included), it may date from the Campanian-Santonian 86–82 m.y.a., diversifying 78–76 m.y.a., while Janssens et al. (2009) date stem group Solanales to 101 m.y.a. and Bremer et al. (2004) to ca. 106 m.y.a.

The current topology of the tree for Solanales, obtained with a large number of genes (17) and more samples of Solanaceae and Convolvulaceae (Soltis et al. 2011), is congruent with prior molecular studies (Soltis et al. 2000; Bremer et al. 2002) but has stronger internal support. The order is characterized by the presence of O-methyl flavonols (flavones), inflorescence terminal, pollen tube usually with callose and persistent calyx in fruit (Stevens 2001 onwards).

Bremer (1996) and Soltis and Soltis (1997) found Montiniaceae, a family from Africa and Madagascar, sister to [Solanaceae + Convolvulaceae]. Soltis et al. (2000) reported strong support for the association of *Montinia* and *Hydrolea* (*Sphenoclea* was not included). Later, in a broader sampling in Montiniaceae, Bremer et al. (2002) provided stronger support, though just above 50%, for the association of Montiniaceae with *Sphenoclea* and *Hydrolea*. Finally, Soltis et al. (2011) found Montiniaceae are sister to Spenocleaceae + Hydroleaceae, and this well-supported clade (BS = 98%) in turn is sister to Convolvulaceae + Solanaceae (BS = 100%).

Classification of Solanaceae

I. **Subfam. Goetzeoideae** Thorne & Reveal (2007).
  (Gen. 1–6).
II. **Subfam. Cestroideae** Burnett (1835).
  1. Tribe Cestreae Dumort. (1829).
    (Gen. 7–9).
  2. Tribe Browallieae Hunz. (1995).
    (Gen. 10–11).
  3. Tribe Salpiglossideae Benth. (1835).
    (Gen. 12–13).
  4. Cestroideae: Incertae sedis:
    (Gen. 14–15: *Protoschwenckia* Soler., *Heteranthia* Nees & Mart.).
III. **Subfam. Nicotianoideae** Miers (1848).
  1. Tribe Anthocercideae G. Don (1838).
    (Gen. 16–22).
  2. Tribe Nicotianeae G. Don
    (Gen. 23: *Nicotiana* L.).
IV. **Subfam. Solanoideae** Burnett (1835).
  1. Tribe Lycieae Lowe (1872).
    (Gen. 24).
  2. Tribe Hyoscyameae Endl. (1839).
    (Gen. 25–31).
  3. Tribe Juanulloeae Hunz. (1977).
    (Gen. 32–35).
  4. Tribe Datureae Dumort. (1829).
    (Gen. 36–37).
  5. Tribe Physalideae Miers (1849).
    a. Subtribe Withaniinae Bohs & Olmstead (2008).
      (Gen. 38–43).
    b. Subtribe Iochrominae Reveal (2012).
      (Gen. 44–49).
    c. Subtribe Physalidinae Reveal (2012).
      (Gen. 50–57).
    d. Physalideae: Incertae sedis:
      (Gen. 58–63: *Capsicophysalis* Averett & M. Martínez, *Cuatresia* Hunz., *Larnax* Miers, *Deprea* Raf., *Schraderanthus* Averett, *Tzeltalia* E. Estrada & M. Martínez).
  6. Tribe Capsiceae Dumort. (1827).
    (Gen. 64–65).
  7. Tribe Solaneae Dumort. (1829).
    (Gen. 66–67).
  8. Solanoideae: Incertae sedis:
    (Gen. 68–80: *Latua* Phil., *Jaborosa* Juss., *Sclerophylax* Miers, *Nolana* L.f., *Exodeconus* Raf., *Nicandra* Adans., *Schultesianthus* Hunz., *Trianaea* Planch. & Linden, *Solandra* Sw., *Mandragora* L., *Nectouxia* Kunth, *Salpichroa* Miers, *Darcyanthus* Hunz.).
V. **Solanaceae: Incertae sedis:**
  (Gen. 81–96: Genera not assigned to a subfamily: *Schizanthus* Ruiz & Pav., *Duckeodendron* Kuhlmann, Tribes not assigned to a subfamily: Tribe Benthamielleae Hunz. (2000): *Pantacantha* Speg., *Combera* Sandw., *Benthamiella* Speg., Tribe Petunieae Horan. (1847): *Fabiana* Ruiz & Pav., *Calibrachoa* La Llave & Lex., *Petunia* Juss., *Brunfelsia* L., *Leptoglossis* Benth., *Nierembergia* Ruiz & Pav., *Bouchetia* Dunal, *Hunzikeria* D'Arcy, *Plowmania* Hunz. & Subils, Tribe Schwenckieae Hunz. (1977): *Schwenckia* L., *Melananthus* Walp.).

Key to the Genera of Solanaceae

1. Fruit a drupe, 1-seeded; embryo U-shaped; endosperm oily; corolla aestivation quincuncial; trees up to 30 m high. Brazil, Central Amazonia **82. *Duckeodendron***
– Fruit a berry, capsule, diclesium, or schizocarp, few- to many-seeded, rarely 1-seeded; embryo straight or slightly curved, vermiform, annular or coiled; endosperm mostly starchy or missing; corolla aestivation various; herbs, shrubs or small trees 2
2. Leaves always entire and coriaceous; corolla orange; fruit fleshy, yellow or orange, mesocarp gelatinous at maturity; embryo straight; endosperm missing 3
– Leaves entire or variously lobed or divided, membranaceous, less commonly coriaceous; corolla variously coloured; fruit fleshy or dry, mesocarp rarely gelatinous at maturity; embryo of various shape; endosperm scanty or abundant 6
3. Corolla funnel-shaped, cup-shaped or tubular-curved, > 10 mm long; ovary bilocular 4
– Corolla campanulate or shortly tubular, < 10 mm long; ovary unilocular 5
4. Corolla funnel-shaped or cup-shaped; calyx and corolla hexamerous; stamens 6, equal. Greater Antilles **6. *Goetzea***
– Corolla tubular-curved; calyx and corolla pentamerous (rarely 4- or 6-merous); stamens 5, unequal. Cuba **5. *Espadaea***

5. Corolla campanulate, lobed nearly to base; style pubescent; stamens inserted at base of corolla. Greater Antilles **3. *Coeloneurum***
– Corolla shortly tubular; style glabrous; stamens inserted at middle of corolla. Cuba **4. *Henoonia***
6. Aestivation valvate, valvate-induplicate, valvate-plicate, valvate-conduplicate, cochlear, cochlear-conduplicate, cochlear-plicate, contorted-induplicate, contorted-conduplicate, or reciprocative; corolla actinomorphic or zygomorphic; anthers mostly dithecal; endosperm starchy; embryo variously curved or almost straight, cotyledons usually up to 2.5–3 times shorter, equal, slightly longer or frequently much longer than rest of embryo 7
– Aestivation valvate-supervolute; corolla actinomorphic; anthers mostly monothecal (dithecal in *Anthocercis* and *Symonanthus*); endosperm oily and starchy (starchy only in *Duboisia*); embryo straight, cotyledons very short (ca. 5–12 times shorter than the rest of the embryo) 92
7. Placentation parietal; style strongly exserted and pubescent at the apex; stigma capitate, very thick. Madagascar **2. *Tsoala***
– Placentation mostly axile (basal in *Melananthus*; apical in *Sclerophylax*); style mostly included, glabrous or pubescent in its basal half; stigma, when capitate, thin or thick 8
8. Aestivation valvate-induplicate (valvate also in some *Jaltomata* spp.); pedicels articulated 9
– Aestivation valvate, valvate-plicate, valvate-conduplicate, cochlear, cochlear-conduplicate, cochlear-plicate, contorted-induplicate, contorted-conduplicate, or reciprocative; pedicels not articulated (articulated in some *Solanum* spp.) 12
9. Ovary with an annular persistent corolla remaining at its base (cyathium); anthers dorsifixed, embryo straight or almost curved 10
– Ovary without cyathium; anthers ventrifixed, embryo strongly curved. Americas **66. *Jaltomata***
10. Seeds long and thin (up to 10 times longer than broad), compressed, winged, imbricate in the fruit; capsules few-seeded (1–8 seeds). South America and Haiti **8. *Sessea***
– Seeds slightly longer than broad, thick, subpolyhedrical, wingless, adjacent in the fruit; few-seeded berry or many-seeded capsule 11
11. Anthers included; ovary stipitate, ovules few (4–20, exceptionally more); berry few-seeded. Mexico, Central America, the Antilles, South America **7. *Cestrum***
– Anthers exserted; ovary sessile, ovules many (more than 50); capsules many-seeded. Chile **9. *Vestia***
12. Seeds not compressed and angular, mostly prismatic, polyhedrical, and cuboidal, more rarely subglobose, subreniform, or long, thin and very narrow (in *Benthamiella* and *Combera* seeds slightly compressed) 13
– Seeds compressed and not angular, mostly discoid or reniform 35
13. Stamen fertile 4 or 2 14
– Stamens fertile 5 (rarely also 3, 2 or 1 in *Benthamiella*) 25
14. Anthers basifixed 15
– Anthers ventrifixed or dorsifixed 17
15. Aestivation cochlear; anthers monothecic; style smooth; pollen grains triporate or 3–4-colpate, rarely 3-colporate. Central and South America **89. *Brunfelsia***
– Aestivation reciprocative or cochlear-conduplicate; anthers dithecic, with partial or complete abortion of one theca in the upper stamen pair; style strongly corrugated; pollen grains 4–5/9-colporate 16
16. Herbs 0.08–1(1.5) m high; corolla hypocrateriform, ventricose, not twisted, aestivation reciprocative; stamens inserted in upper half of corolla, lower posterior pair with filaments usually two times longer than anthers, upper lateral pair with compressed filaments shorter than anther. Americas **10. *Browallia***
– Shrubs 1.5–2 m high; corolla infundibuliform, not ventricose, its lower part spirally twisted, aestivation cochlear-conduplicate; stamens inserted in both halves of the corolla, lower posterior pair with filaments 4–5 times longer than anthers, upper lateral pair with terete filaments as long as anthers or somewhat longer. Colombia, Ecuador, Peru **11. *Streptosolen***
17. Anthers dorsifixed; nectary cushion-shaped. Bolivia, Brazil **14. *Protoschwenckia***
– Anthers ventrifixed; nectary absent or when present annular, sheath-like or invaginate-bilobed 18
18. Nectary absent 19
– Nectary always present 20
19. Corolla infundibuliform, slightly bilabiate; stamens inserted at lower half of corolla tube and not connivent on style; thecae unequal; oil-secreting glandular trichomes inside the corolla missing. Brazil **15. *Heteranthia***
– Corolla hypocrateriform; stamens inserted at apex of corolla tube and generally connivent on style; thecae equal; oil-secreting glandular trichomes inside the corolla present. Mexico, South America **91. *Nierembergia***
20. Corolla zygomorphic, limb with five undivided lobes 21
– Corolla actinomorphic or zygomorphic, limb with five trilobulate or trifid lobes 24
21. Style cylindrical, not broadened distally, lacking lateral expansions; stigma laminar; nectary sheath-like and bilobed; pollen grains in monads or tetrads. Americas **92. *Bouchetia***
– Style broadened distally, with 2 lateral expansions or wings; stigma inconspicuous, linear; nectary annular and prominent or invaginate-bilobed; pollen grains in monads 22

22. Calyx campanulate; corolla 42–50 mm long, funnel-shaped, tube comparatively wide; fertile stamens 4 plus one staminode, thecae distinct; nectary annular; shrubs. Mexico, Guatemala **94. *Plowmania***
- Calyx not campanulate; corolla 9–30 mm long, cylindrical or salver-form, tube narrow; fertile stamens 4 or 2 plus 2–3 staminodes, thecae confluent; nectary invaginate-bilobed; small herbaceous xerophytic perennials or ephemerals 23
23. Corolla ventricose, limb narrow (3–6 mm); calyx cylindrical; fruits fusiform; seeds 0.3–0.5 mm long. Peru, Chile, Argentina **90. *Leptoglossis***
- Corolla salver-form, limb broad (10–23 mm); calyx funnel-shaped, campanulate or obconical; fruits almost subglobose; seeds 1.2–1.5 mm long. USA, Mexico, Venezuela **93. *Hunzikeria***
24. Placentation axile; nectary invaginate-pelviform; aestivation valvate-conduplicate; capsules bilocular, with 6–70 seeds; seeds cuboidal, 0.6–0.8 mm long, testa reticulate. Central and South America **95. *Schwenckia***
- Placentation basal; nectary invaginate-pelviform or slightly bilobulate; aestivation valvate-induplicate; capsules unilocular, 1-seeded; seeds elongated, 1–1.5 mm long, testa smooth. Americas **96. *Melananthus***
25. Fruit a berry or long capsule (35–45 mm long); pollen grains echinate; shrubs or small trees 26
- Fruit a short capsule (3–28 mm long); pollen grains not echinate; usually herbs or small shrubs 27
26. Fruit a thin capsule; seeds very long (30–39 x 2–3.5 mm); embryo straight, endosperm scanty; corolla aestivation contorted-conduplicate. Brazil **1. *Metternichia***
- Fruit a globose berry; seeds slightly longer than broad, ca. 2 mm diam.; embryo slightly curved, endosperm abundant; corolla aestivation cochlear-conduplicate. Chile **68. *Latua***
27. Flowers usually subtended by two opposite bracteoles similar to the leaves; capsules few-seeded (1–5 or up to 7 seeds) 28
- Flowers without a pair of opposite bracteoles; capsules many-seeded 30
28. Herbs; leaves petiolate, sheathless, spathulate, oblong-spathulate, deltoid-ovate or rhomboide-ovate. Argentina, Chile **84. *Combera***
- Plants woody; leaves sessile, sheathed, linear, narrowly elliptic to obovate or linear-subulate 29
29. Low shrubs; leaves not densely imbricate; leaves and calyx teeth rigid and spiny; corolla aestivation valvate-induplicate; seed surface ridged-winged. Argentina **83. *Pantacantha***
- Chamaephytes or lax cushions; leaves densely imbricate; leaves and calyx teeth neither rigid nor spiny; corolla aestivation contorted-conduplicate; seed surface foveolate or reticulate. Argentina, Chile **85. *Benthamiella***
30. Nectary absent; stamens attached to apex of corolla tube and generally connivent on style; oil-secreting glandular trichomes inside the corolla present. Mexico and South America **91. *Nierembergia***
- Nectary always present; stamens attached to lower half of corolla tube (in some *Nicotiana* spp. near apex of corolla tube) and not connivent on the style; oil-secreting glandular trichomes missing 31
31. Anthers dorsifixed; aestivation contorted-conduplicate 32
- Anthers ventrifixed; aestivation cochlear or reciprocative 33
32. Flowers in many-flowered terminal paniculoid inflorescences; glandular trichomes generally with multicellular head (rarely head unicellular), branched trichomes frequent; stamens equal or in two (4 + 1) or three groups (2 + 2 + 1); seeds ca. 100–5000. Americas, Australia, Africa **23. *Nicotiana***
- Flowers solitary or arising from the axil of one leaf; glandular trichomes with unicellular head, branched trichomes absent; stamens equal (only in 2 species) or in two groups (2 + 3); seeds ca. 30–52. Peru, Bolivia, Chile, Argentina **86. *Fabiana***
33. Fertile stamens 5 (2 + 2+ 1); nectary sheathed-bilobed; stigma laminar; pollen grains in monads or tetrads. Americas **92. *Bouchetia***
- Fertile stamens 5 (2 + 2+ 1, rarely 2 + 3); nectary bilobed; stigma capitate, discoid or bilobate; pollen grains always in monads 34
34. Corolla aestivation reciprocative: the anterior induplicative lobe covers the four others which are conduplicate; calyx usually lobed halfway, lobes narrowed towards apex; seeds with straight anticlinal walls. Americas **87. *Calibrachoa***
- Corolla aestivation cochlear; calyx usually deeply lobed, lobes linear or enlarged towards apex; seeds with undulating anticlinal walls. Southern South America **88. *Petunia***
35. Woody epiphytes or occasionally shrubs or trees; embryo almost straight or slightly curved (annular in *Schultesianthus*), cotyledons accumbent or oblique; endosperm comparatively scanty 36
- Herbs, shrubs or trees (rarely vines in *Solanum*); embryo annular, coiled or subcoiled, cotyledons incumbent; endosperm usually abundant 41
36. Aestivation valvate 37
- Aestivation cochlear 38
37. Anthers basifixed; flowers solitary on brachyblasts; cotyledons oblique, shorter and not broader than rest of embryo. Brazil **32. *Dyssochroma***
- Anthers dorsifixed; inflorescences on long peduncles; cotyledons accumbent, as long as (or somewhat shorter or longer) but wider than rest of embryo. Mexico to northern South America **35. *Merinthopodium***
38. Anthers ventrifixed; ovary (4) 5–10-carpellate. Colombia to Peru **75. *Trianaea***

– Anthers dorsi- or basifixed; ovary 2-carpellate 39
39. Flowers zygomorphic, fragrant; calyx thick and coriaceous; gynoecium and androecium usually bent; seeds 5–5.5 mm long; embryo strongly annular. Mexico to Bolivia **74. *Schultesianthus***
– Flowers actinomorphic, not fragrant; calyx membranous; gynoecium and androecium straight; seeds 2.2–3.8 mm long; embryo usually slightly curved 40
40. Anthers dorsifixed; corolla thick and fleshy, tubular. Mexico to Bolivia **33. *Juanulloa***
– Anthers basifixed; corolla membranous, funnel-form or infundibuliform. Panama to Peru **34. *Markea***
41. Fertile stamens five; corolla aestivation cochlear, valvate, valvate-conduplicate or contorted-conduplicate; berry frequently with stone cells, rarely with pyrenes, or fruit a capsule, diclesium or schizocarp 42
– Fertile stamens four or two; corolla aestivation cochlear; fruit a capsule 90
42. Fruit mostly a circumscissile capsule (pyxidium) 43
– Fruit a berry or a non-circumscissile capsule, or diclesium or schizocarp; gynoecium 2–5-carpellate (in *Nolana* up to 30-carpellate) 48
43. Filaments 1.5–4 times longer than anthers, inserted at base of corolla or further up (in some *Hyoscyamus* in upper half of corolla but not near its upper margin); stamens equal or unequal 44
– Filaments shorter than anthers, inserted near upper margin of corolla; stamens equal. China **26. *Przewalskia***
44. Flowers solitary (in *Hyoscyamus* solitary at base and in condensed cymes above); fruiting calyces inflated or not 45
– Inflorescences paniculate; fruiting calyces inflated. Asia **28. *Physochlaina***
45. Flowers pendant on long pedicels; stamens included, filaments inserted at base of corolla; nectary annular, prominent; lobes of fruiting calyx never spine-tipped 46
– Flowers sessile or very shortly pedicellate; stamens usually exserted, filaments inserted at middle of corolla, at a lower level or further up; nectary absent or cryptic; fruiting calyx rigid, lobes spine-tipped. Europe, Asia, Africa **25. *Hyoscyamus***
46. Corolla actinomorphic; filaments straight; calyx lobes usually unequal in size 47
– Corolla zygomorphic (lobes unequal); filaments circinate in bud, curved at anthesis; calyx lobes equal. China **29. *Atropanthe***
47. Corolla and calyx usually almost of the same length; stamens equal, anthers 5–7 mm. Asia **30. *Anisodus***
– Corolla ca. two times longer than calyx; stamens slightly unequal, anthers 2.5–4 mm. Europe and Asia **27. *Scopolia***
48. Gynoecium (3–4)5-carpellate or up to 30-carpellate 49
– Gynoecium bicarpellate (two species of *Jaborosa* with 2–5 carpels) 50
49. Berry enclosed by accrescent, stiff and inflated calyx; flowers actinomorphic; calyx segments auriculate; filaments equal, geniculate at their expanded base; carpels 3–5; annual herbs or subshrubs. Peru to Argentina **73. *Nicandra***
– Mericarps fused or free from each other; flowers commonly zygomorphic; calyx segments not auriculate; filaments unequal (3 + 2), not geniculate nor expanded basally; carpels 2–30; annuals or perennial herbs or weak shrubs. Chile, Peru, Galapagos Islands **71. *Nolana***
50. Nectary with two opposite lobes; pollen grains cryptoaperturate, exine echinate. Europe, Asia **77. *Mandragora***
– Nectary, when present, without lobes; pollen grains usually aperturate (a few species of dioecious *Solanum* with inaperturate monads), exine not echinate 51
51. Filaments straight 52
– Filaments declinate, or incurved towards centre of flower 86
52. Ovary with cyathium; corolla aestivation cochlear or cochlear-plicate. Cosmopolitan **24. *Lycium***
– Ovary without cyathium; corolla aestivation valvate, valvate-plicate, valvate-conduplicate, contorted-conduplicate, contorted-induplicate or contorted-plicate 53
53. Corolla aestivation contorted-conduplicate; anthers basifixed; endosperm with protein and/or oil, lacking starch 54
– Corolla aestivation valvate, valvate-plicate, valvate-conduplicate, contorted-induplicate or contorted-plicate; anthers mostly dorsifixed or dorsi-basifixed, more rarely ventrifixed or basifixed; endosperm starchy 56
54. Fruit a many-seeded capsule or berry; calyx neither accrescent nor thorny; stamens equal; placenta axile; many ovules per locule; embryo coiled; herbs, shrubs or small trees 55
– Fruit a diclesium, generally 2-seeded; calyx accrescent, lobes becoming thorny; stamens unequal; placenta apical; ovules 1–2 per locule; embryo straight or slightly curved; annual or perennial herbs. Argentina, Paraguay, Uruguay **70. *Sclerophylax***
55. Annual herbs or short-lived perennials; branching mostly dichasial; flowers erect; calyx circumscissile forming a reflexed annulus subtending the fruit; ovary partially inferior, tetralocular in lower half, nectary invaginate; capsule dehiscent, mostly with spines or tubercles; seeds with or without a funicular elaiosome. Americas **37. *Datura***
– Shrubs or small trees; branching mostly monochasial; flowers pendant or inclined; calyx not circumscissile, falling off completely or persisting around the fruit; ovary superior, bilocular throughout, nectary annular; berry pendant, unarmed; seeds lacking an elaiosome. South America **36. *Brugmansia***

56. Ovary with basal nectary; filaments usually inserted individually on corolla; anthers opening by longitudinal slits 57
– Ovary without nectary; filaments basally united into a uniform and continuous ring fused at base of corolla; anthers strictly poricidal or opening by terminal pores which sometimes expand into longitudinal slits 85
57. Anthers extrorse, ventrifixed 58
– Anthers introrse, dorsifixed, basifixed or dorsi-basifixed 59
58. Corolla usually glabrous within; calyx usually entire at anthesis; pericarp lacking stone cells. Guatemala to Bolivia **58. *Cuatresia***
– Corolla usually with tufts of dendroid trichomes within; calyx sometimes 5-apiculate, shortly dentate or more frequently divided; pericarp with few stone cells. Europe, Africa, Asia **43. *Withania***
59. Filaments minute, anthers sessile or subsessile; perianth tetramerous; ovary half-inferior. Hawaii **39. *Nothocestrum***
– Filaments at least half as long as anthers; perianth usually pentamerous; ovary superior 60
60. Calyx entire or the five main nerves overtopping the smooth margin as cylindrical or dentate appendages 61
– Calyx 5-divided 63
61. Corolla generally with an inside ring of non-glandular trichomes above the filament insertion; pericarp usually with stone cells. Mexico to Bolivia **50. *Witheringia***
– Corolla glabrous inside or with some tiny glandular trichomes; pericarp lacking stone cells (few in one *Capsicum* species) 62
62. Stapet auriculate; calyx frequently with five or ten appendages or teeth prolonging the nerves; corolla aestivation valvate-plicate; placenta and endocarp never fuse. Southern USA to South America **65. *Capsicum***
– Stapet not auriculate; calyx without appendages, lacking nerve prolongations; corolla aestivation valvate; placenta centrifugally enlarged and fused with endocarp. Asia **40. *Tubocapsicum***
63. Corolla aestivation valvate 64
– Corolla aestivation valvate-induplicate, valvate-plicate or valvate-conduplicate 70
64. Corolla lobes short, tube always longer 65
– Corolla lobes as long as tube or slightly longer 67
65. Corolla urceolate-campanulate or campanulate-infundibuliform; anthers dorsifixed; calyx not accrescent; pericarp lacking stone cells 66
– Corolla generally infundibuliform; anthers basifixed; calyx accrescent, slightly or much-inflated, completely surrounding the berry, stone cells frequent in pericarp. Venezuela to Bolivia **60. *Deprea***
66. Corolla urceolate-campanulate; stamens included; style and anthers pubescent. Africa **38. *Discopodium***
– Corolla campanulate-infundibuliform; stamens exserted; style and anthers glabrous. Mexico to South America **49. *Acnistus***
67. Corolla with a dense ring of trichomes inside; stapet not auriculate, filaments hairy, inserted near middle of corolla. Mexico and Central America **51. *Brachistus***
– Corolla glabrous inside (glabrescent in some *Larnax* species); stapet auriculate, filaments glabrous, inserted at base of corolla 68
68. Herbs or shrubs; calyx membranous, strongly accrescent, tightly enveloping the berry; styles long, always of the same length; filaments and anthers frequently unequal. Central America to Peru **59. *Larnax***
– Shrubs or trees, calyx rather thick, not accrescent or, if accrescent, surrounding the berry completely or partly, but always open upwards; styles heteromorphic; filaments and anthers equal 69
69. Shrubs; calyx almost dialisepalous (tube very short or sometimes almost absent), strongly accrescent. Brazil **41. *Athenaea***
– Shrubs or small trees (up to 8 m high); calyx with evident tube, not accrescent. Brazil, Paraguay, Argentina, Peru, Bolivia **42. *Aureliana***
70. Corolla infundibuliform or tubular-infundibuliform; fruiting calyx not inflated 71
– Corolla rotate or campanulate-rotate; fruiting calyx inflated or not 75
71. Herbs, generally small to very small (3–20 cm high); stamens unequal 72
– Shrubs or trees (up to 10 m, or more); stamens usually equal 73
72. Calyx with equal lobes; corolla 20–40 mm long, 5-angulate; berry lignified at apex or sometimes with stone cells; seeds wingless. Ecuador, Peru, Chile, Argentina **72. *Exodeconus***
– Calyx with unequal lobes; corolla 5–8 mm long, with well-developed lobules; fruit a capsule; seeds with a hyaline peripheral wing. North America **54. *Oryctes***
73. Stapet large, with conspicuous appendages and lateral wings; style subterminal. Colombia to Argentina **44. *Dunalia***
– Stapet not prominent, lacking apical appendages or lateral wings; style terminal 74
74. Young stems densely hairy with deciduous brown specialized trichomes (thickened, multiseriate, bifurcate or not); filaments inserted at base of corolla; berry with pyrenes, pericarp purple to shining black; calyx not accrescent. Venezuela to Bolivia **46. *Saracha***
– Young stems with simple trichomes; filaments inserted between 1/4 and 3/4 of corolla length; berry usually with many stone cells, pericarp yellowish to orange or red, sometimes purple at apex; calyx accrescent. South America **45. *Iochroma***
75. Corolla inside usually with a ring of trichomes above; filaments longer than anthers; stapet not auriculate and not with incipient auricles 76

- Corolla glabrous inside; filaments more or less as long as or slightly longer than anthers; stapet auriculate 84
76. Herbs or weak shrubs; stapet not auriculate; calyx strongly accrescent; style subterminal 77
- Small trees; stapet with incipient auricles; calyx slightly accrescent; style terminal. Argentina, Bolivia **48. *Eriolarynx***
77. Fruiting calyx inflated, urceolate, angled and strongly veined (not angled in *Calliphysalis*), completely closed or with a small apical opening, hiding the berry 78
- Fruiting calyx not inflated, neither angled nor strongly veined, reflexed and not enclosing the berry or enclosing it only partly 81
78. Fruiting calyx coriaceous, partly open at mouth; soft-stemmed reclining shrubs or subshrubs. Mexico **62. *Tzeltalia***
- Fruiting calyx papery, completely closed; herbs, rarely shrubs 79
79. Aerial parts lacking vesiculose water-containing trichomes; flowers solitary or 2–6; corolla usually campanulate-rotate or campanulate, pericarp of uniform thickness, never disintegrating at maturity 80
- Aerial parts with a granular aspect, covered almost completely by vesiculose, whitish, water-containing trichomes; flowers 2–4(5); corolla rotate; pericarp unevenly thickened, disintegrating almost completely in thinnest and weakest parts. USA, Mexico **55. *Quincula***
80. Flowers solitary; fruiting calyx angled, invaginated at base; rhizomatous perennial herbs or annual herbs. Mostly Americas **57. *Physalis***
- Fascicles of 2–6 flowers; fruiting calyx rounded, not invaginated at base; taprooted perennial herbs. USA **53. *Calliphysalis***
81. Inflorescence fasciculate (6–8 flowers) from axils; corolla lobed to divided, 4–5 cm wide; fruiting calyx reflexed under berry at maturity. Mexico, Guatemala, Venezuela **61. *Schraderanthus***
- Flowers 1 or 2 per axil; corolla little or slightly lobed, not divided, 2–4 cm wide; fruiting calyx reflexed or closely appressed and partially or completely enclosing berry at maturity 82
82. Corolla rotate-campanulate, irregular; berry red; fruiting calyx red, reflexed under the berry at maturity; testa tuberculate; erect annual herbs **63. *Capsicophysalis***
- Corolla rotate, actinomorphic; berry green; fruiting calyx green, closely appressed and partially or completely enclosing the berry at maturity; testa rugose-reticulate; erect annual or ascending or spreading perennial herbs 83
83. Corolla 8–9 mm long, aestivation contorted-plicate; pericarp generally dry, lacking stone cells. USA, Mexico **56. *Chamaesaracha***
- Corolla 20–24 mm long, aestivation valvate-plicate; pericarp generally juicy, with stone cells. North America **52. *Leucophysalis***
84. Shrubs or small trees; calyx not accrescent (or very slightly accrescent), never surpassing half of the berry; corolla 5-lobed. Southern South America **47. *Vassobia***
- Herbs; calyx much accrescent, tightly enveloping berry; corolla 5-angulate, lobes inconspicuous. Peru, Bolivia **80. *Darcyanthus***
85. Calyx entire, truncate, with five to ten linear nerve prolongations; anthers strictly poricidal at tips; berry juicy, exceptionally with pyrenes. Americas, Asia **64. *Lycianthes***
- Calyx 5-parted into lobes or segments; anthers poricidal or with a terminal pore expanding into a longitudinal slit; berry juicy or nearly dry, never with pyrenes. Cosmopolitan **67. *Solanum***
86. Corolla zygomorphic or slightly zygomorphic; aestivation cochlear; filaments declinate 87
- Corolla actinomorphic; aestivation valvate, valvate-induplicate or valvate-conduplicate; filaments incurved towards the centre of the flower or curved at least at the apex 88
87. Corolla 10–37 cm long; stamens equal, anthers basifixed; ovary partly inferior tetralocular; scandent shrubs or lianas. Mexico to Brazil **76. *Solandra***
- Corolla 1.9–3.5 cm long; stamens unequal, anthers dorsifixed; ovary superior bilocular; perennial herbs. Europe, Asia, Africa **31. *Atropa***
88. Leaves dentate to pinnatisect, not darkening when dry; flowers 1–7 per axil, or in dense glomerules or fascicles; calyx tube well-developed; ovary 2–5-carpellate; berry globose-depressed; embryo annular, endosperm copious. Peru to Argentina **69. *Jaborosa***
- Leaves entire, darkening when dry; flowers solitary; calyx tube absent or very short; ovary bicarpellate; berry elongated; embryo coiled, endosperm scanty 89
89. Corolla lacking a corona; distal end of filaments broadened or not, lacking a laminar enlargement on the dorsal side of anthers; scandent shrubs, occasionally herbs. Venezuela to Argentina **79. *Salpichroa***
- Corolla with a corona; distal end of filaments broadened into a laminar enlargement on the dorsal side of anthers; herbs. Mexico **78. *Nectouxia***
90. Lateral stamens aborted or with smaller anthers, posterior pair always fertile or with larger anthers; flowers not resupinate 91
- Lateral stamens fertile, posterior pair aborted; flowers resupinate. Chile, Argentina **81. *Schizanthus***
91. Corolla 2.3–5 cm long; anthers dorsifixed, 2–4 mm long. Chile, Argentina **12. *Salpiglossis***
- Corolla 0.6–1.3(–1.5) cm long; anthers basifixed, 0.8–1 mm long. Chile, Argentina **13. *Reyesia***

92. Anthers dithecal; plants dioecious or hermaphrodite 93
- Anthers monothecal; plants hermaphrodite 94
93. Plants hermaphrodite; stamens 4, didynamous or subequal; corolla lobes long and narrow, usually longer or sometimes equal or slightly shorter than tube. Australia **17. *Anthocercis***
- Plants dioecious; stamens (3) 4 or 5, equal; corolla lobes as long as broad, four to six times shorter than tube. Australia **16. *Symonanthus***
94. Fruit a capsule; endosperm oily and starchy; plants usually with dense indumentum of dendritic trichomes 95
- Fruit aberry; endosperm starchy; plants glabrous or glabrate. Australia, New Caledonia **22. *Duboisia***
95. Leafy shrubs; corolla 5–22 mm long 96
- Shrubs appearing leafless, the small leaves early-deciduous; corolla 2.5–4 mm long; stamens 4 (2 + 2) and a sterile staminode usually present. Australia **21. *Crenidium***
96. Stamens 5 97
- Stamens 4, didynamous (sometimes with a staminode present). Australia **19. *Cyphanthera***
97. Stamens unequal (2 long, 3 short); corolla lobes long and narrow, glabrous inside. Australia **18. *Grammosolen***
- Stamens equal; corolla lobes broad and short, densely pubescent inside. Australia **20. *Anthotroche***

I. Subfam. Goetzeoideae Thorne & Reveal (2007).

Trees to shrubs; pollen tricolpate, exine echinate; fruit a capsule pauciseminate or a berry pauciseminate; embryo straight, cotyledons fleshy mostly large; endosperm at most slight.

1. *Metternichia* J.C. Mikan Fig. 61

Metternichia J.C. Mikan, Del. fl. faun. bras. 3, tab. 13 (1823).

Trees. Leaves entire. Flowers 5-merous, solitary or in small fascicles, bracts deciduous; calyx campanulate, zygomorphic; corolla strongly scented, white, long, actinomorphic, infundibuliform, lobes short and wide, aestivation contorted-conduplicate; stamens unequal, 2 + 2 + 1 or 2 + 3, filaments declinate; anthers basifixed; ovary bilocular, ovules 10–16, anatropous; nectary annular. Capsule subcylindric-fusiform, septicidal-loculicidal, 4-valved. Seeds 4–5 per locule, erect, long, very narrow and thin, triquetrous, each angle with a narrow wing, hilum near the base of the ventral face; embryo straight, cotyledons compressed, not wider than the slightly longer radicle; endosperm scanty. $2n = 26$.

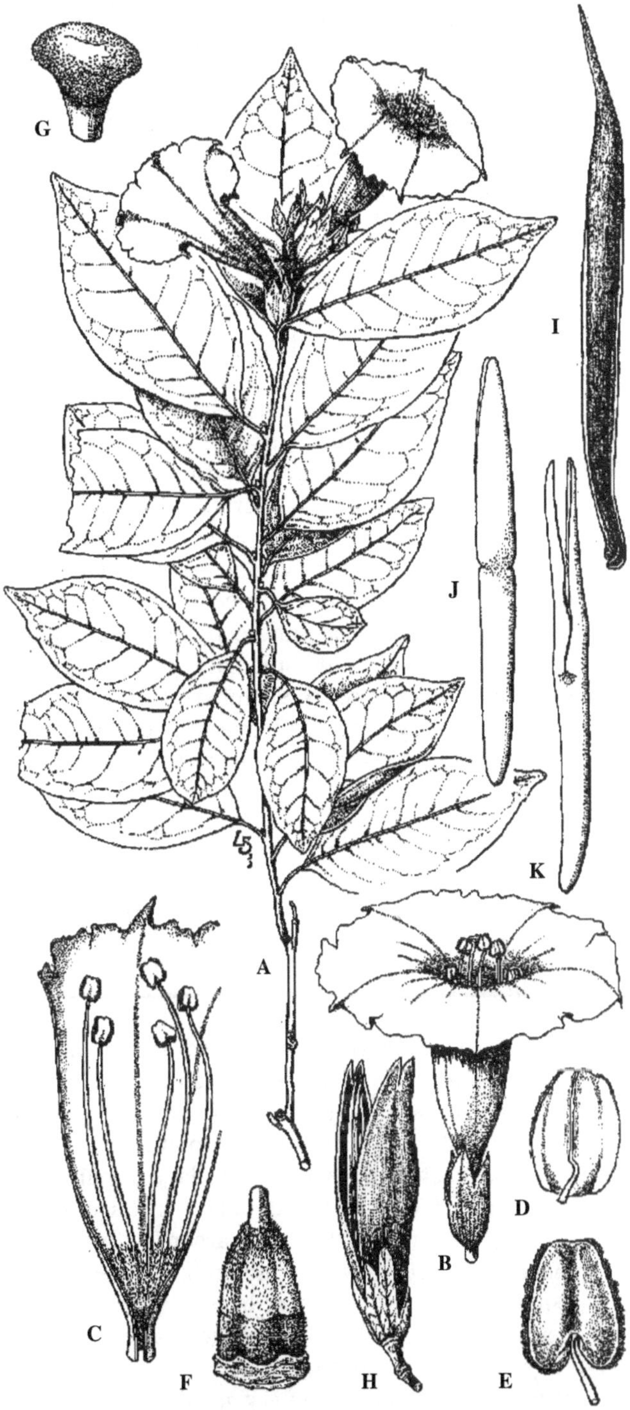

Fig. 61. Solanaceae. *Metternichia principis*. **A** Flowering branch. **B** Flower. **C** Opened corolla. **D** Half an anther showing the curved apex of the filament and its basal insertion. **E** Anther, ventral view. **F** Ovary. **G** Stigma, commissural view. **H** Fruit. **I** Seed. **J** Embryo, front view. **K** Embryo, lateral view. (Hunziker 2001; drawn by L. Sánchez)

A single species from Brazil (Rio de Janeiro state, Bahia, Minas Gerais): *Metternichia principis* J.C. Mikan.

2. ***Tsoala*** Bosser & D'Arcy

Tsoala Bosser & D'Arcy, Bull. Mus. Natl. Hist. Nat., B, Adansonia 14: 8 (1992).

Shrubs or small trees, much branched; young stems and leaves with dense reddish pubescence. Leaves deciduous, entire, petiole canaliculate. Flowers terminal, solitary, 5-merous; calyx lobes subequal, triangular; corolla greenish, hypocrateriform, entirely glabrous, tube thin and very long (up to 15 cm), limb infundibuliform, lobes apiculate; stamens strongly exserted, subequal, filaments glabrous, inserted near apex of tube; anthers basifixed; ovary unilocular or partly bilocular, placentation parietal; nectary annular; style longer than stamens, pubescent at apex, stigma thick, capitate. Capsule apiculate, pericarp green, thin and fragile. Seeds 4–9; embryo thick, straight, cotyledons very short; endosperm abundant.

Monotypic. *Tsoala tubiflora* Bosser & D'Arcy, Madagascar.

3. ***Coeloneurum*** Radlk.

Coeloneurum Radlk., Sitzungsber. Math.-Phys. Cl. Königl. Bayer. Akad. Wiss. München 19: 280 (1890); Liogier, La Flora de la Española VI, vol. 70, ser. Científica 27: 361 (1994).

Shrubs or small trees; stems with dense rusty pubescence. Leaves petiolate, entire, glabrous, leathery. Flowers extra-axillary, solitary, sessile or subsessile; calyx broadly campanulate, shortly 5-lobed, rusty pubescent; corolla campanulate, yellowish or cream, lobed nearly to base, lobes oblong; stamens 5(4), filaments glabrous, inserted basally; ovary unilocular; style sometimes absent, pubescent at base, stigma bilobed. Berry globose, yellow or orange, 1-seeded. Seed large; embryo straight, cotyledons fleshy, thick, radicle tiny; endosperm very scanty.

A single species, *C. ferrugineum* Urb., Hispaniola (Greater Antilles).

4. ***Henoonia*** Griseb.

Henoonia Griseb., Cat. pl. Cub.: 166 (1866); Fuentes Fiallo, Fl. República de Cuba 10(4): 2–14 (2005), rev.

Shrubs; branches greyish, sometimes lateral branches spinescent. Leaves subsessile, leathery, entire, base cordate or wedge-shaped, margins revolute, glabrous and brilliant adaxially, pubescent abaxially. Flowers axillary, solitary or in fascicles, subsessile; calyx cupola-shaped, shortly 5-lobed, pubescent; corolla shortly tubular, deeply lobed, lobes reflexed; stamens 5, filaments glabrous, inserted at middle of corolla tube; anthers dorsifixed; ovary unilocular, uniovulate; style glabrous. Berry globose-apiculate, 1-seeded. Seeds up to 8 mm, broadly ovoid; embryo straight, cotyledons fleshy, thick; endosperm very scanty.

A single species, *H. myrtifolia* Griseb., Cuba.

5. ***Espadaea*** A. Rich.

Espadaea A. Rich., Hist. phys. Cuba, Pl. vasc. 11: 147 (1850); Fuentes, Revista Jard. Bot. Nac. Cuba 3: 51–67 (1982); Fuentes, Fl. República Cuba 10(4): 2–14 (2005), rev.

Trees; stems greyish, rusty pubescent when young. Leaves petiolate, entire, leathery, glabrous and brilliant adaxially, pubescent abaxially. Flowers axillary, solitary, subsessile, (4)5(6)-merous; calyx cup-shaped, shortly dentate; corolla yellow, funnel-shaped, slightly curved, lobes reflexed; stamens 5, unequal, glabrous, strongly exserted, inserted basally; anthers basifixed; ovary tomentose, bilocular, each locule uniovulate; style glabrous. Berry globose-apiculate, orange or yellow, 1–2-seeded. Seeds up to 7.5 mm, broadly ovoid or dorsally flattened; embryo straight, cotyledons fleshy, thick, radicle tiny; endosperm very scanty. $2n = 48$.

A single species, *Espadaea amoena* A. Rich., Cuba. Planted as an ornamental.

6. ***Goetzea*** Wydl.

Goetzea Wydl., Linnaea 5: 423 (1830); Liogier, Descriptive Flora of Puerto Rico and Adjacent Islands IV: 444 (1995).

Trees or shrubs. Leaves shortly petiolate, entire, leathery, broad. Flowers axillary, solitary, long-pedicellate, pendant; calyx campanulate or cup-shaped, 6-lobed; corolla yellow, conspicuously funnel-shaped, 6-lobed, lobes reflexed; stamens 6, filaments glabrous, exserted, inserted basally; anthers pubescent; ovary bilocular, pubescent, ovules few, pendant; style glabrous, stigma bi- or trilobed. Berry ovoid, orange, 1–7-seeded. Seeds large; embryo straight, cotyledons fleshy, thick, radicle tiny; endosperm very scanty.

Two endemic and endangered species in the Greater Antilles: *G. elegans* Wydl. (Puerto Rico) and *G. ekmanii* O.E. Schulz ex O.C. Schmidt (Hispaniola).

II. Subfam. Cestroideae Burnett (1835).

Androecium 4 or 5, often didynamous or heterodynamous, staminode usually present. Cotyledons incumbent; endosperm copious.

II.1. **Tribe Cestreae** Dumort. (1829).

Shrubs. Flowers sessile or with articulate pedicels; corolla aestivation valvate-induplicate, slightly contorted; stamens equal or unequal, filaments straight, generally geniculate at their insertion; ovary with a small annular remainder of the deciduous corolla (cyathium). Embryo straight or slightly curved cotyledons broader than the rest of the embryo.

7. ***Cestrum*** L.

Cestrum L., Sp. pl. 1: 191 (1753); Francey, Candollea 6: 46–398 (1935), 7: 1–132, pl. I–III (1936), rev.; Benítez de Rojas & D'Arcy, Ann. Missouri Bot. Gard. 85(2): 273–351 (1998); reg. rev.; Romanutti & Hunziker, Fl. Fanerog. Argent. 55: 4–12 (1998), reg. rev.; Nee, Solanaceae V. Advances in Taxonomy and Utilization: 109–136 (2001), rev.

Shrubs or trees, rarely vines, some fragrant. Leaves entire, unequal. Inflorescence terminal, monochasial cymes, up to 40-flowered. Bracts deciduous or persistent; calyx campanulate, urceolate, tubular, or cup-shaped; corolla variously coloured, tubular, infundibuliform, hypocrateriform, or slightly urceolate; stamens equal or unequal, geniculate and inserted at different heights in corolla tube, filaments with or without appendages; anthers dorsifixed; ovary stipitate pedicellate, annular nectary prominent; stigma wet, papillate. Berry ovoid or ellipsoid. Seeds subpolyhedrical; cotyledons usually shorter than rest of embryo. $n = 8, 16$; $2n = 16$.

Nearly 150–200 spp. in tropical and subtropical America. Some species cultivated as garden plants (*C. nocturnum* L., etc.), others are poisonous (*C. parqui* L'Hérit., *C. laevigatum* Schltdl.).

Cestrum, *Sessea* and *Vestia* appear to form a monophyletic group (tribe Cestreae) well supported by molecular (Santiago-Valentín and Olmstead 2003; Martins and Barkman 2005; Montero-Castro et al. 2006; Olmstead et al. 2008), morphological (Hunziker 2001) and cytological (Sykorova et al. 2003) characters.

8. ***Sessea*** Ruiz & Pav

Sessea Ruiz & Pav., Fl. peruv. prodr.: 21, tab. 33 (1794); Bitter, Repert. Spec. Nov. Regni Veg. 18 (513/523): 199–225 (1922), rev.; Benítez de Rojas & Nee, Solanaceae V. Advances in Taxonomy and Utilization: 153–159 (2001), reg. rev.
Sesseopsis Hassler (1917).

Shrubs to trees, frequently with dendroid or candelabra hairs. Leaves entire, membranous or subcoriaceous, some looking like stipules. Inflorescence axillary or terminal, mostly several- to many-flowered. Flowers 5-merous, commonly with bracteoles; calyx tubular, lobes equal or unequal; corolla tubular or slightly infundibuliform; stamens geniculate and inserted at different heights, filaments pilose and thickened at insertion; anthers included; ovary pubescent, 4-lobed; stigma lobed, deflexed or capitate. Capsule bivalvate, valves bifid. Seeds 6–19, elongate and imbricate in fruit, with a peripheral narrow hyaline wing. $2n = 16$.

15 spp., Andes of South America.

Montero-Castro et al. (2006) found evidence for the paraphyly of *Sessea* in relation to *Vestia* which needs to be validated with additional sampling and inclusion of related genera.

9. ***Vestia*** Willd.

Vestia Willd., Enum. pl. 1: 208 (1809).

Much-branched shrubs, ill-scented. Leaves entire. Flowers 5-merous, pendant, shortly pedicellate; calyx tubular, minutely dentate, glabrous; corolla greenish yellow, hypocrateriform; stamens strongly exserted, attached slightly above basal fourth of corolla, filaments unequal, thickened at insertion and pubescent from there downwards; anthers basifixed; gynoecium longer than stamens; ovary bilocular, nectary annular;

stigma capitate. Capsule 2-valved, valves bifid. Seeds ca. 50, subpolyhedrical, angulate, wingless; testa tuberculate. $n = 8$; $2n = 16$.

A single species in Chile. *Vestia foetida* (Ruiz & Pav.) Hoffmanns. is a well-known medicinal plant.

II.2. **Tribe Browallieae** Hunz. (1995).

Herbs or small shrubs. Corolla zygomorphic; stamens 4 (anterior aborted), didynamous; pollen grains 4–9-colporate; style geniculate and tortuose at apex, strongly corrugated above; stigma broad, two-horned. Seeds cuboidal or cuboidal-elongated; embryo slightly curved, cotyledons two times shorter than the rest of the embryo.

10. ***Browallia*** L.

Browallia L., Sp. pl. 2: 631 (1753).

Annual herbs. Leaves cordate-ovate to linear-oblong. Inflorescence a raceme or flowers solitary, showy. Calyx actinomorphic; corolla superiorly ventricose, occasionally slightly bilabiate, aestivation reciprocative; upper lateral pair of stamens with filaments flattened, glabrous or with branched trichomes, and anthers with one thecae aborted, lower posterior pair terete and narrow, glabrous or pilose at apex, and anthers with both thecae fertile; ovary glabrous or with rigid trichomes, with cryptic or annular and thick nectary. Capsule globose, elongated or obconic, calyx accrescent and papery. Seeds angulate 10–45, testa reticulate-foveolate. $n = 10, 11, 22, 44$; $2n = 20, 22, 44$.

About eight spp. from southern Arizona (USA) to Andean South America, reaching Bolivia. At least two species, *B. americana* L. and *B. speciosa* Hook., have ornamental value.

A critical treatment of the genus is strongly needed to clarify its obscure taxonomy. *Browallia* and *Streptosolen* form a sister pair well-supported by morphological (Hunziker 2001), palynological (Stafford and Knapp 2006) and molecular (Martins and Barkman 2005; Olmstead et al. 2008) data.

11. ***Streptosolen*** Miers

Streptosolen Miers, Ann. Mag. Nat. Hist. II, 5 (27): 207 (1850).

Evergreen scabrous-pubescent shrubs, densely branched. Leaves ovate, petiolate. Inflorescence subcorymbose cymes. Flowers pedicellate; calyx zygomorphic, tubular, 4–5-lobed; corolla infundibuliform, tube spirally twisted, not ventricose, limb rather narrow, aestivation cochlear-conduplicate; upper lateral pair of stamens with short filaments and small anthers (usually one theca aborted), lower posterior pair with long filaments and larger anthers; ovary pubescent above, nectary annular, thick. Capsule globose to ovate. Seeds 60–80, cuboidal-elongated; testa reticulate. $n = 11, 12$.

A single species, *S. jamesonii* (Benth.) Miers, Colombia, Ecuador and Peru. Ornamental.

II.3. **Tribe Salpiglossideae** Benth. (1835).

Herbs or shrubs. Pubescence of glandular trichomes with multicellular, multiseriate heads and long uniseriate stalk. Flowers mostly solitary on long pedicels; corolla aestivation cochlear; androecium 4–5-merous, stamens fertile didynamous, the posterior pair with larger anthers than the lateral pair; pollen in monads or in tetrads; style expanded or spoon- shaped at apex, stigma linear. Seeds compressed; embryo coiled, cotyledons up to three times shorter than the rest of the embryo.

12. ***Salpiglossis*** Ruiz & Pav. Fig. 62

Salpiglossis Ruiz & Pav., Fl. peruv. prodr.: 94, tab. 19 (1794); Hunziker & Subils, Bot. Mus. Leafl. 27(1–2): 1–43 (1980), rev.

Herbs or shrubs, densely viscid-pubescent; stems leafy or almost without leaves, sometimes xeromorphic (*S. spinescens*) with spine-like processes. Flowers solitary or in panicles, chasmogamous or cleistogamous; perianth 5-merous; calyx actinomorphic, campanulate; corolla zygomorphic, infundibuliform; androecium with one or without staminode (*S. sinuata*); anthers ovoid, dorsifixed; ovary bicarpellate, nectary annular; style hollow, expanded distally. Capsule ovate 2–4-valved. Seeds up to 300 or more, testa reticulate. $n = 22$.

Two spp. endemic to southern Andean South America (Chile and Argentina).

13. ***Reyesia*** Gay

Reyesia Gay, Fl. Chile 4: 418 (1840); Hunziker & Subils, Bot. Mus. Leafl. 27(1–2): 1–43 (1980), rev.

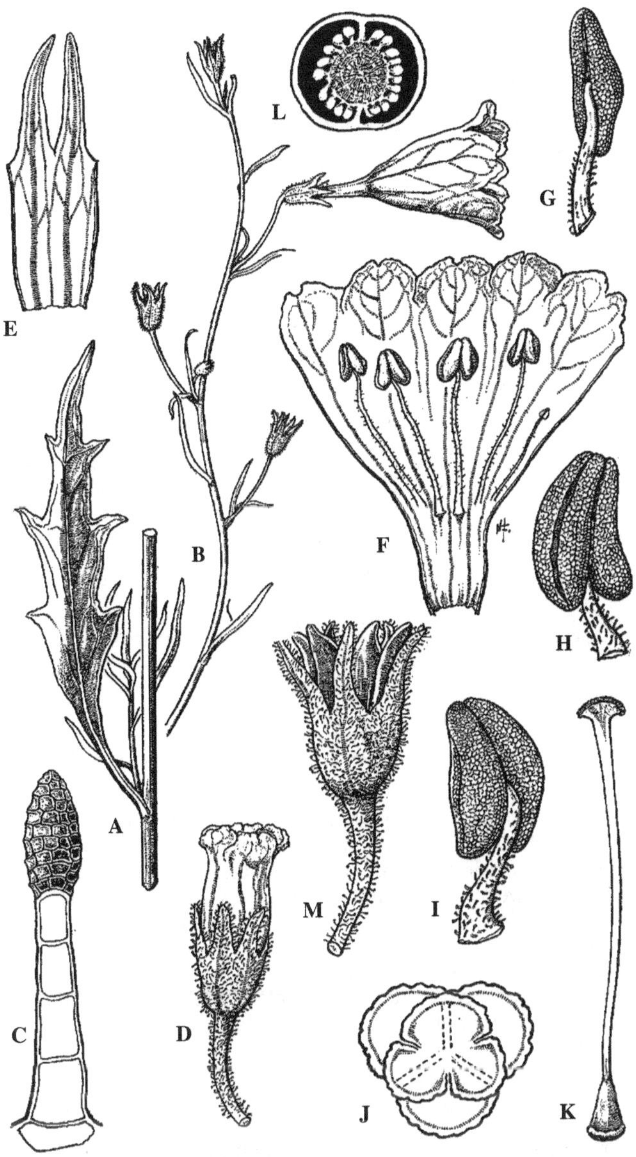

Fig. 62. Solanaceae. *Salpiglossis sinuata*. **A** Basal portion of branch. **B** Flowering branch. **C** Glandular trichome. **D** Flower. **E** Inside view of two calyx segments, showing venation. **F** Opened corolla. **G** Anther, lateral view. **H** Anther, ventral view. **I** Anther, dorsal view. **J** Pollen tetrad. **K** Gynoecium. **L** Transverse section of ovary. **M** Dehiscent fruit. (Hunziker 2001; drawn by N. Flury)

Annual or perennial herbs, dichasially branched, glutinose. Lower leaves entire to pinnatifid or pinnatisect, the upper smaller or linear scales. Flowers solitary, terminal, small; calyx actinomorphic, strongly glanduliferous; corolla zygomorphic, tubular or infundibuliform; the posterior pair of stamens always fertile, the lateral pair with fertile or sterile anthers; anthers basifixed, thecae usually unequal, divaricate; ovary bicarpellate, nectary pelviform, bilobed, style hollow or solid, spoon-shaped distally. Capsule 2–4-valved hidden at the base of the persistent calyx. Seeds (3)20–25, testa reticulate or granulate.

Four spp. in Andean northern Chile, one reaching Argentina.

II.4. **Cestroideae: Incertae sedis**

14. ***Protoschwenckia*** Soler.

Protoschwenckia Soler., Ber. Deutsch. Bot. Ges. 16: 243 (1898); Freire de Carvalho, Sellowia 18(18): 67–72 (1966), reg. rev.
Schwenkiopsis Dammer (1916).

Small shrubs, profusely branched, with non-glandular and glandular trichomes, and occasionally branched trichomes. Leaves entire, ovate, base cordate. Inflorescence paniculate. Flowers pedicellate; calyx actinomorphic, with 5 equal uninerved lobes; corolla actinomorphic, tubular, lobe margins inflexed; aestivation valvate-induplicate; androecium and gynoecium included; stamens 4 (2 + 2), filaments filiform, pubescent from insertion to tube downwards; anthers dorsifixed; ovary with a cushion-shaped nectary; style short, stigma discoidal-capitate. Capsule septicidal-loculicidal, with 2 bifid valves. Seeds ca. 22, subreniform; embryo slightly curved, cotyledons 3 times shorter than rest of embryo.

A single species, *P. mandonii* Soler., Andes of Bolivia, Brazil.

15. ***Heteranthia*** Nees & Mart.

Heteranthia Nees & Mart., Acta Phys.-Med. Acad. Caes. Leop.-Carol. Nat. Cur. 11: 41 (1823).

Perennial herbs. Leaves petiolate, ovate to obovate. Inflorescence racemose. Calyx actinomorphic, 5-lobed; corolla zygomorphic, funnel-shaped, slightly bilabiate, basal tube narrow and short, lobules very short and broad, aestivation unknown; stamens 2 + 2; anthers ventrifixed, thecae unequal; ovary subconical or pyriform, nectary absent; style exserted, curved or almost uncinate. Capsule septicidal, 2-valved; calyx accrescent. Seeds 20–28, cuboidal; embryo straight, very small, cotyledons unequal.

Monotypic, *H. decipiens* Nees & Mart., eastern Brazil.

III. Subfam. Nicotianoideae Miers (1848).

Mostly herbs or shrubs. Calyx non-accrescent or scarcely accrescent; androecium tetramerous (staminode present), or pentamerous (stamens of two lengths). Embryo mostly straight or slightly curved, cotyledons short to very short relative to the rest of the embryo.

III.1. **Tribe Anthocercideae** G. Don (1838).

Shrubs. Calyx non-accrescent; corolla broadly actinomorphic, aestivation valvate-supervolute; anthers ventrifixed, extrorse, attached low in the corolla tube. Mostly capsule. Seeds reniform to elongate, testa reticulate; embryo terete, cotyledons 5–12 times shorter than the rest of the embryo; endosperm with a small oil sector.

16. ***Symonanthus*** Haegi Fig. 63

Symonanthus Haegi, Telopea 2: 175 (1981); Purdie, Symon & Haegi, Fl. Australia 29: 13–17 (1982).

Dioecious or hermaphrodite shrubs, indumentum of glandular and non-glandular branched hairs. Leaves long and narrow, margin revolute. Flowers solitary or in 2–3-flowered cymes; calyx 5-lobed; corolla narrowly tubular or cylindrical-campanulate, upper half funnel-shaped, almost truncate or with 5 equal, very short and broad lobules; staminate flower: stamens equal, 5 or 4 (then staminode sometimes present); anthers dithecal, gynoecium much reduced; pistillate flower: stamens sterile, ovary large, nectary annular, ovules 4–10; style well developed, stigma bilobed. Capsule subglobose, septicidal-loculicidal, 4-valved. Seeds few, reniform; embryo slightly arcuate; endosperm copious. $n = 36$; $2n = 32$.

Two spp., south-western Australia.

17. ***Anthocercis*** Labill.

Anthocercis Labill., Nov. Holl. pl. 2: 19, tab. 158 (1806); Purdie, Symon & Haegi, Fl. Australia 29: 6–13 (1982).

Shrubs, glabrous or pubescent, occasionally viscid, usually with glandular trichomes. Leaves variously shaped, sometimes thick and fleshy or leathery, usually entire, or minutely toothed or

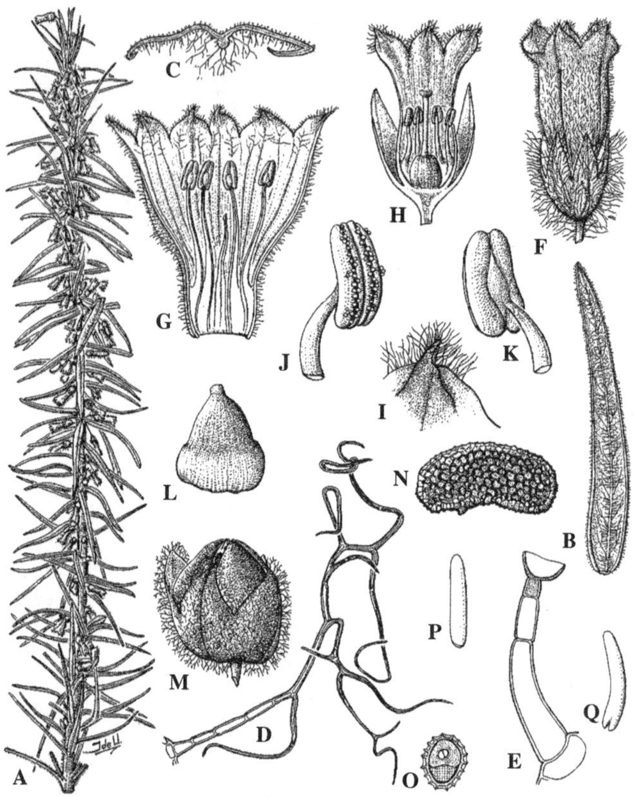

Fig. 63. Solanaceae. *Symonanthus aromaticus*. **A** Flowering branch. **B** Leaf, abaxial view. **C** Transverse section of leaf. **D** Dendritic trichome. **E** Glandular trichome. **F** Staminate flower. **G** Staminate flower, opened corolla. **H** Carpellate flower, longitudinal section. **I** Apex of corolla lobe. **J** Anther, lateral view. **K** Anther, ventral view. **L** Rudimentary ovary of staminate flower. **M** Fruit. **N** Seed. **O** Seed, transverse section (the stippled lower area is oily). **P, Q** Two views of embryo. (Hunziker 2001; drawn by J. de Ugarte)

serrulate-crenulate. Inflorescences cyme-, raceme- or panicle-like. Flowers perfect; calyx actinomorphic, 5-partite; corolla narrowly 5-lobed; stamens 4, didynamous or subequal (sometimes with a staminode); anthers dithecal; nectary annular; stigma bilobed. Capsule pear-shaped or ovoid, septicidal-loculicidal, 2–4-valved. Seeds up to 100, reniform; endosperm copious. $n = 36$.

10 spp., mainly south-western temperate Australia. Some species have occasionally caused poisoning of children or cattle.

18. ***Grammosolen*** Haegi

Grammosolen Haegi, Telopea 2: 178 (1981); Purdie, Symon & Haegi, Fl. Australia 29: 28–29 (1982), rev.

Shrubs; branches and leaves densely tomentose, with dendritic trichomes. Leaves crowded or well-spaced, subsessile. Inflorescences cymose, 1–3-flowered, sometimes forming leafy spikes. Flowers perfect; calyx 5-lobed, densely tomentose; corolla tubular, lobes long and narrow; stamens 5, unequal (2 + 3); anthers monothecal; nectary annular; stigma bilobed. Capsules septicidal-loculicidal, 4-valved. Seeds reniform; endosperm copious. $n = 56$.

Two spp., South Australia.

19. ***Cyphanthera*** Miers

Cyphanthera Miers, Ann. Mag. Nat. Hist. II, 11 (65): 376 (1853); Purdie, Symon & Haegi, Fl. Australia 29: 21–28 (1982).
Anthocercis sect. *Cyphanthera* (Miers) F. Muell. (1861).

Shrubs, usually pubescent, viscid or slightly to densely woolly-tomentose, trichomes glandular or dendritic. Leaves usually sessile, variously shaped. Inflorescence a cyme, or raceme-, panicle- or thyrse-like, or flowers solitary. Flowers perfect; calyx small; corolla funnel-shaped to campanulate; stamens 4 (2 + 2), sometimes also a staminode present; anthers monothecal; nectary annular; stigma slightly bilobed. Capsule septicidal-loculicidal, 4-valved. Seeds ca. 6, reniform. $n = 30$; $2n = 60$.

Nine spp., southern temperate Australia.

A phylogenetic analysis (six spp. sampled) revealed that *Cyphantera* is not monophyletic (García and Olmstead 2003; Clarkson et al. 2004; see comment under *Duboisia*).

20. ***Anthotroche*** Endl.

Anthotroche Endl. in Endl. & Fenzl, Nov. stirp. dec. 1: 6 (1839); Purdie, Symon & Haegi, Fl. Australia 29: 30–34 (1982), rev.

Shrubs with a very dense indumentum of dendritic and glandular trichomes. Leaves shortly petiolate, variously shaped. Flowers solitary or in 2–6-flowered clusters, sometimes forming leafy pseudospikes; calyx actinomorphic; corolla campanulate to rotate, densely pubescent, lobes broad and short; stamens 5, equal, included or exserted, filaments sometimes incurved and pubescent; anthers monothecal, horseshoe-shaped. Capsules ovoid to ellipsoid or globose, septicidal-loculicidal, 4-valved. Seeds reniform; endosperm copious. $n =$ ca. 31.

Four spp., arid inland of south-western Western Australia.

21. ***Crenidium*** Haegi

Crenidium Haegi, Telopea 2: 179. 1981; Purdie, Symon & Haegi, Fl. Australia 29: 34 (1982), rev.

Intricately branched shrubs, leafless except on young twigs; branches spinescent, tomentose with dendritic trichomes. Leaves sessile, margin slightly revolute. Flowers in cymose clusters. Flowers perfect; calyx minutely 5-lobed; corolla narrowly tubular with ovate to broadly ovate and short lobes; stamens 4 (2 + 2 and a staminode usually present), exserted; anthers monothecal; nectary annular; stigma shortly bilobed. Capsules ovoid, septicidal-loculicidal, valves 2, bifid. Seeds 20–100, reniform. $n = 35$.

Monotypic, *C. spinescens* Haegi, inland south-western Western Australia.

22. ***Duboisia*** R. Br.

Duboisia R. Brown, Prodr.: 448 (1810); Purdie, Symon & Haegi, Fl. Australia 29: 17–21 (1982), rev.

Shrubs or trees, glabrescent to glabrous; bark corky and thick; wood extremely light. Leaves narrowly obovate to obovate-elliptic, or linear. Inflorescence paniculate. Flowers perfect; calyx campanulate; corolla campanulate-funnel-shaped; stamens 4 (2 + 2 and sometimes a vestigial staminode); anthers monothecal; nectary annular; stigma bilobed or bifid. Berry globose or subglobose, rarely ellipsoid to ovoid. Seeds 2–12, reniform; endosperm copious, starchy, oily sector absent. $n = 30$; $2n = 60$.

Four spp. in New Caledonia and eastern Australia. Species of *Duboisia* known as "corkwood" are harvested commercially to obtain scopolamine for medicinal purposes, and are toxic to cattle, horses and humans.

Duboisia does not form a monophyletic group (García and Olmstead 2003; Clarkson et al. 2004), and a re-circumscription of this genus and also of *Cyphanthera* (*C. albicans* is resolved in the *Duboisia* clade) is needed.

III.2. **Tribe Nicotianeae** G. Don

Mostly herbs or shrubs. Inflorescence terminal, paniculoid, many-flowered; corolla aestivation contorted-conduplicate; anthers dorsifixed; stamens 5, equal or 4 + 1 or 2 + 2 + 1. Capsule with up to ca. 5000 seeds.

23. ***Nicotiana*** L.

Nicotiana L., Sp. pl. 1: 180 (1753); Goodspeed, The genus *Nicotiana*: 1–536 (1954), rev.; Cocucci & Hunziker, Flora Fanerog. Argent. 89: 3–14 (2005); Knapp, Chase, & Clarkson, Taxon 53: 73–82 (2004), rev.

Herbs to small trees. Basal leaves spathulate, sometimes rosulate, cauline leaves increasingly smaller and similar to inflorescence bracts, margin entire or undulate. Flowers nocturnal, diurnal or cleistogamous; corolla actino- or zygomorphic,$<$90 mm, tubular, salverform, or campanulate, variously coloured, tube straight or curved, sometimes very narrow or broadly inflated, lobes short, often emarginate; stamens included or exserted, sometimes strongly geniculate; anthers circular to elongated. Capsule bivalvate. $2n = 18, 20, 24, 32, 36, 40, 48$. In some species, $2n = 38, 42, 44, 46$.

Near 76 spp., southern South America east of the Andes (40 spp.), Africa (1 sp.), Australia (25 spp.), SW North America (8 spp.).

IV. Subfam. Solanoideae Burnett (1835).

Herbs to small trees; corolla aestivation valvate, cochlear or contorted. Androecium 5-merous (rarely 4-merous). Fruit mostly a berry but also schizocarp and less frequently drupe, diclesium or pixidium. Seeds flattened.

IV.1. **Tribe Lycieae** Lowe (1872).

Shrubs or small trees. Perianth pentamerous, sometimes tetramerous; corolla aestivation cochlear or cochlear-plicate; stamens 5, anthers dorsifixed; ovary 2–4-locular. Fruit a berry or a drupe. Embryo annular or subcoiled, cotyledons incumbent; endosperm copious.

24. ***Lycium*** L.

Lycium L., Sp. pl. 1: 191 (1753); Bernardello, Bol. Acad. Nac. Ci. Republ. Argent. 57: 173–356 (1986), part. rev.; Chiang-Cabrera, Ph.D. diss., The University of Texas at Austin (1981), part. rev.
Grabowskia Schltdl. (1832).
Phrodus Miers (1849).

Mostly shrubs, usually with cauline spines. Flowers axillary, solitary or in clusters; calyx actinomorphic or zygomorphic; corolla tubular or infundibuliform; stamens exserted or included, filaments cylindrical or enlarged at proximal end; thecae free in their basal third or half; ovary bicarpellate, completely bilocular or tetralocular in its upper part, 1–∞ ovules per carpel, nectary annular or invaginate-bilobed, inconspicuous and greenish or conspicuous and red. Berry generally many-seeded, rarely drupaceous with two 1-seeded pyrenes or intermediate with stone cells or sclerenchyma. Seeds discoidal-compressed or ovoid and plano-convex. $n = 12, 18, 24, 36, 48, 60$; $2n = 24, 36, 48$.

Cosmopolitan genus of ca. 92 spp. Recent phylogenetic studies (Levin and Miller 2005; Levin et al. 2007, 2011) have shown that the Chilean monotypic *Phrodus* (*L. bridgesii* (Miers) Levin, Miller & Bernardello) and the few species of the exclusively American genus *Grabowskia*, reduced to one (*L. boerhaviifolium* L.f.), must be included in *Lycium*.

IV.2. **Tribe Hyoscyameae** Endl. (1839).

Perennial herbs. Flowers usually solitary; corolla aestivation cochlear. Fruit mostly a pyxidium many-seeded and hidden in the accrescent calyx. Embryo strongly curved to subcoiled; endosperm rather copious.

25. ***Hyoscyamus*** L.

Hyoscyamus L., Sp. pl. 1: 179 (1753); Hoare & Knapp, Bull. Nat. Hist. Mus. London, Bot. 27: 26 (1997), rev.
Archihyoscyamus Lu (1997).

Herbs with rhizomes or napiform, woody or fleshy roots; glabrous, villose or glandular pubescent. Leaves in a rosette. Lower flowers axillary, solitary, upper ones in condensed inflorescences; calyx variously shaped, densely veined, 5-toothed or 5-lobed; corolla zygomorphic, tube short, expanding upwards, or subrotate, lobes unequal; androecium and gynoecium mainly exserted; filaments unequal; anthers dorsifixed; nectary indistinct; style filiform, stigma capitate. Pyxidium elongate, hidden by the accrescent calyx, calyx teeth spine-tipped and rigid. Seeds 200–500, discoid; embryo strongly curved, cotyledons unequal, as long as or shorter than rest of embryo. $n = 14, 17, 34$; $2n = 28, 34, 68$.

About 17 spp., from Madeira, Canary Islands, and North Africa to India and China.

26. ***Przewalskia*** Maxim.

Przewalskia Maxim., Bull. Acad. Imp. Sci. Saint-Pétersbourg. 27: 507 (1881); Hoare & Knapp, Bull. Nat. Hist. Mus. London, Bot. 27: 27 (1997), rev.

Perennial herbs with fleshy roots; indumentum glandular. Leaves densely aggregated. Inflorescence with clusters of 1–6 axillary flowers; calyx actinomorphic, cylindrical-campanulate; corolla actinomorphic, tubular; stamens equal, exserted, filaments glabrous, attached above; anthers dorsifixed; nectary annular; stigma discoidal-capitate, exserted. Pyxidium globose, included in the much accrescent, urceolate-ovoid and strongly veined calyx, calyptra convex and apiculate. Seeds reniform. $2n = 44$.

A single species, *P. tangutica* Maxim., western China, 3200–5000 m.

27. ***Scopolia*** Jacq.

Scopolia Jacq., Observ. bot.: 32, tab. 20 (1764), nom. cons.; Hoare & Knapp, Bull. Nat. Hist. Mus. London, Bot. 27: 27 (1997), rev.

Perennial herbs with rhizomes. Leaves obovate or oblong. Flowers axillary, solitary; calyx campanulate; corolla campanulate or tubular-campanulate, shallowly 5-lobed, lobes very short and broad; stamens slightly unequal, included, pubescent; anthers probably basifixed; nectary annular; stigma depressed. Pyxidium globose, enclosed in the somewhat enlarged calyx. Seeds discoid; embryo subcoiled, cotyledons slightly shorter than rest of embryo. $n = 24$; $2n = 44, 46, 48, 84, 168$.

Two spp. with disjunct distribution: *S. carniolica* Jacq. in Europe and *S. japonica* Maxim. in East Asia.

Within Hyoscyameae, *Scopolia* is inferred to be paraphyletic in relation to *Przewalskia* (Olmstead et al. 2008).

28. ***Physochlaina*** G. Don

Physochlaina G. Don, Gen. hist. 4: 470 (1838); Hoare & Knapp, Bull. Nat. Hist. Mus. London, Bot. 27: 27 (1997), rev.

Perennial herbs with fleshy, tuberous roots; indumentum non-glandular and glandular. Leaves entire or sinuate-dentate. Inflorescence paniculate, terminal or axillary. Calyx actinomorphic, campanulate, tubular-campanulate or tubular-urceolate, lobes equal; corolla actinomorphic with a basal tube expanding upwards; stamens exserted or included, inserted below middle of corolla; anthers dorsifixed; nectary annular, prominent; stigma discoid. Pyxidium subglobose or elongated included in the much accrescent, inflated, elongated calyx; embryo subcoiled, cotyledons shorter than rest of embryo. $2n = 28, 41, 42, 82$.

About eight spp., Asia.

29. ***Atropanthe*** Pascher

Atropanthe Pascher, Oesterr. Bot. Z. 59(9): 529 (1909); Hoare & Knapp, Bull. Nat. Hist. Mus. London, Bot. 27: 26 (1997), rev.

Perennial herbs with thick rhizomes. Leaves entire, acuminate. Flowers solitary, axillary, pendant; calyx campanulate-urceolate, lobes subequal; corolla slightly zygomorphic, cylindrical-campanulate, lobes unequal, revolute; stamens included, filaments unequal, circinate in bud, curved at anthesis, inserted basally; anthers ventrifixed; nectary annular prominent; stigma bilobed. Pyxidium pendant, small, globose and apiculate, included in the enlarged, urceolate and inflated calyx. Seeds ca. 50, discoid; embryo subcoiled, cotyledons slightly shorter than rest of embryo. $2n = 48$.

Monospecific, *A. sinensis* (Hemsley) Pascher, southern and central China, 1400–3000 m.

30. ***Anisodus*** Link

Anisodus Link ex Sprengel, Syst. veg. 1: 699 (1825); Hoare & Knapp, Bull. Nat. Hist. Mus. London, Bot. 27: 25 (1997), rev.

Cryptophytic herbs, glabrous or pubescent, roots woody, napiform. Leaves entire, sinuose or dentate. Flowers axillary, solitary, on long pedicels; calyx campanulate or urceolate, zygomorphic, irregularly 5-lobed; corolla campanulate or campanulate-urceolate, lobes equal, auriculate, revolute; stamens equal, included, filaments straight, adnate basally to corolla tube; anthers dorsifixed; ovary conical, nectary annular, prominent, thick; stigma discoid. Pyxidium usually globose, 2–4 times shorter than the ribbed calyx. Seeds discoid. $2n = 24, 42, 48, 50$.

Four spp. in the mountains of Nepal, India, Bhutan, and China, 2000–4500 m.

31. ***Atropa*** L.

Atropa L., Sp. pl.: 181 (1753); Hoare & Knapp, Bull. Nat. Hist. Mus. London, Bot. 27(1): 26 (1997), rev.
Pauia Deb & R. M. Dutta (1965).

Perennial herbs, glabrous or glandular-pubescent. Basal leaves alternate, upper geminate and anisophyllous. Flowers solitary, paired or in triplets, pendant or erect; calyx campanulate, tube shorter than lobes; corolla slightly zygomorphic, tubular-campanulate, bell-shaped or infundibuliform, lobes rounded, slightly unequal and curved outwards; androecium and gynoecium included or exserted; filaments unequal, incurved, attached basally; anthers dorsifixed; nectary annular; style somewhat arched, stigma capitate. Berry black and juicy, calyx somewhat enlarged. Seeds ca. 25–160, reniform; embryo annular, cotyledons almost as long as rest of embryo. $n = 36$; $2n = 50, 72, 74$.

Three spp. in Europe, in Asia to Mongolia, and in northern Morocco.

IV.3. **Tribe Juanulloeae** Hunz. (1977).

Mostly epiphytes or lianas, some genera myrmecophilous, others bat-pollinated. Corolla aestivation cochlear or valvate. Fruit a large berry. Embryo straight to annular; cotyledons accumbent or oblique, shorter than the rest of the embryo; endosperm very scarce to copious.

32. ***Dyssochroma*** Miers

Dyssochroma Miers, Ann. Mag. Nat. Hist. II, 4(22): 250 (1849).

Shrubs or climbers, epiphytic or terrestrial; brachyblasts cauliflorous. Leaves coriaceous. Inflorescence terminal, a congested cyme or a single flower. Flower pendant, 5-merous; calyx subcampanulate, coriaceous, deeply divided; corolla thick, shortly cylindrical, expanded upwards into an infundibuliform-campanulate or ventricose part, limb 5-partite, aestivation valvate; stamens straight, attached basally to corolla; anthers basifixed; nectary annular, well developed; style articulate, stigma saddle-shaped or very long and decurrent. Berry ovoid-globose. Seeds boomerang- or rod-shaped; embryo curved, cotyledons oblique, as wide as rest of embryo; endosperm abundant. $2n = 24$.

Two spp. in Brazil.

Knapp et al. (1997) suggested *Dyssochroma* to be congeneric with *Trianaea*.

33. ***Juanulloa*** Ruiz & Pav. Fig. 64

Juanulloa Ruiz & Pav., Prodr. Fl. peruv. prodr.: 27, tab. 4 (1794); Hunziker & Subils, Kurtziana 21: 209–235 (1991), rev.
Laureria Schltdl. (1833).
Sarcophysa Miers (1849).
Ectozoma Miers (1849).
Rahowardiana D'Arcy (1973).

Climbers, shrubs or small trees, epiphytic or terrestrial. Leaves coriaceous or membranous. Inflorescences terminal or lateral, variously modified cymes or raceme-like, usually many-flowered or reduced to 1–2 flowers, inflorescence rhachis short or very long, pendant. Calyx fleshy, 5-lobed, or deeply divided or unlobed and inflated; corolla fleshy, tubular or funnel-form, slightly zygomorphic, aestivation cochlear; stamens generally included, attached at different levels of corolla, pubescent at point of insertion; anthers dorsifixed; nectary annular, well developed; stigma saddle-shaped. Berry ovoid or subglobose, calyx accrescent. Seeds shortly subreniform or boomerang-shaped; embryo curved, cotyledons accumbent, wider than rest of embryo.

About 11 spp. from southern Mexico to Bolivia.

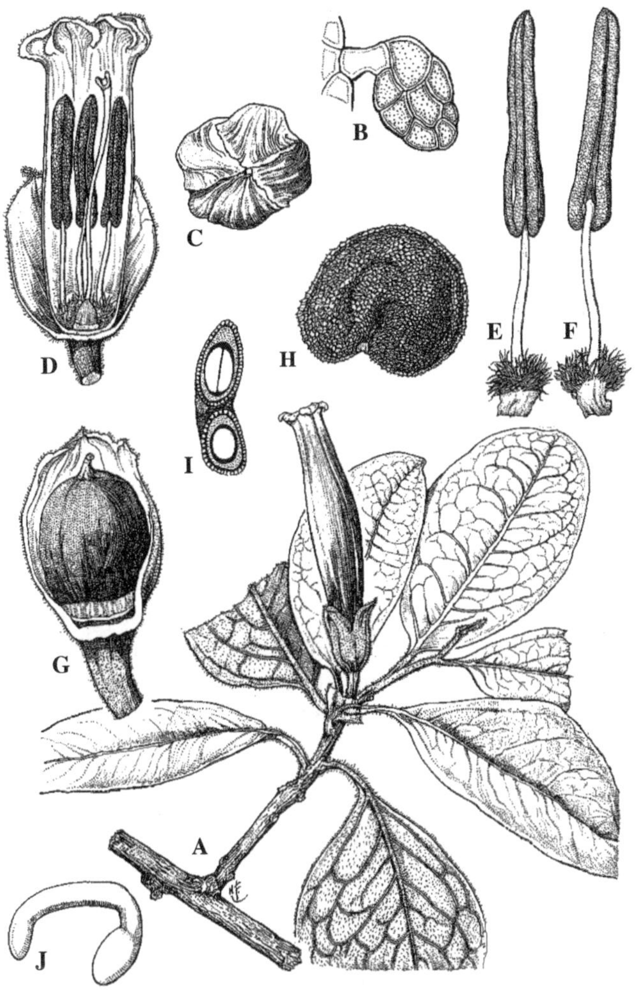

Fig. 64. Solanaceae. *Juanulloa speciosa*. **A** Flowering branch. **B** Glandular trichome. **C** Flower bud, top view showing aestivation. **D** Young flower, longitudinal section. **E** Stamen, ventral view. **F** Stamen, dorsal view. **G** Fruit, accrescent calyx cut longitudinally. **H** Seed. **I** Seed, cross section. **J** Embryo. (Hunziker 2001; drawn by N. Flury)

34. ***Markea*** Rich.

Markea Rich., Actes Soc. Hist. Nat. Paris 1: 107 (1792); Hunziker, Kurtziana 25: 67–113 (1997), rev.
Hawkesiophyton Hunz. (1977).

Epiphytic, shrubs or small trees, frequently with root tuberosities. Leaves in pseudowhorls, coriaceous or membranous. Flowers solitary or in a cyme of 2–17, usually pedicellate, on long and flexible or short and rigid peduncles; calyx tube very short or well developed; corolla thick, funnel-form or infundibuliform, lobes not reflexed at anthesis, aestivation cochlear; stamens adnate basally and pubescent at point of insertion; anthers basifixed, frequently connivent; nectary well developed; stigma capitate, saddle-shaped or with 2 short branches. Berry ellipsoid, calyx accrescent. Seeds 40–100; embryo slightly curved, cotyledons oblique, wider than rest of embryo.

About nine spp. from Panama to Peru.

35. ***Merinthopodium*** Donn. Sm.

Merinthopodium Donn. Sm., Bot. Gaz. 23: 11, pl. I (1897).

Epiphytic shrubs or small trees, rarely terrestrial. Leaves subcoriaceous or membranous, glabrous or minutely pubescent on the main vein beneath. Inflorescence a pendant raceme, peduncles long (12–90 cm), flexible, slender, bearing 1–14 pedicellate flowers. Calyx campanulate, actinomorphic, tube very short; corolla broadly campanulate, lobes short, reflexed at anthesis, aestivation valvate; stamens attached basally; anthers dorsifixed and somewhat versatile, thecae not confluent; nectary annular; style straight, stigma saddle-shaped. Berry ovoid, surrounded by the slightly accrescent calyx. Seeds boomerang-shaped; embryo thick and short, curved, cotyledons accumbent; endosperm rather scarce.

About three spp. from southern Mexico to Venezuela and Colombia. The taxonomy of this genus is obscure.

IV.4. **Tribe Datureae** Dumort. (1829).

Herbs, shrubs or small trees. Flowers solitary, large; corolla aestivation contorted-conduplicate; anthers basifixed, pubescent, dehiscence latrorse or extrorse. Embryo annular; cotyledons shorter than rest of embryo; endosperm scarcely copious, rich in protein.

36. ***Brugmansia*** Pers.

Brugmansia Pers., Syn. pl. 1: 216. 1805; Nee, Solanaceae Newslett. 3 (2): 27–35 (1991); Preissel & Preissel, Engelstrompeten *Brugmansia* und *Datura*: 8–101 (1997), rev.
Methysticodendron R.E. Schult. (1955).

Shrubs or trees. Leaves large, dentate or sinuate-repand. Flowers pendant or inclined, 5-merous;

calyx zygomorphic, unequally toothed, sometimes splitting along one side (spathe-like), falling off after anthesis, or persisting around the fruit; corolla very large, broadly infundibuliform, rarely narrow-infundibuliform to almost cylindrical; stamens inserted near middle of corolla; ovary superior, completely bilocular, nectary annular; stigma long, saddle-shaped. Berry large, pendant. Seeds up to 300, large, testa usually thick, sometimes suberose. $n = 12$; $2n = 24$.

Six spp. in Andean South America (Venezuela to Bolivia), some used as ornamentals or for medicinal and psychotropic preparations, for example: *B. arborea* (L.) Lagerh., *B. sanguinea* (Ruiz & Pav.) D. Don.

37. ***Datura*** L.

Datura L., Sp. pl. 1: 179 (1753); Avery et al.: The genus *Datura*: 1–289 (1959), rev.; Symon & Haegi, in Hawkes et al. (eds.) Solanaceae III. Taxonomy, Chemistry, Evolution: 197–210 (1991); Preissel & Preissel. Engelstrompeten. *Brugmansia* und *Datura*: 102–137 (1997), rev.

Annual herbs or short-lived perennials dichasially branched. Leaves sinuate-dentate or lobed. Flowers erect or oblique; calyx cylindrical, shortly (3)5(6)-lobed, circumscissile after anthesis; corolla broadly funnel-form; stamens included, attached basally; ovary partially inferior, tetralocular in lower half, with small fleshy spines; nectary invaginate-pelviform; stigma bilobed. Capsule erect or deflexed, septifragal with 2–4 valves (*D. ceratocaula* with smooth berry), pericarp with spines, bristles, or blunt tubercles; calyx remnants accrescent and reflexed. Seeds up to ca. 500, sometimes with elaiosome; endosperm with lipid vesicles and rich in protein bodies. $n = 12$, 24; $2n = 24$, 48.

About 9 to 11 spp. from Mexico to southwestern USA. *Datura ferox* L. and *D. stramonium* L. are toxic weeds.

IV.5. **Tribe Physalideae** Miers (1849).

Mostly shrubs, small trees or perennial herbs. Flowers in axillary fascicles; corolla aestivation usually valvate, rarely contorted; stapet frequently present, inconspicuous or conspicuous, with or without basal auricles. Embryo subcoiled to strongly bent.

IV.5.a. **Subtribe Withaniinae** Bohs & Olmstead (2008).

Corolla campanulate, campanulate-urceolate, rotate or salver-form, aestivation valvate; filaments often with conspicuous lateral appendages; calyx often not accrescent.

38. ***Discopodium*** Hochst.

Discopodium Hochst., Flora 2: 22 (1844).

Shrubs or small trees. Leaves frequently anisophyllous. Flowers in fascicles of 5–10; calyx cup-shaped, subtruncate, denticulate; corolla campanulate-urceolate, tube and lobes with dense indumentum of simple trichomes; stamens included, inserted basally; anthers dorsifixed and pubescent on ventral surface, connective thick; nectary annular, protruding, almost as high as the rest of the ovary; style with antrorse trichomes, stigma capitate-discoidal. Berry red, globose; calyx not accrescent. Seeds ca. 20, subreniform, compressed; embryo annular, cotyledons shorter than rest of embryo; endosperm copious.

One sp., *D. penninervium* Hochst., tropical Africa.

39. ***Nothocestrum*** A. Gray

Nothocestrum A. Gray, Proc. Amer. Acad. Arts. 6: 48 (1866).

Shrubs or trees. Leaves often geminate. Inflorescences lateral, with (1)2–3(10) pedicellate and fragrant flowers, on prominent brachyblasts. Flowers 4-merous; calyx unevenly toothed or lobed, or almost bilabiate; corolla salver-form, with 4 well-developed folds; stamens included or exserted; anthers sessile or subsessile, dorsi-basifixed; ovary partly inferior, nectary annular, slightly protruding; style short, stigma saddle-shaped. Berry globose or fusiform, few-seeded, calyx accrescent or not. Seeds subreniform, compressed; embryo annular, cotyledons as long or slightly shorter than rest of embryo; endosperm scanty. $n =$ ca. 24.

A genus of five or six species endemic to the Hawaii Archipelago, some of them federally listed as endangered.

40. ***Tubocapsicum*** (Wettst.) Makino

Tubocapsicum (Wettst.) Makino, Bot. Mag. (Tokyo) 22: 18 (1908); D'Arcy et al., Bot. Bull. Acad. Sin. 42: 67–84 (2001), rev.

Perennial herbs; stems stout, dichasially branched, glabrate. Leaves alternate or anisogeminate. Flowers in fascicles of 2–6. Pedicels obconically thickened at apex; calyx cup-shaped, truncate; corolla broadly campanulate, divided almost to middle or to 2/3 into reflexed segments; stamens basally broad, inserted usually in upper half of corolla; anthers basifixed and versatile; nectary hardly visible; stigma capitate. Berry juicy, red, shining, pericarp lacking stone cells, calyx persistent, hardly accrescent. Seeds ca. 100, suborbiculate, laterally compressed; embryo annular; endosperm rather abundant. $n = 24$.

One species from southern and eastern Asia: *Tubocapsicum anomalum* (Franch. & Sav.) Makino.

41. ***Athenaea*** Sendtn.

Athenaea Sendtn. in Mart., Fl. bras. 10 (fasc. 6): 133 (1846), nom. cons.; Barboza & Hunziker, Bol. Soc. Argent. Bot. 26(1–2): 91–105 (1989), rev.

Shrubs with conspicuous non-glandular or viscid-glandular indumentum. Fascicles of up to 6 flowers, rarely solitary; calyx deeply divided, segments equal or unequal, rarely cordiform; corolla rotate, segments 2–4 times longer than the very short tube, white and with green or red spots inside; stapet prominent with two short lateral auricles; anthers dorsifixed; nectary annular; styles heteromorphic, stigma discoidal-depressed. Berry many-seeded, surrounded completely or in part by the accrescent calyx, pericarp glabrous to pubescent, lacking stone cells. Seeds subreniform, discoid; embryo slightly curved, cotyledons much shorter than rest of embryo; endosperm abundant.

Seven spp., south-eastern Brazil.

42. ***Aureliana*** Sendtn.

Aureliana Sendtn. in Mart., Fl. bras. 10 (fasc. 6): 138 (1846); Hunziker & Barboza, Darwiniana 30 (1–4): 95–113 (1990), rev.

Shrubs or small trees. Leaves geminate or alternate. Flowers axillary, usually in up to 26-flowered fascicles; calyx urceolate in bud, later splitting into 5 unequal and short lobes, or not urceolate and lobules equal; corolla rotate, segments mostly with greenish spots inside; conspicuous stapet bidenticulate, anthers dorsifixed; nectary annular; style heteromorphic, stigma discoidal-depressed. Berry many-seeded, calyx not accrescent, pericarp and placenta with parenchymatic proliferations, pericarp with few small stone cells. Seeds reniform, discoid; embryo subcoiled, cotyledons much shorter than rest of embryo; endosperm abundant. $2n = 24$.

Eight spp. in eastern Brazil, Paraguay, Peru, Bolivia, and north-eastern Argentina.

43. ***Withania*** Pauquy

Withania Pauquy, Belladone: 14, f. 4 a–d (1825), nom. cons.; Hepper, in Hawkes et al. (eds.) Solanaceae III. Taxonomy, Chemistry, Evolution: 211–227 (1991), part. rev.; Zhang, Lu & D'Arcy, Fl. China 17: 309–310 (1994), reg. rev.
Mellisia Hook. f. (1867–1871).
Physaliastrum Makino (1914).
Archiphysalis Kuang (1966).

Shrubs or geophytes. Leaves sometimes opposite and frequently anisogeminate. Flowers usually in groups of 2–6, perfect or functionally unisexual, pendant; calyx segments equal or very unequal; corolla broadly campanulate or somewhat infundibuliform, with dendritic trichomes inside, shortly 5-lobulate and with or without folds; perfect flowers: stamens 5, connivent, equal, usually included, filaments generally expanded into a stapet with 2 inconspicuous lateral extensions fused to the corolla; anthers ventrifixed; nectary annular, somewhat protruding; style articulate; pistillate flowers: androecium rudimentary; staminate flowers: small ovary lacking style and stigma. Berry tightly or loosely enclosed by the accrescent urceolate and sometimes rough and muricate calyx. Seeds up to 150, reniform or discoid; embryo strongly bent. $n = 12, 24, 36$; $2n = 24, 48, 96$.

Almost 20 Old World spp. The taxonomy and phylogeny of the genus need clarification.

IV.5.b. **Subtribe Iochrominae** Reveal (2012).

Corolla tubular, funnel-shaped, salver-form or campanulate, aestivation valvate; stapet usually auriculate, anthers mostly dorsifixed. Berry with stone cells; calyx usually not accrescent.

44. ***Dunalia*** Kunth

Dunalia Kunth in Humb., Bonpl. & Kunth, Nov. gen. sp. 3: 43, tab. 194 (1818), nom. cons.; Hunziker, Bol. Acad. Nac. Ci. Republ. Argent. 41(2): 211–244 (1960), rev.

Shrubs or small trees. Leaves large to small. Inflorescences axillary, sessile, usually 1–5-flowered (20–30-flowered in *D. solanacea*). Flowers hermaphrodite or functionally unisexual; corolla hypocrateriform or infundibuliform, 5-lobulate, limb narrow, lobes broader than long; stamens glabrous, usually equal, inserted basally, stapet with two lateral, narrow and free wings; anthers basi-dorsifixed, fertile ones larger and exserted, sterile ones small and included; fertile ovary with long style and exserted stigma, nectary included. Berry globose, calyx not accrescent, stone cells rarely absent. Seeds ca. 30 or more; embryo annular. $n = 12$.

Five spp., Colombia to central Argentina.

45. ***Iochroma*** Benth.

Iochroma Benth., Bot. Reg. 31, tab. 20 (1845), nom. cons.

Shrubs, exceptionally spiny; young stems with simple trichomes. Leaves petiolate, rarely microphyllous. Flowers axillary, pedicellate, solitary or usually in fascicles of 2–13(–18). Perianth 5-lobulate; calyx urceolate in bud, at anthesis campanulate; corolla funnel-shaped, tubular or salver-form, variously coloured, lobes wide, tube with folds; stamens equal or not, inserted at different level of corolla; ovary bicarpellate (rarely 4-carpellate); stigma shortly saddle-shaped. Berry globose to elongated, with many stone cells, calyx usually accrescent. Seeds up to ca. 250; embryo subcoiled or almost hippocrepiform; endosperm abundant. $n = 12$; $2n = 24$.

Around 26 spp. in western South America, reaching north-western Argentina; many are ornamentals due to their coloured long corollas.

A phylogenetic analysis revealed that *Iochroma* species do not form a monophyletic group (22 spp. sampled); two alternatives have been proposed: to transfer some species to *Acnistus*, or to consider *Acnistus* as a monophyletic group within a monophyletic *Iochroma* s.l. (Smith and Baum 2006).

46. ***Saracha*** Ruiz & Pav.

Saracha Ruiz & Pav., Fl. peruv. prodr.: 31, t. 34 (1794); Álvarez, Systematics of *Saracha* (Solanaceae), Master thesis, University of Missouri, St. Louis, USA: 1–172 (1996), rev.

Shrubs or small trees; young stems densely hairy with deciduous brown trichomes (thickened, multiseriate, bifurcate or not, ending in a uniseriate row of cells). Leaves coriaceous or subcoriaceous. Inflorescences terminal or axillary, 1–3- or up to 15-flowered. Flowers pendant; calyx campanulate; corolla tubular-infundibuliform or broadly campanulate; stamens somewhat expanded or forming a stapet when inserted basally to corolla; anthers basifixed; nectary annular; stigma bilobed. Berry globose, usually with thick stone cells or with one or two many-seeded pyrenes, pericarp thick, coriaceous, calyx not accrescent. Seeds up to 35, embryo strongly bent; endosperm rather abundant. $2n = 24$.

Two spp. from Venezuela to Bolivia, at 2500–3800 m.

Although this genus is morphologically well-defined, its monophyly is still uncertain (Smith and Baum 2006).

47. ***Vassobia*** Rusby

Vassobia Rusby, Bull. New York Bot. Gard. 4: 422 (1927); Hunziker, Kurtziana 17: 91–118 (1984), rev.

Shrubs or small trees, with cauline spines. Leaves large. Inflorescence terminal, a cymose corymb, ca. 50-flowered, or axillary, fasciculate, ca. 8-flowered. Flowers inodorous, 5-merous; calyx urceolate in bud, irregularly lobed at anthesis; corolla rotate or campanulate-rotate, tube with 5 plaits and glabrous inside; stamens equal, stapet auriculate; anthers basi- or dorsi-basifixed; nectary included; stigma bilobate. Berry juicy, pericarp coalescent with placenta, stone cells many, calyx scarcely accrescent. Seeds discoid; embryo annular. $n = 12$.

Two spp., Bolivia, eastern Paraguay, southern Brazil, northern Argentina and north-western Uruguay.

48. ***Eriolarynx*** (Hunz.) Hunz.

Eriolarynx (Hunz.) Hunz., Kurtziana 28: 66, f. 1 (2000). *Vassobia* Rusby sect. *Eriolarynx* Hunziker, Kurtziana 10: 24 (1977), ibid. 17: 91–118 (1984), rev.

Small shrubs or trees. Leaves glabrous to pubescent. Inflorescences axillary, sessile, fasciculate, up to 4-flowered or flowers solitary, pendant. Calyx urceolate in bud, later campanulate, 5-apiculate or shortly 5-lobed; corolla campanulate-rotate or subinfundibuliform-rotate, with a ring of hairs inside, lobes 5, short, aestivation valvate-plicate; stamens unequal, glabrous, stapet with incipient auricles; anthers basi-dorsifixed; nectary slightly protruding; stigma shortly saddle-shaped. Berry juicy, pericarp coalescent with the placenta, stone cells abundant, calyx accrescent. Embryo slightly curved or annular, cotyledons as long as rest of embryo. $n = 12$.

Three spp., one in the Bolivian Andes, the others in north-western Argentina.

Eriolarynx might be monophyletic after inclusion of *Iochroma australe* Griseb. (Smith and Baum 2006).

49. ***Acnistus*** Schott Fig. 65

Acnistus Schott, Wiener Z. Kunst. 1829(4): 1180 (1829); Hunziker, Kurtziana 15: 81–102 (1982), rev.

Shrubs or small evergreen trees; indumentum of simple and branched trichomes. Leaves entire, decurrent. Inflorescences axillary, many-flowered fascicles. Flowers small, fragrant, pedicellate, 5-merous; calyx urceolate in bud, torn into somewhat unequal lobules at anthesis; corolla campanulate-infundibuliform; stamens inserted in lower half of corolla; anthers dorsifixed, exserted; nectary annular, prominent; stigma broad, discoidal-capitate. Berry many-seeded, juicy, stone cells rather scanty; calyx not accrescent. Seeds subreniform; embryo strongly bent. $n = 12$; $2n = 24$.

A single species, *A. arborescens* (L.) Schltdl., southern Mexico to eastern Brazil and Paraguay. A redefinition of *Acnistus* will be necessary if some *Iochroma* spp. are transferred to this genus (Smith and Baum 2006).

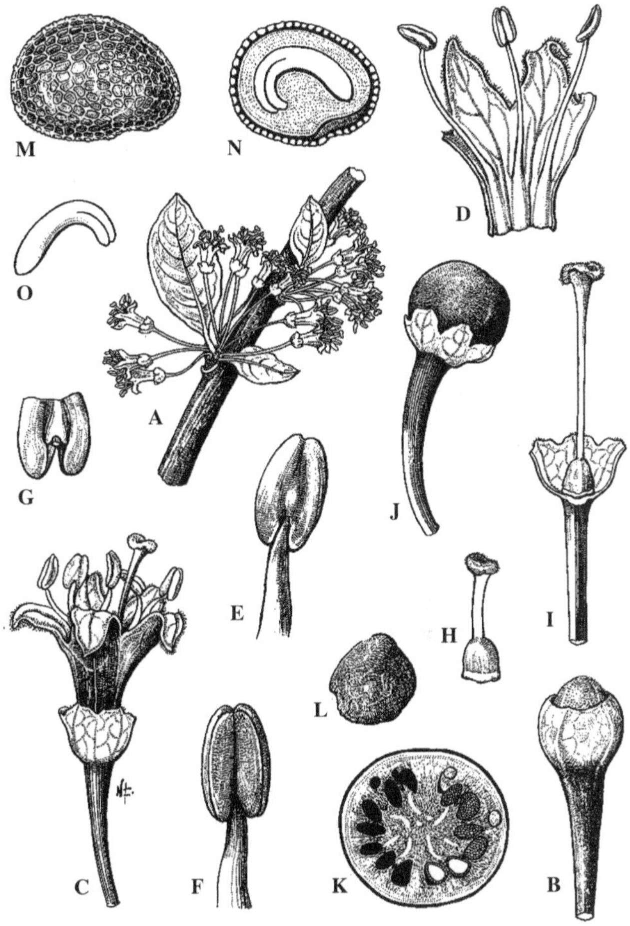

Fig. 65. Solanaceae. *Acnistus arborescens*. **A** Part of branch with flowering brachyblast. **B** Flower bud showing entire calyx. **C** Flower. **D** Inside view of a corolla sector. **E** Anther, dorsal view. **F** Anther, ventral view. **G** Dorsal view of anther base, showing insertion of filament. **H** Gynoecium from a bud. **I** Gynoecium from a flower at anthesis. **J** Fruit. **K** Transverse section of fruit. **L** Stone cell. **M** Seed. **N** Seed, longitudinal section. **O** Embryo. (Hunziker 2001; drawn by N. Flury)

IV.5.c. **Subtribe Physalidinae** Reveal (2012).

Corolla with an internal ring of trichomes, aestivation valvate or contorted. Calyx normally accrescent enveloping partly or entirely the berry.

50. ***Witheringia*** L'Hérit.

Witheringia L'Hérit., Sert. angl.: 1, 33, tab. I (1788); Hunziker, Kurtziana 5: 101–153 (1969), rev.; Sousa-Peña, M.,

Systematics and reproductive biology of the genus *Witheringia* L'Her. (Solanaceae), PhD thesis, University of Connecticut: 1–197 (2001), rev.

Herbs or shrubs; stem glabrous or with simple or branched trichomes. Leaves geminate, unequal in size. Inflorescences axillary, many-flowered, peduncle short or long. Perianth 4–5-merous; calyx cyathiform, truncate; corolla in general 5-parted, tube shorter than elongated lobes, usually with an internal ring of trichomes near stamen insertion, aestivation valvate; stamens equal or not, inserted in upper half of corolla; anthers dorsifixed or dorsi-basifixed; nectary annular; stigma discoidal-depressed or subglobose. Berry red or yellow, stone cells 2–10, calyx sometimes slightly accrescent. Seeds reniform or subreniform; embryo strongly bent. $n = 12$; $2n = 24$.

About 11 spp., southern Mexico to Bolivia.

51. ***Brachistus*** Miers

Brachistus Miers, Ann. Mag. Nat. Hist. II, 3 (16): 262 (1849); Hunziker, Kurtziana 5: 153–163 (1969), rev.

Shrubs or small trees, frequently anisophyllous. Inflorescences axillary, many-flowered fascicles, or flowers solitary, small, odourless. Calyx margin 5-angulate or with 5 triangular teeth; corolla with a campanulate tube and an inner ring of trichomes, segments longer than broad, aestivation valvate; stamens inserted at middle of corolla, stapet not auriculate; anthers dorsi-basifixed; nectary annular. Berry red or orange, many-seeded, calyx strongly or slightly accrescent, pericarp fleshy, stone cells few; embryo annular.

Three spp. in Mexico and Central America. *Brachistus* is a poorly known genus.

52. ***Leucophysalis*** Rydb.

Leucophysalis Rydb., Mem. Torrey Bot. Club 4: 365 (1896).

Annual or perennial herbs, pubescent, anisophyllous. Inflorescences axillary, 1–2-flowered. Flowers showy, pedicels reflexed to pendant; calyx 5-parted, segments unequal; corolla rotate or shallowly bell-shaped, 5-angled or faintly 5-lobuled, inside with hairy ring, aestivation valvate-plicate; stamens equal or unequal, inserted basally; anthers basifixed or dorsi-basifixed; nectary annular; stigma very shortly saddle-shaped. Berry green, ovoid or globose, pericarp with abundant stone cells, calyx accrescent, tightly or loosely appressed to berry, partially or completely enclosing it. Seeds many, discoid or reniform; embryo subcoiled or annular. $n = 12$; $2n = 24$.

Two spp., Canada and USA.

53. ***Calliphysalis*** Whitson

Calliphysalis Whitson, Rhodora 114: 137 (2012).

Erect perennial herbs with taproot, viscid pubescent. Leaves alternate and geminate, long petiolate. Inflorescence axillary in fascicle of 2–6 flowers. Flowers pendant, actinomorphic, 5-merous. Calyx campanulate, shallowly 5-lobed; corolla broadly campanulate, with five patches of olive-green to mustard-yellow spots at the interior base; stamens inserted at the corolla base, its maturation staggered; anthers basifixed; ovary with many ovules, nectary annular, stigma capitate. Berry mustard-yellow, entirely hidden by the non-angled and scarcely papery calyx. Seeds discoid to reniform; stone cells numerous. $n = 12$.

Monotypic, *C. carpenteri* (Riddell) Whitson, south-eastern USA.

54. ***Oryctes*** S. Watson

Oryctes S. Watson, Botany (fortieth parallel): 274, pl. 28, f. 5–10 (1871).

Very small viscid-pruinose annuals. Leaves entire, slightly undulate. Flowers in fascicles of 2–5 in axils, on short peduncles. Calyx zygomorphic, campanulate, with unequal segments; corolla tubular, purple, lobes short, deltoid, slightly unequal, as long as broad, aestivation valvate-conduplicate; androecium heteromorphic: 2 longest stamens exserted or equalling the corolla, filaments inserted basally; anthers basifixed, thecae slightly unequal; nectary included; stigma discoidal-depressed. Capsule almost as long as accrescent, faintly veined calyx. Seeds strongly compressed, with a hyaline peripheral wing; embryo coiled, cotyledons incumbent; endosperm scanty.

Monotypic, *O. nevadensis* S. Watson, western North America.

55. ***Quincula*** Raf.

Quincula Raf., Atl. J. 1: 145 (1832); Barboza, Kurtziana 28 (1): 69–79 (2000), rev.

Herbs with dense indumentum of water-containing, vesicular, whitish trichomes. Leaves pinnatifid or pinnatipartite. Flowers axillary, up to five. Calyx campanulate; corolla rotate, lobes inconspicuous, inside with five pads of branched trichomes; filaments inserted basally, anthers basiventrifixed; nectary inconspicuous; style curved. Berry with few seeds, completely enveloped by the inflated calyx with invaginated base, pericarp thin, thickness uneven, easily breaking into irregular pieces. Seed margin clearly irregular or somewhat crenate; embryo coiled, endosperm abundant. $n = 11, 12$; $2n = 20, 22, 44$.

One species, *Q. lobata* (Torrey) Raf., southwestern USA to northern Mexico.

56. ***Chamaesaracha*** (A. Gray) Benth. & Hook.

Chamaesaracha (A. Gray) Benth. & Hook., Gen. pl. 2: 891 (1876); Averett, Rhodora 75: 325–365 (1973), rev.

Perennial ascending or spreading herbs, glabrous to variously pubescent. Leaves variously shaped. Flowers axillary 1–2, 5-merous. Calyx lobed; corolla rotate, slightly lobed, with white tomentose pads in the throat, aestivation contorted-plicate; stamens inserted basally; anthers dorsifixed; nectary strongly developed; stigma bilobed. Fruit a dry, green berry, tightly and partially enclosed by the accrescent calyx, pericarp thin. Seeds up to 40, testa rugose-reticulate; embryo subcoiled. $n = 12, 18, 24, 36$.

Small, mainly Mesoamerican genus with ca. 10 spp.

57. ***Physalis*** L.

Physalis L., Sp. pl. 1: 182 (1753); Benítez de Rojas & Magallanes, Acta Bot. Venez. 21(2): 11–42 (1998), reg. rev.; Martínez, Anales Inst. Biol. Univ. Nac. Autón. México, Bot. 69(2): 71–117 (1998), part. rev.
Margaranthus Schltdl. (1838).

Herbs, rarely shrubs; taproot thick or roots gemmiferous; stems occasionally viscid; indumentum varied, of simple or branched, non-glandular or glandular trichomes. Leaves entire or pinnatifid. Flowers solitary, pendant. Calyx invaginate; corolla rotate or campanulate-rotate, rarely urceolate, aestivation contorted-plicate, in general with a dense hairy ring inside; anthers basifixed, connective broad; nectary included; stigma shortly saddle-shaped. Berry many-seeded, completely enclosed by inflated calyx, reticulate-veined, strongly urceolate. Seeds enveloped by parenchymatic cells of placental and pericarpic origin; embryo subcoiled or coiled. $n = 12, 24, 36$; $2n = 24, 48, 72$.

About 90 spp., mostly American (Mexico, USA and Central America, reaching South America), except the Eurasian *P. alkekengi* L.

A recent molecular study (Whitson and Manos 2005) revealed paraphyly of *Physalis* in its current circumscription which calls for taxonomic changes to correct the paraphyly.

IV.5.d. **Physalideae: Incertae sedis**

58. ***Cuatresia*** Hunz.

Cuatresia Hunz., Kurtziana 10: 15, f. 4 (1977); Hunziker, Opera Botanica 92: 73–82 (1987), rev.

Shrubs or small trees. Leaves geminate, strongly anisophyllous. Flowers 2–45, in sessile or subsessile axillary inflorescences or on long peduncle; calyx cup-shaped or urceolate, margin almost entire at anthesis, or obscurely divided at fruiting stage; corolla usually glabrous within, campanulate or infundibuliform, 5-lobed or 5-parted, aestivation valvate-induplicate; stamens equal; anthers ventrifixed; nectary included or slightly prominent, stigma shortly saddle-shaped. Berry juicy, pericarp lacking stone cells, calyx not accrescent or investing the entire berry. Seeds reniform, discoid; embryo strongly curved; endosperm very abundant.

About 15 spp., Guatemala to eastern Bolivia.

59. ***Larnax*** Miers

Larnax Miers, Ann. Mag. Nat. Hist., ser. II, 4: 37 (1849); Sawyer, Monogr. Syst. Bot. 104: 259–285 (2005); Leiva & Barboza, Arnaldoa 16: 29–36 (2009), key.

Herbs, shrubs or small trees. Inflorescence axillary, fascicles up to 30-flowered; calyx shortly apiculate or toothed, exceptionally with long segments; corolla mostly stellate, variously coloured, lobes 2–4 times longer than tube, aestivation valvate; androecium heterodynamous or homodynamous, filaments broadened at their insertion into a conspicuous or inconspicuous stapet with or without auricles; anthers basifixed; nectary conspicuous; stigma discoid, saddle-shaped or bilobed. Berry completely enveloped by the accrescent calyx, calyx tightly appressed to berry or inflated and loosely surrounding it. Seeds up to ca. 150, reniform; embryo strongly bent, cotyledons sometimes very short.

About 32 spp., Central America to Peru. A taxonomic revision is needed.

60. ***Deprea*** Raf.

Deprea Raf., Sylva tellur.: 57 (1838); Barboza & Hunziker, Kurtziana 23: 45–65 (1994), rev., Sawyer, Monogr. Syst. Bot. 104: 259–285 (2005).

Herbs or shrubs. Leaves rarely geminate. Flowers generally in fascicles of 1–7, perfect or exceptionally functionally unisexual. Calyx tube 2.5–3 times longer than unequal lobes; corolla narrowly campanulate, funnel-shaped or urceolate, lobes always shorter than tube, aestivation valvate; filament base almost always with an inconspicuous stapet; anthers basifixed; nectary conspicuous; stigma discoid or saddle-shaped. Berry subglobose or ellipsoidal, loosely or tightly enveloped by the strongly accrescent calyx. Seeds subreniform; embryo strongly bent, cotyledons shorter than rest of embryo.

Ten spp., Venezuela to Bolivia.

61. ***Schraderanthus*** Averett

Schraderanthus Averett, Phytologia 91: 54 (2009).

Erect herbs or soft-wooded shrubs, viscid, glandular-pubescent throughout. Leaves geminate, anisophyllous. Inflorescences axillary, fasciculate, 6–8(–10)-flowered. Calyx deeply 5-parted, exceeding length of corolla, segments acute; corolla rotate and reflexed, 5-lobed to parted, inside with a hairy ring and with greenish spots in the throat, aestivation valvate-plicate; stamens unequal, exserted, attached basally to corolla, anthers bluish, dorsi-basifixed. Berry globose, fleshy, red to orange-red, calyx accrescent, broadly campanulate, exceeding berry but becoming reflexed at maturity. Seeds ca. 50–75, reniform; embryo subcoiled, endosperm abundant. $2n = 24$ (sub nom. *Leucophysalis viscosa*).

Monotypic, *S. viscosus* (Schrad.) Averett, Mexico, Guatemala, Venezuela.

62. ***Tzeltalia*** E. Estrada & M. Martínez

Tzeltalia E. Estrada & M. Martínez, Brittonia 50: 289 (1998).

Weak shrubs; stem fistulose, glabrous to densely villose. Leaves alternate or geminate, petiole winged. Inflorescence axillary (1)2–7-flowered fascicles. Calyx thin-chartaceous, glabrous at anthesis, truncate or lobed; corolla rotate, whitish or cream-white usually with 5 purplish-green or olive green spots near the base of the lobes, inside throat and base of lobes with 5 more or less confluent pads of branched trichomes surrounding and above the filaments insertion, lobes 2–3 times longer than tube; filaments blue or purplish; anthers subdorsifixed. Berry greenish, calyx accrescent, open at mouth and not invaginate at base, strongly 5- to 10-angled and strongly reticulate-veined. Seeds reniform, alveolate.

Three spp., southern Mexico and Guatemala.

63. ***Capsicophysalis*** (Bitter) Averett & M. Martínez

Capsicophysalis (Bitter) Averett & M. Martínez, J. Bot. Res. Inst. Texas 38(1): 72 (2009).

Annual or weak perennial herbs. Leaves petiolate, thin to membranous. Flowers axillary 1–2. Calyx campanulate, rounded at the base; corolla campanulate-subrotate, unequally 5-lobed, irregular, white, yellowish or yellow-green with villous pads in the throat; filaments equal, anthers basifixed. Berry fleshy, shiny red or orange-red, calyx accrescent, at first loosely investing the berry, then splitting and becoming reflexed below the berry. Seeds discoidal, testa tuberculate with rod-like projections.

One species, *C. potosina* (B.L. Rob. & Greenm.) Averett & M. Martínez, southern Mexico to Guatemala and Honduras.

IV.6. **Tribe Capsiceae** Dumort. (1827).

Shrubs, trees, or perennial herbs. Flowers solitary or in fascicle pauciflowered. Calyx entire, truncate, with 0–10 linear nerve prolongations (teeth); corolla rotate or stellate, rarely campanulate, aestivation valvate. Calyx not accrescent. Endosperm copious.

64. ***Lycianthes*** (Dunal) Hassl.

Lycianthes (Dunal) Hassl., Ann. Conserv. Jard. Bot. Genève 20: 180 (1917), nom. cons.; Bitter, Abh. Naturwiss. Verein Bremen 24(2): 292–520 (1919), rev.; Barboza & Hunziker, Darwiniana 31(1–4): 17–34 (1992), reg. rev.; Benítez de Rojas & D'Arcy, Ann. Missouri Bot. Gard. 84: 167–200 (1997), reg. rev.; Dean, Bot. J. Linnean Soc. 145: 385–424 (2004), part. rev.

Shrubs, trees, vines or perennial herbs. Leaves various. Flowers axillary or extra-axillary; calyx with (0)5–10 linear teeth; corolla rotate, apically 5-angled or deeply divided; androecium usually heterodynamous, filaments fused in a ring adnate to the basal edge of corolla; anthers basifixed, tightly coherent or connate in a column, dehiscing with a terminal pore; nectary absent; stigma discoid or saddle-shaped. Berry sometimes with stone cells or occasionally with 1–2-seeded pyrenes. Seeds discoid or subreniform; embryo coiled, cotyledons as long as rest of embryo. $n = 12$; $2n = 24$.

American-Asian genus of ca. 150 spp. or more, mostly neotropical.

65. ***Capsicum*** L.

Capsicum L., Sp. pl.1: 188 (1753); Hunziker, Darwiniana 36: 201–203 (1998), reg. rev.; Barboza et al., Syst. Bot. 36 (3): 768–781 (2011), key.

Shrubs, herbs or trees. Leaves entire, alternate or geminate. Flowers axillary, solitary or in fascicles of 2–9 (18), pedicels usually geniculate distally. Calyx in general with (0)5–10 linear-compressed teeth; corolla rotate, stellate, campanulate, or campanulate-urceolate, lobes usually shorter than rest of corolla, aestivation valvate, tube with plaits; stamens usually equal, filament basis with two lateral, short and thick auricles fused to the corolla (except *C. chacoense*); anthers dorsifixed; nectary included; style sometimes heteromorphic, stigma bilobed. Berry usually with capsaicinoids in the epidermis of the septum and adjacent areas and consequently hot. Seeds many; embryo annular. $n = 12, 13$; $2n = 24, 26$.

About 32 spp., southern USA to central Argentina. The cultivated peppers belong to five spp., the most important is *C. annuum* L. with many cultivars.

IV.7. **Tribe Solaneae** Dumort. (1829).

Herbs, shrubs or trees. Corolla rotate, campanulate, stellate, infundibuliform, tubular, urceolate; aestivation valvate; stamen usually equal in length, markedly enlarged at base where adnate to base of corolla; anthers dehiscing longitudinally or by pores then expanding into a longitudinal slit; stigma capitate or bilobed; calyx slightly or strongly accrescent.

66. ***Jaltomata*** Schltdl.

Jaltomata Schltdl., IndexSem. (Halle): 8 (1838); Mione, Anderson & Nee, Brittonia 45: 138–145 (1993), part. rev.; Mione, Leiva G. & Yacher, Taxon 17: 49–58 (2007).
Hebecladus Miers (1845).

Perennial herbs or shrubs. Leaves entire, repand or toothed. Inflorescence axillary or terminal dichasia or umbellate, pedunculate; flowers solitary, 2–3 or up to 12–41, pendant, usually protogynous; pedicels basally articulated. Calyx 5-lobed; corolla rotate, campanulate, urceolate, crateriform or tubular, 5-lobed, or lobes alternating with lobules totalling 10, aestivation valvate or valvate-induplicate; androecium usually homomorphic, filaments enlarged at insertion, stapet usually swollen and hairy; anthers ventrifixed; nectary well developed, nectar orange, yellowish, or red. Berry frequently edible, many-seeded, calyx strongly accrescent, rotate to reflexed, never enclosing the berry. Seeds small; embryo strongly bent; endosperm abundant. $n = 12$; $2n = 24, 48$.

Ca. 60 spp. in two centres of diversity: southwestern USA and Mexico (*J. antillana* (Krug & Urban) D'Arcy in Greater Antilles), and western South America from Venezuela and Colombia to Bolivia (*J. werfii* D'Arcy in Galápagos).

67. *Solanum* L.

Solanum L., Sp. pl. 1: 184 (1753); Hawkes & Hjerting, The potatoes of Argentina, Brazil, Paraguay and Uruguay: 1–525 (1969), reg. rev.; Morton, A revision of the Argentine species of *Solanum*: 1–260 (1976), reg. rev.; Whalen, Gentes Herb. 12: 179–282 (1984); Francisco-Ortega et al., Pl. Syst. Evol. 185: 189–205 (1993), rev.; Bohs, Fl. Neotrop. Monogr. 63: 1–175 (1994), part. rev.; Knapp, Fl. Neotrop. Monogr. 84: 1–404 (2002), part. rev.; Spooner et al., Syst. Bot. Monogr. 68: 1–209 (2004), reg. rev.; Peralta, Spooner & Knapp, Syst. Bot. Monogr. 84: 1–186 (2008), part. rev.
Lycopersicon Mill. (1754).
Melongena Mill. (1754).
Bassovia Aubl. (1775).
Triguera Cav. (1786), nom. cons.
Androcera Nutt. (1818).
Cyphomandra Sendtn. (1845).
Normania Lowe (1872).
Pheliandra Werderm. (1940).

Herbs to trees, rarely vines, sometimes tuber-bearing, glabrous or with a profuse variety of trichomes, sometimes prickly. Leaves alternate or paired, frequently unequal, simple or compound, sometimes with pseudostipules. Inflorescences terminal, axillary or extra-axillary, cymes branched or unbranched, in racemose, paniculate or scorpioid structures. Flowers sessile or on pedicels, articulated or not, actinomorphic or zygomorphic, usually perfect, rarely unisexual, rarely dioecious. Perianth 4–5-merous, occasionally 4–6(10)-merous; calyx cyathiform or variously divided; corolla rotate, campanulate-urceolate, stellate, infundibuliform or urceolate, aestivation valvate or valvate-induplicate; stamens equal or unequal, filaments basally fused in a ring adnate to corolla tube, sometimes also into an upper ring free from the corolla; anthers equal or unequal, connivent or not, basifixed to basi-ventrifixed, exclusively poricidal or with a terminal pore expanding into a longitudinal slit, or anthers opening introrsely with longitudinal slits, sometimes with a thick glandular excrescence at the base or with a prolonged and sterile apical sector; nectary absent. Fruit a berry, sometimes dry and dehiscent, stone cells usually present, calyx slightly or strongly accrescent. Seeds often flattened; embryo coiled, endosperm abundant. $n = 11, 15$; $2n = 24, 48, 72, 96$.

Cosmopolitan; ca. 1400 spp., mainly American. Genus of worldwide economic importance (see section "Economic Importance").

Phylogenetic studies in this genus have included increasingly more species in recent years. Currently, 13 major clades are recognized (Bohs 2005; Weese and Bohs 2007; Stern et al. 2011).

IV.8. Solanoideae: Incertae sedis

68. *Latua* Phil.

Latua Phil., Bot. Zeitung (Berlin) 16 (33): 241 (1858).

Shrubs or trees with heteroblastic growth and cauline spines. Leaves fasciculate on short shoots or scattered on long shoots. Flowers solitary, pedicels tomentose, arising in the axil of a spine from an involucre of scales; calyx campanulate, lobes equal; corolla actinomorphic, densely pilose throughout, aestivation cochlear-conduplicate; stamens 5, filaments unequal, adnate basally to corolla tube; anthers basifixed; nectary annular. Berry globose, yellow or orange-yellow, calyx accrescent, splitting irregularly. Seeds numerous, thick; embryo bent, cotyledons shorter than rest of embryo; endosperm abundant. $2n = 24$.

A single species, *L. pubiflora* (Griseb.) Baill., coastal mountains of southern Chile.

69. *Jaborosa* Juss.

Jaborosa Juss., Gen. pl.: 125 (1789); Barboza & Hunziker, Kurtziana 19: 77–153 (1987), rev.
Trechonaetes Miers (1845).

Perennial herbs, frequently with gemmiferous roots and rhizomes. Leaves entire to pinnatisect. Flowers in 1–7-flowered fascicules or in dense glomerules (ca. 100 or more flowers). Calyx campanulate, 5-lobed; corolla tube cylindrical, rotate, campanulate-rotate, urceolate or hypocrateriform, pubescent inside, aestivation valvate-induplicate or valvate-conduplicate; stamens 5, usually included, inserted at different levels on corolla, bent towards the centre of the flower, tip inflexed; ovary bicarpellate, rarely 3–5-carpellate, nectary annular; stigma lobulate or branched. Berry globose-depressed, calyx accrescent. Seeds up to ca. 90; embryo annular; endosperm copious. $n = 12$; $2n = 24$.

22 spp., southern Peru to Argentina.

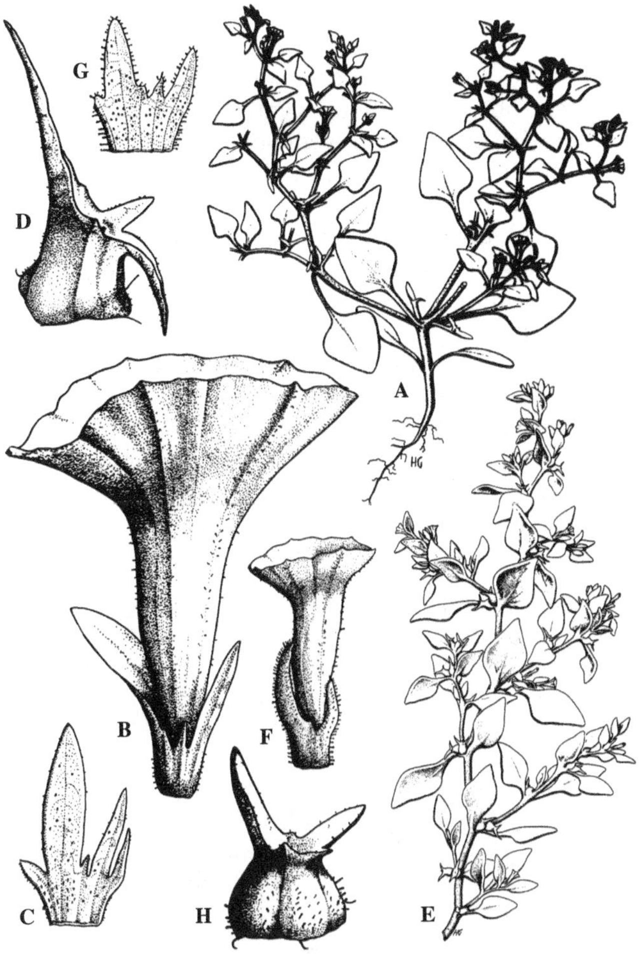

Fig. 66. Solanaceae. **A–D** *Sclerophylax arnottii*. **A** Flowering branch. **B** Flower. **C** Calyx. **D** Fruiting calyx. **E–H** *Sclerophylax ruiz-lealii*. **E** Flowering branch. **F** Flower. **G** Calyx. **H** Fruiting calyx. (Di Fulvio 1961; drawn by H. García)

70. ***Sclerophylax*** Miers Fig. 66

Sclerophylax Miers, London J. Bot. 7: 18 (1848); Di Fulvio, Kurtziana 1: 9–103 (1961), rev.

Branched annual or perennial herbs, indumentum eglandular or glandular or plants subglabrous. Two entire leaves per node, generally not opposite, often anisophyllous. Flowers solitary (rarely paired) in leaf axils, 5-merous; sepals united at base, unequal, accrescent with lobes becoming more or less thorny in fruit; corolla white or violet, infundibuliform, slightly zygomorphic, deciduous, lobes inconspicuous, aestivation contorted-conduplicate; stamens included; anthers dorsifixed; ovary bicarpellate, bilocular; placentation apical, ovules apotropous, pendant, 1–2 per locule; style terminal, stigma simple. Fruit dry, indehiscent, generally one seed per locule, pericarp membranous included in the accrescent and sclerified calyx (diclesium). Seeds with aleuron, exarillate; embryo straight or curved; endosperm oily. $n = 12$; $2n = 24$.

14 spp. mainly in Argentina, also in Uruguay and Paraguay.

71. ***Nolana*** L.f. Fig. 67

Nolana L.f., Dec. pl. horti upsal.1: 3 (1762).

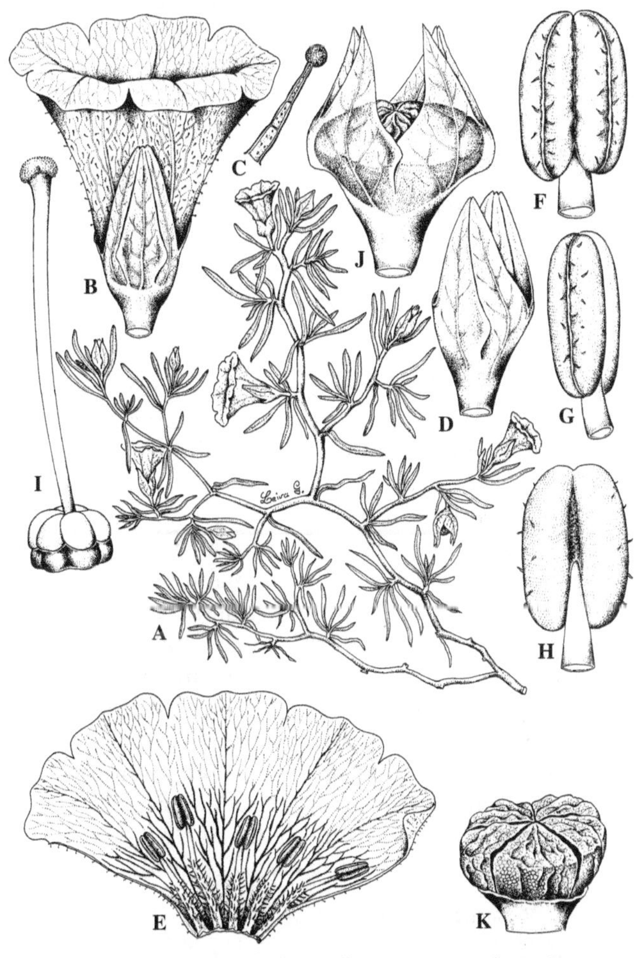

Fig. 67. Solanaceae. *Nolana lezamae*. **A** Flowering branch. **B** Flower. **C** Floral trichome. **D** Calyx with connate lobes. **E** Dissected corolla. **F** Anther, ventral view. **G** Anther, lateral view. **H** Anther, dorsal view. **I** Gynoecium. **J** Calyx expanding with mericarps. **K** Mericarps. (Drawn by S. Leiva García)

Annuals to shrubs, sometimes with basal rosettes from taproots. Leaves succulent, terete to laminar. Flowers 5-merous; calyx regular, actinomorphic, to bilabiate, zygomorphic, enclosing the ovary; corolla regular, or more commonly irregular to weakly zygomorphic, infundibuliform, campanulate or tubular, more rarely suburceolate or salver-form; stamens adnate basally to the corolla tube, often pubescent, unequal (3 long, 2 short); ovary 5-carpellate. Fruit a schizocarp, mericarps 2–30, highly variable in shape, from polygons to spheres, and external sculpturing smooth to rugulose. Seeds with copious oily endosperm. $2n = 24$.

89 species in Chile, Peru and one in the Galapagos Islands (Ecuador). The genus has been traditionally divided into subgenera *Alona* and *Nolana*; however, recent studies have shown that additional monophyletic groups are in need of formal classification (Dillon, pers. obs.).

72. ***Exodeconus*** Raf.

Exodeconus Raf., Sylva tellur.: 57 (1838); Axelius, Pl. Syst. Evol. 193: 153–172 (1994), rev.
Cacabus Bernh. (1839).

Small annual viscid herbs. Leaves entire to sinuate-dentate. Flowers solitary, 5-merous. Calyx tubular, infundibuliform, or cyathiform to urceolate, deeply divided; corolla tubular-campanulate or infundibuliform with a very narrow base, aestivation contorted-induplicate; stamens included, usually heterodynamous, filaments unequal, inserted basally; anthers basifixed; nectary annular, inconspicuous; style terminal, stigma much depressed. Berry sometimes sclerified apically to in upper half, base with sclereids and stone cells, calyx accrescent, innervation very conspicuous. Seeds up to ca. 300; embryo bent, cotyledons shorter than rest of embryo. $2n = 24$.

Six spp. mostly in southern Ecuador, Peru, northern Chile and northern Argentina.

73. ***Nicandra*** Adans.

Nicandra Adans., Fam. pl. 2: 219 (1763), nom. cons.; Carrizo García, Flora Fanerog. Argent. 88: 3–4 (2005), reg. rev.
Physalodes Boehm. (1760), nom. rej.

Annual herbs; stems hollow. Leaves generally sinuate. Flowers extra-axillary, solitary, 5-merous. Calyx aestivation reduplicate-valvate, segments auriculate or cordate at base and fused almost to middle; corolla broadly campanulate, aestivation cochlear-plicate; stamens geniculate, incurved; anthers basifixed, connivent; ovary 3–5-carpellate, nectary included, crenulate; stigma capitate-lobed. Berry pendant, pericarp almost dry, semitranslucent, easily ruptured, stone cells present, calyx strongly accrescent, chartaceous, inflated, with conspicous venation. Seeds discoidal-reniform; embryo coiled, cotyledons shorter than rest of embryo; endosperm abundant. $n = 10$; $2n = 19, 20, 21, 40$.

Three spp., two endemic to Peru and *N. physalodes* (L.) Gaertn. from Peru to northern Argentina; this species is an ornamental and ruderal in tropical and subtropical areas throughout the world.

74. ***Schultesianthus*** Hunz.

Schultesianthus Hunz., Kurtziana 10: 35, f. 9 (1977); Bernardello & Hunziker, in Hawkes et al. (eds.) Solanaceae III. Taxonomy, Chemistry, Evolution: 1–17 (1991), rev.

Climbing epiphytes or scandent shrubs. Leaves coriaceous, variously shaped, abundantly punctate below. Inflorescence terminal cymes, 2–10-flowered, occasionally reduced to a single flower. Flowers showy, fragrant, 5-merous; calyx usually actinomorphic, lobes thick, coriaceous; corolla zygomorphic, funnel-form to strongly infundibuliform, lobes auriculate, aestivation cochlear; stamens declinate or straight, inserted basally and pubescent at point of insertion; anthers basifixed; nectary annular, slightly exserted; style declinate or straight. Berry globose or ovoid; calyx accrescent, woody. Seeds ca. 30, reniform, large; embryo strongly annular, cotyledons accumbent, as broad as long and not wider than rest of embryo; endosperm scarce.

Eight spp., southern Mexico to Bolivia.

Knapp et al. (1997) defined the genus more broadly and included three species recognized under *Markea* by Hunziker (1997, 2001).

75. ***Trianaea*** Planch. & Linden

Trianaea Planch. & Linden in Linden, Cat. pl. exot.: 717 (1853).

Shrubs or lianas, usually epiphytic, sometimes with a woody tuberosity, exceptionally small trees. Leaves thick, shiny. Flowers terminal or axillary, solitary or in groups of 2–3, showy, 5- or 4-merous, pendant. Calyx campanulate; corolla campanulate, large and fleshy, tube short, lobules rounded, reflexed, aestivation cochlear; stamens strongly geniculate at base and attached basally; anthers ventrifixed; ovary 4- or 5-carpellate, nectary prominent; stigma capitate to clavate. Berry large (up to 8 cm diam.), surrounded by the accrescent calyx. Seeds reniform, elongated, thin; embryo curved, cotyledons oblique, longer than broad; endosperm rather abundant.

About six spp., north-western South America from Colombia to northern Peru.

76. ***Solandra*** Sw.

Solandra Sw., Kongl. Vetensk. Acad. Nya Handl. 8: 302, tab. XI (1787), nom. cons.; Bernardello & Hunziker, Nordic J. Bot. 7: 639–652 (1987), rev.

Climbing epiphytic shrubs or lianas or small trees, glabrous or pubescent. Leaves coriaceous or subcoriaceous, lustrous. Flowers solitary, terminal, usually zygomorphic; calyx tubular-campanulate; corolla large, limb short, tube cylindrical basally and expanded apically, aestivation cochlear; stamens equal, declinate, attached near middle of corolla or more basally; anthers basifixed; ovary bicarpellate, tetralocular, ∞-ovuled, partly inferior, nectary annular; style declinate, stigma discoidal-capitate. Berry ∞-seeded. Seeds discoid or reniform, compressed; embryo subcoiled, cotyledons incumbent or slightly oblique, shorter than rest of embryo; endosperm abundant. $n = 12$; $2n = 24$.

10 spp., Mexico and West Indies to Peru, Bolivia, and Brazil; some species are widely cultivated as ornamentals, such as *S. grandiflora* Sw., *S. longiflora* Tussac, and *S. maxima* (Sessé & Mociño) P.S. Green.

77. ***Mandragora*** L.

Mandragora L., Sp. pl. 1: 181 (1753); Ungricht, Knapp & Press, Bull. Nat. Hist. Mus. London, Bot. 28: 17–40 (1998), rev.

Perennial rosette-forming herbs, with fleshy, thick taproots. Leaves rugose-bullate abaxially. Flowers 5(6)-merous; calyx deeply lobed; corolla campanulate, divided to almost half its length or more, aestivation cochlear; stamens homomorphic, incurved towards the centre of the flower and attached basally, filaments with a conspicuous basal tuft of trichomes; anthers dorsifixed, connective thick and fleshy; nectary prominent, two-lobed; stigma capitate. Berry globose, with a pleasant odour at maturity, later odour heavy and disagreeable, calyx accrescent. Seeds up to 50, reniform, large; embryo coiled, cotyledons unequal, longer than rest of embryo; endosperm scanty. $2n = 24, 48, 84, 96$.

Three spp., disjunct in Mediterranean Europe, central Asia, and the Sino-Himalaya region.

78. ***Nectouxia*** Kunth

Nectouxia Kunth in Humb., Bonpl. & Kunth, Nov. gen. sp. 3: 10, tab. 193 (1818).

Hemicryptophytic herbs, somewhat fetid; rhizome fleshy with minute cataphylls; stems glandular-pubescent. Leaves cordate. Flowers solitary, pendant. Calyx divided almost to base into equal segments, densely hirsute; corolla yellow or yellow-green, hypocrateriform, aestivation valvate-induplicate, with a ring-like corona at junction of tube and limb; stamens included or subincluded, inserted distally, thecae dorsally with a laminar enlargement of the filament apex; nectary rather prominent; stigma discoid. Berry usually apiculate. Seeds up to 10; embryo coiled; endosperm rather scanty.

One species, *N. formosa* Kunth, Mexico.

79. ***Salpichroa*** Miers

Salpichroa Miers, London J. Bot. 4: 324 (1845); Keel, PhD thesis, University of New York: 1–196. University Microfilms International, Michigan, USA (1984), rev.; Barboza & Hunziker, Fl. Fanerog. Argent. 54: 19–24 (1998), reg. rev.

Scandent, pendant, prostrate or straggling shrubs. Leaves solitary or geminate. Flowers solitary, rarely in pairs, pendant. Calyx 5-parted almost to base or with a tube as long as segments; corolla cylindrical, urceolate, or urceolate in basal half and cylindrical upwards, aestivation valvate or slightly valvate-induplicate, inside normally glabrous (not *S. origanifolia*); stamens inserted at various levels in upper half of corolla, filament apex usually spathulate; anthers dorsifixed, frequently connivent; nectary prominent; stigma discoid. Berry elongated. Seeds up to 28; embryo subcoiled; endosperm rather scanty. $n = 12$; $2n = 24$.

16 spp., Venezuela to north-western Argentina and Chile; *S. origanifolia* (Lam.) Baillon naturalized in the USA, Europe, North Africa, and Australia.

80. ***Darcyanthus*** Hunz. Fig. 68

Darcyanthus Hunz., Bol. Soc. Argent. Bot. 35: 345 (2000).

Annual sticky herbs, indumentum of long glandular multicellular trichomes. Inflorescences fasciculate, sessile, flowers 7–10, pedicellate, 5-merous. Calyx teeth triangular, unequal, shorter than tube; corolla delicate, rotate or rotate-campanulate, angulate, cream with 5 purple spots within, tube very short; stapet conspicuous, with lateral outgrowths fused to corolla, filaments glabrous, anthers dorsi-basifixed; nectary included, stigma discoidal-capitate. Berry globose, red, pericarp thin, not fleshy, with stone cells, tightly surrounded by accrescent and thin calyx. Seeds small; embryo annular.

Monotypic, *D. spruceanus* (Hunz.) Hunz., oriental rainforests of Peru and Bolivia.

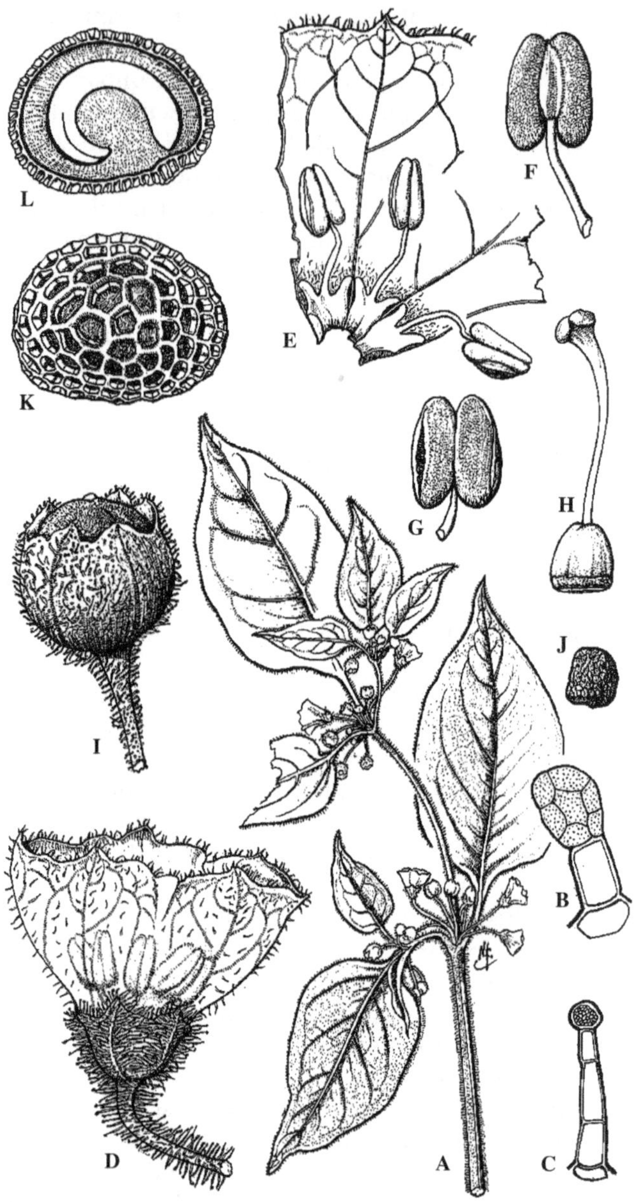

Fig. 68. Solanaceae. *Darcyanthus spruceanus.* **A** Flowering branch. **B, C** Glandular trichomes. **D** Flower. **E** Inside view of a corolla sector. **F** Anther, dorsal view. **G** Anther, ventral view. **H** Gynoecium. **I** Fruit. **J** Stone cell. **K** Seed. **L** Seed, longitudinal section. (Hunziker 2001; drawn by N. Flury)

V. Solanaceae: Incertae sedis

Genera not assigned to subfamily

81. ***Schizanthus*** Ruiz & Pav.

Schizanthus Ruiz & Pav., Fl. peruv. prodr.: 6 (1794); Grau and Gronbach, Mitt. Bot. Staatssamml. München 20: 111–203 (1984), rev.; Hunziker and Cosa, Fl. Fanerog. Argent. 74: 3–6 (2001), reg. rev.

Annual or biennial herbs, sometimes woody at base, usually sticky, with non-glandular unicellular trichomes and glandular shaggy hairs. Leaves entire, slightly serrate, lobed to pinnatisect. Inflorescences terminal. Flowers 5-merous, resupinate (twist of ca. 72° at anthesis); calyx segments slightly unequal, linear or spathulate, tube almost absent; corolla zygomorphic, papilionate (segments unequal, deeply laciniate: one anterior, two lateral and the two posterior ones fused forming a keel), aestivation cochlear; 2 fertile stamens in lateral position, and 3 staminodes; anthers dorsifixed, dehiscing explosively; ovary with annular nectary; style filiform, stigma inconspicuous, lacking papillae. Capsule septicidal, 2-valved. Seeds compressed, up to 40, ellipsoidal or reniform; embryo arcuate, cotyledons incumbent, clearly shorter than rest of embryo; endosperm nuclear, copious. $n = 10$.

A genus of 12 spp., Chile and Argentina (2). Three spp. (*S. pinnatus* Ruiz & Pav., *S. retusus* Hook. and *S. grahamii* Gillies) and a few hybrid strains (*S.* x *wisetonensis* Hort., *S. hybridus* Hort., etc.) are cultivated as ornamentals.

82. ***Duckeodendron*** Kuhlmann

Duckeodendron Kuhlmann, Arch. Jard. Bot. Rio de Janeiro 4: 361, tab. 31 h–l (1925); Martins et al. in Ribeiro et al., Flora da Reserva Ducke 583–587 (1999); Nee in Smith et al., Flowering Plants of the Neotropics: 133–134 (2004).

Trees ca. 30 m high. Leaves entire, coriaceous, pubescent abaxially; petiole canaliculate, with dense ochre indumentum. Inflorescence terminal, cymose, ca. 15-flowered, brownish-pubescent. Flowers subtended by one tiny pilose bract, pedicels articulated; calyx aestivation quincuncial, lobes broad, as long as tube; corolla greenish, tubular, slightly infundibuliform, pubescent inside, lobes rounded, aestivation quincuncial; stamens equal, inserted near middle of corolla; anthers exserted, dorsi-basifixed; ovary bicarpellate, each locule uniovulate; style strongly exserted, stigma minutely bilobed. Drupe 1-seeded. Seed ovoid; embryo U-shaped, cotyledons extremely short; endosperm oily.

Monotypic, central Amazonia (Manaus, Brazil): *D. cestroides* Kuhlmann.

Tribes not assigned to subfamily

Tribe Benthamielleae Hunz. (2000).

Chamaephytes, small shrubs or perennial microphyllous herbs. Leaves sometimes rigid and spinescent. Flowers subtended by two opposite bracteoles. Stamens usually 5 (rarely 1–3), anthers dorsifixed. Capsule pauciseminate. Embryo curved; endosperm copious.

83. ***Pantacantha*** Speg.

Pantacantha Speg., Anales Soc. Ci. Argent. 53: 171 (1902); Rossow, Fl. Patag. 6: 278–279 (1999), rev.; Hunziker, Fl. Fanerog. Argent. 64: 9–10 (2000), rev.

Low shrubs; stems branched, densely leafy. Leaves sessile, sheathing, linear-subulate to narrow-elliptic, rigid, spinescent. Flowers 5-merous, axillary, actinomorphic, solitary, subtended by two leaf-like bracteoles; calyx campanulate or tubular, teeth equal or subequal, rigid, spiny; corolla greenish-yellow or yellow, cylindrical or slightly infundibuliform, aestivation valvate-induplicate; stamens inserted near base of corolla, filaments equal, pubescent; ovary with prominent annular nectary, ovules ca. 20 per locule; stigma discoid, somewhat bilobate. Capsule septicidal or partly loculicidal. Seeds (1)2–3, angulate-subreniform, strongly ridged or winged; embryo curved, cotyledons shorter than rest of embryo. $2n = 22$.

One species, *P. ameghinoi* Speg., western Argentina.

84. ***Combera*** Sandw.

Combera Sandw. in Hook., Icon. pl. 34, t. 3325 (1939); Rossow, Fl. Patag. 6: 235–237 (1999) reg. rev.; Hunziker, Fl. Fanerog. Argent. 64 (9): 10–11 (2000), reg. rev.

Small, erect perennial herbs; stems densely leafy, partly subterranean. Leaves petiolate, sheathless, spathulate to deltate-ovate or rhombic-ovate, fleshy, closely imbricate. Flowers clustered among the uppermost leaves. Calyx actinomorphic to slightly zygomorphic, 5-dentate, with long unequal teeth; corolla actinomorphic, infundibuliform, aestivation contorted-conduplicate; stamens 5, filaments equal, attached at lower half of corolla; ovary with an included annular nectary, ovules 4–10 per locule. Capsule septicidal or partly loculicidal. Seeds up to 7, embryo curved. $2n = 22$.

Two spp., Patagonia (Argentina and Chile).

85. ***Benthamiella*** Speg.

Benthamiella Speg., Anales Soc. Ci. Argent. 15: 109 (1883); Arroyo, Bot. Not. 133: 67–76 (1980), rev.; Arroyo-Leuenberger, Fl. Fanerog. Argent. 64: 4–9 (2000), reg. rev.
Saccardophytum Speg. (1902).

Pulvinate chamaephytes or lax cushions; stems branched, densely leafy. Leaves sessile, sheathed, membranous or fleshy, closely imbricate. Flowers normally at the end of branches. Calyx actinomorphic, campanulate or tubular, with equal or subequal, acute or truncate lobes; corolla whitish or yellow, actinomorphic, infundibuliform or subcampanulate; aestivation contorted-conduplicate; fertile stamens 1, 2 or 5, staminodes sometimes present, filaments equal or not, attached at 1 or 2 levels on the upper or the lower half of corolla tube; ovules 4–10 per locule. Capsule septicidal or partly loculicidal. Seeds 1–4, embryo curved cotyledons shorter than rest of embryo. $2n = 22$.

12 spp., southern Patagonia (Argentina and Chile).

Tribe Petunieae Horan. (1847).

Herbs or shrubs. Corolla commonly zygomorphic, funnel-shaped or salver-form, aestivation cochlear, reciprocative or less frequently contorted; androecium usually heterodynamous, stamens fertile 5 or 4 or 2, staminodes present.

86. ***Fabiana*** Ruiz & Pav.

Fabiana Ruiz & Pav., Fl. peruv. prodr.: 18, pl. 34 (1794); Barboza & Hunziker, Kurtziana 22: 109–153 (1993), rev.

Chamaephytes or microphyllous shrubs, of homoblastic or heteroblastic growth; stems and leaves with dense resiniferous indumentum. Leaves sessile, imbricate, fasciculate or rosulate. Flowers solitary, 5-merous, actinomorphic; calyx lobes shorter than tube; corolla whitish, rarely lilac or bluish, funnel-shaped or salverform, aestivation contorted-conduplicate; stamens homodynamous or heterodynamous (2 + 3), filaments usually geniculate; anthers dorsifixed, thecae confluent or free; ovary with a prominent invaginate-bilobed nectary; stigma capitate or saddle-shaped. Capsule septicidal, 2-valved. Seeds ca. 50, polyhedric, angular; embryo straight or slightly curved. $n = 9$; $2n = 18$.

15 spp., western South America, usually Andes. *Fabiana imbricata* Ruiz & Pav. is an ornamental species and a medicinal plant in Argentina and Chile.

87. ***Calibrachoa*** La Llave & Lex.

Calibrachoa La Llave & Lex., Nov. veg. descr. 2: 3 (1825). *Leptophragma* Benth., in Dunal, De Candolle, Prodr. 13 (1): 578 (1852).

Small shrubs, more rarely annual herbs, generally viscid-pubescent; stem usually with brachyblasts. Leaves variously shaped. Flowers in monochasia, apparently solitary and arising from a pair of subopposite equal bracts; pedicels mostly deflected at fruiting time; calyx usually lobed halfway, lobes narrowing towards apex; corolla slightly zygomorphic, funnel-shaped, aestivation reciprocative (the anterior induplicative lobe covers the four others, which are conduplicate); stamens 2 + 2 + 1, filaments sometimes strongly geniculate; anthers ventrifixed; nectary bilobed. Capsules septicidal, 2-valved. Seeds up to 60, reticulate-foveate with straight thick-walled ridges. $n = 9$; $2n = 18$.

26 spp., south-eastern and southern Brazil, Paraguay, Uruguay, and Argentina, *C. parviflora* (Juss.) D'Arcy disjunct between South and North America.

88. ***Petunia*** Juss.

Petunia Juss., Ann. Mus. Hist. Nat. 2: 215, t. 47 (1803). *Stimoryne* Raf., Fl. Tellur. 3: 76 (1857).

Annual herbs, rarely chamaephytes (*P. patagonica*), viscid-pubescent; brachyblasts absent. Leaves variously shaped. Inflorescence sympodial; flowers in monochasia arising from a pair of subopposite equal leaves. Calyx usually deeply lobed, lobes linear or enlarged towards the apex; corolla actinomorphic or slightly zygomorphic, tube funnel-shaped, salverform or cylindric, aestivation cochlear; stamens usually 2 + 2 + 1; anthers ventrifixed; nectary bilobed. Capsules septicidal, many-seeded, 2-valved, fruiting pedicels usually thin and deflexed. Seeds < 1 mm, reticulate-foveate with wavy thick-walled ridges. $n = 7$; $2n = 14$.

14 spp., Brazil, Bolivia, Uruguay, Paraguay, Argentina. *Petunia* x *hybrida* (Hook.) Vilm., a hybrid between *P. integrifolia* (Hook.) Schinz & Thell. and *P. axillaris* (Lam.) Britton, Stern & Poggenb., is among the most popular annual bedding plants.

89. ***Brunfelsia*** L.

Brunfelsia L., Sp. pl. 1: 191 (1753); Plowman, Fieldiana, Bot. 39: 1–135 (1998), rev.
Franciscea Pohl (1826).

Evergreen shrubs or trees. Leaves entire, petioles glabrous to pilose. Flowers in terminal cymes. Calyx actinomorphic or slightly zygomorphic, lobes subequal; corolla zygomorphic, hypocrateriform, aestivation cochlear, lobes unequal; stamens 4, didynamous, upper lateral pair with long filaments and anthers occasionally scarcely exserted, lower posterior pair shorter, with anthers always included; anthers basifixed and monothecous; ovary bicarpellate, nectary annular; style curved at apex. Capsule bivalved or fruit subbaccate. Seeds 2–40, of different form; embryo straight or slightly curved, cotyledons incumbent, shorter than or as long as rest of embryo. $n = 11$; $2n = 22$.

Around 50 spp., Central America and the Antilles to South America. Some spp. are widely cultivated as garden plants; a few others have medicinal properties, or are hallucinogenic or toxic species.

90. ***Leptoglossis*** Benth.

Leptoglossis Benth., Bot. Voy. Sulphur: 143 (1845); Hunziker & Subils, Bot. Mus. Leafl. 27: 1–43 (1980), rev.

Xerophytic annual or perennial herbs. Leaves narrow or linear, rarely dimorphic. Calyx actinomorphic, cylindrical, 5-cleft (in *L. albiflora* zygomorphic, 7–10-cleft); corolla zygomorphic, ventricose, aestivation cochlear; fertile stamens 4 (2 + 2) or 2 (the remaining aborted), filaments inserted basally or halfway to the enlarged distal part of corolla tube; anthers circular, thecae always confluent; nectary invaginate-bilobed, sometimes pelviform; style with two lateral expansions above. Capsules septicidal, fusiform. Seeds ca. 100, polyhedric; embryo slightly curved, cotyledons shorter than rest of embryo. $n = 10$; $2n = 20$.

Seven spp., one in Argentina and six in coastal Peru and Chile.

91. ***Nierembergia*** Ruiz & Pav.

Nierembergia Ruiz & Pav., Fl. peruv. prodr.: 23 (1794); Cocucci & Hunziker, Fl. Fanerog. Argent. 15: 1–15 (1995), reg. rev.

Herbs, sometimes rhizomatous or tuberous, or shrubs. Leaves linear to circular. Flowers solitary or in lax cymes, pedicels short. Calyx funnel-shaped or bell-shaped with 10 longitudinal veins; corolla zygomorphic, tube filiform, abruptly expanded into funnel-, bell-, or cup-shaped limb, inside with elaiophores, aestivation cochlear; stamens 5 or 3 + 2 (occasionally 4), attached to the mouth of tube; anthers ventrifixed; nectary absent; stigma, when crescent-shaped, embracing the stamens. Capsule septicidal-loculicidal, 4-valved. Seeds 20–100, polyhedric or trigonal; embryo slightly curved, cotyledons slightly longer than rest of embryo. $n = 8, 9, 24$; $2n = 16, 18$.

20 spp., disjunct between South America and Mexico.

Two main clades within the genus were identified (Tate et al. 2009): one composed primarily of herbaceous species with broad stigmas and the other containing woody species with crescent-shaped stigmas.

92. ***Bouchetia*** Dunal

Bouchetia Dunal in DC., Prodr. 13(1): 589 (1852); Hunziker & Subils, Bol. Soc. Argent. Bot. 22 (1–4): 275–295 (1983), rev.

Rhizomatous or root sprouting, decumbent herbs. Basal leaves oblong-spathulate or rhombic-ovate, upper leaves narrowly elliptic. Flowers solitary, axillary. Calyx tubular-campanulate, corolla zygomorphic, cylindrical below, funnel-shaped above, aestivation cochlear; stamens fertile 5 (2 + 2 + 1) or 4 (2 + 2, one staminode present or absent); anthers ventrifixed; ovary and style compressed, nectary sheath-like and bilobed. Capsules septicidal-loculicidal, 4-valved. Seeds 50–60, polyhedric; embryo straight or slightly curved, cotyledons shorter than rest of embryo. $n = 8$.

One species in South America and two in North America.

93. ***Hunzikeria*** D'Arcy

Hunzikeria D'Arcy, Phytologia 34: 283 (1976); D'Arcy, Ann. Missouri Bot. Gard. 65: 705–708 (1978), rev.

Small herbs up to 25 cm, usually much branched from a perennial root. Leaves ovate-elliptic. Flowers solitary or paired. Calyx funnel-shaped, campanulate or obconic; corolla salver-form, ventricose, aestivation cochlear; fertile stamens 4 (staminode sometimes present), didynamous, anthers ventrifixed, equal or heteromorphic; style with two lateral expansions above, nectary invaginate-bilobed. Capsule subglobose, septicidal-loculicidal, 4-valved. Seeds 7–40, reniform; embryo straight or slightly curved, cotyledons shorter than rest of embryo. $n = 16$.

Three spp., south-western USA, Mexico, Venezuela.

94. ***Plowmania*** Hunz. & Subils

Plowmania Hunz. & Subils, Kurtziana 18: 127, f. 1–3 (1986).

Plants up to 1.5 m high, usually scandent. Leaves ovate-elliptic. Flowers 5-merous; calyx campanulate; corolla brightly coloured, zygomorphic, funnel-shaped, slightly ventricose, aestivation cochlear; stamens 4 (sometimes with a vestigial fifth stamen), didynamous, the lateral longer, posterior shorter; anthers ventrifixed; ovary with annular prominent nectary; style with two lateral expansions above. Capsule apparently septicidal, 2-valved, pericarp coriaceous. Mature seeds unknown.

Monotypic, *P. nyctaginoides* (Standl.) Hunz. & Subils, Mexico (Chiapas) and Guatemala.

Tribe Schwenckieae Hunz. (1977).

Herbs or small shrubs. Corolla with 5 trilobulate or trifid lobes; anthers ventrifixed; stamens fertile 2–4. Embryo tiny, cotyledons 2–3 times shorter than the rest of the embryo.

95. ***Schwenckia*** L.

Schwenckia L., Gen. pl. 6: 577 (1764); Freire de Carvalho, Rodriguesia 44: 307–524 (1978), reg. rev.

Annual or perennial herbs, sometimes shrubs. Leaves sessile to petiolate. Inflorescences racemose or paniculate, axillary or terminal. Calyx tubular to campanulate, actinomorphic or slightly zygomorphic; corolla actino- or zygomorphic, 5-lobate, each lobe trilobulate or trifid, aestivation valvate-conduplicate; 2 stamens + 3 staminodes, or 2 + 2 fertile stamens; ovary bilocular, nectary invaginate-pelviform; style smooth, stigma discoid. Capsule commonly globose. Seeds up to 70, cuboidal; embryo straight; endosperm copious. $n = 12$.

Around 25 spp., Central America and Antilles to north-eastern Argentina.

96. ***Melananthus*** Walp.

Melananthus Walp., Bot. Zeitung (Berlin) 8: 788 (1850); Freire de Carvalho, Sellowia 18 (18): 51–66 (1966), reg. rev.
Microschwenckia Benth. (1882).

Tiny herbs, or small shrubs; indumentum glandular. Leaves sessile or subsessile. Inflorescences spiciform, terminal. Calyx zygomorphic, segments 5, unequal, triangular-elongate or linear; corolla zygomorphic or actinomorphic, tubular, straight or slightly curved, tube very narrow, cylindrical, aestivation valvate-induplicate; stamens 2 + 2; ovary unilocular (unicarpellate?), uniovulate , nectary invaginate-pelviform or slightly bilobed. Capsule zygomorphic, septicidal, valves 2, unequal. Seed 1, smooth; embryo straight or slightly curved; endosperm scanty.

Five or six spp., Cuba, Brazil, Mexico, Guatemala, Belize, Venezuela, Argentina.

While this account was in press, two genera have been sunk into synonymy: *Larnax* was included in *Deprea* (Deanna et al. 2015) and *Aureliana* in *Athenaea* (Zamberlan et al. 2015; Knapp et al. 2015).

Selected Bibliography

Acosta, M.C., Bernardello, G., Guerra, M., Moscone, E.A. 2005. Karyotype analysis in several South American species of *Solanum* and *Lycianthes rantonnei* (Solanaceae). Taxon 54: 713–723.

Anderson, G.J. 1979. Dioecious *Solanum* species of hermaphroditic origin is an example of broad convergence. Nature 282: 836–838.

Anderson, G.J., Symon, D.E. 1985. Extrafloral nectaries in *Solanum*. Biotropica 17: 40–45.

Anderson, G.J., Symon, D.E. 1989. Functional dioecy and andromonoecy in *Solanum*. Evolution 43: 204–219.

Anderson, G.J., Bernardello, G., Opel, M., Santos Guerra, A., Anderson, M. 2006. Reproductive biology of the dioecious Canary Islands endemic *Withania aristata* (Solanaceae). Amer. J. Bot. 93: 1295–1305.

Ando, T., Saito N., Tatsuzawa, F., Kakefuda, T., Yamakage, K., Ohtani, E., Koshi-ishi, M., Matsusake, Y., Kokubun, H., Watanabe, H., Tsukamoto, T. Ueda, Y., Hashimoto, G., Marchesi, E., Asakura, K., Hara, R., Seki, H. 1999. Floral anthocyanins in wild taxa of *Petunia* (Solanaceae). Biochem. Syst. Ecol. 27: 623–650.

Ando, T., Takahashi, M., Nakajima, T., Toya, Y., Watanabe, H., Kokubun, H., Tatsuzawa, F. 2004. Delphinidin accumulation is associated with abnormal flower development in petunias. Phytochemistry 65: 2219–2227

APG III (Angiosperm Phylogeny Group III) 2009. An update of the Angiosperm Phylogeny Group classification for the orders and families of flowering plants: APG III. Bot. J. Linn. Soc. 161: 105–121.

Araújo, N.I., Miranda Coelho, V.P., Agra, M.F. 2010. Estudo farmacobotânico comparativo de folhas de *Solanum crinitum* Lam., *Solanum gomphodes* Dunal e *Solanum lycocarpum* A. St. Hil., Solanaceae. Rev. Bras. Farmacog. 20: 666–674.

Axelius, B. 1992. Testa patterns in some species of *Physalis* L. and some other genera in the tribe Solaneae (Solanaceae). Int. J. Plant Sci. 153: 488–502.

Badr, A., Khalifa, S.F., Aboel-Atta, A.I., Abou-El-Enain, M. M. 1997. Chromosomal criteria and taxonomic relationships in the Solanaceae. Cytologia 62: 103–113.

Bañares, A., Blanca, G., Güemes, J., Moreno, J.C., Ortiz, S. (eds.) 2004. Atlas y libro rojo de la Flora Vascular amenazada de España. Madrid: Dirección General de Conservación de la Naturaleza.

Banks, P. 1984. A new diploid chromosome number for tomato (*Lycopersicon esculentum*)? Can. J. Genet. Cytol. 26: 636–639.

Barboza, G.E. 1986. Estudios palinológicos en *Jaborosa* Juss. y *Trechonaetes* Miers (Solanaceae). Bol. Acad. Nac. Ci. Rep. Argent. 57: 357–376.

Barboza, G.E. 1989. Sobre la naturaleza tricelular de los granos de polen en la tribu Jaboroseae (Solanaceae). Kurtziana 20: 139–145.

Barboza, G.E. 1991. El sistema reproductivo en *Jaborosa* (Solanaceae). I. Esporogénesis, gametogénesis y fecundación. Kurtziana 21: 39–79.

Barboza, G.E., Carrizo García, C. 2005. Sobre el estado actual de la embriología en Solanaceae. Memoria del XVI Congreso Venezolano de Botánica, Saber Supl. 17: 461–470.

Barboza, G.E., Hunziker A.T. 1987. Estudios sobre Solanaceae XXV. Revisión de *Jaborosa*. Kurtziana 19: 77–153.

Barboza, G.E., Chiarini, F.E., Stehmann, J.R. 2010. Real identity of *Witheringia sellowiana* (Solanaceae), typification and chromosome number. Syst. Bot. 35 (2): 420–424.

Barboza, G.E., Agra, M.F., Romero, M.V., Scaldaferro, M. A., Moscone, E.A. 2011. New endemic species of *Capsicum* (Solanaceae) from the Brazilian caatinga: Comparison with the re-circumscribed *C. parvifolium*. Syst. Bot. 36 (3): 768–781.

Benítez de Rojas, C. 2007. Anatomy of the foliar epidermis of the *Solanum nudum* species group of *Solanum* sect. *Geminata* (Solanaceae). Acta Hort. (ISHS) 745: 287–293.

Benítez de Rojas, C., Jáuregui, D. 2005. Morphology of foliar trichomes in species of *Sessea* (Solanaceae). Monogr. Syst. Bot. Missouri Bot. Gard. 104: 321–338.

Bennett, M.D., Bhandol, P., Leitch, I.J. 2000. Nuclear DNA amounts in Angiosperms and their modern uses – 807 new estimates. Ann. Bot. 86: 859–909

Berg, C., Greilhuber, J. 1993. Cold-sensitive chromosome regions and heterochromatin in *Cestrum aurantiacum* (Solanaceae). Plant Syst. Evol. 185: 259–273.

Bernardello, L.M. 1983. Estudios en *Lycium* (Solanaceae). III. Estructura y desarrollo de fruto y semilla en *Lycium* y *Grabowskia*. Bol. Soc. Argent. Bot. 22: 147–176.

Bernardello, L.M. 1987. Comparative floral morphology in Lycieae (Solanaceae). Brittonia 39: 112–129.

Bernardello, G. 2007. A systematic survey of floral nectaries. In: Nicolson, S.W., Nepi, M., Pacini, E. (eds.) Nectaries and nectar. Dordrecht: Springer, pp. 19–128.

Bernardello, L.M., Anderson, G.J. 1990. Karyotypic studies in *Solanum* section *Basarthrum* (Solanaceae). Amer. J. Bot. 77: 420–431.

Bernardello, L.M., Luján, M.C. 1997. Pollen morphology of tribe Lycieae: *Grabowskia*, *Lycium*, *Phrodus*. Rev. Paleobot. Palynol. 96: 305–315.

Bessis, J., Guyot, M. 1979. An attempt to use stomatal characters in systematic and phylogenetic studies of Solanaceae. In: Hawkes, J.G., Lester, R.N., Skelding, A.D. (eds.) The Biology and Taxonomy of the Solanaceae. Linn. Soc. Symp. Ser. 7: 321-326. London: Academic Press.

Bohs, L. 2005. Major clades in *Solanum* based on ndhF sequence data. Monogr. Syst. Bot. Missouri Bot. Gard. 104: 27–49.

Bondeson, W.E. 1986. Gynoecial morphology and funicular germination plugs in the Nolanaceae. Nord. J. Bot. 6: 183–198.

Botha, C.J., Schultz, R.A., Van Der Lugt, J.J., Retief, E., Labuschagne, L. 1999. Neurotoxicity in calves induced by the plant, *Nierembergia hippomanica* Miers var. *violacea* Millán in South Africa. Onderstepoort J. Veterin. Res. 66: 237–244.

Bremer, B. 1996. Phylogenetic studies within Rubiaceae and relationships to other families based on molecular data. Op. Bot. Belgica 7: 33–50.

Bremer, B., Bremer, K., Heidari, N., Erixon, P., Olmstead, R.G., Anderberg, A.A., Källersjö, M., Barkhordarian, E. 2002. Phylogenetics of asterids based on 3 coding and 3 non-coding chloroplast DNA markers and the utility of non-coding DNA at higher taxonomic levels. Mol. Phylogenet. Evol. 24: 274–301.

Bremer, K., Friis, E.M., Bremer, B. 2004. Molecular phylogenetic dating of asterid flowering plants shows Early Cretaceous diversification. Syst. Biol. 53: 496–505.

Brouwer, Y.M., Clifford, H.T. 1990. An annotated list of domatia-bearing species. Notes Jodrell Lab. 12: 1–50.

Brown, C.R. 2005. Antioxidants in potato. Amer. J. Potato Res. 82: 163–172.

Buschi, C.A., Pomilio, A.M. 1987. Pyrole-3-carbamidine: a letal principle from *Nierembergia hippomanica*. Phytochemistry 26: 863–865.

Cabrera, V.A., Dottori, N., Cosa, M.T. 2010. Germinación, éxito reproductivo y fenología reproductiva de *Solanum chenopodioides* (Solanaceae). Bol. Soc. Argent. Bot. 45:7–80.

Carlquist, S. 1987. Wood anatomy of Nolanaceae. Aliso 11: 463–471.

Carlquist, S. 1988. Wood anatomy and relationships of Duckeodendraceae and Goetzeaceae. I.A.W.A. Bull., N.S. 9: 3–12.

Carlquist, S. 1992. Wood anatomy of Solanaceae: a survey. Allertonia 6: 279–326.

Carrizo García, C. 2002a. Estudios morfológicos del androceo y el gineceo en Solanaceae. Derivaciones taxonómicas y filogenéticas. Ph.D. Thesis, University of Córdoba, Argentina.

Carrizo García, C. 2002b.An approach to the diversity of endothecial thickenings in Solanaceae. Flora 197: 214–223.

Carrizo García, C., Matesevach, M., Barboza, G.E. 2008. Features related to anther opening in *Solanum* (Solanaceae). Bot. J. Linn. Soc. 158: 344–354.

Chase, M.W., Knapp, S., Cox, A.V., Clarkson, J.J., Butsko, Y., Joseph, J., Savolainen, V., Parokonny, A.S. 2003. Molecular systematics, GISH and the origin of hybrid taxa in *Nicotiana* (Solanaceae). Ann. Bot. 92: 107–127.

Chen, L.-X., He, H., Qiu, F. 2011. Natural withanolides: An overview. Nat. Prod. Rep. 28: 705–740.

Chiarini, F., Barboza, G.E. 2008. Karyological studies in Argentinean species of *Jaborosa* (Solanaceae). Bot. J. Linn. Soc. 156: 467–478.

Chiarini, F., Moreno, N., Barboza, G.E., Bernardello, G. 2010. Karyotype characterization of Andean Solanoideae (Solanaceae). Caryologia 63(3): 278–291.

Child, A. 1979. A review of branching patterns in the Solanaceae. In: Hawkes, J.G., Lester, R.N., Skelding, A.D. (eds.) The Biology and Taxonomy of the Solanaceae. Linn. Soc. Symp., Ser. 7: 345–356. London: Academic Press.

Clarkson, J.J., Knapp, S., García, V.F., Olmstead, R.G., Leitch, A.R., Chase, M.W. 2004. Phylogenetic relationships in *Nicotiana* (Solanaceae) inferred from multiple plastid DNA regions. Mol. Phylogenet. Evol. 33: 75–90.

Cocucci, A.A. 1999. Evolutionary radiation in neotropical Solanaceae. In: Nee, M., Symon, D.E., Lester, R.N., Jessop, J.P. (eds.) Solanaceae IV. Advances in biology and utilization. Richmond: Royal Botanic Gardens, Kew, pp. 9–22.

Cosa, M.T. 1989. Multiplicación vegetativa en Solanaceae: *Nierembergia stricta*, *N. aristata*, *Bouchetia anomala* y *Leptoglossis linifolia*. Kurtziana 20: 147–167.

Cosa, M.T. 1991. Estudio morfo-anatómico de órganos vegetativos en Cestroideae (Solanaceae). I. Tribu Nicotianeae. Kurtziana 21: 111–152.

Cosa, M.T. 1993. Estudio morfo-anatómico de órganos vegetativos en Cestroideae (Solanaceae). II. Tribu Salpiglossideae. Kurtziana 22: 47–72.

Cosa, M.T. 1994. Estudio morfoanatómico de órganos vegetativos en Cestroideae (Solanaceae). III. Tribu Schwenckieae. Kurtziana 23: 9–25.

Cosa, M.T., Bruno, G., Dottori, N. 1998. Anatomía de los órganos vegetativos en *Solanum juvenale* y su comparación con *S. elaeagnifolium* (Solanaceae). Anales Inst. Biol. Univ. Nac. Autón. México, Bot. 69: 9–22.

Cosa, M.T., Dottori, N., Bruno, G. 2000. Propagación y anatomía de órganos vegetativos en *Solanum hieronymi* (Solanaceae). Kurtziana 28: 211–220.

Costa, E.F., Portiansky, E.L., Massone, A.R., Marino, F.P., Idiart, J.R., Gimeno, E.J. 1998. Alteraciones en la diferenciación y proliferación celular cutánea en bovinos, inducidas por Hipervitaminosis D de origen vegetal. Analecta Veterin. 18: 7–13.

Cramer, M.J., Mesquita, C.G., Bentos, T.V., Moser, B., Williamson, G.B. 2007. Forest fragmentation reduces seed dispersal of *Duckeodendron cestroides*, a Central Amazon endemic. Biotropica 39: 709–718.

D'Arcy, W.G. 1979. The classification of the Solanaceae. In: Hawkes, J.G., Lester, R.N., Skelding, A.D. (eds.) The Biology and Taxonomy of the Solanaceae. Linn. Soc. Symp., Ser. 7: 3–48. London: Academic Press.

D'Arcy, W.G. 1991. The Solanaceae since 1976, with a review of its biogeography. In: Hawkes, J.G., Lester, R., Nee, M., Estrada, E. (eds.) Solanaceae III. Taxonomy, Chemistry, Evolution. Royal Botanic Gardens, Kew, pp. 75–137.

D'Arcy, W.G., Keating, R.C., Buchmann, S. 1996. The calcium oxalate package or so-called resorption tissue in some angiosperm anthers. In: D'Arcy, W.G, Keating, R.C. (eds.) The anther: form, function and phylogeny. Cambridge: Cambridge University Press, pp. 159–192.

D'Arcy, W.G., Keating, R.C., Zhang, Z.-Y., Peng, C.-I. 2001. The genus *Tubocapsicum* (Solanaceae). Bot. Bull. Acad. Sin. 42: 67–84.

Danert, S. 1958. Die Verzweigung der Solanaceen im reproduktiven Bereich. Abh. Deutsch. Akad. Wiss. Berlin. Math.-Naturwiss. Kl. 1957(6): 1–184.

Danert, S. 1967. Die Verzweigung als infragenerisches Gruppenmerkmal in der Gattung *Solanum* L. Kulturpflanze 15: 275–292.

Daunay, M.-C., Janick, J., Laterrot, H. 2007. Iconography of the Solanaceae from Antiquity to the XVIIth century: a rich source of information on genetic diversity and uses. Acta Hort. (ISHS) 745: 59–88.

Daunay, M.-C., Lester, R.N., Gebhardt, Ch. Hennart, J.W., Jahn, M., Frary, A., Doganlar, S. 2001. Genetic resources of eggplant (*Solanum melongena* L.) and applied species: a new challenge for molecular geneticists and eggplant breeders. In: van der Berg, R.G., Barendse, G.W.M., van der Weerden, G.M., Mariani, C. (eds.) Solanaceae V: Advances in taxonomy and utilization. Nijmegen: Nijmegen University Press, pp. 251–274.

Davis, G. 1966. Systematic embryology of the Angiosperms. New York: Wiley.

Deanna, R., Leiva González, S., Barboza, G.E. 2015. Changes in the circumscription of *Deprea* (Physalideae, Solanaceae): Thirty two new combinations. PhytoKeys 46: 73–87.

Dharamadhaj, P., Prakash, N. 1978. Development of the anther and ovule in *Capsicum* L. Austr. J. Bot. 26: 433–439.

Di Fulvio, T.E. 1961. El género *Sclerophylax* (Solanaceae): estudo anatómico, embriológico y cariológico con especial referencia a la taxonomía. Kurtziana 1: 9–103.

Di Fulvio, T.E. 1971. Morfología floral de *Nolana paradoxa* (Nolanaceae), con especial referencia a la organización del gineceo. Kurtziana 6: 41–51.

Diez, M.J., Ferguson, I.K. 1984. Pollen morphology of *Mandragora autumnalis* Bertol. (Solanaceae). Pollen et Spores 26: 151–160.

Dillon, M.O. 2005. Solanaceae of the Lomas formations of Coastal Peru and Chile. Monogr. Syst. Bot. Missouri Bot. Gard. 104: 131–155.

Dottori, N., Bruno, G., Cosa de Gastiazoro, M.T. 2000. Anatomía comparada de las plántulas de *Solanum elaeagnifolium* y *S. juvenale* (Solanaceae). Bol. Soc. Argent. Bot. 35(1–2): 115–123.

Douglas, A.C., Freyre, R. 2010. Floral development, stigma receptivity and pollen viability in eight *Nolana* (Solanaceae) species. Euphytica 174: 105–117.

Eich, E. 2008. Solanaceae and Convolvulaceae: Secondary Metabolites. Berlin, Heidelberg: Springer.

Eichler, A.W. 1875. Blütendiagramme. I. Leipzig: Wilhelm Engelmann.

Eichler, A.W. 1878. Blütendiagramme. II. Leipzig: Wilhelm Engelmann.

Evans, W.C. 1979. Tropane alkaloids of the Solanaceae. In: Hawkes, J.G., Lester, R.N., Skelding, A.D. (eds.) The Biology and Taxonomy of the Solanaceae. Linn. Soc. Symp., Ser. 7: 241–254. London: Academic Press.

Facciuto, G., Pannunzio, M.J., Coviella, M.A., Soto, S., Hagiwara, J.C., Borja, M. 2006. Characterization of the ornamental value of *Calibrachoa* spp. native to Argentina. Acta Hort. (ISHS) 714: 37–42.

Farruggia, F., Bohs, L. 2010. Two new South American species of *Solanum* section *Crinitum* (Solanaceae). PhytoKeys 1: 67–77.

Fay, M.F., Olmstead, R.G., Richardson, J.E., Santiago, E., Prance, G.T., Chase, M.W. 1998. Molecular data support the inclusion of *Duckeodendron cestroides* in Solanaceae. Kew Bull. 53: 203–212.

Figueroa, S.D., Dottori, N., Cosa, M.T. 2008. Anatomía de órganos vegetativos en *Solanum chenopodioides* (Solanaceae). Arnaldoa 15(2): 247–254.

Freire de Carvalho, L., Machado, R.D., Bovini, M.G. 1999. Seed coat micromorphology of Brazilian species of *Schwenckia*. In: Nee, M., Symon, D.E., Lester, R.N., Jessop, J.P. (eds.) Solanaceae IV. Advances in biology and utilization. Royal Botanic Gardens, Kew, pp. 23–32.

Freitas da Silva, M., Braga Lisbôa, P.L., Lobato Lisbôa, R. C. 1977. Nomes vulgares de plantas amazônicas. Manaus: INPA.

Freyre, R., Douglas, A., Dillon, M.O. 2005. Artificial hybridizations in five species of Chilean *Nolana* (Solanaceae). HortScience 40: 532–536.

Gallo, G.G. 1987. Plantas tóxicas para el ganado en el Cono Sur de América. Buenos Aires: Ed. Hemisferio Sur.

García, V.F., Olmstead, R.G. 2003. Phylogenetics of Tribe Anthocercideae (Solanaceae) based on ndhF and trnL/F sequence data. Syst. Bot. 28: 609–615.

Gentry, J.L. 1979. Pollen morphology of the Salpiglossoideae (Solanaceae). In: Hawkes, J.G., Lester, R.N., Skelding, A.D. (eds.) The Biology and Taxonomy of the Solanaceae. Linn. Soc. Symp., Ser. 7: 327–332. London: Academic Press.

Gentry, J.L. 1986. Pollen morphology in the Cestreae (Solanaceae). In: D'Arcy, W.G. (ed.) Solanaceae: Biology and Systematics. New York: Columbia Press, pp. 138–158.

Giacomin, L., Stehmann, J.R. 2011. A new heterandrous species of *Solanum* section *Gonatotrichum* Bitter (Solanaceae) from Bahia, Brazil. PhytoKeys 7: 1–9.

Goldraij, A, Kondo, K., Lee, C.B., Hancock, C.N., Sivaguru, M., Vazquez-Santana, S., Kim, S., Phillips, T.E., Cruz-Garcia, F., McClure, B. 2006. Compartmentalization of S-RNase and HT-B degradation in self-incompatible *Nicotiana*. Nature 439: 805–810.

Goodspeed, T.H. 1947. Maturation of the gametes and fertilization in *Nicotiana*. Madroño 9: 110–120.

Granados-Tochoy, J.C., Knapp, S., Orozco, C.I. 2007. *Solanum humboldtianum* (Solanaceae): An endangered new species from Colombia rediscovered 200 years after its first collection. Syst. Bot. 32: 200–207.

Hadid, M., Cosa, M.T., Dottori, N., Liscovsky, I. 2007. Anatomía de la raíz de *Solanum sisymbriifolium* (Solanaceae). Latin Amer. J. Pharm. 26(1): 10–14.

Haegi, L. 1981. A conspectus of the tribe Anthocercideae. Telopea 2: 173–180.

Hartl, D. 1963. Das Placentoid der Pollensäcke, ein Merkmal der Tubifloren. Ber. Deutsch. Bot. Ges. 76: 70–72.

Hawkes, J.G. 1990. The potato. Evolution, biodiversity and genetic resources. Washington, D.C.: Smithsonian Institution Press.

Hawkes, J.G. 1999. The economic importance of the family Solanaceae. In: Nee, M., Symon, D.E., Lester, R.N., Jessop, J.P. (eds.) Solanaceae IV. Advances in biology and utilization. Royal Botanic Gardens, Kew, pp. 1–8.

Heiser, Ch.B. 2001. Interspecific hybridization and the improvement of the naranjilla (*Solanum quitoense*). In: van der Berg, R.G., Barendse, G.W.M., van der Weerden, G.M., Mariani, C. (eds.) Solanaceae V: Advances in taxonomy and utilization. Nijmegen: Nijmegen University Press, pp. 307–310.

Hepper, F.N. 1991. Old World *Withania* (Solanaceae): a taxonomic review and key to the species. In: Hawkes, J.G., Lester, R., Nee, M., Estrada, E. (eds.) Solanaceae III. Taxonomy, Chemistry, Evolution. Royal Botanic Gardens, Kew, pp. 211–227.

Heslop-Harrison, Y., Shivanna, K.R. 1977. The receptive surface of the Angiosperm stigma. Ann. Bot. 41: 1233–1258.

Hijmans, R.J., Spooner, D.M., Salas, A.R., Guarino L., de la Cruz, J. 2002. Atlas of wild potatoes. Rome: International Plant Genetic Resources Institute.

Holm, L., Doll, J., Holm, E., Pancho, J., Herberger, J. 1997. World weeds. Natural histories and distribution. New York: Wiley.

Huber, K.A. 1980. Morphologische und entwicklungsgeschichtliche Untersuchungen an Blüten und Blütenständen von Solanaceen und von *Nolana paradoxa* Lindl. (Nolanaceae). Diss. Bot. 55: 1–252.

Hunziker, A.T. 1979. South American Solanaceae: a synoptic survey. In: Hawkes, J.G., Lester, R.N., Skelding, A.D. (eds.) The Biology and Taxonomy of the Solanaceae. Linn. Soc. Symp., Ser. 7: 49–85. London: Academic Press.

Hunziker, A.T. 1997. Estudios sobre Solanaceae. XLIII. Revisión de las especies de *Markea*. Kurtziana 25: 67–113.

Hunziker, A.T. 2001. Genera Solanacearum. The genera of Solanaceae illustrated, arranged according to a new system. Ruggell: A.R.G. Gantner Verlag K.-G.

Ichiyanagi, T., Kashiwada, Y., Shida, Y., Ikeshiro, Y. Kaneyuki, T. Konishi, T. 2005. Nasunin from eggplant consists of cis-trans iosmers of delphinidin 3-[4-(*p*-coumaroyl)-L-rhamnosyl(1 → 6)glucopyranoside]-5-glucopyranoside. J. Agric. Food Chem. 54: 530–5312.

IUCN 2007. IUCN Red List of Threatened Species. www.iucnredlist.org. Accessed 15 September 2007.

Jagannadham, D. 1988. Embryological studies in *Nicotiana tabacum* and *N. rustica*. J. Indian Bot. Soc. 67: 135–141.

Janssens, S.B., Knox, E.B., Huysmans, S., Smets, E.F., Merckx, V.S.F. 2009. Rapid radiation of *Impatiens* (Balsaminaceae) during Pliocene and Pleistocene: Results of a global climate change. Mol. Phylogenet. Evol. 52: 806–824.

Japan Tobacco Inc. 1994. The genus *Nicotiana* Illustrated. Tokyo: Japan Tobacco Inc.

Jones, R.N., Rees, H. 1982. B chromosomes. London: Academic Press.

Kenton, A., Parokonny, A.S., Gleba, Y.Y., Bennett, M.D. 1993. Characterization of the *Nicotiana tabacum* L. genome by molecular cytogenetics. Mol. Gen. Genet. 240: 159–169.

Knapp, S. 2002a. Floral diversity and evolution in the Solanaceae. In: Cronk, Q.C.B., Bateman, R.M., Hawkins, J.A. (eds.) Developmental Genetics and Plant Evolution. London: Taylor & Francis; pp. 267–297.

Knapp, S. 2002b. Tobacco to tomatoes: a phylogenetic perspective on fruit diversity in the Solanaceae. J. Exp. Bot. 53: 2001–2022.

Knapp, S. 2007. *Solanum coalitum* (Solanaceae), a new endemic species from southern Ecuador. Novon 17 (2): 212–216.

Knapp, S. 2010a. On 'various contrivances': pollination, phylogeny and flower form in the Solanaceae. Philos. Trans. Royal Soc. B Biol. Sci. 365(1539): 449–460.

Knapp, S. 2010b. Four new vining species of *Solanum* (Dulcamaroid Clade) from montane habitats in Tropical America. PLoS ONE 5(5): e10502. doi:10.1371/journal.pone.0010502.

Knapp, S. 2010c. New species of *Solanum* (Solanaceae) from Peru and Ecuador. PhytoKeys 1: 33–51.

Knapp, S., Nee, M. 2009. *Solanum anomalostemon* (Solanaceae), an endangered new species from southern Peru with unusual anther morphology. Novon 19: 178–181.

Knapp, S., Persson, V., Blackmore, S. 1997. A phylogenetic conspectus of the tribe Juanulloeae. Ann. Missouri Bot. Gard. 84: 67–89.

Knapp, S., Persson, V., Blackmore, S. 1998. Pollen morphology and functional dioecy in *Solanum* (Solanaceae). Plant Syst. Evol. 210: 113–139.

Knapp, S., Stafford, P., Persson, V. 2000. Pollen morphology in the Anthocercideae (Solanaceae). Kurtziana 28: 7–18.

Knapp, S., Barboza, G.E., Romero, M.V., Vignoli-Silva, M., Giacomin, L.L., Stehmann, J.R. 2015. Identification and lectotypification of the Solanaceae from Vellozo's Flora Fluminensis. Taxon 64(4): 822–836.

Konar, R.N., Linskens, H.F. 1966a. The morphology and anatomy of the stigma of *Petunia hybrida*. Planta 71: 356–371.

Konar, R.N., Linskens, H.F. 1966b. Physiology and biochemistry of the stigmatic fluid of *Petunia hybrida*. Planta 71: 372–387.

Lachman, J., Hamouz, K. 2005. Red and purple potatoes as a significant antioxidant source in human nutrition-a review. Plant Soil Environ. 51: 477–482.

Las Peñas, M. L., Chiarini, F.E., Bernardello, G., Benítez de Rojas, C. 2006. Karyotypes of some species of *Cestrum*, *Sessea*, and *Vestia* (tribe Cestreae, Solanaceae). Caryologia 59(2): 131–137.

Lester, R.N., Francisco Ortega, J., Al-Ani, M. 1999. Convergent evolution of heterandry (unequal stamens) in *Solanum*, proved by sporoderm SEM. In: Nee, M., Symon, D.E., Lester, R.N., Jessop, J.P. (eds.) Solanaceae IV. Advances in biology and utilization. Royal Botanic Gardens, Kew, pp. 51–69.

Levin, R.A., Miller, J.S. 2005. Relationships within tribe Lycieae (Solanaceae): paraphyly of *Lycium* and multiple origins of gender dimorphism. Amer. J. Bot. 92: 2044–2053.

Levin, R.A., Shak, J.R., Miller, J.S., Bernardello, G., Venter, A.M. 2007. Evolutionary relationships in tribe Lycieae (Solanaceae). Acta Hort. (ISHS) 745: 225–240.

Levin, R.A., Bernardello, G., Whiting, C., Miller, J.S. 2011. A new generic circumscription in tribe Lycieae (Solanaceae). Taxon 60: 681–690.

Levine, D.A., Anderson, G.J. 1986. Evolution of dioecy in an American *Solanum*. In: D'Arcy, W.G. (ed.) Solanaceae: Biology and Systematics. New York: Columbia Press, pp. 264–273.

Lim, K.Y., Matyasek, R., Lichtenstein, C.P., Leitch, A.R. 2000. Molecular cytogenetic analyses and phylogenetic studies in the *Nicotiana* section *Tomentosae*. Chromosoma 109: 245–258.

Lippman, Z.B., Cohen, O., Alvarez, J.P., Abu-Abied, M., Pekker, I., Paran, I., Eshed, Y., Zamir, D. 2008. The making of a compound inflorescence in tomato and related nightshades. PLoS Biol. 6: 2424–2435.

Liscovsky, I.J. 2007. Estudio comparado de los órganos vegetativos y reproductivos en Cestreae G. Don (Solanaceae) y su implicancia taxonómica en el sistema de la familia. Ph.D. Thesis, University of Córdoba, Argentina.

Liscovsky, I.J., Cosa, M.T. 2005. Anatomía comparativa de hoja y tallo en los representantes de Cestreae G. Don (Solanaceae) de Argentina. Gayana Bot. 62: 33–43.

Liscovsky, I.J., Cosa, M.T., Dottori, N. 2001. Crecimiento secundario atípico del tallo de *Datura ferox* (Solanaceae). Kurtziana 29: 7–13.

Liscovsky, I.J., Cosa, M.T., Dottori, N. 2002. Estudio anatómico de órganos vegetativos en representantes de Datureae (Solanaceae). Bol. Soc. Argent. Bot. 37: 171–180.

Lu, A.-M., Zhang, Z.-Y., Chen, Z.-D., Pan, K.-Y. 1999. Embryology and adaptive ecology in the genus *Przewlaskia*. In: Nee, M., Symon, D.E., Lester, R.N., Jessop, J.P. (eds.) Solanaceae IV. Advances in biology and utilization. Richmond: Royal Botanic Gardens, Kew, pp. 71–80.

Machado, A.S., Cosa, M.T., Dottori, N. 2008. Desarrollo de fruto, semilla y germinación en *Nicotiana glauca* Graham (Solanaceae). Arnaldoa 15: 227–235.

Machado, A.S., Cosa, M.T., Barboza, G.E. 2011. Anatomía de fruto y semilla de *Duckeodendron cestroides*, una enigmática especie de Solanaceae. Bol. Soc. Argent. Bot. 46 (suppl.): 106.

Manirakiza, P., Covaci, A., Schepens, P. 2003. Pungency principles in *Capsicum* - analytical determinations and toxicology. In: Krishna De, A. (ed.) *Capsicum*: The genus *Capsicum*. London: Taylor and Francis, pp. 71–86.

Martine, C.T., Vanderpool, D., Anderson, G.J., Les, D.H. 2006. Phylogenetic relationships of andromonoecious and dioecious Australian species of *Solanum* subgenus *Leptostemonum* section *Melongena*: Inferences from ITS sequence data. Syst. Bot. 31: 410–420.

Martins, T.R., Barkman, T. 2005. Reconstruction of Solanaceae phylogeny using the nuclear gene SAMT. Syst. Bot. 30: 435–447.

McClure, B.A., Franklin-Tong, V. 2006. Gametophytic self-incompatibility: understanding the cellular mechanisms involved in "self" pollen tube inhibition. Planta 224: 233–245.

Mentz, L.A., de Oliveira, P.L. 2004. *Solanum* (Solanaceae) na região Sul do Brasil. Pesquisas, Bot. 54: 9–327.

Mentz, L., de Oliveira, P.L., da Silva, M.V. 2000. Tipologia dos tricomas das espécies do gênero *Solanum* (Solanaceae) na Região Sul do Brasil. Iheringia Bot. 54: 75–106.

Metcalfe, C.R., Chalk, I. 1950. Anatomy of the dicotyledons. 2 Vols. [Solanaceae: pp. 965–978]. Oxford: Clarendon Press.

Miller, J.S., Diggle, P.K. 2003. Diversification of andromonoecy in *Solanum* section *Lasiocarpa* (Solanaceae): The roles of phenotypic plasticity and architecture. Amer. J. Bot. 90: 707–715.

Miller, J.S., Venable, D.L. 2002. The transition to gender dimorphism on an evolutionary background of self-incompatibility: An example from *Lycium* (Solanaceae). Amer. J. Bot. 89: 1907–1915.

Minne, L., Spies, J.J., Venter, A.J.T., Venter, A.M. 1994. Breeding systems in some representatives of the genus *Lycium* (Solanaceae). Bothalia 24: 107–110.

Misico, R.I., Oberti, J.C., Nicotra, V., Barboza, G.E., Gil, R. R., Burton, G. 2011. Withanolides and related steroids. In: Kinghorn, A.D. et al. (eds.) Progress in the Chemistry of Organic Natural Products 94: 127–229.

Modilewski, J. 1935. Cytological investigation of the genus *Nicotiana*. Cytology and embryology of the amphidiploids *Nicotiana ditagla*. Vseukr. Akad. Nauk. Inst. Bot. Zh. 7: 7–29.

Mohan, K., Singh, B. 1969. Morphological studies in Solanaceae. III. Structure and development of seed of *Solanum torvum* Sw. J. Indian Bot. Soc. 48: 338–345.

Moncada, M., Fuentes, V.R. 1993. Palinología de Goetzeaceae. Rev. Jard. Bot. Nac. Univ. Habana 12: 75–79.

Montero-Castro, J.C., Delgado-Salinas, A., De Luna, E., Eguiarte, L.E. 2006. Phylogenetic analysis of *Cestrum* section *Habrothamnus* (Solanaceae) based on plastid and nuclear DNA sequences. Syst. Bot. 31: 843–850.

Montes Hernández, S., Aguirre Rivera, J.R. 1992. Tomate de cáscara (*Physalis philadelphica*). In: Hernández, B., León, J.E. (eds.) Cultivos marginados, otra perspectiva de 1942. Colección FAO: Producción y protección vegetal No. 26, Roma, pp. 115–120.

Morton, J.F. 1987. Cape Gooseberry. In: Morton, J.F. (ed.) Fruits of warm climates. Miami: University of Miami, Media Incorporated, pp. 430–434.

Moscone, E.A. 1989. Estudios citotaxonómicos en las tribus Solaneae y Nicotianeae (Solanaceae) de América del Sur. Ph.D. Thesis, University of Córdoba, Argentina.

Moscone, E.A. 1992. Estudios de cromosomas meióticos en Solanaceae de Argentina. Darwiniana 31: 261–297.

Moscone, E.A., Lambrou, M., Hunziker, A.T., Ehrendorfer, F. 1993. Giemsa C-banded karyotypes in *Capsicum* (Solanaceae). Plant Syst. Evol. 186: 213–229.

Moscone, E.A., Loidl, J., Ehrendorfer, F., Hunziker, A.T. 1995. Analysis of active nucleolus organizing regions in *Capsicum* (Solanaceae) by silver staining. Amer. J. Bot. 82: 276–287.

Moscone, E.A., Scaldaferro, M.A., Gabriele, M., Cecchini, N.M., Sánchez García, Y., Daviña, J.R., Ducasse, D.A., Barboza, G.E., Ehrendorfer, F. 2007. The evolution in chili peppers (*Capsicum*-Solanaceae), a view from the cytogenetics. Acta Hort. (ISHS) 745: 137–170.

Moscone, E.A., Scaldaferro, M.A., Grabiele, M., Romero, M.V., Debat, H., Seijo, J.G., Acosta, M.C., Daviña, J. R., Barboza, G.E., Ducasse, D.A. 2011. Genomic characterization of the chili peppers (*Capsicum*, Solanaceae) germplasm by classical and molecular cytogenetics. In: Physical Mapping Technologies for the Identification and Characterization of Mutated Genes to Crop Quality, IAEA TECDOC 1664: 97–104. International Atomic Energy Agency, Wien, Austria.

Murad, L., Lim, K.Y., Christopodulou, V., Matyasek, R., Lichtenstein, C.P., Kovarik, A., Leitch, A.R. 2002. The origin of tobacco's T genome is traced to a particular lineage within *Nicotiana tomentosiformis* (Solanaceae). Amer. J. Bot. 89: 921–928.

Nanetti, A. 1912. Sulle probabili cause della parthenocarpia del *Solanum muricatum* Ait. Nuovo Giorn. Bot. Ital. 19: 91–111.

Newbigin, E., Anderson, M., Clarke A. 1999. Molecular basis of self-incompatibility in the Solanaceae. In: Nee, M., Symon, D.E., Lester, R.N., Jessop, J.P. (eds.) Solanaceae IV. Advances in biology and utilization. Richmond: Royal Botanic Gardens, Kew, pp. 189–195.

Nurit-Silva, K., Agra, M.F. 2011. Leaf epidermal characters of *Solanum* sect. *Polytrichum* (Solanaceae) as taxonomic evidence. Microsc. Res. Tech. 74: 1186–1191.

Nurit-Silva, K., Basílio, I.J.L., Agra, M.F. 2007. Estudo farmacobotânico comparativo entre *Solanum paniculatum* L. e *Solanum rhytidoandrum* Sendtn. Rev. Bras. Bioci. 5: 243–245.

Nurit-Silva, K., Costa-Silva, R., Coelho, V.P.M., Agra, M.F. 2011. A pharmacobotanical study of vegetative organs of *Solanum torvum*. Braz. J. Pharmacog. 21 (4): 568–574.

Olmstead, R., Bohs, L. 2007. A summary of molecular systematic research in Solanaceae: 1982–2006. Acta Hort. (ISHS) 745: 255–268.

Olmstead, R.G., Palmer, J.D. 1992. A chloroplast DNA phylogeny of the Solanaceae: subfamilial relationships and character evolution. Ann. Missouri Bot. Gard. 79: 346–360.

Olmstead, R.G., Palmer, J.D. 1997. *Solanum*: implications for phylogeny, classification, and biogeography from cpDNA restriction site variation. Syst. Bot. 22: 19–29.

Olmstead, R.G., Sweere, J.A., Spangler, R.E., Bohs, L., Palmer, J.D. 1999. Phylogeny and provisional classification of the Solanaceae based on chloroplast DNA. In: Nee, M., Symon, D.E., Lester, R.N., Jessop, J.P. (eds.) Solanaceae IV. Advances in biology and utilization. Richmond: Royal Botanic Gardens, Kew, pp. 111–137.

Olmstead, R.G., Bohs, L., Migid, H.A., Santiago-Valentin, E., García, V.F., Collier, S.M. 2008. A molecular phylogeny of the Solanaceae. Taxon 57(4): 1159–1181.

Parokonny, A.S., Kenton, A.Y. 1995. Comparative physical mapping and evolution of the *Nicotiana tabacum* L. karyotype. In: Brandham, P.E., Bennett, M.D. (eds.) Kew Chromosome Conference IV. Richmond: Royal Botanic Gardens, Kew, pp. 301–320.

Patel, R.C., Inamdar, J.A. 1971. Structure and ontogeny of stomata in some Polemoniales. Ann. Bot. 35(2): 389–409.

Pavari, F. 1957. Ricerche embriologiche e cariologiche su *Cestrum elegans* L. (Solanaceae). Caryologia 9: 437–452.

Perícola, N., Dottori, N., Cosa, M.T., Hadid, M., Bruno, G. 2004. Estudio morfoanatómico de plántula de *Solanum palinacanthum*. Bol. Soc. Argent. Bot. 39(1-2): 67–73.

Persson, V., Knapp, S., Blackmore, S. 1994. Pollen morphology and systematics of tribe Juanulloeae A.T. Hunziker. Rev. Paleobot. Palynol. 83: 1–30.

Persson, V., Knapp, S., Blackmore, S. 1999. Pollen morphology and the phylogenetic analysis of *Datura* and *Brugmansia*. In: Nee, M., Symon, D.E., Lester, R.N., Jessop, J.P. (eds.) Solanaceae IV. Advances in biology and utilization. Richmond: Royal Botanic Gardens, Kew, pp. 171–187.

Prasad, T., Singh D. 1978. Gamethophytes and seed development in *Nicandra physalodes* Gaertn. J. Indian Bot. Soc. 57: 76–83.

Pringle, G.J., Murray, B.G. 1993. Karyotypes and C-banding patterns in species of *Cyphomandra* Mart. ex Sendtner (Solanaceae). Bot. J. Linn. Soc. 111: 331–342.

Prohens, J., Nuez, F. 2000. The Tamarillo (*Cyphomandra betacea*): A review of a promising small fruit crop. Small Fruits Rev. 1: 43–68.

Prohens, J., Nuez, F. 2001. The use of wild species in the improvement of pepino (*Solanum muricatum*). In: van der Berg, R.G., Barendse, G.W.M., van der Weerden, G.M., Mariani, C. (eds.) Solanaceae V: Advances in taxonomy and utilization. Nijmegen: Nijmegen University Press; pp. 297–306.

Queensland Nature Conservation Act 1992. Office of the Queensland Parliamentary Counsel Warning.

Raven, P.H. 1975. The bases of angiosperm phylogeny: cytology. Ann. Missouri Bot. Gard. 62: 724–764.

Record, S.J., Hess, R.W. 1942. Timbers of the New Worlds. New Haven: Yale University Press.

Reis dos, C., Das Graças Sajo, M., Stehmann, J.R. 2002. Leaf structure and taxonomy of *Petunia* and *Calibrachoa* (Solanaceae). Brazil. Arch. Biol. Technol. 45: 59–66.

Ricci, M. 2006. Conservation status and ex situ cultivation efforts of endemic flora of the Juan Fernández Archipelago. Biodivers. Conserv. 15: 3111–3130.

Ritterbusch, A. 1976. Die Organopoiëse der Blüte von *Calceolaria tripartita* R. et P. (Scrophulariaceae). Bot. Jahrb. Syst. 95: 267–320.

Robyns, W. 1931. L'organisation florale des Solanacées zygomorphes. Memoires de l'Académie Royale de Belgique, Cl. Sci. 11: 1–84.

Rodríguez, I.G. 1998. Anatomía floral comparada en la Tribu Solanaeae (Solanaceae). Ph.D. Thesis, University of Córdoba, Argentina.

Rodríguez, I.G. 2000. Anatomía floral de *Deprea orinocensis*, *Physalis minima* y *P. viscosa* (Solanaceae: Solaneae). Kurtziana 28: 19–34.

Roe, K.E. 1971. Terminology of hairs in the genus *Solanum*. Taxon 20: 501–508.

Rozen, J.G. Jr. 2003. A new tribe, genus, and species of South American panurgine bee (Andrenidae, Panurginae), oligolectic on *Nolana* (Nolanaceae). In: G.A. R. Melo & I. Alves-dos-Santos, Apoidea Neotropica: Homenagem aos 90 Anos de Jesus Santiago Moure. Criciúma: Editora UNESC, pp. 93–106.

Samuelsson, G. 1913. Studien über die Entwicklungsgeschichte der Blüten einiger Bicornes-typen. Svensk Bot. Tidskr. 7: 97–188.

Santiago-Valentin, E., Olmstead, R.G. 2003. Phylogenetic of the Antillean Goetzeoideae (Solanaceae) and their relationships within the Solanaceae based on chloroplast and ITS DNA sequence data. Syst. Bot. 28: 452–460.

Sarmento da Silva, T.M., de Carvalho M.G., Braz-Filho, R., Agra, M.F. 2003. Occurrence of flavones, flavonols and its glycosides in species of the genus *Solanum* (Solanaceae). Quim. Nova 26(4): 517–522.

Sawyer, N.W. 2005. Systematics of *Deprea* and *Larnax* (Solanaceae) based on morphological evidence. Monogr. Syst. Bot. Missouri Bot. Gard. 104: 259–285.

Sawyer, N.W., Anderson, G.J. 2000. Dioecy in South American *Deprea* (Solanaceae). Biotropica 32: 291–298.

Schultes, R.E. 1979. Solanaceous hallucinogens and their role in the development of New World cultures. In: Hawkes, J.G., Lester, R.N., Skelding, A.D. (eds.) The Biology and Taxonomy of the Solanaceae. Linn. Soc. Symp., Ser. 7: 137–160. London: Academic Press

Seithe, A. 1962. Die Haararten der Gattung *Solanum* L. und ihre taxonomische Verwertung. Bot. Jahrb. Syst. 81: 262–335.

Sharma, M. 1974. Contribution to the palinotaxonomy of genus *Solanum*. J. Palynol. 10(1): 51–68.

Sharma, R.C., Raghuvanshi, R.K., Singh, D. 1987. Embryological studies in *Atropa* L. and *Hyoscyamus* L. J. Indian Bot. Soc. 66: 311–316.

Siddiqui, S.A., Khan, F.A. 1986. Megasporangium and development of female gametophyte in some *Solanum* species. J. Indian Bot. Soc. 65 (Suppl.): 72.

Siddiqui, S.A., Khan, F.A. 1988. Ontogeny and dehiscence of anther in Solanaceae. Bull. Soc. Bot. France 135, Lettr. Bot. 2: 101–109.

Sink, K.C. 1984. *Petunia*. Berlin: Springer.

Smith, S.D., Baum, D.A. 2006. Phylogenetics of the florally diverse Andean clade Iochrominae (Solanaceae). Amer. J. Bot. 93: 1140–1153.

Solano, J., Anabalón, L., Hauenstein, E. 2011. Experiencia de conservación ex-situ de *Solanum fernandezianum* (Solanaceae). Gayana Bot 68(1): 102–105.

Soltis, D.E., Soltis, P.S. 1997. Phylogenetic relationships in Saxifragaceae sensu lato: A comparison of topologies based on 18S rDNA and rbcL sequences. Amer. J. Bot. 84: 504–522.

Soltis, D.E., Soltis, P.S., Chase, M.W., Mort, M.E., Albach, D.C., Zanis, M., Savolainen, V., Hahn, W.H., Hoot, S. B., Fay, M.F., Axtell, M., Swensen, S.M., Prince, L.M., Kress, W.J., Nixon, K.C., Farris, J.S. 2000. Angiosperm phylogeny inferred from 18S rDNA, rbcL, and atpB sequences. Bot. J. Linn. Soc. 133: 381–461.

Soltis, D.E., Smith, S.A., Cellinese, N., Wurdack, K.J., Tank, D.C. , Brockington, S.F., Refulio-Rodriguez, N.F., Walker, J.B., Moore, M.J., Carlsward, B.S., Bell, C.D., Latvis, M., Crawley, S., Black, C., Diouf, D., Xi, Z., Rushworth, C.A., Gitzendanner, M.A., Sytsma, K. J., Qiu, Y.-L., Hilu, K.W., Davis, C.C., Sanderson, M. J., Beaman, R.S., Olmstead, R.G., Judd, W.S., Donoghue, M.J., Soltis, P.S. 2011. Angiosperm phylogeny: 17 genes, 640 taxa. Amer. J. Bot. 98: 704–730.

Stafford, P., Knapp, S. 2006. Pollen morphology and systematics of the zygomorphic-flowered nightshades

(Solanaceae: Salpiglossideae sensu D'Arcy, 1978 and Cestroideae sensu D'Arcy, 1991, pro parte): a review. Syst. Biodivers. 4: 173–201.

Stehmann, J.R., Greppi, J.A. 2011. Two new species of *Calibrachoa* (Solanaceae) from subtropical South America. Phytotaxa 25: 31–38.

Stern, S., Bohs, L. 2012. An explosive innovation: Phylogenetic relationships of *Solanum* section *Gonatotrichum* (Solanaceae). PhytoKeys 8: 89–98.

Stern, S., Agra, M.F., Bohs, L. 2011. Molecular delimitation of clades within New World species of the "spiny solanums" (*Solanum* subg. *Leptostemonum*). Taxon 60(5): 1429–1441.

Stevens, P.F. 2001 onwards. Angiosperm Phylogeny Website. Version 12, July 2012 [and more or less continuously updated since]: http://www.mobot.org/MOBOT/research/APweb/.

Stiefkens, L., Bernardello, G. 2000. Karyotypes and DNA content in diploid and polyploid *Lycium* (Solanaceae). Bol. Soc. Argent. Bot. 35: 237–244.

Stuessy, T.F., Swenson, U., Crawford, D.J., Anderson, G., Silva, M. 1998. Plant conservation in the Juan Fernandez Archipelago, Chile. Aliso 16: 89–98.

Svensson, H.G. 1926. Zytologische-embryologische Solanaceenstudien. I. Über die Samenentwicklung von *Hyoscyamus niger* L. Svensk Bot. Tidskr. 20: 420–434.

Sykorova, E., Lim, K.Y., Chase, M.W., Knapp, S., Leitch, I. J., Leitch, A.R., Fajkus, J. 2003. The absence of Arabidopsis-type telomeres in *Cestrum* and closely related genera *Vestia* and *Sessea* (Solanaceae): first evidence from eudicots. Plant J. 34: 283–291.

Tago-Nakazawa, M., Dillon, M.O. 1999. Biogeografía y evolución en el clado *Nolana* (Nolaneae-Solanaceae). Arnaldoa 6: 81–116.

Tanksley, S.D. 2004. The genetic, developmental, and molecular bases of fruit size and shape variation in tomato. Plant Cell 16 (Suppl. 2004): 181–189.

Tate, J.A., Acosta, M.C., McDill, J., Moscone, E.A., Simpson, B.B., Cocucci, A.A. 2009. Phylogeny and character evolution in *Nierembergia* (Solanaceae): molecular, morphological, and cytogenetic evidence. Syst. Bot. 34: 198–206.

Threatened Species Protection Act 1995. http://www.austlii.edu.au/au/legis/tas/consol_act/tspa1995305/.

Troll, W. 1969. Die Infloreszenzen. Typologie und Stellung im Aufbau des Vegetationskörpers. Vol. 2(1). Stuttgart: Springer.

Tu, T., Volis, S., Dillon, M.O., Sunand, H., Wen, J. 2010. Dispersals of Hyoscyameae and Mandragoreae (Solanaceae) from the New World to Eurasia in the early Miocene and their biogeographic diversification within Eurasia. Mol. Syst. Evol. 57: 1226–1237.

Vaknin, H., Bar-Akiva, A., Ovadia, R., Nissim-Levi, A., Forer, I., Weiss, D., Oren-Shamir, M. 2005. Active anthocyanin degradation in *Brunfelsia calycina* (yesterday-today-tomorrow) flowers. Planta 222: 19–26

Wang, X., Hughes, A.L., Tsukamoto, T., Ando, T., Kao, T. H. 2001. Evidence that intragenic recombination contributes to allelic diversity of the S-RNase gene at the self-incompatibility (S) locus in *Petunia inflata*. Plant Physiol. 125: 1012–1022.

Wesse, T.L., Bohs, L. 2007. A three-gene phylogeny of the genus *Solanum*. Syst. Bot. 32: 445–463.

Whalen, M.D., Costich, D.E. 1986. Andromonoecy in *Solanum*. In: D'Arcy, W.G. (ed.) Solanaceae Biology and Systematics. New York: Columbia University Press; pp. 284–302.

Whitson, M., Manos, P.S. 2005. Untangling *Physalis* (Solanaceae) from the Physaloids: a two-gene phylogeny of the Physalinae. Syst. Bot. 30: 216–230.

Wikström, N., Savolainen, V., Chase, M.W. 2001. Evolution of the angiosperms: Calibrating the family tree. Proc. Roy. Soc. B 268: 2211–2220.

Wollenweber, E., Dörsan, M., Dörr, M., Roitman, J.N., Valant-Vetschera, K.M. 2005. Chemodiversity of surface flavonoids in Solanaceae. Z. Naturf., C 60: 661–670.

Yuan, Y., Zhang, Z.-Y, Olmstead, R.G. 2006. A retroposon insertion to the waxy gene: Defining monophyly of the tribe Hyoscyameae (Solanaceae), and revealing the allopolyploid origin of *Atropa belladonna*. Mol. Biol. Evol. 23: 2263–2267.

Zamberlan, P.M., Rodrigues, I.M.C., Mäder, G., Castro, L., Stehmann, J.R., Bonatto, S.L., Freitas, L.B. 2015. Re-evaluation of the generic status of *Athenaea* and *Aureliana* (Withaniinae, Solanaceae) based on molecular phylogeny and morphology of the calyx. Bot. J. Linnean Soc. 177 (3): 322–334.

Sphenocleaceae

Sphenocleaceae (Lindley) Baskerville, Affin. Pl.: 110 (1839), nom. cons.
Sphenocleoideae Lindley (1836).
Sphenocleeae A. Gray (1878).

T.G. Lammers

Erect annual glabrous herbs, up to 1.5 m tall. Roots fibrous. Stems fistulose, fleshy or spongy, freely branching. Leaves alternate, simple, exstipulate, petiolate; lamina pinnately veined, margin entire. Inflorescence a dense terminal pedunculate acropetal conico-cylindric spike. Flowers small, tetracyclic, perfect, radially symmetrical, subtended by an inconspicuous spathulate bract and a pair of linear bracteoles. Calyx synsepalous, adnate to the ovary, forming a hypanthium; lobes 5, imbricate, persistent. Corolla sympetalous, radially symmetric, white or pale yellow, urceolate-campanulate, caducous; lobes 5, imbricate. Stamens 5, alternating with the corolla lobes; filaments distinct, adnate to the middle of the corolla tube, slightly dilated at base; anthers tetrasporangiate, dithecal, basifixed, distinct. Nectary 0. Gynoecium syncarpous, 2-locular; ovary half-inferior; ovules numerous; placentae axile, spongy; style simple, short or almost wanting; stigma capitate. Fruit a circumscissile capsule, enclosed by the accrescent calyx lobes. Seeds small, oblong, yellowish brown, shining, numerous, exarillate, testa striate; embryo straight; endosperm very scanty or lacking.

Sphenocleaceae comprise but a single genus with a single species, indigenous to the tropics and subtropics of the Old World and naturalized in the Americas.

Vegetative Morphology and Anatomy. The stem is aerenchymatous, with a hollow pith at maturity, surrounded by well-developed air canals in the cortex; these continue into the petiole. The cortical air-canals are separated from one another by single or double rows of parenchyma cells, which contain numerous clustered crystals of calcium oxalate; such crystals may be found in other parenchyma cells as well. Xylem and phloem form a continuous cylinder traversed by medullary rays one or two cells wide. Perforation plates of vessel elements are simple. Lacticifers typical of Campanulaceae, where the genus was formerly included, are lacking. Thin-walled sclerenchyma fibers are found in the pericycle. Leaves are dorsiventral, with one or two palisade layers in the mesophyll. Stomata are tetracytic and chiefly abaxial.

Floral Structure and Anatomy. Proterandry and specialized mechanisms of secondary pollen presentation typical of Campanulaceae are lacking. Floral aestivation is imbricate, in contrast to the valvate Campanulaceae. Petal venation is simple, with just two veins arising near the base of the midrib. The stamens are distinct and inserted on the short corolla tube near the middle. The style is very short and thick, or nearly lacking, and lacks the pollen-collecting hairs typical of Campanulaceae (Fig. 69).

Embryology. Anther wall development follows the dicotyledonous pattern, forming a single middle layer. Fibrous thickenings are found in the endothecium. The tapetum is of the glandular (secretory) type, and the tapetal cells are binucleate. Ovules are anatropous, unitegmic, and tenuinucellar. Embryo sac formation is monosporic and of the Polygonum type. Upon coming in contact with the embryo sac, the inner layer of the integument develops as an endothelium (integumentary tapetum). Embryogenesis follows the Solanad pattern. Endosperm formation is cellular ab initio, but endosperm is absent or nearly so from mature seeds. Four-celled haustoria form at both the chalazal and micropylar ends and are equally aggressive. No hypostase forms.

J.W. Kadereit and V. Bittrich (eds.), *Flowering Plants. Eudicots*, The Families and Genera of Vascular Plants 14,
DOI 10.1007/978-3-319-28534-4_30,

Fig. 69. Sphenocleaceae. *Sphenoclea zeylanica*. **A** Flowering plant. **B** Flower. **C** Gynoecium. **D** Petals and stamens spread out. **E** Gynoecium, vertical section. (From Takhtajan 1981)

Pollen Morphology. Pollen grains are subprolate, tricolporate, trinucleate when shed, and have a smooth surface from which small rounded protrusions arise; exine layering is obscure (Dunbar 1975).

Karyology. Counts of $2n = 24$, 32, 40, 42, 48 have been reported for *Sphenoclea zeylanica* (Lammers 1992), but the data are too few for any definite conclusions regarding karyological variation in the genus.

Pollination and Reproductive Systems. Nothing is known of the reproductive biology of this family, aside from its clear lack of any mechanism of secondary pollen presentation.

Phytochemistry. Little is known of the secondary chemistry of the family, aside from its lack of alkaloids, the production of cyclic thiosulfinates and secoiridoids, and the storage of carbohydrate as inulin (Porembski and Koch 1999; Hiraia et al. 2000).

Affinities. Until recently, almost all authors pointed to a relationship between *Sphenoclea* and Campanulaceae. The genus was assigned to Campanuloideae by Bentham (1876), Hutchinson (1973), and Dahlgren (1980). Schönland (1889) segregated it as its own tribe within Campanulaceae, the Sphenocleeae, while Thorne (1976) and Takhtajan (1980) did the same at subfamilial rank (Sphenocleoideae). However, the genus differs from Campanulaceae in its lack of articulated lacticifers and proterandrous secondary pollen presentation, and in its possession of tetracytic stomata, imbricate calyx and corolla lobes, capitate stigma, and exalbuminous seeds. For these reasons, many authors (Wagenitz 1964; Dahlgren 1975; Cronquist 1981; Lammers 1992; Thorne 1992; Takhtajan 1997) excluded the genus from Campanulaceae as the monotypic Sphenocleaceae. Of these authors, most still considered the family to be allied to Campanulaceae, assigning it to the order Campanulales. This hypothesis was supported by a phylogenetic analysis based on morphological data (Gustafsson and Bremer 1995), in which *Sphenoclea* emerged as the sister-group of Stylidiaceae, which together formed the sister-group of Campanulaceae.

Airy Shaw (1948, 1973), however, questioned the campanulalean affinity of *Sphenoclea*, suggesting (largely on the basis of similarities in habit) that the family was allied to Phytolaccaceae. That family (a member of Caryophyllales) differs from Sphenocleaceae not only in numerous floral features, but also in embryological, anatomical, and phytochemical features (Monod 1980; Rosatti 1986).

The family here is assigned to Solanales in accordance with the results of phylogenetic analyses of molecular data (Cosner et al. 1994; Soltis et al. 2011), which placed *Sphenoclea* as a member of a clade comprising *Hydrolea* (Hydroleaceae) and *Montinia* (Montiniaceae). None of the known features of Sphenocleaceae is inconsistent with this assignment, although *Sphenoclea* is unusual among Solanales in lacking a nectary.

Distribution and Habitat. The family comprises a single genus, monotypic *Sphenoclea*. Originally native to the tropics and subtropics of the Old World, *Sphenoclea zeylanica* has spread to the Western Hemisphere via human agency in historic times. It is found growing in a variety of low-elevation wet habitats, in both seasonally dry and continuously rainy climates.

Economic Importance. *Sphenoclea zeylanica* is a common weed of rice paddies. An exudate from living roots is an effective control for the rice root nematode, *Hirschmaniella oryzae* (Mohandas et al. 1981). In Java, the leaves are sold in markets under the name *goendapadi*, to be eaten as a vegetable with rice (Airy Shaw 1948).

One genus:

Sphenoclea Gaertner Fig. 69

Sphenoclea Gaertner, Fruct. Sem. Pl. 1: 113 (1788), nom. cons.

See family description.

One species, *Sphenoclea zeylanica* Gaertner (including *S. dalzielii* N. E. Br. of tropical western Africa).

Selected Bibliography

Airy Shaw, H.K. 1948. Sphenocleaceae. In: van Steenis, C. G.G.J. (ed.) Flora Males. I, 4(1). Djakarta: Noordhoff-Kolff, pp. 27–28.

Airy Shaw, H.K. 1973. A dictionary of the flowering plants and ferns, ed. 8. Cambridge: Cambridge University Press.

d'Almeida, J.F.R. 1934. On the development of secondary aerating tissues in the stem of *Sphenoclea zeylanica* Gaertn. Proc. Indian Sci. Congr. Assoc. 1934: 314.

Bentham, G. 1876. Campanulaceae. In: Bentham, G., Hooker, J.D., Genera plantarum 2: 541–564. London: Reeve.

Cosner, M.E., Jansen, R.K., Lammers, T.G. 1994. Phylogenetic relationships in the Campanulales based on *rbc*L sequences. Plant Syst. Evol. 190: 79–95.

Cronquist, A. 1981. An integrated system of classification of flowering plants. New York: Columbia University Press.

Dahlgren, R.M.T. 1975. A system of classification of the angiosperms to be used to demonstrate the distribution of characters. Bot. Not. 128: 119–147.

Dahlgren, R.M.T. 1980. A revised system of classification of the angiosperms. Bot. J. Linn. Soc. 80: 91–124.

Dunbar, A. 1975. On pollen of Campanulaceae and related families with special reference to the surface ultrastructure II. Campanulaceae subfam. Cyphioideae and subfam. Lobelioideae; Goodeniaceae; Sphenocleaceae. Bot. Not. 128: 102–118.

Erbar, C. 1995. On the floral development of *Sphenoclea zeylanica* (Sphenocleaceae, Campanulales) - SEM investigations on herbarium material. Bot. Jahrb. Syst. 117: 469–493.

Gupta, D.P. 1959. Vascular anatomy of the flower of *Sphenoclea zeylanica* Gaertn. and some other related species. Proc. Natl. Inst. Sci. India 25B: 55–64.

Gustafsson, M.H.G., Bremer, K. 1995. Morphology and phylogenetic interrelationships of the Asteraceae, Calyceraceae, Campanulaceae, Goodeniaceae, and related families (Asterales). Amer. J. Bot. 82: 250–265.

Hiraia, N., Sakashita, S., Sano, T., Inoue, T., Ohigashi, H., Premasthira, C., Asakawa, Y., Harada, J., Fujii, Y. 2000. Allelochemicals of the tropical weed *Sphenoclea zeylanica*. Phytochemistry 55: 131–140.

Hutchinson, J. 1973. The families of flowering plants, ed. 3. Oxford: Clarendon Press.

Kausik, S.B., Subramanyam, K. 1946. A contribution to the life-history of *Sphenoclea zeylanica* Gaertn. Proc. Indian Acad. Sci. 23B: 274–280.

Lammers, T.G. 1992. Circumscription and phylogeny of the Campanulales. Ann. Missouri Bot. Gard. 79: 388–413.

Mohandas, C., Rao, Y.S., Sahu, S.C. 1981. Cultural control of rice root nematodes (*Hirschmanniella* spp.) with *Sphenoclea zeylanica*. Proc. Indian Acad. Sci., Animal Sci. 90: 373–376.

Monod, T. 1980. A propos du *Sphenoclea zeylanica* (Sphenocleaceae). Adansonia II, 20: 147–164.

Porembski, S., Koch, M. 1999. Inulin occurrence in *Sphenoclea* (Sphenocleaceae). Plant Biol. 1: 288–289.

Rosatti, T.J. 1986. The genera of Sphenocleaceae and Campanulaceae in the southeastern United States. J. Arnold Arb. 67: 1–64.

Schönland, S. 1889. Campanulaceae. In: Natürl. Pflanzenfamilien IV(5). Leipzig: W. Engelmann, pp. 40–70.

Soltis, D.E., Smith, S.A., Cellinese, N., Wurdack, K.J., Tank, D.C., Brockington, S.F., Refulio-Rodriguez, N. F., Walker, J.B., Moore, M.J., Carlsward, B.S., Bell, C. D., Latvis, M., Crawley, S., Black, C., Diouf, D., Xi, Z. X., Rushworth, C.A., Gitzendanner, M.A., Sytsma, K. J., Qiu, Y.L., Hilu, K.W., Davis, C.C., Sanderson, M.J., Beaman, R.S., Olmstead, R.G., Judd, W.S., Donoghue, M.J., Soltis, P.S. 2011. Angiosperm phylogeny: 17 genes, 640 taxa. Amer. J. Bot. 98: 704–730.

Subramanyam, K. 1950. A contribution to our knowledge of the systematic position of the Sphenocleaceae. Proc. Indian Acad. Sci. 31(B): 60–65.

Takhtajan, A. 1980. Outline of the classification of flowering plants (Magnoliophyta). Bot. Rev. 46: 225–359.

Takhtajan, A. 1981. Plant Life, vol. 5(2). Moskau: Proswjestschenije (in Russian).

Takhtajan, A. 1997. Diversity and classification of flowering plants. New York: Columbia University Press.

Thorne, R.F. 1976. A phylogenetic classification of the Angiospermae. In: Hecht, M.K., Steere, W.C., Wallace, B. (eds.) Evolutionary Biology 9. New York: Plenum Press, pp. 35–106.

Thorne, R.F. 1992. Classification and geography of the flowering plants. Bot. Rev. 58: 225–348.

Wagenitz, G. 1964. Reihe Campanulales. In: Melchior, H. (ed.) A. Engler's Syllabus der Pflanzenfamilien, ed. 12, 2. Berlin: Bornträger, pp. 478–497.

Wilbur, R.L. 1976. Flora of Panama, Campanulaceae. Ann. Missouri Bot. Gard. 63: 593–655.

Sphenostemonaceae

Sphenostemonaceae P. Royen & Airy Shaw, Kew Bull. 27: 325 (1972).

M. Thiv

Evergreen shrubs or trees up to 25 high. Leaves alternate, opposite to subopposite, or pseudo-whorled, petiolate, exstipulate (present, minute?), simple, coriaceous, entire to glandular-dentate, penninerved. Inflorescences axillary or terminal racemes, bracteate at base, floral bracts caducous. Flowers hermaphrodite or rarely unisexual, actinomorphic; perianth composed of 4–6 distinct, imbricate, caducous sepals, and 4–6 distinct, imbricate, fleshy, caducous petals, or apetalous; stamens 4–6 or 8–13, distinct, in one whorl, antesepalous, filaments short or almost absent, anthers introrse; disk absent, ovary superior, sessile, bicarpellate, bilocular, fusiform to cylindric, style absent, stigma slightly bipartite, cup-shaped; ovules one per locule, pendant, anatropous, apotropic, unitegmic. Fruit a berry, sometimes asymmetric, crowned by the stigma. Seed 1 or 2, non-arillate; embryo surrounded by chalazal envelope with copious, ruminate endosperm.

A monogeneric family with nine species in New Caledonia, Northern Queensland, New Guinea, the Moluccas and Celebes.

Vegetative Morphology. The leaf venation of *Sphenostemon* is characteristic. The secondary veins are brochidodromous with marginal arches developed apically. Tertiary veins are weakly transverse (Dickison and Baas 1977).

Vegetative Anatomy. The leaf, bark and wood anatomy of *Sphenostemon* was studied in detail by Baas (1975) and also partly by Metcalfe (1956). Leaves and stems are usually glabrous, although unicellular hairs can occur. Stomata are anomocytic with 4-6 neighbouring cells. Crystals as styloids (up to 350 x 30 μm) are present in the mesophyll and the phloem of the major vascular bundles. Nodes are trilacunar. The wood is unspecialized with scalariform vessel perforations, long fibre-tracheids, and heterogeneous rays. It is diffuse-porous and growth rings are absent. Pronounced scalariform vessel pits are purported to be typical for *Sphenostemon*. Within the genus two distinct anatomical leaf groups were found. In contrast to the remainder, the New Caledonian species (sect. *Sphenostemon*) are characterised by a multilayered epidermis and a closed vascular system in the petiole and midrib.

Flower Structure (Fig. 70). Flowers of *Sphenostemon* show some variation in sex, number of petals and stamens, and stamen morphology. Flowers are usually hermaphrodite, only a single New Caledonian species (*S. balansae*) has unisexual flowers. The number of petals and stamens and stamen morphology correlate with the infrageneric classification. Whereas four to six petals and stamens occur in sect. *Sphenostemon*, sect. *Apetalae* is characterised by the absence of petals and the presence of six or 8–13 stamens. Bailey (1956) described two different types of stamens. Usually, the thecae are located laterally on a thick connective in sect. *Apetalae* or ± ventrally on thick petaloid stamens in sect. *Sphenostemon*. In *S. pachycladus* of the latter section, the fibrous endothecium is restricted to the regions of the partly merged pollen-sacs (Solnzewa 1987). Endress (2008) described the irregularly cuneate stamens of sect. *Apetalae* which are densely crowded in bud and have an imprinted shape resulting from pressure in the bud.

Embryology. Only a few embryological data are available. According to Bremer et al. (2001), its

J.W. Kadereit and V. Bittrich (eds.), *Flowering Plants. Eudicots*, The Families and Genera of Vascular Plants 14,
DOI 10.1007/978-3-319-28534-4_31, © Springer International Publishing Switzerland 2016

Fig. 70. Sphenostemonaceae. A–E *Sphenostemon comptonii.* A Flowering branch. B Floral diagram. C Anthetic flower. D Stamens. E Ovary in longitudinal section. (Baker 1921, courtesy Missouri Botanical Garden)

embryo sac development follows the *Polygonum* type. According to Doweld (2000), ovules are crassinucellate.

Pollen Morphology. According to Erdtman (1952), Bailey (1956), Lobreau-Callen (1977) and Jérémie (1997), pollen of Sphenostemonaceae is isopolar, subcircular (sect. *Apetalae*) or sub-tri-angular to hexagonal (sect. *Sphenostemon*), between 10–15(−30) µm in diameter and tri- or tetraporate or rarely inaperturate. The exine structure is tectate-reticulate or finely verrucate.

Karyology. Chromosome numbers of $2n \approx 26$ have been reported by Carr and McPherson (1986).

Reproductive Systems. Monoecious or dioecious unisexual flowers have been reported for *S. balansae* (Jérémie 1997).

Fruits and Seeds. The fruits of *Sphenostemon* were mostly described as a drupe with a pyrene. The origin of the sclerified tissue, however, remained unclear for a long time. Bailey (1956) recognized that it is not part of the pericarp, but of the seed, being derived from the testa. This was later reviewed by Steenis (1986) who concluded that the sclerified tissue is derived from the chalaza, which completely surrounds the seed during development. Recent karpological studies showed that the fruits are berries with a single-layered exocarp, a many-layered parenchymatous mesocarp with sporadic small lysigenous cavities and groups of asterosclerids which do not form a stone, and a 2–3-layered parenchymal endocarp (Doweld 2000; Savinov 2003).

Dispersal. The fleshy fruits indicate zoochory.

Phytochemistry. Only the absence of cephalaroside in *Sphenostemon* has been reported (Bremer et al. 2001).

Distribution and Habitats. Species of *Sphenostemon* grow in montane rainforests. Both sections have different areas of distribution. Whereas sect. *Sphenostemon* is restricted to New Caledonia, sect. *Apetalae* occurs in Northern Queensland and East Malesia (Steenis 1986). The age of this disjunction between the sections is unknown. Still, it must be younger than the stem node age of Sphenostemonaceae which may date back to the Cretaceous-Cenomanian (Magallón and Castillo 2009).

Affinities. The systematic relationships of Sphenostemonaceae have been controversially debated in the past and have been summarised by Bernardi (1964) and Baas (1975). Affinities have been postulated to Aquifoliaceae (Baillon 1875; Steenis 1955; Baas 1975), Icacinaceae (Baas 1975; Savinov 2003), Theaceae and Ochnaceae (Loesener 1942), Dichapetalaceae (Lobreau 1969), Guttiferae (Steenis 1952) and to the basal angiosperm family Trimeniaceae (Gibbs 1917). Most later authors followed Airy Shaw (1972) by giving *Sphenostemon* family rank. The analysis of *rbc*L sequences indicated that *Sphenostemon* is part of the Dipsacales of the euasterids II, where it is sister to *Paracryphia* (Savolainen et al. 2000). Such a relationship was already assumed by Dickison and Baas (1977) and Cameron (2002) based on similarities in vegetative anatomy. In accordance with these results, APG III (2009) and Tank and Donoghue (2010) proposed to sink Sphenostemonaceae together with Quintiniaceae into Paracryphiaceae of Paracryphiales (Campanulidae) as sister to Dipsacales.

A single genus:

Sphenostemon **Baill.** Fig. 70

Sphenostemon Baill., Bull. Mens. Soc. Linn. Paris 1: 53 (1875); Steenis, Flora Males. 10: 146 (1986); Jérémie in Morat, Flore Nouvelle-Calédonie 21: 3 (1997).
Nouhuysia Lauterb. (1912).
Idenburgia Gibbs (1917).

Characters as for the family.

Nine spp., of which six belong to sect. *Sphenostemon* and three to sect. *Apetalae* (Steenis 1986).

Selected Bibliography

Airy Shaw, H.K. 1972. Validation of the family name Sphenostemonaceae. Kew Bull. 27: 325–326.

APG III (Angiosperm Phylogeny Group) 2009. An update of the Angiosperm Phylogeny Group classification for the orders and families of flowering plants: APG III. Bot. J. Linn. Soc. 161: 105–121.

Baas, P. 1975. Vegetative anatomy and the affinities of Aquifoliaceae, *Sphenostemon*, *Phelline*, and *Oncotheca*. Blumea 22: 311–407.

Bailey, I.W. 1956. The relationship between *Sphenostemon* of New Caledonia and *Nouhuysia* of New Guinea. J. Arnold Arbor. 37: 360–365.

Baillon, H. 1875. Sur le nouveau genre *Sphenostemon*. Bull. Mens. Soc. Linn. Paris 1: 53.

Baker, E.G. 1921. *Sphenostemon comptonii*. In: Rendle, A. B. et al., Systematic account of the plants collected in New Caledonia and the Isle of Pines by Prof. R. H. Compton, M.A., in 1914. Part 1, Flowering Plants (Angiospermae). J. Linn Soc., Bot 45: 287, pl. 16.

Bernardi, L. 1964. La position systématique du genre *Sphenostemon* Baillon sensu van Steenis. Candollea 19: 199–205.

Bremer, K., Backlund, A., Sennblad, B. et al. 2001. A phylogenetic analysis of 100+ genera and 50+ families of euasterids based on morphological and molecular data with notes on possible higher level morphological synapomorphies. Plant Syst. Evol. 229: 137–169.

Cameron, K.M. 2002. On the phylogenetic position of the New Caledonian endemic families Paracryphiaceae, Oncothecaceae, and Strasburgeriaceae: a comparison of molecules and morphology. Bot. Rev. 68: 428–443.

Carr, G.D., McPherson, G. 1986. Chromosome Numbers of New Caledonian plants. Ann. Missouri Bot. Gard. 73: 486–489.

Dickison, W.C., Baas, P. 1977. The morphology and relationships of *Paracryphia* (Paracryphiaceae). Blumea 23: 417–438.

Doweld, A.B. 2000. Sphenostemonaceae. In: Takhtajan, A., Anatomia seminum comparativa. Rosidae I 5: 118–119. St. Petersburg: Nauka.

Endress, P.K. 2008. The whole and the parts: relationships between floral architecture and floral organ shape, and their repercussions on the interpretation of fragmentary floral fossils. Ann. Missouri Bot. Gard. 95: 101–120.

Erdtman, G. 1952. Pollen Morphology and Plant Taxonomy. Angiosperms. Stockholm: Almqvist & Wiksell.

Gibbs, L.S. 1917. A contribution to the phytogeography and flora of the Arfak mountains. London: Taylor & Francis, p. 136.

Jérémie, J. 1997. Sphenostemonaceae. In: Morat, P. (ed.) Flore de la Nouvelle-Calédonie 21: 3–21.

Lobreau, D. 1969. Les limites de l'ordre Celastrales d'après le pollen. Pollen et Spores 11: 499–555.

Lobreau-Callen, D. 1977. Les pollens des Célastrales. Sphenostemonaceae. Mém. Trav. Inst. Montpellier 3: 51–54.

Loesener, T. 1942. Aquifoliaceae. In: Engler, A., Prantl, K., Die natürlichen Pflanzenfamilien, 2nd edn, 20b: 6–86.

Magallón, S., Castillo, A. 2009. Angiosperm diversification through time. Amer. J. Bot. 96: 349–365.

Metcalfe, C.R. 1956. The taxonomic affinities of *Sphenostemon* in the light of the anatomy of its stem and leaf. Kew Bull. 11: 249–253.

Savinov, I.A. 2003 Comparative carpology of the genus *Sphenostemon* (Sphenostemonaceae) in the context of its taxonomy and phylogeny. Bot. Zh. 88: 5–16.

Savolainen, V., Fay, M.F., Albach, D.C. et al. 2000. Phylogeny of the eudicots: a nearly complete familial analysis based on rbcL gene sequences. Kew Bull. 55: 257–309.

Solnzewa, M.P. 1987. Sphenostemonaceae. In: Batygina, T.B, Yakovlev, M.S. (eds.) Comparative embryology of flowering plants. Davidiaceae-Asteraceae. P. 47. Leningrad: Nauka, p. 391.

Steenis, C.G.G.J. van 1952. Reduction of two endemic monotypic Papuan genera. Act. Bot. Neerl. 1: 94–98.

Steenis, C.G.G.J. van 1955. Some notes on the flora of New Caledonia and reduction of *Nouhuysia* to *Sphenostemon*. Svensk Bot. Tidskr. 49: 19–23.

Steenis, C.G.G.J. van 1986. Sphenostemonaceae. Flora Males. I, 10: 145–149.

Tank, D.C., Donoghue, M.J. 2010. Phylogeny and phylogenetic nomenclature of the Campanulidae based on an expanded sample of genes and taxa. Syst. Bot. 35: 425–441.

Stemonuraceae

Stemonuraceae Kårehed, Amer. J. Bot. 88: 2267 (2001).

M.J. Potgieter, M. Schori, and T.M.A. Utteridge

Evergreen trees or shrubs. Lower branches self-pruning and leaving conspicuous scars in some genera. Buttresses absent (except *Cantleya*), stilt-roots and pneumatophores sometimes present (*Stemonurus*). Leaves simple, alternate, sometimes appearing distichous, entire, penninerved (rarely triplinerved), exstipulate. Inflorescences axillary, terminal or leaf-opposed, occasionally ramiflorous, cymose, rarely reduced to 1 or few flowers. Flowers bisexual or functionally unisexual, actinomorphic, usually 5-merous, usually small (3–6 mm), but *Whitmorea* up to 1.5 cm. Calyx persistent, articulated with the pedicel, small, generally cupular, sepals 4–5, connate, calyx rim entire, apiculate or shortly lobed. Petals 4–5(7), minute to conspicuous, free (coherent), sometimes united in *Gomphandra*, often inflexed at apex, lobes valvate, midrib usually thickened ventrally. Stamens isomerous, alternipetalous, filaments filiform (rarely) or broad and flattened, glabrous or with papillose to clavate hairs ventrally, dorsally, and/or at apex of connective; anthers bisporangiate, basifixed, introrse or extrorse (*Grisollea*), longitudinally dehiscent. Staminodes in pistillate flowers present or absent (*Grisollea*), resembling stamens or much smaller (*Codiocarpus*) but pollen sacs sterile. Disk absent or present, cupular, hippocrepiform, or squamiform and unilateral, lobed or entire, persistent in fruit. Ovary superior, carpel apparently 1 but likely pseudomonomerous, unilocular, often with exterior lateral appendage above the disk, ovules 2, apical, pendulous, anatropous, unitegmic. Stigma on short style or more often sessile, peltate, subcapitate, or punctiform (only in bisexual flowers), entire to lobed, crenate, or ridged, often eccentric, persistent in fruit. Ovary rudiment in staminate flowers usually smaller than ovary and occasionally with 2 ovules which do not develop, disk sometimes present, stigma absent or much reduced. Fruit a drupe of various shape, in most genera laterally compressed, when compressed with conspicuous lateral appendage which develops from the ovary wall (except *Grisollea*). Mesocarp divided into two parts, outer mesocarp usually fleshy, thin, variously coloured, inner mesocarp of sclerified bundles forming prominent longitudinal ridges visible in dry fruit. Endocarp thin to thick, of sclerified cells arranged perpendicular to the inner mesocarp. Seed 1, testa vascularised with post-chalazal or pachychalazal bundles. Embryo small.

Twelve genera with about 90 species, distributed mainly in the tropics of the Old World, rapidly decreasing in number towards the subtropics.

Vegetative Morphology. Members of the family are small to medium-sized trees or shrubs. Some species have buttresses (*Cantleya*, *Stemonurus*) or pneumatophores (*Stemonurus*). Branchlets and petioles are puberulous or glabrous. Bark smooth, hooped. Terminal buds abruptly contracted, often resinous in *Stemonurus* and *Whitmorea*. Some genera with lower branches self-pruning, leaving scars. Leaves simple, petiolate, alternate. Leaves mostly oblong-elliptic, sometimes lanceolate. Leaf bases typically obtuse. Leaf margin entire, usually flat, sometimes strongly (e.g. *Gastrolepis alticola*, Munzinger et al. 2008) to slightly revolute (e.g. *Irvingbaileya australis*, Howard 1943). Texture ranges from thinly chartaceous to thickly coriaceous. Leaves penninerved with the secondary veins in 2–10 pairs, rarely triplinerved. Tertiary veins typically obscure. Domatia absent (Kårehed 2001). Leaves often tubercled on the abaxial surface in *Gomphandra* (Sleumer 1972).

J.W. Kadereit and V. Bittrich (eds.), *Flowering Plants. Eudicots*, The Families and Genera of Vascular Plants 14,
DOI 10.1007/978-3-319-28534-4_32, © Springer International Publishing Switzerland 2016

Vegetative Anatomy. Gelatinisation of the upper leaf epidermis in *Gomphandra* (Solereder 1908). Stomata anisocytic (*Discophora*, *Gomphandra*, *Irvingbaileya*, *Lasianthera* and *Medusanthera*), cyclocytic (*Cantleya*, *Gastrolepis* and *Hartleya*) or transitional (*Codiocarpus*, *Grisollea*, *Stemonurus* and *Whitmorea*). Stomata in *Gomphandra* occasionally heliocytic (van Staveren and Baas 1973). A peristomal rim may be present (*Codiocarpus*) or absent (*Cantleya*, *Gastrolepis*, *Gomphandra*, *Medusanthera* and *Whitmorea*). In *Stemonurus* stomatal guard cells with or without distinct peristomal rim. Polar T-pieces present (*Cantleya*, *Codiocarpus*, *Gastrolepis* and *Stemonurus*), absent (*Medusanthera* and *Whitmorea*) or variable in *Gomphandra* (van Staveren and Baas 1973). Stomatal guard cells without inner stomatal ledges. Subsidiary cells partly to almost completely submersed in *Cantleya* and *Stemonurus*, but not to hardly submersed in *Codiocarpus*, *Gastrolepis* and *Gomphandra*. The stomatal index can be extremely variable within species (van Staveren and Baas 1973; Baas 1974). Various types of hairs are found in Stemonuraceae and include, amongst others, normal hairs (hairs with an elongate or tapering body: *Discophora* and *Medusanthera*), cylindrical hairs (on stamens of most genera), and what Heintzelman and Howard (1948) named 'icacinaceous hairs' (hairs with a short stalk and a body at right angles to this stalk), which occur in *Cantleya*, *Codiocarpus*, *Gastrolepis* and *Grisollea*. Thin-based hairs (long, tapering hairs with a thick wall and restricted lumen in the body and a very thin-walled base) are present in some species of *Cantleya*, *Codiocarpus* and *Medusanthera*. Druses are abundant in all genera, except *Gomphandra*, *Irvingbaileya* and *Lasianthera* (Heintzelman and Howard 1948). Rhombic crystals are less common but found additionally in *Discophora*, *Gastrolepis*, *Lasianthera*, *Medusanthera* and *Stemonurus*. *Cantleya corniculata* and *Stemonurus malaccensis* contain tannin crystals which are scattered in the mesophyll and ground tissue of the midrib (Teo and Haron 1999). Crystal sand in *Cantleya*, *Discophora*, *Medusanthera* and *Stemonurus*. Growth ring boundaries generally absent, indistinct in *Hartleya* and *Stemonurus*. Wood diffuse-porous (Lens et al. 2008). The vessels range from (4)7–45(50)/mm^2 (Lens et al. 2008) and are solitary in *Cantleya* and *Lasianthera*, solitary and in radial and/or tangential multiples of 2–4 in *Discophora*, *Gastrolepis*, *Medusanthera* and *Grisollea thomassetii*. In *Stemonurus* the vessels may be solitary or in clusters of 3–7 cells. Perforation plates always simple (*Cantleya*) to usually simple (*Codiocarpus*, *Discophora*, *Gomphandra*, *Medusanthera* and *Stemonurus*) or equally mixed with scalariform perforations (*Hartleya*); scalariform perforations with (1)3–18(64) bars, occasionally double simple perforations in *Discophora* (Lens et al. 2008). The inter-vascular and vessel-parenchyma pits are numerous and relatively small in *Stemonurus* (Bailey and Howard 1941a). Intervessel pits 5–35 μm in horizontal diameter, non-vestured, arrangement usually opposite or scalariform, less commonly alternate (*Discophora*, *Lasianthera*). Scalariform vessel-ray pitting with reduced pit borders in species of *Cantleya*, *Discophora*, *Gastrolepis*, *Hartleya*, *Lasianthera*, *Medusanthera* and *Stemonurus*. In some species of *Stemonurus*, the lumina of the imperforate tracheary elements are much reduced or are occluded by excessive thickening of the secondary walls (Bailey and Howard 1941b). *Discophora*, *Gastrolepis* and *Cantleya* tend to have enlarged pits in the lateral walls of the vessels. Tangential diameter of vessels (30)45–130(2,000) μm (Lens et al. 2008). Wall thickenings, tyloses, interxylary phloem, and tracheids are all absent. Fibres non-septate, thick- to thin-walled, with distinctly bordered pits (1,100)2,000–3,650(4,500) μm long, pit borders 4–9 μm in horizontal diameter (Lens et al. 2008). Axial parenchyma typically diffuse-in-aggregates and scanty paratracheal in *Cantleya* and *Gastrolepis*. Paratracheal parenchyma dominant in species of *Gastrolepis*, *Lasianthera*, *Stemonurus* and *Whitmorea*. Uniseriate rays generally scarce to completely absent (*Medusanthera*), 0–5 rays/mm, (50)210–759(1,550) μm high, consisting of square to procumbent ray cells. Multiseriate rays generally 2–7(17)-seriate, (300)835–2,810(4,400) μm high in most genera, but much higher in *Gomphandra* and *Medusanthera*, indistinct sheath cells sometimes present in *Discophora*, *Gomphandra*, *Hartleya*, *Lasianthera* and *Stemonurus*; thick-walled sclereids in rays of *Medusanthera*.

Inflorescence Structure. Inflorescences of Stemonuraceae are basically cymose but may appear

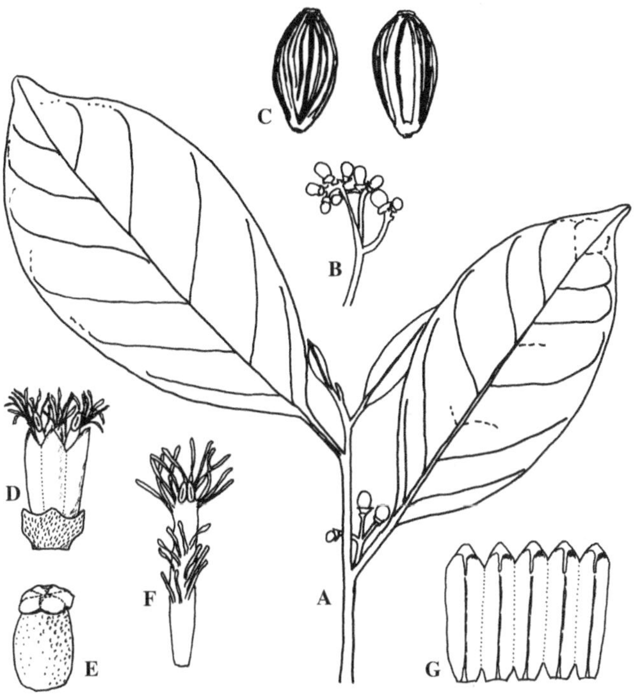

Fig. 71. Stemonuraceae. *Gomphandra*. **A** Twig with pistillate inflorescence. **B** Staminate inflorescence. **C** Fruit, dorsal and ventral views. **D** Staminate flower. **E** Ovary with sessile stigma. **F** Stamen, ventral view. **G** Corolla, showing inflexed petal apices. (Illustrations by M. Schori)

capitate or umbelliform, or flowers may be solitary. Scorpioid cymes are present in *Stemonurus* and *Gomphandra*. The typical structure is of several branched cymes borne on a common peduncle, each cyme with 1–3 orders of branching. Inflorescences may be axillary, leaf-opposed, terminal or ramiflorous. Bracts are present but are often deciduous shortly after anthesis, though persistent in some species of *Stemonurus*. Occasionally a small leaf will be present in the inflorescence (*Gomphandra*). Flowers (sub)sessile or on short pedicels, in 1-flowered inflorescences the peduncle serves as pedicel. In dioecious taxa, staminate inflorescences tend to have significantly more flowers and are more highly branched than pistillate inflorescences.

Flower Structure (Fig. 71). Flowers are small and often clustered. The calyx is usually cupular, entire to 4–5-lobed, radially symmetrical but sometimes split irregularly. The corolla is radially symmetrical, with 4–5 free to coherent or connate petals. In most genera, the petals have a thickened midrib and an inflexed tip. Stamen number is equal to petal number, although extra stamens of irregular length have rarely been observed (*Gomphandra*). The stamens are usually exserted and have clavate hairs; they are often showier than the petals. The filaments (most genera) are flattened and dilated towards the apex. The anthers are basifixed, longitudinally dehiscent, and introrse (extrorse in *Grisollea*). Fresh pollen is usually white. Staminodes in pistillate flowers (absent in *Grisollea*) resemble the stamens but are often shorter and less pubescent and have pollen sacs which are brownish and devoid of pollen. Nectariferous tissue is present in most genera (Schori, unpublished data) but is not always clearly differentiated. Some genera have a morphologically discrete disk at the base of the ovary or ovary rudiment (in staminate flowers), while others have a swollen but relatively undifferentiated region characterised by nectary slits. The disk may be annular, squamiform, or hippocrepiform. The ovary is superior, approximately equal in length to the stamens or petals, apparently 1-carpellate, unilocular, with two pendent apical ovules, only one of which develops. The stigma is sessile or on a short style, and is capitate, lobed or punctiform. In several genera it is as broad as the ovary (accrescent in *Gomphandra*). The ovary rudiment in staminate flowers is minute and much shorter than the stamens or petals, conical to cylindrical, often apparently embedded in a disk, with a very reduced or absent stigma. Ovules have been reported from ovary rudiments, but there is no evidence that they mature.

Embryology. Embryology has been investigated in *Stemonurus* (Mauritzon 1936) and *Gomphandra* (Fagerlind 1945; Padmanabhan 1961). The tapetum is secretory and cells of the inner layer are multinucleate. Cytokinesis is simultaneous after meiosis, and the pollen grains are 2-celled when shed. The ovules are anatropous, unitegmic and crassinucellate. The vascular bundle in the chalaza extends up to the top of the single integument. In the linear tetrad of megaspores, the chalazal megaspore develops into a Polygonum-type embryo sac while the other megaspores disintegrate. The synergids are pyriform and hooked, with the filiform apparatus. The polar nuclei fuse at the time of fertilisation and the 3 antipodal cells are ephemeral. Endosperm development is nuclear. Embryogeny is unknown.

Pollen Morphology. Pollen grains are oblate to markedly oblate (Dahl 1952; Erdtman 1966) to oblately spheroidal in *Codiocarpus*. Grains are typically triporate and triangular in polar view, but additional pores (4–5, rarely up to 9) have been recorded for most genera (Schori and Furness 2014). Their size varies from 6 x 19 μm in *Stemonurus* to 21 x 32 μm in *Gomphandra*. Polar to equatorial quota $\leq$0.9. Pollen grains commonly binucleate. *Grisollea* pollen has an echinulate exine (Rasoarimalala et al. 1982) which is also microperforate, as do *Codiocarpus*, *Discophora* and *Medusanthera*. Exines in other genera vary from striate (*Gomphandra*, *Whitmorea*) to rugulate (*Cantleya*, *Gastrolepis*, *Gomphandra*), fossulate-verrucate (*Stemonurus*), microgemmate (*Irvingbaileya*), or microechinate at the poles (*Lasianthera*) (Schori and Furness 2014). A distinct annulus is present in four genera, and seven genera have a vestibulum at the pores (Schori and Furness 2014). Irregular grains are common in certain species of *Gomphandra* and *Gastrolepis*. Pollen sterility (Schori and Furness 2014) has been recorded for both normal and irregular grains in *Gomphandra*.

Karyology. A chromosome number of n = 22 was reported for *Stemonurus* cf. *ammui* (Oginuma et al. 1998).

Pollination and Reproductive Systems. The flowers of Stemonuraceae are generally small to minute (*Whitmorea* excepted) and white, green, yellow or sometimes red (*Stemonurus*). Herbarium labels variously report certain species as having fragrant flowers (*Codiocarpus*, *Gomphandra*, *Grisollea*, *Medusanthera*, *Stemonurus*). Photographs of Stemonuraceae indicate that flowers are diurnal, though not all species or genera have been observed in flower. Individual flower duration is not known, but most herbarium specimens indicate a flowering time of one to several weeks for inflorescences or individual trees. Specimens of widely collected taxa provide evidence for blooming periods which range from one to several months, and some species may flower at any time of the year. Pollen is abundant in *Gomphandra* (Schori, pers. obs.), and it is possible that the clavate hairs characterising Stemonuraceae aid in pollen transfer and/or pollinator attraction. Although a disk is present in some genera, it is not known whether it has a role in pollination. Pollinators are not documented for any species.

Stemonuraceae have two main reproductive systems. Several genera have bisexual flowers (*Cantleya*, *Gastrolepis*, *Lasianthera*, *Stemonurus*, *Whitmorea*), although a unisexual flower is occasionally encountered. The other genera, with the exception of *Discophora*, are functionally dioecious. *Discophora* has both bisexual and staminate flowers. Ovules have been reported from ovary rudiments in staminate flowers, but there is no evidence that ovary rudiments ever develop into fruit. Pistillate flowers produce staminodes which are often quite similar to stamens but have sterile anthers. A pair of pistillate trees of *Gomphandra luzoniensis* in Hawaii produces fruit (but perhaps not viable seed), and parthenocarpy is known from *G. papuana*. One species of *Gastrolepis* may be agamospermous, given its extremely irregular pollen (Schori and Furness 2014). There is some evidence that species of *Gomphandra* may hybridise.

Fruit and Seed. The fruits of Stemonuraceae are indehiscent drupes with a thin, fleshy to leathery mesocarp, its inner part consisting of one to several layers of sclerotic bundles. The endocarp is formed of sclerified cells which enclose the seed and are perpendicular to the bundles of the mesocarp. Druses, crystal sand, rhombics and mucilage idioblasts may be present in the mesocarp. *Stemonurus* and *Whitmorea* have relatively large, bicoloured fruits. Parthenocarpy is present in *Gomphandra papuana*, often producing fruits which have no sclerified inner mesocarp, endocarp or seed. The drupe of *Cantleya* has an invaginated fruit wall. *Gomphandra* has a prominent terminal stigma and a feature which has been referred to as a 'sulcus complex' (Schori and Utteridge 2010)—a prominent pair of longitudinal bundles and grooves flanking a smaller central bundle. All other genera have a laterally compressed drupe and, of these, only *Grisollea* lacks a fleshy appendage. The fleshy appendage develops from the ovary wall, in a position which correlates with the placement of the sulcus complex in *Gomphandra*. The appendage can become quite large and may be larger than the main part of the fruit. Appendage colour ranges from white to pink and generally contrasts with the main fruit body. Seeds have a soft, vascularised testa. Most genera appear to have a prominent chalazal and post-chalazal bundle which can be adnate to the endocarp. At least two genera (*Codiocarpus* and *Medusanthera*) are evidently pachychalazal (Schori 2010).

Dispersal. Dispersal may occur by several routes. Fruits may drop off the parent tree and germinate in place. The fruits of mangrove species (some *Stemonurus*) may disperse via water. The most common mode of dispersal is by mammals or birds. Red leaf monkeys (Supriatna et al. 1986; Davies 1991), fruit pigeons (Frith et al. 1976), birds of paradise (LeCroy et al. 1980), civet cats and cassowaries have all been reported as dispersers.

Phytochemistry. According to Kaplan et al. (1991), both *Cantleya* and *Lasianthera* contain the same carbocyclic or similar iridoid constituents. Trunk bark of *Cantleya* contains cantleyoside, a glucoside of the ester of secologanic acid with loganin. Only cantleyine is isolated when the extract is treated with ammonia.

Subdivisions and Relationships Within the Family. No molecular phylogeny for the family has yet been published.

Affinities. The family was formerly included in Icacinaceae, but as early as 1941 it had emerged as a distinct group on evidence of wood anatomy (Bailey and Howard 1941b). Recent morphological and molecular analyses have supported the distinctness of the group and have found it be widely separated from Icacinaceae sensu stricto. Accordingly, Stemonuraceae were described as a new family by Kårehed (2001). For further discussion, see Kårehed (2002). Molecular data place Stemonuraceae in Aquifoliales as sister to Cardiopteridaceae (Kårehed 2001; Winkworth et al. 2008). Morphologically, the family is not especially distinctive, but terminal bud shape, branching pattern, usually cymose inflorescences, filaments with clavate hairs, the unusual fruits, and the vascularised testa are reliable characters for its identification.

Distribution and Habitats. The family occurs in all major tropical wet forest regions. *Lasianthera* (1 sp.) is found in Guineo-Congolan Africa, *Grisollea* (3 spp.) in Madagascar, the Seychelles and the Comoros, *Irvingbaileya* (1 sp.) in Queensland, and *Discophora* (2–3 spp.) from Panama to Brazil, while the other 8 genera grow in tropical and subtropical Asia. *Gomphandra* (~60 spp.) grows from India to China and the Solomons; *Stemonurus* (14 spp.) from the Andamans and Sri Lanka to the Solomons; *Medusanthera* (10 spp.) from Malaya to Samoa; and *Cantleya*, *Codiocarpus*, *Gastrolepis*, *Hartleya* and *Whitmorea* each have 1–2 spp. with narrower distributions. Evergreen tropical rainforest is the typical habitat, though certain species occur in coastal forests or in mangrove swamps.

Palaeobotany. Fossil pollen of *Stemonurus* has been reported from a Holocene peat and a Miocene coal deposit in Borneo (Anderson and Muller 1975). Wood pieces taken from the Deccan Intertrappean beds (Upper Cretaceous to Lower Tertiary) of Kachchh, Gujarat, Western India were identified as *Gomphandroxylon samnapurensis* (Guleria and Srivastava 2001). Suparan et al. (2001) found *Stemonurus* pollen to be present at a depth of approximately 400 m at the Kau Plain in Halmahera, Indonesia. Pollen of *Lasianthera* was found in Upper Quaternary sediments from Lake Barombi in western Cameroon (Maley and Brenac 1998). Silicified wood of *Icacinoxylon cantleyoides* and *Icacinoxylon crassiradiatum*, related to *Cantleya*, has been found in southern Germany and dated to the Late Ottnangian to Middle Badenian period (Böhme et al. 2007).

Economic Importance. The Malesian *Cantleya corniculata* produces highly prized timber often used in house- or ship-building or as a substitute for sandalwood. *Codiocarpus*, *Gomphandra*, *Stemonurus* and *Whitmorea* have minor local uses in construction. The wood of *Lasianthera africana* is used for large pestles, and the leaves are eaten as a vegetable (Akpanabiatu et al. 1998). This species is also used to treat hypertension and asthma in Nigeria (Ajibesin et al. 2008), and stomach aches in Cameroon (Sandberg et al. 2005).

Key to the Genera

1. Flowers bisexual 2
– Flowers unisexual 7
2. Twig tips usually resinous, fruit usually bicoloured, always uncompressed laterally and without appendage 3
– Twigs not resinous, fruit either not bicoloured or if bicoloured then laterally compressed with an appendage 4
3. Calyx less than 5 mm wide, stamens exserted, stigma punctiform; widespread throughout Malesia and Melanesia **11. *Stemonurus***
– Calyx at least 5 mm wide, stamens included, stigma invaginated; restricted to Solomon Islands **12. *Whitmorea***
4. Inflorescence bracts not imbricate, fruit laterally compressed, appendage present; extra-Malesian 5
– Inflorescence with many imbricate bracts, fruit not laterally compressed, appendage absent; Malesia **1. *Cantleya***
5. Twigs stout, inflorescences not umbellate 6
– Twigs slender, inflorescences umbellate; wet tropical Africa **9. *Lasianthera***
6. Stigma papillose, anthers pubescent ventrally but hairs not attached to a flap of tissue; New Caledonia **4. *Gastrolepis***
– Stigma not papillose, ventral anther pubescence attached to a flap of tissue; Central and South America **3. *Discophora***
7. Staminate material (unknown for *Hartleya*) 8
– Pistillate material (including fruits) 13
8. Stamens flattened and dilated 9
– Stamens filiform; Andamans and Philippines **2. *Codiocarpus***
9. Anther dehiscence introrse, petals visible without magnification 10
– Anther dehiscence extrorse, petals vestigial; Madagascar, Seychelles, Comoros **6. *Grisollea***
10. Petals free at anthesis 11
– Petals connate at anthesis; widespread in Southeast Asia to Melanesia **5. *Gomphandra***
11. Ovary rudiment without a lateral basal swelling, pollen not echinulate 12
– Ovary rudiment with a lateral basal swelling, pollen echinulate; Malesia and Melanesia **10. *Medusanthera***
12. Ovary rudiment embedded in a discrete cupular disk, pollen with no discernible exine ornamentation (light microscopy); northern Queensland **8. *Irvingbaileya***
– Ovary rudiment swollen at base or not but without a discrete disk, pollen ornamentation various but discernible; widespread in Southeast Asia to Melanesia, Queensland **5. *Gomphandra***
13. Petals well-developed at maturity, stigma without projections, fruit, when laterally compressed, bearing an appendage 14
– Petals vestigial, stigma with a ring of small projections, fruit laterally compressed but without an appendage; Madagascar, Seychelles, Comoros **6. *Grisollea***
14. Ovary and fruit with discrete lateral disk 15
– Ovary and fruit without discrete lateral disk 17
15. Lateral disk squamiform, flowers small but more than 3 mm across, fruits green with white or pink appendage 16
– Lateral disk reniform, flowers minute (2–3 mm across), fruits black with white or yellow appendage; Central and South America **3. *Discophora***
16. Lateral disk trilobed, dry fruit with more than 12 closely spaced, evenly prominent ridges on side without appendage, not winged; Papua New Guinea **7. *Hartleya***
– Lateral disk entire, dry fruit with a few unequal ridges, flattened and almost winged; Queensland **8. *Irvingbaileya***
17. Petals free, staminodes rudimentary, much shorter than ovary, fruit white with two prominent ribs under the appendage; Andamans and Philippines **2. *Codiocarpus***
– Petals free to connate, staminodes well developed, as long as ovary, fruit colour various, appendage absent or if present without prominent ribs 18
18. Petals connate (rarely free), ovary generally not curved, fruit uncompressed, one colour, without appendage; widespread in Southeast Asia and Melanesia **5. *Gomphandra***
– Petals free, ovary generally strongly curved in sicco, fruit laterally compressed, bicoloured, with appendage; Malesia and Melanesia **10. *Medusanthera***

Genera of Stemonuraceae

1. *Cantleya* Ridl.

Cantleya Ridl., Fl. Malay. Penins. 1: 436 (1922); R.A. Howard, J. Arnold Arbor. 21: 479, t. 2, f. 1–7 (1940), rev.; Sleumer, Fl. Malesiana 7(1): 51 (1972), rev.

Trees. Leaves ovate-elliptic, thinly coriaceous. Inflorescences cymose, axillary and solitary, peduncle with minute imbricate bracts. Flowers bisexual, (4–)-5-merous. Sepals cupular, lobed. Petals free, probably coherent. Stamens free; filaments flattened and dilated upwards, with clavate-penicillate hairs. Ovary elongate-conical, stigma capitate, slightly lobed. Drupe fusiform, inner mesocarp thickly sclerified, fruit wall invaginated on one side.

One sp. (*C. corniculata* (Becc.) R.A. Howard), Malaysia, Indonesia (Sumatra, Kalimantan).

2. *Codiocarpus* R.A. Howard

Codiocarpus R.A. Howard, Brittonia 5: 53, f. 2 & 3 e–f (1943); Sleumer, Blumea 17: 188 (1969), rev.

Trees. Leaves oblong-elliptic, chartaceous to thinly coriaceous, venation prominent. Inflorescences cymose, multiply branched, axillary or often ramiflorous. Flowers unisexual, 5(−6)-merous. Sepals fused. Petals free, recurved. Staminate flowers: stamens free, conspicuous, filaments filiform; ovary rudiment small, globose with a short style. Pistillate flowers: sepals and petals shorter than in male flowers, staminodes weakly developed, shorter than ovary; ovary sub-cylindrical, stigma capitate. Drupe oblongoid, compressed, slightly curved, with several prominent longitudinal ridges on the convex surface, and a longitudinal, median furrow covered by the fleshy appendage on the concave surface.

One to three spp., Philippines, India (Andaman and Nicobar Is.), one collection from Aru Islands.

3. *Discophora* Miers

Discophora Miers, Ann. Mag. Nat. Hist. II, 10: 118 (1852); Duno de Stefano, Candollea 63: 177–187 (2008), rev.
Kummeria Mart. ex Engl. (1872).

Shrubs or trees. Leaves oblong-elliptic, coriaceous, venation prominent below. Inflorescences thyrsoid, axillary, supra-axillary or cauliflorous. Flowers minute, bisexual or unisexual (staminate), 5-merous. Sepals cupular, apiculate. Petals free. Staminate flowers: stamens free, filaments flattened, with an inconspicuous adaxial swelling with tangled clavate hairs below the anthers. Ovary rudiment minute, depressed, apparently embedded in an encircling disk, style apiculate. Bisexual flowers: stamens as in staminate flowers, ovary cylindrical, disk reniform, subtending lateral appendage, stigma sessile, capitate. Drupe compressed, slightly curved, black, with 3–7 longitudinal ridges on the convex side, and a white fleshy appendage on the concave side.

Two to three spp., Central America, N South America to Ecuador, Peru, Brazil, Bolivia.

4. *Gastrolepis* Tiegh.

Gastrolepis Tiegh., Bull. Soc. France 44: 115 (1897).

Trees. Leaves ovate-elliptic to obovate, coriaceous. Inflorescences of clustered flowers in cymose panicles, axillary or terminal. Flowers bisexual, 5-merous. Calyx cupular, lobed. Petals free at apex, coherent below. Stamens free, filaments flattened, with clavate hairs on both sides below the anthers. Disk squamiform, unilateral, subtending the lateral appendage. Ovary asymmetrically curved, stigma papillose. Drupe compressed, slightly curved, with multiple low longitudinal ridges on the convex side, and a fleshy appendage on the concave side.

Two spp., New Caledonia.

5. *Gomphandra* Wall. ex Lindl. Fig. 71

Gomphandra Wall. ex Lindl., Nat. Syst. Bot. ed. 2: 439 (1836); Sleumer, Blumea 17: 189 (1969), rev.
Stemonurus Blume p.p. (1825–1826).
Platea Miers (1852) non Blume.
Lasianthera Miq. (1856) non P. Beauv.

Shrubs to medium-sized trees. Leaves chartaceous to coriaceous. Inflorescences axillary, leaf-opposed, or terminal. Flowers unisexual. Calyx cupular, entire to minutely apiculate or lobed. Petals 4–5, free to connate. Staminate flowers: stamens 4–5, generally exserted, filaments flattened and dilated upwards, glabrous or more often pubescent ventrally, dorsally, and/or at the connective; ovary rudiment small, cylindrical to ovoid-conical, possibly embedded in a disk. Pistillate flowers: staminodes resembling stamens, sometimes less pubescent; ovary cylindrical, sometimes gibbous at base but without lateral appendage, stigma sessile, capitate, often lobed or ridged, accrescent. Fruit not compressed, drupe shape various, flesh thin, inner mesocarp usually with 8–13 longitudinal ridges, prominent sulcus complex on one side (stigma often displaced towards sulcus).

About 60 spp., tropical Asia to NE Australia. Certain species of *Gomphandra* with free petals or with very immature buds may be extremely difficult to distinguish from *Medusanthera*. Aspects of leaf venation, calyx pubescence, bract position, and inflorescence branching often must be used to separate specimens of *M. gracilis* and *M. laxiflora* from *Gomphandra* in regions where the genera co-occur.

6. ***Grisollea*** Baill.

Grisollea Baill., Adansonia 4: 217, t. 3 & 4 (1863–64).

Small trees. Leaves oblong-elliptic to obovate, thinly to thickly coriaceous. Inflorescences axillary or terminal cymose panicles. Flowers unisexual, 5-merous, minute. Calyx deeply lobed. Disk absent. Staminate flowers: petals minute, not inflexed at tip, adherent to calyx at base; anthers extrorse; ovary rudiment sub-conical or globular. Pistillate flowers: petals minute, not inflexed at tip, adherent to calyx at base; staminodes absent; ovary cylindrical, curved, pubescent, with minute projections at the centre and around the edges of the stigma. Drupe elliptic, compressed, with several longitudinal ridges on the convex side and a sulcus on the concave side.

Three spp., Madagascar, Seychelles, Comoros.

7. ***Hartleya*** Sleumer

Hartleya Sleumer, Blumea 17: 218 (1969).

Trees. Leaves ovate-elliptic to obovate, coriaceous. Inflorescences cymose, axillary. Flowers unisexual, 4–5-merous. Staminate flowers unknown. Pistillate flowers: calyx cupular, lobed. Petals free. Staminodes pubescent dorsally and ventrally below the anthers. Disk squamiform, subtly trilobed, subtending ovary appendage. Ovary with a lateral appendage, scarcely differentiated in flower; stigma sessile, subcapitate, papillose, displaced towards ovary appendage. Drupe curved, somewhat compressed, ovate-acuminate, with numerous prominent nerves on the convex side, and a short rib from the apex to the appendage on the concave side, appendage attached to a thickened pad, disk evident below the appendage.

One sp. (*H. inopinata* Sleumer), New Guinea.

8. ***Irvingbaileya*** R.A. Howard

Irvingbaileya R.A. Howard, Brittonia 5: 50 (1943).

Trees. Leaves elliptic, firmly chartaceous. Inflorescences axillary cymes. Flowers unisexual, 5-merous. Calyx cupular, shallowly lobed. Petals free. Staminate flowers: stamens free, filaments flattened, villous above middle. Ovary rudiment conical, immersed in a cupular disk. Pistillate flowers: staminodes present. Ovary asymmetric, with a disk at base on concave side. Drupe asymmetric, calyx displaced towards concave side, ellipsoid or ovoid, flattened and winged, with multiple longitudinal ridges on the concave side, with a much larger white fleshy lateral appendage on the convex side.

One sp. (*I. australis* (C.T. White) R.A. Howard), NE tropical Australia.

9. ***Lasianthera*** P. Beauv.

Lasianthera P. Beauv., Fl. Oware and Benin 1: 85, t. 51 (1806).

Trees or shrubs. Leaves lanceolate to ovate, elliptic, or obovate, often caudate, chartaceous. Inflorescences umbels, opposite the leaves. Flowers bisexual, 5-merous. Calyx cupular, lobed. Petals connate, apices free. Filaments free, with short straight hairs on an adaxial flap below the anthers and long tangled hairs abaxially at the connective. Disk unilateral, fleshy, covering half of the ovary in flower, overlapping the lateral appendage. Ovary and fruit with a lateral appendage. Drupe laterally compressed, oblong-elliptic, flattened, longitudinally ribbed.

One sp. (*L. africana* P. Beauv.) with two varieties, tropical Africa.

10. ***Medusanthera*** Seem.

Medusanthera Seem., J. Bot. 2: 74 (1864); R.A. Howard, Lloydia 6: 133 (1943), rev.; Sleumer, Blumea 17: 266 (1969), rev.
Tylecarpus Engl. (1893).

Trees. Leaves various in shape, chartaceous to coriaceous. Inflorescences axillary panicles. Flowers unisexual, 5-merous. Calyx cupular, minutely toothed or lobed. Petals free. Staminate flowers: stamens free, filaments flattened and fleshy, persistently clavate-hairy to barbate in the upper part; ovary rudimentary, with a distinct lateral swelling. Pistillate flowers: staminodes present; ovary sub-cylindrical, base with a lateral swelling. Drupe oblongoid to ellipsoid, or sub-ovoid, laterally compressed, the concave side with a fleshy appendage and 1–3(−5) prominent longitudinal ridges on the convex side.

Ten spp., Fiji, Samoa, Malesia (Malaysia, Sumatra, the Philippines, and Sulawesi to the Carolines, Bismarcks, Solomons). *Medusanthera gracilis* and *M. laxiflora* may be difficult to distinguish from certain species of *Gomphandra* if only staminate or very immature floral material is available. See notes under *Gomphandra*.

11. ***Stemonurus*** Blume

Stemonurus Blume, Bijdr. Fl. Ned. Ind. 648 (1826); Becc., Malesia 1: 108–111 (1877); Sleumer, Blumea 17: 255 (1969), rev.
Urandra Thwaites (1855).
Lasianthera Miq. (1856) non P. Beauv.

Trees, some species with buttresses or pneumatophores. Terminal buds abruptly contracted, resinous. Leaves lanceolate to elliptic, ovate, or obovate, often stiffly coriaceous. Inflorescences axillary, either compound scorpioid cymes or umbelliform. Flowers bisexual, sessile, (4–)5-merous. Calyx cupular, almost entire to shallowly lobed, persistent in fruit. Petals connate at base, coherent into a tube with the sutures evident. Stamens 5, connate with the petals at base only, filaments fleshy, flat, pubescent below the anther and long-penicillate distally on the abaxial side. Disk cupular, often lobed. Ovary conical. Style very short, stigma punctiform. Drupe ovoid, ellipsoid or oblongoid-fusiform, mesocarp bicoloured (except *S. corrugatus*).

Approximately 14 spp., Andamans, Vietnam (Cochinchina and Annam), Sri Lanka, Malesia.

12. ***Whitmorea*** Sleumer

Whitmorea Sleumer, Blumea 17: 263 (1969).

Trees. Terminal buds resinous. Leaves elliptic, coriaceous. Flowers bisexual, (4–)5-merous. Inflorescence axillary, of 3–4 umbelliform scorpioid cymes in a cross-like arrangement with a common peduncle. Calyx cupular, entire. Petals generally connate at base. Stamens fused to the petals; filaments trapezoidal in cross-section, fleshy, with subclavate hairs dorsally below the anthers. Ovary ovoid-cylindrical, apex truncate, invaginated, stigma within invagination. Drupe oblong-ellipsoid, bicoloured.

One sp. (*W. grandiflora* Sleumer), Solomon Is.

Selected Bibliography

Ajibesin, K.K., Ekpoa, B.A., Bala, D.N., Essien, E.E., Adesanya, S.A. 2008. Ethnobotanical survey of Akwa Ibom State of Nigeria. J. Ethnopharmacol. 115: 387–408.

Akpanabiatu, M.I., Bassey, N.B., Udosen, E.O., Eyong, E. U. 1998. Evaluation of some minerals and toxicants in some Nigerian soup meals. J. Food Composition Analysis 11: 292–297.

Anderson, J.A.R., Muller, J. 1975. Palynological study of a Holocene peat and a Miocene coal deposit from NW Borneo. Rev. Palaeobot. Palynol. 19: 291–351.

Baas, P. 1974. Stomatal types in Icacinaceae. Additional observations on genera outside Malesia. Acta Bot. Neerl. 23: 193–200.

Bailey, I.W., Howard, R.A. 1941a. The comparative morphology of the Icacinaceae. I. Anatomy of the node and internode. J. Arnold Arbor. 22: 125–131.

Bailey, I.W., Howard, R.A. 1941b. The comparative morphology of the Icacinaceae. II. Vessels. J. Arnold Arbor. 22: 171–187.

Bailey, I.W., Howard, R.A. 1941c. The comparative morphology of the Icacinaceae. III. Imperforate tracheary elements and xylem parenchyma. J. Arnold Arbor. 22: 432–442.

Böhme, M., Bruch, A.A., Selmeier, A. 2007. The reconstruction of Early and Middle Miocene climate and vegetation in Southern Germany as determined from the fossil wood flora. Palaeogeogr. Palaeoclimatol. Palaeoecol. 253: 91–114.

Dahl, A.O. 1952. The comparative morphology of the Icacinaceae. 4. The pollen. J. Arnold Arbor. 33: 252–291.

Davies, G. 1991. Seed-eating by red leaf monkeys (*Presbytis rubicunda*) in dipterocarp forest of Northern Borneo. Int. J. Primatol. 12: 119–144.

Duno de Stefano, R. 2008. El género *Discophora* Miers (Stemonuraceae) en el Neotrópico. Candollea 63: 177–187.

Erdtman, G. 1966. Pollen morphology and plant taxonomy. Angiosperms. New York: Hafner.

Fagerlind, F. 1945. Bau des Gynoeceums, der Samenlage und des Embryosackes bei einigen Repräsentanten der Familie Icacinaceae. Svensk Bot. Tidskr. 39: 346–364.

Frith, H.J., Crome, F.H.J., Wolfe, T.O. 1976. Food of fruit-pigeons in New Guinea. The Emu 76: 49–58.

Guleria, J.S., Srivastava, R. 2001. Fossil dicotyledonous woods from the Deccan Intertrappean beds of Kachchh, Gujarat, Western India. Palaeontographica Abt. B 257: 17–33.

Heintzelman, C.E., Howard, R.A. 1948. The comparative morphology of the Icacinaceae. V. The pubescence and the crystals. Am. J. Bot. 35: 42–52.

Howard, R.A. 1943. Studies of the Icacinaceae. VI. *Irving-baileya* and *Codiocarpus*, two new genera of the Icacineae. Brittonia 5: 47–57.

Kaplan, M.A.C., Ribeiro, J., Gottlieb, O.R. 1991. Chemo-geographical evolution of terpenoids in Icacinaceae. Phytochemistry 30: 2671–2676.

Kårehed, J. 2001. Multiple origin of the tropical forest tree family Icacinaceae. Am. J. Bot. 88: 2259–2274.

Kårehed, J. 2002. Not just hollies – the expansion of Aquifoliales. In: Evolutionary Studies in Asterids emphasising Euasterids II. Uppsala, Sweden, Acta Univ. Uppsala, thesis 761.

LeCroy, M., Kulupi, A., Peckover, W.S. 1980. Goldie's bird of paradise: display, natural history and traditional relationships of people to the bird. The Wildon Bull. 92: 289–301.

Lens, F., Kårehed, J., Baas, P., Jansen, S., Rabaey, D., Huysmans, S., Hamann, T., Smets, E. 2008. The wood anatomy of the polyphyletic Icacinaceae s.l., and their relationships within asterids. Taxon 57: 525–552.

Maley, J., Brenac, P. 1998. Vegetation dynamics, palaeoenvironments and climatic changes in the forests of western Cameroon during the last 28,000 years B.P. Rev. Palaeobot. Palynol. 99: 157–187.

Mauritzon, J. 1936. Embryologische Angaben über Stackhousiaceae, Hippocrateaceae und Icacinaceae. Svensk Bot. Tidskr. 30: 541–550.

Munzinger, J., McPherson, G., Lowry II, P.P. 2008. A second species in the endemic New Caledonian genus *Gastrolepis* (Stemonuraceae) and its implications for the conservation status of high-altitude maquis vegetation: coherent application of the IUCN Red List criteria is urgently needed in New Caledonia. Bot. J. Linn. Soc. 157: 775–783.

Oginuma, K., Kiaptranis, R., Damas, K., Tobe, H. 1998. A cytological study of some plants from Papua New Guinea. Acta Phytotaxon. Geobot. 49: 105–114.

Padmanabhan, D. 1961. A contribution to the embryology of *Gomphandra polymorpha*. Proc. Natl. Inst. Sci. India, Part B, Biol. Sci. 27: 389–398.

Rasoarimalala, L., Albers, F., Straka, H. 1982. Palynologia Madagassica et Mascarenica. Pollen et Spores 24: 65–91.

Sandberg, F., Perera-Ivarsson, P., Rushdey El-Seedi, H.R. 2005. A Swedish collection of medicinal plants from Cameroon. J. Ethnopharmacol. 102: 336–343.

Schori, M. 2010. A systematic revision of *Gomphandra* (Stemonuraceae). Ph.D. dissertation, Ohio University, Athens, OH.

Schori, M., Furness, C.A. 2014. Pollen diversity in Aquifoliales. Bot. J. Linn. Soc. 175: 169–190.

Schori, M., Utteridge, T.M.A. 2010. Three new species and a new name in Southeast Asian *Gomphandra* (Stemonuraceae/Icacinaceae s.l.). Blumea 55: 189–195.

Sleumer, H. 1969. Materials towards the knowledge of the Icacinaceae of Asia, Malesia, and adjacent areas. Blumea 17: 181–264.

Sleumer, H. 1972. Icacinaceae. In: van Steenis, C.G.G.J. (ed.) Flora Malesiana 7: 1–87. Leyden: Noordhoff.

Solereder, H. 1908. Systematic anatomy of the dicotyledons. Oxford: Clarendon Press.

Suparan, P., Dam, R.A.C., van der Kaars, S., Wong, T.E. 2001. Late Quaternary tropical lowland environments on Halmahera, Indonesia. Palaeogeogr. Palaeoclimatol. Palaeoecol. 17: 229–258.

Supriatna, J., Manullang, B.O., Soekara, E. 1986. Group composition, home range, and diet of the maroon leaf monkey (*Presbytis rubicunda*) at Tanjung Puting Reserve, Central Kalimantan, Indonesia. Primates 27: 185–190.

Teo, S.P., Haron, N.W. 1999. Anatomical studies in West Malaysian Icacinaceae. Austr. Syst. Bot. 11: 729–738.

Utteridge, T.M.A., Schori, M. 2009. *Stemonurus corrugatus* (Stemonuraceae, Icacinaceae s.l.) a new species from Sarawak. Kew Bull. 64: 327–331.

van Staveren, M.G.C., Baas, P. 1973. Epidermal leaf characters of the Malesian Icacinaceae. Acta Bot. Neerl. 22: 329–359.

Winkworth, R.C., Lundberg, J., Donoghue, M. J. 2008. Toward a resolution of Campanulid phylogeny, with special reference to the placement of Dipsacales. Taxon 57: 53–65.

Tribelaceae

Tribelaceae (Engl.) Airy Shaw, Kew Bull. 18: 269 (1965).
Tribeleae Engl. in Engler & Prantl, Nat. Pflanzenfam., 2nd edn, 18a: 111 (1930) ('Tribeliseae').

A. Backlund

Small creeping subshrubs, roots developing from the nodes; entire plant glabrous. Leaves alternate, pinnately veined with sheathing leaf-bases, simple, glaucescent, leaf margins entire but conspicuously tridentate at leaf tip. Flowers solitary, terminal, bisexual, ± actinomorphic, pentamerous; sepals basally fused, imbricate in bud, persistent; petals appearing distinct, imbricate or slightly contorted in bud; stamens oppositisepalous, distinct and isomerous with the petals, anthers basifixed, dithecal and tetrasporangiate; ovary epigynous, of three carpels, syncarpous, trilocular, style simple, stigma tri-lobed; fertile ovules numerous, anatropous, unitegmic, tenuinucellate. Fruit a many-seeded, ovate, loculicidal leathery capsule, seeds remaining attached to fruit axis after dehiscence. Endosperm nuclear, seeds small, rounded to obovate, testa black.

Tribelaceae comprise only one monotypic genus in southern Chile and Argentina.

Vegetative Morphology. Small creeping or incidentally epiphytic subshrubs with semi-erect, slender and densely leafy twigs, reaching up to 0.3 m in height. Roots developing from the nodes. Leaves simple, glaucescent, up to 1.5 cm long, alternating, leaf-bases sheathing. Leaf venation pinnate but sparse, leaf margins entire but conspicuously tridentate at leaf tip. The entire plant is glabrous.

Vegetative Anatomy and Ultrastructure. Very little is known about the anatomy and ultrastructure of *Tribeles*. Calcium oxalate crystals occur as druses, abundantly in leaves, flowers and wood. Wood fibres are non-septate and of the tracheid type.

Flower Structure and Floral Anatomy (Fig. 72). Flowers solitary, terminal, with one of the petals abaxially oriented. Sepals green with one central prominent vein. Petals white or pinkish, their apex slightly inflexed or clawed, veins prominent, branched. Floral nectaries have not been observed. Anthers stout, basifixed and non-versatile, extrorse, opening with longitudinal slits, staminodia absent. Crystal druses prominent in anther walls. Gynoecium syncarpous, of three carpels, trilocular, epigynous. Each locule with several ovules with axile placentation, placenta non-intrusive. Style simple, stigma trilobate.

Fruit and Seed. Fruit a many-seeded, ovate, loculicidal leathery capsule, brownish, ca. 8–10 mm long with persistent remains of calyx and style. Seeds small, ca. 1 mm, black, rounded to obovate, with a black and slightly rugose testa of hexangular cells.

Embryology. Endosperm formation is nuclear, resulting in a nuclear, oily, copious endosperm. Embryogenesis has not been studied in *Tribeles*. The dicotyledonous embryo is straight, small, but well developed, without chlorophyll detected.

Pollen Morphology. Pollen grains spherical, tricolporate, semi-tectate, with a well developed layer of branched columellae.

Affinities. Traditionally *Tribeles* has been treated as a member of Escalloniaceae, or placed in subfamily Escallonioideae of Saxifragaceae (Engler 1930). Based largely on nucleotide sequence information, a new interpretation of the former Escalloniaceae/Escallonioideae sensu Engler has

J.W. Kadereit and V. Bittrich (eds.), *Flowering Plants. Eudicots*, The Families and Genera of Vascular Plants 14,
DOI 10.1007/978-3-319-28534-4_33, © Springer International Publishing Switzerland 2016

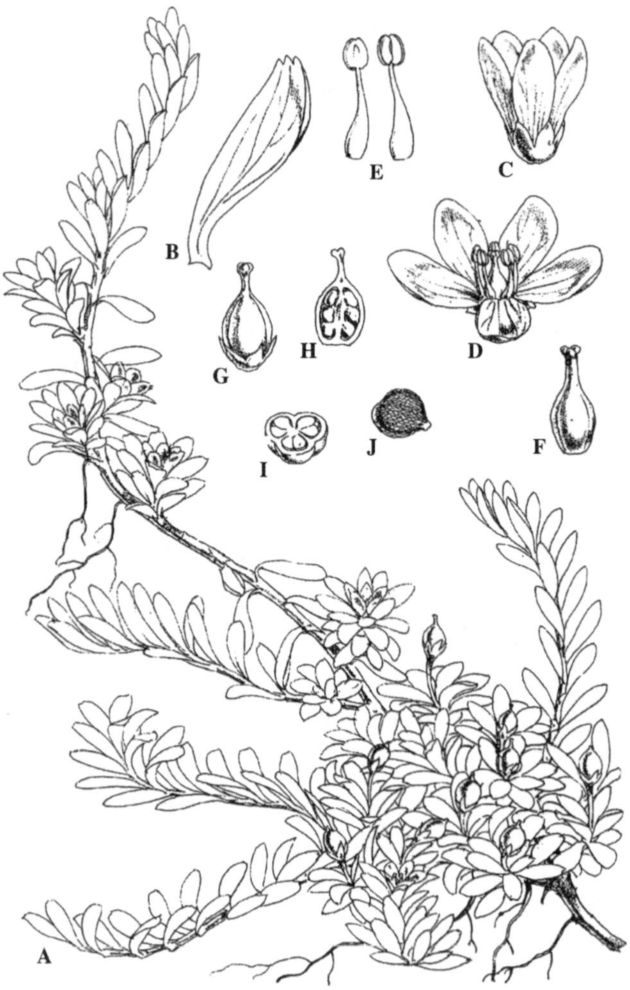

Fig. 72. Tribelaceae. **A–J** *Tribeles australis*. **A** Habit. **B** Leaf. **C, D** Flowers. **E** Stamens. **F** Gynoecium. **G** Young fruit. **H** Longitudinal section of young fruit. **I** Cross section of young fruit. **J** Immature seed. (Hooker 1871 [as *Chalepoa magellanica*], courtesy Missouri Botanical Garden)

emerged. In this recent view a number of genera of the former Escalloniaceae are shown to occupy systematic positions at different places in the Asteridae, even if not as the traditionally known Escalloniaceae. One of these is *Tribeles*, which in some studies shows affinities to *Escallonia*, *Quintinia*, *Eremosyne* and *Forgesia* (e.g. Gustafsson et al. 1996; Backlund and Bremer 1997; Savolainen et al. 2000; Bremer et al. 2001; Winkworth et al. 2008). In others, Escalloniaceae s.nov. also include *Valdivia* and *Polyosma* and are even recognized at ordinal level as Escalloniales, which were found to be the closest relative of the Asterales albeit with low support (e.g. Soltis et al. 2011). APG III (2009) includes Eremosynaceae, Polyosmaceae and Tribelaceae in Escalloniaceae s.l., which thus represents the only family of order Escalloniales.

Distribution and Habitats. The single species of *Tribeles*, *T. australis* Phil., is found in humid heathlands and grasslands on the western and eastern slope of the southern Andes from Tierra de Fuego to 40°S at an elevation from sea level in the south up to 1200 m at its northern border.

One genus:

1. ***Tribeles*** Phil. Fig. 72

Tribeles Phil., Anales de la Universidad de Chile, Santiago 23: 382 (1863).
Chalepoa Hook. f. (1871).

Characters as for family.

A single sp., *T. australis* Phil., in the southern Andes.

Selected Bibliography

Airy Shaw, H.K. 1966. Tribelaceae (Engl.). In: Airy Shaw, H.K. (ed.) Willis's A dictionary of the flowering plants and ferns. Cambridge: Cambridge University Press.

APG III 2009. An update of the Angiosperm Phylogeny Group classification for the orders and families of flowering plants: APG III. Bot. J. Linn. Soc. 161: 105–121.

Backlund, A., Bremer, B. 1997. Phylogeny of the Asteridae s.str. based on *rbc*L sequences, with particular reference to the Dipsacales. Plant Syst. Evol. 207: 225–254.

Bremer, K., Backlund, A., Sennblad, B., Swenson, U., Andreasen, K., Hjertson, M., Lundberg, J., Backlund, M., Bremer, B. 2001. A phylogenetic analysis of 100+ genera and 50+ families of euasterids based on morphological and molecular data. Plant Syst. Evol. 229: 137–169

Corner, E.J.H. 1976. The seeds of dicotyledons, vols. 1 and 2. Cambridge: Cambridge University Press.

Engler, A. 1930. Saxifragaceae. In: Engler, A., Die natürlichen Pflanzenfamilien, 2nd ed. 18a. Leipzig: W. Engelmann, pp. 74–226.

Gustafsson, M.H.G., Backlund, A., Bremer, B. 1996. Phylogeny of the Asterales sensu lato based on *rbc*L sequences with particular reference to the Goodeniaceae. Plant Syst. Evol. 199: 217–242.

Hooker, J.D. 1871. *Chalepoa magellanica*, Hook. f. Hooker's Icones Plantarum 11: 65–66, t. 1082.

Philippi, R.A., 1863. Nuevas Plantas Chilenas. Anales Universidad de Chile, Santiago 23: 382.

Philipson, W.R. 1974. Ovular morphology and the major classification of the dicotyledons. Bot. J. Linn. Soc. 68: 89–108.

Savolainen, V., Fay, M.F., Albach, D.C., Backlund, A., van der Bank, M., Cameron, K.M., Johnson, S.A., Lledó, M.D., Pintaud, J.-C., Powell, M., Sheahan, M.C., Soltis, D.E., Soltis, P.S., Weston, P., Whitten, W.M., Wurdack, K.J., Chase, M.W. 2000. Phylogeny of the eudicots: a nearly complete familial analysis based on *rbc*L gene sequences. Kew Bull. 55: 257–309.

Soltis, D.E., Smith, S.A., Cellinese, N., Wurdack, K.J., Tank, D.C., Brockington, S.F., Refulio-Rodriguez, N. F., Walker, J.B., Moore, M.J., Carlsward, B.S., Bell, C. D., Latvis, M., Crawley, S., Black, C., Diouf, D., Xi, Z., Rushworth, C.A., Gitzendanner, M.A., Sytsma, K.J., Qiu, Y.L., Hilu, K.W., Davis, C.C., Sanderson, M.J., Beaman, R.S., Olmstead, R.G., Judd, W.S., Donoghue, M.J., Soltis, P.S. 2011. Angiosperm phylogeny: 17 genes, 640 taxa. Amer. J. Bot. 98: 704–730.

Winkworth, R.C., Lundberg, J., Donoghue, M.J. 2008. Towards a resolution of campanulid phylogeny, with special reference to the placement of Dipsacales. Taxon 57: 53–65.

Vahliaceae

Vahliaceae Dandy in Hutch., Fam. Fl. Pl., ed. 2: 461 (1959).

M. Thiv

Annual or perennial, erect, branched herbs or subshrubs, glabrous or pubescent; nodes unilacunar. Leaves opposite, sessile to subsessile, exstipulate, simple, ovate to linear, entire. Inflorescences axillary, of paired, small hermaphrodite, actinomorphic flowers. Perianth with distinct pentamerous calyx and corolla; calyx lobes valvate; petals free, valvate, obovate-spathulate, entire to crenate-dentate, sometimes hairy at the base inside; stamens 5, inserted on the margin of an epigynous disk, antesepalous, filaments free, subulate, glabrous or sparsely hairy, sometimes with scale-like appendage at the base; anthers 2-thecous, introrse, dorsifixed; ovary inferior, campanulate to subglobose, 2–3-carpellate, syncarpous, unilocular, with 2–3 distinct, glabrous or sparsely hairy stylodia, stigmata capitate; ovules numerous, attached to two pendulous placentas, anatropous, bitegmic, tenuinucellate. Fruit a subglobose or obovoid capsule, apically with 2–3 valves. All floral parts except anthers persistent. Seeds numerous, minute, oblong.

A monogeneric family with five species in Africa, Iraq, Iran, Vietnam and the Indian subcontinent.

Vegetative Anatomy. The leaf and stem anatomy of *Vahlia* was studied by Inamdar and Patel (1971). They reported on the development of anomocytic, anisocytic and paracytic stomata, and of simple uniseriate filiform and capitate filiform glandular trichomes, which are found on the leaves, stems and calyces. They emphasised the distinctiveness in details of the structure and ontogeny of stomata in *Vahlia* from Saxifragaceae s.l., which supports a treatment as an independent family. The distribution and anatomy of trichomes was studied by Al-Shammary and Gornall (1994). Sieve element plastids, however, correspond to the S-type, also found in the Saxifragaceae s.l. (Behnke 1988). Druses in the cortical parenchyma and piths and resinous substances in intercellular canals in the cortex were observed. The cork initially arises in the epidermis. The wood is formed by evenly distributed vessels with simple perforation plates, uniseriate rays, libriform fibres, fibre-tracheids and vascular tracheids. Growth rings and axial parenchyma are absent (Gornall and Al-Shammary 1998).

Embryology. The embryology of *Vahlia* was described by Raghavan and Srinivasan (1942) and Saxena (1973). Characteristic are the pendant placentas (Fig. 73C), otherwise rarely observed in angiosperms, a cellular endosperm and a differentiated endothelium. Embryo-sac development follows the *Polygonum* type. The embryogeny of *Vahlia* differs from many Saxifragaceae s.l. and conforms to the Crucifer type (Saxena 1973). Plurinucleate tapetal cells are the result of repeated divisions of their nuclei (Raghavan and Srinivasan 1942).

Pollen Morphology. The palynology of *Vahlia* was studied by Hideux and Ferguson (1976) and Polevova (2006). The pollen is released in monads. The tricolporate, tectate-columellate grains are of oval shape with a complete and striate tectum.

Karyology. Base numbers of $x = 6, 7$ and 9 and karyotypes were reported for *Vahlia* by Hamel (1953) and Rangaswami Ayyangar (1971).

Pollination. The protogynous flowers of *Vahlia* (Engler 1930) with white or yellow petals are entomophilous. Visits of bees (*Afranthidium* Michener, Megachilidae) have been reported by Gess and Gess (2007).

J.W. Kadereit and V. Bittrich (eds.), *Flowering Plants. Eudicots*, The Families and Genera of Vascular Plants 14,
DOI 10.1007/978-3-319-28534-4_34, © Springer International Publishing Switzerland 2016

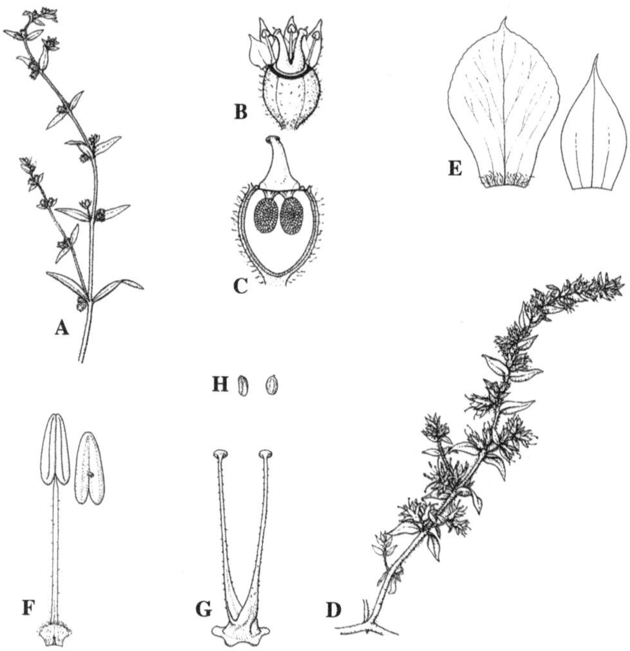

Fig. 73. Vahliaceae. **A–C** *Vahlia digyna.* **A** Habit. **B** Flower. **C** Pendant placentas. **D–H** *Vahlia somaliensis.* **D** Habit. **E** Petal and sepal. **F** Stamen. **G** Stylodia. **H** Seeds. (Bridson 1975, with permission of the Royal Botanic Gardens, Kew.)

Fruit and Seed. The septicidal capsules of *Vahlia* dehisce by two or three valves between the bases of the styles (Bridson 1975). The seed structure was described by Krach (1976). Seeds are oblong, ovoid to cylindrical or spindle-shaped and characterised by longitudinal ridges. Exotestal cells are elongate, while endotestal cells have U-shaped lignified cell wall thickenings (Nemirowich-Danchenko 1996). The endosperm contains storage lipids and aleuron grains.

Phytochemistry. Iridioids are present in Vahliaceae according to Erbar and Leins (2011). Majinda et al. (1997) detected kaempferol, quercetin, afzelin, astragalin, quercitrin, isoquercitrin, rutin, gallic acid, chiro-inositol, dulcitol and a specific biflavonoid in *Vahlia capensis.*

Family Status and Affinities. Traditionally, *Vahlia* has been regarded as part of Saxifragaceae (e.g. Engler 1930). It was transferred into the newly created monogeneric Vahliaceae by Dandy (1959), belonging to Saxifragales. A close relationship to Saxifragales, however, was later ruled out by several molecular analyses. All of them placed *Vahlia* in the Euasterid I clade (Morgan and Soltis 1993; Savolainen et al. 2000; Albach et al. 2001; Bremer et al. 2001), which corresponds to the Lamiidae of APG III (2009). Whereas such placement would be contradicted by the bitegmic ovules, it gains support by the presence of iridoids and tenuinucellate ovules. Nonetheless, the exact relationship remains controversial because it occupied different positions in the various molecular studies. At present, the best hypothesis may be a position as sister group of a clade formed by Boraginales, Lamiales, Solanales and Gentianales (Soltis et al. 2011).

Palaeobotany. Fossil flowers from the Upper Cretaceous (Late Santonian–Early Campanian) in Sweden described by Friis and Skarby (1982) as *Scandianthus* were found to strongly resemble those of *Vahlia* (Dahlgren 1983). Similarities concern the pentamerous perianth and the inferior, bicarpellate, unilocular ovary with two pendant placentae. The two taxa differ mainly in the number of stamens. Therefore, many authors postulate close affinities between Vahliaceae and *Scandianthus* (Morgan and Soltis 1993; Friis et al. 2006; Martínez-Millán 2010). This would be in accordance with an estimated stem node age of Vahliaceae in the Cretaceous (Bell et al. 2010).

Distribution and Habitats. Vahliaceae are a palaeotropical family with its centre of diversity in Africa where all five species occur. The distribution of three of these species extends to Iraq, Iran, Vietnam and the Indian subcontinent (for distribution map, see Bridson 1975). These intercontinental disjunctions are likely the result of recent long-distance dispersal of the numerous small seeds of the genus. The species mostly grow in semiarid areas, along sandy river banks, in cultivated land, grassland and woodland.

A single genus:

Vahlia Thunb. Fig. 73

Vahlia Thunb., Nova Gen. 2: 36 (1782), nom. cons.; Bridson, Kew Bull. 30:163–182 (1975), rev.
Bistella Adans., Fam. Pl. 2: 226 (1763), nom. rej.

Characters as for the family. Five species.

Selected Bibliography

Al-Shammary, K.I.A., Gornall, R.J. 1994. Trichome anatomy of the Saxifragaceae s.l. from the southern hemisphere. Bot. J. Linn. Soc. 114: 99–131.

Albach, D.C., Soltis, P.S., Soltis, D.E., Olmstead, R.G. 2001. Phylogenetic analysis of asterids based on sequences of four genes. Ann. Missouri Bot. Gard. 83: 163–212.

APG III (Angiosperm Phylogeny Group) 2009. An update of the Angiosperm Phylogeny Group classification for the orders and families of flowering plants: APG III. Bot. J. Linn. Soc. 161: 105–121.

Behnke, H.-D. 1988. Sieve element plastids and systematic relationships of Rhizophoraceae, Anisophyllaceae, and allied groups. Ann. Missouri Bot. Gard. 75: 1387–1409.

Bell, C.D., Soltis, D.E., Soltis, P.S. 2010. The age and diversification of the angiosperms re-revisited. Amer. J. Bot. 92: 284–296.

Bremer, K., Backlund, A., Sennblad, B. et al. 2001. A phylogenetic analysis of 100+ genera and 50+ families of euasterids based on morphological and molecular data with notes on possible higher level morphological synapomorphies. Plant Syst. Evol. 229: 137–169.

Bridson, D.M. 1975. A revision of the family Vahliaceae. Kew Bull. 30: 163–182.

Dahlgren, R. 1983. General aspects of angiosperm evolution and macrosystematics. Nordic J. Bot. 3: 119–149.

Dandy, J.E. 1959. Vahliaceae. In: Hutchinson, J., The Families of Flowering Plants, ed. 2, 1: 1–510.

Engler, H. 1930. Saxifragaceae. In: Engler, A., Prantl, K. Die natürlichen Pflanzenfamilien, 2nd edn, 18a: 74–226.

Erbar, C., Leins, P. 2011. Synopsis of some important, non-DNA character states in the asterids with special reference to sympetaly. Plant Div. Evol. 129: 93–123.

Friis, E.M., Skarby, A. 1982. *Scandianthus* gen. nov., angiosperm flowers of Saxifragalean affinities from the Upper Cretaceous of southern Sweden. Ann. Bot. 50: 569–583.

Friis, E.M., Pedersen, K.R., Crane, P.R. 2006. Cretaceous angiosperm flowers: Innovation and evolution in plant reproduction. Palaeogeogr. Palaeoclimatol. Palaeoecol. 232: 251–293.

Gess, S.K., Gess, F.W. 2007. Notes on nesting and flower visiting of some Anthidiine bees (Hymenoptera: Megachilidae: Megachilinae: Anthidiini) in Southern Africa. J. Hymenopt. Res. 16: 30–50.

Gornall, R.J., Al-Shammary, K.I.A. 1998. Vahliaceae. In: Cutler, D.F., Gregory, M. (eds.) Anatomy of the Dicotyledons 4: 232–235. Oxford: Clarendon Press.

Hamel, J.L. 1953. Contribution à l'étude cyto-taxinomique des Saxifragacées. Rev. Cytol. Biol. Vég. 14: 113–313.

Hideaux, M.J., Ferguson, I.K. 1976. The stereostructure of the exine and its evolutionary significance in Saxifragaceae sensu lato. In: Ferguson, I.K., Muller, J. (eds.) The evolutionary significance of the exine. London: Academic Press.

Inamdar, J.A., Patel, R.C. 1971. Structure and development of trichomes, stomata and systematic position of *Vahlia digyna* (Retz.) O. K. Acta Bot. Acad. Sci. Hung. 17: 361–369.

Krach, J.E. 1976. Samenanatomie der Rosifloren I. Die Samen der Saxifragaceae. Bot. Jahrb. Syst. 97: 1–60.

Majinda, R.R., Motswaledi, M., Waigh, R.D., Waterman, P.G. 1997. Phenolic and antibacterial constituents of *Vahlia capensis*. Planta Med. 63(3): 268–270.

Martínez-Millán, M. 2010. Fossil record and age of the Asteridae. Bot. Rev. 76: 83–135.

Morgan, D.R., Soltis, D.E. 1993. Phylogenetic relationships among members of Saxifragaceae sensu lato based on *rbc*L sequence data. Ann. Missouri Bot. Gard. 80: 631–660.

Nemirowich-Danchenko, E.N. 1996. Vahliaceae. In: Danilova, M., Anatomia seminum comparativa. Rosidae I. Vol. 5: 73–76. St. Petersburg: Mir et Semja.

Polevova, S.V. 2006. Review of the sporoderm ultrastructure of members of the Asterales. Paleontol. J. 40 Suppl. 5: S656–S663.

Raghavan, T.S., Srinivasan, V.K. 1942. A contribution to the life-history of *Vahlia viscosa* Roxb. and *Vahlia oldenlandioides* Roxb. Proc. Indian Acad. Sci. 15: 83–105.

Rangaswami Ayyangar, K. 1971. Cytology of *Vahlia oldenlandioides*. Proc. Indian Sci. Congr. Assoc. 58: 469.

Savolainen, V., Fay, M.F., Albach, D.C. et al. 2000. Phylogeny of the eudicots: a nearly complete familial analysis based on rbcL gene sequences. Kew Bull. 55: 257–309.

Saxena, N.P. 1973. Studies in the family Saxifragaceae VIII. Floral anatomy, seed development and systematics of *Bistella* Adans. J. Indian Bot. Soc. 52: 244–251.

Soltis, D.E., Smith, S.A., Cellinese, N. et al. 2011. Angiosperm phylogeny: 17 genes, 640 taxa. Amer. J. Bot. 98: 704–730.

Valerianaceae

Valerianaceae Batsch, Tab. Affin. Regni Veg.: 227 (1802), nom. cons.

F. WEBERLING AND V. BITTRICH

Annual to perennial, often fetid herbs, subshrubs, sometimes shrubs, rarely ligneous or herbaceous vines; root often a taproot, often slender, sometimes a napiform or fusiform turnip, the stem-borne roots sometimes tuberous; plants often semirosulate, sometimes rosulate; stem often fistulous. Leaves mostly decussate, rarely 3-verticillate, sometimes alternate, rarely distichous or tristichous, entire, pinnatifid or pinnate, exstipulate, rarely ericoid or carnose, glabrous or with simple and glandular hairs. Inflorescence usually bracteate, a dense thyrse or pleiothyrse without terminal flower, basimesotonic or acrotonic and pseudodichotomous, sometimes capitate by condensation of the flower-bearing branches; in some genera supernumerary bracts occur below the flowers. Flowers epigynous, bisexual or unisexual, then plants dioecious, polygamous or often gynodioecious, basically pentamerous, but altered by reduction of organs, usually irregularly zygomorphic; calyx rarely regularly 5-lobed, mostly inconspicuous or strongly reduced, or often divided into up to 30 segments, inrolled during anthesis; corolla sympetalous, funnelform, infundibuliform, salverform, rotate or campanulate, tube often saccate at base and nectariferous inside, sometimes spurred, limb with 3 or 5 more or less unequal lobes, often bilabiate; stamens 1–4(5), usually inserted in the corolla tube, alternating with the corolla lobes, anthers tetrasporangiate or rarely bisporangiate; ovary inferior, 3-locular, one locule with 1 pendent ovule, the other two sterile, often very small; style 1, filiform with a two- or three-lobed stigma. Fruit an achene with a persistent, usually accrescent calyx, sterile locules persistent, sometimes inflated, or apparently absent; the calyx mostly developing into a plumose or awned pappus (segments inrolled in flower) or a solid rim, or consisting of 5 regular teeth or fewer or more irregular lobes, sometimes 1–3 stout horns or a small or larger number of spines, sometimes regularly 6-lobed and coroniform, rarely inflated. Mature seed usually without endosperm, rarely endosperm copious; embryo large, straight.

The family comprises 5 genera (with *Valeriana* s.l. including various Andean segregate genera) and about 400 species, distributed mainly in the northern hemisphere and the mountainous regions of South America, absent from Australia and Polynesia. *Triplostegia* is treated in this volume under Dipsacaceae.

VEGETATIVE MORPHOLOGY. Mostly annual or perennial herbs are half-rosette plants (sometimes subrosulate) with terminal inflorescences, often with a taproot. The innovation shoots arise from the rosulate zone or the preceding parts, which are more or less lignified. The main stem remains extremely short in several high-elevation Andean species of *Valeriana* s.l. (incl. *Stangea*, *Belonanthus*, *Aretiastrum* and *Phyllactis*), some of which form large and dense globe-shaped cushions by periodical acrotonic ramification of the rosettes (Fig. 74). Some species of *Valeriana* grow as small shrubs, a few species as woody lianas (*V. clematitis*) or herbaceous climbers (*V. scandens*).

Leaves are usually decussate, in some cases, however, starting from a distichous condition (*Valeriana officinalis*). In some species of *Valeriana* s.l. (*Stangea*) a transition from an oblique decussate phyllotaxis in the first nodes into spirodecussate arrangement and higher degrees of dispersion in the inflorescence takes place (Rauh and Weberling 1960). In a similar way phyllotaxis changes in some other high Andean

Weberling (deceased).

J.W. Kadereit and V. Bittrich (eds.), *Flowering Plants. Eudicots*, The Families and Genera of Vascular Plants 14,
DOI 10.1007/978-3-319-28534-4_35,

species of *Valeriana* (*V. pycnantha*, *V. globularis*, a.o.) from orthodistichy in the first nodes to oblique decussation and to a tristichous arrangement. In other species of *Valeriana* s.l. (*Phyllactis*), the position of the first leaves is orthostichous changing to spirodistichous in later leaves and finally to a spiral arrangement. Also in *Valeriana* s.l. (*Belonanthus*), foliage leaves show a divergence of 3/8. In species of *Patrinia* an adnation

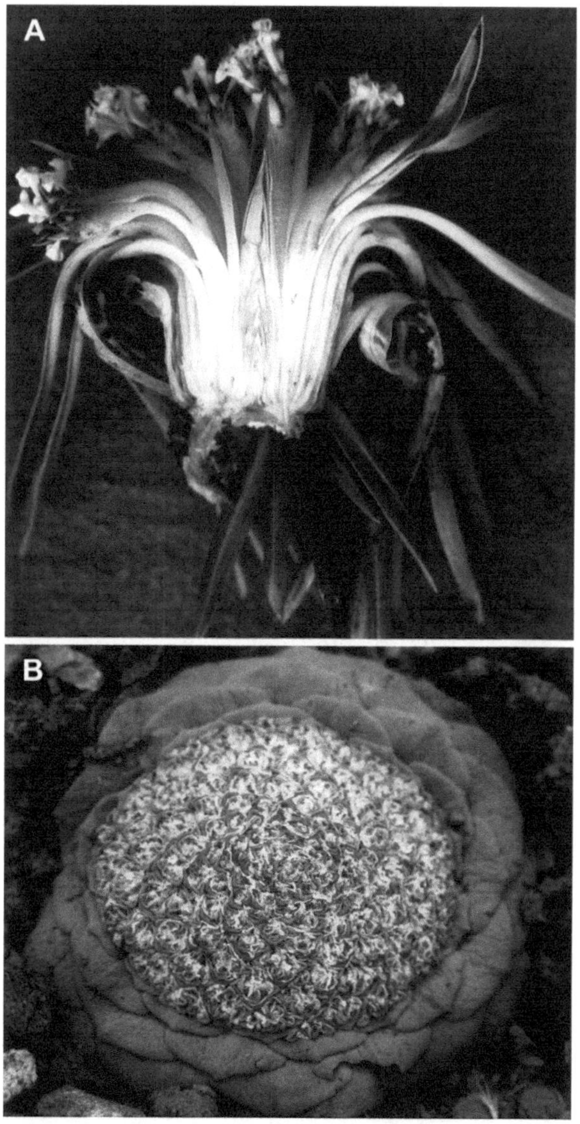

Fig. 74. Valerianaceae. **A** *Valeriana rigida*, rosette in longitudinal section (Cahuish-Tunnel, central Peru, 4178 m elevation). **B** *Valeriana henrici*, fruiting plant (Paso Ticlio, central Peru, 4843 m). **C, D** *Valeriana aschersoniana*. **C** Vegetative shoot. **D** Cushion with patches of snow (Valle Puypuy, central Peru, ca. 4400 m)

Fig. 74. (continued)

between the leaf bases and the cortex of the internode below can be observed.

Vegetative Anatomy. Simple hairs, unicellular or sometimes transversely septate glandular hairs with unicellular or multicellular stalks and small multicellular heads are common. The cortex lacks sclerenchymatic elements. In young shoots a distinct endodermis consisting of large cells is visible (Solereder 1899). Nodal anatomy was studied by Neubauer (1978). The nodes are usually 3-lacunar; *Valeriana aschersoniana* has unilacunar, a few species, like *V. polystachya* and *V. tiliaefolia*, have 5-lacunar nodes, and *V. aliariifolia* has 5-, 6-, or 7-lacunar nodes. As in several genera of Dipsacaceae and *Morina* of Morinaceae, the outer branches of the bifurcating lateral traces of opposite leaves join in the nodal cortex, forming a bridge. From the apex of such a bridge two bundles are leaving, one entering each petiole; thus, in case of a 3-lacunar node, a total of 5 traces enter the base of each petiole. In *Valeriana celtica* and *V. saxatilis* an anomalous vascularization was found. In all species investigated, well-developed transfer cells are present in the primary xylem and the primary phloem of the shoot. They were frequently found in nodes, but also in internodes (Lörcher and Hildenbrand 1987). The derivatives of the procambium and the first cambial derivatives are able to form wall protuberances, persisting especially in the region of the pit membranes (vestured pits). The vessels of the metaxylem and secondary xylem show a continuous transition in their characters (length, pitting, perforation, vestures) and often have scalariform perforation plates. In the secondary xylem of all species investigated except *Patrinia*, tracheids are missing. Fibres and/or fibre tracheids occur. The vessels even of the secondary xylem are mostly short and extremely small (mean tangential diameter <25 μm). They are arranged in long radial multiples, with simple oblique perforations and moderately large alternating intervascular pits. The pitting of a vessel often shows a transition from elongated pits in scalariform arrangement to alternating shorter circular pits. Exclusively alternate pitting in all vessels of the secondary xylem was only observed in a few species of *Valeriana*. Rays usually are 4–6 cells wide and consist almost entirely of square upright cells. In herbaceous species xylem rays usually are absent. The axial xylem parenchyma usually is scanty and paratracheal, in some species vasicentric or confluent. Phloem fibres were found in some woody species. In high Andean species of *Valeriana* s.l. (incl. *Stangea*, *Belonanthus*, *Phyllactis* and *Aretiastrum*), the xylem is interrupted by broad parenchymatic rays. As a consequence, the cross section of vascular bundles of older stems and rhizomes has an almost star-like or deeply lobed shape. Lörcher (1990) interpreted the stem anatomy of these species as showing anomalous secondary growth. However, transversal sections in the branching zone of cushion-like plants or the basal parts of (semi) rosulate plants revealed that the structures described as anomalous secondary growth by Lörcher (1990) represent lateral branches embedded in a common cortex (T. Stützel, pers. comm.). In some species of *Patrinia* and *Valeriana* tannins were found in the parenchyma cells of the cortex. In some high-elevation Andean *Valeriana* species, oil cells are frequently present in the whole shoot. In the above-ground shoot of *Valeriana officinalis* and some other species, numerous lipid droplets occur in the transfer cells.

In allorhizous species of *Valeriana*, especially alpine species with taproots, secondary growth has been observed in the primary root. At the onset of secondary growth, a periderm arises from the outer layers of the cortex. The cortex can persist for a long time. However, in *Centranthus*, *Valerianella*, *Patrinia*, a.o., in which the roots show strong secondary growth, the primary cortex is lost and a periderm develops from the pericycle. Somewhat elongated oil cells with suberized walls occur in the outer part of the cortex and cork of *Centranthus*, *Patrinia*, *Valeriana* and *Valerianella*. The stem-borne roots of *Valeriana officinalis* only have a primary structure, and the exodermis persists after degeneration of the rhizodermis.

The structure of the foliage leaves normally is dorsiventral; in leaves of somewhat xerophytic structure—e.g. the ericoid leaves of *Valeriana* (*Aretiastrum*) *aschersoniana* and the small succulent leaves of *Valeriana* (*Aretiastrum*) *magellanicum*—the lamina can be aequifacial, the stomata being sunken. Stomata are of the ranunculaceous type, sometimes paired (*Valerianella*). In some species they occur on both sides of the leaves, otherwise the leaves are hypostomatic. Hydathodes have been reported for the leaf teeth of *Valeriana sambucifolia*. No crystals have been recorded in leaves except in *Patrinia*. In *Valerianella* (incl. *Fedia*) crystals of calcium oxalate were found in certain layers of the pericarp, called "Cristarque" by Vidal (1903, with reference to Van Tieghem; for further literature, see Ernet 1978).

Inflorescences. Detailed studies were undertaken by Weberling (1961, 1966). Usually the inflorescences are terminal and the mode of ramification is thyrsic. Generally the cymose partial inflorescences of the thyrses are arranged in opposite pairs (Fig. 75B). Exceptions are found in some species of *Valeriana* with alternate leaves. The thyrsic mode of ramification can be repeated by branches preceding the thyrsic region of the main stem, leading to di- or pleiothyrsic ramification. Comparative studies revealed that, despite the lack of a terminal flower, the inflorescences are principally monotelic and can be considered

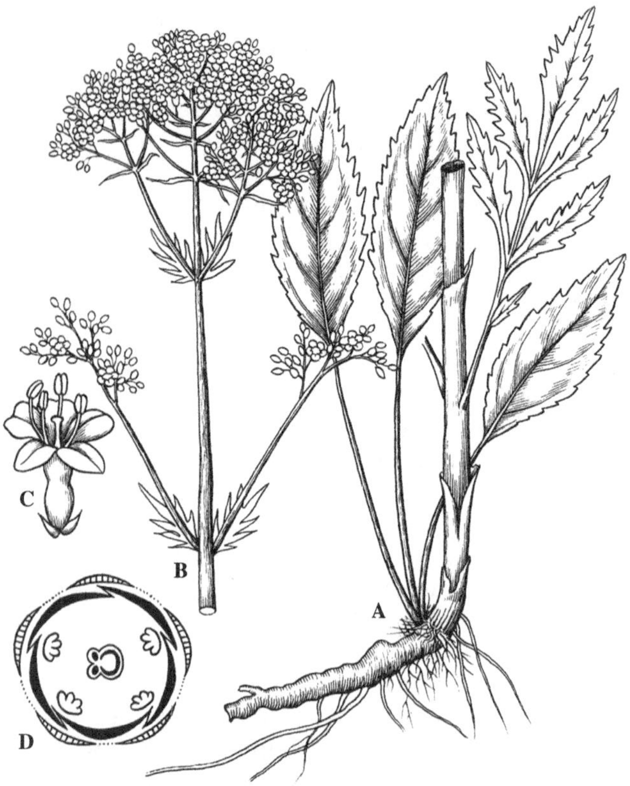

Fig. 75. Valerianaceae. *Patrinia scabiosifolia*. **A** Basal part of plant. **B** Inflorescence. **C** Flower. **D** Floral diagram. (after Steward 1958, courtesy Missouri Bot. Garden)

truncate monotelic synflorescences (Weberling 1961), which can be designated as truncate thyrsoids or pleiothyrsoids. This can be explained by the fact that the uppermost paracladia "arise so early and involve so much of the apical meristem that no residuum of the apex remains between them to continue the main axis" (Philipson 1947). In the extremely acrotonic inflorescences of *Valerianella* (incl. *Fedia*) this happens during the initiation of the only, but strongly developed pair of cymose paracladia. These continue this mode of ramification, and the suppression of the terminal flower continues during their cymose branching until, in the last ramifications, a terminal flower is formed and branching ends in the formation of one or two cincinni. The same acrotonic development occurs in the cymose paracladia of some *Valeriana* species (*V. alliariaefolia*, a.o.). The dichasial ramification in the cymose partial paracladia can continue over a long period, producing extended dichasial flowering systems, which show the regular formation of prophylls. Beyond this, in most species of *Patrinia* and in some of *Valeriana* s.l. (*Stangea*) sterile supernumerary bracts occur above the branches arising from the prophyllar axils (Weberling 1961). In many species of *Patrinia* three supernumerary bracts are united to form a scarious three-lobed or cordate fruit wing (Weberling 1961, 1966; see also Hofmann and Göttmann 1990). Differences in the shape of inflorescences often result from 1) the different number of cymose and thyrsoidal paracladia and the intensity of their ramification, 2) the different growth of the internodes, either intensive elongation of all internodes leading to loosely branched systems (*Valeriana effusa*), or contraction of the entire inflorescence or parts of it (*Valeriana capitata*, *V. nivalis*) or, when restricted to the cymose paracladia, to spike-like forms (*Valeriana celtica*, *V. dinorrhiza*), 3) the mesotonic or acrotonic support of the development of the paracladia, in the latter case often leading to an extremely vigorous development of the only pair of cymose paracladia and dichotomous-like branching being characteristic for *Valerianella* and some species of *Valeriana*.

In some species of *Valeriana* (e.g. *V. macrorhiza*, *V. carnosa*, *V. clarioniaefolia*), the apex of the strong and extremely short primary shoot remains permanently indeterminate and forms foliage leaves. After giving rise to a series of 'sterile' leaves, it produces 'fertile' leaves which have thyrsoidal paracladia in their axils. Then the shoot again returns to vegetative growth. This alternation even takes place in the dense cushions of rosettes by rhythmic production of new, at first vegetative branches which arise from the lower, lignified parts of the rosulate branches (*Valeriana* s.l.: *Phyllactis pulvinata*). In some rosette-forming species of *Valeriana* (e.g. *V. henrici*, cf. Eriksen 1989b, fig. 5), the terminal inflorescence forms a broad head by intensive primary growth of the pith of the short inflorescence axis.

Flower Morphology. The calyx is variously developed, accrescent or not, rarely absent. Following other authors, Eichler (1875) confirmed that the feathery pappus present in the majority of Valerianaceae is equivalent to a pentamerous calyx. It develops from 5–30 sections inrolled in flower. The gamopetalous, mostly five-lobed, in some species of *Valeriana* s.l. (*Phyllactis*, *Belonanthus*, *Aretiastrum*) three-lobed corolla is subactinomorphic to more or less zygomorphic and often bulged, saccate or spurred at the base (Fig. 75C). Inside the corolla tube, it contains a nectary formed by a cushion of one-celled epidermal hairs. This is situated at the base of the tube or somewhat higher or in the gibba or spur which then serves both as nectary and nectar container (*Patrinia*, *Valeriana*, *Centranthus*). In the upper part of the petal tube, strong, usually single lateral veins occur between adjacent petals ("transpetal veins"), typical also for Dipsacaceae (Gustafsson 1995). In *Valeriana* the nectary is separated from the corolla tube by a septum, in *Centranthus* the septum is separating the nectary-bearing spur (see Weberling 1970). The zygomorphic form of the 5- or rarely 3-lobed corolla and the position of its lobes, the position of the 1–5 stamens, the three-carpellate structure of the gynoecium in combination with the position of the bulge or spur and the separating septum result in an asymmetric structure of the flower. Form and size of the corolla, and the length of the corolla tube can be very different; in some species of *Valeriana* or *Valerianella* (sect. *Siphonella*), a long tube can be formed, whereas in others, especially many species of *Valerianella*, the tube remains very short. The size of the corolla can even vary within a species.

In the Eurasian, African and North and Central American taxa the anthers are tetrasporangiate, whereas all South American taxa of *Valeriana*, except *Valeriana chaerophylloides* and *Valeriana* sect. *Porteria*, have bisporangiate anthers (Eriksen 1989a), possible related to the development of smaller, selfing flowers (Moore and Donoghue 2007). The ovules are anatropous, solitary, unitegmic and tenuinucellate. As reported by Eichler (1875), only the carpel facing the β-prophyll of the inferior three-carpellate ovary is fertile, the other two carpels are sterile and pressed to the side of the α-prophyll. Mostly the sterile locules are more or less degenerated; however, in *Nardostachys*, *Patrinia*, some species of *Valeriana* (e.g. *V. saliunca*, *V. hyalinorhiza*, *V. chaerophylloides*) and *Valerianella*, the empty locules are inflated, appearing externally as bulges. Asplund (1920) found that the sterile locules of *Patrinia*, *Valerianella* and *Valeriana* can contain one or two more or less reduced ovules (see also Wichura 1846; Buchenau 1854/55; Payer 1857; Eckardt 1937).

EMBRYOLOGY. The tapetum of the Valerianaceae is periplasmodial, and its nuclei can be highly polyploid (*Valeriana officinalis*, Skalinska 1958). The pollen grains are 3-nucleate. The raphe of the anatropous pendulous ovules is lateral. The ovules are unitegmic and tenuinucellate. There is a tendency to the formation of several (2–5) archesporial cells, appearing before the initiation of the integument. As in most sympetalous families, the megaspore mother cell develops without separation of parietal cells. The tetrad of megaspores resulting from the meiotic division of the megaspore mother cell is often irregularly T-shaped, the lowest of the cells giving rise to the embryo sac. Megasporogenesis follows the monosporic type (Polygonum type). After its completion the antipodial cells undergo mitotic divisions of their nuclei, thus becoming bi- to polynucleate, and persist during early fertilization stages. The embryo sac is enclosed by an endothelium developing from the inner epidermis of the integument. With the possible exception of species of *Patrinia*, the endosperm, which develops according to the nuclear type, is almost completely used up by the developing embryo (Asplund 1920; Peng et al. 1995). Thus, an endosperm is lacking in the mature fruit.

POLLEN MORPHOLOGY. The family is relatively stenopalynous (Patel and Skvarla 1979). The pollen grains are 3-zonocolpate to 3-colpo-roidate, more or less spheroidal (suboblate to subprolate), the longest axis up to 34–67 μm long, their size often depending on the degree of polyploidy. According to the literature, there is no endexine in Valerianaceae. However, during the development of the pollen wall, lamellated structures can be seen by SEM on the inner side of the exine before the formation of the intine. The "white lines" occurring at this stage were interpreted as an indication for the existence of an endexine (Rowley 1981; Guédès 1982). The colpi are bounded by a margin encircled by a band which appears bright in the light microscope (halo) and is formed by a columella-free area of sexine and a thin discontinuous nexine. While the margin is nearly exclusively formed by the tectum, this is absent on the surface of the colpus, the bacula on the surface of the aperture being homologous to columellae (as assumed already by Wagenitz 1956). Only few differences are found in supratectate structures. The exine surface is usually echinate, in *Stangea* and *Valeriana pygmaea* baculate (*Stangea* type, Blankenhorn 1978), in *Centranthus* spinulose (*Centranthus* type). *Valerianella*, *Plectritis* (*Valerianella* type), *Aretiastrum*, *Belonanthus* and some species of *Centranthus* are exclusively micro-echinate or granulate. In *Patrinia* and *Nardostachys* the individual spines are supported by dome-shaped portions of the exine (*Patrinia* type), "very large verrucae" in the sense of Clarke (1978). All these types are found in *Valeriana*. According to the results of Blankenhorn (1978) and Clarke (1978), pollen morphology suggests the subdivision of the family into the basal Patrinieae (*Nardostachys* and *Patrinia*, a group now known to be paraphyletic) and the other genera. *Nardostachys* and *Patrinia* are characterized by the dome-shaped soccles of the exine bearing the individual spines (*Patrinia* type of Blankenhorn 1978), or by "thick walled grains with verrucae on the sexine and in which the columellae are not differentiated at the poles" (Clarke 1978). Within the rest of the family there "is a tendency for these characters to become progressively reduced". In *Valeriana* s. l., Blankenhorn (1978) found a wide range of forms. *Valerianella* (incl. *Fedia*), *Valeriana* spp. (*Plectritis*) and *Centranthus* have similar pollen with very reduced supratectate structures (*Valerianella* type) and are "characterized by the differentiation of the columellae at the poles, showing an increase of length and being fused in their lower parts and branched above" (Clarke 1978). In a comparative study of pollen wall structure by means of light microscopy, SEM and TEM of longitudinal and cross sections, Reese-Krug et al. (2001) could distinguish four different pollen "types": 1. *Patrinia* type, with a transitional *Valeriana coarctata* subtype, somewhat corresponding to the *Patrinia* type of Blankenhorn (1978), 2. *Valeriana palmeri* type, with a waved or folded and partially extremely thickened tectum forming arched sculptures, 3. *Valerianella* type, corresponding to the *Valerianella* type distinguished by Blankenhorn (1978), and 4. *Stangea* type, characterizing the genus *Stangea* (= *Valeriana* s.l.), from which Blankenhorn (1978) derived the name, but also including the pollen grains of some high Andean species of *Valeriana*. These "types", however, represent slightly distinct reference points within a more or less continuous spectrum of forms. In contrast to the data in the literature, the occurrence of an endexine in pollen of Valerianaceae could be ascertained.

Karyology. The chromosome numbers often form polyploid series based on $x = 7, 8, 9, 12$, unusual numbers being $x = 11$, $2n = 22$ in *Patrinia* (up to $2n = 88$) and $x = 13$, $2n = 26$ in *Nardostachys*. In *Valerianella* (incl. *Fedia*), the predominant chromosome numbers are $n = 8$, $2n = 16, 32$ (Xena de Enrech et al. 1991), but also $n = 7$, $2n = 14$ or $n = 15, 44, 45$ have been reported (Ernet in Weberling 1970). In *Valeriana* the basic numbers reported are $x = 7, 8, 11, 12$. Besides diploids, there are many tetraploid and octoploid taxa, sometimes as infraspecific polyploid races. In the Eurasian taxa (*Valeriana officinalis* group, a.o.) the series are based on $x = 7$. This also applies to the South African *Valeriana capensis* ($2n = 14$) and to the tropical American *V. scandens* ($2n = 28, 56$). In the Eurasian mountain species as well as all the high-elevation Andean species (incl. the segregate genera *Astrephia*, *Stangea*, *Aretiastrum*, *Phyllactis* and *Belonanthus*) investigated, the chromosome number is $2n = 32$. The same applies to *Centranthus*. For further literature, see Engel (1976). Hidalgo et al. (2010) optimized chromosome numbers on a Valerianaceae phylogeny. According to their results, a clade of *Valerianella* (incl. *Fedia*), *Centranthus* and most *Valeriana* spp. has $x = 8$ with further reduction to $x = 7$ in a subclade of *Valeriana* p.p. The three basal clades of the family have higher base numbers: *Patrinia* $x = 11$, *Nardostachys* $x = 13$, and a small clade of *Valeriana* p.p. (see below under "Subdivision of and relationship within the family") has $x = 11$. The counts (IPCN reports online 2012) for the 3 species in this clade, however, are somewhat ambiguous: *V. celtica* (3 different subspecies) $2n = 24, 48, 96$, *V. hardwickii* $n = 8$, $2n = 16, 28$, and *V. saxatilis* $2n = 22, 24$.

Fruit and Seeds. The fruit is an achene, i.e. a dry indehiscent monospermous fruit derived from an inferior ovary (Wagenitz 1976). The evolution of fruit and seed characters based on a phylogeny of the family was recently investigated by Jacobs et al. (2010). The achenes of *Nardostachys* and *Patrinia* have a persistent and regularly five-lobed calyx. In most species of *Valeriana*, all species of *Centranthus* and one species of *Valerianella* (*V. hirsutissimia*; Richardson 1975), a pappus consisting of 5 or up to 30 plumose rays is formed during maturation of the fruit. Achenes with pappus appear in the Valerianaceae in the distal clade of *Valeriana* + *Centranthus* (Jacobs et al. 2010) but also in the more basal clade composed of few *Valeriana* species (see Hidalgo et al. 2010). Thus, either the pappus evolved once at the base of Dipsacaceae and Valerianaceae and was reduced again in *Patrinia*, *Nardostachys* and *Valerianella*, or the pappus evolved independently in Dipsacaceae and once or twice in the Valerianaceae. In several species of *Valeriana* s.l. (incl. *Phyllactis*, *Belonanthus*, *Aretiastrum*, *Plectritis*, *Stangea*), a solid ring-shaped rim is formed instead of a pappus. In some taxa (*Nardostachys*, *Patrinia*, several species of *Valeriana*, e.g. *V. saliunca*, *V. hyalinorrhiza* and *Valerianella* spp.), the two sterile locules are persistent and even inflated. Jacobs et al. (2010), on the other hand, found the sterile locules to be absent in the mature fruits of *Centranthus* and the *Valeriana* (incl. *Plectritis*) species studied by them. In several species of *Valerianella* (*V. patellaria*, *V. amblyotis*, a.o.), the sterile locules form wing-like projections which, however, are not homologous with the wings found in some species of *Valeriana* s.l. (*Plectritis*; Morey 1962). *Plectritis* (*Valeriana* s.l.) has dimorphic fruits, individuals of the same population may have wings as outgrowths of the fruit wall of the fertile locule or not (Dempster 1958). For *P. brachystemon*, Ganders et al. (1977) reported that "the polymorphism is controlled by a single locus with the allele for winged fruits exhibiting complete dominance". Heterocarpy as well as achene polymorphism are found in some species of *Valerianella* (Ernet 1978; Mathez and Xena de Enrech 1985a, 1985b). The different development of fruits in the same individual depends on the position in the inflorescence: in the proximal parts of the inflorescence, the sterile locules of the achenes are poorly developed but they are well developed in the distal parts; intermediate types occur as well. The achenes of the distal parts of the inflorescences are deciduous and can be dimorphic or more rarely trimorphic within the same population regarding development of the calyx and the sterile locules. Genetic models for the observed fruit polymorphism were developed by Eggers Ware (1983) and Xena de Enrech and Mathez (1998). *Valeriana ozarkana* (incl. *V. bushii*) shows simple Mendelian inheritance, with the *ozarkana* (winged) fruit-type allele dominant over the *bushii* (fusiform) allele (Eggers Ware 1983). Fruit polymorphism in *Valeriana* (species of the former genus *Plectritis*) and in *Valerianella* evolved independently according to their morphology and the family phylogeny (Jacobs et al. 2010). See following section for elaiosomes.

The seed is enclosed in the indehiscent fruit. It basically lacks endosperm in the core valerians (probably a synapomorphy for this clade), but the seeds of *Patrinia* (basal branch in the family phylogeny) have copious endosperm and also a smaller embryo than in the other genera. No data about the endosperm are available for *Nardostachys*. In the chalazal region of *Centranthus* an unlignified "hypostasis" occurs, apparently missing in *Valeriana*; the embryo is sometimes green, and contains oil but no starch (Netolitzky 1926).

Pollination and Dispersal. Reproductive biology of *Nardostachys jatamansi* was studied by Chauhan et al. (2008), who observed bees, flies and butterflies as flower visitors. The flowers are protogynous and autogamous, although fruit set is higher under open-pollination conditions. The fragrant and nectariferous flowers of some *Valerianella* species, especially those with large funnel- or salverform corollas, are pollinated by insects. Flies, beetles, Hymenoptera and even Lepidoptera are known as flower visitors. Predominant autogamy is found in numerous other *Valerianella* species with small funnel-shaped flowers (Ernet 1977). The flowers of *Valeriana*, although highly derived in their morphology by being asymmetric, gibbous or even spurred, are regarded to be only slightly specialized. Since the corolla tubes of most flowers are short, the nectar is accessible even for insects with a short proboscis. The scent of the flowers is characterized as paraffinoid. Predominantly Diptera, but also Coleoptera, Hymenoptera (bees), Lepidoptera (in *Valeriana dioica*) and Nematocera (*V. montana*, *V. tripteris*) have been observed as visitors. *Valeriana officinalis* is reported to be strictly allogamous (Skalinska

1947). The long-spurred flowers of most *Centranthus* species are accessible only to insects with a long proboscis (Lepidoptera, mainly Papilionoidea, Hymenoptera with long proboscis). The flowers of all species investigated are distinctly protandrous.

Achenes with a pappus are adapted to wind dispersal, but also can attach to the fur or feathers of passing animals. Many species of *Valerianella* have fruits with clinging devices (uncinate aristae, horns on calyx teeth, uncinate hairs). Furthermore, in this genus larger calyx lobes or wings may serve for anemochory, and inflated sterile locules or calyces (*Valerianella*) and air-filled tissues as floating organs. The supernumerary bracts of many species of *Patrinia* become wing-like in fruit, and the wings of the fruits of some *Valeriana* spp. probably enhance wind dispersal. Tumble weeds are found in some *Valerianella* species. In some species of *Valeriana* (species formerly included in *Astrephia*) and *Valerianella* (species formerly included in *Fedia*), the inflorescence axis and nodes begin to swell after anthesis. When the resulting tension is released by, for example, touch, the fruits are expelled (Eriksen 1989a). Sernander (1906) and Bresinsky (1963) found pericarp swellings below the exocarp between the walls of the sterile locules of *Valerianella* (*Fedia*), consisting of a small-celled tissue containing lipids. Both authors interpreted them as elaiosomes. Sernander (1906) already observed that the lipid containing tissue in the ventral pericarp of the "elaiosomic" fruits (Bresinsky 1963) is macroscopically visible as a white basal swelling, and only this form of fruit is myrmecochorous. Around the zone of abscission of the fruits of several species of *Valerianella*, Ernet (1978) also found swellings of the pericarp formed of cells with an oily liquid, and identified them as elaiosomes. Sernander (1906) observed myrmecochory in the field. In many species the viability of the seed extends over several years. Germination, especially of mountain species, is often somewhat delayed and was found to take place in light and to be enhanced by frost.

Phytochemistry. The principal compounds responsible for the long-known healing properties of Valerianaceae, especially the sedative effects, have been revealed as a specific and otherwise unknown group of iridoid substances named valeriana-epoxy-triesters, for convenience abbreviated valepotriates (Thies 1966, 1968; Thies and Funke 1966). Meanwhile, a large series of valepotriates has been detected which differ by the presence of one or two double bonds in the ring structures of the cyclopenta-[c]-pyran skeleton and diverse residues at five different positions (for further information incl. biosynthesis, methods of extraction and analysis, literature, see Backlund and Moritz 1998). Variation in distribution and concentration of different valepotriates often are characteristic for taxa. In general they indicate a "chemosystematic difference" between South American and Eurasian taxa (Schild and Seitz 1971). In contradiction to older investigations, valepotriates have also been found in *Patrinia* and *Nardostachys* (see Backlund and Moritz 1998). Until now, no traces of valepotriates could be detected in taxa outside the Valerianaceae, suggesting the uniqueness of valepotriates for this family. However, since *Triplostegia* is now considered a basal branch in the Dipsacaceae, it cannot be excluded that in the Dipsacaceae the valepotriates might have been reduced secondarily. The characteristic smell of most plants of the family is due to the degradation product isovaleric acid. For further information and literature, see Backlund and Moritz (1998). During the search for the principal compounds responsible for the pharmacological effects, several iridoid-like alkaloids as well as a number of sesquiterpenes and monoterpenes and derivatives have been identified. In the roots of *Patrinia* glycosidically bound saponines have been found—e.g. patrinin being separated from patrigenin (Matveeva and Abubakirov 1964).

Useful taxonomic data were provided by comparative analyses of flavonoids in foliage leaves, which revealed their distinctive chemistry for *Nardostachys* and *Patrinia* (occurrence of isorhamnetin, luteolin and acacetin), and yielded new perspectives for the classification of *Valerianella* (Greger and Ernet 1971, 1973). The essential oils of the roots and rhizomes of various *Valeriana* species used for the preparation of perfumes were studied in detail. The oil of *V. hardwickii* contains α-patchoulene, patchoulol, viridiflorol and α-cadinol, besides various unknown compounds (Novak et al. 2000), that of *V. hardwickii* var. *arnottiana* had as major compounds valeracetate, bornyl acetate, methyl linoleate, cuparene and α-cedrene (Sati and Mathela 2004); in *V. wallichii* Bos et al. (1997) found an unidentified sesquiterpene hydrocarbon, α-santalene, *ar*-curcumene and xanthorrhizol in European material, while in the material from Nepal patchoulol was the major compound. Similarly, Tanaka and Komatsu (2008) studied the volatile components of rhizomes of *Nardostachys jatamansi* and identified β-maaliene, 9-aristolene, calarene and patchoulol, aromadendrene, cube-11-ene, epi-α-selinene, spirojatamol and valeranone.

Subdivision of and Relationship Within the Family. The Patrinieae, with *Patrinia* and *Nardostachys*, were usually regarded as a natural group because of their regularly 5-lobed calyx, usually 4 stamens and well-developed sterile locules in the gynoecium. However, these characters are probably plesiomorphic. Evidence obtained from the analyses of DNA sequence data appears somewhat contradictory. From combined *rcb*L and *ndh*F sequences, Pyck et al. (2002) concluded "that the traditionally circumscribed Patrinieae can no longer be recognized", whereas "parsimony analysis based on a morphological data set supported a monophyletic Patrinieae". Hidalgo et al. (2004) deduced from a cladistic analysis of DNA sequence data (*mat*K) that tribe Patriniae is monophyletic (with 94 % bootstrap support) and that tribe Valerianeae also forms a clade. Bell and Donoghue (2005), on the other hand, concluded that Patrinieae are paraphyletic, with *Patrinia* being sister to the rest of the family and *Nardostachys* the next basal branch (100 % bootstrap support; see also Jacobs et al. 2011). In consequence, a formal infrafamilial classification was abandoned in this treatment.

Although diverging from earlier results of Backlund and Donoghue (in Backlund 1996), recent data (Hidalgo et al. 2004; Bell and Donoghue 2005; Jacobs et al. 2011) suggest that *Triplostegia* needs to be removed from Valerianaceae as the genus is more closely related to the Dipsacaceae. In this volume it is thus included in a monogeneric subfamily Triplostegioideae of Dipsacaceae. Hofmann and Göttmann (1990) already suggested that, after the exclusion of *Triplostegia* from the Valerianaceae, this family is very homogeneous. Differing from a prior statement (Bell et al. 2001), Bell and Donoghue (2005), based on a much improved data set, stated that their new data clearly indicate that *Plectritis* is nested within South American *Valeriana*. Hidalgo et al.'s (2004) data also support this view, and this genus is therefore merged with *Valeriana*. It has long been discussed whether *Aretiastrum*, *Astrephia*, *Belonanthus*, *Phuodendron*, *Phyllactis* and *Stangea* should also be included into *Valeriana* (Borsini 1944; Eriksen 1989a). Eriksen (1989a) regarded the reduction of the number of anther locules in most South American *Valeriana* species as a strong argument supporting a wider circumscription of the genus. Whereas all Asian, European, African and North American species of the genus *Valeriana* have tetrasporangiate anthers, those of nearly all South American species, except *Astrephia* and the Venezuelan sect. *Porteria* of *Valeriana*, are bisporangiate (cf. Eriksen 1989a). Moreover, in Central America and the West Indies both southern species with bisporangiate and northern species with tetrasporangiate anthers occur. Eriksen (1989a) suggested that the South American species with bisporangiate anthers represent "an independent evolutionary line" derived from North American *Valeriana*. She therefore regarded the South American species group, including the former genera *Phyllactis*, *Aretiastrum*, *Belonanthus*, and *Stangea* with bisporangiate anthers, as a subgenus of *Valeriana* (subgen. *Phyllactis* (Pers.) Borsini emend. Eriksen). Indeed, bisporangiate anthers seem to be a synapomorphy of a large clade of mainly Central and South American species (cf. cladogram in Bell and Donoghue 2005). This clade also includes species with tetrasporangiate anthers, however, and the data suggest three independent reversals to the tetrasporangiate condition (the former genera *Astrephia* and *Plectritis* and the former sect. *Porteria* of *Valeriana*). This clade is deeply nested within *Valeriana* (Bell and Donoghue 2005), and its formal recognition would necessitate a formal subdivision of the remainder of the genus into at least three monophyletic taxa. Furthermore, in the most recent phylogenetic analyses (Hidalgo et al. 2004, 2010; Bell and Donoghue 2005), three Eurasian species of *Valeriana* surprisingly appeared in positions clearly outside of *Valeriana* even in the wider circumscription proposed by Eriksen (1989a, 1989b) and also adopted in the present treatment. Thus, *V. celtica*, *V. hardwickii* and *V.*

saxatilis appeared as basal branches in the Valerianeae clade and *V. longiflora* as sister to *Centranthus*. This would suggest an even wider circumscription of a monophyletic *Valeriana* with the inclusion of *Centranthus* and *Valerianella* (incl. *Fedia*). It seems that species like *V. celtica*, etc. should be included in a new genus (these species also show a different chromosome base number than the Valerianeae clade, Hidalgo et al. 2010). *Valeriana longiflora* could be included in *Centranthus* (already Willkomm 1851, when describing *V. longiflora* as new, mentioned similarities in the corolla morphology with *Centranthus*) but, according to the phylogeny of Hidalgo et al. (2010), *Valeriana* appears still paraphyletic with the exclusion of *Centranthus*. More data are needed, however, and we here abstain from the radical solution to include *Centranthus* and *Valerianella* in *Valeriana*. A new infrageneric classification of *Valeriana* at the current state of knowledge is certainly premature. The DNA sequence data also indicate an inclusion of *Fedia* into *Valerianella*. Already Höck (1882) concluded that morphological characters pointed to a derivation of *Fedia* from *Valerianella* series *Locustae*.

Affinities. It is generally agreed that the family is to be placed in the Dipsacales together with Caprifoliaceae (incl. the now sometimes separated Diervillaceae and Linnaeaceae), Morinaceae and Dipsacaceae. This is based on morphological and anatomical studies (see Weberling 1966) as well as on molecular data (e.g. Backlund 1996; Bell et al. 2001; Winkworth et al. 2008; Jacobs et al. 2011). Molecular data supported the idea that Adoxaceae s.l. also belong into the order as a basal branch. Besides vegetative characters (predominantly decussate and often pinnate leaves with often amplexicaulous bases), the deep-seated cork cambium, perulate buds, a ± inferior ovary, a calyx persistent in fruit, and a vascularized testa might be synapomorphic for the order (e.g. version 2008 of Stevens 2001 onwards). A close relationship of Valerianaceae to Diervillaceae, Caprifoliaceae, Linnaeaceae, Morinaceae and Dipsacaceae is also shown by general trends in flower and inflorescence morphology—e.g. the shift in several lineages from monotelic over truncated monotelic to polytelic inflorescences (Weberling 1961); the supernumerary sterile bracts found in many Caprifoliaceae and some Valerianaceae may be seen as a tendency which has led to the formation of an epicalyx as a constant feature in Dipsacaceae. The reduction of flower parts also is a prominent feature, which is also expressed by a decrease of carpel number, carpel fertility and the number of ovules to finally one fertile carpel with a single pendulous ovule. The nectaries described are a common character of Caprifoliaceae, Diervillaceae, Valerianaceae, Morinaceae and Dipsacaceae (Wagenitz and Laing 1984), as are monosymmetric flowers, larger calyx lobes, tubular corollas, elongate styles and capitate stigmas (e.g. Howarth and Donoghue 2005). A strong similarity between Valerianaceae and Dipsacaceae is reported especially in shoot (Metcalfe and Chalk 1983, II, 197f.) and trichome anatomy (Weberling 1977), and the strong and mostly single transpetal veins (Gustafsson 1995). Recent studies using DNA sequence data also confirm a sister-group relationship of Dipsacaceae (incl. *Triplostegia*) and Valerianaceae (e.g. Hidalgo et al. 2004; Winkworth et al. 2008). APG III (2009) unites Caprifoliaceae s.s., Morinaceae, Dipsacaceae and Valerinaceae into Caprifoliaceae s.l., which is sister to Adoxaceae s.l.

Distribution and Habitats. The family comprises five genera and about 400 species distributed mainly in the northern hemisphere and the mountainous regions of western South America, but absent from Australia and Polynesia. Important data about the historical biogeography have been contributed by Bell and Donoghue (2005) and Bell et al. (2012), based on phylogenetic studies using sequences of several cpDNA markers and the nuclear ribosomal ITS. The crown Valerianaceae may be about 50–60 million years old. It is now well supported that the family probably originated in the eastern Himalaya and the mountainous regions of SW China, corresponding to the present range of *Nardostachys* and many species of *Patrinia*. Both genera form the first subsequent splits in the family phylogeny (Pyck et al. 2002; Bell and Donoghue 2005; Hidalgo et al. 2010). *Patrinia* is distributed eastwards towards Taiwan and Japan. From the Himalaya region, dispersal to Europe and the New World took place. *Valeriana* in South America is not monophyletic, and probably four colonization events occurred from the north (Bell et al. 2012). Eriksen (1989a) hypothesized that the first colonization event occurred only after the closure of the Isthmus of Panama, but the data of Bell and Donoghue (2005) and Bell et al. (2012) rather support that independent colonization events occurred earlier (about 12–20 Ma) via long-distance dispersal. For some widespread species such as *Valeriana scandens*, however, colonization via the Isthmus of Panamá seems also possible. Within New World *Valeriana* s.l., a monophyletic group of mainly high Andean species is supported by the data, although weakly. This clade contains several small segregate genera such as *Phyllactis* and *Stangea*. For one group of *Valeriana*, including *Plectritis*, dispersal from western South America to western North America may have occurred. It seems that the newly formed habitats in the Andes with many unoccupied niches allowed a very rapid diversification with the formation of an estimated 0.8–1.34 species/million years (Weberling 1968; Bell and Donoghue 2005; Moore and Donoghue 2007). Diversification seems to be more due to the colonization of new habitats, especially in mountainous regions, than to morphological innovations like pappose fruits (Moore and Donoghue 2007; Bell et al. 2012).

Many of the Eurasian *Valeriana* species, which all are perennials and often have vigorous cryptocorms (*V. tuberosa* with bulbs), occur in mountainous regions, often in alpine meadows and rocky heaths, in crevices (*V. saxatilis*, *V. tripteris*) or more or less creeping on gravelly soil (*V. supina*, *V. montana*, *V. montana*); others persist in half-shade in deciduous or coniferous forests. Most occur in moist habitats. In the winter rain zones around the Mediterranean Sea, *Valerianella* and *Centranthus* have many species well adapted to these climatic conditions in life cycle and dispersal strategies. *Valerianella* is distributed around the Mediterranean Sea, but extends into SW Asia towards Afghanistan. In North America *Valerianella* is represented with about ten or more species, a few of them (*V. locusta*, *V. carinata*) naturalized from Europe. Two species (*V. longiflora*, *V. nuttallii*) are endemic to Arkansas. In North America *Valeriana* comprises about 11 species and the tropical American *V. scandens* (sect. *Sorbifoliae*), which is also found in the southern, tropical part of Florida. In Mexico, Central America and the West Indies the genus is represented by 20 more species, most of them perennials, but also several annuals (e.g. *V. apiifolia*, *V. sorbifolia*) with fusiform or subnapiform taproots, which still enable the plants to ripen their fruits at the beginning of a dry period when these plants die. There are also some lianas like *V. clematitis*, while *V. scandens* f. *scandens* and f. *candolleana* are completely herbaceous. Many of the high Andean species of *Valeriana* are growing as semirosulate or pulvinate plants in the Puna at altitudes of up to nearly 5000 m. Here, they have to survive recurrent night frosts of sometimes -1 to -4.5 °C, with the risk of frost drought because the upper layers of soil often remain frozen for many hours. In the páramos of Colombia, Ecuador and northern Peru, associations of ligneous, shrubby or even more or less arborescent species occur. On the other hand, especially in Chile, a considerable number of annual species is found. Some species, e.g. *V. polystachya* and *V. salicifolia*, grow in swampy areas.

Parasites. Virus diseases (e.g. cucumber mosaic virus) have been reported for diverse species of *Valerianella* (especially *V. locusta*) used as "corn salad". Parasitic fungi such as downy mildew (*Peronospora valerianellae*) and powdery mildew (*Erysiphe polyphaga*) cause mycoses in *Valerianella*. In *Valerianella* aecidiospores of *Puccinia* species have also been found. *Erysiphe valerianae* often causes considerable damage in cultures of *Valeriana officinalis* and also attacks other species of this genus. The roots of *Valeriana officinalis* are damaged by *Phoma*. Beyond these, many other fungi (Uromyces, *Puccinia valerianae*, *P. commutata*, *Colleotrichum valerianae*, *Ramularia valerianae*) are known as parasites of *Valeriana* species. Among the animal parasites, *Macrophaga albicincta*, the "valerian wasp", can cause complete defoliation of *Valeriana* cultures. Damages are also caused by aphids, gall midges (*Dasyneura valerianae*) and gall mites (*Eriophyes macrotuberculatus*), and *Contarinia crispans* often causes deformations in leaves, inflorescences and flowers (virescence). Destruction by feeding is caused by caterpillars of several butterfly species, especially *Hepiolus humuli*, and a species of *Tephroclystia* has been named *T.*

valerianata because of its association with *Valeriana*. There are also miners among the Lepidoptera, Diptera and Coleoptera.

Economic Importance. Since ancient times rhizomes of *Valeriana*, especially *V. officinalis* and *V. phu*, have been widely used in Europe for sedative purposes. The same applies to *V. wallichii*, *V. hardwickii*, *V. jatamansi* in southern Asia, and to *V. edulis* and other species as well as *Stangea henrici* in Central and Andean America. Rhizomes of *Nardostachys jatamansi*, the "Spike Nard" or spikenard, have been used as a perfume drug since the time of Hammurabi (1792–1750 B.C.), and were transported as precious ointment from the high Himalaya to Mesopotamia. This is mentioned in the bible (Song of Solomon *1*, 12, *4*, 13–14, in Mark *14*, 3 and John *12*, 3 in the report of the anointment of Christ in Bethania). From India to China the drug still serves as a stimulant in the ayurvedic medicine. It is now included in CITES Appendix II. *Valeriana celtica* was used as a perfume drug under the name spica "nardi" or "nardus celtica" by the Romans and is still in use. In Europe and the Mediterranean, species of *Valerianella* (*V. locusta*, *V. cornucopiae*, a.o.) are widely consumed as "corn salad", "lamb's lettuce", "mâche", "Feldsalat", and the same applies to *Valeriana edulis* in Mexico and the S.E. United States. Some species of *Centranthus* (*C. ruber*, *C. macrosiphon*) and *Patrinia* as well as *Valerianella cornucopiae* serve as ornamental plants.

Key to the Genera of Valerianaceae

1. Stamen 1; achene pappose; corolla red, lilac or white, never yellow **4. *Centranthus***
– Stamens 2–5 (rarely one, then corolla yellow); achenes pappose or not 2
2. Calyx lobes 5, not developing into a pappus (Asia) 3
– Calyx inconspicuous in flower, developing into a pappus or forming 1 or several teeth of various shape at maturity, or forming a solid ring only (Asia, Europe, Africa, Americas) 4
3. Rosette plants, not terminating with an inflorescence, with spathulate-lanceolate leaves with axillary partial inflorescences; the 5 calyx lobes well developed in flower, accrescent in fruit; corolla white or pink, stamens 4 **2. *Nardostachys***
– Semirosulate plants with dentate, pinnatifid or pinnate leaves, terminating with an inflorescence; the 5 calyx lobes smaller than 1 mm, not accrescent in the fruit; corolla yellow (in *P. villosa* white to pink); stamens (1–) 3–4(5) **1. *Patrinia***
4. Annuals, appearing dichotomously branched; stamens 2–3; achenes very rarely with pappus (*V. hirsutissimia*) **3. *Valerianella***
– Annuals or perennials, not appearing dichotomously branched (except for a few species with extremely acrotonic inflorescences, mostly stunted forms), life form various (herbs, cushion and rosette plants, vines, shrubs, small trees); stamens 3; achenes pappose or epappose **5. *Valeriana***

Genera of Valerianaceae

Patrinia-Nardostachys grade

Calyx lobes 5, persistent, but not forming a pappus. Stamens mostly 4 (rarely 1–3 or 5), sterile locules of the fruit well developed. Perennial herbs.

1. ***Patrinia*** Juss. Fig. 75

Patrinia Juss., Ann. Mus. Natl. Hist. Nat. 10: 311 (1807), nom. cons.; Hong Deyuan et al., Flora of China 19 (2010).

Rhizomatous herbs with subentire to pinnatifid leaves and corymbiform pleiothyrsic inflorescences. Flowers in many species preceded by supernumerary bracts which are concrescent and form a scarious three-lobed or cordate fruit wing; calyx lobes small, ovate, not accrescent in fruit; corolla tube short, with subequal lobes, obscurely gibbous near the base, yellow or (*P. villosa* Juss.) white to pink; stamens (1–3)4(5); stigma capitate. Achene ovoid, obovoid or oblongoid, crowned by the persistent calyx. $2n = 22$, 44, 88.

About 25 species from Central Asia and the Himalaya to E Asia. Sect. *Centrotrinia* Maxim., ca. six species, sect. *Patrinia* ("*Palaeopatrinia*" Höck), about 17 species, sect. *Monandropatrinia* Höck, ca. two species.

2. ***Nardostachys*** DC.

Nardostachys DC., Prodr. 4: 624 (1830); Mém. Valér.: 4, t. 2 (1832); Weberling, Bot. Jahrb. Syst. 99: 188–221 (1978), monogr.; Hong Deyuan et al., Flora of China 19 (2010).

Rhizomatous herbs of very different size, subrosulate, with entire, elongate, spathulate-lanceolate leaves, first leaves distichous, later leaves decussate; rhizome covered with the reticulated remains of marcescent leaves. Inflorescences originating from the axils of living or dead basal leaves, pleiothyrsic, flowers condensed in a terminal and one or two pairs of lateral heads, the lateral heads originating in the axils of spathulate or ovate cauline leaves; bracts lanceolate to narrowly lanceolate. Calyx with 5 upright lobes, accrescent in fruit; corolla 3,5–13(−24) mm long, tubular-campanulate, lobes subequal, tube with a slight gibbous protrusion with nectary, white or pink; stamens 4, inserted at the base of the corolla, mostly somewhat exserted, long-haired in the lower half of the filaments; style exserted, stigma capitate; ovary hairy or glabrous. Achene somewhat compressed, elongate-ovate with two longitudinal swellings (the sterile locules), crowned by cartilaginous calyx, the 5 calyx lobes united at the bases by a low fringe. $2n = 26$.

One or two species, from the W Himalaya to SW China.

Valeriana clade

Perennial or annual herbs, subshrubs, shrubs, lianas. Calyx mostly not fully developed before fruit ripens, then regularly or irregularly lobed, inflated, forming 1 to 3 stout horns or several spines, a regularly 6-lobed crown, a spreading star, a cartilaginous cup or a solid rim only, or up to 30 segments often developing into a plumose or awned pappus.

3. *Valerianella* Mill.

Valerianella Mill., Gard. Dict. abr. ed. 4 (1754); Krok, Kgl. Svensk. Vet.-Akad. Handl., N.F. 5(1): 1–105 (1864), rev.; Ernet, Plant Syst. Evol. 127: 243–276 (1977a), 128: 1–22 (1977b), 130: 85–126 (1978), morph.; Xena de Enrech & Mathez, Naturalia monspeliensia, sér. Bot., Fasc. 54: 3–77 (1989), rev. of *Fedia*; Devesa & Martínez, Flora Ibérica 15: 233–258 (2007).
Fedia Gaertn. (1791), nom. cons., non Adanson (1763).
Pseudobetckea (Höck) Lincz. (1958).
Siphonella (Torrey & A. Gray) J.K. Small (1903).

Annuals, appearing dichotomously branched, due to the acrotonic development of the inflorescence. Flowers usually small, in dense cymosely branched partial inflorescences; in some species the prophylls within the monochasial (cincinnus-like) parts at maturity form rows of spreading sclerified horns. Calyx limb inconspicuous or not developed in flower; corolla funnelform or in some species salverform with a long tube, subactinomorphic or zygomorphic, bluish, white, red or pink; stamens 2–3, two sometimes united; stigma bifid or trifid. Achenes displaying a large variety of forms, sometimes heterocarpic, sterile locules inflated to inconspicuous, calyx transformed into regular or irregular lobes, sometimes 1–3 stout horns or a small or larger number of spines, a regularly 6-lobed crown, a spreading star, a cartilaginous cup, or a solid rim only, calyx and sterile locules can be inflated in fruit; in some species the distal deciduous achenes are dimorphic (rarely trimorphic) within the same population. $n = 8$, $2n = 16$, 32, but also $n = 7$, $2n = 14$, or $n = 15$, 44, 45.

60–70 species, mainly in SW Asia, the Mediterranean and Europe, but also in North America, rare in South America.

Phylogenetic studies (Hidalgo et al. 2004; Bell and Donoghue 2005) suggest paraphyly for *Valerianella* after exclusion of *Fedia*; thus, the latter genus is here included in *Valerianella* s.l.

4. *Centranthus* DC.

Centranthus DC. in Lam. & DC., Fl. franç. (ed. 3) 4: 238 (1805); Richardson, Bot. J. Linn. Soc. 71: 211–234 (1975), rev.; Martinez & Devesa, Flora Ibérica 15: 223–233 (2007).

Annuals or rhizomatous perennials, glabrous, usually glaucous, leaves simple or divided. Inflorescences pleiothyrsic, the thyrsic parts with numerous cymose partial inflorescences which end in many-flowered cincinni. Flowers perfect; calyx divided into up to 25 segments, inrolled in flower, forming a plumose pappus in fruit; corolla with 5 unequal lobes, infundibuliform or with a long cylindrical tube, gibbous near the middle of the tube or spurred near the base and with an internal longitudinal membrane reaching from the insertion of the spur to the mouth of the corolla tube, red, lilac or white; stamen 1; stigma subclavate to trifid. Sterile locules of the achene reduced. $2n = 32$.

Ten species and several subspecies. Southern Europe, Mediterranean and SW Asia. *Centranthus ruber* (L.) DC. is planted as an ornamental and today occurs as a garden escape outside the original distribution area.

5. ***Valeriana*** L. Fig. 74

Valeriana L., Sp. Pl. 31 (1753); Weberling, Bot. Jahrb. Syst. 79: 394–404 (1960), tax.; Borsini, Lilloa 32: 369–374 (1966), tax.; Hong Deyuan et al., Flora of China 19 (2010), reg. rev.; Borsini, Lilloa 8: 353–377 (1942), reg. rev.; Larsen, Nord. J. Bot. 6: 427–446 (1986), rev. *Phyllactis*; Lörcher, Flora 184: 231–254 (1990), morph. *Phyllactis*; Meyer, Ann. Miss. Bot. Gard. 38: 377–503 (1951), reg. rev.; Morey, thesis, Stanford Univ. (1962), biosyst. *Plectritis*; Weberling & Uhlarz, Plant Syst. Evol. 127(4): 217–242 (1977), morph. *Aretiastrum*; Weberling & Weberling, Akad. Wiss. Lit. Mainz, Trop. subtrop. Pflanzenwelt 36 (1981), morph. anat. *Belonanthus*; Weberling, Abh. Akad. Wiss. Lit. Mainz, math.-naturw. Kl. 2 (2001), morph. *Stangea*; Weberling & Stützel, Wulfenia 13: 193–205 (2006), morph. anat. *Aretiastrum*; Hidalgo et al., J. Bot. (2010), karyol., phylog.
Aretiastrum (DC.) Spach (1841).
Astrephia Dufr. (1811).
Belonanthus Graebn. (1906).
Phuodendron (Graebn.) Dalla Torre & Harms (1905).
Phyllactis Pers. (1805).
Plectritis (Lindl.) DC. (1831).
Porteria Hook. (1851).
Stangea Graebn. (1906).

Annual or perennial rhizomatous herbs, often semirosulate or rosulate, subshrubs or shrubs, small trees or lianas; innovation shoots of perennial species arising from the axils of basal leaves. Leaves entire, pinnatifid or pinnate, herbaceous, coriaceous or rarely fleshy. Inflorescences pleiothyrsic, of very diverse shape, supernumerary bracts on pedicel sometimes present. Flowers bisexual or unisexual, then plants dioecious, polygamous or often gynodioecious; calyx either forming a solid ring, not accrescent in fruit, or up to 30 segments inrolled in the flower and expanding into a plumose pappus in fruit; corolla with (3–)5 mostly unequal lobes, infundibuliform, salverform or tubular, mostly slightly gibbous near the base, rarely spurred, white, yellow, pink or red; stamens 3(4), anthers tetrasporangiate or bisporangiate; stigma 2(3)-lobed. Achenes of different shape, in some species di- or trimorphic, usually somewhat compressed, elongate-ovate, rarely winged, pappose or epappose, sterile locules usually small, sometimes well developed. $n = 8$, $2n = 14, 16, 18, 22, 24, 28, 32, 48, 54, 64, 96$.

About 270 species in Eurasia, North and South America and the southern part of Africa. High Andean species were included in various segregated genera, but these were shown to be deeply nested with *Valeriana* s.s. and belong to a clade of mainly high Andean species ("páramo clade" of Bell and Donoghue 2005). Even in the wider sense as accepted here, the genus is currently still paraphyletic or polyphyletic with exclusion of *Centranthus* and even *Valerianella* (incl. *Fedia*). Further studies on critical species are needed (see Hidalgo et al. 2010 and sections "Karyology" and "Subdivision of and relationship within the family").

Selected Bibliography

APG III 2009. An update of the Angiosperm Phylogeny Group classification for the orders and families of flowering plants: APG III. Bot. J. Linn. Soc. 161: 105–121.

Ascherson, P. 1881. Subflorale Axen als Flugapparate. Jahrb. Königl. Bot. Gart. Berlin 1: 318–336, pl. VI.

Asplund, E. 1920. Studien über die Entwicklungsgeschichte der Blüte einiger Valerianaceen. Kgl. Svenska Vetenskapsakad. Handl. 61, Nr. 3. Stockholm.

Backlund, A. 1996. Phylogeny of the Dipsacales. Comprehensive Summaries of Uppsala Diss., Fac. of Science and Technol. 243. Acta Univ. Upsaliensis.

Backlund, A., Moritz, T. 1998. Phylogenetic implications of an expanded valepotriate distribution in the Valerianaceae. Biochem. Syst. Ecol. 26: 309–335.

Backlund, A., Gustafsson, H.G., Bremer, B. 1997. Phylogeny of the Asteridae s.str. based on *rbc*L sequences, with particular reference to the Dipsacales. In: Backlund, A., 1996 (see preceding reference).

Barrie, F. 1990. A systematic study of the Mexican and Central American species of Valeriana (Valerianaceae). PhD Thesis, Univ. of Texas at Austin.

Bell, C.D., Donoghue, M.J. 2005. Phylogeny and biogeography of Valerianaceae (Dipsacales) with special reference to the South American valerians. Organisms Diversity Evol. 5(2): 147–159.

Bell, C.D., Edwards, E.J., Kim, S-T., Donoghue, M.J. 2001. Dipsacales phylogeny based on chloroplast DNA sequences. Harvard Papers Bot. 6: 481–499.

Bell, C.D., Kutschker, A., Arroyo, M.T.K. 2012. Phylogeny and diversification of Valerianaceae (Dipsacales) in the southern Andes. Mol. Phylogenet. Evol. 63: 724–737.

Blankenhorn, B. 1978. Pollenmorphologisch-systematische Untersuchungen an Valerianaceae. Bot. Jahrb. Syst. 99: 108–138.

Boissier, E. 1875. Valerianaceae. In: Flora orientalis 3: 84–113. London: H. Georg.

Borsini, O.E. 1944. Valerianaceae. In: Descole, H.R. (ed.) Genera et species plantarum Argentinarum II: 275–382. Buenos Aires: Guillermo Kraft.

Bos, R., Woerdenbag, H.J., Hendriks, H., Smit, H.F., Wikstrom, H.V., Scheffer, J.J. 1997. Composition of the essential oil from roots and rhizomes of *Valeriana wallichii* DC. Flav. Fragr. J. 12: 123–31.

Bresinsky, A. 1963. Bau, Entwicklungsgeschichte und Inhaltsstoffe der Elaiosomen. Studien zur myrmekochoren Verbreitung von Samen und Früchten. Bibliotheca Botanica (Hrsg. H. Melchior) Heft 29. Stuttgart: Schweizerbarth.

Buchenau, F. 1854/55. Ueber die Blüthenentwickelung einiger Dipsaceen, Valerianeen und Compositen. Abh. Senckenb. Naturf. Ges. 1854.

Chauhan, R.S., Kaul, M.K., Kumar, A., Nautiyal, M.C. 2008. Pollination behaviour of *Nardostachys jatamansi*: an endangered medicinal and aromatic herb. Sci. Hortic. 117: 78–81.

Clarke, G.C.S. 1978. Pollen morphology and generic relationships in the Valerianaceae. Grana 17: 61–75.

Clarke, G.C.S., Jones, M. 1977. The Northwest European Pollen Flora, 16. Valerianaceae. Rev. Palaeobot. Palynol. 24: 155–179.

Davis, G.L. 1966. Systematic Embryology of the Angiosperms. New York: Wiley.

Dempster, L.T. 1958. Dimorphism in the fruits of *Plectritis* and its taxonomic implications. Brittonia 10: 14–28.

Dyal, S.C. 1941. Studies in the Family *Valerianaceae*. Thesis, Cornell Univ., Ithaca.

Eckardt, T. 1937. Untersuchungen über Morphologie, Entwicklungsgeschichte und systematische Bedeutung des pseudomonomeren Gynoeceums. Nova Acta Leop. N. F. 5, Nr. 26.

Eggers Ware, D.M. 1969. A revision of *Valerianella* in North America. Thesis, Vanderbilt Univ., Nashville.

Eggers Ware, D.M. 1983. Genetic fruit polymorphism in North American *Valerianella* (Valerianaceae) and its taxonomic implications. Syst. Bot. 8(1): 33–44.

Eichler, A.W. 1875. Blüthendiagramme. Teil I. Leipzig (Reprint Eppenheim: O. Koeltz 1954).

Engel, K. 1976. Beiträge zur Systematik der Valerianaceae unter besonderer Berücksichtigung cytosystematischer Ergebnisse. Diss. Univ. Giessen, Fachber. Biologie.

Eriksen, B. 1989a. Notes on generic and infrageneric delimitation in the Valerianaceae. Nord. J. Bot. 9: 179–187.

Eriksen, B. 1989b. Valerianaceae. In: Harling, G., Andersson, L. (eds.) Flora of Ecuador 34. Nordic Journal of Botany.

Ernet, D. 1972. Valerianaceae. In: Löve, A. (ed.) IOPB Chromosome Number Reports XXXVII. Taxon 21: 495–496.

Ernet, D. 1977. Blütenbau und Fortpflanzungsbiologie von *Valerianella* und *Fedia* (*Valerianaceae*). Plant Syst. Evol. 128: 1–22.

Ernet, D. 1978. Fruchtbau und Verbreitungsbiologie von *Valerianella* und *Fedia* (*Valerianaceae*). Plant Syst. Evol. 130: 85–126.

Ganders, F., Carey, K., Griffiths, A.J.F. 1977. Outcrossing rates in natural populations of *Plectritis brachystemon* (Valerianaceae). Can. J. Bot. 55(15): 2070–2074.

Graebner, P. 1906a. Valerianaceae andinae. Bot. Jahrb. Syst. 37: 436–451.

Graebner, P. 1906b. Die Gattungen der natürlichen Familie der Valerianaceae. Bot. Jahrb. Syst. 37: 464–480.

Gray, A. 1862. Proc. Amer. Acad. Sci. 5: 322.

Greger, H., Ernet, D. 1971. Flavonoide und Systematik der Valerianaceae. Naturwissenschaften 58: 416–417.

Greger, H., Ernet, D. 1973. Flavonoid-Muster, Systematik und Evolution bei *Valerianella*. Phytochemistry 12: 1693–1699.

Guédès, M. 1982. Exine stratification, ectexine structure and angiosperm evolution. Grana 21: 161–170.

Gustafsson, M.H.G. 1995. Petal venation in the Asterales and related orders. Bot. J. Linn. Soc. 118: 1–18

Handel-Mazzetti, H. 1934. Symbolae Sinicae. VII/4. Anthophyta, Valerianaceae, pp. 1052–1056. Wien.

Hidalgo, O., Garnatje, T., Susanna, A., Mathez, J.L. 2004. Phylogeny of Valerianaceae based on matK and ITS markers, with reference to matK individual polymorphism. Ann. Bot. 93: 283–293.

Hidalgo, O., Mathez, J., Garcia, S., Garnatje, T., Pellicer, J., Vallès, J. 2010. Genome size study in the Valerianaceae: First results and new hypotheses. J. Bot. 2010, 797246, 19 pp.

Höck, F. 1882. Beiträge zur Morphologie, Gruppirung und geographischen Verbreitung der Valerianaceen. Bot. Jahrb. Syst. 3: 1–73.

Höck, F. 1891. Valerianaceae. In: Engler & Prantl (eds.) Die natürlichen Pflanzenfamilien IV, 4, pp. 172–182. Leipzig: W. Engelmann.

Hofmann, U., Göttmann, J. 1990. *Morina* L. und *Triplostegia* Wall. ex DC. im Vergleich mit Valerianaceae und Dipsacaceae. Bot. Jahrb. Syst.111: 499–553.

Howarth, D.G., Donoghue, M.J. 2005. Duplications in cyc-like genes from Dipsacales correlate with floral form. Int. J. Plant Sci. 166(3): 357–370.

Index to plant chromosome numbers (IPCN) 1979 onwards. P. Goldblatt & D.E. Johnson (eds.) Missouri Botanical Garden, St. Louis. Accessed online June 2012. http://www.tropicos.org/Project/IPCN

Jacobs, B., Bell, C., Smets, E. 2010. Fruits and seeds of the *Valeriana* clade (Dipsacales): Diversity and evolution. Int. J. Plant Sci. 171: 421–434.

Jacobs, B., Geuten, K., Pyck, N., Huysmans, S., Jansen, S., Smets, E. 2011. Unraveling the phylogeny of *Heptacodium* and *Zabelia* (Caprifoliaceae): An interdisciplinary approach. Syst. Bot. 36: 231–252.

Killip, E.P. 1937. Valerianaceae. In: Macbride, E.P., Flora of Peru VI, 2. Fieldiana, Bot. Ser. 13: 287–321.

Lörcher, H. 1990. Achsenverdickung und Sprossanatomie bei Valerianaceae. Akad. Wiss. Lit. Mainz, Trop. u. subtrop. Pflanzenwelt 74.

Lörcher, H., Hildenbrand, M. 1987. Transferzellen im Xylem der Valerianaceae. Plant Syst. Evol. 156: 69–89.

Mathez, J. 1984. Introduction à une révision du genre *Fedia* Gaertn. emend. Moench. Mém. Soc. Brot. 27: 129–175.

Mathez, J., Xena de Enrech, N. 1985a. Le polymorphisme génétique de la morphologie des fruits du genre *Fedia* Gaertn. (Valerianaceae). I. Détermination du mécanisme de contrôle génétique chez les espèces *F.*

cornucopiae (L.) Gaertn. et *F. graciliflora* Fisch. & Meyer. Candollea 40: 425–434.
Mathez, J., Xena de Enrech, N. 1985b. Heterocarpy, fruit polymorphism and discriminating dissemination in the genus *Fedia* (*Valerianaceae*). In: Jacquard, P. et al. (eds.) NATO ASI Series G5, Genetic Differentiation and Dispersal in Plants. Berlin, Heidelberg: Springer.
Matveeva, A.V., Abubakirov, N.K. 1964. Saponins of *Patrinia intermedia*. Khimicheskii Zhurnal 8(5): 43–46.
Metcalfe, C., Chalk, L. 1983. Anatomy of the Dicotyledons, 2nd ed., II, p. 197 f. Oxford: Clarendon Press.
Moore, B.R., Donoghue, M.J. 2007. Correlates of diversification in the plant clade Dipsacales: Geographic movement and evolutionary innovations. Amer. Natur. 170: S28–S55.
Morey, D.H. Jr. 1962. The Biosystematics of the genus *Plectritis*. Thesis, Stanford Univ. (University Microfilms, Inc.), Ann Arbor, MI.
Netolitzky, F. 1926. Anatomie der Angiospermensamen. In: K. Linsbauer, Handbuch der Pflanzenanatomie II, X. Berlin: Gebr. Borntraeger.
Neubauer, H.F. 1978 (publ. 1979). On nodal anatomy and petiole vascularization of some Valerianaceae and Dipsacaceae. Phytomorphology 28: 431–436.
Novak, J., Novak, S., Bitsch, C., Franz, C.M. 2000. Essential oil composition of underground parts of *Valeriana celtica* ssp. from Austria and Italy. Flavour Fragrance J. 15(1): 40–42.
Patel, N.D., Skvarla, J.J. 1979. Valerianaceae pollen morphology. Pollen et Spores 21: 81–103.
Payer, J.B. 1857. Traité d'organogénie comparée de la fleur, pl. 130, 132. Paris.
Peng, C.I., Tobe H., Takahashi, M. 1995. Reproductive morphology and relationships of *Triplostegia* (Dipsacales). Bot. Jahrb. Syst. 116: 505–516.
Philipson, W.R. 1947. Studies in the development of the inflorescence. III. The thyrse of *Valeriana officinalis* L. Ann. Bot. 11: 409–416.
Pyck, N., Van Lysebetten, A., Stessens, J., Smets, S. 2002. The phylogeny of *Patrinia* sensu Graebner (Valerianaceae) revisited: additional evidence from *ndhF* sequence data. Plant Syst. Evol. 233: 29–46.
Rauh, W., Weberling, F. 1960. Morphologische und anatomische Untersuchungen an der Valerianaceengattung *Stangea* Graebner. Abh. Akad. Wiss. Lit. Mainz, math.-naturw. Kl. 1959, Nr. 10.
Reese-Krug, H., Meyer, E., Hildenbrand, M., Weberling, F. 2001. Palynological investigations in Valerianaceae. Some elementary aspects and problems. Wulfenia 8: 61–80.
Richardson, I.B.K. 1975. A revision of the genus *Centranthus* DC. (Valerianaceae). Bot. J. Linn. Soc. 71: 211–234.
Rowley, J. 1981. Pollen wall characters with emphasis upon applicability. Nord. J. Bot. 1: 357–380.
Sati, S., Mathela, C.S. 2004. Essential oil composition of *Valeriana hardwickii* var. *arnottiana* from the Himalayas. Flavour Fragrance J. 20(3): 299–301.
Seitz, W. 1969. Zytotaxonomische Studien an Valerianaceen I. Ber. Deutsch.Bot. Ges. 82: 651–655.
Sernander, R. 1906. Entwurf einer Monographie der europäischen Myrmekochoren. Kgl. Svensk. Vet.-Akad. Handl. 41(7): 1–410.
Schild, W., Seitz, W. 1971. Studien an Valerianaceen. 2. Valepotriate in südamerikanischen Arten. Ber. Deutsch. Bot. Ges. 84: 225–241.
Skalinska, M. 1947. Polyploidy in *Valeriana officinalis* Linn. in relation to its ecology and distribution. J. Linn. Soc. Bot. 53: 159–186.
Skalinska, M. 1958. Studies in the karyological differentiation of the tapetum in *Valeriana officinalis* L. Acta Biol. Crac. 1: 45–54.
Solereder, H. 1899. Sytematische Anatomie der Dicotyledonen. Stuttgart: Ferdinand Enke.
Stevens, P.F. 2001 onwards. Angiosperm Phylogeny Website. Version 9, June 2008. http://www.mobot.org/MOBOT/research/APweb/.
Steward, A.N. 1958. Manual of vascular plants of the lower Yangtze Valley, China. Corvallis: Oregon State College.
Tanaka, K., Komatsu, K. 2008. Comparative study on volatile components of *Nardostachys* rhizome. J. Nat. Med. 62: 112–116.
Thies, P.W. 1966. Über die Wirkstoffe des Baldrians. 2. Zur Konstitution der Isovaleriansäureester Valepotriat, Acetoxylvalepotriat und Dihydrovalepotriat. Tetrahedron Lett. 11: 1163–1170.
Thies, P.W. 1968. Die Konstitution der Valepotriate. Mitteilung über die Wirkstoffe des Baldrians. Tetrahedron 24: 313–347.
Thies, P.W., Funke, S. 1966. Nachweis und Isolierung von sedativ wirksamen Isovaleriansäureestern aus Wurzeln und Rhizomen von verschiedenen *Valeriana*- und Kentranthus-Arten. Tetrahedron Lett. 11: 1155–1162.
Vidal, L. 1903. Contribution à l'anatomie des Valérianacées. Ann. Univ. Grenoble 15: 561–605.
Wagenitz, G. 1956. Pollenmorphologie der mitteleuropäischen Valerianaceen. Flora 143: 473–485.
Wagenitz, G. 1976. Was ist eine Achäne? Zur Geschichte eines karpologischen Begriffs. Candollea 31: 79–85.
Wagenitz, G., Laing, B. 1984. Die Nektarien der Dipsacales und ihre systematische Bedeutung. Bot. Jahrb. Syst. 104: 483–507.
Weberling, F. 1961. Die Infloreszenzen der Valerianaceen und ihre systematische Bedeutung. Abh. Akad. Wiss. Lit. Mainz, math.-naturw. Kl. Nr. 5.
Weberling, F. 1966. Ordnung Dipsacales, general introduction. In: Hegi, G. (ed.) Flora von Mitteleuropa VI/2, 2. Aufl. (ed. G. Wagenitz). München: Carl Hanser.
Weberling, F. 1968. Zum Florenaustausch zwischen Nord- und Südamerika. Bot. Jahrb. Syst. 88: 366–381.
Weberling, F. 1970. Valerianaceae. In: Hegi, G. (ed.) Flora von Mitteleuropa VI/2, 2. Aufl. (ed. G. Wagenitz). München: Carl Hanser.
Weberling, F. 1977. Vergleichende und entwicklungsgeschichtliche Untersuchungen über die Haarformen der Dipsacales. Beitr. Biol. Pflanzen 53: 61–89.
Weberling, F. 1983. Valerianaceae. In: Löve, A. (ed.) IOPB Chromosome number reports LXXX. Taxon 32: 504.
Weberling, D., Weberling, F., Necker, C. 1976. Valerianaceae. In: Löve, A. (ed.) IOPB Chromosome number reports LIII, 498. Taxon 25: 483–500.

Wichura, M. 1846. Beiträge zur Lehre von der Blatt- und Knospenstellung. III. Ueber den Blüthenbau der Valerianeen. Flora (1846): 241–247, Fig. VI.

Willkomm, H.M. 1851. *Valeriana longiflora* n.sp. Flora 34: 733.

Winkworth, R.C., Lundberg, J., Donoghue, M.J. 2008. Toward a resolution of Campanulid phylogeny, with special reference to the placement of Dipsacales. Taxon 57(1): 53–65.

Xena de Enrech, N., Mathez, J. 1998. Genetic control of fruit polymorphism in the genus *Fedia* (Valerianaceae) in the light of dimorphic and trimorphic populations of *F. pallens*. Plant Syst. Evol. 210: 199–210.

Xena de Enrech, N., Cardona, A., Mathez, J. 1991. Estudio citotaxonómico del Género *Fedia* Gaertn. (*Valerianaceae*). An. Jard. Bot. Madrid 48(2): 157–169.

Xena de Enrech, N., Lebreton, Ph., Mathez, J. 1996. Los Flavonoides del género *Fedia* Gaertn. (*Valerianaceae*) y su interpretación quimotaxonómica. An. Jard. Bot. Madrid 54: 327–335.

Wellstediaceae

Wellstediaceae (Pilg.) Novák in S. Prát (ed.), Rostlinopis 9: 530 (1943).
Boraginaceae subfam. Wellstedioideae Pilg., Bot. Jahrb. Syst. 46: 558 (1912).

H.H. Hilger and M. Weigend

Annual herbs, sometimes with basal rosette of large leaves, or more often subperennial to perennial shrublets; branches and leaves often more or less distichous and congested; stems lignified, strong taproot present; indumentum sericeous, of unicellular, densely scabrid trichomes, usually acroscopically appressed. Leaves alternate, exstipulate, shortly petiolate or subsessile, entire, linear to (ob)ovate-elliptical, narrowed at base, apex acute, acuminate or apiculate, indistinctly veined, sometimes primary vein visible below, rarely venation pinnate with ascending secondary veins. Inflorescences bracteose, of dense scorpioid monochasia or dichasia, rarely reduced to single flowers, morphologically terminal but overtopped by and sometimes partly fused with uppermost lateral shoot; when fused, flowers arranged in alternate rows on this shoot. Flowers erect, tetramerous, actinomorphic, bisexual, hypogynous; calyx divided nearly to base, calyx lobes equal, narrowly triangular to narrowly ovate or linear, pubescent or sericeous, slightly enlarging in fruit; corolla pink, white or yellowish, sympetalous, hypocrateriform with very short tube and half-erect to spreading lobes, tube inside with 4 protrusions often fused into a distinct rim, lobes triangular-ovate, margin entire or denticulate to erose; stamens 4, alternating with petal lobes, filaments equal, shorter or longer than anthers, anthers included or exserted; nectary unknown; ovary bicarpellate, style terminal with two very short stigmatic lobes. Fruit a bivalved capsule. Seeds one, rarely two, asymmetrically ovoid in lateral view, strongly laterally compressed, pubescent often with ring of longer hairs near funicular pole, exendospermous.

One genus of six species with a disjunct distribution between the Horn of Africa region (five species in Ethiopia, Somalia to Socotra, Yemen) and western southern Africa (one species in NW South Africa and Namibia).

Vegetative Morphology and Anatomy (Fig. 76). *Wellstedia* is usually described as a dwarf shrub, with rosette leaves rarely present. Branches are decumbent, ascending or rarely erect and usually densely congested. All axes are very hard and likely lignified. The leaves are more or less distinctly distichous. Little is known about the life history of *Wellstedia*, but it likely germinates during the rainy season and continues to grow, flower and fruit as long as sufficient moisture is available. It may thus behave as an annual or ephemeral plant in some places. Branches incurve when the plants dry out completely and spread out again when the plant is wetted (at least in *W. dinteri*). Plants are up to ca. 40 cm tall, but mostly smaller. All parts of the plants, apart from the inside of the flower, are densely sericeous with a dense cover of long, unicellular, usually densely scabrid, rarely smooth trichomes. Cystolithic foot cells are absent. The presence of multicellular peltate glands reported by Hunt (1969) could not be verified by Thulin and Johansson (1996). Shorter, smooth unicellular trichomes are sometimes present. Leaves are narrowly obovate to nearly linear in most taxa, and differ characteristically in size, indumentum and shape between species. Only one species, *W. laciniata*, has distinctly ovate-elliptical leaves with well-visible secondary veins and a distinct petiole. Leaves are amphistomatic.

Inflorescence and Flower (Fig. 76). The inflorescence is extremely congested and consists of terminal mono- or dichasia which soon become overtopped by a branch arising from the axil of the uppermost foliage leaf below inflorescence and is then displaced into a lateral position on

J.W. Kadereit and V. Bittrich (eds.), *Flowering Plants. Eudicots*, The Families and Genera of Vascular Plants 14,
DOI 10.1007/978-3-319-28534-4_36, © Springer International Publishing Switzerland 2016

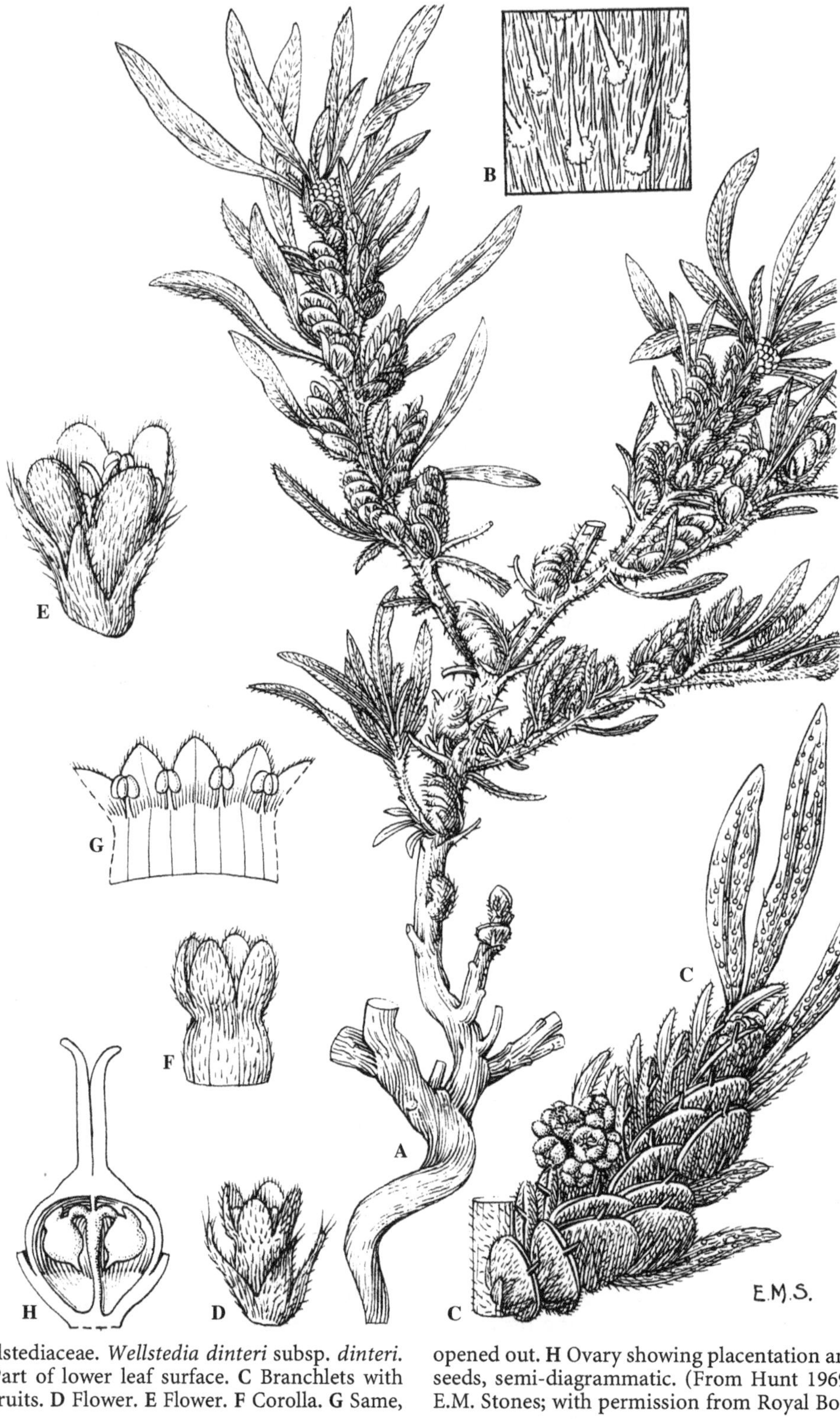

Fig. 76. Wellstediaceae. *Wellstedia dinteri* subsp. *dinteri*. **A** Habit. **B** Part of lower leaf surface. **C** Branchlets with flowers and fruits. **D** Flower. **E** Flower. **F** Corolla. **G** Same, opened out. **H** Ovary showing placentation and developing seeds, semi-diagrammatic. (From Hunt 1969; artwork by E.M. Stones; with permission from Royal Bot. Gds Kew)

the upper side of the usually more or less plagiotropic or ascending branches. The inflorescence axis itself is then often fused with the overtopping branch, with flowers arising apparently directly from a vegetative branch in two alternating rows. Narrowly ovate to linear bracts are usually present. The flowers are pink, white or yellowish and tetramerous. The calyx is divided nearly to the base with 4 usually more or less triangular, densely sericeous lobes. The corolla tube is very short, cylindrical, and contains 8 vascular bundles. The tube has inside four horizontal protrusions near or below the points of filament insertion. These may be homologous with the faucal scales of Boraginaceae s.str. Corolla lobes are imbricate in bud, (triangular-)ovate, abaxially pubescent and adaxially glabrous, 3-veined and equal. Filaments are inserted near the middle or near the apex of the corolla tube and may be very short (shorter than the anthers) or distinctly longer, far exceeding the anthers. Filaments are usually slightly bent inwards, positioning the anthers in the opening of the corolla tube. Anthers are dorsifixed with introrse dehiscence. The ovary is superior, broadly sessile on the receptacle and widely ovoid at anthesis. It is incompletely bilocular, at least apically densely pubescent, and crowned with a terminal style with a shortly bifid stigma. The two united carpels are partly or largely separated by a thin, hyaline septum. The septum may be present only at the base of the fruit or almost reach the style base. Placentation is subapical with one ovule in each locule or, more commonly, one locule is sterile. Ovules are anatropous and initially pendent and epitropous.

Fruit and Seed. The fruit is wedge-shaped, rhomboidal or obcordate in lateral view and laterally strongly compressed. It contains mostly one, rarely two seeds (*W. dinteri*). The style may be overtopped by the lateral parts of the growing capsule. Depending on the extent of the septum, the ovule will shift from a pendulous to an erect position at maturity. Dehiscence is loculicidal, with the sharp edge of the capsule opening from apex to base. At least in *W. dinteri* fruits are hygrochasious and open only when wetted, releasing the seeds after rare rains. The seeds are laterally strongly compressed, (obliquely) ovoid in lateral view and densely pubescent. Trichomes are short at the chalazal pole and gradually longer towards the funicular pole, ending in a distinct ring of long trichomes near the funicular pole. The slightly curved funicular pole itself is glabrous. The short funicle remains attached to the mature seed. The seed is completely filled by the large embryo and is exalbuminous. The embryo has accumbent cotyledons and a superior radicle. Germination is only known for *W. dinteri*, where it is phanerocotylar with widely ovate, densely pubescent cotyledons.

Pollen Morphology. Pollen grains are isopolar, oblate to subprolate and 12–25 × 8–15 μm in size. Grains are tricolporate, with colpi 11–14 μm long and 1–3 μm wide, and have a granular colpus membrane. Pseudocolpi are sometimes present. The tectum is perforate to reticulate, usually coarsely so at the centre of the mesocolpia with lumina smaller towards apocolpia and colpus margin (Thulin and Johansson 1996; Retief and van Wyk 2008).

Dispersal. Plants apparently die after fruiting, remaining in the soil with incurved twigs. After wetting the congested and incurved twigs spread out and the capsules open and release the seeds. The characteristic pubescent seeds may also be dispersed by wind after release. Fruiting plants are also said to form tumbleweeds by breaking off at the base and rolling over the ground.

Affinities. Morphologically, *Wellstedia* is quite isolated in Boraginales with its peculiar vegetative morphology, tetramerous flowers, bilaterally compressed, capsular, 1–2-seeded fruit and unique seed morphology. Balfour (1888) placed it as a "genus anomalum" at the end of the Gamopetalae, and both Gürke (1893) and Pilger (1912) assigned it to Boraginaceae s.l. as subfam. Wellstedioideae. Family rank was first suggested by Novák (1943) and has been followed by, for example, Friedrich-Holzhammer (1967). Because of its ecology and habit, an affinity to Ehretiaceae (*Tiquilia* and *Coldenia*, the latter also with tetramerous flowers) was then assumed. However, all Ehretiaceae have fruits separating into mericarpids. Pollen morphology has been used to argue for an affinity with Hydrophyllaceae (Constance and Chuang 1982; Thulin and Johansson 1996) and with *Codon* (Retief et al. 2001), here treated as a separate family Codonaceae. However, pollen morphology itself is not particularly distinct, with the subprolate, reticulate, tricolporate pollen representing one of the most common pollen types in Eudicots. Regarding its specific characters in capsule, flower, seed or vegetative morphology, *Wellstedia* is quite dissimilar from both Codonaceae and Hydrophyllaceae. Recent molecular studies based on four chloroplast markers (Weigend et al. 2013, 2014) could finally clarify the correct placement for Wellstediaceae as sister to Boraginaceae s.str., a clade which is in turn sister to Codonaceae and represents one of the two major clades of Boraginales. The relationship to Hydrophyllaceae and Ehretiaceae is quite remote. The peculiar flowers and capsular fruit for Wellstediaceae clearly warrant separation from Boraginaceae (with four nutlets) and recognition as a distinct, monogeneric family.

Distribution and Habitats. *Wellstedia* is disjunct between NE Africa and Socotra (5 spp.) and Namibia and the extreme north of the Cape Province of South Africa (1 sp.), a distribution pattern shared by many arid zone taxa. Plants grow in arid habitats with open vegetation. *W. socotrana* is listed as threatened on the IUCN Red List.

One genus:

Wellstedia Balf. f. Fig. 76

Wellstedia Balf. f., Proc. Roy. Soc. Edinb. 12: 407 (1884); Friedrich-Holzhammer, Prodromus einer Flora von Südwestafrika, Lieferung 13 (1967); Retief & van Wyk, Bothalia 38: 57–63 (2008); Thulin, Flora of Somalia 3: 31–60 (2006).

Characters as for family.

Species are well circumscribed (Thulin and Johansson 1996; Thulin 2006). Thulin and Johansson (1996) assume a basal position of the southwest African *W. dinteri* based on morphology, but molecular data are not yet available.

Selected Bibliography

Balfour, I.B. 1888. Botany of Socotra. Edinburgh: R. Grant, pp. 1–446.

Constance, L., Chuang, T.I. 1982. SEM survey of pollen morphology and classification in Hydrophyllaceae (waterleaf family). Amer. J. Bot. 69: 40–53.

Friedrich-Holzhammer, M. 1967. 121. Wellstediaceae. In: Merxmüller, H. (ed.) Prodromus einer Flora von Südwestafrika, Lieferung 13. Lehre: Cramer, p. 1

Gürke, M. 1893. Borraginaceae (Asperifoliaceae). In: Engler, A., Prantl, K. (eds.) Die natürlichen Pflanzenfamilien, IV, 3a. Leipzig: Engelmann, pp. 71–131.

Hunt, D.R. 1969. *Wellstedia socotrana, W. dinteri* var. *dinteri, W. dinteri* var. *gracilior*. Hook. Ic. Pl. 37: tab. 3665–3667.

Novák, F.A. 1943. Wellstediaceae. Prát, Rostlinopsis 9: 530.

Pilger, R. 1912. Die Gattung *Wellstedia* in Südwestafrika. Bot. Jahrb. Syst. 46: 558–561.

Retief, W., van Wyk, A.E. 2008. The genus *Wellstedia* (Boraginaceae: Wellstedioideae) in southern Africa. Bothalia 38: 57–63.

Retief, E., van Wyk, A.E., Van der Merwe, C.F. 2001. Tapetal orbicules in southern African Boraginaceae. Proc. Micros. Soc. Southern Africa 31: 56.

Thulin, M. 2006. Boraginaceae. In: Thulin, M. (ed.) Flora of Somalia 3. Angiospermae (Ericaceae-Asteraceae). Kew: Royal Botanic Gardens, pp. 31–60.

Thulin, M., Johansson, A.N.B. 1996. Taxonomy and biogeography of the anomalous genus *Wellstedia*. In: Van der Maesen, L.J.G., van der Burgt, X.M., van Medenbach de Rooy, J.M. (eds.) The Biodiversity of African Plants. Proc. XIV AETFAT Congr. Dordrecht: Kluwer, pp. 73–86.

Weigend, M., Luebert, F., Selvi, F., Brokamp, G., Hilger, H. H. 2013. Multiple origins for Hounds tongues (*Cynoglossum* L.) and Navel seeds (*Omphalodes* Mill.) – the phylogeny of the borage family (Boraginaceae s.str.). Mol. Phylogenet. Evol. 68: 604–618.

Weigend, M., Luebert, F., Gottschling, M., Couvreur, T.L. P., Hilger, H.H., Miller, J. 2014. From capsules to nutlets—phylogenetic relationships in the Boraginales. Cladistics 30(5): 508–518.

Index

References to accepted names in **bold**-faced print, to synonyms in upright print, to illustrations in *italics*.

J.W. Kadereit and V. Bittrich (eds.), *Flowering Plants. Eudicots*, The Families and Genera of Vascular Plants 14,
DOI 10.1007/978-3-319-28534-4, © Springer International Publishing Switzerland 2016

Volumes published in this series

Volume I	Pteridophytes and Gymnosperms Edited by K.U. Kramer and P.S. Green (1990) Date of publication: 28.9.1990
Volume II	Flowering Plants. Dicotyledons. Magnoliid, Hamamelid and Caryophyllid Families Edited by K. Kubitzki, J.G. Rohwer, and V. Bittrich (1993) Date of publication: 28.7.1993
Volume III	Flowering Plants. Monocotyledons: Lilianae (except Orchidaceae) Edited by K. Kubitzki (1998) Date of publication: 27.8.1998
Volume IV	Flowering Plants. Monocotyledons: Alismatanae and Commelinanae (except Gramineae) Edited by K. Kubitzki (1998) Date of publication: 27.8.1998
Volume V	Flowering Plants. Dicotyledons: Malvales, Capparales and Non-betalain Caryophyllales Edited by K. Kubitzki and C. Bayer (2003) Date of publication: 12.9.2002
Volume VI	Flowering Plants. Dicotyledons: Celastrales, Oxalidales, Rosales, Cornales, Ericales Edited by K. Kubitzki (2004) Date of publication: 21.1.2004
Volume VII	Flowering Plants. Dicotyledons: Lamiales (except Acanthaceae including Avicenniaceae) Edited by J.W. Kadereit (2004) Date of publication: 13.4.2004
Volume VIII	Flowering Plants. Eudicots: Asterales Edited by J.W. Kadereit and C. Jeffrey (2007) Date of publication: 6.12.2006
Volume IX	Flowering Plants. Eudicots: Berberidopsidales, Buxales, Crossosomatales, Fabales p.p., Geraniales, Gunnerales, Myrtales p.p., Proteales, Saxifragales, Vitales, Zygophyllales, Clusiaceae Alliance, Passifloraceae Alliance, Dilleniaceae, Huaceae, Picramniaceae, Sabiaceae Edited by K. Kubitzki (2007) Date of publication: 6.12.2006
Volume X	Flowering Plants. Eudicots: Sapindales, Cucurbitales, Myrtaceae Edited by K. Kubitzki (2011) Dates of publication: 10.12.2010 (online); 10.1.2011 (print)
Volume XI	Flowering Plants. Eudicots: Malpighiales Edited by K. Kubitzki (2014) Dates of publication: 27.11.2013 (online); 17.12.2013 (print)

(continued)

Volume XII	Flowering Plants. Eudicots: Santalales, Balanophorales By J. Kuijt and B. Hansen (2015) Dates of publication: 21.10.2014 (online); 18.11.2014 (print)
Volume XIII	Flowering Plants. Monocots: Poaceae By E.A. Kellogg (2015) Dates of publication: 18.5.2015 (online); 12.6.2015 (print)
Volume XIV	Flowering Plants. Eudicots: Aquifoliales, Boraginales, Bruniales, Dipsacales, Escalloniales, Garryales, Paracryphiales, Solanales (except Convolvulaceae), Icacinaceae, Metteniusaceae, Vahliaceae Edited by J.W. Kadereit and V. Bittrich (2016)

Kolibri 360 GmbH
Ferdinand-Jühlke-Straße 7
99095 Erfurt, Deutschland
produktsicherheit@kolibri360.de